Global Distribution of the Major Biomes

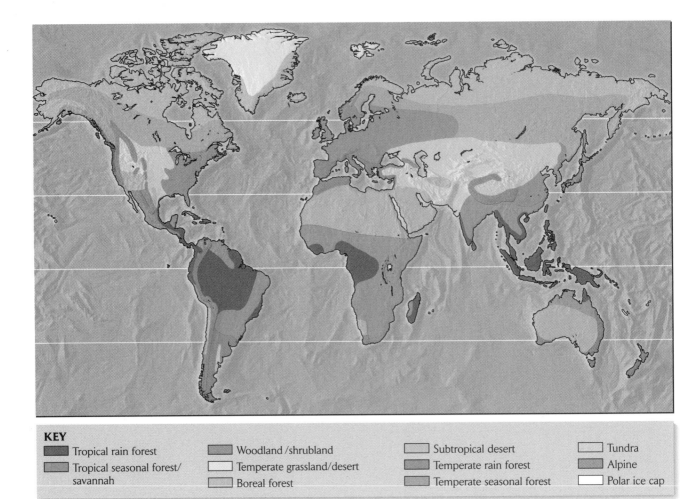

KEY

- Tropical rain forest
- Tropical seasonal forest/ savannah
- Woodland /shrubland
- Temperate grassland/desert
- Boreal forest
- Subtropical desert
- Temperate rain forest
- Temperate seasonal forest
- Tundra
- Alpine
- Polar ice cap

How to Use This Book and Its E-Study Center at
http://www.whfreeman.com/ricklefs

The E-Study Center has been developed with *The Economy of Nature, Fifth Edition* to enhance and enrich the ecological concepts discussed in the text. The E-Study Center offers study aids, Web activities, Living Graphs, and additional material organized by chapter and content type. The following chart shows how the book and the E-Study Center work together.

What you see in the book

Help on the Web icons in the textbook indicate that useful tutorials are found at the E-Study Center.

> **HELP ON THE WEB** *Go to Living Graphs at http://www.whfreeman. com/ricklefs* and use the interactive tutorial to better understand the derivation and properties of the Lotka–Volterra model.

More on the Web icons in the text proper indicate that additional enrichment essays can be found at the E-Study Center.

> **MORE ON THE WEB** *Alarm calls as altruistic behaviors.* Belding's ground squirrels give alarm calls warning of predators more often in the presence of close relatives.

More on the Web icons in the *Practicing Ecology* question sets indicate URLs that provide additional information needed to answer the questions.

> **MORE ON THE WEB** 3. Visit the *Burgess Shale Fossils* Web site through *Practicing Ecology on the Web* at *http:// www.whfreeman.com/ricklefs.* The Burgess Shale is located in Yoho National Park in the Rocky Mountains of British Columbia, Canada. What is it about this locality that accounts for its unusual assemblage of fossils? What can we learn about historical patterns of biodiversity from the Burgess Shale?

What you find at the E-Study Center
http://www.whfreeman.com/ricklefs

Living Graph interactive tutorials provide the opportunity to work with important ecology models such as exponential and geometric growth, the logistic equation, and the Lotka–Volterra predator–prey model. Answers to the exercises at the end of each tutorial can be sent directly to the instructor.

More on the Web enrichment essays provide further depth for selected topics.

Practicing Ecology on the Web research links cited in each end-of-chapter *Practicing Ecology* case study are housed here. The E-Study Center also provides a place to answer the *Practicing Ecology* questions and the means to send them directly to the instructor.

The **E-Study Center** also provides the following study aids:

Q & A Online Self-Tests provide practice tests. Each test answer is accompanied by a text page number that links the test question to the appropriate material in the text. Self-tests can also be sent to the instructor.

Chapter Outlines can serve as templates for class notes as well as for exam preparation.

Flashcards can be used to review key vocabulary terms.

Web Links reference interesting Web sites related to topics covered in the text.

The Economy of Nature

The Economy of Nature

Fifth Edition

ROBERT E. RICKLEFS

University of Missouri–St. Louis

 W. H. FREEMAN AND COMPANY

NEW YORK

Executive Editor: Sara Tenney

Marketing Director: John Britch

Senior Development Editor: Randi Rossignol

Associate Editor for Media and Supplements: Joy Hilgendorf

Project Editor: Diane C. Davis

Copy Editor: Norma Roche

Production Editors: Nancy Brooks/Penelope Hull

Cover and Text Designer: Victoria Tomaselli

Photo Editor: Jennifer MacMillan

Photo Researcher: Tobi Zausner

Illustration Coordinators: Shawn Churchman/Bill Page

Illustration: J. B. Woolsey Associates

Production Coordinator: Susan Wein

Composition: Sheridan Sellers/W. H. Freeman Electronic Publishing Center

Manufacturing: R R Donnelley & Sons Company

Cover Photo: Pat O'Hara/Tony Stone

Back Cover Inset Photo: A. Bazzaz

Library of Congress Cataloging-in-Publication Data

Ricklefs, Robert E.

The economy of nature/Robert E. Ricklefs—5th ed.

p. cm.

ISBN 0-7167-3883-X (pbk.)

1. Ecology. I. Title.

QH541.R54 2000

577—dc21

00-011831

Printed in the United States of America

Second printing, 2001

W. H. Freeman and Company

41 Madison Avenue, New York, NY 10010

Houndmills, Basingstoke, RG21 6XS, England

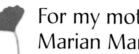 For my mother,
Marian Markarian Ricklefs

ABOUT THE AUTHOR

Robert E. Ricklefs is Curators' Professor of Biology at the University of Missouri–St. Louis, where he joined the faculty in 1995 after 27 years in the Biology Department at the University of Pennsylvania. Professor Ricklefs is a native of California and holds an undergraduate degree from Stanford University and a Ph.D. from the University of Pennsylvania. His interests include the energetics of reproduction in birds, evolutionary differentiation of life histories, biogeography, and the historical development of biological communities, including the generation and maintenance of large-scale patterns of biodiversity. His research has taken him to a wide variety of habitats from the lowland tropics to seabird islands in Antarctica. Professor Ricklefs is a Fellow of the American Association for the Advancement of Science and the American Academy of Arts and Sciences. He is also the coauthor of *Ecology,* now in its fourth edition, and *Aging: A Natural History,* both published by W. H. Freeman and Company.

Brief Contents

Detailed Contents

Student E-Study Center

http://www.whfreeman.com.ricklefs

In addition to finding practice tests and study aids for each chapter at the new Student E-Study Center, you will find Living Graph interactive tutorials and text enrichment topics.

LIVING GRAPHS

HELP ON THE WEB | *Help on the Web* icons throughout the text indicate that the Student E-Study Center at *http://www.whfreeman.com/ricklefs* includes interactive tutorials to help students master key ecological equations. *Living Graphs* are found on the E-Study Center for the following chapter topics:

ENRICHMENT TOPICS

MORE ON THE WEB | *More on the Web* icons throughout the text direct the student to go to the E-Study Center at *http://www.whfreeman.com/ricklefs* for chapter enrichment topics that complement and enhance the coverage found in the text or for Web links that will help in answering the end-of-chapter Practicing Ecology Web research questions. The following enrichment topics are found at the E-Study Center.

Preface

The Enduring Vision and Organization

Throughout this textbook you will see a high value placed on three tenets of teaching introductory ecology:

First, a solid grounding in natural history. The more we know about habitats and their resident organisms, the better we are able to generalize.

Second, an appreciation of the organism as the fundamental unit of ecology. The structure and dynamics of populations, communities, and ecosystems express the activities of, and interactions among, the organisms they comprise. That a population of insects increases to outbreak proportions depends on the fecundity and survival of individuals in the population, and the fecundity and survival reflect, in turn, the interaction of the individual with resources, predators, and physical conditions of the environment. Similarly, the regeneration of nutrients within an ecosystem depends in large part on the activities of individual microorganisms that make waste products of their own feeding and metabolism available as resources to plants.

Third, the central position of evolutionary thinking in the study of ecology. The qualities of all ecological systems express the evolutionary adaptations of their component species. It is impossible to understand how ecological systems develop, function, and respond to perturbation without understanding the evolutionary dynamics of populations.

I have therefore organized the book so that students are immediately introduced to the physical environment and the ways in which organisms adapt to their surroundings in Chapters 1–4. The book then introduces the concept of the biome to illustrate the diversity of habitats on earth in Chapter 5. Chapters 6–8 discuss how energy and elements move among the different components of the biosphere, emphasizing the interconnectedness of life and the physical world. The book then treats processes at the level of the organism in Chapters 9–12: these chapters focus on adaptation and the fundamental trade-offs that organisms make. This is fol-lowed by discussions of population structure and dynamics in Chapters 13–16 and of interactions between populations of different species, including predator–prey interactions and competition, in Chapters 17–20. Chapters 21–24, which discuss the organization and regulation of ecological communities, emphasize how ecological systems are continually challenged and maintained and how the dynamics of such systems shape patterns of biodiversity over the globe. Chapters 25 and 26 complete the book with an overview of environmental problems and some of their solutions.

What Is New to This Edition

NEW MATERIAL

• *Updated coverage.* This new edition updates much of the material in the text to reflect the most recent results and thinking in ecology. For example, this edition features new research on the fitness advantage of sexual reproduction in the presence of parasite infection (Chapter 11), the use of stable isotopes to estimate ocean temperatures during the Ice Ages (Chapter 4), the analy-

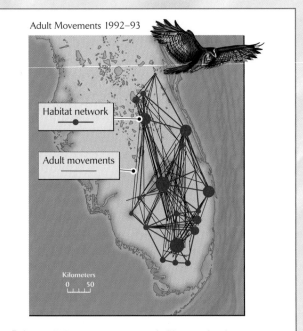

Figure 13.8 Movement among habitat patches integrates a population.

sis of forest dynamics as a way of understanding tropical tree diversity (Chapter 23), sampling populations at different scales to determine the dimensions of population processes (Chapter 13), species sorting through environmental filters to form local communities from the regional species pool (Chapter 23) and evolution of beak size in finches in response to changing food resources resulting from El Niño events (Chapter 16.)

• *Earlier and enhanced coverage of evolution.* In response to a trend to teach more evolution as part of the introductory ecology course, the new edition covers the basic concepts of Darwinian evolution in Chapter 9, "Adaptation to Life in Varying Environments." This chapter cements the relationship among natural selection, adaptation, and evolution. This material is presented as the force behind much of organismal and population ecology. Further treatment of evolution is integrated throughout the remainder of the text as in previous editions.

ECOLOGISTS IN THE FIELD

Predator avoidance and growth performance in frog larvae

Staying out of harm's way can depress growth rates when predator avoidance limits prey individuals to poor feeding areas. The effects of predation risk on the growth of tadpoles have been demonstrated in laboratory and field experiments on bullfrogs (*Rana catesbiana*) by Rick Relyea and Earl Werner at the University of Michigan. They conducted laboratory experiments in which newly hatched tadpoles were placed in aquaria with caged ... poles reduced their acti...

• *More field ecology.* Students appreciate concepts if they see how they are developed and how they are applied. With this in mind, the coverage of the practice of ecology is highlighted and much increased in the Fifth Edition. This new approach is showcased in two ways:

PRACTICING ECOLOGY
CHECK YOUR KNOWLEDGE

Variation over Space and Time

As we have discussed throughout this chapter, variations in the physical environment play an important role in ecological interactions that determine the distributions and abundances of organisms. The way in which individual organisms respond to changes in climate is critical to their success in surviving and reproducing. Therefore, it is important to understand how past and present patterns of climatic variability affect plants and animals so that we can predict the probable effects of future changes in the environment.

Ecologists in the Field case studies describe both classic and current field and laboratory research. There are more than 30 *Ecologist in the Field* case studies throughout the text. For example, Chapter 17, "Predation and Herbivory," includes two *Ecologists in the Field* sections. One features a comparative analysis and a foraging-time model that address the greater relative size of prey hunted by larger predators; the other features an experimental analysis of the effect of predators on the behavior and growth of frog larvae.

Practicing Ecology synthesis problems appear at the end of every chapter. Each *Practicing Ecology* ties chapter concepts together within the context of an ecological study. Students can then check their understanding of the concepts and the research protocol underlying the study,

analyze the data obtained in the research, and finally do guided research on their own through the *E-Study Center* at *www.whfreeman.com/ricklefs*. For example, the *Practicing Ecology* at the end of Chapter 4, "Variations in the Physical Environment," focuses on the effect of climate warming on the timing of activities in animals and plants and the potential that seasonal shifts in activity have for disrupting population and ecosystem processes.

• *More qualitative coverage.* This edition focuses more on the qualitative treatment of mathematical models. As well, mathematical expressions are now presented in a more accessible stepwise fashion and are accompanied by explanatory text. Students receive additional guidance in learning mathematical models when they visit the *E-Study Center* at *www.whfreeman.com/ricklefs.* Here they will find *Living Graphs* that are computerized simulations of key mathematical models such as the Lotka–Volterra equation and the logistic equation. The *Living Graph* simulations are indicated in the body of the text by *Help on the Web* icons.

Much of the math that was considered by instructors and students to be beyond the scope of this course has been moved out of the text and onto the book's *E-Study Center.* These topics can be used for course enrichment. *More on the Web* icons within the body of the text call out these enrichment topics.

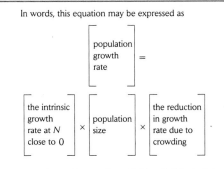

In words, this equation may be expressed as

$$\left[\begin{array}{c}\text{population}\\\text{growth}\\\text{rate}\end{array}\right] = $$

$$\left[\begin{array}{c}\text{the intrinsic}\\\text{growth}\\\text{rate at } N\\\text{close to 0}\end{array}\right] \times \left[\begin{array}{c}\text{population}\\\text{size}\end{array}\right] \times \left[\begin{array}{c}\text{the reduction}\\\text{in growth}\\\text{rate due to}\\\text{crowding}\end{array}\right].$$

According to this equation, which is called the **logistic equation,** the exponential rate of increase decreases as a linear function of the size of a population. Such a decrease reasonably approximated the data for the population of the United States (❚ Figure 14.15).

HELP ON THE WEB ❚ *Go to Living Graphs at http://www.whfreeman.com/ ricklefs.* Use the interactive tutorial to see how the logistic equation represents a population that always tends toward an equilibrium carrying capacity.

ENHANCED STUDENT PEDAGOGY

• *New art program.* The new edition includes 175 new concept-building art pieces, 20 new maps, 35 new climatographs, and 70 new illustrations depicting experiments. This is 50% more art than in the last edition. New art was added where it would help elucidate basic concepts and mathematical models. Graphs have been simplified so that experimental data are much more accessible. Many graphs have also been enhanced to show where the experiment was done as well as to illustrate the experimental organism. Every illustration was examined to ensure that it would help students read and understand ecological concepts, graphs and models. *The Economy of Nature* has always included superb photographs and we continue this tradition.

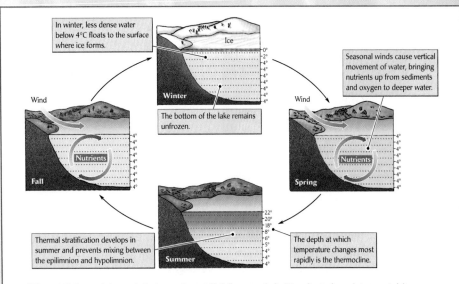

❚ Figure 4.13 Seasonal changes in the temperature profile influence vertical mixing of water layers in temperate lakes.

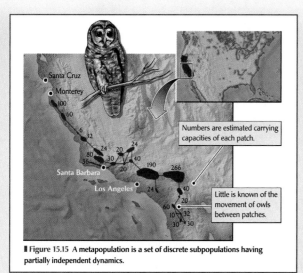

Numbers are estimated carrying capacities of each patch.

Little is known of the movement of owls between patches.

▌ Figure 15.15 A metapopulation is a set of discrete subpopulations having partially independent dynamics.

• **New and innovative art pedagogy.** The art program also features devices that help the student comprehend the art as well as make connections from the art back to the text. Annotations in the illustrations now lead students through the art, connecting the data and the concepts. When helpful, numbers indicate the steps of an ecological process or model. Full-sentence figure titles at the start of every figure legend state the take-home message of each illustration and photograph. Maps have been added throughout. And finally, figure locator icons (▌) help students move between the art and text.

• **New chapter structure.** A new chapter format is designed to make the material more manageable for the reader. Within chapters, full-sentence headings state clearly the purpose of each section. A list of the full-sentence headings appears at the beginning of each chapter as an overview of the key concepts found in the chapter. Sections are now divided into subsections to make reading assignments easier for the instructor and more manageable for the student.

Climate is the major determinant of plant distribution

• **New E-Study Center** at *www.whfreeman.com/ricklefs*. The *E-Study Center* is a component of the text's dedicated Web site, providing a place for students to enhance, test, and expand their grasp of the material. The *E-Study Center* is designed to help instructors and students use the Web as part of the learning environment in their ecology course. It offers the following:

Simulations. *Living Graph* interactive tutorials help students master important ecology equations such as exponential and geometric growth, the logistic equation, and the Lotka–Volterra predator–prey model. *Help on the Web* icons in the textbook tell

HELP ON THE WEB

students when to go to the Web to find these simulations. Answers to exercises at the end of each tutorial can be sent directly to the instructor for a grade.

Practice Tests help students prepare for tests. Text-page references for each test answer serve as a quick guide for students to re-read material in the text after they've taken a practice test.

MORE ON THE WEB

Text Enrichment Topics and Web Research Links. *More on the Web* icons in the textbook tell students to go to the Web for additional topics that enhance those in the text.

Practicing Ecology **Web Research Links** cited in each end-of-chapter *Practicing Ecology* synthesis problem are at the *E-Study Center*, providing easy access to Web sites mentioned in the text.

SUPPLEMENTS FOR INSTRUCTORS

Presentation tools

NEW Text Images and PowerPoint Presentation Slides by Thomas Wentworth, North Carolina State University. Available in the following formats:

• Dual Platform CD-ROM, ISBN 0-7167-3982-8

• *www.whfreeman.com/ricklefs*

All the text images are available in JPEG and PICT formats for use with any presentation software. PowerPoint Presentation slides written especially for this edition are also available. With these tools, instructors can quickly create classroom presentations that integrate the lecture outlines and figures, as well as content saved locally or found on the Internet.

MORE Overhead Transparency Set

ISBN: 0-7167-3983-6

Contains 300 illustrations from the text especially formatted for maximum readability in large lecture halls. The transparencies follow standards that were tested and set in a university auditorium (1,000 seats, 8 ft × 10 ft screen). Labels are enlarged and boldfaced, complicated illustrations are split into multiple pages, and colors have been chosen with optimal projection in mind.

Course management tools

NEW WebCT e-Learning

• *www.whfreeman.com/ricklefs*

All the supplements are available formatted for downloading to a WebCT course Web site.

NEW Instructor's Resource Web Site

• *www.whfreeman.com/ricklefs*

The password-protected instructor side of the Ricklefs site includes access to all the images from the text in JPEG and PICT format; PowerPoint lecture outline slides; the test bank files; online quizzing reports; the list of overhead transparencies; and the downloadable WebCT cartridge.

Assessment tools

NEW Test Bank by Thomas Wentworth, North Carolina State University. Available in the following formats:

• Dual Platform CD-ROM, ISBN 0-7167-3986-0

• Printed, ISBN 0-7167-3985-2

• *www.whfreeman.com/ricklefs*

The new test bank offers hundreds of multiple-choice, fill-in-the-blank, and essay questions. Each chapter also contains at least five interrelated application and integration questions based on actual experiments and data or on hypothetical situations. The easy-to-use CD version includes Windows and Mac versions on a single disc in a format that lets the instructor add, edit, and re-sequence questions.

PDF files of the test bank are available on the password-protected side of the Ricklefs Web site. As well, the CD-ROM and Web site include an electronic test manager powered by Diploma from the Brownstone Research Group. With Diploma, instructors can create their own examinations from the test bank, with the flexibility to add, edit, and re-sequence questions to suit their needs. Instructors can also create and administer secure exams over a network and over the Internet, with questions incorporating multimedia and interactive exercises. The Diploma Software lets instructors restrict tests to specific computers or time blocks, and it includes an impressive suite of gradebook and result-analysis features.

NEW Practice Tests by Thomas Wentworth, North Carolina State University. Powered by Question Mark.

• *www.whfreeman.com/ricklefs*

This feature allows instructors to quiz students online easily and securely using prewritten multiple-choice questions for each text chapter (not from the test bank. Students receive instant feedback and can take the quizzes more than once. Instructors can view results by quiz, student, or question, or can get weekly results via e-mail. Question Mark's Perception is a widely used Web-based quizzing and assessment tool with full multimedia capabilities.

Acknowledgments

The process of making a book requires the cooperation and common vision of many people, but there are rare times when this process has a special quality of personal interaction that brings unusual pleasure to an author and produces a book of unusual quality for the reader. Such was the case for the Fifth Edition of *The Economy of Nature.* Executive Editor, Sara Tenney; Senior Development Editor, Randi Rossignol; and Art Developer, John Woolsey, have each left a personal imprint on this book through their own sensitivities to and understanding of books and readers, and through the synergism developed over many long and stimulating hours of going through text and figures together. Sara provided the overall style and direction for the new edition, guiding the process when it was getting off track but always giving support and enthusiasm. John, with whom I have worked on book projects for almost 30 years, has a unique and unfailing sense of design that combines visual appeal with getting information across clearly. Through her astute editing, Randi has given the book its strong organization and lively and straightforward language, and, through her constant encouragement and gentle prodding, has sustained my interest and enthusiasm and taught me much about writing.

I am grateful for the hard work and professionalism of Victoria Tomaselli, Assistant Art Director; Jennifer MacMillan, Photo Editor; Joy Hilgendorf, Associate Editor; Diane Davis, Project Editor; Norma Roche, Copy Editor; and Nancy Brooks and Penelope Hull, Production Editors. With great proficiency, Sheridan Sellers, W. H. Freeman's Electronic Publishing Center's Manager; Susan Wein, Production Coordinator; Bill Page, Senior Illustration Coordinator; and Shawn Churchman, Illustration Coordinator, managed much of the book's production.

Of particular importance to me are colleagues who read the manuscript and provided useful advice and guidance. These reviewers have my gratitude: Peter Alpert, University of Massachusetts, Amherst; David M. Armstrong, University of Colorado; Stephen G. Bousquin, Oklahoma State University; Martin S. Cohen, University of Hartford; Mark D. Decker, University of Minnesota; Evan H. DeLucia, University of Illinois; George F. Estabrook, University of Michigan; Paul W. Ewald, Amherst College; Lloyd C. Fitzpatrick, University of North Texas; Bradford A. Hawkins, University of California, Irvine; Lauraine Hawkins, Pennsylvania State University, Mont Alto; Stephen B. Heard, University of Iowa; Robert D. Holt, University of Kansas; Keith T. Killingbeck, University of Rhode Island; Douglas W. Larson, University of Guelph; Michael E. Loik, University of California, Santa Cruz; James B. McGraw, West Virginia University; Joseph F. Merritt, Carnegie Museum of Natural History; Jon C. Pigage, University of Colorado, Colorado Springs; John M. Pleasants, Iowa State University; Willem M. Roosenburg, Ohio University; Sallie Sheldon, Middlebury College; Steve R. Simcik, Texas A&M University; Robert J. Steidl, University of Arizona; Alan E. Stiven, University of North Carolina, Chapel Hill; Irwin A. Ungar, Ohio University; Thomas Wentworth, North Carolina State University; Peter Wetherwax, University of Oregon; Loreen A. Woolstenhulme, Brigham Young University.

Additionally, instructors met with the editors at a focus group held at the Ecological Society of America conference in August 1999 in Spokane, Washington. Their ideas and feedback were instrumental in the plans for the student and instructor supplements. These participants have my gratitude: Robert Christopherson, American River College; Andy Guss, Utah State University; Gregg Hartvigsen, SUNY Geneseo; Jack Hayes, Paine College; Elizabeth Newell, Hobart and William Smith Colleges; Douglas Slack, Texas A&M University; Thomas Wentworth, North Carolina State University.

CHAPTER 1

Introduction

Ecological systems can be as small as individual organisms or as large as the entire biosphere

Ecologists study nature from several different perspectives

Plants, animals, and microorganisms play different roles in ecological systems

The habitat defines an organism's place in nature; the niche defines its functional role

All ecological systems and processes have characteristic scales in time and space

Ecological systems are governed by general physical and biological principles

Ecologists study the natural world by observation and experimentation

Humans are a prominent part of the biosphere

Human impacts on the natural world have increasingly become a focus of ecology

* The Horse Chestnut tree (*Aesculus hippocastanum*) is native to Asia and Northern Greece, but it is now cultivated in many areas of Europe and North America.

In his book *Uncommon Ground*, William Cronon challenges two common perceptions about nature and about humankind's relationship to nature. The first is the idea that nature tends toward a self-restoring equilibrium when left alone, a notion referred to as "the balance of nature." The second is the idea that in the absence of human interference, nature exists in a pristine state. Ecological studies present scientific evidence both for and against the idea of balance in nature and show how humans have influenced ecological systems. However, Cronon goes beyond these issues to address the cultural foundations of the way we view our own relationship with nature. He advances the idea that the conservation movement and, to some extent, the scientific field of ecology regard pristine nature as an absolute against which there can be no appeal. The unspoiled Amazon rain forest, for example, is likened by many to the Garden of Eden before Adam and Eve, which embodies complete good and also the temptations of complete evil. Cronon suggests that, in the minds of some people, the extinction of species brings out our own deep fear of losing paradise or of having to face the reality of our imperfect world.

Ecological studies paint a different picture. They show historical variation in nature and demonstrate that the pervasive influence of human activities extends to the most remote regions of the earth. These findings challenge the notion of a pristine, balanced environment. Paradise never did exist, at least not in human experience. Where we humans fit in a less than perfect world is a judgment each of you must make, guided by your own sense of values and moral beliefs. Regardless of your own stand, it will be more useful to you and to humankind in general if your judgment is informed by a scientific understanding of how natural systems work and the ways in which humans are a part of the natural world. The purpose of *The Economy of Nature* is to help you achieve that understanding.

The English word *ecology* is taken from the Greek *oikos*, meaning "house," and thus refers to our immediate surroundings, or environment. In 1870, the German zoologist Ernst Haeckel gave the word a broader meaning:

> *By ecology, we mean the body of knowledge concerning the economy of nature—the investigation of the total relations of the animal both to its organic and to its inorganic environment; including above all, its friendly and inimical relation with those animals and plants with which it comes directly or indirectly into contact—in a word, ecology is the study of all the complex interrelationships referred to by Darwin as the conditions of the struggle for existence.*

Thus, ecology is the science by which we study how organisms (animals, plants, and microbes) interact in and with the natural world.

The word *ecology* came into general use only in the late 1800s, when European and American scientists began to call themselves ecologists. The first societies and journals devoted to ecology appeared in the early decades of the twentieth century. Since that time, ecology has undergone immense growth and diversification, and professional ecologists now number in the tens of thousands. The science of ecology has produced an immense body of knowledge about the world around us. At the same time, rapid growth of the human population and its increasing technology and materialism have greatly accelerated the deterioration of the earth's environment. As a result, ecological understanding is now needed more than ever to learn the best policies for managing the watersheds, agricultural lands, wetlands, and other areas—what are generally called environmental support systems—upon which humanity depends for food, water, protection against natural catastrophes, and public health. Ecologists provide that understanding through studies of population regulation by predators, the influence of soil fertility on plant growth, the evolutionary responses of microbes to environmental contaminants, the spread of organisms over the surface of the earth, and a multitude of similar issues. Management of biotic resources in a way that sustains a reasonable quality of human life depends on the wise application of ecological principles to solve or prevent environmental problems and to inform our economic, political, and social thought and practice.

This chapter will start you on the road to ecological thinking. We shall first discuss several vantage points from which ecological knowledge and insight can be viewed—for example, as different levels of complexity, varieties of organisms, types of habitat, and scales in time and space.

We shall see how we can regard many different entities as ecological systems, by which we mean any organism, assemblage of organisms, or complex of organisms and their surroundings, united by some form of regular interaction or dependence of parts of the system on one another. Although the extent and complexity of ecological systems vary from a single microbe to the entire biosphere blanketing the surface of the earth, all ecological systems obey similar principles. Some of the most important of these principles concern physical and chemical attributes of ecological systems, regulation of structure and function in ecological systems, and evolutionary change. Applying these principles to environmental issues can help us to meet the challenge of maintaining a supportive environment for natural systems—and for ourselves—in the face of increasing ecological stresses.

Ecological systems can be as small as individual organisms or as large as the entire biosphere

An ecological system may be an organism, a population, an assemblage of populations living together (often called a community), an ecosystem, or the entire biosphere of the earth. Each smaller ecological system is a subset of the next larger one, and so the different types of ecological systems form a hierarchy of size. This arrangement is shown diagrammatically in ▮ Figure 1.1, which represents the idea that a population is made up of many individual organisms, a community comprises many interacting populations, an ecosystem represents the linkage of many communities through their use of energy and nutrient resources, and the biosphere encompasses all the ecosystems on earth.

The **organism** is the most fundamental unit of ecology, the elemental ecological system. No smaller unit in biology, such as the organ, cell, or molecule, has a separate life in the environment (although, in the case of single-celled protists and bacteria, cell and organism are synonymous). Every organism is bounded by a membrane or other covering across which it exchanges energy and materials with its surroundings. This boundary separates the "internal" processes and structures of the ecological system—in this case an organism—from the "external" resources and conditions of the surroundings.

---

* The Maidenhair tree (*Ginkgo biloba*) is the sole survivor of a genus that has changed little over the past 150 million years. Discovered in temple gardens in China, the Maidenhair tree now grows in many countries.

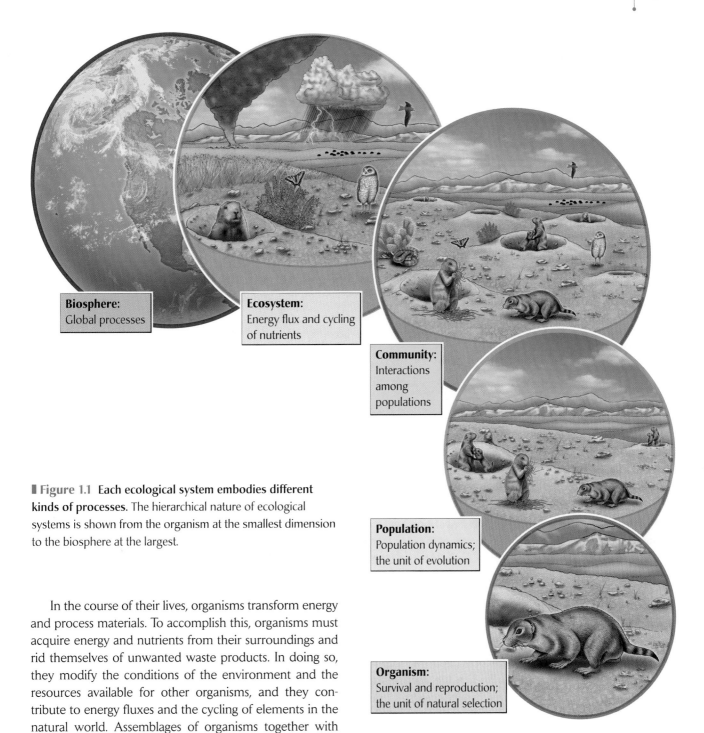

Biosphere:
Global processes

Ecosystem:
Energy flux and cycling
of nutrients

Community:
Interactions
among
populations

Population:
Population dynamics;
the unit of evolution

Organism:
Survival and reproduction;
the unit of natural selection

Figure 1.1 Each ecological system embodies different kinds of processes. The hierarchical nature of ecological systems is shown from the organism at the smallest dimension to the biosphere at the largest.

In the course of their lives, organisms transform energy and process materials. To accomplish this, organisms must acquire energy and nutrients from their surroundings and rid themselves of unwanted waste products. In doing so, they modify the conditions of the environment and the resources available for other organisms, and they contribute to energy fluxes and the cycling of elements in the natural world. Assemblages of organisms together with their physical and chemical environments make up an **ecosystem.** Ecosystems are large and complex ecological systems, sometimes including many thousands of different kinds of organisms living in a great variety of individual surroundings. A warbler flitting among the leaves overhead searching for caterpillars and a bacterium decomposing the organic soil underfoot are both part of the same forest ecosystem. We may speak of a forest ecosystem, a prairie ecosystem, and an estuarine ecosystem as distinct units because relatively little energy and few substances are exchanged between these units compared with the innumerable transformations going on within each of them. Hence the analogy with the organism of having "internal" processes and exchange with the "external" surroundings holds, allowing us to treat both organism and ecosystem as ecological systems.

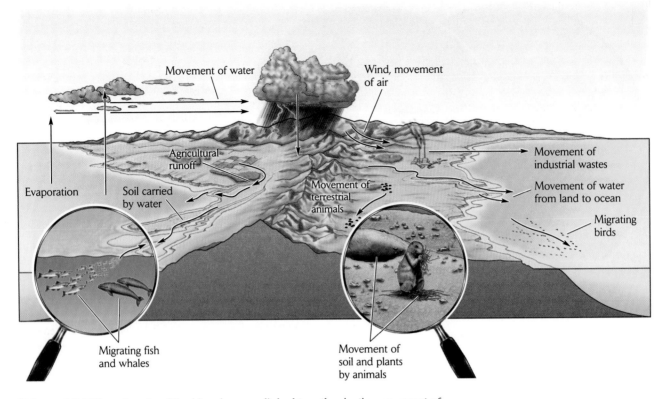

❙ Figure 1.2 Different parts of the biosphere are linked together by the movement of air, water, and organisms.

Ultimately, all ecosystems are linked together in a single **biosphere** that includes all the environments and organisms of the earth. The far-flung parts of the biosphere are linked together by the energy and nutrients carried by currents of wind and water and the movements of organisms. Water flowing from a headwater to an estuary connects the terrestrial and aquatic ecosystems of the watershed to those of the marine realm (❙ Figure 1.2). The migrations of gray whales link the ecosystems of the Bering Sea and the Gulf of California because feeding conditions in the Bering Sea influence the numbers of migrating whales and their reproductive success in the Gulf of California. This in turn determines the effect of the whale population on ecosystems in the breeding area. The importance of movement of materials between ecosystems within the biosphere is underscored by the global consequences of human activities. For example, industrial and agricultural wastes spread far from their points of origin, affecting all regions of the earth.

The biosphere is the ultimate ecological system. All that is external to the biosphere is the sunlight striking the surface of the earth and the black coldness of space. Except for the energy arriving from the sun and heat lost to the depths of space, all the transformations of the biosphere are internal. We have all the materials that we will

ever have; our wastes have nowhere to go and must be recycled within the biosphere.

The concepts of ecosystems and the biosphere emphasize the transformation of energy and the synthesis and degradation of materials—ecological systems as physical machines and chemical laboratories. Another perspective emphasizes the uniquely biological properties of ecological systems that are embodied in the dynamics of populations. A **population** consists of many organisms of the same kind living together. Populations differ from organisms in that they are potentially immortal, their numbers being maintained over time by the births of new individuals that replace those that die. Populations also have properties, such as geographic boundaries, densities (number of individuals per unit of area), and variations in size or composition (for example, evolutionary responses to environmental change and periodic cycles of numbers) that are not exhibited by individual organisms.

Many populations of different kinds living in the same place make up an ecological **community.** The populations within a community interact in various ways. For example, many species are predators that eat other kinds of organisms; almost all are themselves prey. Some, such as bees and the plants whose flowers they pollinate, and many microbes living together with plants and animals, enter into cooperative arrangements in which both parties benefit from the interaction. All these interactions influence the numbers of individuals in populations. Unlike organisms but like ecosystems, communities have no rigidly defined boundaries; no perceptible skin separates a community from what surrounds it. The interconnectedness of ecological systems means that interactions between populations spread across the globe as individuals and materials move between habitats and regions. Thus, the community is an abstraction representing a level of organization rather than a discrete unit of structure in ecology.

Ecologists study nature from several different perspectives

Each different level of the hierarchy of ecological systems has unique structures and processes. Therefore, each level has given rise to a different approach to the study of ecology. Of course, all the approaches have intersections. Within these areas of overlap ecologists may bring several perspectives to the study of particular ecological problems, as shown by the simple diagram in ❙ Figure 1.3.

The organism approach to ecology emphasizes the way in which an individual's form, physiology, and behavior help it to survive in its environment. This approach also seeks to understand why the distribution of each type of organism is limited to some environments and not others, and why related organisms living in different environments have different characteristic appearances. For example, as we shall see later in this book, the dominant plants of warm, moist environments are trees, while regions with cool, wet winters and hot, dry summers typically support shrubs with small, tough leaves.

Ecologists who use the organism approach are often interested in studying the adaptations of organisms to their environments. Adaptations are modifications of structure and function that better suit an organism for life in its environment: for example, enhanced kidney function to conserve water in deserts; cryptic coloration to avoid detection by predators; flowers shaped to be probed by certain kinds of pollinators. Adaptations are the result of evolutionary change by natural selection. Because evolution occurs through the replacement of one type of organism by another within a population, the study of adaptation represents a point of overlap between the organism and population approaches to ecology.

The population approach is concerned with numbers of individuals and their variations through time, including evolutionary changes within populations. Changes in numbers reflect births and deaths within a population.

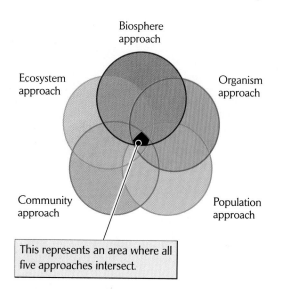

This represents an area where all five approaches intersect.

❙ **Figure 1.3** **There are five approaches to the study of ecology.** Although each approach addresses a different level of the hierarchy of ecological systems, they are portrayed on a single plane of scientific inquiry, each approach interacting with the others to varying degrees.

These may be influenced by physical conditions of the environment—temperature and the availability of water, for example. In the process of evolution, genetic mutations may alter birth and death rates, new lineages of individuals may become common within a population and the overall genetic makeup of the population may change. Other types of organisms, such as food items, pathogens, and predators, also influence the births and deaths of individuals within a population. In some cases, such interactions can produce dramatic oscillations of population size or less predictable population changes. Interactions between different kinds of organisms are the common ground of the population and community approaches.

The community approach to ecology is concerned with understanding the diversity and relative abundances of different kinds of organisms living together in the same place. The community approach focuses on interactions between populations, which both promote and limit the coexistence of species. These interactions include feeding relationships, which are responsible for the movement of energy and materials through the ecosystem, providing a link between the community and ecosystem approaches.

The ecosystem approach to ecology describes organisms and their activities in terms of common "currencies," principally amounts of energy and various chemical elements essential to life, such as oxygen, carbon, nitrogen, phosphorus, and sulfur. The study of ecosystems deals with movements of energy and materials within the

▌Figure 1.4 Ocean currents and winds carry moisture and heat over the earth. This satellite image of the North Atlantic Ocean during the first week of June, 1984, shows the Gulf Stream moving along the coast of Florida and breaking up into large eddies as it begins to cross the Atlantic toward Northern Europe. Warm water is indicated by red and cold water by green or blue, and then by red at the top of the picture. Courtesy of Otis Brown, Robert Evans, and Mark Carle, University of Miami Rosenstiel School of Marine and Atmospheric Science.

environment, and how these movements are influenced by climate and other physical factors in the environment. Ecosystem function results from the activities of organisms as well as from physical and chemical transformations in the soil, atmosphere, and water. Thus, the activities of organisms as different as bacteria and birds can be compared by describing the energy transformations of a population in such units as watts per square meter of habitat. However, in spite of their commonalties, community and ecosystem approaches to ecology provide different ways of looking at the natural world. We may speak of a forest ecosystem, or we may speak of the community of animals and plants that live in the forest, using different jargon and referring to different facets of the same ecological system.

Concerned with one extreme of the spectrum of ecological systems, the biosphere approach to ecology tackles the movements of air and water, and the energy and chemical elements they contain, over the surface of the earth (▌Figure 1.4). Ocean currents and winds carry the heat and moisture that define the climates at each point on earth, which govern the distributions of organisms, the dynamics of populations, the composition of communities,

and the productivity of ecosystems. Understanding natural variations in climate, such as El Niño, and such human-abetted changes as the formation of the ozone hole over the Antarctic and the conversion of grazing lands to desert over much of Africa is also an important goal of the biosphere approach to ecology.

Plants, animals, and microorganisms play different roles in ecological systems

The characteristics that distinguish plants, animals, fungi, protists, and bacteria (prokaryotes) have important implications for the way we study and come to understand nature. Different kinds of organisms have different functions in natural systems (▌Figure 1.5). The largest and most conspicuous forms of life, plants and animals, perform a large share of the energy transformations within the biosphere, but no more so than the countless unseen bacteria in soils, water, and sediments.

Moreover, plants and animals are relatively recent developments in the long history of evolution on earth. Early ecosystems were dominated by bacteria of various forms, which not only modified the biosphere, making it possible for more complex forms of life to exist, but were also the ancestors of all other forms of life. Photosynthetic bacteria in some of the earliest ecosystems produced oxygen as a by-product when they assimilated carbon dioxide. The resulting increase in oxygen concentration in the atmosphere and oceans (▌Figure 1.6) eventually permitted the evolution of complex, mobile life forms with high metabolic requirements, which have dominated the earth for the last 500 million years. As these new forms of life evolved, however, the more primitive types survived because their unique biochemical capabilities allowed them to use resources and tolerate ecological conditions not accessible to their more complex descendants. Indeed, the characteristics of modern ecosystems depend on the activities of many varied forms of life, with each major group filling a unique and necessary role in the biosphere.

Plants use the energy of sunlight to produce organic matter

All ecological systems depend on transformations of energy. For most systems, the ultimate source of that energy is sunlight. On land, plants use the energy of sunlight to synthesize organic molecules from carbon dioxide and water. Most plants have structures with large exposed surfaces—their leaves—to capture the energy of sunlight. Their leaves are thin because surface area for light capture

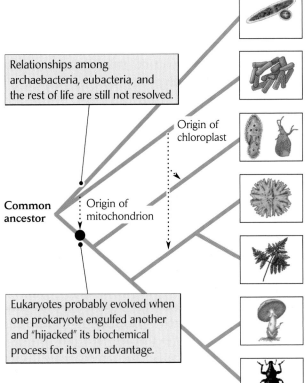

Archaebacteria
Simple prokaryotic organisms lacking an organized nucleus and other cell organelles. Adapted to living in extreme conditions of high salt concentration, high temperature, and pH (both acid and alkaline).

Eubacteria
Like archaebacteria, simple prokaryotic organisms having a wide variety of biochemical reactions of ecological importance in cycling elements through the ecosystem. Many forms are symbiotic or parasitic.

Various protists
An extremely diverse group of mostly single-celled eukaryotes–organisms having nuclear membranes and other cell organelles–ranging from slime molds and protozoans to the photosynthetic red, brown, and green algae.

Green algae
One of the lines of photosynthetic protists, which are responsible for most of the biological production in aquatic systems, and thought to have been the ancestor of green plants.

Green plants
Complex, primarily terrestrial photosynthetic (photoautotrophic) organisms responsible for fixing most of the organic carbon in the biosphere.

Fungi
Primarily terrestrial heterotrophic organisms of great importance in recycling plant detritus in ecosystems. Many forms are pathogenic and others form important symbioses (lichens, mycorrhizae).

Animals
Aquatic and terrestrial heterotrophic organisms that feed on other forms of life or their remains. Complexity and mobility have led to remarkable diversification of animal life.

Relationships among archaebacteria, eubacteria, and the rest of life are still not resolved.

Origin of chloroplast

Common ancestor

Origin of mitochondrion

Eukaryotes probably evolved when one prokaryote engulfed another and "hijacked" its biochemical process for its own advantage.

❚ Figure 1.5 Different organisms have different functions in natural systems. The major divisions of life and their evolutionary relationships are shown by the branching pattern at left.

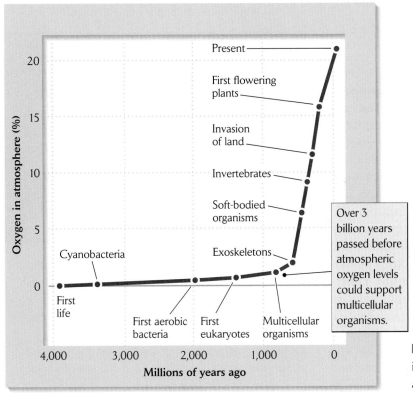

❚ Figure 1.6 The concentration of oxygen in the atmosphere has increased since the appearance of life on earth.

Figure 1.7 Epiphytic air plants form entire ecosystems. These plants grow high above the ground on the limbs of trees in tropical rain forests. Photo by R. E. Ricklefs.

is more important than bulk. Rigid stems support their aboveground parts. To obtain carbon, plants take up gaseous carbon dioxide from the atmosphere. At the same time, they lose prodigious amounts of water by evaporation from their leaf tissues to the atmosphere. Thus, plants need a steady supply of water to replace water lost during photosynthesis. Not surprisingly, most plants are firmly rooted in the ground, in constant touch with water in the soil. Those that are not, such as orchids and other tropical "air plants" (epiphytes), can be actively photosynthetic only in humid environments bathed in cloud mists (**Figure 1.7**).

Animals feed on other organisms or their remains

The organic carbon produced by photosynthesis provides food, either directly or indirectly, for the rest of the ecological community. Some animals consume plants; some consume animals that have eaten plants; others consume the dead remains of plants or animals.

Animals and plants differ in many important ways besides their sources of energy (**Figure 1.8**). Animals, like plants, need large surfaces for exchanging substances with their environments. However, because they don't need to capture light as an energy source, their exchange surfaces can be enclosed within the body. A modest pair of human lungs has a surface area of about 100 square meters, which is half the size of a tennis court. The gut also presents a large surface across which nutrients are assimilated into the body. For example, the intestine of a bird the size of a robin is about 30 centimeters long and has an absorptive surface area of over 200 square centimeters, or about half the size of this page. By internalizing their exchange

surfaces, animals can achieve bulk and streamlined body shapes, and they can develop the skeletal and muscular systems that make mobility possible. In terrestrial environments, the internalized surfaces of animals also lose less water by evaporation than do the exposed leaves of plants, and so land animals don't need to be continuously supplied with water.

Fungi are highly effective decomposers

The fungi assume unique roles in the ecosystem because of their distinctive growth form. Most fungi, like plants and animals, are multicellular organisms (except for yeasts and their relatives). But unlike plants and animals, the fungus grows from a microscopic spore without passing through an embryonic stage. Most fungal organisms are made up of threadlike structures called hyphae that are only a single cell in diameter. These hyphae may form a loose network, which can invade plant or animal tissues or dead leaves and wood on the soil surface, or grow together into the reproductive structures that we recognize as mushrooms (**Figure 1.9**). Because fungi can penetrate deeply, they readily decompose dead plant material, eventually making many of the nutrients contained in it available to other organisms. Fungi digest their foods externally, secreting acids and enzymes into their immediate surroundings, cutting through dead wood and dissolving recalcitrant nutrients from soil minerals. Fungi are the primary agents

Figure 1.8 Plants derive their energy from sunlight, and animals derive their energy from plants. A mammal browsing on vegetation on a savanna in eastern Africa emphasizes the fundamental differences between plants, which assimilate the energy of sunlight and convert atmospheric carbon dioxide to organic carbon compounds, and animals, which derive their energy ultimately from the production of plants. Photo by R. E. Ricklefs.

▌Figure 1.9 Fungi are effective decomposers of wood and other dead organic material. Mushrooms are fruiting bodies produced by the larger masses of threadlike hyphae, shown here in an exposed fungus that had developed in decaying leaf litter. The fruiting bodies extend up and to the right in this photograph. Photo by Larry Jon Friesen/Saturdaze.

of rot—unpleasant to our senses and sensibilities, perhaps, but very important to ecosystem function.

Protists are the single-celled ancestors of more complex life forms

The protists are a highly diverse group of mostly single-celled eukaryotic organisms that includes the algae, slime molds, and protozoans. There are a bewildering variety of protists filling almost every ecological role. For instance, algae, including diatoms, are the primary photosynthetic organisms in most aquatic systems. Algae can form large plantlike structures—some seaweeds can be up to 100 meters in length (see, for example, Figure 1.23)—but their cells are not organized into the specialized tissues and organs that one sees in plants.

The other members of this group are not photosynthetic. Foraminifera and radiolarians are protozoans that feed on tiny particles of organic matter or absorb small dissolved organic molecules, and that secrete shells of calcite or silicate. Some of the ciliate protozoa are effective predators—on other microorganisms, of course.

Bacteria have a wide variety of biochemical mechanisms for energy transformations

Bacteria, or prokaryotes, are the biochemical specialists of the ecosystem. Each bacterium consists of a simple, single cell, lacking a nucleus and chromosomes to organize its

DNA (▌Figure 1.10). Nonetheless, the enormous range of metabolic capabilities of bacteria enables them to accomplish many unique biochemical transformations. Some bacteria can assimilate molecular nitrogen (N_2, the common form found in the atmosphere), which they use to synthesize proteins and nucleic acids. Others can use inorganic compounds such as hydrogen sulfide (H_2S) as sources of energy. Plants, animals, fungi, and most protists cannot accomplish these feats. Furthermore, many bacteria live under anaerobic conditions (lacking free oxygen) in mucky soils and sediments, where their metabolic activities regenerate nutrients and make them available for plants. We will have much more to say about the special place of microorganisms in the functioning of the ecosystem.

Many types of organisms cooperate in nature

Because each type of organism is specialized to a particular way of life, it is not surprising that there are many different types of organisms that live together in close association, forming a **symbiosis.** In such relationships, each partner provides something that the other lacks. Some familiar examples

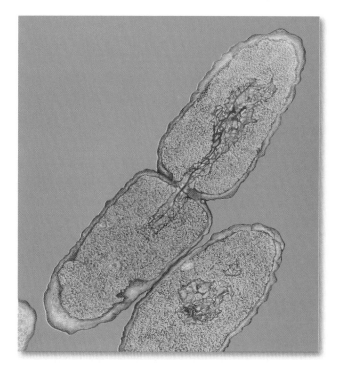

▌Figure 1.10 Bacteria are distinguished by their simple structure. They lack intracellular membranes and organelles. This *Salmonella typhimurium* bacterium, which is a gut parasite of many animals, was caught in the act of dividing. The stringy orange-colored material in the center of the cells is the DNA. The magnification is about ×15,000. Photo by Kari Lounatmaa/Science Photo Library/PhotoResearchers.

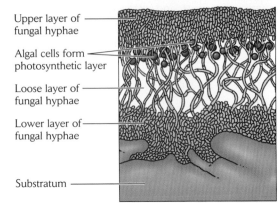

Upper layer of fungal hyphae

Algal cells form photosynthetic layer

Loose layer of fungal hyphae

Lower layer of fungal hyphae

Substratum

Figure 1.11 A lichen is a symbiotic association of a fungus and a green alga. Photo by R. E. Ricklefs.

include lichens, which comprise a fungus and an alga in one organism (**Figure 1.11**); bacteria that ferment plant material in the guts of cows; protozoans that digest wood in the guts of termites; fungi associated with the roots of plants that help them to extract mineral nutrients from the soil in return for carbohydrate energy from the plant; photosynthetic algae in the flesh of corals and giant clams; and nitrogen-fixing bacteria in the root nodules of legumes. The specialized organelles so characteristic of the eukaryotic cell—chloroplasts for photosynthesis, mitochondria for various oxidative energy transformations—originated as symbiotic prokaryotes (bacteria) living within the cytoplasm of host cells.

The habitat defines an organism's place in nature; the niche defines its functional role

Ecologists who apply the organism approach have found it useful to distinguish between where an organism lives and what it does. The **habitat** of an organism is the place, or physical setting, in which it lives. Habitats are characterized by their conspicuous physical features, often including the predominant form of plant life or, sometimes, animal life (**Figure 1.12**). Thus, we speak of forest habitats, desert habitats, and coral reef habitats. Ecologists have devoted

(a)   (b)   (c)

(d)

Figure 1.12 Terrestrial habitats are distinguished by their dominant vegetation. (a) Warm temperatures and abundant rainfall support the highest levels of biological production and diversity of life on earth in the tropical rain forest. In tropical seasonal forest habitats (b), trees lose their leaves during the pronounced dry season to avoid water stress. Tropical grasslands (c), which develop where rainfall is sparse, nonetheless support vast herds of grazing herbivores during the productive rainy season. (d) Freezing temperatures on the Antarctic ice cap preclude all life except for occasional bacteria in crevices of sun-warmed exposed rock. Photos by R. E. Ricklefs.

(a)

(b)

(c)

(d)

▌Figure 1.13 Each species has a distinctive niche. Four species of anole lizards occupy different niches in woodland habitat on the islands of Hispaniola and Jamaica in the Greater Antilles. (a) *Anolis insolitus*, a twig anole; (b) *A. garmani*, a crown giant; (c) *A. chlorocyanus*, a trunk-crown anole; (d) *A. cybotes*, a trunk-ground anole. Courtesy of Jonathan B. Losos.

much effort to classifying habitats. For example, ecologists distinguish terrestrial and aquatic habitats; among aquatic habitats, freshwater and marine; among marine habitats, ocean and estuary; among ocean habitats, benthic (on or within the ocean bottom) or pelagic (in the open sea). However, as such classifications become more complex, they ultimately break down, because habitat types overlap broadly and absolute distinctions between them rarely exist. The idea of habitat nonetheless is useful because it emphasizes the variety of conditions to which organisms are exposed. Inhabitants of abyssal ocean depths and tropical rain forest canopies experience vastly different conditions of light, pressure, temperature, oxygen concentration, moisture, viscosity, and salts, not to mention food resources and enemies.

An organism's **niche** represents the ranges of conditions that it can tolerate and the ways of life that it pursues—that is, its role in the ecological system. An important principle of ecology is that each species has a distinct niche

(▌Figure 1.13). No two species are exactly the same, because each has distinctive attributes of form and function that determine the conditions it can tolerate, how it feeds, and how it escapes its enemies.

The variety of habitats holds the key to much of the diversity of living organisms. No one organism can live under all the conditions on earth; each must specialize with respect to both the range of habitats within which it can live and the niche that it can occupy within a habitat.

All ecological systems and processes have characteristic scales in time and space

We have glimpsed the great variety of conditions on earth. Most things that we can measure in the environment, such as air temperature or the number of individuals in a population per unit of area, vary from one place to the next

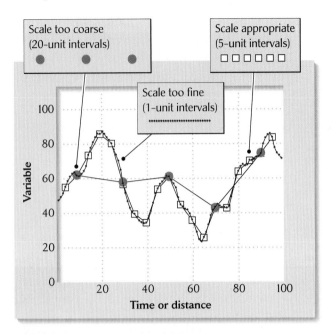

Figure 1.14 Patterns of variation have different scales in time and space. If the environment is measured at a scale that is too coarse, details of the pattern are missed. If the scale is too fine, the additional detail does not help to define the pattern.

and from one moment to the next. As a result, each measurement exhibits highs and lows, and the intervals or distances between successive highs or successive lows are separated by short or long intervals in time or distances in space. Yet variation in each measurement exhibits a characteristic **scale,** which is the dimension in time or space over which variation is perceived. It is important to select the appropriate scale of measurement to match the scale of variation of an ecological pattern in either time or space (**Figure 1.14**). For example, over time, the temperature of the air can plunge dramatically in a matter of hours as a cold front passes through a region, whereas a particular area of the ocean may require weeks or months to cool the same amount. Hours, weeks, months, and years are typical time scales of ecological patterns and processes. Millimeters, meters, and kilometers are typical ecological spatial scales.

Temporal variation

We perceive temporal variation as our environment changes over time, for example, with the alternation of day and night and the seasonal progression of temperature and precipitation. Superimposed on these more or less predictable cycles are irregular and unpredictable variations. Winter weather is generally cold and wet, but the weather at any particular time cannot be predicted much

in advance; it varies perceptibly over intervals of a few hours or days with the passage of cold fronts and other atmospheric phenomena. Some irregularities in conditions, such as a string of especially wet or dry years, occur over longer periods. Other events of great local ecological consequence, such as fires and tornadoes, strike a particular place only at very long intervals.

In general, the more extreme the condition, the less frequent it is. However, both the severity and the frequency of events are relative measures, depending on the organism that experiences them. Fire may touch an individual tree many times but skip dozens of generations of an insect population. How organisms and populations respond to change in their environments depends on how often it occurs.

Scales of temporal variation may be determined by intrinsic properties of systems as well as by variation in external factors. For example, in pine woodlands, the probability of a destructive fire increases with time since the last such event. As litter and other fuels accumulate, they produce a characteristic fire cycle for a particular habitat. Similarly, the rapid spread of contagious disease through a population often depends on the accumulation of young, non-immunized individuals following the last epidemic (**Figure 1.15**).

Spatial variation

The environment also differs from place to place. Variations in climate, topography, and soil type cause large-scale heterogeneity (across meters to hundreds of kilometers; see the variation in water temperature in the western Atlantic Ocean illustrated in Figure 1.4). At smaller scales, heterogeneity is generated by the structures of plants, the activities of animals, and the content of soils. A particular scale of spatial variation may be important to one animal and not to another. The difference between the top and the underside of a leaf is important to an aphid, but not to a moose, which happily eats the whole leaf, aphid and all.

As an individual moves through an environment that varies in space, it encounters environmental variation as a sequence in time. In other words, a moving individual perceives spatial variation as temporal variation. The faster an individual moves, and the smaller the scale of spatial variation, the more quickly it encounters new environmental conditions and the shorter the temporal scale of variation. This applies to plants as well as to animals. Roots growing through soil may encounter new conditions if the scale of spatial variation in soil characteristics is small enough. Wind and animals disperse seeds, which may land in a variety of habitats depending on how far they travel relative to the scale of spatial variation in the habitat.

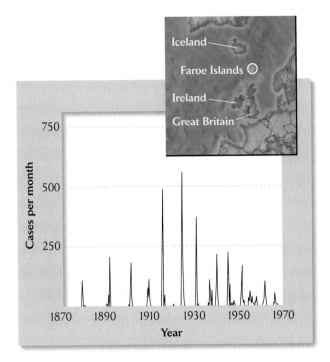

Figure 1.15 More extreme conditions generally occur less frequently than less extreme conditions. This is true even for outbreaks of contagious diseases. The number of cases of whooping cough in the population of the Faroe Islands from 1881 through 1969 indicates that the size of an epidemic outbreak is larger the longer the interval since the previous outbreak. This pattern occurs because the number of previously unexposed, hence susceptible, children increases with time in a population. After C. J. Rhodes et al., *Proceedings of the Royal Society of London* B 264:1639–1646 (1997).

devastation inflicted by hurricanes over periods of days or weeks. In the oceans, at one extreme, small eddy currents may last only a few days; at the other extreme, ocean gyres (circulating currents encompassing entire ocean basins) are stable over millennia.

Compared with marine and, especially, atmospheric phenomena, variations in landforms have very long temporal scales at a particular spatial scale. The reason is simple: landforms are determined by underlying topography and geology, which are transformed at a snail's pace by such processes as mountain building, volcanic eruptions, erosion, and even continental drift. In contrast, spatial

Spatial and temporal dimensions in ecology are correlated

With regard to ecologically important phenomena, duration in time usually increases with the size of the area affected (Figure 1.16). For example, tornadoes last only a few minutes and affect small areas compared with the

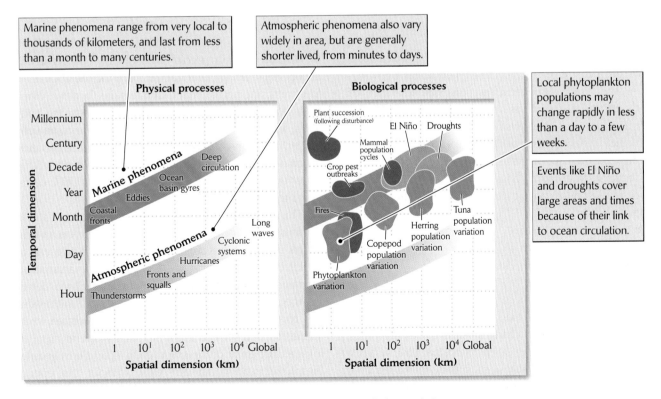

Figure 1.16 Ecological events often show temporal and spatial correlation. Variation in atmospheric and marine systems shows that an event's duration in time usually increases with the size of the area affected. After J. H. Steele, *J. Theor. Biol.* 153:425–436 (1991).

heterogeneity in the open ocean results from physical processes in water, which are obviously more changeable than those on the land. Because air is even more fluid than water, atmospheric processes have very short periods at a given spatial scale, as shown in Figure 1.16.

A principle related to the space–time correlation states that the frequency of a phenomenon is generally inversely related to its spatial dimension or local severity. Thus tornadoes and hurricanes occur at longer intervals, on average, than do winter storms. The frequency of forest fires or brush fires is inversely related to the area they burn. In addition, such disturbances create patches of habitat in various stages of ecological development, or succession, thereby contributing to the spatial heterogeneity of the environment on many scales of time and space.

Ecological systems are governed by general physical and biological principles

We can cope more easily with the complexity of ecological systems when we understand that they are all governed by a small number of basic principles. A brief consideration of four of these principles will illustrate the underlying unity of ecology.

Ecological systems are physical entities

Life builds upon the physical properties and chemical reactions of matter. The diffusion of oxygen across body surfaces, the rates of chemical reactions, the resistance of vessels to the flow of fluids, and the transmission of nerve impulses all obey the physical laws of thermodynamics. Biological systems are powerless to alter these fundamental physical qualities of matter and energy, but, within the broad limits imposed by physical constraints, life can pursue many options, and it has done so with astounding invention.

Ecological systems exist in dynamic steady states

Whether we focus on the organism, the population, the ecosystem, or the entire biosphere, each of these ecological entities continually exchanges matter and energy with its surroundings (▌Figure 1.17). That ecological systems remain more or less unchanged implies that gains and losses are more or less balanced. This is the essence of a dynamic steady state: a system exchanges energy or materials with its surroundings but nonetheless maintains its characteristics unchanged. A warm-blooded animal con-

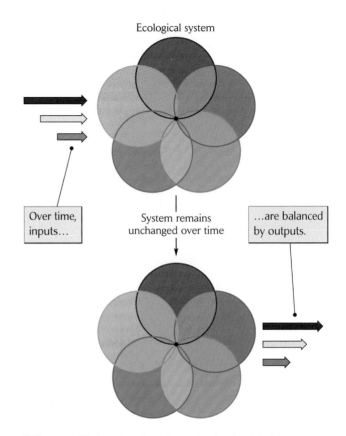

▌**Figure 1.17 A system in a dynamic steady state** has continual inputs and outputs but remains unchanged over time.

tinually loses heat to a cold environment. This loss is balanced, however, by heat gained from the metabolism of foodstuffs, and body temperature remains constant. When gains do not add up to losses for some reason, the body cools. Similarly, the proteins of our bodies are continually broken down and replaced by newly synthesized proteins. Much of the material in the bodies we all carried around a year ago has been replaced, although we still look pretty much the same.

This idea of maintaining a steady state in the face of continuous flux of materials and energy between an ecological system and its surroundings applies to all levels of ecological organization. For the individual, assimilated food and energy must balance their metabolic breakdown. For the population, gains and losses are births and deaths. The diversity of a biological community decreases when species become extinct, and it increases when new species invade the habitat of the community. Ecosystems and the biosphere itself could not exist without the energy received from the sun, yet this gain is balanced by heat energy radiated at infrared wavelengths back out into space. How the steady states of ecological systems are maintained and regulated is one of the most important questions posed by ecologists, one to which we will return frequently throughout this book.

The maintenance of living systems requires expenditure of energy

Because life is so special, being composed of molecules that are rare or nonexistent in the nonliving world, living organisms exist out of equilibrium with the physical environment. What the organism loses to its surroundings, however, is not returned by the environment for free. If it were, life would be the equivalent of a perpetual motion machine. The organism must procure energy or materials to replace its losses. To do this, it must expend energy. Thus, energy lost as heat and motion must be replaced by metabolizing food, which the organism captures and assimilates at a cost. The price of maintaining a living system as a dynamic steady state is energy.

Ecological systems undergo evolutionary change over time

The history of life on earth has shown that the attributes of organisms change over time. Such change is referred to as **evolution.** Although the physical and chemical properties of matter and energy are immutable, what living systems do with matter and energy is as variable as all the forms of organisms that have existed in the past, are alive today, or might evolve in the future. The structures and functions of organisms are products of evolutionary change in populations in response to the features of the environment with which individual organisms must contend. Such features include both the physical conditions that prevail and the various other kinds of organisms with which each population interacts. For example, animals that have visually hunting predators are often colored in such a way that they blend in with their backgrounds and escape notice (**Figure 1.18**). Many plants that grow in hot, dry climates have thick, waxy cuticles that reduce the loss of water by evaporation across their leaf surfaces. These attributes of structure and function that suit an organism to the conditions of its environment are called **adaptations.**

The close correspondence between organism and environment is no accident. It derives from a fundamental and unique principle of biological systems: **natural selection.** Only those individuals that are well suited to their environments survive and produce offspring. The inherited traits that they pass on to their progeny are preserved. Unsuccessful individuals do not survive, or they produce fewer offspring, and their less suitable traits therefore disappear from the population as a whole. Charles Darwin was the first to recognize that this process allowed populations

(a)

(b)

Figure 1.18 Adaptations to environmental conditions help organisms survive. The cryptic colorations of (a) a Costa Rican mantid and (b) a North American tree frog blend in with the backgrounds of these animals and reduce their risk of being seen by visually hunting predators. Photo (a) by Michael Fogden/DRK; photo (b) by David Northcott/DRK.

to respond, over periods of many generations, to changes in their environments. A wonderful thing about natural selection and evolution is that as each species changes, new possibilities for further change are opened up for itself and for other species with which it interacts. In this way, the complexity of ecological communities and ecosystems builds upon, and is fostered by, existing complexity. An important goal of ecology as a science is to understand how ecological systems came into being and how they function in their environmental settings.

Ecologists study the natural world by observation and experimentation

Like other scientists, ecologists apply many methods to learn about nature. Most of these methods reflect three facets of scientific investigation: (1) observation and description, (2) development of hypotheses or explanations, and (3) testing of these hypotheses, often with experiments.

Most research programs begin with a set of facts about nature that invite explanation. Usually these facts describe a consistent pattern. For example, measurements of rainfall and plant growth over several years might reveal a correlation between precipitation and plant production. To cite another example, exploration during the nineteenth century established that the number of animal and plant species in tropical regions greatly exceeds that in temperate regions. Recognition of this relationship between biodiversity and latitude grew out of comparisons of the accumulated observations of many scientists until it became confirmed as a general pattern. Like the relationship between rainfall and plant growth, this pattern invites explanation. Because many explanations are plausible, it is necessary to conduct experiments or other kinds of investigations to determine which explanations best account for the facts.

Hypotheses are ideas about how a system works—that is, they are explanations. If correct, a hypothesis may help us to understand the cause of an observed pattern. Suppose we observe that male frogs sing on warm nights after periods of rain. If a reasonable amount of observation produces few exceptions to this pattern, it may be regarded as a generalization that enables us to predict the behavior of frogs from the weather. Having established the existence of such a pattern, we may wish to understand it better. For example, we may wish to explain how a frog responds to temperature and rainfall; we may also wish to explain why a frog responds the way it does. The "how"

part of this particular phenomenon involves details of sensory perception, the interplay between environmental stimuli and hormonal status, and neuromotor effectors—in other words, it involves physiological processes. The "why" question deals with the costs and benefits of the behavior to the individual; it is more ecological and evolutionary in nature. If we suspect that male frogs sing in order to attract females, we may entertain the idea that males sing after rains because that is when females look for mates. If frogs chorused at other times, they might attract few mates (low benefit) but still expose themselves to predation or other risks (high cost) in the attempt. We now have generated a number of hypotheses about how frogs behave: (1) singing by the males attracts females and leads to mating; (2) females actively search for males only after rains; (3) singing imposes a cost, which compels males to save their singing for times when it will do them the most good.

If we are to convince ourselves that a hypothesis is valid, we must put it to the test. Only rarely can a particular idea be proved beyond a doubt, but our confidence increases the more we explore the implications of a hypothesis and find it to be consistent with the facts. If our second hypothesis about frog singing were true, we would expect to observe more receptive females on nights following rains than on nights following fair weather. This is a **prediction,** which is a statement that follows logically from a hypothesis. If observations of the activity of females confirm this prediction, then the hypothesis is strengthened; if not, the hypothesis is weakened, or perhaps it may be rejected altogether.

The strongest tests of hypotheses are often the outcomes of **experiments,** in which one or a small number of variables are manipulated independently of others to reveal their particular effects. In the frog example, one test of our second hypothesis would be to determine whether mating success is lower when a male sings after fair weather than it is when a male sings after rains. Unfortunately, males normally don't sing unless it rains. Perhaps, by some suitable manipulation, we could trick a male into singing on the "wrong" night. This would be a good experiment if we could make frogs sing without altering other aspects of their behavior. To eliminate all variables except singing, we would have to ascertain that silent frogs tested on the same night didn't also attract females. Such a treatment, which reproduces all aspects of an experiment except the variable of interest, is called an experimental **control.** Another experiment that comes to mind would be to record the songs of male frogs on a tape, play them through speakers on different nights, and tally the numbers of females that are attracted to the calls after periods of rain versus fair weather.

Tests of hypotheses generate new information that often initiates additional rounds of hypothesis formation and testing. For example, if we find that female frogs are more active after rainy weather, we have discovered a new pattern that invites explanation. In this way scientific discovery builds upon itself, generating a rich understanding of the workings of natural systems.

ECOLOGISTS IN THE FIELD

An experimental test of a hypothesis

To illustrate how ecologists use experiments to test a hypothesis, we'll dissect a field study into its basic components. This study was conducted by Robert Marquis and Chris Whelan of the University of Missouri at St. Louis.

Observation: In spite of a large variety of potential herbivores, only a small proportion of the leaf area of a forest is consumed during the growing season.

Observation: Birds eat insects.

Hypothesis: Bird predation on insect herbivores reduces the amount of leaf area consumed.

Experimental test: Exclude birds from foliage by constructing bird-proof cages (▌ Figure 1.19) that allow insects to pass freely.

Control: Obtain data for trees without exclusion cages paired with experimental trees to account for spatial and temporal variation in insect or bird populations.

Control for experimental effects: Because the exclusion cages might have other effects on the foliage (shading, for example), enclose some trees within incomplete cages that allow birds access to the foliage.

▌ **Figure 1.19 Experiments are the strongest tests of hypotheses.** A cage has been placed around a white oak sapling to exclude bird predators that would otherwise consume caterpillars. Courtesy of C. Whelan.

Marquis and Whelan found that where birds were excluded, the numbers of insects recorded on foliage increased by 70%, and the percentage of leaf area missing at the end of the growing season increased from 22% to 35%. These findings led them to conclude that avian predators reduce the abundance of insect herbivores as well as the damage caused by herbivores to trees. The findings also led to another question: Will the decreases in bird populations caused by fragmentation of forests in the eastern United States and elsewhere result in increased insect damage to forests?

Although the ways of acquiring scientific knowledge appear to be straightforward, many pitfalls exist. For example, a correlation between variables does not imply a causal relationship; the mechanism of causation must be determined independently by suitable investigation. In addition, many hypotheses cannot be tested by experimental methods because the scales of the relevant processes are too large. These limitations become particularly critical with patterns that have evolved over long periods and with systems such as entire populations or ecosystems that are too large for practical manipulation.

Different hypotheses may explain a particular observation equally well, and one must make predictions that distinguish among the alternatives. The observation that biodiversity decreases at higher latitudes has stimulated many explanations. As one travels north from the equator, average temperature and precipitation decrease, light intensity and biological production decrease, and seasonality and other environmental variation increases. Each of these factors could interact with biological systems in ways that could affect the number of species that can coexist in a locality, and dozens of hypotheses based on these factors have been proposed. Isolating the effect of each factor has proved difficult because each tends to vary in parallel with the others.

Faced with these difficulties, ecologists have resorted to several alternative approaches to hypothesis testing. One of these is the **microcosm** experiment, which attempts to replicate the essential features of a system in a simplified laboratory or field setting (▌ Figure 1.20). Thus, an aquarium with five animal species may behave like the more complex natural system in a pond, or even like ecological systems more generally; if so, experimental manipulations of the microcosm may yield results that can be generalized to the larger system. The hypothesis that diversity decreases as environmental variation increases might be approached in a microcosm experiment by determining whether variations in temperature, light, acidity, or nutrient resources cause species to disappear from the

Figure 1.20 Microcosm experiments are designed to replicate the essential features of an ecological system. Communities of freshwater invertebrates are housed in cattle tanks at the Kellogg Biological Station of Michigan State University. Numerous tanks are used for replicates of different experimental treatments. Photo by R. E. Ricklefs.

system. Of course, it is a long stretch to generalize from an aquarium to a "real" ecological system, but if variation consistently resulted in a loss of species in a variety of microcosms, the hypothesis would be strengthened.

Another approach is to construct a mathematical model of a complex system, in which the investigator represents the system as a set of equations. These equations portray our understanding of how the system works in the sense that they describe the relationship of each of the system's components to other components and to outside influences. A mathematical model is a hypothesis; it provides an explanation of the observed structure and functioning of a system. Models can be tested by comparing the predictions they yield with what is actually observed. Most models make predictions about attributes of the system that have not been measured or about the response of the system to perturbation. Whether these predictions are consistent with observations determines whether the hypothesis on which they are based is supported or rejected. For example, detailed models have been developed to describe the spread of communicable diseases (see Figure 1.15). These models include such factors as the proportion of the population that is susceptible, exposed, infectious, and recovered (and thus resistant because of acquired immunity) as well as rates of transmission and the virulence of the disease organism.

At a larger scale, ecologists have created global carbon balance models to investigate the effect of the burning of fossil fuels on the carbon dioxide content of the atmosphere. Understanding this relationship is critically important to managing the environment of the earth. Global carbon balance models include, among other factors, equations for the uptake of carbon dioxide by plants and the solution of carbon dioxide in the oceans. However, the output of early versions of these models failed to match the observed data, specifically by overestimating the annual increase in atmospheric carbon dioxide. The real world evidently contains carbon dioxide "sinks" that remove the gas from the atmosphere but were not represented in the model. This discrepancy has caused ecosystem modelers to look more closely at processes such as the regeneration of forests and the movement of carbon dioxide across the air–water interface. Such processes contribute to more refined descriptions of the functioning of the biosphere—models that will provide more accurate predictions of the future of atmospheric change.

Humans are a prominent part of the biosphere

Why do we do all this? The wonders of the natural world summon our natural curiosity about life and our desire to understand our surroundings. For many of us, our curiosity about nature and the challenges of taking a scientific approach to its study are reason enough. In addition, however, an understanding of nature is becoming more and more urgent as the growing human population stresses the capacity of natural systems to maintain their structure and functioning. Environments that human activities either dominate or have produced—including our urban and suburban living places, our agricultural breadbaskets, our recreation areas, tree farms, and fisheries—are also ecological systems. The welfare of humanity depends on maintaining the functioning of these systems, whether they are natural or artificial. Virtually all of the earth's surface is, or soon will be, strongly influenced by people, if not fully under their control. Already, humans usurp more than 40% of the biological productivity of the biosphere. We cannot take this responsibility lightly.

The human population has recently passed the 6 billion mark, and it consumes energy and resources, and produces wastes, far in excess of needs dictated by biological metabolism. This has caused two related problems of global dimensions. The first is the impact of human activity on natural systems, including the disruption of ecological processes and the extermination of species. The second is the steady deterioration of humankind's own environment as we push the limits of what ecological systems can sustain. Understanding ecological principles is a necessary step in dealing with these problems. Two examples drive this point home.

Introduction of the Nile perch into Lake Victoria

During the 1950s and early 1960s, the Nile perch was introduced into Lake Victoria, a large, shallow lake straddling the equator in East Africa. This was done with the well-intentioned purpose of providing additional food for the people living in the area and additional income from export of the surplus catch (▮Figure 1.21). However, because basic ecological principles were ignored, the introduction ended up destroying most of the lake's traditional fishery. Until the introduction of the Nile perch, Lake Victoria supported a sustainable catch of a variety of local fishes, mostly of species belonging to the family Cichlidae, which feed primarily on detritus, plants, and small animals. Nile perch are very large, and they eat vast quantities of other fish: the smaller cichlids, in this instance. However, because energy is lost with each step in the food chain, predatory fish cannot be harvested at as high a rate as their prey species. Furthermore, the perch was alien to Lake Victoria, and the local cichlids had no innate behaviors to help them escape predation. Inevitably, the perch annihilated the cichlid populations, driving many unique species to extinction, destroying the native fishery, and severely reducing its own food supply. Consequently, the perch's voracious habits among defenseless prey brought about its own demise as an exploitable fish species and completely changed the Lake Victoria ecosystem. Introduction of the Nile perch had secondary consequences for the terrestrial ecosystems surrounding the lake as well. The flesh of the perch is oily and must be preserved by smoking rather than sun-drying, so local forests have been cut rapidly for firewood.

To be sure, the native fishery was already precariously close to being overexploited as a result of the burgeoning local human population and the use of advanced, nontraditional fishing technologies. However, an appropriate solution to these problems would have been better management of the cichlids and development of food sources other than fish, not the introduction of an efficient predator upon them.

The California sea otter

Half a world away from Lake Victoria, efforts to save the sea otter along the coast of California illustrate the intricate intermingling of ecology and other human concerns (▮Figure 1.22). The sea otter was once widely distributed around the North Pacific Rim from Japan to Baja California. In the 1700s and 1800s, intense hunting for otter pelts reduced the population to near extinction. Predictably, the fur industry collapsed as it overexploited its economic base. Subsequent protection enabled the California sea otter population to increase to several thousand individuals by the 1990s, well above the danger level. The sea otter's success, alas, has irked some California fishers, who claim that the otters—which do not need commercial fishing licenses—drastically reduce stocks of valuable abalone, sea urchins, and spiny lobsters. Matters deteriorated at one point to the marine equivalent of a range war between the fishing industry and conservationists, with the otter caught in the line of fire, often fatally.

Ironically, the otters benefit a different commercial marine enterprise, the harvesting of kelps, which are large seaweeds used in making fertilizer. Kelps grow in shallow water in stands called kelp forests, which provide refuge

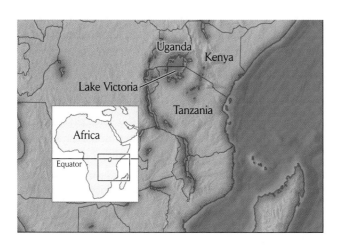

▮Figure 1.21 **The introduction of a new species to an ecosystem can have drastic effects.** Nile perch were introduced into Lake Victoria in the 1950s to improve the local fishery, but have driven many endemic native fishes to extinction and completely changed the ecosystem of the lake. Courtesy of Tim Baily/The African Angler and Joe Bucher Tackle Company.

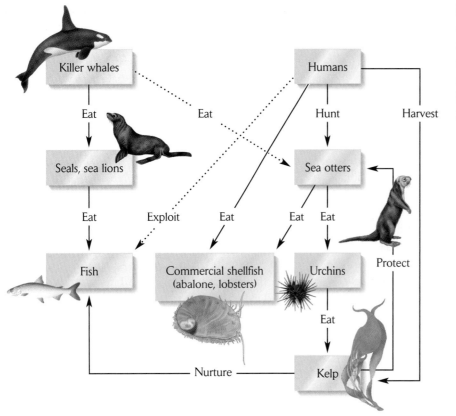

▮ **Figure 1.22 Human activities have complex effects on ecosystems.** Several components of the otter–urchin–kelp ecosystem are altered when humans reduce otter populations by hunting them. After J. A. Estes et al., *Science* 282:473–476 (1998).

and feeding grounds for larval fish (▮Figure 1.23). Kelps are also grazed by sea urchins, which, when abundant, can denude an area. The sea otter is a principal predator of sea urchins. When the expanding otter population spread into new areas, urchin populations were controlled, allowing kelp forests to regrow.

Elsewhere, where other factors are at work, the otter population is declining. In a report published in 1998 in the journal *Science,* J. A. Estes and his colleagues at the University of California at Santa Cruz showed that populations of otters in the vicinity of the Aleutian Islands, Alaska, have declined precipitately during the 1990s. The reason? Killer whales, which previously had not preyed on otters, have been coming close to shore and taking large numbers of otters. A predictable result of the change in otter populations has been a dramatic increase in urchins and decimation of kelps in affected areas. Why have killer whales changed their predatory behavior recently? Estes points out that populations of the principal prey of killer whales—seals and sea lions—collapsed during the same period, perhaps inducing the whales to seek alternative food sources. Why did the seals and sea lions decline? One can only speculate at this point. However, intense human fisheries have reduced fish stocks exploited by the seals to levels low enough to seriously affect their populations.

Human impacts on the natural world have increasingly become a focus of ecology

Although the plight of endangered species may arouse us emotionally, ecologists increasingly realize that the only effective means of preserving and using natural resources is through the conservation of entire ecological systems and the management of broad-scale ecological processes. Individual species, including the ones that humans rely on for food and other products, themselves depend on the maintenance of ecological support systems. We have already seen how predators such as the Nile perch and the sea otter may assume key roles in the functioning of natural systems, for better or worse, depending on the circumstances. By manipulating populations of such important species, humans can change the composition of biological communities and influence the functioning of entire ecosystems. When our interference in ecological systems is localized and focused on one or a few species, it is possible to manage the situation once the basic problem is understood. Unfortunately, much of our influence on the environment results from multiple, widespread impacts

Figure 1.23 The integrity of the kelp forest habitat depends on the presence of sea otters. The kelp forest provides feeding grounds and refuge for many species of fish and invertebrates. The otters eat sea urchins that would otherwise destroy young kelp. Photo by E. Hanauer, courtesy of Paul Dayton.

that are more difficult for scientists to characterize and for legislative and regulatory bodies to control. For this reason, a sound scientific understanding of environmental problems is a necessary prerequisite to action.

The daily newspapers are filled with environmental problems: disappearing tropical forests, the ozone hole, depleted fish stocks, global warming. Wars create staggering environmental catastrophe as well as human tragedy. But there are success stories as well. Many developed countries, including the United States and most European nations, have made great strides in cleaning up their rivers, lakes, and air. Fish are once again migrating up the major rivers of North America and Europe to spawn. Acid rain has decreased, thanks to changes in the combustion of fossil fuels. The release of chlorofluorocarbons, which damage the ozone layer that shields the earth's surface from ultraviolet radiation, has decreased dramatically. The possibility of global warming caused by increased atmospheric carbon dioxide has set off an international research effort and provoked global concern. Conservation efforts, including the breeding of endangered species in captivity, have saved some animals, such as the California condor, from certain extinction. They have also heightened public awareness of environmental issues, and they have sometimes sparked public controversy. However, without public concern and understanding, political action is impossible.

Particularly encouraging is the growing level of international cooperation exemplified in such organizations as the International Union for the Conservation of Nature (IUCN) and the World Wildlife Fund (WWF). In addition, the nations of the world have reached several important agreements for the protection of wildlife and nature. One of these is the Convention on International Trade in Endangered Species (CITES), which makes it illegal to transport endangered species or their products (hides, feathers, and ivory, for example) across international boundaries, depriving poachers of markets. A second important agreement is the Rio Convention on Biodiversity, which recognizes the proprietary interest of countries in their own biological heritage and guarantees fees and royalties for the exploitation of local plants and animals for uses such as pharmaceutical products.

These successes have been based on increased scientific knowledge of the natural world. Understanding ecology will not by itself solve our environmental problems in all their political, economic, and social dimensions. However, as we contemplate the need for global management of natural systems, our success will hinge on our understanding of their structure and functioning—an understanding that depends on knowing the principles of ecology.

 Summary

*

1. Ecology is the scientific study of the natural environment and of the relationships of organisms to one another and to their surroundings.

2. Organism, population, community, ecosystem, and biosphere represent levels of organization of ecological structure and functioning. They form a hierarchy of progressively more complex entities.

3. Ecologists use several different approaches to study nature, focusing on the interactions of organisms with their environments, the resulting transformations of energy and chemical elements in ecosystems and the biosphere, the dynamics of individual populations, and the interactions of populations within ecological communities.

4. Different kinds of organisms play different roles in the functioning of ecosystems. Plants and algae fix the energy of sunlight; animals and protozoa consume biological forms of energy. Fungi are able to penetrate soil and dead plant material and so play an important role in breaking down biological materials and regenerating nutrients in the ecosystem. Bacteria are biochemical specialists, able to accomplish such transformations as the biological assimila-

* Sword ferns have been a prominent part of the earth's vegetation for millions of years.

tion of nitrogen and the use of hydrogen sulfide as an energy source, both of which are essential components of ecosystem function.

5. An individual's habitat is the place in which that individual lives. The habitat concept emphasizes the structure of the environment. An individual's niche is the ranges of conditions that it can tolerate and the ways of life that it can pursue—that is, its functional role in the natural system.

6. Ecological processes and structures have characteristic scales of time and space. In general, scales of patterns and processes in time and in space are correlated.

7. The variety and complexity of ecological systems are understandable in terms of a small number of basic ecological principles. Among these is the idea that ecological systems are physical entities and function within the physical and chemical constraints governing energy transformations. Furthermore, all ecological systems exchange materials and energy with their surroundings. When inputs and outputs are balanced, the system is said to be in a dynamic steady state.

8. All ecological systems are subject to evolutionary change, which results from the differential survival and reproduction, within populations, of individuals that exhibit different genetically determined traits. As a result of natural selection, organisms exhibit adaptations of structure and function that suit them to the conditions of their environments.

9. Ecologists employ a variety of techniques to study natural systems. The most important of these are observation, development of hypotheses to explain observations, and testing of hypotheses to confirm the predictions they generate. Experiments are an important tool in testing hypotheses. When natural systems do not lend themselves readily to experimentation, ecologists may work with microcosms or mathematical models of systems.

10. Humans play a dominant role in the functioning of the biosphere, and human activities have created an environmental crisis of global proportions. Solving our acute environmental problems will require the intelligent application of general principles of ecology within the framework of political, economic, and social action.

PRACTICING ECOLOGY

CHECK YOUR KNOWLEDGE

Who, How, and Why

Who practices ecology? Of course, the simple answer is ecologists. Ecologists are people who use scientific methods to learn how organisms interact with their environments. The information that ecologists gather while practicing ecology is important for understanding how natural systems work—and fail to work—under various stresses. Human activities are increasingly threatening the ability of our planet to support life, and ecologists are at the forefront of research intended to understand what these threats mean for the future of life on earth.

How do you practice ecology? Practicing ecology requires an ability to recognize patterns and processes in nature and an ability to design and perform experiments (often in remote settings under harsh conditions). Practicing ecology also means training young students to become the ecologists of the future. And it requires the communication of results to other scientists working on similar problems. Increasingly, ecologists are finding that practicing ecology also means communicating the implications of their results to the media and elected officials. Throughout this book, "Practicing Ecology" will examine the how, what, and why of the hands-on research that ecologists are conducting at the start of a new millennium.

Why does one become an ecologist? If you conducted a survey, the reasons would probably vary widely. Some ecologists probably had their interest sparked as children during an annual family camping trip. Others may have been inspired by meeting a field biologist, or were influenced by a professor in college. Many would say that they were motivated to study ecology by reading Rachel Carson's *Silent Spring* or some other book with a powerful message relating ecology to environmental problems and solutions.

CHECK YOUR KNOWLEDGE

1. Survey the ecologists in your college or university, or at a local field biology station. You might consider seeking them out via their Web pages. Where did your professors go to college and graduate school? If you contact them in person, you might ask them what specific influential observation, book, or inspiring person motivated them to practice ecology.

★ MORE ON THE WEB **2.** Go to *Practicing Ecology on the Web* through *The Economy of Nature* Web site at *http://www.whfreeman.com/ricklefs* and read about how and why the author of the text and the authors of its supplements became ecologists.

How does society benefit from the study of ecology? As we have noted, humans are a part of the earth's ecosys-

---

* "More on the Web" icons appear throughout the text. These icons indicate that the book's Web site, *www.whfreeman.com/ricklefs*, contains additional detail on this topic.

tems, and our activities are taking an increasing toll on the earth's diversity of life. This diversity underlies the life-support systems of Planet Earth. Therefore, it is incumbent upon humans to understand the effects they are having on ecosystems through population growth, pollution, habitat destruction, and introductions of invasive species (among other things).

CHECK YOUR KNOWLEDGE

MORE ON THE WEB

3. Go to *http://www.whfreeman.com/ricklefs* and read a press release from the World Wildlife Fund about global warming and its impact on human health. Then read the following articles from three news sources on the same topic. How do the news articles differ in their coverage and detail? Do reporters articulate what the results of the study mean in terms of economic, social, or political policy?

 Suggested Readings

Barel, C. D. N., et al. 1985. Destruction of fisheries in Africa's lakes. *Nature* 315:19–20. (Introduction of the Nile perch into Lake Victoria.)

Bartholomew, G. A. 1986. The role of natural history in contemporary biology. *BioScience* 36:324–329. (An influential biologist's personal approach to scientific investigation.)

Booth, W. 1988. Reintroducing a political animal. *Science* 241:156–158. (The ecological role of sea otters in kelp communities.)

Cohn, J. P. 1998. Understanding sea otters. *BioScience* 48(3):151–155.

Cronon, W. (ed.). 1996. *Uncommon Ground: Rethinking the Human Place in Nature.* W. W. Norton, New York.

Estes, J. A., M. T. Tinker, T. M. Williams, and D. F. Doak. 1998. Killer whale predation on sea otters linking oceanic and nearshore systems. *Science* 282:473–476.

Franklin, J. F., C. S. Bledsoe, and J. T. Callahan. 1990. Contributions of the long-term ecological research program. *BioScience* 40:509–523.

Goldschmidt, T., F. Witte, and J. Wanink. 1993. Cascading effects of the introduced Nile perch on the detritivorous/phytoplanktivorous species in the sublittoral areas of Lake Victoria. *Conservation Biology* 7:686–700.

Harley, J. L. 1972. Fungi in ecosystems. *Journal of Animal Ecology* 41:1–16.

Leibold, M. A. 1995. The niche concept revisited—mechanistic models and community context. *Ecology* 76:1371–1382.

Margulis, L., D. Chase, and R. Guerrero. 1986. Microbial communities. *BioScience* 36:160–170.

Margulis, L., and K. V. Schwartz. 1998. *Five Kingdoms: An Illustrated Guide to the Phyla of Life on Earth,* 3d ed. W. H. Freeman, New York.

Marquis, R. J., and C. Whelan. 1994. Insectivorous birds increase growth of white oak through consumption of leaf-chewing insects. *Ecology* 75:2007–2014.

McIntosh, R. P. 1985. *The Background of Ecology: Concept and Theory.* Cambridge University Press, New York. (The most complete history of the development of the science of ecology.)

Nichols, F. H., J. E. Cloern, S. N. Luoma, and D. H. Peterson. 1986. The modification of an estuary. *Science* 231:567–573. (An illustration of how the multiplication of technology and population has transformed San Francisco Bay into an ecological crisis.)

Price, P. W. 1996. Empirical research and factually based theory: What are their roles in entomology? *American Entomologist* 42(2):209–214. (Thoughts about the study of ecological systems by an eminent insect ecologist.)

Sinclair, A. R. E., and J. M. Frywell. 1985. The Sahel of Africa: Ecology of a disaster. *Canadian Journal of Zoology* 63:987–994. (Intense grazing and collection of firewood has caused deforestation and habitat degradation over large areas of sub-Saharan Africa.)

Urban, D. L., R. V. O'Neill, and H. H. Shugart, Jr. 1987. Landscape ecology. *BioScience* 37:119–127. (The scale of habitat patchiness within an area influences population processes and the distributions of organisms.)

CHAPTER 2

The Physical Environment

Water has many properties favorable for the maintenance of life

All natural waters contain dissolved substances

The concentration of hydrogen ions profoundly affects ecological systems

Carbon and oxygen are intimately involved in biological energy transformations

The availability of inorganic nutrients influences the abundance of life

Light is the primary source of energy for the biosphere

The thermal environment provides several avenues of heat gain and loss

Organisms must cope with temperature extremes

Organisms use many physical stimuli to sense the environment

Those of you who saw the movie *Titanic* will recall the terrible loss of human life caused by freezing in the cold arctic waters. One might wonder how blood and body tissues could freeze solid in perfectly liquid water. The answer is that dissolved substances depress the freezing temperature of water and other liquids. While pure water freezes at 0°C, seawater, which contains about 3.5% dissolved salts, freezes at –1.9°C, or almost 2°C colder. The blood and body tissues of most vertebrates, including humans, contain less than half the salt content of seawater, and thus freeze at a higher temperature than the freezing point of the ocean. This was a terrible problem for victims of the *Titanic* disaster. It is also a problem for fish living in polar seas.

Two questions come to mind: First, why don't polar fish have high salt levels in their blood and tissues? Second, how can these fish survive at such low temperatures? Polar fish do not use salts to keep their body fluids from freezing because salts interfere with many biochemical processes. Maintaining a low-salt internal environment allows more rapid and efficient metabolism and movement. Instead, antarctic fish have circumvented their resulting susceptibility to freezing by raising their blood and tissue levels of such compounds as glycerol–common drugstore glycerin–which lower the freezing temperature of their body fluids, but do not severely disrupt functioning. As we look at these and similar questions in this chapter, we will see that, although physical properties of the environment and of biological materials constrain life, they also provide solutions to many of its problems.

We often speak of the living and the nonliving as opposites. But although we can easily distinguish these two great realms of the natural world, they do not exist in isolation from each other. Life depends on the physical world. Living beings also affect the physical world: soils, the atmosphere, lakes and oceans, and many sedimentary rocks owe their properties in part to the activities of plants and animals.

Although distinct from physical systems, life forms nonetheless function within limits set by physical laws. The physical world provides the context for life, but also constrains its expression. Biological systems must use energy to counteract the physical forces of gravity, heat flow, diffusion, and chemical reaction. A bird in flight constantly expends energy to maintain itself aloft against the pull of gravity. Therefore, life exists out of equilibrium with the physical world.

The ability to act against external physical forces distinguishes the living from the nonliving. A bird in flight supremely expresses this quality, but plants also perform work when they absorb soil minerals into their roots and synthesize the highly complex carbohydrates and proteins that make up their structure. Like internal combustion engines, organisms transform energy to perform work. An automobile engine burns gasoline chemically, and it transmits power from the cylinder to the tires mechanically. When an organism metabolizes carbohydrates to provide the energy to move its appendages, it follows related chemical and mechanical principles.

Above all, unlike physical systems, living organisms have a purposeful existence. Their structures, physiology, and behavior are directed toward procuring energy and resources and producing offspring. Certainly life is constrained by physics and chemistry, just as architecture is constrained by the properties of building materials. However, as in biological systems, the purpose of the design of a building is unrelated to, and transcends, the qualities of bricks and mortar.

In the final analysis, life is a special part of the physical world, but it exists in a state of constant tension with its physical surroundings. Organisms ultimately receive their energy from sunlight and their nutrients from the soil and water, and they must tolerate extremes of temperature, moisture, salinity, and the other physical factors of their surroundings. The heat and dryness of deserts exclude most species, just as the bitter cold of polar regions discourages all but the most hardy. But we need not search so far as such extreme conditions for evidence of the tension between the physical and biological realms. The form and function of all plants and animals have evolved partly in response to conditions prevailing in the physical world.

In this chapter, we shall explore those attributes of the physical environment that are most consequential for life. Because life processes take place in an aqueous environment, and because water makes up the largest part of all organisms, water seems a logical place to start.

Water has many properties favorable for the maintenance of life

Water is abundant over most of the earth's surface, and within the temperature range usually encountered there, it is liquid. Water also is a powerful solvent. Consequently, water is an excellent medium for the chemical processes of living systems. It is hard to imagine life having any other basis than water. No other common substance at the surface of the earth is liquid, and this property is necessary for life as we know it. Movement by living organisms depends on the fluidity of water. The high concentrations of molecules necessary for rapid chemical reactions depend on the density of water. Try to imagine life based on a rigid solid or a thin gas.

Thermal properties of water

Water stays liquid over a broad range of temperatures because it resists changes in temperature. In addition, water conducts heat rapidly, which tends to spread heat evenly throughout a body of water. Thus, the temperature of water changes slowly, even when heat is removed or added rapidly, as can happen at the air–water interface or at an organism's surface. Water also resists change of state between solid (ice), liquid, and gaseous (water vapor) phases. Over 500 times as much energy must be added to evaporate a quantity of water as to raise its temperature by 1°C! Freezing requires the removal of 80 times as much heat as that needed to lower the temperature of the same quantity of water by 1°C. This property helps to keep large bodies of water from freezing solid during winter.

Another curious, but fortunate, thermal property of water is that, whereas most substances become denser at colder temperatures, water becomes less dense as it cools below 4°C. Water also expands and becomes even less dense upon freezing. Consequently, ice floats (■ Figure 2.1), which not only makes ice skating possible, but also prevents the bottoms of lakes and oceans from freezing and enables aquatic plants and animals to find refuge there in winter.

Figure 2.1 Water becomes less dense as it freezes, so ice floats. But because the density of ice is 0.92 g per cm³ (not very different from that of liquid water, which is 1 g per cm³), more than 90% of the bulk of this antarctic iceberg lies below the surface. Photo by R. E. Ricklefs.

The buoyancy and viscosity of water

Water both supports and constrains life. Because water is dense (800 times denser than air), it provides considerable support for organisms, which, after all, are themselves mostly water. But water is also viscous, meaning that it resists flow or the movement of a body through it. High density and viscosity tend to retard motion. This is an example of physical properties creating a favorable environment for life, but at the same time placing limits on its development.

Organisms often deal with these limits by taking advantage of the physical properties of natural substances or exploiting physical principles. For example, animals and plants contain bone, proteins, dissolved salts, and other materials that are denser than salt water and much denser than fresh water. These materials would cause aquatic organisms to sink were it not for a variety of mechanisms that reduce their density or retard their rate of sinking. Many species of fish have a gas-filled swim bladder, a mechanism that takes advantage of the low density of gases. The swim bladder's size can be adjusted to make the density of the body equal to that of the surrounding water. Some large kelps have gas-filled bulbs that float their leaves to the sunlit surface waters (see Figure 1.23).

Many of the microscopic, unicellular algae that float in great numbers in the surface waters of lakes and oceans (phytoplankton) contain droplets of oil that compensate for their natural tendency to sink (**Figure 2.2**). Most fats and oils have densities between 0.90 and 0.93 grams per cubic centimeter (90–93% of the density of pure water). Accumulated lipids also enhance the buoyancy of fish and aquatic organisms.

Trimmed-down skeletons, reduced musculature, and perhaps even the decreased salt concentrations of their body fluids further lighten the bodies of aquatic organisms. Unlike bony fishes, sharks and rays lack a swim bladder. But they also lack bony skeletons, and the absence of heavy mineral salts in their skeletons partly compensates for the swim bladder's absence. Calcium carbonate and calcium phosphate, the principal components of mineralized bone, have densities approximately three times that of water. The density of the cartilage skeleton of sharks and rays is much less—close to that of water.

While the high viscosity of water hampers the movement of some marine organisms, others use that property to avoid sinking. Many tiny marine animals have evolved long, filamentous appendages that retard sinking (**Figure 2.3**),

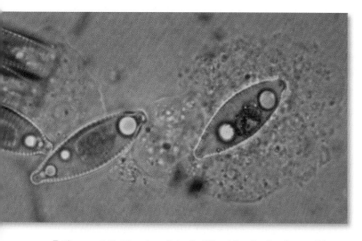

Figure 2.2 The droplet of oil in this algal cell provides buoyancy. Photo by Larry Jon Friesen/Saturdaze.

Figure 2.3 Filamentous and feathery projections from the body of a tropical marine planktonic crustacean retard sinking. Photo by Image Quest 3-D.

Figure 2.4 Streamlined shapes reduce the drag of water. The sleek bodies of barracudas allow them to swim rapidly with a low expenditure of energy. Photo by Larry Jon Friesen/Saturdaze.

just as a parachute slows the fall of a body through air. In contrast, fast-moving aquatic animals have evolved stream-lined shapes that reduce the drag encountered in moving through a dense and viscous medium. Barracuda and other swift fishes of the open ocean closely approach the shape of a body ideally proportioned for moving quickly through water (**Figure 2.4**).

All natural waters contain dissolved substances

Water has an impressive capacity to dissolve various substances, making them accessible to living systems and providing a medium within which they can react to form new compounds. Water is a formidable solvent because water molecules are strongly attracted to many solids. Some solid compounds consist of electrically charged atoms or groups of atoms called **ions.** For example, common table salt, sodium chloride (NaCl), contains positively charged sodium ions (Na^+) and negatively charged chlorine ions (Cl^-) arranged in close proximity in a crystal lattice. In water, however, the charged sodium and chlorine ions are powerfully attracted by water molecules, which themselves have both positive and negative charges. These forces of attraction are stronger than the forces that hold salt crystals together, with the result that they readily separate into their component ions when surrounded by water molecules—another way of saying that the salt dissolves.

The powerful solvent properties of water are responsible for most of the minerals in streams, rivers, lakes, and oceans. Water vapor in the atmosphere condenses to form clouds and, eventually, precipitation (rain, snow, and so on). At this point, the water is nearly pure, except for dissolved atmospheric gases, principally nitrogen, oxygen, and carbon dioxide. Rainwater acquires some minerals from dust particles and droplets of ocean spray in the atmosphere as it falls, and picks up more as it flows over and under the ground. Surface waters, such as streams and

rivers, pick up additional minerals from the substrates through which they flow. The water in most lakes and rivers contains 0.01–0.02% dissolved minerals, which is far less than the average salt concentration of the oceans (3.4% by weight), in which salts and other minerals have accumulated over several billion years.

The minerals dissolved in fresh water and salt water differ in composition as well as in quantity. Seawater abounds in sodium (Na^+) and chlorine (Cl^-) ions and contains significant amounts of magnesium (Mg^{2+}) and sulfate (SO_4^{2-}) ions. Fresh water contains a greater variety of ions, but calcium (Ca^{2+}), bicarbonate (HCO_3^-), and sulfate ions tend to predominate. The concentrations of minerals in fresh water reflect the composition of and solubilities of materials in the rock and soil that the water flows through. Limestone consists primarily of calcium carbonate ($CaCO_3$), and waters flowing through regions with limestone bedrock have high concentrations of calcium and bicarbonate ions (**Figure 2.5**). Granite is composed of minerals that lack calcium and resist dissolution; waters flowing through granitic areas contain few dissolved substances.

The ocean functions like a large still, concentrating minerals as mineral-laden water arrives via streams and rivers and as pure water evaporates from its surface. Here the concentrations of some elements, particularly calcium, reach limits set by the maximum solubility of the compounds they form. In the oceans, calcium readily forms calcium carbonate, which is not very soluble in water. It dissolves only to the extent of 0.014 grams per liter of water, or 14 milligrams per liter (mg per L). Its concentration in the oceans reached this level eons ago, so excess calcium ions washing into the oceans each year precipitate to form limestone sediments (**Figure 2.6**). At the other extreme, the solubility of sodium compounds, such as sodium chloride (360 g per L) and sodium bicarbonate (69 g per L), is very high, far exceeding the concentration of sodium in seawater (10 g per L). Most of the sodium chloride washing into ocean basins remains dissolved, and so its concentration in seawater has increased greatly over geologic time.

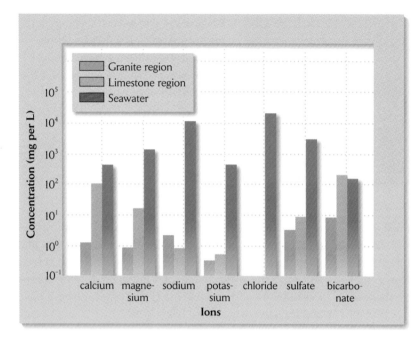

▌Figure 2.5 Minerals have accumulated in the oceans for eons. The average concentrations of minerals in streams, whether in areas having granitic or limestone bedrock, are much lower than in the oceans. The exception is bicarbonate ion, which has limited solubility in water and precipitates out of the water column in the ocean.

Aquatic life requires many mineral elements. Among the most important of these are nitrogen and phosphorus, which are needed to make amino acids, nucleic acids, and other important biological molecules. Nitrogen enters bodies of fresh water in relative abundance in the runoff from surrounding terrestrial ecosystems. Typical concentrations in fresh water are about 0.40 mg per L, mostly in the form of nitrate (NO_3^-) and dissolved organic nitrogen compounds, with smaller amounts of ammonium (NH_4^+).

In contrast, most of the phosphorus in fresh water readily forms chemical complexes with iron and precipitates out of the system, typically leaving about 0.01 mg per L in solution as phosphate (PO_4^{3-}). As a result, phosphorus, rather than nitrogen, usually limits plant growth in freshwater systems. The situation is reversed in the oceans, where phosphorus concentrations are typically higher (0.01–0.1 mg per L) than nitrogen concentrations (often less than 0.01 mg per L).

▌Figure 2.6 The limestone sediments that form many mountains represent calcium carbonate precipitated out of solution in shallow seas. Photo by Larry Jon Friesen/Saturdaze.

The concentration of hydrogen ions profoundly affects ecological systems

Hydrogen ions (H^+) deserve special mention because they are extremely reactive. In high concentrations, they affect the activities of most enzymes and have other, generally negative consequences for life processes. They also play a crucial role in dissolving minerals from rock and soil.

The concentration of hydrogen ions in a solution is referred to as **acidity.** Acidity is commonly measured on a scale of pH, which is the negative of the common logarithm of hydrogen ion concentration, measured in moles per liter (▌Figure 2.7). In pure water at any given time, a small fraction of the water molecules (H_2O) are dissociated into their hydrogen (H^+) and hydroxide (OH^-) ions. The pH of pure water, which is defined as neutral pH, is 7, which means that the concentration of hydrogen ions is 10^{-7} (0.0000001) moles per liter, or one ten-millionth of a gram per kilogram of water. In contrast, strong acids,

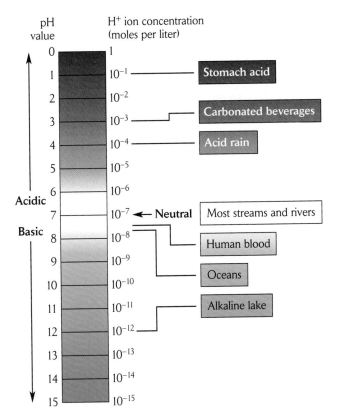

■ Figure 2.7 The pH scale of hydrogen ion concentration extends from 0 (highly acidic) to 15 (highly alkaline).

such as sulfuric acid (H_2SO_4) and hydrochloric acid (HCl), dissociate almost completely when dissolved in water. At high concentrations, such acids can produce pH values approaching 0—that is, 1 mole of H^+ per liter. The acid in your stomach has a pH of 1. Most natural waters contain weak acids, such as carbonic acid (H_2CO_3) and various organic acids, and tend to have pH values close to neutral. Some natural waters are somewhat basic, or alkaline (pH > 7), having an excess of OH^- over H^+. The normal range of pH in natural waters is between 6 and 9, although small ponds and streams in regions with acid rainfall, or which are polluted by sulfuric acid draining out of coal mining wastes, can reach pH values as low as 4.

Hydrogen ions, because of their high reactivity, dissolve minerals from rock and soil. For example, in the presence of hydrogen ions, the calcium carbonate that makes up limestone dissolves readily, according to the chemical equation

$$H^+ + CaCO_3 \rightarrow Ca^{2+} + HCO_3^-.$$

As you can see, this chemical reaction removes hydrogen ions from the water and thus increases its pH. Consequently, water in limestone areas contains abundant calcium ions, which make it "hard," and relatively few hydro-

gen ions, resulting in a somewhat alkaline pH (that is, pH > 7). Where limestone is absent, water contains few calcium ions and is "soft." Also, carbonic acid (H_2CO_3), formed when atmospheric CO_2 dissolves in water, tends to accumulate, and this lowers the pH of the water. Calcium ions are important to life processes, and their presence in high concentrations is vital to organisms, such as snails, that form shells made of calcium carbonate. Indeed, mollusks are less abundant and diverse in streams and lakes with soft water than in those with hard water. Thus, hydrogen ions are essential for making certain nutrients available for life processes. However, this same capacity of hydrogen ions also helps to dissolve highly toxic heavy metals, such as arsenic, cadmium, and mercury, that are detrimental to life.

Hydrogen ions react strongly with living matter as well as with rock and soil. Most organisms keep the pH of their blood and cells close to a neutral pH. We humans maintain the pH of our blood between 7.3 and 7.5. Some microorganisms are more tolerant of high acidity. For example, some photosynthetic cyanobacteria can function at a pH as low as 4. Other kinds of bacteria tolerate acidity down to almost pH 0, but do so by maintaining their internal pH in the range of 6 to 7 at great metabolic cost.

Carbon and oxygen are intimately involved in biological energy transformations

Organisms are composed of carbohydrates, lipids, proteins, and other biological molecules. These compounds contain energy in the form of chemical bonds, primarily between carbon atoms. The energy in these bonds can be released for use by the organism through reactions that break the bonds. In biological systems, one of the most common of these reactions is the **oxidation** of organic forms of carbon (■ Figure 2.8). Oxidation decreases the chemical energy potential of the carbon atom, and the released energy can be used for other biochemical work, such as building cell membranes. The opposite of oxidation is

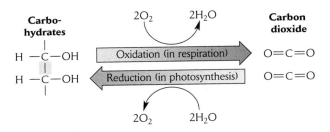

■ Figure 2.8 Oxidation and reduction change the chemical state and energy level of carbon.

reduction. The reduction of carbon increases the energy potential of the atom and allows it to react with other carbon atoms or nitrogen atoms to form organic molecules. Of course, the process of reducing carbon itself requires a source of energy.

Animals and most microorganisms obtain carbon that is already reduced in their food. Such organisms are referred to as **heterotrophs** because they obtain their energy by consuming (hence "troph") other (hence "hetero") organisms. Plants, algae, and many bacteria assimilate oxidized forms of carbon, particularly carbon dioxide, which they reduce chemically using other forms of energy. (The term **assimilation** refers specifically to the incorporation of energy or matter into the tissues of living organisms.) Such organisms are referred to as **autotrophs,** literally "self-feeders." Autotrophs derive the energy they need to reduce carbon from sunlight (photoautotrophs) or, as in the case of some bacteria, from other chemical reduction reactions (chemoautotrophs), as we shall see in a later chapter.

Photosynthesis and respiration

The ultimate source of carbon for making organic molecules is carbon dioxide (CO_2), which is an inorganic oxidized form of carbon present in the atmosphere and dissolved in water. During **photosynthesis,** plants reduce the carbon atom in carbon dioxide using energy from light. All organisms, including plants, undo the results of photosynthesis by oxidizing organic carbon back to carbon dioxide; this process is known as **respiration.** The oxidation of carbon during respiration releases energy, and organisms can harness a portion of this energy to synthesize proteins, maintain cellular ion concentrations, and move; the rest escapes as heat.

Photosynthesis and respiration involve the complementary reduction and oxidation of carbon and oxygen:

$$\text{energy} + 6CO_2 + 6H_2O \rightleftharpoons C_6H_{12}O_6 + 6O_2.$$

Notice that as carbon is reduced during photosynthesis, oxygen is oxidized from its form in water to its molecular form, O_2. This molecular oxygen is found as a gas in the atmosphere and dissolved in water. In its reduced state, oxygen has an excess of electrons and readily forms water (H_2O) in combination with positively charged hydrogen ions. During respiration, oxygen is reduced to form water, and carbon is oxidized to form carbon dioxide. Because less energy is needed to reduce oxygen than to reduce carbon, the oxidation of carbon releases more energy than the reduction of oxygen requires. Therefore, the coupling of the oxidation of carbon with the reduction of oxygen results in a net release of energy, which the organism can use to perform other work.

The limited availability of inorganic carbon

Plants assimilate more carbon through photosynthesis than they oxidize by way of respiration (otherwise they would not grow), so they require an external source of carbon. The only practical source of nonbiological carbon, CO_2, has an extremely low concentration in the atmosphere (about 0.03%). Carbon dioxide enters plant cells because there is a higher concentration of CO_2 in the atmosphere than there is in the cells, where CO_2 is continually used up by photosynthesis. However, the atmosphere-to-plant difference in the concentration of CO_2 is much, much less than the plant-to-atmosphere difference in the concentration of water vapor, which drives water out of plant cells into the surrounding air. This makes water conservation a problem for terrestrial plants, especially in arid environments. Plants evaporate 500 g of water from their leaves, more or less, for every gram of carbon they assimilate (■ Figure 2.9).

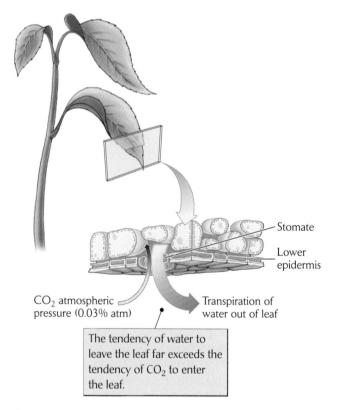

CO₂ atmospheric pressure (0.03% atm)

Transpiration of water out of leaf

Stomate

Lower epidermis

The tendency of water to leave the leaf far exceeds the tendency of CO_2 to enter the leaf.

■ **Figure 2.9 Gas exchange occurs across the surface of a leaf.** Schematic cross section of the lower portion of a leaf, showing the slow diffusion of carbon dioxide into the leaf compared with the evaporation of water from the leaf surface to the surrounding air. The lower epidermis of the leaf is relatively impermeable to water; gas exchange occurs primarily through pores (stomates) on the undersurface of the leaf. Because the plant uses carbon dioxide in photosynthesis, the concentration of that gas remains lower in the leaf than in the surrounding air.

Getting enough carbon poses different problems for aquatic plants. The solubility of carbon dioxide in fresh water is about 0.0003 liters per liter of water, which is 0.03% by volume, or about the same as its concentration in the atmosphere. When carbon dioxide dissolves in water, however, most of the molecules form carbonic acid (H_2CO_3). Depending on how acid the water is, carbonic acid molecules dissociate into bicarbonate ions (HCO_3^-) and carbonate ions (CO_3^{2-}). Within the range of acidity that is typical of most fresh and salt water (pH values between 6 and 9), the more common form is bicarbonate, which dissolves readily in water. As bicarbonate forms, carbon dioxide is removed from solution, and more of the gas can then enter into solution from the atmosphere:

$$CO_2 + H_2O \rightarrow H_2CO_3 \rightarrow H^+ + HCO_3^-.$$

This process continues until the concentration of bicarbonate ions is equivalent to 0.03–0.06 liters of carbon dioxide gas per liter of water (3–6%), more than 100 times the concentration of carbon dioxide in air (∎Figure 2.10). Thus, bicarbonate ions provide a large reservoir of inorganic carbon in aquatic systems.

Carbon dioxide diffuses slowly through water

Inorganic carbon is abundant in water, to be sure. But if that carbon doesn't move rapidly, plants don't have access to an abundant supply. The rate of diffusion of carbon dioxide through unstirred water is about 10,000 times less than it is in air, and the larger bicarbonate ions diffuse even more

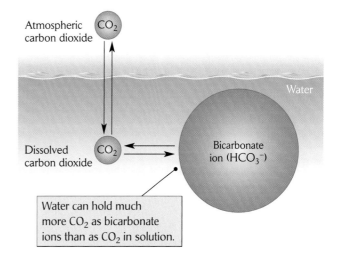

Water can hold much more CO_2 as bicarbonate ions than as CO_2 in solution.

∎ **Figure 2.10 Carbon dioxide dissolved in water** exists in equilibrium with a larger concentration of bicarbonate ion.

slowly. Every surface of an aquatic plant, alga, or microbe has a **boundary layer** of unstirred water, which may range from as little as 10 micrometers (μm) for single-celled algae in turbulent waters to 500 μm (0.5 mm) for a large aquatic plant in stagnant water (∎Figure 2.11). Thus, in spite of the high concentration of bicarbonate ions in the water surrounding these organisms, photosynthesis may nonetheless be limited by a diffusion barrier of still water at the surface of the organism.

Both carbon dioxide and bicarbonate ions enter the cells of aquatic plants. Once inside the cells, bicarbonate ions can be used directly as a source of carbon for

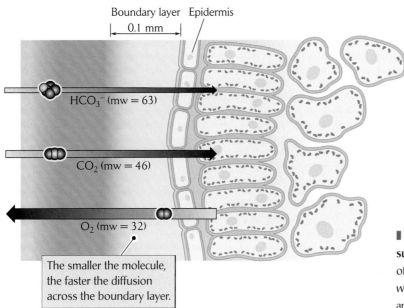

The smaller the molecule, the faster the diffusion across the boundary layer.

∎ **Figure 2.11 The boundary layer at the surface of an aquatic plant** retards the exchange of gases between a leaf and the surrounding water (mw = molecular weight). After H. B. A. Prins and J. T. M. Elzenga, *Aquatic Botany* 34:59–83 (1989).

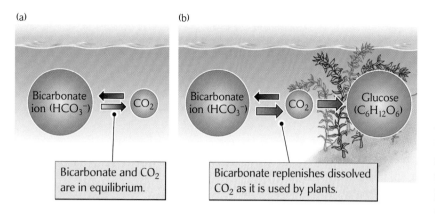

(a)

Bicarbonate ion (HCO_3^-) ⇌ CO_2

Bicarbonate and CO_2 are in equilibrium.

(b)

Bicarbonate ion (HCO_3^-) ⇌ CO_2 → Glucose ($C_6H_{12}O_6$)

Bicarbonate replenishes dissolved CO_2 as it is used by plants.

Figure 2.12 Bicarbonate ion is a source of carbon dioxide in aquatic systems. When aquatic plants and algae deplete carbon dioxide in their immediate vicinity, it is replenished from the pool of bicarbonate ions. (a) The equilibrium between bicarbonate and dissolved carbon dioxide in water. (b) When plants and algae remove carbon dioxide from the water during photosynthesis, the reduced CO_2 concentration causes bicarbonate to release additional CO_2 into solution ($H^+ + HCO_3^- \rightarrow H_2O + CO_2$).

photosynthesis, although at only 10–40% of the efficiency of utilizing carbon dioxide. As carbon dioxide itself is taken up during photosynthesis, and thereby depleted within cells, bicarbonate ions associate once more with hydrogen ions to produce more carbon dioxide:

$$H^+ + HCO_3^- \rightarrow CO_2 + H_2O$$

(**Figure 2.12**). Bicarbonate ions and carbon dioxide exist in a chemical equilibrium, which represents the balance achieved between H^+ and HCO_3^-, on one hand, and CO_2 and H_2O on the other.

Oxygen is scarce in water

The low solubility of oxygen in water often limits the metabolism of animals in aquatic habitats. This limitation is compounded by the vastly lower rate of diffusion of oxygen in water than in air. Compared with its concentration of 0.21 liters per liter (21% by volume) in the atmosphere, the solubility of oxygen in water reaches a maximum (at 0°C in fresh water) of 0.01 liters per liter (1%). Furthermore, below the limit of light penetration in deep bodies of water and in waterlogged sediments and soils, no oxygen is produced by photosynthesis. Therefore, as animals and microbes use oxygen to metabolize organic materials, such habitats may become severely depleted of dissolved oxygen. Habitats, such as deeper layers of water in lakes and mucky sediments of marshes, that are devoid of oxygen are referred to as **anaerobic** or **anoxic** habitats. Such conditions pose problems for terrestrial plants, whose roots need oxygen for respiration. Many plants that live in waterlogged habitats have special vascular tissues that conduct air directly from the atmosphere to the roots. The roots of cypress trees and many mangroves grow vertical extensions that project above the anoxic soil and conduct oxygen directly from the atmosphere to the roots (**Figure 2.13**).

The availability of inorganic nutrients influences the abundance of life

Organisms are composed of a variety of chemical elements. After hydrogen, carbon, and oxygen, which are the elements in carbohydrates, those required in greatest quantity are nitrogen, phosphorus, sulfur, potassium, calcium, magnesium, and iron. The functions of these elements in biological systems are outlined in Table 2.1. Certain organisms need other elements in abundance as well. For example, diatoms construct their glassy shells of silicates (**Figure 2.14**); tunicates, which are sessile marine inverte-

Figure 2.13 The knees of bald cypress trees conduct air from the atmosphere to roots when a swamp is flooded and the waterlogged sediments contain little or no free oxygen. Photo by R. E. Ricklefs.

Table 2.1   Major nutrients required by organisms, and some of their primary functions

Element	Function
Nitrogen (N)	Structural component of proteins and nucleic acids
Phosphorus (P)	Structural component of nucleic acids, phospholipids, and bone
Sulfur (S)	Structural component of many proteins
Potassium (K)	Major solute in animal cells
Calcium (Ca)	Structural component of bone and of material between woody plant cells; regulator of cell permeability
Magnesium (Mg)	Structural component of chlorophyll; involved in the function of many enzymes
Iron (Fe)	Structural component of hemoglobin and many enzymes
Sodium (Na)	Major solute in extracellular fluids of animals

brates, accumulate vanadium in high concentrations, possibly as a defense against predators; nitrogen-fixing bacteria require molybdenum as a part of the key enzyme in nitrogen assimilation.

The scarcity (relative to need) of inorganic nutrients often limits plant growth. Plants acquire mineral nutrients—other than oxygen, carbon, and some nitrogen—as ions from water in the soil around their roots. Nitrogen exists in soil as ammonium (NH_4^+) and nitrate ions (NO_3^-), phosphorus as phosphate ions (PO_4^{3-}), calcium and potassium as their elemental ions Ca^{2+} and K^+, and so on. The availability of these elements varies with their chemical form in the soil and with temperature, acidity, and the presence of other ions. Phosphorus, in particular,

(a)

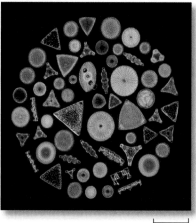

0.3 mm

(b)

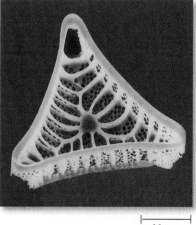

30 µm

(c)

5 µm

(d)

Figure 2.14 Diatoms are photosynthetic protists. The outer shells of diatoms are made up of silicates. (a) A selection of marine diatoms as seen through a light microscope. (b), (c) Scanning election micrographs of *Entogonia* and *Cyclotella*, respectively. (d) The stalked diatom *Licmophora*. Photo (a) by M.I. Walker/Photo Researchers; (b) F. Rossi; (c) Ann Smith/Photo Researchers; and (d) Biophoto Associates/Science Source/Photo Researchers.

often limits plant production in terrestrial environments; even when it is abundant, most of the compounds it forms in the soil do not dissolve easily. We shall have much more to say about nutrient uptake by plants in later chapters.

Light is the primary source of energy for the biosphere

Green plants, algae, and some bacteria absorb light and assimilate its energy by photosynthesis, but not all the light striking the earth's surface can be used in this way. Rainbows and prisms show that light consists of a spectrum of wavelengths that we perceive as different colors. Actually, visible light represents only a small part of the spectrum of electromagnetic radiation, which extends from gamma rays (the shortest wavelengths) to radio waves (the longest). Wavelengths are usually expressed in nanometers (nm: one-billionth of a meter). The visible portion of the spectrum, which corresponds to the wavelengths of light suitable for photosynthesis, ranges between about 400 nm (violet) and 700 nm (red). This range is called the photosynthetically active region (PAR) of the spectrum. Light of wavelengths shorter than 400 nm makes up the **ultraviolet** part of the spectrum; light of wavelengths longer than 700 nm is

called infrared. Infrared radiation is perceived primarily as heat. The energy intensity of light varies inversely with its wavelength: shorter-wavelength blue light has a higher energy level than longer-wavelength red light.

Ozone and ultraviolet radiation

Starting in the visible portion of the spectrum and moving toward shorter wavelengths, one encounters ultraviolet radiation and high-energy X rays. Because of its high energy level, ultraviolet light can damage exposed cells and tissues. Fortunately, the earth's atmosphere is completely transparent only to the visible range of the spectrum. As light passes through the atmosphere, most of its ultraviolet components are absorbed, primarily by a molecular form of oxygen known as ozone (O_3) that occurs in the upper atmosphere. The atmosphere thus shields life at the earth's surface from the most damaging wavelengths of light (❚Figure 2.15).

Certain pollutants in the atmosphere, particularly the chlorofluorocarbons (CFCs) formerly used as refrigerants and as propellants in aerosol cans, chemically destroy ozone in the upper atmosphere. This degradation has produced "ozone holes"—areas of low ozone concentration in the upper atmosphere—over some parts of the earth, particularly at high latitudes. Consequently, the danger of tis-

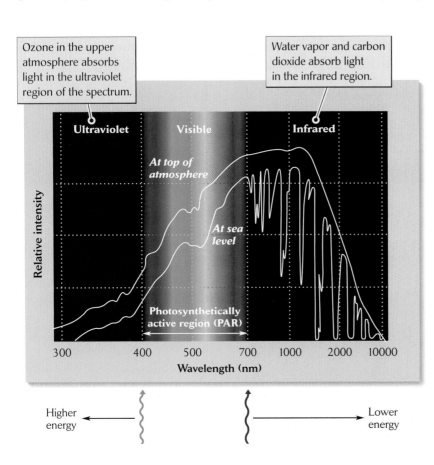

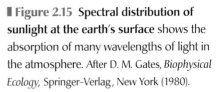

❚ **Figure 2.15 Spectral distribution of sunlight at the earth's surface** shows the absorption of many wavelengths of light in the atmosphere. After D. M. Gates, *Biophysical Ecology,* Springer–Verlag, New York (1980).

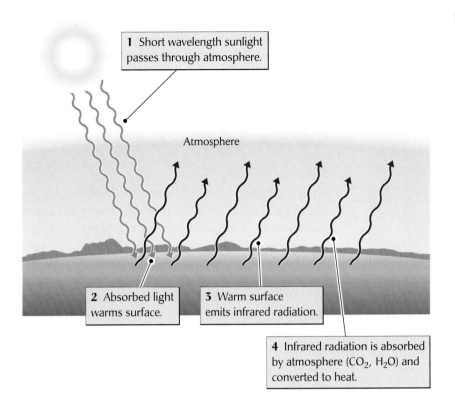

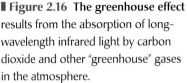

Figure 2.16 The greenhouse effect results from the absorption of long-wavelength infrared light by carbon dioxide and other "greenhouse" gases in the atmosphere.

1 Short wavelength sunlight passes through atmosphere.

Atmosphere

2 Absorbed light warms surface.

3 Warm surface emits infrared radiation.

4 Infrared radiation is absorbed by atmosphere (CO_2, H_2O) and converted to heat.

sue damage from ultraviolet radiation has increased for humans and most likely for other forms of life. Concern over the increasing size of ozone holes has led to strict controls over the manufacture and release into the atmosphere of substances such as CFCs.

Infrared light and the greenhouse effect

Toward the other end of the spectrum from ultraviolet light, one passes through the infrared region to extremely long-wavelength, low-energy radiation such as radio waves. The presence of water vapor, carbon dioxide, methane, and other gases in the atmosphere makes it relatively opaque to infrared light. These gases absorb much of the infrared portion of sunlight, and this absorbed energy contributes to the warming of air. More importantly, because of its infrared opacity, the atmosphere also absorbs radiation from the surface of the earth. Most of the energy in the visible portion of the solar spectrum that reaches the earth's surface is absorbed by vegetation, soil, and surface waters and converted to heat energy. This heat is then radiated from the warmed surface of the earth back toward space as low-intensity infrared radiation. Much of this radiation is absorbed by the atmosphere, which thereby acts as a blanket covering the earth and keeping its surface warm. Because this warming effect resembles the manner

in which glass keeps a greenhouse warm, it is called the **greenhouse effect** (**Figure 2.16**). Eventually, this absorbed energy reaches the upper levels of the atmosphere and is lost to space, but at a much slower rate than would occur in the absence of the infrared-opaque components of air—the so-called greenhouse gases. Overall, the greenhouse effect greatly benefits life by maintaining temperatures on earth within a favorable range. However, our addition of carbon dioxide to the atmosphere by clearing forests and burning fossil fuels has intensified the greenhouse effect, and the surface of the earth is becoming warmer.

The absorption spectra of plants

Vision and the photochemical conversion of light energy to chemical energy by photosynthetic organisms occur primarily within that portion of the solar spectrum at the earth's surface that contains the greatest amount of energy. The absorption of radiant energy depends on the nature of the absorbing substance. Water only weakly absorbs light in the visible region of the spectrum; therefore, a glass of water appears colorless. Dyes and pigments strongly absorb some wavelengths in the visible region, reflecting or transmitting light of the color that identifies them. Leaves contain several kinds of pigments, particularly chlorophyll (green) and carotenoids (yellow), that absorb light and

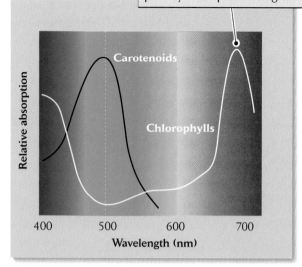

Chlorophyll absorbs mostly in the violet and red wavelengths. What is not absorbed—green and yellow—is reflected back, which is why photosynthetic plants look green.

Carotenoids

Chlorophylls

Relative absorption

400   500   600   700
Wavelength (nm)

▌ **Figure 2.17 Two groups of photosynthetic pigments**—chlorophylls and carotenoids—absorb different wavelengths of light. After R. Emerson and C. M. Lewis, *J. Gen. Physiol.* 25: 579–595 (1942).

harness its energy (▌ Figure 2.17). Carotenoids, which give carrots their orange color, absorb primarily blue and green light and reflect light in the yellow and orange regions of the spectrum. Chlorophyll absorbs red and violet light while reflecting green and blue.

The transparency of a glass of water is deceptive. Although it appears colorless in small quantities, water absorbs or scatters enough light to limit the depth of the sunlit zone of the sea. In pure seawater, the energy content of light in the visible part of the spectrum diminishes to 50% of the surface value at a depth of 10 meters, and it drops to less than 7% within 100 meters. Moreover, water absorbs longer wavelengths more strongly than shorter ones; most of the infrared radiation disappears within the topmost meter of water. Short wavelengths (violet and blue) tend to scatter when they strike water molecules, so they too fail to penetrate deeply. Because of the absorption and scattering of light by water, green light predominates with increasing depth.

The photosynthetic pigments of aquatic algae parallel this spectral shift. Algae near the surface of the oceans, such as the green sea lettuce (*Ulva*), which grows in shallow water along rocky coasts, have pigments resembling those of terrestrial plants and best absorb blue and red light and reflect green light. The deep-water red alga *Porphyra* has additional pigments that enable it to use green light more effectively in photosynthesis (▌ Figure 2.18).

The absorption of light by water limits the depth at which aquatic photosynthetic organisms can exist. The narrow zone close to the surface where there is sufficient light for photosynthesis is called the **euphotic** zone. The lower limit of the euphotic zone, where the assimilation of

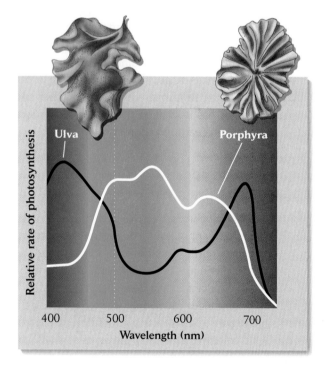

▌ **Figure 2.18 Relative rates of photosynthesis** by the green alga *Ulva* and the red alga *Porphyra* (*right*) differ as a function of the color of light. After F. T. Haxo and L. R. Blinks, *J. Gen. Physiol.* 33:389–422 (1950). Photo by Larry Jon Friesen/Saturdaze.

energy by photosynthesis just balances the release of energy by respiration, may lie 100 meters below the surface in some exceptionally clear lakes and oceans, but this is a rare condition. In productive waters with dense phytoplankton, or in waters turbid with suspended silt particles, the euphotic zone may be as shallow as 1 meter.

Light intensity

Ecologists measure the intensity of light as the energy content of the light from the photosynthetically active region of the spectrum striking a unit of surface area per unit of time. Light intensity is sometimes referred to as **radiant flux,** which is commonly expressed as watts per square meter (W per m²). The watt, which is the familiar unit used to rate the power consumption of light bulbs and appliances, is equal to one joule of energy per second.

A flat surface above the atmosphere of the earth directly facing the sun would receive approximately 1,400 W per m². This intensity of solar radiation—the energy reaching the outer limit of the atmosphere—is called the **solar constant.** In reality, the average intensity of light at any area on the surface of the earth is far less. Nighttime periods without light, the low incidence of light early and late in the day and at high latitudes, absorption of light by the atmosphere, and reflection of light by clouds all diminish light intensity at the earth's surface.

The thermal environment provides several avenues of heat gain and loss

Most of the solar radiation absorbed by water, soil, plants, and animals is converted to heat. Each object and each organism on earth continually exchanges heat with its surroundings. When the temperature of the environment exceeds that of an organism, the organism gains heat and becomes warmer. When the environment is cooler, the organism loses heat to the environment and cools. An individual organism's heat budget includes several avenues of heat gain and heat loss (**Figure 2.19**).

Radiation is the absorption or emission of electromagnetic energy. Sources of radiation in the environment include the sun, the sky (scattered light), and the landscape (which radiates heat it has absorbed from the sun). How rapidly an object loses energy by radiation depends on the temperature of the radiating surface. The relationship is nonintuitive in that radiation increases with the *fourth* power of absolute temperature (K). (Absolute zero, that is, 0 degrees Kelvin—0 K—is equal to −273°C.) Accordingly, a small mammal with a skin temperature of 37°C (310 K) radiates heat 30% more rapidly than a lizard of similar size with a skin temperature of 17°C (290 K). At night, objects

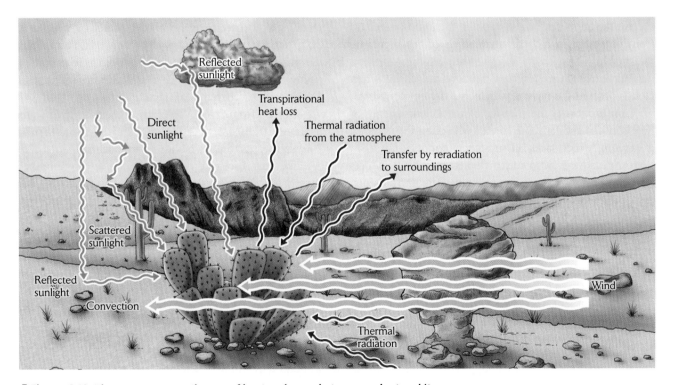

Figure 2.19 There are many pathways of heat exchange between a plant and its environment. After D. M. Gates, *Biophysical Ecology,* Springer-Verlag, New York (1980).

Figure 2.20 Thermal images of Canada geese show variation in thermal radiation over different parts of the body. It is clear that the geese lose more heat across their necks and legs than from their well-insulated bodies. Courtesy of R. Boonstra, from R. Boonstra et al., *J. Field Ornithol.* 66:192–198 (1995).

that have warmed in the sunlight radiate their stored heat to colder parts of the environment and, eventually, to space. The bodies of organisms, especially warm-blooded birds and mammals, often are the "brightest" objects emitting infrared radiation in the night (**Figure 2.20**). Because we are so much hotter than the black void of space (which has a temperature close to 0 K), we radiate tremendous quantities of energy to the clear night sky. We can also receive radiation from water vapor in the atmosphere and from vegetation, which balances some of our nighttime radiation loss. That is why, at a given temperature, one feels warmer at night in a humid environment, particularly when clouds obscure the night sky, than in a dry environment.

Conduction is the transfer of the kinetic energy of heat between substances in contact with one another. Thus, a vacuum, which lacks all substance, conducts no heat. Water, because it is so much denser than air, conducts heat more than 20 times faster than air. The rate at which heat passes between an organism and its surroundings depends on the insulating value of the organism's surface (its resistance to heat transfer), its surface area, and the temperature difference between the organism and its surroundings. An organism can either gain or lose heat by conduction, depending on its temperature relative to that of the environment. That is why lizards often lie flat on hot rocks, warming their bodies by conduction.

Convection is the movement of heat in liquids and gases of different temperatures, particularly over surfaces across which heat is transferred by conduction. Air conducts heat poorly. In still air, a boundary layer of air forms over a surface, just as a boundary layer of still water forms over the surfaces of aquatic plants. A warm organism tends to warm its boundary layer to the temperature of its own body, effectively insulating itself against heat loss. A current of air flowing past a surface tends to disrupt the boundary layer and to increase the rate of heat exchange by conduction (**Figure 2.21**). This convection of heat away from the body surface is the basis of the "wind chill factor" we hear about in winter on the evening weather report. On a cold day, air movement makes you feel as cold as you would on an even colder windless day. For example, a wind blowing 32 km per hour at an air temperature of −7°C has the cooling power of still air at −23°C.

Evaporation also affects the movement of heat. The evaporation of 1 g of water from the body surface removes 2.43 kilojoules (kJ) of heat when the temperature of the surface is 30°C. As plants and animals exchange gases with the environment, some water evaporates from their exposed surfaces. In plants, the evaporation of water from the surface of a leaf is referred to as **transpiration.** The rate of heat loss from a surface by evaporation and transpiration depends on the permeability of the surface to water, the relative temperatures of the surface and the air, and the vapor pressure of the atmosphere. Vapor pressure is a measure of the capacity of the atmosphere to hold water. Vapor pressure is expressed in atmospheres (the pressure of the atmosphere at the surface of the earth), and represents the fractional weight of water vapor in saturated air. At 30°C, the vapor pressure of water is 0.042 atm, meaning that the air can hold 4.2% water by weight. At 20°C, the vapor pressure of water is only 0.023 atm. Thus, the water-holding capacity of air varies with its temperature, nearly doubling over each 10°C increase in temperature. Consequently, when the temperature of air saturated with water drops from 30° to 20°C, its capacity to hold water decreases from 4.2 to 2.3%, and the difference—almost 2%—condenses to form clouds or precipitation.

Like heat, moisture can be trapped in the boundary layer of air that forms above surfaces. Convection tends to disrupt boundary layers and therefore increases evapora-

Heated boundary layer

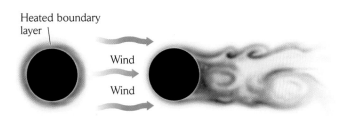

Wind

Wind

Figure 2.21 The boundary layer that develops around an object in a still fluid is disrupted by movement of the fluid.

tive heat loss as well as conductive heat loss. Because warm air can hold more water than cold air, it has greater potential for evaporating water. In hot climates, water evaporating from the skin and respiratory surfaces cools many animals. For warm-blooded animals in cold climates, evaporation can become a problem as cold inhaled air containing little water warms when it encounters the body and respiratory surfaces, thereby speeding evaporation. We see evidence of such water loss on winter days when water evaporated from the warm surfaces of our lungs condenses as our breath mingles with the cold atmosphere.

All of the gains and losses of heat by an organism constitute its **heat budget.** The heat budget of an organism can be expressed by a simple equation that relates the rate of change in its heat content to the rates of gains and losses through radiation, conduction, convection, and evaporation, plus the internal heat it generates by metabolizing foods:

$$\text{change in heat content} = \text{metabolism} - \text{evaporation} \pm \text{radiation} \pm \text{conduction} \pm \text{convection.}$$

Because radiation, conduction, and convection can either add to or remove heat from the organism, a plus/minus (±) sign precedes these terms. When gains and losses are perfectly balanced, the change in heat content is zero.

Because evaporation and metabolism influence body heat, the heat budget is connected to the organism's budgets of water, food, and salts, as illustrated in ▌Figure 2.22. Food is the source of metabolically produced heat, and it also contains water and salts. Evaporative heat loss requires a source of free water, which can be obtained by drinking (where water is available) and produced by the metabolism of organic compounds (as we saw in the chemical formula for respiration above). For example, the metabolism of a gram of fat produces 1.07 grams of water.

ECOLOGISTS IN THE FIELD

Keeping cool on tropical islands

Organisms that live in the open in tropical climates gain a tremendous amount of heat by radiation from the sun and by conduction and convection from the hot environment in which they live. In many circumstances, this heat can be gotten rid of by evaporation of water from body surfaces. Humans certainly sweat when they are hot. Many animals also lose heat by evaporation of water from the respiratory surfaces of the lungs. Where water is scarce, however, evaporation is not an option, and animals generally find ways of avoiding intense sunlight. So here is an interesting puzzle: Sev-

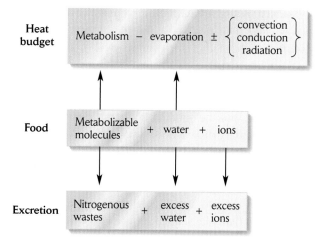

▌**Figure 2.22 The heat, water, energy, and salt budgets of organisms** are coupled by diet, evaporative water loss, and excretion.

eral species of seabirds nest on bare sand on small coral atolls in the Tropics, such as Tern Island in the northwestern Hawaiian Islands. These birds are exposed to punishing levels of solar radiation during the middle of the day. In this harsh environment, the sooty tern nests on the surface of the sand in full sunlight (▌Figure 2.23), while another species of similar size and coloration, the wedge-tailed shearwater, builds its nests in deep burrows beneath the surface of the sand.

▌**Figure 2.23 A sooty tern can tolerate a hot environment** because its food contains abundant water for evaporation cooling. This bird is sitting on its egg in the hot sun on Christmas Island, located on the equator in the central Pacific Ocean. Photo by R. E. Ricklefs.

Seabird biologist Paul Sievert wondered why the two species place their nests so differently. The conventional wisdom had been that shearwaters nest in burrows to avoid predators such as frigatebirds, which, ever watchful, swoop down to snatch unattended eggs and chicks. By chance, however, the density of shearwaters on Tern Island was so great, and the sand so hard to dig through, that many shearwaters nested on the surface out of desperation. These birds had very low nesting success because they were forced to abandon their eggs under the intense solar radiation. If the eggs weren't lost to frigatebirds, they heated up in the sun, and the developing embryos died. Sievert found that if he shaded surface nests with plywood A-frames, the shearwaters were able to reproduce successfully because the adults could remain on their eggs throughout the middle of the day (▌Figure 2.24).

This simple experiment demonstrated the importance of the thermal environment for shearwaters, but it did not explain how sooty terns could nest on the surface in the same environment in full sun. The key to this puzzle lies in the diets and feeding regimes of the two bird species. Sooty terns feed on fish and squid obtained in areas close to their nesting sites. The male and female sooty terns alternate incubation duty, and neither one stays at the nest for more than a day or two at a time. Shearwaters have a diet similar to that of terns, but they feed hundreds of kilometers from their nesting sites.

They digest most of what they eat while foraging at sea, and convert the surplus energy to fat deposits to use during their weeklong spells tending the egg. Thus, sooty terns come back to their nests from the sea with a stomach full of water-laden food, which provides a reservoir of free water for evaporative heat loss. Shearwaters have plenty of fat for a prolonged fast, but fat contains little water, and even the water produced by fat metabolism is insufficient to dissipate the heat load absorbed under full sunlight. So why don't shearwaters drink the abundant seawater all about them? Seawater contains so much salt that they would lose, through their kidneys and salt-secreting glands, as much water as they consumed. Water, water, everywhere, nor any drop to drink!

Organisms must cope with temperature extremes

Unlike birds and mammals, most organisms do not regulate their body temperatures, which vary in parallel with the temperature of their surroundings. Most life processes occur only within the range of temperature at which water is liquid: 0°–100°C at the earth's surface. Relatively few plants and animals can survive body temperatures above 45°C, which defines the upper limit of the physiological

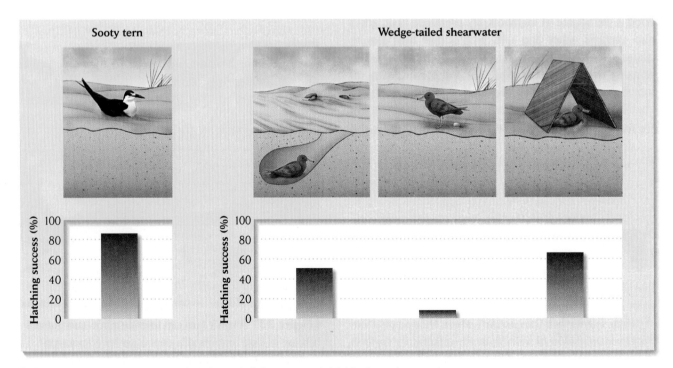

▌**Figure 2.24 Hatching success of wedge-tailed shearwaters is highly dependent on the thermal environment.** Individuals protected from the sun in burrows or provided with artificial shade have higher success than do those nesting in the open. Data courtesy of Paul Sievert.

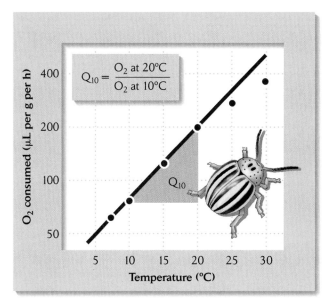

Figure 2.25 Oxygen consumption increases as a function of temperature. These data are for the Colorado potato beetle. After K. Marzusch, *Zeitschr. Vergl. Physiol.* 34:75–92 (1952).

range for most eukaryotic organisms. Clearly, temperature influences life processes. How do organisms cope with temperatures outside the physiological range?

Tolerance of heat

Much of the influence of temperature on life processes results from the way in which heat affects organic molecules. Heat imparts a high kinetic energy to living systems, and causes biological molecules to move and change shape at a high rate. By increasing the rate of movement of molecules, heat also accelerates chemical reactions. The rates of most biological processes increase between two and four times for each 10°C rise in temperature throughout the physiological range (**Figure 2.25**). Higher temperatures mean that organisms can develop more rapidly, swim, run, and fly faster, and digest and assimilate more food. Thus, increasing temperature has a positive effect on biological productivity. But high temperatures also have a depressing effect on life processes. Proteins and other biological molecules become less stable at higher temperatures and may not function properly or retain their structure.

Proteins are long chains of amino acids joined together. The functioning of any particular protein depends on the complicated folding of the amino acid chain. The shape of the protein is maintained by forces of attraction between certain types of amino acids strategically placed in the sequence of the chain. The molecular motion caused

by heat tends to open up, or denature, the structure of proteins. Existence at high temperatures requires that proteins and other biological structures have strong forces of attraction within and between molecules to resist being literally shaken apart. The proteins of thermophilic ("heat-loving") bacteria have higher proportions of amino acids that form strong bonds among one another than do those of other, heat-intolerant organisms. Consequently, some photosynthetic cyanobacteria tolerate temperatures as high as 75°C, and some archaebacteria can live in hot springs at temperatures up to 110°C.

Temperature affects other biological compounds as well. The physical properties of fats and oils, which are major components of cell membranes and constitute the energy reserves of animals, depend on temperature. When cold, fats become stiff (picture in your mind the fat on a piece of meat taken from the refrigerator); when warm, they become fluid.

Tolerance of freezing

Temperatures on the earth's surface rarely exceed 50°C, except in hot springs and at the soil surface in hot deserts. However, temperatures below the freezing point of water occur commonly, particularly on the land and in small ponds, which may freeze solid during winter. When living cells freeze, the crystal structure of ice disrupts most life processes and may damage delicate cell structures, eventually causing death. Many organisms successfully cope with freezing temperatures, either by maintaining their body temperatures above the freezing point of water or by activating chemical mechanisms that enable them to resist freezing or tolerate its effects.

Dissolved substances that interfere with the formation of ice can depress the freezing point of water below 0°C. Plants and animals take advantage of this physical property to lower the freezing point of their body fluids. As we saw above, salt decreases the freezing point of water. But protein structure and function are also disrupted by high salt concentrations, and so using salts for this purpose is physiologically impractical. Instead, the body fluids of many marine organisms contain high concentrations of dissolved glycerol, which is a three-carbon alcohol that is used to form the backbone of triglyceride fats. A 10% glycerol solution lowers the freezing point of water by about 2.3°C without severely affecting biochemical processes. Glycoproteins, the class of proteins that contain one or more carbohydrates, also may be used to lower the freezing temperature. Such antifreeze-like compounds in their tissues allow fish in antarctic regions to remain active in seawater that is colder than the normal freezing point of the blood of fish inhabiting temperate or tropical seas

(■ Figure 2.26). Terrestrial invertebrates also use the antifreeze approach, and their body fluids may contain up to 30% glycerol, in extreme cases, as winter approaches.

Supercooling provides a second physical solution to the problem of freezing. Under certain circumstances, fluids can cool below the freezing point without ice crystals developing. Ice generally forms around some object, called a seed, which can be a small ice crystal or other particle. In the absence of seeds, pure water may cool more than 20°C below its melting point without freezing. Supercool-

ing has been recorded to −8°C in reptiles and to −18°C in invertebrates. Glycoproteins in the blood of these cold-adapted animals impede ice formation by coating developing crystals, which would otherwise act as seeds.

As you can see, many types of organisms use a variety of physical mechanisms to cope with physical stresses in the environment, emphasizing the general principle that organisms are physical systems, albeit very special ones. This point is further emphasized by the ways in which organisms use physical principles to sense their environments.

Organisms use many physical stimuli to sense the environment

To function in a complex and changing environment, organisms must be able to sense environmental change, detect and locate objects, and navigate the landscape. A predator must find its food before it can eat it. Salmon must recognize the proper river at the end of their spawning migration. Plants must sense the changing seasons to flower at the right time. The senses an organism uses generally match the types of physical stimuli available in the environment and the ways in which the organism relates to it (plants, for example, don't need the acute vision that some predators have).

Sensing electromagnetic radiation

That so many organisms rely on vision to sense the environment is not surprising, considering the high energy levels available in the visible portion of the spectrum and the fact that light travels in a straight line, allowing accurate location and resolution of objects. We ourselves primarily use vision to locate food, particularly as it is now displayed on the shelves of supermarkets. Yet our visual acuity is rather pathetic compared with that of hawks, and many insects and birds can perceive ultraviolet light, which is invisible to us (■ Figure 2.27). Insects also can detect rapid movement, such as that of wings beating 300 times per second; we humans cannot distinguish individual movie frames flickering at even 30 times per second. Thus, different organisms use the available visual information to different extents.

Some animals that are active at night, when visible light levels are too low to be used effectively, rely on other kinds of radiation. Among the more unusual sensory organs are the pit organs of the pit vipers, a group of reptiles that includes the rattlesnakes. The pit organs, located on each side of the head in front of the eyes (■ Figure 2.28), detect the infrared (heat) radiation given off by the warm

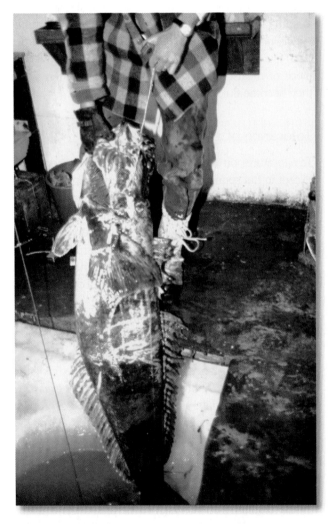

■ **Figure 2.26 Glycoproteins act as biological antifreeze in the antarctic cod.** Blood and tissues are prevented from freezing by the accumulation of high concentrations of glycoprotein, which lower the freezing point of body fluids to below the minimum temperature of seawater (−1.8°C) and prevent ice crystal formation. This fish is being pulled through a hole in the ice near McMurdo Station, Antarctica. Note the rich blood supply to its gills exposed in this photograph. Photo by John Bortniak, courtesy of NOAA.

(a)

(b)

Figure 2.27 Many organisms use signals that are "visible" only in ultraviolet light.
(a) The human eye sees this Yellow Daisy in reflected light in the range of 400–700 nanometers. (b) Bees see a different pattern in the same flower in light from the ultraviolet portion of the spectrum (300–400 nanometers). The light flecks on the petals are pollen grains. Photos by Leonard Lessin/Photo Researchers.

bodies of potential prey (see Figure 2.20). Pit vipers are so sensitive to infrared radiation that they can detect small rodents several feet away in less than a second. Moreover, because the pits are directionally sensitive, these snakes can locate warm objects precisely enough to strike them. Even plants make use of light stimuli to measure the length of the day as a cue for the seasonal changes that regulate flowering or initiate dormancy.

Sensing sound

What we perceive as sounds are pressure waves in air created by movements and impacts of objects, vibrating objects, or even turbulence in air flowing around objects.

Figure 2.28 The pit organs of the rattlesnake, located between the eye and the nostril, are used for detection of infrared radiation from potential prey. Courtesy of R. B. Suter.

Pressure waves are propagated in all directions like ripples on the surface of water. This makes sounds easy to detect but difficult to localize. The energy of pressure waves also decreases with distance from the source, effectively limiting the range of detection. Nonetheless, sound may warn of an approaching predator, regardless of the direction from which it comes. Some nocturnal predators can sense the direction of a sound source by the different times of arrival of sound waves at each ear. When the source of a sound is directly ahead, pulses of sound pressure arrive at each ear simultaneously. When the head is turned relative to a sound's point of origin, peaks and troughs of pressure reach the ears at different times, partially canceling rather than reinforcing each other. Directional sensitivity of hearing is greatest when the distance between the ears is about the same as the distance between sound waves. High-pitched sounds have shorter wavelengths and therefore are more useful sources of information for smaller animals. In fact, the ranges of sound frequencies that an animal can make and hear both vary inversely with size, just as a bass fiddle produces lower sounds than a smaller violin. Owls have such sensitive and directionally informative hearing that they can locate mice and other prey by the sounds they make as they move through the habitat. Their ability to pinpoint sound is further aided by the asymmetrical shapes of their outer ears, which deaden sound arriving from certain angles relative to sounds received from other directions.

Bats can use sound to find their way around the environment and locate prey in the absence of environmental sounds, because they produce the sound themselves by means of a biological sonar system. Bats emit very loud, high-pitched pulses of sound—generally above the range of human hearing—and sense the echoes that bounce back from objects in the environment, including such prey as moths in flight. The sound has to be produced in pulses so that the bat can listen for the fainter echoes during the quiet intervals between pulses. As a bat closes in on its prey, it emits pulses more frequently to increase the rate of incoming information. A bat can direct most of the energy in its echolocation sound directly in front of its line of flight, which increases the effective range of the sonar. The bat's hearing is also highly directional because its large external ears can channel faint returning echoes to the eardrum (❙ Figure 2.29).

One of the most distinctive uses of sound is the long-distance communication of whales. The high density of water is ideal for propagating sounds, especially those having very low frequencies. Some large whales produce extremely loud, deep sounds close to the lower limit of detection by humans (about 20 cycles per second). These sounds travel hundreds of kilometers and presumably allow widely dispersed groups of whales to communicate with one another. What they talk about isn't known.

Sensing odors

Smell is the detection of molecules diffusing through air or water. This source of information has properties that differ considerably from those of electromagnetic waves (sight) and pressure waves in air or water (hearing). Odors carried by air and water currents are difficult to localize. However, because odors are persistent, the presence of a substance can be detected long after its source has disappeared. Once an odor is detected, an organism can move upstream in the direction of the current to locate the source of the smelly molecules. This is the basis for a great deal of chemical communication, including the production of volatile mate attractants by many insects and the fragrances that many plants use to attract pollinators. Some predators follow trails of volatile chemicals to locate potential prey or other food sources. Snakes sense chemicals by flicking their tongues against the ground as they move and transferring chemicals that adhere to the tongue to sensitive organs of smell located in the roof of the mouth. The forked tongues of snakes and other reptiles allow them to simultaneously test for odors to the left and to the right to determine the correct direction of travel.

Sensing electric fields

A few aquatic animals have developed the sensory ability to detect electric fields. Some species of electric fish continuously discharge electricity from specialized muscle organs, creating a weak electric field around them. Nearby objects distort the field, and these changes are picked up by receptors on the surface of the fish. As one might expect, the production and sensation of electric fields are most highly developed in fishes that inhabit murky waters where visibility is poor. The long, flat snout of the paddlefish, which lives in silt-laden rivers, is highly sensitive to electrical disturbances produced by small prey organisms (❙ Figure 2.30). Because its snout projects so far in front of its mouth, the paddlefish has time to capture the small zooplankton it detects as they float past. Some species use electric signals to communicate, and the electric ray *Torpedo* employs powerful electric currents (up to 50 volts at several amperes) to defend itself and to kill prey.

❙ **Figure 2.29 The head of a leaf-nose bat** is adapted for the production and detection of sonar signals. Courtesy of A. Guillen.

Figure 2.30 The long rostrum of the paddlefish is packed with sensory organs. This juvenile paddlefish can detect weak electrical signals emitted by its tiny crustacean prey. In an experimental chamber, the wires produce electric pulses that mimic the signal given off by natural prey, tricking the paddlefish to home in on them. Courtesy of L. Wilkens.

Sensing physical contact

In contrast to the magnificent senses of many organisms, others perceive their surroundings only dimly and rely on chance to bump into things. In rivers where visibility is poor, bottom-dwelling species such as catfish use elongated fins and barbels around the mouth as sensitive touch and taste receptors. Even with long barbels and bristles, the tactile sense has a very short range. Nonetheless, touch can provide a tremendous amount of information not available through other senses because of the textural and structural richness of the environment.

We have barely touched, so to speak, on the ways in which organisms perceive and find their way around their environments. Sensory modalities are limited by the availability of information that can be interpreted to reveal pattern and change in the environment, and that information is a feature of the physical environment. The use of information to sense the environment reminds us once again that organisms, above all, are physical systems and, as such, must obey physical laws and operate within limits set by the physical environment.

 ## Summary

1. Water is the basic medium of life. It is liquid within the range of temperatures encountered over most of the earth, and it has an immense capacity to dissolve inorganic compounds. These properties, and its abundance at the earth's surface, make water an ideal medium for living systems.

2. Water conducts heat rapidly and resists changes in temperature, and temperatures are therefore relatively evenly distributed throughout bodies of water and organisms.

3. Water is denser than air and provides more buoyancy, but it is also more viscous and therefore impedes movement.

4. All natural waters contain dissolved substances picked up in the atmosphere or from soils and rocks through which water flows. In limestone-rich areas, streams and lakes have abundant calcium (Ca^{2+}) and bicarbonate (HCO_3^-) ions. In the oceans, sodium (Na^+), chlorine (Cl^-), and sulfate (SO_4^{2-}) ions predominate owing to their greater solubility.

5. Acidity refers to the concentration of hydrogen (H^+) ions and is expressed as pH. Most natural waters have pH values between 6 (slightly acid) and 9 (slightly alkaline). Some organisms can tolerate high acidity (low pH) in the environment, but they maintain their internal environments between pH 6 and 7, or close to neutral, because high concentrations of the strongly reactive hydrogen ions disrupt biological processes.

6. Biological energy transformations are based largely on the chemistry of carbon and oxygen. Energy assimilated during photosynthesis is chemically reduced from its low-energy state in carbon dioxide to its high-energy state in carbohydrates. In a coupled reaction, oxygen is oxidized from its form in water to molecular oxygen. The energy stored in carbohydrates is released by the oxidation of carbon to carbon dioxide (respiration).

7. Carbon dioxide is scarce in the atmosphere (0.03%), but is more abundantly distributed in aquatic systems, where it forms soluble bicarbonate ions. The availability of carbon dioxide in aquatic systems is limited by the rate of diffusion of the gas through water, especially through still boundary layers of unmixed water that adhere to the surfaces of plants.

8. Oxygen, abundant in the atmosphere, is relatively scarce in water, where its solubility and rate of diffusion are low. Oxygen may be depleted by bacterial respiration of organic matter (producing anoxic conditions) in environments where it cannot be replenished by photosynthesis.

9. Organisms require many elements to build necessary biological structures and maintain life processes. The availability of these elements varies tremendously among environments. The scarcity (relative to need) of nitrogen and phosphorus often limits plant growth.

10. Most of the energy for life ultimately comes from sunlight. Solar radiation varies over a spectrum of wavelengths. Short wavelengths (ultraviolet) are absorbed in the atmosphere by ozone; carbon dioxide and water vapor absorb long wavelengths. Absorption in the atmosphere of infrared light radiated by the earth is known as the greenhouse effect and is responsible for the moderate temperatures at the surface of the earth. Plants extract energy primarily in the high-intensity, short-wavelength portion of the spectrum, which roughly coincides with visible light.

11. Light is attenuated by water. The depth of the euphotic zone, at the bottom of which photosynthesis balances respiration, varies from 100 meters in clear waters to a few tens of centimeters in turbid or polluted water.

12. Radiation, conduction, convection, and evaporation determine the thermal environment of organisms, especially in terrestrial habitats. In still air, organisms are surrounded by boundary layers, which impede exchange of heat and water vapor with the environment. The heat budget of an organism is intimately tied to the metabolism of food and regulation of water and salts.

13. Higher temperatures generally increase the rate of biological processes by a factor of about 2 to 4 for each 10°C. Higher thermal energy also causes proteins and other biological molecules to unfold and lose their function, setting an upper limit to temperature tolerance.

14. Most organisms cannot survive temperatures much greater than 45°C, but thermophilic bacteria grow in hot springs at up to 110°C. They tolerate such temperatures because their proteins are chemically designed to generate strong forces of attraction to hold molecules together.

15. Organisms in cold environments withstand freezing temperatures by metabolically maintaining elevated body temperatures, by lowering the freezing point of their body fluids with glycerol or glycoproteins, or by supercooling their body fluids.

16. The senses of organisms depend on the availability of information in the physical environment. The nature of this information—whether it consists of light waves, sound (pressure) waves, or volatile or dissolved molecules—determines how, and how well, the organism can detect and localize sources of information.

PRACTICING ECOLOGY

CHECK YOUR KNOWLEDGE

The Future Physical Environment

Organisms exhibit a certain geographic distribution in part because of their adaptations to the physical environment. Temperature, precipitation, mineral nutrients, light intensity, day length, oxygen content, and pH all limit where an organism can live and reproduce successfully. In most cases, the importance of these factors is species-specific; that is, the relative influence of each of these factors differs among species.

Many physical factors in the biosphere are expected to change as a direct or indirect consequence of human activities. One such factor is the concentration of CO_2 in the atmosphere. Scientists are concerned that an increase in atmospheric CO_2 could enhance the greenhouse effect enough to cause rapid global warming. In the late 1950s, Dr. Charles Keeling began recording atmospheric CO_2 concentrations atop Mauna Loa, an extinct volcano, in Hawaii. He wanted to test the hypothesis that CO_2 was increasing as a result of emissions from the burning of fossil fuels and the clearing of forests. At the time Keeling began his study, scientists did not have accurate long-term measurements of atmospheric CO_2. Mauna Loa was an ideal place to conduct the study because Hawaii is remote from concentrated sources of human CO_2 production, and measurements made at high altitude there would provide an estimate of average global effects of CO_2 emissions on the atmosphere.

Keeling observed that the average atmospheric CO_2 concentration was increasing dramatically (▌Figure 2.31). At the start of his observations in 1958, the CO_2 concentration was about 315 parts per million (ppm; 315 CO_2 molecules per million molecules of air), and had increased to about 352 ppm by 1990. By January 2000, the value had reached 370 ppm. Most scientists agree that it will continue to increase if humans continue to burn fossil fuel and clear forests at current rates.

CHECK YOUR KNOWLEDGE

1. Use the data in the essay and in Figure 2.31 to determine the percentage increase in CO_2 concentrations from 1958 to 2000.

MORE ON THE WEB **2.** Visit the *Carbon Dioxide Information Analysis Center* Web site that reports on Dr. Keeling's research through *Practicing Ecology on the Web* at *http://www.whfreeman.com/ricklefs*. How is the concentration of atmospheric CO_2 measured?

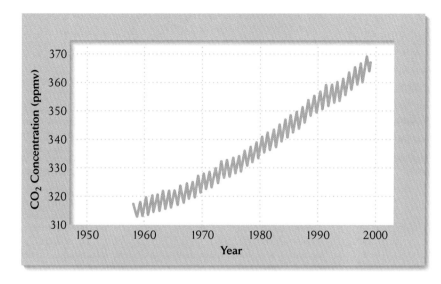

■ **Figure 2.31 CO$_2$ concentrations in the atmosphere have changed over time.** These measurements were recorded at Mauna Loa, Hawaii, where the effects of nearby human activities are minimal. CO$_2$ concentration in the Northern Hemisphere is lower during summer, when plant photosynthesis removes carbon from the atmosphere, and higher during winter, when respiration exceeds plant production. Courtesy of C. D. Keeling and T. P. Whorf, Carbon Dioxide Information Analysis Center, Oak Ridge National Laboratory, U.S. Department of Energy (1999).

As we have already seen, CO$_2$ acts as a greenhouse gas. Predictions about how much the earth will warm up as a result of these increases vary considerably (the accuracy of such predictions is expected to increase as computing power increases). Currently, it is expected that the average temperature of the earth will increase between 1.5° and 3°C (and maybe more) by the year 2050 (assuming that emissions rates stay at 1990s levels). Note that this estimate represents the average temperature over the entire planet, meaning that warming may be greater or less depending on where you are. For example, arctic ecosystems could experience an increase of 5°C.

What does global warming mean for the physical conditions that dictate the geographic distributions of organisms? Certainly, the distributions of many species are expected to shift slowly poleward and toward higher elevations. Dr. Robert L. Peters has predicted that if the climate warms by 3°C, species distributions will shift up mountainsides by about 500 m (■ Figure 2.32). Similarly, the northern limits of species ranges are expected to move poleward by about 500 km.

But are these predictions realistic? In natural systems, species will experience fragmentation of their ranges, and the rate at which suitable environments are displaced will determine whether species will be able to shift fast enough to keep up with their preferred ranges. For humans, as agricultural zones shift and as sea levels rise due to melting ice caps there may be catastrophic displacement of populations. We know that the earth has been warm before, much warmer than it is at present or is likely to be in the near future. However, the predicted changes will require that plants and animals alter their interactions at a rate that is 10 to 100 times faster than they have ever had to in response to past climate changes.

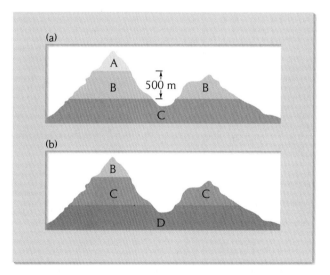

■ **Figure 2.32 The elevational distributions of species are likely to change in response to climatic warming.** (a) Ranges of species A, B, and C before climatic warming. (b) Species A has gone extinct, B has moved to a smaller habitat at a higher elevation, C's habitat has become fragmented onto two portions of the mountain, and D has moved into new habitat at higher elevations. From R. L. Peters, in R. L. Peters and T. E. Lovejoy (eds.), *Global Warming and Biological Diversity*, Yale University Press, New Haven, CT (1992), pp. 15–30.

CHECK YOUR KNOWLEDGE

3. Why are scientists concerned about rising levels of CO_2?

MORE
ON THE
WEB

4. The mission of the Environmental Protection Agency is to "protect human health and to safeguard the natural environment – air, water, and land – upon which life depends." Go to the global warming page of their site through *Practicing Ecology on the Web* at *http://www.whfreeman.com/ricklefs*. How do greenhouse gases combine to cause the greenhouse effect?

Suggested Readings

Brock, T. D. 1985. Life at high temperatures. *Science* 230:132–138. (Bacteria living close to the boiling point of water in Yellowstone hot springs.)

Fenchel, T., and B. J. Finlay. 1994. The evolution of life without oxygen. *American Scientist* 82:22–29.

Gates, D. M. 1965. Energy, plants, and ecology. *Ecology* 46:1–13.

Gates, D. M. 1971. *Man and His Environment: Climate.* Harper & Row, New York. (The thermal environment of a familiar organism.)

Hochachka, P. W., and G. N. Somero. 1984. *Biochemical Adaptation.* Princeton University Press, Princeton, NJ.

Keeling, G. D., and T. P. Whorf. 1996. Atmospheric CO_2 records from sites in the SIO area sampling network. In *Trends: A Compendium of Data on Global Change.* Carbon Dioxide Information Center, Oak Ridge National Laboratory, Oak Ridge, TN.

Knoll, A. H. 1991. End of the Proterozoic eon. *Scientific American* 265:64–73. (The role of primitive organisms in modifying the early environment of the earth.)

Peters, R. L. 1992. Conservation of biological diversity in the face of climate change. In R. L. Peters and T. E. Lovejoy (eds.), *Global Warming and Biological Diversity,* pp. 15–30. Yale University Press, New Haven, CT.

Schwenk, K. 1994. Why snakes have forked tongues. *Science* 263:1573–1577.

Vogel, S. 1981. *Life in Moving Fluids: The Physical Biology of Flow.* Princeton University Press, Princeton, NJ.

CHAPTER 3

Adaptation to Aquatic and Terrestrial Environments

Availability of water depends on the physical structure of soil

Plants obtain water from the soil by the osmotic potential of their root cells

Forces generated by transpiration help to move water from roots to leaves

Adaptations to arid environments control loss of water from leaves

Plants obtain mineral nutrients from soil water

Photosynthesis varies with levels of light

Plants modify photosynthesis in environments with high water stress

Salt balance and water balance go hand in hand

Animals excrete excess nitrogen in the form of small organic molecules

Water conservation mechanisms are important in hot environments

Organisms maintain a constant internal environment

Large animals deliver oxygen to their tissues through circulatory systems

Countercurrent circulation increases transfer of heat and substances between fluids

Each organism functions best under a restricted range of conditions

Sperm whales routinely dive to depths of 500 meters and occasionally as deep as 2 kilometers, staying below the surface for more than an hour. The reason sperm whales and other deep divers take the plunge is clear: They are after food. Fish, squid, krill, and other prey are abundant at these depths. How the whales manage such feats of diving is another story, which demonstrates the extremes to which some organisms have adapted to their environments.

Like all mammals, sperm whales must breathe air. Indeed, all diving mammals, and diving birds as well, are ultimately limited when under water by the need for oxygen to sustain their metabolism. As you would expect, a diver begins its descent with a large supply of oxygen stored in its body. What may come as a surprise, however, is that very little of this oxygen resides in the lungs. The Weddell seal, a native of antarctic waters and an excellent diver (■ Figure 3.1), carries an average of 87 milliliters of oxygen per kilogram of body mass when it submerges. Only 5% is in the lungs. Two-thirds is bound to hemoglobin in the blood, and the rest is bound to a similar molecule, myoglobin, in the muscles. In contrast, humans have only 20 ml per kg oxygen at the beginning of a dive, a quarter of which is in the lungs and only 15% in the muscles. While under water, deep-diving mammals shut down their metabolism considerably by blocking blood flow to nonvital organs such as the skin, viscera, lungs, kidneys, and muscles (which have their own oxygen supply bound to myoglobin), and keep blood flowing primarily to the brain and heart. Consequently, the temperature of all but a few key organs drops, heart rate slows, and demand for oxygen drops to a minimum.

The adaptations of diving mammals show how the structure and functioning of an organism is adapted to the particular environment it lives in. All organisms are constrained by their physical environments, whether water, light, or nutrients are in short supply or conditions of

Figure 3.1 The Weddell seal is an excellent diver. This seal, though clumsy on land, can dive to over 500 meters depth and remain submerged for up to an hour and 20 minutes. Photo by R. E. Ricklefs.

temperature, acidity, or salinity are stressful. In this chapter, we shall explore the various ways in which both aquatic and terrestrial plants and animals cope with their environments. Because adaptations to the physical environment result in specialized anatomy and physiology, this discussion will also help us to understand why the distributions of plants and animals are restricted to particular environments. After all, whales can't do much out of water!

As we saw in the previous chapter, the physical environment includes many factors that are important to the well-being of living organisms. Each type of organism is adapted to function best within a rather narrow range of physical conditions. Extending these limits exacts costs, requiring mechanisms that allow the individual either to tolerate more extreme conditions or to maintain its internal environment within a favorable range.

In this chapter, we shall explore a variety of the adaptations that animals and plants have evolved so that they can function well in their physical environments. Many of these adaptations enable plants and animals to control the movement of heat and various substances across their surfaces. By regulating the exchange of heat and substances with the physical environment, organisms can maintain their own internal environments in a state that is favorable for their life processes while obtaining necessary resources and ridding themselves of unnecessary, or even dangerous, waste products of their metabolism.

The mechanisms by which organisms interact with their physical environments help us to understand why

organisms are specialized to rather narrow ranges of environmental conditions, particularly those organisms living under extreme conditions. This specialization provides a basis for understanding the ecological distributions of populations and why certain adaptations of morphology and physiology are associated with certain physical conditions.

We shall begin this exploration by considering how plants cope with the limited availability of water in their environment. Plants need water in prodigious quantities because they lose so much by evaporation from their leaves while taking up carbon dioxide from the atmosphere. The ability of plants to obtain water from the soil is determined by the physical properties of soil and of water and by the way differences in solute concentrations cause water to move.

Availability of water depends on the physical structure of soil

Most terrestrial plants obtain the water they need from the soil. The amount of water in soil and its availability to plants varies with the physical structure of the soil. Water is sticky. The capacity of water molecules to cling to one another (the basis for surface tension) and to the surfaces of soil particles (a tendency known as capillary attraction) is responsible for the retention of water in soil. The more surface area it has, the more water a soil can hold.

Soils consist of grains of clay, silt, and sand, as well as particles of organic material, in varying proportions. Because the total surface area of particles in a given volume of soil increases as their size decreases, soils with abundant clay (particles less than 0.002 mm in diameter) and silt (0.002–0.05 mm) hold more water than coarse sands (> 0.05 mm), through which water drains quickly (**Figure 3.2**).

Plant roots easily take up water that clings loosely to soil particles. But close to the surfaces of soil particles, water adheres tightly by more powerful forces of attraction. The strength of these forces is called the **water potential** of the soil. Because the physical matrix of the soil generates these forces, they are often referred to as the **matric potential** of the soil. Soil scientists quantify soil water potential in terms of pressure. In the International System of Units (see Appendix A), the unit of pressure is the pascal (Pa) or the megapascal (MPa), where 1 MPa equals 1 million Pa. However, in this book, we shall use the more familiar unit of the atmospheric pressure at the surface of the earth. One atmosphere (1 atm) is equal to 101,325 Pa, or 0.1 MPa. By convention, water potentials

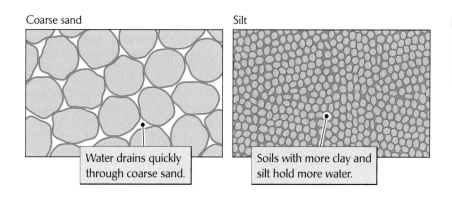

Coarse sand   Silt

Water drains quickly
through coarse sand.

Soils with more clay and
silt hold more water.

Figure 3.2 Finer-grained soils hold more water. Soils with large grain sizes have large spaces between them that are not filled completely at field capacity.

are expressed as negative values because potentials measure the capacity of one substance to extract water from another. Larger negative values indicate greater water potential. Water moves in the direction of the lower (more negative) water potential—that is, toward the substance with the strongest attraction for water.

Matric potential is greatest at the surface of soil particles and decreases with distance from them. Water held by a matric potential of less than about −0.1 atm drains out of the soil under the pull of gravity and joins the groundwater in the crevices of the bedrock below. This applies to water in the interstices between large soil particles, which is generally more than 0.005 mm from their surfaces. The amount of water held against gravity by water potential more negative than −0.1 atm is called the **field capacity** of the soil. Imagine a particle of silt with a diameter of 0.01 mm enlarged to the size of this page (×25,000); the film of water held at field capacity by the matric potential would be as thick as half the width of the page.

As soils dry out, the remaining water is held with increasingly stronger force, on average, because a greater proportion of the water lies close to the surfaces of soil particles. Soils having water potentials as great as −100 atm are very dry. Most plants can extract water held with water potentials as great as −15 atm. At more negative water potentials, plants wilt, even though some water still remains in the soil. Thus, ecologists refer to a water potential of −15 atm as the **wilting coefficient** or **wilting point** of the soil.

Plants obtain water from the soil by the osmotic potential of their root cells

Water in the environment and in the bodies of organisms has many substances dissolved in it. Dissolved substances, called **solutes,** influence the diffusion of water molecules. Plants take advantage of the tendency of water to move

from regions of low solute concentration (high water concentration) to regions of high solute concentration. Ions and other solutes diffuse through water in the opposite direction—from regions of high concentration to regions of low concentration. This movement of ions and water molecules tends to even out concentrations of solutes within a volume of water. Water also moves readily across most biological membranes. Thus, when a cell maintains a high concentration of ions and other solutes, water tends to move from the surrounding environment into the cell. This process is called **osmosis.** The force with which a solution attracts water by osmosis is known as its **osmotic potential.** Like the water potential of soil, osmotic potential is expressed as a pressure; specifically, as the pressure that would be required to keep water from diffusing into the solution (**Figure 3.3**). It is the osmotic potential in the roots of trees that causes water to enter the roots from the soil against the attraction of soil particles.

If the solute responsible for the osmotic potential of a solution also can diffuse across cell membranes, then its concentration within cells and its concentration in the surrounding water will eventually come into equilibrium. At this point, the osmotic potentials of the cell and its surroundings will be equal, and there will be no net movement of water across the cell membrane. This equalization of osmotic potential can be prevented by two mechanisms. First, a membrane can be **semipermeable,** meaning that some small molecules and ions can diffuse across it, but larger ones cannot. Many carbohydrates and most proteins are too large to pass through the pores of a cell membrane. Membranes may also transport ions and small molecules actively against a diffusion gradient to maintain their concentrations within the cell. However, this **active transport** requires expenditure of considerable energy.

The osmotic potential generated by an aqueous solution depends on the concentration of the solute. More specifically, it depends on the number of solute molecules or ions per volume of solution. Thus, a given mass of a

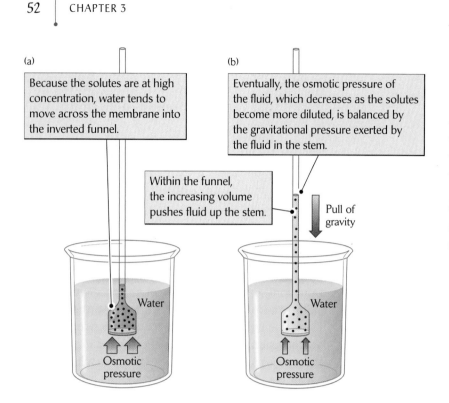

(a)

Because the solutes are at high concentration, water tends to move across the membrane into the inverted funnel.

Within the funnel, the increasing volume pushes fluid up the stem.

Water

Osmotic pressure

(b)

Eventually, the osmotic pressure of the fluid, which decreases as the solutes become more diluted, is balanced by the gravitational pressure exerted by the fluid in the stem.

Pull of gravity

Water

Osmotic pressure

▌ **Figure 3.3 Solutes enclosed within a membrane permeable to water develop an osmotic potential.** (a) Because the solutes are at high concentration, water tends to move across the membrane into the inverted funnel. (b) Within the funnel, the increasing volume of water pushes the fluid up the stem. Eventually, the osmotic pressure of the fluid, which decreases as the solutes become more diluted, is balanced by the gravitational pressure exerted by the fluid in the stem.

small solute generates greater osmotic potential than the same mass of a larger molecule. You will remember from your introductory chemistry course that the concentration of molecules in solution is expressed in terms of gram molecular weights, or moles, per liter. For example, the sugar glucose ($C_6H_{12}O_6$) has a molecular weight of 180, and so a 1 molar solution of glucose contains 180 grams of glucose per liter of water.

A molar concentration of a substance in solution creates an osmotic potential of −21 atmospheres. Thus, the water potential of the root hair cells of plants with a wilting point of −15 atm is equivalent to the osmotic potential of an approximately 0.7 molar solution. Plants growing in deserts and salty environments have been shown to increase (make more negative) the water potential of their roots to as much as −60 atm by increasing the concentrations of amino acids, carbohydrates, or organic acids in their root cells. They pay a high metabolic price, however, to maintain such concentrations of dissolved substances.

Forces generated by transpiration help to move water from roots to leaves

Osmotic potential draws water from the soil into the cells of plant roots. But how does the water get from the roots to the leaves? Plants conduct water to their leaves through xylem elements, which are the empty remains of xylem cells in the cores of roots and stems, connected end-to-end to form the equivalent of water pipes. For water to flow into these elements, their water potential must be more negative than that in the living cells of roots, into which water enters from the soil. Then, for water to move through the xylem from roots to leaves, the water potential of the leaves must exceed that of the roots enough to draw water upward against the osmotic potential of the root cells, the pull of gravity, and the resistance of the xylem elements.

Leaves generate water potential when water evaporates from leaf cell surfaces into the atmosphere, a process known as **transpiration.** Dry air at 20°C has a water potential of −1,332 atm. Thus, even with high relative humidity, air has more than enough water potential to pull water through the roots, xylem, and leaves. Transpiration creates a continuous gradient of more negative water potential from the soil to the air.

The water potential that is produced in leaf cells by transpiration has led to the **tension–cohesion theory** of water movement (▌ Figure 3.4). This theory states that the force required to move water within xylem elements is generated when water moves from the vessels in the vascular tissues of leaves to leaf cells to replace transpiration losses. This force, which represents a water potential as high as −20 to −50 atm, is transmitted all the way to the roots of the plants through the water column in the xylem elements

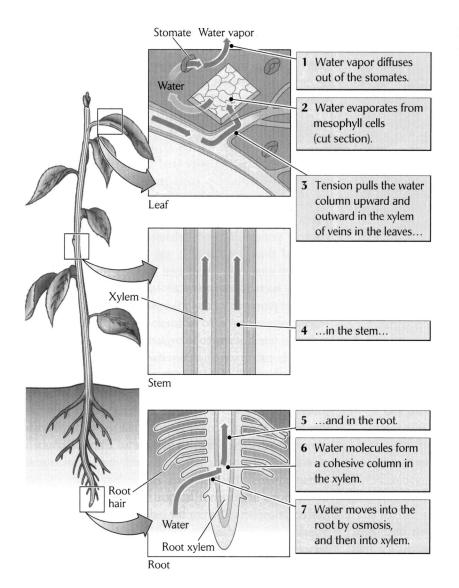

Stomate Water vapor

1 Water vapor diffuses out of the stomates.

Water

2 Water evaporates from mesophyll cells (cut section).

3 Tension pulls the water column upward and outward in the xylem of veins in the leaves...

Leaf

Xylem

4 ...in the stem...

Stem

5 ...and in the root.

6 Water molecules form a cohesive column in the xylem.

Root hair

7 Water moves into the root by osmosis, and then into xylem.

Water

Root xylem

Root

Figure 3.4 According to the tension–cohesion theory of water movement, the water potential that moves water from the roots to the leaves of a plant is generated by transpiration.

---

of the stem. Thus, water is literally pulled from the roots by the force generated from transpiring leaves. The water potential required to overcome soil water potential and draw water against the force of gravity to the top of a tall tree represents a force equivalent to many atmospheres.

Adaptations to arid environments control loss of water from leaves

For plants that live in dry habitats, where water is limiting, the rate of photosynthesis represents a balance between the need to acquire carbon dioxide and the need to conserve water. Openings at the leaf surfaces, called **stomates** (**Figure 3.5**), are the point of entry for CO_2, and also allow water to escape to the atmosphere by transpiration. Plants control water loss by closing their stomates. As leaf water potential decreases, the cells bordering the stomates collapse slightly and close the openings. This prevents further water loss, and also prevents carbon dioxide from entering the leaf.

The loss of water that accompanies the uptake of CO_2 presents a problem for plants in hot climates (see Figure 2.9). And because the vapor pressure of water increases with temperature, heat magnifies the problem of water loss. Heat- and drought-adapted plants have anatomic and physiologic modifications that reduce transpiration across plant surfaces, reduce heat loads, and enable plants to tolerate high temperatures. When plants absorb sunlight, they heat up. Plants can minimize overheating by increasing their surface area for heat dissipation and by protecting

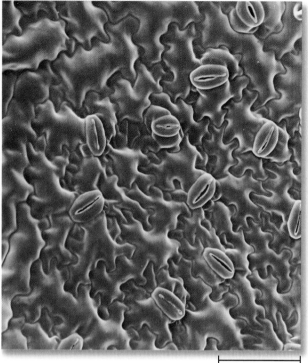

50 μm

Figure 3.5 Plants control water loss by opening and closing stomates on the leaf surface. Courtesy of Michele McCauley, from P. H. Raven, R. F. Evert, and S. E. Eichorn, *Biology of Plants,* 6th ed., W. H. Freeman and Company and Worth Publishers, New York (1999), p. 630.

their surfaces from direct sunlight with dense hairs and spines (**Figure 3.6**). Spines and hairs also produce a still boundary layer of air that traps moisture and reduces evaporation. Because thick boundary layers retard heat loss as well, hairs are prevalent in arid environments that

are cool, but less so in hot deserts. Plants may further reduce transpiration by covering their surfaces with a thick, waxy cuticle that is impervious to water and by recessing the stomates in deep pits, often themselves filled with hairs (**Figure 3.7**).

Plants obtain mineral nutrients from soil water

Plants acquire mineral nutrients—primarily nitrogen, phosphorus, potassium, and calcium—from dissolved ionic forms of these elements in soil water. Their uptake of abundant elements whose ions diffuse rapidly in the soil solution, such as calcium (Ca^{2+}) and magnesium (Mg^{2+}), is limited primarily by the absorptive capacity of the roots. Plants compensate for low levels of a nutrient in the soil by active transport or by increasing root growth.

In laboratory experiments, barley and beet roots took up phosphorus by diffusion when its concentration in the water surrounding the roots exceeded a critical level, approximately 0.2–0.5 millimolar (mM) (1 millimole equals one-thousandth of a mole), which is slightly higher than the concentration of phosphorus in root tissues. However, at soil concentrations below 0.2–0.5 mM, the rate of phosphorus uptake by diffusion would have been too low to meet the plant's needs, and the roots actively transported phosphorus across their surfaces. Active transport requires the expenditure of energy as the root tissue moves ions against a concentration gradient.

Plants may also respond to a scarcity of soil nutrients by increasing the extent of the root system. When plants increase their root growth, they do so at the expense of

(a)   (b)

Figure 3.6 Spines and hairs help plants adapt to heat and drought. (a) Cross section and (b) surface view of the leaf of the desert perennial herb *Enceliopsis argophylla*, which uses this strategy. Courtesy of J. R. Ehleringer. From H. R. Ehleringer, in E. Rodrigues, P. Healy, and I. Mehta (eds.), *Biology and Chemistry of Plant Trichomes,* Plenum Press, New York (1984), pp. 113–132.

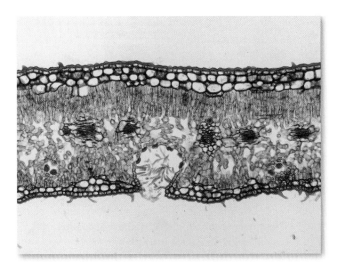

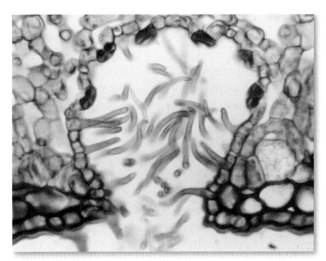

❚ Figure 3.7 Oleander, a drought-resistant plant, reduces water loss by placing its stomates in hair-filled pits on the leaf's undersurface. The hairs reduce water loss by slowing air movement and trapping water. The photo on the right, showing the pit in detail, is magnified about 400 times. The dark red staining cells in the pit lining are the guard cells surrounding the stomate openings. Photos by Jack M. Bostrack/Visuals Unlimited.

shoot growth (❚ Figure 3.8). This strategy balances the nutrient requirements of the plant with nutrient availability by reducing the nutrient demand created by the leaves, increasing the absorptive surface area of the root system,

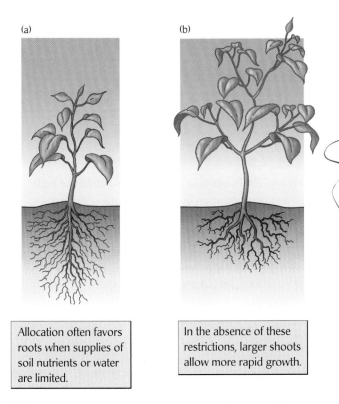

(a) Allocation often favors roots when supplies of soil nutrients or water are limited.

(b) In the absence of these restrictions, larger shoots allow more rapid growth.

❚ Figure 3.8 Plants respond to scarcity of nutrients by allocation of root or shoot growth.

and sending roots into new areas of soil from which the plant has not already removed scarce minerals.

Species adapted to nutrient-poor soils may also cope with low nutrient availability by establishing symbiotic relationships with fungi, which enhance mineral absorption, and by growing slowly and retaining leaves for long periods, thereby reducing nutrient demand. Such species typically cannot respond to artificially increased nutrient levels by increasing their growth rates. Instead, their roots absorb more nutrients than the plant requires and store them for subsequent use when the soil nutrient availability declines.

ECOLOGISTS IN THE FIELD

Effects of patchy soil nutrients on plant growth

Like most of the environment, soils are heterogeneous, often at very small scales. The nutrient supply in a small patch of soil depends on recent inputs from decaying wood, fruits, leaves, feces, urine, and bodies of animals as well as disturbances, such as trampling and burrowing, and local depletion of nutrients by plant roots. In one study in a sagebrush steppe habitat near Logan, Utah, pairs of soil samples only 12.5 cm apart varied by a factor as high as 12 in available nitrogen. Samples 3 cm apart varied on average by a factor of almost 3. How do plants respond to so much variation in their immediate environments? Do plants grow better or worse in highly heterogeneous soils than in uniform soils? Do all plants respond in the same way to this variation?

To answer these questions, James Cahill and Brenda Casper, at the University of Pennsylvania in Philadelphia, grew two common herbaceous plants individually in pots having either patchy or uniform distributions of nutrients in the soil. The plants were ragweed (*Ambrosia artemesiifolia*) and pokeweed (*Phytolacca americana*). Nutrients were added to the soil in the form of 75 ml of dried cow manure, which was either mixed uniformly throughout the soil (homogeneous treatment) or concentrated in a 1.5 cm diameter hole approximately 5 cm from the center of the pot (heterogeneous treatment). The plants were watered on a regular basis. After 9 weeks, Cahill and Casper terminated the experiments and determined the dry masses of the root and shoot portions of the plants. Roots were removed from the soil by washing them over a fine sieve. In both species, shoot biomass was greater in the heterogeneous soil than in the homogeneous soil (▌Figure 3.9). Although the total root biomass did not differ between treatments for either species, both grew higher densities of roots in the cow manure patches in the heterogeneous treatment. Because of the high concentration of nutrients in these patches, the rate of nutrient uptake increased enough to support faster growth aboveground.

It was surprising, therefore, that when ragweed and pokeweed were grown together in the same pot, heterogeneous soils did not stimulate greater shoot growth in either species. One explanation is that both plants grew roots into the nutrient patch and divided the nutrients between them, causing nutrient levels in the patch to drop to levels too low to stimulate aboveground growth. These experiments show that plants can respond to variations in nutrient levels in the soil, but they also suggest that fine-scale heterogeneity may not be so important in natural environments, where the root responses of plants quickly exhaust patches of concentrated nutrients. Thus, by seeking out and utilizing nutrient patches, plants tend to increase the uniformity of the soil on a local scale.

Photosynthesis varies with levels of light

The rate of photosynthesis in plants varies in direct proportion to the amount of light under low light intensities. With brighter light, however, the rate of photosynthesis increases more slowly or levels off as intensity increases. The response of photosynthesis to light intensity has two reference points (▌Figure 3.10). The first, called the **compensation point,** is the level of light intensity at which photosynthetic assimilation of energy just balances plant

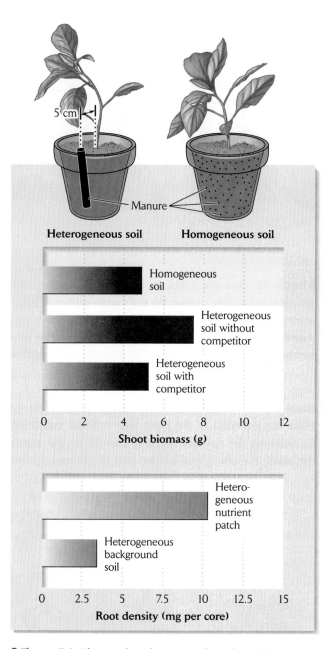

▌**Figure 3.9 Plants take advantage of patches of high nutrient concentration in heterogeneous soils.** Shoot growth of ragweed and pokeweed in homogeneous and heterogeneous soils without neighbors (*above*) and with neighbors (*below*). In heterogeneous soils, both species grew more roots in the high-nutrient patch than in the background soil. From J. F. Cahill and B. B. Casper, *Annals of Botany* 83: 471–478 (1999).

respiration. Above the compensation point, the energy balance of the plant is positive; below the compensation point, the energy balance is negative. The second reference point is the **saturation point,** above which the rate

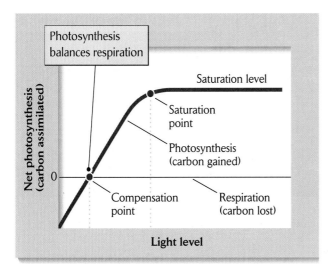

Figure 3.10 Photosynthesis increases asymptotically with increasing light intensity. The compensation point is the level of light at which photosynthesis (measured here by carbon dioxide assimilation) balances respiration (black line), and the saturation point is where photosynthesis levels off. After M. G. Barbour, J. H. Burk, and W. D. Pitts, *Terrestrial Plant Ecology*, Benjamin/Cummings, Menlo Park, CA (1980).

of photosynthesis no longer responds to increasing light intensity because the photosynthetic pigments are saturated with light. Among terrestrial plants, the compensation points of species that normally grow in full sunlight (a maximum of about 500 W per m^2) occur between 1 and 2 W per m^2. The saturation points of such species usually are reached between 30 and 40 W per m^2—less than a tenth of the energy level of bright, direct sunlight. As one might expect, the compensation and saturation points of plants that typically grow in shade occur at lower light intensities.

Plants modify photosynthesis in environments with high water stress

C_3 photosynthesis

For most plants, the first step in photosynthesis is the conversion of CO_2 into an organic molecule called 3PG, or phosphoglycerate (**Figure 3.11a**). We can represent this step as

$$CO_2 + RuBP \rightarrow 2 \ 3PG,$$
$$\text{1 carbon} \quad \text{5 carbons} \quad \text{3 carbons}$$

where RuBP (ribulose bisphosphate) is a five-carbon compound. Because the product of this step is a three-carbon compound, biologists call this pathway **C_3 photosynthesis.** 3PG then enters what is known as the Calvin–Benson cycle, which regenerates one molecule of RuBP while making one carbon atom available to synthesize glucose. All of these processes occur in the mesophyll cells of the leaves.

The enzyme responsible for the assimilation of carbon, RuBP carboxylase, has a low affinity for carbon dioxide. Consequently, at the low concentration of CO_2 found in the atmosphere and the resulting low concentration in the mesophyll cells, plants assimilate carbon inefficiently. To achieve high rates of carbon assimilation, plants must pack their mesophyll cells with large amounts of RuBP carboxylase—up to 30% of the dry weight of leaf tissue. However, this enzyme also catalyzes the reverse reaction,

$$2 \ 3PG \rightarrow CO_2 + RuBP,$$

in the presence of high oxygen and low carbon dioxide concentrations, especially at elevated leaf temperatures. This reaction partially undoes what RuBP carboxylase accomplishes when it assimilates CO_2, making photosynthesis inefficient and self-limiting. Carbon assimilation therefore tends to inhibit itself as levels of CO_2 decline and levels of oxygen produced by photosynthesis increase in the leaves. Plants could lessen this condition by keeping their stomates open—but that, of course, would lead to high water loss.

C_4 photosynthesis

Raising the concentration of CO_2 and reducing the concentration of O_2 in the leaf tissue could solve the problem of water loss created by the inefficiency of C_3 photosynthesis. Many plants in hot climates modify C_3 photosynthesis by using a different initial step in the assimilation of carbon dioxide as well as by spatially separating that initial assimilation step from the Calvin–Benson cycle within the leaf. Biologists call this modification **C_4 photosynthesis** because the assimilation of CO_2 initially results in a four-carbon compound:

$$CO_2 + PEP \rightarrow OAA,$$
$$\text{1 carbon} \quad \text{3 carbons} \quad \text{4 carbons}$$

where PEP (phosphoenol pyruvate) contains three carbons and OAA (oxaloacetic acid) contains four. The assimilatory reaction is catalyzed by the enzyme PEP carboxylase, which, unlike RuBP carboxylase, has a high affinity for CO_2. Assimilation occurs in the mesophyll cells

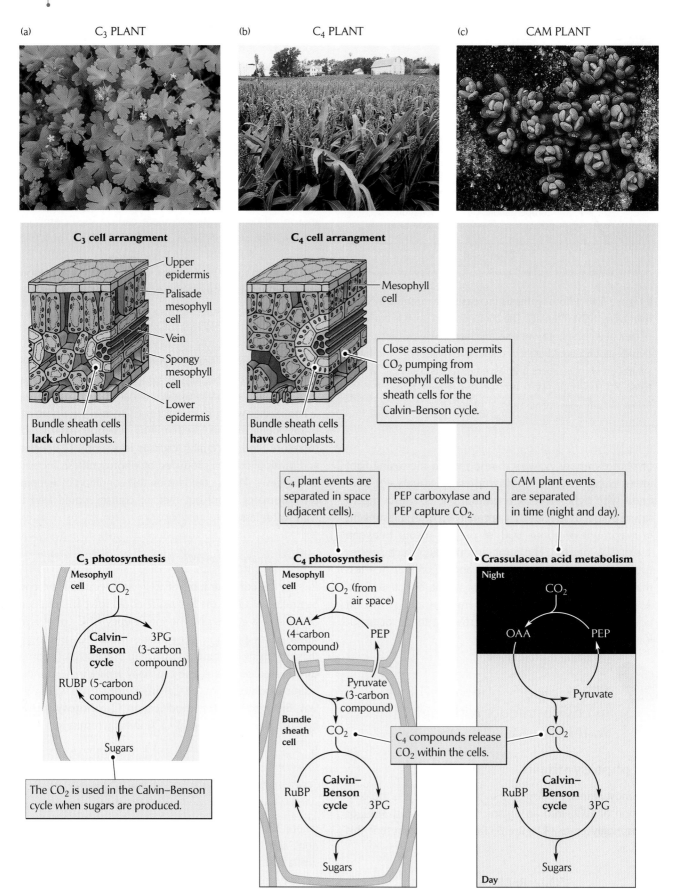

(a)   C$_3$ PLANT

(b)   C$_4$ PLANT

(c)   CAM PLANT

C$_3$ cell arrangment

Upper epidermis

Palisade mesophyll cell

Vein

Spongy mesophyll cell

Lower epidermis

Bundle sheath cells **lack** chloroplasts.

C$_4$ cell arrangment

Mesophyll cell

Close association permits CO$_2$ pumping from mesophyll cells to bundle sheath cells for the Calvin–Benson cycle.

Bundle sheath cells **have** chloroplasts.

C$_4$ plant events are separated in space (adjacent cells).

PEP carboxylase and PEP capture CO$_2$.

CAM plant events are separated in time (night and day).

C$_3$ photosynthesis

Mesophyll cell

CO$_2$

Calvin–Benson cycle

3PG (3-carbon compound)

RUBP (5-carbon compound)

Sugars

The CO$_2$ is used in the Calvin–Benson cycle when sugars are produced.

C$_4$ photosynthesis

Mesophyll cell

CO$_2$ (from air space)

OAA (4-carbon compound)

PEP

Pyruvate (3-carbon compound)

Bundle sheath cell

CO$_2$

Calvin–Benson cycle

RuBP   3PG

Sugars

C$_4$ compounds release CO$_2$ within the cells.

Crassulacean acid metabolism

Night

CO$_2$

OAA   PEP

Pyruvate

CO$_2$

Calvin–Benson cycle

RuBP   3PG

Sugars

Day

❚ Figure 3.11 The process of photosynthesis is modified in plants in water-stressed habitats. (a) A C_3 plant, the wild dovefoot geranium (*Geranium molle*); (b) A C_4 plant, cultivated sorghum (*Sorghum vulgare*); (c) a CAM plant, the Sierra sedum (*Sedum obtusatum*). Below the photos are idealized cross sections of a leaf, illustrating the locations of chloroplasts (small dark green dots) in each plant type. At the bottom, the major steps of the Calvin–Benson cycle are shown for each plant type. Photo (a) by Bert Kragas/Visuals Unlimited; photo (b) by John Spragens, Jr.; photo (c) by John Gerlach/DRK Photo.

of the leaf, but in most C_4 plants, photosynthesis (including the Calvin–Benson cycle) takes place in specialized cells surrounding the leaf veins, called bundle sheath cells (❚ Figure 3.11b). Oxaloacetic acid diffuses into the bundle sheath cells, where it is converted to malic acid, which then breaks down to produce CO_2 and pyruvate, a three-carbon compound. The CO_2 enters the Calvin–Benson cycle, just as it does in C_3 plants. The pyruvate moves back into the mesophyll cells, where enzymes convert it to PEP. This strategy solves the problem of creating high concentrations of CO_2 in leaf tissue. Because the bundle sheath cells are removed from the surface of the leaf, oxygen concentrations also are reduced.

C_4 photosynthesis confers an advantage because CO_2 can be concentrated within the bundle sheath cells to a level that far exceeds its equilibrium established by diffusion from the atmosphere. At this higher concentration, the Calvin–Benson cycle operates more efficiently. Also, because the enzyme PEP carboxylase has a high affinity for CO_2, it can bind CO_2 at a lower concentration in the cell, thereby allowing the plant to open its stomates less and reduce water loss. The disadvantage of C_4 carbon assimilation is that less leaf tissue is devoted to photosynthesis itself, thereby reducing the maximum potential photosynthetic rate. Consequently, C_3 photosynthesis is favored in cool climates with abundant soil water.

Carbon assimilation in CAM plants

Certain succulent plants in water-stressed environments use the same biochemical pathways as C_4 plants, but segregate CO_2 assimilation and the Calvin–Benson cycle between day and night. The discovery of this arrangement in plants of the family Crassulaceae (the stonecrop family; sedum is one example) and the initial assimilation and storage of CO_2 as four-carbon organic acids (malic acid and OAA) led to the name **crassulacean acid metabolism,** or **CAM.**

CAM plants open their stomates for gas exchange during the cool desert night, at which time transpiration of

water is minimal. CAM plants initially assimilate CO_2 in the form of four-carbon OAA and malic acid, which the leaf tissues store in high concentrations in vacuoles within the cells (❚ Figure 3.11c). During the day, the stomates close, and the stored organic acids are gradually recycled to release CO_2 to the Calvin–Benson cycle. The assimilation of CO_2 and the regeneration of PEP are regulated by different enzymes that have different temperature optima. CAM photosynthesis results in extremely high water use efficiencies and enables some types of plants to exist in habitats too hot and dry for other, more conventional species.

Salt balance and water balance go hand in hand

The water balance of aquatic animals is closely tied to the concentrations of salts and other solutes in their body tissues and in the environment. The osmotic potential of seawater is about −12 atmospheres, and that of fresh water is close to zero. The body fluids of vertebrate animals, which have an osmotic potential of about −3 to −5 atm (30–40% of that of seawater), occupy an intermediate position. Thus, the tissues of freshwater fish have higher concentrations of salts than the surrounding water. Such organisms, which are **hyperosmotic,** tend to gain water from and lose solutes to their surroundings. Marine fish, which have lower concentrations of salts than the surrounding seawater, are referred to as **hypo-osmotic.** They tend to gain solutes and lose water. Fish solve these osmotic problems by using active transport mechanisms to pump ions in one direction or the other across various body surfaces (skin, kidney tubules, and gills), expending considerable energy in the process.

Ion retention is critical to freshwater organisms

Freshwater fish continuously gain water by osmosis across the surfaces of the mouth and gills, which are the most

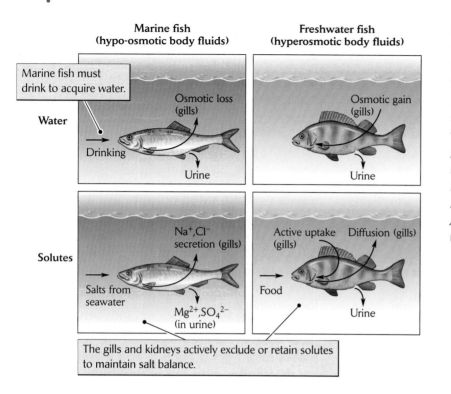

**Marine fish
(hypo-osmotic body fluids)**

**Freshwater fish
(hyperosmotic body fluids)**

Marine fish must drink to acquire water.

Water

Osmotic loss (gills)

Drinking

Urine

Osmotic gain (gills)

Urine

Solutes

Na⁺,Cl⁻ secretion (gills)

Salts from seawater

Mg^{2+}, SO_4^{2-} (in urine)

Active uptake (gills)

Diffusion (gills)

Food

Urine

The gills and kidneys actively exclude or retain solutes to maintain salt balance.

Figure 3.12 Pathways of exchange of water and solutes differ between marine and freshwater fish. The body fluids of marine fish are hypo-osmotic (have a lower salt concentration than the surrounding water), whereas those of freshwater fish are hyperosmotic. The gills and kidneys actively exclude or retain solutes to maintain salt balance. Marine fish must drink to acquire water. After K. Schmidt-Nielsen, *Animal Physiology: Adaptation and Environment*, Cambridge University Press, Cambridge (1975).

permeable of their tissues that are exposed to the surroundings, and in their food (**Figure 3.12**). To counter this influx, fish eliminate excess water in their urine. If fish did not also selectively retain dissolved ions, however, they would soon become lifeless bags of water. The kidneys of freshwater fish retain salts by actively removing ions from the urine and infusing them back into the bloodstream. In addition, the gills can selectively absorb ions from the surrounding water and release them into the bloodstream.

Water retention is critical to marine organisms

Marine fish are surrounded by water with a salt concentration higher than that of their bodies. As a result, they tend to lose water to the surrounding seawater and must drink seawater to replace it (see Figure 3.12). The salt that comes in with the water and with food, as well as that which diffuses in across body surfaces, must be excreted at great metabolic cost from the gills and kidneys.

Some sharks and rays have found a solution to the problem of water flux. Sharks retain urea [CO(NH₂)₂]—a common nitrogenous waste product of metabolism in vertebrates—in the bloodstream instead of excreting it from the body in the urine. The urea raises the osmotic potential of the blood to the level of seawater without any increase in the concentration of sodium and chloride ions (**Figure 3.13**). Consequently, there is no net movement of

water across a shark's surfaces. This makes it much easier for sharks to regulate the flux of ions such as sodium because they do not have to drink extra salt-laden water to replace water lost by osmosis. The fact that freshwater species of sharks and rays do not accumulate urea in their blood emphasizes the importance of urea for osmoregulation in marine members of this group.

The small copepod *Tigriopus* takes an approach to water balance similar to that of sharks. *Tigriopus* lives in pools high in the splash zone along rocky coasts (**Figure 3.14**). These pools receive seawater infrequently from the splash of high waves, and as the water evaporates, the salt concentration rises to high levels. A heavy rainfall, on the other hand, can rapidly lower the salt concentration in these pools. Ron Burton, at the Scripps Institute of Oceanography, has shown that *Tigriopus*, like sharks, manages its water balance by changing the osmotic potential of its body fluids. When salt concentration is high, it synthesizes large quantities of certain amino acids such as alanine and proline. These small molecules increase the osmotic potential of the body to match that of the habitat, without the deleterious physiological consequences of high levels of salt. This response to salt stress is costly, however. In a laboratory experiment, individual *Tigriopus* were switched from 50% seawater to 100% seawater, as might happen when waves at high tide filled a pool previously flushed with rainwater. In response to this change, the respiration rate of the copepods initially declined, owing to the initial

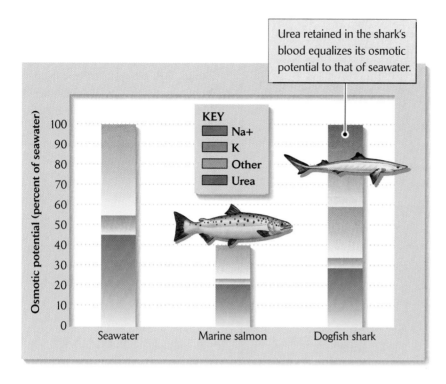

Urea retained in the shark's blood equalizes its osmotic potential to that of seawater.

Figure 3.13 Sodium, potassium, urea, and other solutes (mostly chloride ions) contribute differently to the osmotic potential of seawater and of the body fluids of marine fish and sharks. From data in K. Schmidt-Nielsen, *Animal Physiology: Adaptation and Environment* (5th ed.), Cambridge University Press, London and New York (1997), Table 8.6.

salt stress, and then increased as they synthesized alanine and proline to restore their water balance (**Figure 3.15**). When salinity was suddenly decreased from 100% of seawater down to 50% of seawater, their respiration rate immediately increased as excess free amino acids were rapidly degraded and metabolized.

Certain environments pose special osmotic problems. Aquatic environments with salt concentrations greater than that of seawater occur in some landlocked basins, particularly in dry regions where evaporation considerably exceeds precipitation. The Great Salt Lake (20% salt, or

about six times saltier than normal seawater) in Utah and the Dead Sea (23% salt), lying between Israel and Jordan, are well-known examples of such hypersaline environments. The osmotic potentials of these bodies of water—well in excess of −100 atm—would shrivel most organisms. However, a few aquatic creatures, such as brine shrimp (*Artemia*), can survive in salt water concentrated to the point of crystallization (300 grams per liter, or 30%). Brine shrimp excrete salt at a prodigious rate, and a high energy cost, to keep the salts in their body fluids less concentrated than those in their surroundings.

(a)

(b)

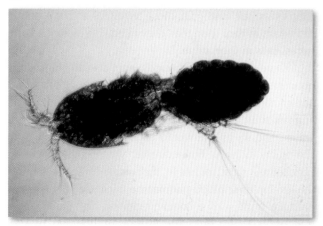

Figure 3.14 The tiny copepod *Tigriopus* lives in splash pools high in the rocky intertidal zone in California. Photo (a) courtesy of Ron Burton; photo (b) by R. E. Ricklefs.

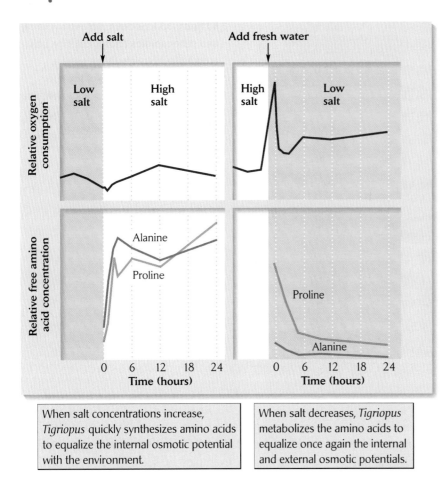

Figure 3.15 *Tigriopus* alters free amino acid levels and metabolic rate in response to hyperosmotic stress (50% to 100% seawater, *left*) and hypo-osmotic stress (100% to 50% seawater, *right*). From E. M. Goolish and R. S. Burton, *Functional Ecology* 3:81–89 (1989).

When salt concentrations increase, *Tigriopus* quickly synthesizes amino acids to equalize the internal osmotic potential with the environment.

When salt decreases, *Tigriopus* metabolizes the amino acids to equalize once again the internal and external osmotic potentials.

Water balance and salt balance in terrestrial organisms

Terrestrial plants transpire hundreds of grams of water for every gram of dry matter they accumulate in tissue growth, and they inevitably take up salts along with the water that passes into their roots. In saline environments, plants actively pump excess salts back into the soil across their root surfaces, which therefore function as the plant's "kidneys." Mangrove plants grow on coastal mudflats that are inundated daily by high tides (**Figure 3.16**). Not only does this habitat impose a high salt load, but the high osmotic potential of the root environment also makes it difficult for the roots to take up water. To counter this problem, many mangrove plants have high levels of organic solutes, such as the amino acids proline and glycinebetaine and the six-carbon sugar sorbitol, in their roots and leaves to increase their osmotic potential. In addition, they have salt glands that secrete salt by active transport to the exterior surface of the leaves. The roots of many species exclude salts, apparently by means of semipermeable membranes that do not allow the salts to enter. Mangrove plants further reduce salt loads by decreasing the transpiration of water from their leaves.

Because many of these adaptations resemble those of plants from environments where water is scarce, the mangrove habitat has been referred to as an osmotic desert.

Because they are not immersed continuously in fresh water, terrestrial animals have little trouble retaining ions. They acquire the mineral ions they need in the water they drink and the food they eat. Lack of sodium in some areas forces animals to obtain salt directly from such mineral sources as salt licks, but most terrestrial animals obtain more salts in their food than they need. They eliminate the excess salts in their urine.

Where fresh water abounds, animals can drink large quantities of water to flush out salts that would otherwise accumulate in the body. Where water is scarce, however, animals must produce a concentrated urine to conserve water. And so, as one would expect, desert animals have champion kidneys. For example, whereas humans can concentrate most solutes in their urine to about 4 times the level in their blood plasma, the kangaroo rat's kidneys produce urine with a concentration as high as 14 times that of its blood, and the Australian hopping mouse, another desert-adapted species, produces urine that has 25 times the solute concentration of its blood. Because

(a)

(b)

❚ Figure 3.16 Mangrove plants have adaptations for coping with a high salt load.
(a) The roots of mangrove vegetation are immersed in salt water at high tide. Some species exclude salt from their roots. (b) Specialized glands in the leaves of the button mangrove, *Conocarpus erecta*, excrete salt, which precipitates on their outer surfaces. Photos by R. E. Ricklefs.

sodium and chloride ions are part of the mechanism by which the kidney retains water, the kidney is not particularly effective at excreting these ions. Hence, many organisms lacking access to fresh water have specialized salt-secreting organs. Birds and reptiles have "salt glands," which are modified tear glands located in the orbit of the eye and which are capable of secreting a concentrated salt solution. These glands are especially well developed in species that feed on marine organisms and receive high salt loads in their diets.

The following experiment shows the relative importance of the salt gland in ridding the body of excess sodium ions. A gull was given 134 ml of seawater containing 63 millimoles (mmol) of sodium—a concentration of 470 mM. After 3 hours, it had excreted 47.3 mmol of sodium in 56.3 ml of water from the salt gland, at a concentration of 800 mM, nearly twice the concentration of salt in seawater. At the same time, only 4.4 mmol of

sodium were excreted by the kidneys, in 75.2 ml of water at an average concentration of 59 mM. Without the salt gland, the gull could not rid itself of salt in its diet without losing too much water.

Animals excrete excess nitrogen in the form of small organic molecules

Most carnivores, whether they eat crustaceans, fish, insects, or mammals, consume excess nitrogen in their diets. This nitrogen, which is part of the proteins and nucleic acids of their prey, must be eliminated from the body when these compounds are metabolized. Most aquatic animals produce the simple metabolic by-product ammonia (NH_3). Although ammonia is mildly poisonous to tissues, aquatic animals eliminate it rapidly in a copious, dilute urine, or directly across the body surface, before it reaches a dangerous concentration within the body. Terrestrial animals cannot afford to use large quantities of water to excrete nitrogen. Instead, they produce metabolic by-products that are less toxic than ammonia and which can therefore accumulate to higher levels in the blood and urine without danger. In mammals, this waste product is urea [$CO(NH_2)_2$], the same substance that sharks produce and retain to achieve osmotic balance in marine environments. Because urea dissolves in water, excreting it requires some urinary water loss—how much depends on the concentrating power of the kidneys. Birds and reptiles have carried adaptation to terrestrial life one step further: they excrete nitrogen in the form of uric acid ($C_5H_4N_4O_3$), which crystallizes out of solution and can then be excreted as a highly concentrated paste in the urine. Although excreting urea and uric acid saves water, there is also a cost, which is the energy lost in the organic carbon used to form these compounds. For each atom of nitrogen excreted, 0.5 and 1.25 atoms of organic carbon are lost in urea and uric acid respectively.

Water conservation mechanisms are important in hot environments

When air and substrate temperatures approach or exceed the maximum tolerable body temperature, animals can dissipate heat only by evaporating water from their skin and respiratory surfaces. In deserts, the scarcity of water makes evaporative heat loss a costly mechanism; animals often must reduce their activity, seek cool microclimates, or undertake seasonal migrations to cooler regions. Many desert plants orient their leaves in such a way as to avoid the direct rays of the sun; others shed their leaves and become inactive during periods of combined heat and water stress.

Among mammals, the kangaroo rat is well suited to life in a nearly waterless environment (■ Figure 3.17). Its large intestine resorbs water from waste material so efficiently that it produces virtually dry feces. Kangaroo rats also recover much of the water that evaporates from their lungs by condensation in their enlarged nasal passages. When the kangaroo rat inhales dry air, moisture in its nasal passages evaporates, cooling the nose and saturating the inhaled air with water. When moist air is exhaled from the lungs, much of the water vapor condenses on the cool nasal surfaces. By alternating condensation with evaporation during breathing, the kangaroo rat reduces its respiratory water loss. The cold, moist nose of a dog serves the same function.

■ **Figure 3.17 The behavior and physiological features of the kangaroo rat adapt it exquisitely to desert environments.** This banner-tailed kangaroo rat (*Dipodomys spectabilis*) of the American Southwest is active at night when the air is cool. The darkness also affords some safety from predators. Photo by Marty Cordano/DRK Photo.

Kangaroo rats avoid the desert's greatest heat by feeding only at night, when it is cooler; during the blistering heat of the day, kangaroo rats remain comfortably below ground in their cool, humid burrows. In sharp contrast, ground squirrels remain active during the day. They also conserve water by restricting evaporative cooling. As you would expect, their body temperatures rise when they forage above ground, exposed to the hot sun. How do they manage? Before their body temperatures become dangerously high, they return to their burrows, where they cool down by conduction and radiation rather than by evaporation. By shuttling back and forth between their burrows and the surface, ground squirrels extend their activity into the heat of the day and pay a relatively small price in water loss.

Like that of the ground squirrel, the camel's body temperature rises in the heat of the day—by as much as 6°C. Large body size gives the camel a distinct advantage, however. Because of its low surface-to-volume ratio, the camel heats up so slowly that it can remain in the sun most of the day. It dumps excess heat at night by conduction and radiation to the cooler surroundings.

Organisms maintain a constant internal environment

Homeostasis is an organism's ability to maintain constant internal conditions in the face of a varying external environment. All organisms exhibit homeostasis to some degree, as we have seen in the case of water and salt balance, although the occurrence and effectiveness of homeostatic mechanisms vary. Regardless of how organisms regulate their internal environments, all homeostatic systems exhibit **negative feedback,** meaning that when the system deviates from its norm, or desired state, internal response mechanisms act to restore that state. Those of you who use thermostats to regulate temperature in your homes can readily understand how a negative feedback system works. When the house is cold, a temperature-sensitive switch turns on a heater, which restores the temperature to its desired setting. Homeostatic mechanisms in animals and plants work in much the same way (❚ Figure 3.18).

Most mammals and birds maintain their body temperatures between 36°C and 41°C, even though the temperature of their surroundings may vary from −50°C to +50°C. Such temperature regulation, which is referred to as **homeothermy** (the Greek root *homos* means "same"), creates constant temperature (homeothermic) conditions within cells, under which biochemical processes can proceed efficiently. Cold-blooded, or **poikilothermic,** organisms, such as frogs and grasshoppers, conform to the external temperature (the Greek root *poikilo* means "varying"). Thus, frogs cannot function at either high or low temperature extremes, so they are active only within a narrow part of the range of environmental conditions over which mammals and birds thrive.

Many so-called cold-blooded organisms, including reptiles, insects, and plants, adjust their heat balance behaviorally simply by moving into or out of shade, by changing their orientation with respect to the sun, or by adjusting their contact with warm substrates. When horned lizards are cold, they lie flat against the ground and gain heat by conduction from the sun-warmed surface. When hot, they decrease their exposure to the surface by standing erect upon their legs. Basking behavior is widespread among reptiles and insects, which can use it effectively to regulate their body temperatures within a narrow range. Indeed, their temperatures may rise considerably above that of surrounding air, well into the range of the "warm-blooded" birds and mammals. Because their source of heat lies outside the body, biologists refer to these animals as **ectotherms** (external heat); animals that generate their

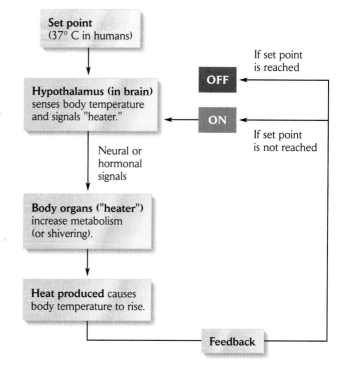

❚ **Figure 3.18 The essential features of a negative feedback system include sensors and switches.** The hypothalamus, like a thermostat, compares body temperature with a set point; when the two differ, it signals the effector organs to bring body condition back into line with the set point.

body heat internally are referred to as **endotherms** (internal heat).

Homeostasis is costly

Sustaining internal conditions that differ significantly from the external environment requires work and energy. Let us consider the costs to birds and mammals of maintaining constant high body temperatures in cold environments. As air temperature decreases, the gradient (difference) between internal and external environments increases. Heat is lost across body surfaces in direct proportion to this gradient. An animal that maintains its body temperature at 40°C loses heat twice as fast at an ambient (surrounding) temperature of 20°C (a gradient of 20°C) as at an ambient temperature of 30°C (a gradient of only 10°C). To maintain a constant body temperature, endothermic organisms replace heat lost to their environment by generating heat metabolically. Thus, the rate of metabolism required to maintain body temperature increases in direct proportion to the difference between body and ambient temperature, all other things being equal.

Limits to homeothermy

An organism's ability to sustain a high body temperature while exposed to extremely low ambient temperatures is limited. Over the short term, the physiological capacity to generate heat limits heat production, and therefore defines the coldest temperature that a homeotherm can withstand. Over the long term, a homeotherm is limited by its ability to gather food or metabolize nutrients to satisfy the energy requirements of generating heat. The maximum rate at which an organism can perform work, even during the most strenuous exercise, generally does not exceed ten to fifteen times its minimum, or basal, level of metabolism. Such high rates of metabolism, typical of a bird in flight, are rarely maintained for more than a few minutes or hours at a time. Over the course of a day, few organisms—even migrating birds—expend energy at a rate exceeding four or five times the basal metabolic rate.

When the environment becomes so cold that heat loss exceeds an organism's physiological capacity to produce heat, body temperature begins to drop, a condition that is fatal to most homeotherms. The lowest environmental temperatures that homeotherms can survive for long periods often depend on their ability to gather food, rather than on their ability to assimilate and metabolize the energy in food. At low temperatures, animals may starve rather than freeze to death when they metabolize food energy to maintain body temperature more rapidly than they can gather food.

Partial homeostasis

All birds and mammals generate heat metabolically to regulate body temperature, but many cold-blooded species also become endothermic or partially endothermic at times. Pythons, for example, maintain high body temperatures while incubating eggs. Some large fishes, such as the tuna, maintain temperatures up to 40°C in the center of their metabolically active muscles; swordfish employ special metabolic heaters, derived from muscle tissue, to keep their brains warm. Large moths and bees often require a preflight warm-up period during which the flight muscles shiver to generate heat. Even some plants, notably philodendron and skunk cabbage, use metabolic heat production to raise the temperature of their floral structures. In this case, the high temperatures volatilize chemicals used to attract insect pollinators.

Because they are small, hummingbirds have a large exposed surface area relative to their mass, and consequently lose heat rapidly relative to the amount of tissue that is available to produce it. As a result, hummingbirds must sustain very high metabolic rates to maintain their at-

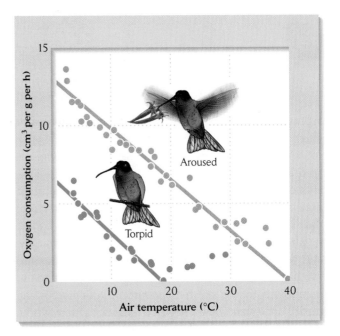

▌**Figure 3.19 Hummingbirds maintain a constant low body temperature when in torpor.** Energy metabolism (measured by oxygen consumption) increases with decreasing air temperature in *Eulampis jugularis* during periods of both torpor and normal arousal. The bird regulates its body temperature in each case, but at different set points. After F. R. Hainsworth and L. L. Wolf, *Science* 168:368–369 (1970).

rest body temperatures near 40°C. Species living in cool climates would starve overnight if they did not enter **torpor,** a voluntary, reversible condition of lowered body temperature and inactivity. The West Indian hummingbird, *Eulampis jugularis,* drops its body temperature to between 18°C and 20°C when resting at night. It does not cease to regulate its body temperature; it merely changes the setting on its thermostat to reduce the difference between ambient and body temperature, and thereby reduces the energy expenditure needed to maintain its temperature at the regulated set point (▌Figure 3.19).

Large animals deliver oxygen to their tissues through circulatory systems

Most animals release the chemical energy contained in organic compounds primarily by respiration. Because oxygen plays such an important role in this process, low availability of oxygen can restrict metabolic activity. Oxygen availability is a particular problem in aquatic habitats, where its solubility is low and its diffusion is slow. Even ter-

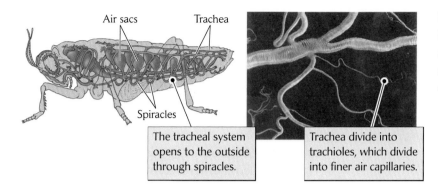

Air sacs Trachea

Spiracles

The tracheal system opens to the outside through spiracles.

Trachea divide into trachioles, which divide into finer air capillaries.

▌Figure 3.20 Insects get oxygen to their body tissues through the tracheal system, a system of branching pipes through which air can move. Scanning electron micrograph courtesy of Thomas Eisner, Cornell University.

restrial organisms, which breathe an atmosphere containing abundant oxygen, must get oxygen through an aqueous medium to all the tissues of the body.

Active organisms require an abundant supply of oxygen for cell respiration. Diffusion can satisfy the oxygen needs of tiny aquatic organisms, but the centers of organisms larger than about 2 mm in diameter are too far from the external environment for diffusion to ensure a rapid supply of oxygen. Tissue metabolism consumes diffusing oxygen before it has gone much farther than a millimeter. Insects have solved this problem with systems of branching pipes (tracheae) that carry air directly to the tissues (▌Figure 3.20). Other animals have blood circulatory systems to distribute oxygen from the respiratory surfaces to the body.

Complex protein molecules such as hemoglobin and hemocyanin, to which oxygen molecules readily attach,

increase the oxygen-carrying capacity of the blood of most animals. While blood plasma can carry only a small amount of oxygen in solution (about 1% by volume), whole blood can transport up to 50 times more oxygen bound to these oxygen-carrying molecules. In most animals utilizing hemoglobin, including all vertebrates, this protein is packed densely into red blood cells, whose color is that of hemoglobin. When oxygen binds to hemoglobin—four molecules of oxygen can bind to each molecule of hemoglobin—it comes out of solution in the blood plasma, making room for the diffusion of more oxygen into the blood from the lungs or gills. In the body tissues, where oxygen concentrations are low, the binding process reverses, and oxygen is released (▌Figure 3.21).

These adaptations for procuring oxygen illustrate a set of solutions to the problems that organisms confront at the

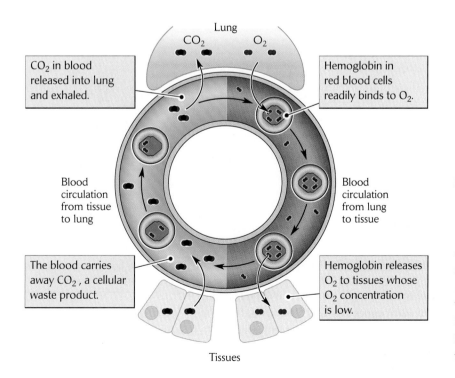

Lung

CO_2 O_2

CO_2 in blood released into lung and exhaled.

Hemoglobin in red blood cells readily binds to O_2.

Blood circulation from tissue to lung

Blood circulation from lung to tissue

The blood carries away CO_2, a cellular waste product.

Hemoglobin releases O_2 to tissues whose O_2 concentration is low.

Tissues

▌Figure 3.21 Most circulatory systems have oxygen-binding proteins. The binding of oxygen by hemoglobin in the red blood cells lowers the concentration of dissolved oxygen in the blood plasma and speeds the diffusion of oxygen from the lungs into the bloodstream. In the body's tissues, where the concentration of oxygen is low, the process is reversed, and oxygen is offloaded from hemoglobin and diffuses toward regions of high metabolic rate.

interface between themselves and their environments. The hemoglobin molecule, which must bind oxygen efficiently in the lungs but release it easily in the tissues, demonstrates that adaptation often requires compromise. No hemoglobin molecule designed only for maximum efficiency of oxygen binding could also release oxygen easily where it was needed, and vice versa. Therefore, the hemoglobin molecules that have evolved represent a compromise between these two functions.

Countercurrent circulation increases transfer of heat and substances between fluids

Solutes diffuse from regions of high concentration to regions of low concentration. With time, this movement equalizes concentrations, and the net movement of solutes stops. Heat is conducted from hotter to cooler substances. Eventually temperatures equalize, and net movement of heat comes to a standstill. Thus, diffusion reduces the efficiency of gas exchange and the transport of ions in excretory organs, and conduction works against the retention of heat within the body. How do organisms counteract these problems?

Such problems have been solved in many cases by a simple and effective arrangement of vessels carrying moving fluids called **countercurrent circulation**. This mechanism is illustrated by the structure of fish gills, which causes water and blood to flow in opposite directions (■Figure 3.22). In a countercurrent system, blood moving in one direction continually encounters water flowing past in the opposite direction. This water has progressively greater oxygen concentrations because it has traveled a progressively shorter distance along the gill lamella. This arrangement maintains a large gradient of oxygen concen-

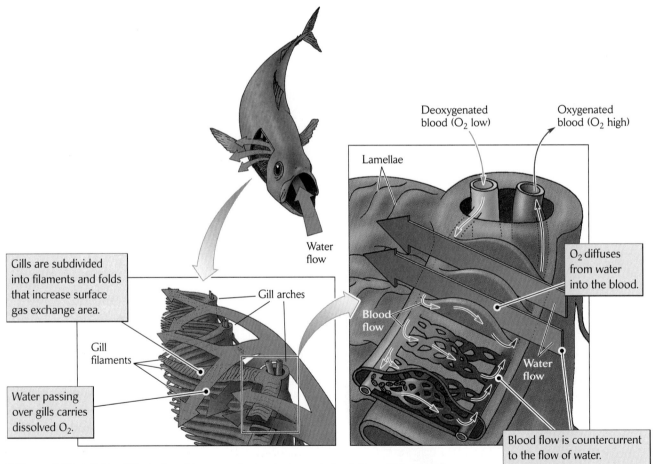

Deoxygenated blood (O$_2$ low)

Oxygenated blood (O$_2$ high)

Lamellae

Water flow

Gill arches

Gills are subdivided into filaments and folds that increase surface gas exchange area.

Gill filaments

Water passing over gills carries dissolved O$_2$.

Blood flow

O$_2$ diffuses from water into the blood.

Water flow

Blood flow is countercurrent to the flow of water.

■**Figure 3.22 A fish's gill is designed to promote countercurrent circulation of blood and water.** The gill consists of several gill arches, each of which carries two rows of filaments. The filaments have thin lamellae (leaflike structures) oriented parallel to the flow of water through the gill. Within the lamellae, blood flows in the direction opposite to the movement of water past the surface, establishing a countercurrent circulation. From D. J. Randall, *Am. Zool.* 8:179–189 (1968).

tration between the blood plasma and the surrounding water, so that the oxygen from the water diffuses readily into the blood. With this mechanism, the oxygen concentration in the blood can approach the concentration in the surrounding water.

The countercurrent circulation principle appears frequently in adaptations that increase the flux of heat or materials between fluids. Among terrestrial organisms, birds have a unique lung structure that, unlike that of the lungs of mammals, results in a one-way flow of air opposite to the flow of blood. This adaptation allows birds, with lungs whose weight and volume are small, to achieve the high rates of oxygen delivery required by their active lives. The extremities of some birds and mammals have countercurrent blood circulation to reduce loss of heat to the surrounding environment. Because the legs and feet of most birds do not have feathers, they would be major avenues of heat loss in cold regions were they not held at a lower temperature than the rest of the body (Figure 3.23). Gulls conserve heat by using countercurrent heat exchange in their legs. Warm blood in arteries leading to the feet cools as it passes close to veins that return cold blood to the body. In this way, heat is transferred from arterial to venous blood and transported back into the body rather than lost to the environment. Tuna use the same principle to retain heat in the active swimming muscles close to the body core, a strategy that allows them to swim very fast and pursue smaller fish as prey even in cold oceans.

Each organism functions best under a restricted range of conditions

Each organism generally has a narrow range of environmental conditions to which it is best suited, which define its **optimum.** The optimum is subject to natural selection, which acts on variations in the properties of enzymes and lipids, the structures of cells and tissues, and the form of the body to enable the organism to function well under the particular conditions of its environment.

Adaptation to different environmental conditions often involves shifts in the optima of biochemical processes. For example, several enzymes of the halophilic (salt-loving) bacterium *Halobacterium salinarium* exhibit peaks of activity at higher salt concentrations than the same enzymes of species adapted to low salt concentrations (Figure 3.24). However, although the optimum salt concentration for *H. salinarium* is very high, not all of its individual enzymes have high salt optima. The structures of some enzymes make it impossible for them to function well at high salt concentrations, and the organism just has to make do.

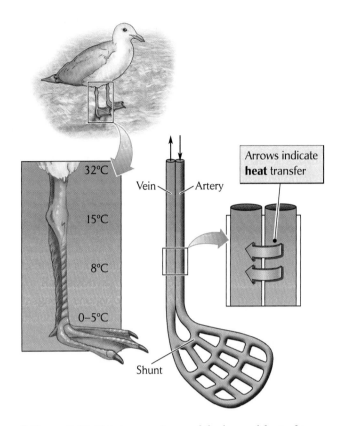

 Figure 3.23 **Skin temperatures of the leg and foot of a gull standing on ice show that heat is retained in the body.** The anatomic arrangement of blood vessels and countercurrent heat exchange between arterial and venous blood are diagrammed at right. A shunt between the artery and vein in the leg allows blood vessels in the feet to constrict, thereby reducing blood flow and heat loss further, with no increase in blood pressure. After L. Irving, *Sci. Am.* 214:93–101 (1966); K. Schmidt-Nielsen, *Animal Physiology*, Cambridge University Press, New York (1975).

Organisms sometimes accommodate predictable changes in environmental conditions by having more than one form of an enzyme or structural molecule, each of which functions best within a different range of conditions. The rainbow trout, for example, experiences low temperatures in its native habitat during winter, when water temperatures may drop close to the freezing point, and much higher temperatures in summer. These seasonal changes in temperature are predictable, and the trout responds by producing different forms of many enzymes in winter and summer. One of these enzymes is acetylcholinesterase, which plays an important role in degrading neurotransmitters and ensuring proper functioning of the nervous system. The affinity of this enzyme for its substrate, acetylcholine, is a good measure of enzyme function.

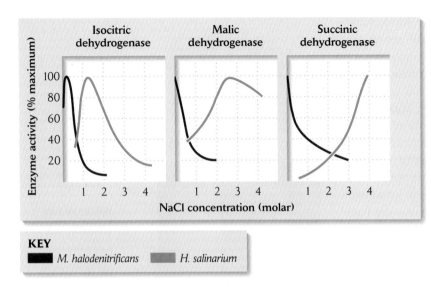

Figure 3.24 Optimal conditions for biological molecules reflect the prevalent conditions of their environment. Activity levels of selected enzymes found in the halophilic bacterium *Halobacterium salinarium*, and of the same enzymes in bacteria that cannot tolerate high salinities, at various salt concentrations. (The concentration of salt in seawater is about 0.6 molar.) After H. Larsen, in I. C. Gunsalus and R. Y. Stanier (eds.), *The Bacteria: A Treatise on Structure and Function*, Vol. 4, *The Physiology of Growth*, Academic Press, New York (1962), pp. 297–342.

Substrate affinity in the winter form of the enzyme is high between 0°C and 10°C, but drops rapidly at higher temperatures. Substrate affinity of the summer form of the enzyme is low at 10°C, rises to a peak between 15°C and 20°C, and drops slowly at higher temperatures (**Figure 3.25**). Which form of the enzyme a trout produces at a particular time depends directly on the temperature of the water. When trout are maintained at 2°C, they produce the winter form; at 17°C, they produce only the summer form.

Measurements of enzyme activities are made in the laboratory from extracts of plant and animal tissues. The values obtained in such studies reveal the performance of a single type of molecule under highly controlled conditions. The performance of the whole organism, however, depends on the integration of many biochemical processes that together must be adjusted to conditions that the organism experiences. Comparisons of the function of organisms from different environments make the best test of how well performance matches environment. For example, many fish in the frigid oceans surrounding Antarctica swim actively and consume oxygen at a rate comparable to fish living among tropical coral reefs. Put a tropical fish in cold water, however, and it becomes sluggish and soon dies; conversely, antarctic fish cannot tolerate temperatures warmer than 5°–10°C.

How can fish from cold environments swim as actively as fish from the Tropics? Swimming depends on a series of biochemical transformations, most of which are catalyzed by enzymes. Because most of these transformations occur more rapidly at high temperatures than at low temperatures, cold-adapted organisms must either have more of the substrate for a biochemical reaction, more of the enzyme that catalyzes the reaction, or a qualitative change in the enzyme itself. As we have seen, a particular enzyme obtained from a variety of organisms may exhibit different

catalytic properties when tested over ranges of temperature, pH, salt concentration, and substrate abundance (see Figures 3.24 and 3.25).

This picture of metabolic compromise is greatly simplified, of course. Adapting to changes in the environment requires adjustment of whole metabolic pathways, which may involve changes in enzyme structure or concentration or the use of alternative metabolic pathways. Indeed, the examples of adaptation to the physical environment con-

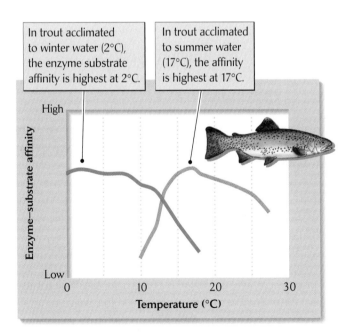

In trout acclimated to winter water (2°C), the enzyme substrate affinity is highest at 2°C.

In trout acclimated to summer water (17°C), the affinity is highest at 17°C.

Figure 3.25 Optimum conditions respond to changes in the environment. The relative substrate affinity of the enzyme acetylcholinesterase differs in trout raised at winter and at summer temperatures. Data from J. Baldwin and P. W. Hochachka, *Biochemical Journal* 116:883–887 (1970).

sidered in this chapter emphasize the unity of the organism and the interrelatedness of all facets of organism structure and function.

Summary

1. The mechanisms by which organisms interact with their physical environments help us to understand why organisms are specialized to narrow ranges of conditions and how adaptations of morphology and physiology are associated with certain conditions.

2. Because water clings tightly to soil surfaces, its availability depends in part on the physical structure of soil. Soils having a high proportion of small clay particles hold water more strongly than do sandy soils. The force by which soil holds water is called the water potential, or matric potential, of the soil. Most plants cannot remove water from soil when the water potential is more negative than −15 atmospheres of pressure. This is referred to as the wilting point of soil.

3. Plants extract water from soils by using solutes to generate high osmotic potentials in their root cells. According to the cohesion–tension theory, water is drawn from the roots to the leaves by the gradient in water potentials generated by transpiration—the evaporation of water from leaf surfaces.

4. Heat- and drought-adapted plants cut down on transpiration by reducing heat loads, using hairs on leaf surfaces to establish boundary layers of humid air, and waterproofing leaf surfaces with waxy cuticles.

5. Plants obtain mineral nutrients from the soil by passive diffusion when nutrients are abundant and by active uptake when they are scarce. When the concentrations of soil nutrients are low, plants can increase their total root surface area at the expense of shoot growth.

6. Photosynthesis varies in proportion to light intensity at low light levels. The level of light below which plant respiration exceeds photosynthesis is called the compensation point. Above the saturation point, usually 10–20% of direct sunlight, photosynthesis levels off.

7. During photosynthesis, most plants assimilate carbon through a reaction (the C_3 pathway) catalyzed by the enzyme RuBP carboxylase. This enzyme has a low affinity for carbon dioxide and brings about oxidation at high temperatures, resulting in a low efficiency of carbon assimilation. Plants adapted to high temperatures interpose a more efficient (C_4) carbon assimilation step, which is spatially separated from the C_3 reactions in the leaf. CAM plants

separate carbon assimilation and the Calvin–Benson cycle reactions into nighttime and daytime phases.

8. To maintain salt balance and water balance, freshwater organisms, which are hyperosmotic, retain salts while excreting the water that continuously diffuses into their bodies. Marine organisms, which are hypo-osmotic, actively exclude salts. Some marine organisms increase the level of solutes, such as urea and amino acids, in their body fluids to match the osmotic potential of seawater and thus reduce the movement of water out of their bodies.

9. Terrestrial organisms reduce water loss in part by concentrating salts and nitrogenous waste products in their urine or by excreting them through salt glands.

10. Nitrogenous waste products of protein metabolism are excreted as ammonia by most aquatic organisms, as urea by mammals, and as uric acid by birds and reptiles. Because uric acid crystallizes out of solution, birds and reptiles may excrete it at high concentrations and thereby gain considerable economy of water use.

11. Water stress increases with temperature. In dry environments, evaporative cooling is impractical, and animals must use other strategies to get rid of excess heat, such as seeking cool microclimates during the hottest period of the day.

12. Maintenance of constant internal conditions, called homeostasis, depends on negative feedback responses. Organisms sense changes in their internal environment and respond in such a manner as to return those conditions to a set point.

13. Homeostasis requires energy when a gradient between internal and external conditions must be maintained. For example, endotherms must generate heat metabolically to balance loss of heat to their cooler surroundings.

14. Oxygen diffuses too slowly to reach tissues more than about a millimeter from an organism's surface. Large animals overcome this problem either by conducting air directly to the tissues via a multibranched tracheal system (as in insects) or by transporting oxygen dissolved in circulating fluids throughout the body. Oxygen-binding proteins, such as hemoglobin, compensate for the low solubility of oxygen in water.

15. The uptake of oxygen by aquatic organisms is greatly facilitated by countercurrent circulation of blood through the gills in a direction opposite to that of water flowing over the gill surfaces. In this way, countercurrent circulation maintains high gradients of oxygen concentration, and the blood can achieve nearly the oxygen concentration of the surrounding water. Countercurrent arrangements are also used to retain heat within the body.

16. Most organisms function best within a narrow range of environmental conditions. These optima may be shifted by evolution to match more closely the environmental conditions within which the organism lives. This is often accomplished by altering the structure and quantity of enzymes responsible for controlling metabolic processes.

17. Overall, adaptation to the physical environment depends on reaching compromises between opposing functions to increase both the individual's chances of survival and its productivity in a particular environment.

PRACTICING ECOLOGY

CHECK YOUR KNOWLEDGE

Adaptations and Conservation

Throughout this chapter, we have seen many examples of how plants and animals have evolved to survive and reproduce in a variety of habitats. Deserts are particularly harsh terrestrial environments because of low, sporadic rainfall, high temperatures in summer (and in some deserts, low temperatures at night or in winter), high rates of evaporation, and strong winds. Desert tortoises are among the many animal species that are highly adapted to deserts. They lead solitary lives for most of the year. They spend a portion of the day in underground burrows to prevent overheating in summer and freezing in winter, and to reduce the amount of water they need for cooling. Desert tortoises start to reproduce at 12 to 20 years of age, when they lay about 4 to 6 eggs one or two times per year, depending, of course, on the ambient conditions of the desert.

Dr. Ken Nagy of the University of California, Los Angeles, and his colleagues have studied the water and nutrient balances of desert animals, including the desert tortoise *Gopherus agassizi*. Not surprisingly, water is of primary importance for the survival of desert tortoises. When rain does fall in the desert, the tortoises dig depressions in the soil and drink the rain they collect. They can also resorb water from their bladders during dry times. Desert tortoise reproduction is highest in years when there is enough winter rain for annual plants, which are important food resources for tortoises, to grow. Cattle grazing on desert lands are affecting the abundances of annual plants and, therefore, the desert tortoises.

CHECK YOUR KNOWLEDGE

1. Why is it important for desert tortoises to conserve their energy?

2. Go to Ken Nagy's Web page from *Practicing Ecology on the Web* at *http://www.whfreeman.com/ricklefs*. How did Nagy and his colleagues measure the amount of water used by desert animals?

Dr. Park Nobel (also of UCLA) and his colleagues have studied the adaptations of desert cacti to the hot, dry conditions of the desert Southwest, and in particular the functions of their spines. Cactus spines are leaves that have been modified over millions of years to serve several purposes, such as protection from herbivory, shading the stem to prevent overheating in summer, and insulating sensitive growth regions to protect them from freezing in winter. The combination of their unusual stem shapes (for water storage), the intricate patterns of their spines, and their bright flowers in red, orange, yellow, pink, and purple make cacti beautiful plants to observe in nature. Indeed, the remarkable appearance of cacti makes their growth and cultivation a popular hobby.

CHECK YOUR KNOWLEDGE

3. How are cacti and desert tortoises similar in their adaptations to the desert environment?
4. How do they differ?

 Suggested Readings

Cahill, J. F., Jr., and B. B. Casper. 1999. Growth consequences of soil nutrient heterogeneity for two old-field herbs, *Ambrosia artemesiifolia* and *Phytolacca americana*, grown individually and in combination. *Annals of Botany* 83:471–478.

Canny, M. J. 1998. Transporting water in plants. *American Scientist* 86:152–159.

Chapin, F. S. III. 1991. Integrated responses of plants to stress. *BioScience* 41:29–36.

Ehleringer, J. R., R. F. Sage, L. B. Flanagan, and R. W. Pearcy. 1991. Climate change and the evolution of C₄ photosynthesis. *Trends in Ecology and Evolution* 6:95–99.

Feldman, L. J. 1988. The habits of roots. *BioScience* 38:612–618.

Hochachka, P. W., and G. N. Somero. 1984. *Biochemical Adaptation.* Princeton University Press, Princeton, NJ.

Karov, A. 1991. Chemical cryoprotection of metazoan cells. *BioScience* 41:155–160.

Keeley, J. E. 1990. Photosynthetic pathways in freshwater aquatic plants. *Trends in Ecology and Evolution* 5(10): 330–333.

Kooyman, G. L., and P. J. Ponganis. 1997. The challenges of diving to depth. *American Scientist* 85:530–539.

Lee, R. E., Jr. 1989. Insect cold-hardiness: To freeze or not to freeze. *BioScience* 39:308–313.

Schmidt-Nielsen, K. 1998. *Animal Physiology: Adaptations and Environment,* 5th ed. Cambridge University Press, London and New York.

Vogel, S. 1988. *Life's Devices.* Princeton University Press, Princeton, NJ.

CHAPTER 4

Variations in the Physical Environment

Global patterns in temperature and precipitation are established by the energy of solar radiation

Ocean currents redistribute heat and moisture

Seasonal variation in climate is caused by the movement of the sun's zenith

Temperature and winds drive seasonal cycles in temperate lakes

Climate sustains irregular fluctuations

Topographic and geologic features cause local variation in climate

Climate and the underlying bedrock determine the diversification of soils

Few people make important decisions based on the evening weather report. Forecasting weather is notoriously difficult because of the irregular and unpredictable changes in the causes of weather patterns. On a global scale, one of the most dramatic weather conditions is the so-called El Niño, which is associated with periodic changes in air pressure patterns over the central and western Pacific Ocean. The cause of these changes is poorly understood, but their effects have been experienced, for better or worse, by most of the human population. For example, the 1991–92 El Niño event, one of the strongest on record, was accompanied by the worst drought of the twentieth century in Africa, which was followed by poor crop production and widespread starvation. The event also brought extreme dryness to many areas of tropical South America and Australasia. Droughts and hot temperatures in Australia reduced populations of red kangaroos to less than half their pre-El Niño levels. Outside the Tropics and Subtropics, El Niño events tend to increase, rather than decrease, precipitation, boosting the production of natural and agricultural systems, but also causing flooding. The most recent El Niño, in 1997–98, was blamed for 23,000 deaths—mostly from famine—and $33 billion in damages to crops and property worldwide.

Fluctuations in climate, whether local or affecting most of the globe, are one manifestation of variation in the earth's environment. Much of this variation can be traced to changes in incident radiation from the sun or to spatial patterns established by the shapes and positions of the earth's ocean basins, continents, and mountain ranges. On top of these predictable variations, physical and biological processes themselves establish new patterns of variation as the outcome of unpredictable interactions among their components. Ecologists strive to understand both the origin of variation in climate and its consequences for ecological systems.

The physical environment varies widely over the surface of the earth. Conditions of temperature, light, substrate, moisture, salinity, soil nutrients, and other factors have shaped the distributions and adaptations of plants, animals, and microbes. The earth has many distinct climate zones whose extents are broadly determined by the intensity of solar radiation and the redistribution of heat and moisture by wind and water currents. Within climate zones, such geologic factors as topography and composition of bedrock further differentiate the environment on a finer spatial scale. This chapter explores some important patterns of variation in the physical environment that underlie diversity in the biological components of ecosystems.

The surface of the earth, its waters, and the atmosphere above it make up a giant heat-transforming machine. Climatic patterns originate as the earth absorbs the energy in sunlight. As its surface varies from bare rock to forested soil, open ocean, and frozen lake, its ability to absorb sunlight varies as well, thus creating differential heating and cooling. The heat energy absorbed by the earth eventually radiates back into space, after undergoing further transformations that perform the work of evaporating water and driving the circulation of the atmosphere and oceans. All these factors have created a great variety of physical conditions that, in turn, have fostered the diversification of ecosystems.

Global patterns in temperature and precipitation are established by the energy of solar radiation

The earth's climate tends to be cold and dry toward the poles and hot and wet toward the equator. Although there are many exceptions to this general rule, climate does exhibit broadly defined patterns. The primary cause of global variation in climate is the greater intensity of sunlight at the equator than at higher latitudes. This is a simple consequence of the angle of the sun relative to the surface of the earth at different latitudes (▌Figure 4.1). The sun warms the atmosphere, oceans, and land most when it lies directly overhead. A beam of sunlight spreads over a greater area when the sun approaches the horizon, and it also travels a longer path through the atmosphere, where much of its energy either is reflected or is absorbed by the atmosphere and reradiated into space as heat. The sun's highest position each day varies from directly overhead in the Tropics to near the horizon in polar regions; thus the warming effect of the sun diminishes from the equator to the poles.

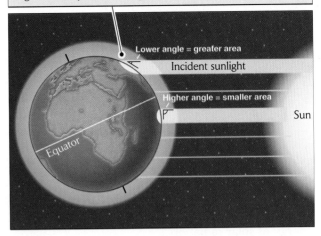

At higher latitudes, light strikes the earth's surface at a lower angle, and is spread over a greater area.

▌**Figure 4.1 The warming effect of the sun is greatest at the equator.** The sun is closer to the perpendicular at the equator and shines directly down at the surface of the earth during the middle of the day.

Patterns of change in temperature and precipitation are as important to biological systems as long-term averages. Periodic cycles in climate follow astronomical cycles: the rotation of the earth upon its axis causes daily periodicity; the revolution of the moon around the earth creates lunar cycles in the amplitude of the tides; and the revolution of the earth around the sun brings seasonal change.

The equator is tilted 23½° with respect to the path the earth follows in its orbit around the sun. Therefore, the Northern Hemisphere receives more solar energy than the Southern Hemisphere during the northern summer, and less during the northern winter (▌Figure 4.2). The seasonal range in temperature increases with distance from the equator, especially in the Northern Hemisphere, where there is less area of ocean to moderate temperature changes (▌Figure 4.3). At high latitudes in the Northern Hemisphere, mean monthly temperatures vary by an average of 30°C, with extremes of more than 50°C annually. For example, at 60°N, the average coldest month is −12°C and the average warmest month is 16°C, a difference of 28°C. Average temperatures of the warmest and coldest months in the Tropics are much higher, and they differ by as little as 2°C or 3°C.

The tilt of the earth also results in a seasonal shift in the area near the equator that receives the greatest amount of sunlight. This area lies at the **solar equator,** which is the parallel of latitude lying directly under the sun's zenith.

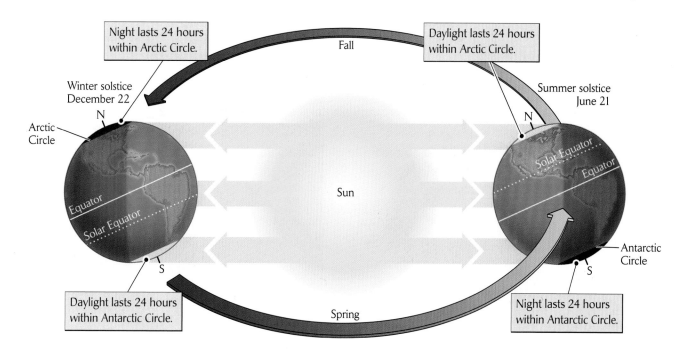

Figure 4.2 The orientation of the earth's axis relative to the sun changes between winter and summer, causing seasonal variation in climate. The position of the solar equator at the summer and winter solstices is indicated.

The solar equator reaches 23½° N on June 21 and 23½° S on December 21.

Hadley cells

Warming air expands, becomes less dense, and tends to rise. As air heats up, its ability to hold water vapor increases, and evaporation quickens. As we saw in Chapter 2, the rate of evaporation from a wet surface nearly doubles with each 10°C rise in temperature. The heat of the sun warms a mass of air in the Tropics, which rises and eventually spreads to the north and south in the upper layers of the atmosphere. This air is replaced from below by surface-level air from subtropical latitudes. The rising tropical air mass cools as it radiates heat back into space. By the time this air has extended to about 30° north and

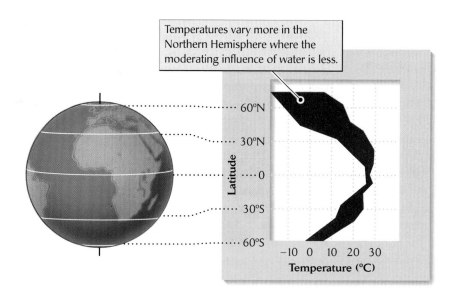

Figure 4.3 Annual range of mean monthly temperatures (red area) is greatest at high latitude in the Northern Hemisphere.

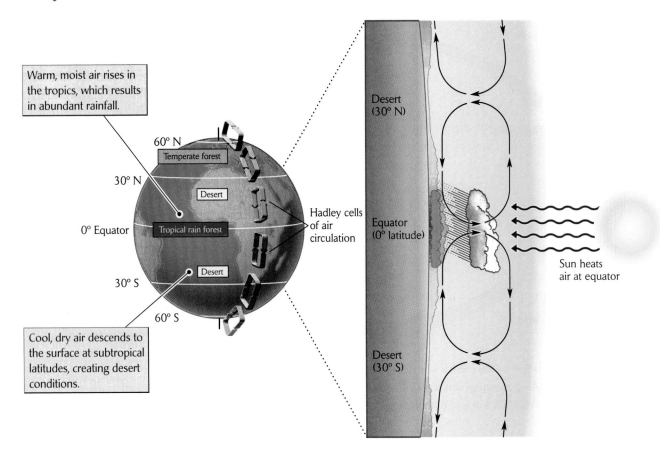

Warm, moist air rises in the tropics, which results in abundant rainfall.

60° N
Temperate forest
30° N
Desert
0° Equator
Tropical rain forest
Hadley cells of air circulation
Desert
30° S
60° S

Cool, dry air descends to the surface at subtropical latitudes, creating desert conditions.

Desert (30° N)

Equator (0° latitude)

Sun heats air at equator

Desert (30° S)

▌Figure 4.4 Differential warming of the earth's surface creates Hadley cells.
Warm, moist air rises in the Tropics, and cool, dry air moves toward the Tropics from subtropical latitudes to replace it. The intertropical convergence is the latitudinal belt at the solar equator within which surface winds converge from the north and south.

south of the equator, it has become dense enough to sink back to the earth's surface and spread out to the north and south, thus completing a cycling of air within the atmosphere (▌Figure 4.4). This type of circulation pattern is called a **Hadley cell.**

One Hadley cell forms around the earth immediately to the north of the equator and another to the south, like a pair of giant waistbands girdling the earth. The sinking air of the tropical Hadley cells drives secondary Hadley cells in temperate regions, which circulate in the opposite direction. The circulation of Hadley cells in temperate latitudes (roughly 30°–60° north and south of the equator) causes air to rise at about 60°N and 60°S, which in turn leads to the formation of polar Hadley cells. All this circulation of air is driven by the differential heating of the atmosphere with respect to latitude. The Hadley cells are linked together by the rising or falling air at the northern and southern edges of the cells. Thus, the movement of air in each Hadley cell helps to drive the circulation of the adjacent cells.

The intertropical convergence and subtropical high pressure belt

The region within which surface currents of air from the northern and southern Subtropics meet near the equator and begin to rise under the warming influence of the sun is referred to as the **intertropical convergence.** As the moisture-laden tropical air rises and begins to cool within the convergence area, the moisture condenses to form clouds and precipitation. Thus, the Tropics are wet not because there is more water at tropical latitudes than elsewhere, but because water cycles more rapidly through the tropical atmosphere. The heating effect of the sun causes water to evaporate and warmed air masses to rise; the cooling of the air as it rises and expands causes precipitation because cold air has a lower capacity to hold water.

The air mass moving high in the atmosphere to the north and south, away from the intertropical convergence, has already lost much of its water to precipitation in the Tropics. Because this air has cooled, it becomes denser and

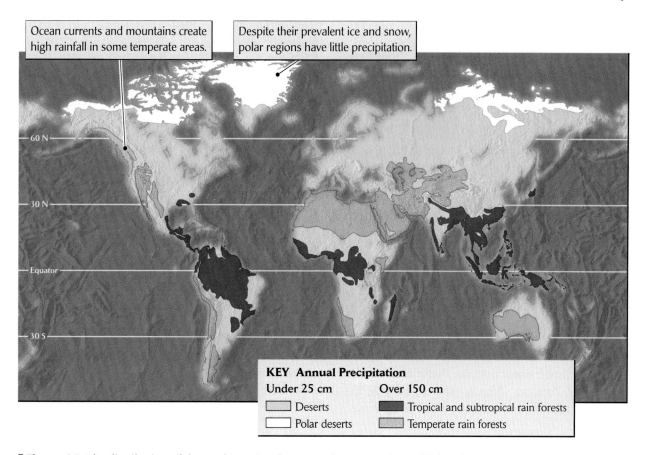

Ocean currents and mountains create high rainfall in some temperate areas.

Despite their prevalent ice and snow, polar regions have little precipitation.

KEY Annual Precipitation

Under 25 cm

☐ Deserts

☐ Polar deserts

Over 150 cm

■ Tropical and subtropical rain forests

▨ Temperate rain forests

▌**Figure 4.5 The distribution of the earth's major deserts and wet areas is established by the major latitudinal climate belts.** Wet areas in western North America, Chile, and New Zealand are temperate rain forests. Dry areas at high latitude are polar deserts.

begins to sink. This descending mass of heavy air creates a high atmospheric pressure, and so these regions north and south of the equator are known as the **subtropical high pressure belts.** As the air sinks and begins to warm again at subtropical latitudes, its capacity to evaporate and hold water increases. As it descends to ground level and spreads to the north and south, it draws moisture from the land, creating zones of arid climate centered at latitudes of about 30° north and south of the equator (▌Figure 4.5). The great deserts of the world—the Arabian, Sahara, Kalahari, and Namib of Africa; the Atacama of South America; the Mojave, Sonoran, and Chihuahuan of North America; and the Australian—all fall within subtropical high pressure belts.

Surface winds and rain shadows

The rotation of the earth deflects the surface flows in the Hadley cells because the rotational speed of the earth is higher close to the equator than it is at higher latitudes.

Consequently, surface flows are shifted to the west in the Tropics, where air moves away from the equator, and to the east in the middle latitudes, where air moves toward the equator (▌Figure 4.6). The resulting wind patterns, known as the trade winds and the westerlies, respectively, help to distribute water vapor through the atmosphere.

The positions of continental landmasses exert a secondary effect on the global pattern of precipitation. At any given latitude, rain falls more plentifully in the Southern Hemisphere because oceans and lakes cover a greater proportion of its surface (81% compared with 61% of the Northern Hemisphere). Water evaporates more readily from exposed surfaces of water than from soil and vegetation. For the same reason, the interior of a continent usually experiences less precipitation than its coasts, simply because it lies farther from the major site of water evaporation, the surface of the ocean. Furthermore, coastal (maritime) climates vary less than interior (continental) climates because the heat storage capacity of ocean waters reduces temperature fluctuations. For example, the hottest and coldest

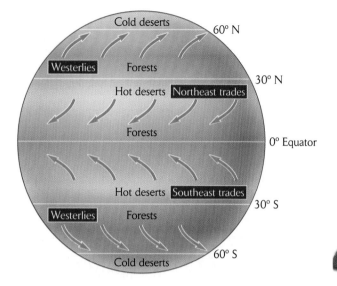

Figure 4.6 The earth's rotation deflects the air movement of Hadley cells to create global wind patterns.

Washington; 26°C at Helena, Montana; and 33°C at Bismarck, North Dakota.

Global wind patterns interact with other features of the landscape to create precipitation. Mountains force air upward, causing it to cool and lose its moisture as precipitation on the windward side of a mountain range. As the dry air descends the leeward slope and travels across the lowlands beyond, it picks up moisture and creates arid environments called **rain shadows** (▌Figure 4.7). The Great Basin deserts of the western United States and the Gobi Desert of Asia lie in the rain shadows of extensive mountain ranges.

Ocean currents redistribute heat and moisture

The physical conditions of the oceans, like those of the atmosphere, are complex. Variation in marine conditions is caused by winds, which propel the major surface currents of the oceans, and by the underlying topography of the ocean basin. In addition, deep currents are established by

mean monthly temperatures near the Pacific coast of the United States at Portland, Oregon, differ by only 16°C. Farther inland, this range increases to 18°C at Spokane,

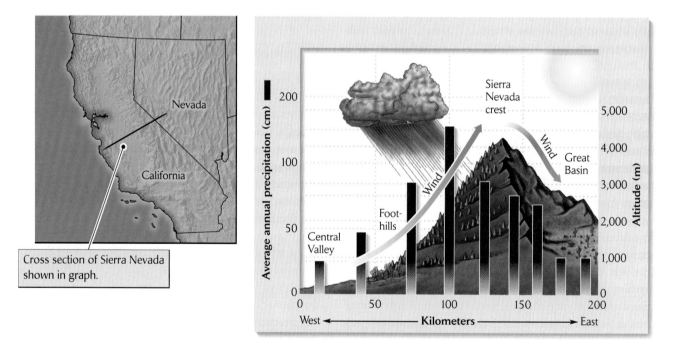

Figure 4.7 Mountain ranges influence local precipitation patterns. In the Sierra Nevada of California, the wind comes predominantly from the west across the Central Valley of California. As moisture-laden air is deflected upward by the mountains, it cools and its moisture condenses, resulting in heavy precipitation on the western slope. As the air rushes down the eastern slope, it warms and begins to pick up moisture, creating arid conditions in the Great Basin. After E. R. Pianka, *Evolutionary Ecology*, 4th ed., Harper & Row, New York (1988).

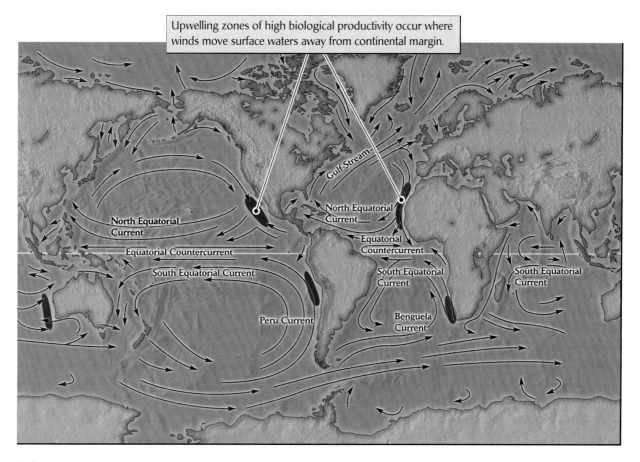

Upwelling zones of high biological productivity occur where winds move surface waters away from continental margin.

❚ Figure 4.8 **The major ocean currents are driven by winds and the earth's rotation.**
After A. C. Duxbury, *The Earth and Its Oceans*, Addison-Wesley, Reading, MA (1971).

differences in the density of ocean water caused by variations in temperature and salinity. In large ocean basins, cold water circulates toward the Tropics along the western coasts of the continents, and warm water circulates toward temperate latitudes along the eastern coasts of the continents (❚ Figure 4.8). The cold Peru Current of the eastern Pacific Ocean, which moves northward from the Antarctic Ocean along the coasts of Chile and Peru, creates cool, dry environments along the west coast of South America, in the rain shadow of the Andes Mountains, all the way to the equator. As a result, the coasts of northern Chile and Peru have some of the driest deserts on earth. Conversely, the warm Gulf Stream, emanating from the Gulf of Mexico, carries a mild climate far to the north into western Europe and the British Isles (see Figure 1.4).

Any upward movement of ocean water is referred to as **upwelling.** Upwelling occurs wherever surface currents diverge, as in the western tropical Pacific Ocean. As surface currents move apart, they tend to draw water upward from deeper layers. Strong upwelling zones are also established on the western coasts of continents where surface

currents move toward the equator. A curious consequence of the rotation of the earth is the deflection of these currents away from the continental margins, which is aided by winds. As this water moves away from the continents, it is replaced by water from greater depths. Because deep water tends to be rich in nutrients, upwelling zones are often regions of high biological productivity. The most famous of these support the rich fisheries of the Benguela Current along the western coast of southern Africa and the Peru Current along the western coast of South America.

Seasonal variation in climate is caused by the movement of the sun's zenith

Within the Tropics, the seasonal northward and southward movement of the solar equator determines the seasonality of rainfall. The intertropical convergence follows the solar equator, producing a moving belt of rainfall. Therefore,

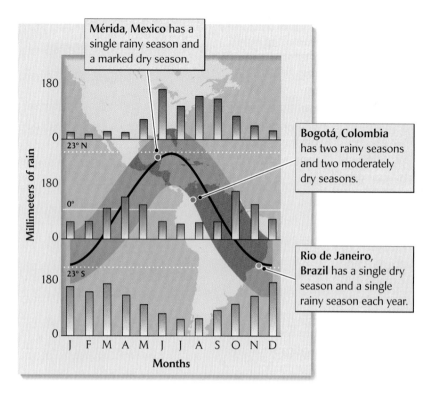

Mérida, Mexico has a single rainy season and a marked dry season.

Bogotá, Colombia has two rainy seasons and two moderately dry seasons.

Rio de Janeiro, Brazil has a single dry season and a single rainy season each year.

180
0
23° N
Millimeters of rain
180
0
0°
180
0
23° S
J F M A M J J A S O N D
Months

Figure 4.9 The movement of the solar equator affects precipitation patterns. The seasonal latitudinal movement of the solar equator (see Figure 4.2) results in two seasons of heavy precipitation at the equator and a single wet season alternating with a pronounced dry season at the edges of the Tropics.

seasonality of rainfall is most pronounced in broad latitudinal belts lying about 20° north and south of the equator.

Mérida, located on Mexico's Yucatán Peninsula, lies about 20° north of the equator. The intertropical convergence reaches Mérida only during the Northern Hemisphere summer months, which are the rainy season for that region (■ Figure 4.9). During the winter, the intertropical convergence lies far to the south of Mérida, and the local climate comes under the influence of the subtropical band of high pressure. Rio de Janeiro, at the same latitude as Mérida, but to the south of the equator, has its rainy season during the Northern Hemisphere winter, roughly six months after Mérida. Close to the equator at Bogotá, Colombia, the intertropical convergence passes overhead twice each year at the time of the equinoxes, resulting in two rainy seasons with peak rainfall in April and October. Thus, as the seasons change, tropical regions alternately come under the influence of the intertropical convergence, which brings heavy rains, and subtropical high-pressure belts, which bring clear skies.

Panama lies at 10°N and, like Mérida, it has a dry and windy winter and a humid, rainy summer. Panama's climate is wetter on the northern (Caribbean) side of the isthmus—the direction from which the prevailing trade winds come—than on the southern (Pacific) side; mountains intercept moisture coming from the Caribbean side and produce a rain shadow (■ Figure 4.10). The Pacific low-

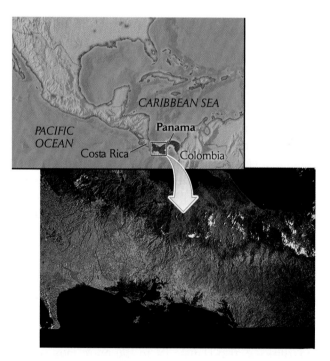

Figure 4.10 Trade winds create a rain shadow in Central America. This false-color satellite image of western Panama during the dry season shows heavy forest (brown) to the north of the continental divide, where the prevailing winds blow humid air from the Caribbean Sea. On the Pacific side of the isthmus, the green color indicates pasture and dry forest. Courtesy of Marcos A. Guerra, Smithsonian Tropical Research Institute.

Figure 4.11 Many trees shed their leaves during the dry season, as these trees on the Pacific slope of Panama have done. Photo by R. E. Ricklefs.

lands are so dry during the winter months that most trees lose their leaves (**Figure 4.11**). Tinder-dry forests and bare branches contrast sharply with the wet, lush, more typically tropical forest that flourishes during the wet season.

Farther to the north, outside the Tropics, climates come under the influence of the westerly winds that blow at middle latitudes. Here, temperatures, as well as rainfall, vary between winter and summer. The difference in climate between tropical and temperate regions can be illustrated by temperature and rainfall graphs of three locations in northern Mexico and the southwestern United States (**Figure 4.12**). At 30°N in the Chihuahuan Desert of central Mexico, rainfall comes only during the summer, when the solar equator reaches its northward limit. During the rest of the year this region falls within the dry subtropical high-pressure belt. Summer rainfall extends north into the Sonoran Desert of southern Arizona and New Mexico at 32°N. This area also receives moisture during the winter from the Pacific Ocean, carried by the southwesterly winds emanating from the subtropical high-pressure belt farther south. Thus, the Sonoran Desert experiences both a winter and a summer peak of rainfall.

Southern California, at the same latitude, lies to the west of the summer rainfall belt and has a winter-rainfall, summer-drought climate, often referred to as a **Mediterranean climate.** Named for the Mediterranean region of

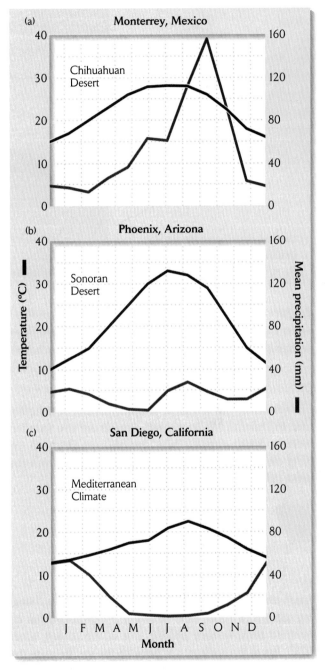

Figure 4.12 Seasonal climatic patterns differ between subtropical localities. (a) The Chihuahuan Desert in central Mexico has a summer rainy season. (b) The Sonoran Desert has a combined climatic pattern, with rainfall in summer and winter. (c) The Pacific coast and the Mojave Desert have a winter rain and summer drought (Mediterranean) climatic pattern.

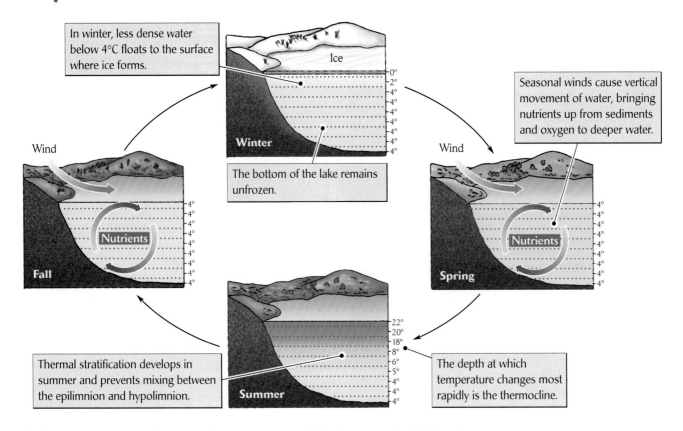

In winter, less dense water below 4°C floats to the surface where ice forms.

The bottom of the lake remains unfrozen.

Seasonal winds cause vertical movement of water, bringing nutrients up from sediments and oxygen to deeper water.

The depth at which temperature changes most rapidly is the thermocline.

Thermal stratification develops in summer and prevents mixing between the epilimnion and hypolimnion.

▌Figure 4.13 Seasonal changes in the temperature profile influence vertical mixing of water layers in temperate lakes. Vertical mixing is aided by wind-driven currents when the temperature of water is uniform from the surface to the bottom of the lake.

Europe, which has the same seasonal pattern of temperature and rainfall, Mediterranean climates are also found in western South Africa, Chile, and Western Australia, all lying along the western sides of continents at about the same latitude north or south of the equator.

Temperature and winds drive seasonal cycles in temperate lakes

Small temperate zone lakes respond quickly to the changing seasons (▌Figure 4.13). In winter, a typical lake has an inverted **temperature profile;** that is, the coldest water (0°C) lies at the surface, just beneath the ice. Because the density of water increases between the freezing point and 4°C, the warmer water within this range sinks, and the temperature increases to as much as 4°C toward the bottom of the lake. In early spring, the sun warms the lake surface gradually. But until the surface temperature exceeds 4°C, the sun-warmed surface water tends to sink

into the cooler layers immediately below. This vertical mixing distributes heat throughout the water column from the surface to the bottom, resulting in a uniform temperature profile. Winds cause deep vertical movement of water, or **spring overturn,** in early spring, bringing nutrients to the surface from the bottom sediments and bringing oxygen from the surface to the depths.

Later in spring and in early summer, as the sun rises higher each day and the air above the lake warms, surface layers of water gain heat faster than deeper layers, creating a zone of rapid temperature change at intermediate depth, called the **thermocline.** Once the thermocline is well established, water does not mix across it. Now, the warmer, less dense surface water literally floats on the cooler, denser water below, a condition known as **stratification.** The depth of the thermocline varies with local winds and with the depth and turbidity of the lake. It may occur anywhere between 5 and 20 m below the surface; lakes less than 5 m deep usually lack stratification.

The thermocline demarcates an upper layer of warm water called the **epilimnion** and a deeper layer of cold water called the **hypolimnion.** Most of the primary production of the lake occurs in the epilimnion, where sunlight is most intense. Oxygen produced by photosynthesis supplements oxygen entering the lake at its surface, keeping the epilimnion well aerated and thus suitable for animal life. Plants and algae in the epilimnion, however, often deplete the supply of dissolved mineral nutrients and, in doing so, curtail their own production. The thermocline isolates the hypolimnion from the surface of the lake, so animals and bacteria that remain below the thermocline, where there is little or no photosynthesis, deplete the water of oxygen, creating anaerobic conditions. Thus, during late summer, the productivity of temperate lakes may become severely depressed, as nutrients needed to support plant growth are lacking in surface waters and oxygen needed to support animal life is lacking in the depths.

During the fall, the surface layers of the lake cool more rapidly than the deeper layers, become denser than the underlying water, and begin to sink. This vertical mixing, called **fall overturn,** persists into late fall, until the temperature at the lake surface drops below 4°C and winter stratification ensues. Fall overturn speeds the movement of oxygen to deep waters and pushes nutrients to the surface. In lakes where the hypolimnion becomes warm by midsummer, deep vertical mixing may take place in late summer, while temperatures remain favorable for plant growth. The resulting infusion of nutrients into surface waters often causes an explosion in the population of phytoplankton—the **fall (autumn) bloom.** In deep, cold lakes, vertical mixing does not penetrate to all depths until late fall or early winter, when water temperatures are too cold to support phytoplankton growth.

Climate sustains irregular fluctuations

Most aspects of climate seem unpredictable. Everyone knows that weather is difficult to forecast far in advance. We often remark that a certain year was particularly dry or cold compared with others. The flooding in the Mississippi Valley and the increased intensity of hurricanes along the east coast of the United States in recent years drive home the capriciousness of nature. Such extreme conditions occur infrequently, but they may affect organisms disproportionately. The rich Peruvian fishing industry thrives on the abundant fish in the nutrient-rich waters of the Peru Current, as do some of the world's largest seabird colonies (▌Figure 4.14). The Peru Current is a mass of cold water that flows up the western coast of South America and finally veers offshore at Ecuador, toward the Galápagos archipelago. North of this point, warm, tropical inshore waters prevail along the coast. Each year, a warm countercurrent known as El Niño ("little boy" in Spanish, a name referring to the Christ child because this countercurrent appears around Christmastime) moves down the coast toward Peru. In some years, it flows strongly enough and far enough to force the cold Peru Current offshore, taking with it the food supply of millions of birds.

During "normal" years between El Niño "events," a steady wind blows across the equatorial central Pacific Ocean from an area of high atmospheric pressure centered over Tahiti to an area of low pressure centered over Darwin, Australia. An El Niño event appears to be triggered

▌**Figure 4.14 Upwelling currents often support high biological productivity.** Like its counterpart Peru Current off the western coast of South America, the Benguela Current off the western coast of South Africa has a zone of upwelling and supports an important fishery. The Cape gannets in this dense nesting colony feed on small fish in the adjacent cold, nutrient-rich waters. Accumulated guano is occasionally scraped off the rocks during the nonbreeding season and used for fertilizer. Photo by R. E. Ricklefs.

(a)

Low pressure over Tahiti and high pressure over Darwin, Australia

Increased cloud formation

Weak trade winds allow warmer waters to move east

The subtropical jet stream carries moisture east

Decreased development of Atlantic hurricanes

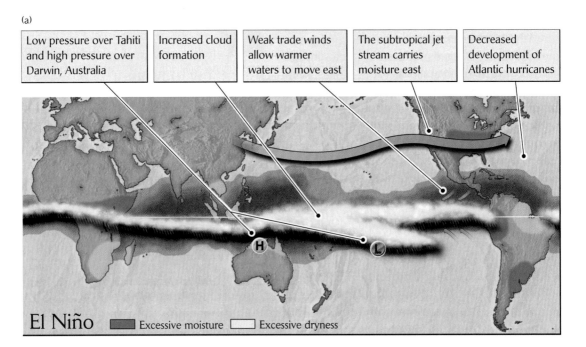

El Niño — Excessive moisture — Excessive dryness

(b)

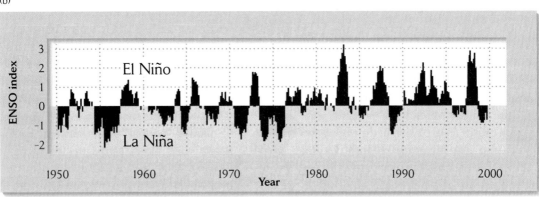

(c)

High pressure over Tahiti and low pressure over Darwin, Australia

Increased cloud formation

Strong trade winds push warmer waters west

The subtropical jet stream is separated and weakened

Increased development of Atlantic hurricanes

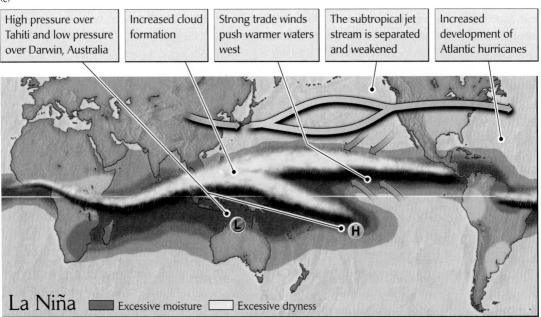

La Niña — Excessive moisture — Excessive dryness

▌**Figure 4.15 El Niño–Southern Oscillation (ENSO) events result in dramatic climatic changes.** (a) A map of the changes that occur during ENSO events. (b) ENSO events are marked by large positive anomalies in the ENSO index, which is correlated with sea surface temperature in South American coastal waters. (c) A map of the changes that occur during La Niña events. (a) and (c) from C. Suplee, *National Geographic Magazine* 195:73–95 (1999); (b) courtesy of the NOAA Climate Diagnostics Center.

by a reversal of these pressure areas (the so-called Southern Oscillation) and of the winds that flow between them (▌Figure 4.15a). As a result, the westward-flowing equatorial currents stop or even reverse, upwelling off the coast of South America weakens or ceases, and warm water—the El Niño current—piles up along the coast of South America. Historical records of atmospheric pressure at Tahiti and Darwin, and of sea surface temperatures on the Peruvian coast, reveal pronounced ENSO (El Niño–Southern Oscillation) events at irregular intervals of 2 to 10 years (▌Figure 4.15b).

El Niño events are often followed by La Niña, a period of strong trade winds that accentuate normal oceanic and upwelling currents and bring extreme weather of a different sort than El Niño to much of the world (▌Figure 4.15c). La Niña is characterized by heavy rainfall in many regions of the Tropics, drought in north-temperate regions, and an increase in hurricane activity in the North Atlantic Ocean.

The climatic and oceanographic effects of an ENSO event extend over much of the world, affecting ecosystems in such distant areas as India, South Africa, Brazil, and western Canada. A record ENSO event in 1982–83 disrupted fisheries and destroyed kelp beds in California, caused reproductive failure of seabirds in the central Pacific Ocean, and resulted in widespread mortality of coral in Panama. Precipitation was also dramatically affected in many terrestrial ecosystems. The deserts of northern Chile, normally the driest place on earth, received their first recorded rainfall in over a century.

The 1982–83 ENSO event drew the world's attention to the far-reaching effects of oceanographic and climatic changes in many parts of the world. For example, data from Zimbabwe for the period 1970–1993 show striking variation in yields of maize. As one might expect, these variations in yield were correlated with variations in rainfall, but, more surprisingly, they were also correlated with sea surface temperatures in the eastern tropical Pacific Ocean (▌Figure 4.16). One can see the far-reaching effects of the 1982–83 and 1991–92 El Niño events in these data.

During the 1991–92 El Niño, rainfall was so high in the Great Basin of the western United States that runoff nearly doubled the volume of water in Great Salt Lake. This reduced the lake's salinity from the usual 100 grams of salt per liter (about 3 times that of seawater) to 50 g per L, which caused marked changes in the ecosystem of the lake. Reduced salinity allowed predaceous corixid bugs to move into shallower parts of the lake. The corixids ate the brine shrimp *Artemia*, which graze on algae and normally dominate the ecosystem. With brine shrimp numbers reduced, algae increased dramatically, turning the lake into the aquatic equivalent of a lawn.

Some of the most striking effects of El Niño events are evident in the Galápagos archipelago, whose islands straddle the equator some 1,000 km off the west coast of Ecuador. The climate of the Galápagos is strongly influenced by the Peru Current, which brings cold water and periods of extreme dryness to the islands. When the Peru Current fails during El Niño conditions, warm water pervades the archipelago, triggering a drastic deterioration of local coldwater fish stocks and bringing extraordinary amounts of precipitation. Thus, El Niño conditions cause populations of seabirds and sea lions that rely on abundant fish to crash. On land, the heavy rains result in luxuriant growth of vegetation and abundant insects and seeds for the populations of birds and reptiles that rely on these

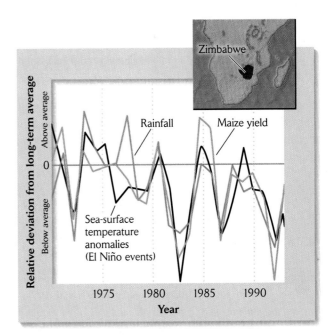

▌**Figure 4.16 ENSO events have far-reaching effects.** Deviations from the long-term average in rainfall and maize production in Zimbabwe are correlated with sea surface temperatures in the eastern equatorial Pacific Ocean. From M. A. Cane, G. Eshel, and R. W. Buckland, *Nature* 370:204–205 (1994).

❚ Figure 4.17 Heavy rains during El Niño events support lush plant growth in the Galápagos archipelago. These photographs show a hillside on Tower Island (Isla Genovesa) at the end of a normal dry season in January 1982 (*left*) and in the middle of a strong El Niño event in March 1983 (*right*). The most important difference is the dramatic increase in understory shrubs and vines. The larger *Bursera* trees are not affected by the exceptionally wet conditions. Courtesy of Robert L. Curry, Villanova University.

foods (❚ Figure 4.17). As we shall see later in the book, this seesawing between scarcity and abundance has important consequences for the population dynamics and evolution of organisms in the Galápagos archipelago.

ECOLOGISTS IN THE FIELD

A half-million-year climate record

Humans have kept records of climate systematically for about 200 years and sporadically for several hundred years before that. Variation in the thickness of the growth rings in trees extends records of climate in some regions—at least from a tree's point of view—back to several thousand years. Whether a climatic record encompasses decades, centuries, or millennia, one sees both regular cycles of climate and irregular fluctuations. But what about longer periods? We know from geological evidence that the Northern Hemisphere has undergone multiple glacial cycles during the past mil-

lion years, and that these cycles reflect broader patterns of global climatic change. Later in this book, we shall consider the evidence linking glacial cycles to changes in biological communities throughout the world. Here we will consider physical evidence of changes in temperature over the past half-million years.

The sediments of the ocean basins consist largely of the calcium carbonate shells of small protists known as foraminifera (❚ Figure 4.18). The shells of these long-dead creatures are tiny permanent thermometers that provide a long-term record of temperature fluctuations. The foraminifera have made this possible by incorporating oxygen in the form of carbonate into their shells. Most of the oxygen in the biosphere has an atomic weight of 16, and is referred to as the form, or isotope, ^{16}O. Oxygen also occurs in an isotope with two additional neutrons, which has an atomic weight of 18. Oxygen-18, or ^{18}O, is relatively rare, making up only 0.2% of the total in the biosphere. The rate of incorporation of the heavier ^{18}O into calcium carbonate shells is influenced by temperature. Thus, the proportion of

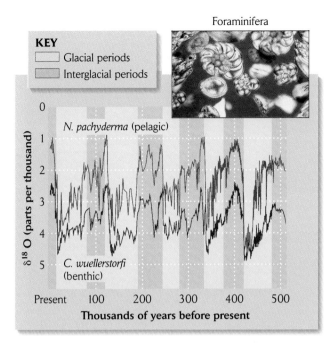

Foraminifera

KEY
☐ Glacial periods
☐ Interglacial periods

N. pachyderma (pelagic)

C. wuellerstorfi (benthic)

δ¹⁸O (parts per thousand)

Present   100   200   300   400   500

Thousands of years before present

■ **Figure 4.18 Variations in marine temperatures are recorded by foraminifera in ocean-floor sediments.** Variations in oxygen isotopes in the shells of foraminifera in sediments from the North Atlantic Ocean during the last 500,000 years. The $\delta^{18}O$ value becomes more negative the colder the temperature of the water in which the foraminifera lived. The record clearly shows five warm interglacial periods separated by cold glacial periods. From J. F. McManus, D. W. Oppo, and J. L. Cullen, *Science* 283:971–975 (1999). *Inset:* Shells of several species of the single-celled protists. Photo by Charles Gellis/ Photo Researchers.

^{18}O in shells decreases by approximately one part per thousand ($\delta^{18}O$) for each 4°C decrease in temperature.

Jerry McManus and his colleagues at the Woods Hole Oceanographic Institute analyzed a 65-meter-long core of sediments taken from the North Atlantic Ocean northwest of Ireland. The record of $\delta^{18}O$ values from the sediment core is shown in ■ Figure 4.18. As one might expect, temperatures indicated by the shells of the surface-dwelling foraminiferan *Neogloboquadrina pachyderma* are several degrees higher than those indicated by the bottom-dwelling *Cibicidoides wuellerstorfi*. (Sorry, they don't have common names.) Both species, however, exhibit 100,000-year-long cycles of temperature, corresponding to glacial and interglacial climatic cycles. Temperature changes at the bottom of the ocean parallel those at the surface, confirming that no place on earth escapes variations of climate. Superimposed on the long-term temperature cycles are numerous variations of shorter duration. These correspond to a wide range of global climatic patterns resulting from periodic variations in the distance of the earth from the sun.

Topographic and geologic features cause local variation in climate

Topography and geology can modify the environment on a local scale within regions of otherwise uniform climate. In hilly areas, the slope of the land and its exposure to the sun influence the temperature and moisture content of the soil. Soils on steep slopes drain well, often causing drought stress for plants on the hillside at the same time that water saturates the soils of nearby lowlands. In arid regions, stream bottomlands and seasonally dry riverbeds may support well-developed **riparian** forests, which accentuate the contrasting bleakness of the surrounding desert. In the Northern Hemisphere, south-facing slopes directly face the sun, whose warmth and drying power limit vegetation to shrubby, drought-resistant **xeric** forms. The adjacent north-facing slopes remain relatively cool and wet and harbor moisture-requiring **mesic** vegetation (■ Figure 4.19).

Air temperature decreases with altitude by about 6°–10°C for each 1,000 m increase in elevation, depending on the region. This decrease in temperature, which is caused by the expansion of air in the lower atmospheric pressures at higher altitudes, is referred to as **adiabatic cooling.** Climb high enough, even in the Tropics, and you will encounter freezing temperatures and perpetual snow. Where the temperature at sea level averages 30°C, freezing temperatures are reached at about 5,000 m, the approximate altitude of the snow line on tropical mountains.

■ **Figure 4.19 Topography can modify the environment on a local scale.** Exposure influences the vegetation growing on a series of mountain ridges near Aspen, Colorado. The cool and moist north-facing slopes permit the development of spruce forest. Shrubby, drought-resistant vegetation grows on the south-facing slopes. Photo by R. E. Ricklefs.

Hudsonian zone

Alpine zone

Upper Sonoran zone

Transition zone

Lower Sonoran zone

Upper Sonoran zone

In north-temperate latitudes, a 6°C drop in temperature with each 1,000 m of altitude corresponds to the temperature change encountered over an 800-km increase in latitude. In many respects, the climate and vegetation of high altitudes resemble those of sea-level localities at higher latitudes. But despite their similarities, alpine environments usually vary less from season to season than their low-elevation counterparts at higher latitudes. Temperatures in tropical montane environments remain nearly constant, and some of these areas remain frost-free over the year, which makes it possible for many tropical plants and animals to live in the cool environments found there.

In the mountains of the southwestern United States, changes in plant communities with elevation result in more or less distinct belts of vegetation, which the nineteenth-century naturalist C. H. Merriam referred to as **life zones.** Merriam's scheme of classification included five broad zones, which he named, from low to high elevation (or from south to north), the Lower Sonoran, Upper Sonoran, Transition, Canadian (or Hudsonian), and Alpine (or Arctic-Alpine) (❙ Figure 4.20). At low elevations, one encounters a cactus and desert shrub association characteristic of the Sonoran Desert of northern Mexico and southern Arizona. In the riparian forests along streambeds, the plants and ani-

Figure 4.20 Vegetation changes at increasing elevation in the mountains of Arizona. At the lowest elevations (bottom photographs), the Lower Sonoran zone supports mostly saguaro cactus, small desert trees such as paloverde and mesquite, numerous annual and perennial herbs, and small succulent cacti. Agave and grasses are conspicuous elements of the Upper Sonoran zone, and oaks appear toward its upper edge. At higher elevations, large trees predominate: ponderosa pine in the Transition zone, spruce and fir in the Hudsonian zone. These gradually give way to bushes, willows, herbs, and lichens in the Alpine zone above the tree line. Photo by Tom Bean/DRK Photo.

mals have a distinctly tropical flavor. Many hummingbirds and flycatchers, ring-tailed cats, jaguars, and peccaries make their only temperate zone appearances in this area. In the Alpine zone, 2,600 m higher, one finds a landscape resembling the tundra of northern Canada and Alaska. Thus, by climbing 2,600 m, one experiences changes in climate and vegetation that would occur in the course of a journey to the north of 2,000 km or more at sea level.

Climate and the underlying bedrock determine the diversification of soils

Climate affects the distributions of plants and animals indirectly through its influence on the development of soil, which provides the substrate within which plant roots grow and many animals burrow. The characteristics of soil determine its ability to hold water and to make available the minerals required for plant growth. Thus, its variation provides a key to understanding the distributions of plant species and the productivity of biological communities. **Soil** defies simple definition, but we may describe it as the layer of chemically and biologically altered material that overlies rock or other unaltered materials at the surface of the earth. It includes minerals derived from the parent rock, modified minerals formed anew within the soil, organic material contributed by plants, air and water within the pores of the soil, living roots of plants, microorganisms, and the larger worms and arthropods that make the soil their home. Where a recent road cut or excavation exposes soil in cross section, one often notices distinct layers, which are called **horizons** (**Figure 4.21**). A generalized, and somewhat simplified, soil profile has four major divisions, the O, A, B, and C horizons. The A horizon has two subdivisions, A_1 and A_2 (Table 4.1). Five factors determine the characteristics of soils: climate, parent material (underlying rock), vegetation, local topography, and, to some extent, age. Soil horizons reveal the decreasing influence of climatic and biotic factors with increasing depth.

(a)   (b)

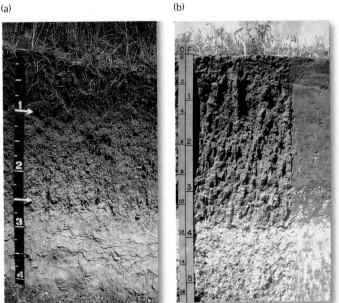

Figure 4.21 Soil profiles may show distinct layers, or horizons. (a) This profile of a prairie soil from Nebraska is weathered to a depth of about 3 feet (0.9 m), where the subsoil contacts the parent material, which consists of loosely aggregated, calcium-rich, wind-deposited sediments (loess). A_1 and A_2 horizons are not clearly distinguished. The B horizon (between the arrows) contains less organic material. Rainfall in Nebraska is sufficient to leach readily soluble ions completely from the soil; hence there are no B layers of redeposition. The C horizon is light-colored and has been leached of some of its calcium. (b) A prairie soil from Texas. The A layer is only about 6 inches thick. The B layer extends down to the bottom of the dark layer, which represents organic material redeposited from the A layer above. Considerable calcium has been redeposited at the base of the B layer and in the C layer below. Courtesy of the U.S. Department of Agriculture, Soil Conservation Service.

Table 4.1   Characteristics of the major soil horizons

Soil horizon	Characteristics
O	Primarily dead organic litter. Most soil organisms inhabit this layer.
A_1	A layer rich in humus, consisting of partly decomposed organic material mixed with mineral soil.
A_2	A region of extensive leaching of minerals from the soil. Because minerals are dissolved by water—that is to say, mobilized—in this layer, plant roots are concentrated here.
B	A region of little organic material, whose chemical composition resembles that of the underlying rock. Clay minerals and oxides of aluminum and iron leached out of the overlying A_2 horizon are some times deposited here.
C	Primarily weakly altered material, similar to the parent rock. Calcium and magnesium carbonates accumulate in this layer, especially in dry regions, sometimes forming hard, impenetrable layers or "pans."

Soils exist in a dynamic state, changing as they develop on newly exposed rock. And even after soils achieve stable properties, they remain in a constant state of flux. Groundwater removes some substances; other materials enter the soil from vegetation, in precipitation, as dust from above, and from the rock below. Where little rain falls, the parent material decomposes slowly, and plant production adds little organic detritus to the soil. Thus arid regions typically have shallow soils, with bedrock lying close to the surface. Soils may not form at all where decomposed bedrock and detritus erode as rapidly as they form. Soil development also stops short on alluvial deposits, where fresh layers of silt deposited each year by floodwaters bury older material.

At the other extreme, soil formation proceeds rapidly in parts of the humid Tropics, where chemical alteration of parent material may extend to depths of 100 meters. Most soils of temperate zones are intermediate in depth, extending to an average of about 1 meter. The variety of soil types, their characteristics, and their distributions are presented in Table 4.2.

Weathering

Weathering—the physical and chemical alteration of rock material near the earth's surface—occurs wherever surface water penetrates. The repeated freezing and thawing of

---

Table 4.2   Soil types, their characteristics, and their distribution

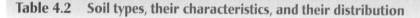

KEY

Alfisols – Moist, moderately weathered mineral soils

Aridosols – Dry mineral soils with little leaching and accumulations of calcium carbonate

Entisols – Recent mineral soils lacking development of soil horizons

Histosols – Organic soils of peat bogs; mucks

Inceptisols – Young, weakly weathered soils

Mollisols – Well-developed soils high in organic matter and calcium; very productive

Oxisols/ Andisols – Deeply weathered, lateritic soils of moist tropics (**not represented in continental United States**)

Spodosols – Acid, podsolized soils of moist, often cool climates with shallow leached horizon and a deeper layer of deposition

Ultisols – Highly weathered soils of warm, moist climates with abundant iron oxides

Vertisols – High content of swelling-type clays developing deep cracks in dry seasons

water in crevices physically breaks rock into smaller pieces and exposes a greater surface area to chemical action. Initial chemical alteration of the rock occurs when water dissolves some of its more soluble minerals, especially sodium chloride (NaCl) and calcium sulfate (CaSO$_4$). Other materials, such as the oxides of titanium, aluminum, iron, and silicon, dissolve less readily.

The weathering of granite exemplifies some basic processes of soil formation. The minerals responsible for the grainy texture of granite—feldspar, mica, and quartz—consist of various combinations of oxides of aluminum, iron, silicon, magnesium, calcium, and potassium, along with other, less abundant compounds. The key to weathering is the displacement of certain elements in these minerals—notably calcium, magnesium, sodium, and potassium—by hydrogen ions, followed by the reorganization of the remaining oxides into new minerals. This chemical process provides the basic structure of soil.

Feldspar and mica grains consist of aluminosilicates of potassium, magnesium, and iron. Hydrogen ions percolating through granite displace potassium and magnesium ions, and the remaining iron, aluminum, and silicon form new, insoluble materials, particularly clay particles. These particles are important to the water-holding and nutrient-holding capacity of soils. Quartz, a type of silica (SiO$_2$), is relatively insoluble and therefore remains more or less unaltered in the soil as grains of sand. Different changes in chemical composition as granite weathers from rock to soil in different climate regions show that weathering is most severe under tropical conditions of high temperature and rainfall (▌Figure 4.22).

Where do the hydrogen ions involved in weathering come from? They derive from two sources. One of these is the carbonic acid that forms when carbon dioxide dissolves in rainwater (see Chapter 2). In regions not affected by acidic pollution, concentrations of hydrogen ions in rainwater produce a pH of about 5. The other source of hydrogen ions is the oxidation of organic material in the soil itself. The metabolism of carbohydrate, for example, produces carbon dioxide, and dissociation of the resulting carbonic acid generates additional hydrogen ions. In the Hubbard Brook Forest of New Hampshire, which is a particularly well studied watershed, these internal processes account for about 30% of the hydrogen ions needed for the weathering of bedrock; the remainder come from precipitation. In the Tropics, however, internal sources of hydrogen ions assume greater importance and may lead to more rapid weathering.

Podsolization occurs in acid soils

Under mild, temperate conditions of temperature and precipitation, sand grains and clay particles resist weathering

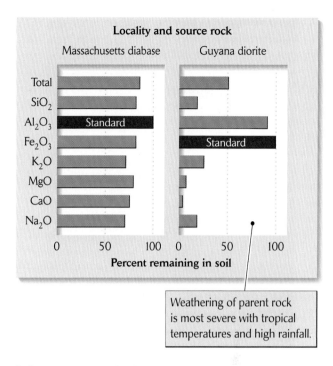

Weathering of parent rock is most severe with tropical temperatures and high rainfall.

▌**Figure 4.22 Weathering is more severe in tropical than in temperate climates.** Differential weathering results in differential removal of minerals from granitic rocks in Massachusetts and in Guyana. Values are compared with those of the mineral (aluminum or iron oxide) assumed to be the most stable component of the soil in its region (labeled Standard). After E. W. Russell, *Soil Conditions and Plant Growth,* 9th ed., Wiley, New York (1961).

and form stable components of the soil. In acid soils, however, clay particles, which retain nutrients in the soil, break down in the A horizon, and their soluble ions are transported downward and deposited in lower horizons. This process, known as **podsolization,** reduces the fertility of the upper layers of the soil.

Acid soils occur primarily in cold regions where needle-leaved trees dominate the forests. The slow microbial decomposition of plant litter shed by spruce and fir trees produces organic acids. In addition, rainfall usually exceeds evaporation in regions of podsolization. Under these moist conditions, because water continuously moves downward through the soil profile, little clay-forming material is transported upward from the weathered bedrock below.

In North America, podsolization advances farthest under spruce and fir forests in New England and the Great Lakes region and across a wide belt of southern and western Canada. A typical profile of a highly podsolized soil (▌Figure 4.23) reveals striking bands corresponding to the regions of leaching and redeposition. The topmost layer of

Figure 4.23 Podsolized soils have reduced fertility. This profile of a podsolized soil in northern Michigan shows strong leaching of the A horizon. The light-colored A₂ horizon and the dark-colored B₁ horizon immediately below it form distinct bands. Compare the general absence of roots in the A₂ horizon with their presence in the lower B₁ horizon. Photo by R. E. Ricklefs.

the profile (A_1) is dark and rich in organic matter. It is underlain by a light-colored horizon (A_2) that has been leached of most of its clay content. As a result, the A_2 horizon consists mainly of sandy skeletal material that holds neither water nor nutrients well. One usually finds a dark band immediately below the A_2 horizon. This is the uppermost layer of the B horizon, where iron and aluminum oxides are redeposited. Other, more mobile minerals may accumulate to some extent in lower parts of the B horizon, which then grades almost imperceptibly into a C horizon and the parent material.

Laterization occurs in warm, wet climates

Soils weather to great depths in the warm, wet climates of many tropical and subtropical regions. One of the most conspicuous features of weathering under these conditions is the breakdown of clay particles, which results in the leaching of silica from the soil, leaving oxides of iron and aluminum to predominate in the soil profile. This process is called **laterization,** and the iron and aluminum oxides give lateritic soils their characteristic reddish coloration (**Figure 4.24**). Even though the rapid decomposi-

(a)

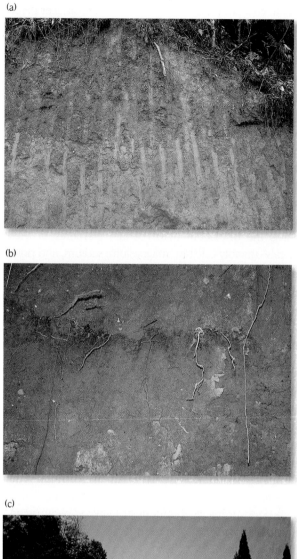

(b)

(c)

Figure 4.24 Laterized soils are leached of clay and hold few nutrients. (a) A fresh road cut in the Amazon basin in Ecuador shows a typical laterized soil profile. (b) Notice the roots at the top of a layer of redeposition of organic matter in the soil profile. (c) Highly oxidized and deeply weathered soils are also found in the southeastern United States, as in this deeply eroded area of West Tennessee. Photos (a) and (b) by R. E. Ricklefs; photo (c) courtesy of the U.S. Department of Agriculture, Soil Conservation Service.

tion of organic material in tropical soils contributes an abundance of hydrogen ions, these are quickly neutralized by the bases formed by the breakdown of clay minerals; consequently, lateritic soils usually are not acidic. Laterization is enhanced in certain soils that develop on parent material deficient in quartz (SiO_2) but rich in iron and magnesium (basalt, for example); these soils contain little clay to begin with because they lack silicon. Regardless of the parent material, weathering reaches deepest and laterization proceeds farthest on low-lying soils, such as those of the Amazon basin, where highly weathered surface layers are not eroded away and the soil profiles are very old.

One of the consequences of laterization in many parts of the Tropics is that the capacity of the soil to hold nutrients is very poor. Without clay and humus particles to hold mineral nutrients, they are readily leached out of the soil. Where soils are weathered deeply, new minerals formed by the decomposition of the parent material are simply too far from the surface layers of the soil to contribute to soil fertility. Besides, heavy rainfall keeps water moving down through the soil profile, preventing the upward movement of nutrients. In general, the deeper the ultimate sources of nutrients in the unaltered bedrock, the poorer the surface layers. Rich soils do, however, develop in many tropical regions, particularly in mountainous areas where erosion continually removes nutrient-depleted surface layers of soil, and in volcanic areas where the parent material of ash and lava is often rich in such nutrients as potassium.

Soil formation emphasizes the role of the physical environment, particularly climate, geology, and landforms, in creating the tremendous variety of environments for life that exist at the surface of the earth and in its waters. In the next chapter, we shall see how this variety affects the distribution of life forms and the appearance of biological communities.

ECOLOGISTS IN THE FIELD

Which came first, the soil or the forest?

When the glaciers retreated from most of Europe and North America, beginning about 18,000 years before the present (BP), dramatic changes in vegetation and soils moved across the landscape. In central Europe, cold, dry steppes were replaced by coniferous forests and then by the deciduous forests that occur throughout the region today. At about the same time as the coniferous–deciduous forest transition, there was a change from strongly podsolized soils to richer brown forest soils. But, as British ecologist Kathy Willis and her colleagues asked, "Which changed first? Did climatic warming result in a

transformation from one soil type to another, which in turn resulted in a change in forest composition, or did the vegetation change first and subsequently alter the soil?"

The answer, at least for one area in northeastern Hungary, comes from a core of sediments removed from small, shallow Lake Kis-Mohos Tó. Lake sediments preserve a record of the local conditions over time. Pollen grains become trapped in the sediments (❚ Figure 4.25), as do minerals

❚ **Figure 4.25 Pollen grains from different types of plants have distinctive surface patterns** that allow them to be recognized in lake sediments. These scanning electron micrographs depict pollen grains of three subtropical plants from North America: *Ipomaea arborescens*, *Ceiba pentandra*, and *Agave palmeri*, magnified approximately 500 times. Courtesy of Norman Hodgkin.

carried in water draining from soils surrounding the lake. The pollen and minerals tell the story of changes in vegetation and soils.

What does the sediment core from Lake Kis-Mohos Tó reveal? First, the pollen record tells us that the local forest changed from coniferous to deciduous in a few centuries. You can see in ▌Figure 4.26 that spruce, pine, and birch, trees typical of boreal forests, abruptly disappeared from the region about 9,500 years BP, and were just as quickly replaced by an oak–hornbeam deciduous forest. Up until this transition, most of the sediment in the lake was inorganic, suggesting that the area was cold and unproductive.

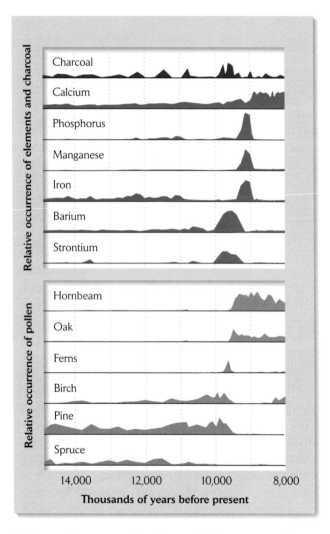

▌**Figure 4.26 Layers of sediments in lakes preserve the history of environmental change in the surrounding watershed.** This core from Lake Kis-Mohos Tó in Hungary shows the replacement of needle-leaved forest by broad-leaved deciduous forest and accompanying changes in soils about 10,000 years ago. From K. J. Willis et al., *Ecology* 78(3):740–750 (1997).

Abundant aluminum, potassium, and magnesium in the sediment core suggested rapid breakdown and leaching of clay components of the surrounding soils, typical of a heavily podsolized area. The first indication of change was a release of large amounts of strontium and barium into the lake. Spruce trees preferentially take up these elements from the soil instead of calcium. The strontium and barium are deposited in the spruce needles and then accumulate in the thick layer of litter on the forest floor. Willis and her colleagues interpreted the release of these elements into ground and surface water flowing into Lake Kis-Mohos Tó as resulting from a rapid breakdown of the spruce litter.

What triggered this breakdown? It is difficult to know with certainty, but again the sediment core provides a clue in the contemporaneous increase in charcoal particles entering the lake. Climate modelers suggest that central Europe experienced a warm, dry period between 10,000 and 9,000 BP. This climate may have promoted natural fires that burned away the litter layers of the coniferous forests. This period also marks the disappearance of spruce from the area. Spruce requires cooler and moister conditions than pines can tolerate. The appearance of charcoal in the sediment core is also associated with a spike of fern spores, which is a sure sign of frequent fires. Ferns colonize burned areas quickly and produce luxuriant growth for a few years after a fire has swept through a forest (▌Figure 4.27). The fires mark the transition from coniferous to deciduous forest because pines disappear and are replaced by oaks at this time. After broad-leaved deciduous trees became established, large quantities of iron, magnesium, and phosphorus were released into the lake during another brief period. This represents a period of leaching of these elements under the still acid conditions of the forest soils, probably accompanied by a transient reduction in soil fertility. The final phase of the transition is marked by an increase in calcium in the sediment core. Calcium is not particularly abundant in the rock underlying the region, but deciduous trees like oaks preferentially take up calcium from the soil and begin to enrich the calcium content of the top layers of soil through annual leaf fall.

So, which changed first, the soil or the forest? Clearly, the forest retained its acidic, podsolized nature until well after the establishment of deciduous vegetation, so apparently the vegetation change caused the soil change in this case, illustrating the contribution of vegetation to the development of soil. The change in the vegetation itself was evidently sparked, so to speak, by warmer and drier climates, which were less favorable for spruce and fostered fires that created openings in the pine forests. These openings allowed oak and other broad-leaved species to invade.

Figure 4.27 Ferns grow abundantly in recently burned areas, such as this aspen forest in northern Michigan. Photo by R. E. Ricklefs.

 Summary

1. Global patterns of temperature and rainfall result from differential input of solar radiation in different regions and from the redistribution of heat energy by winds and ocean currents. Prominent features of terrestrial climates include a band of warm, moist climate over the equator and bands of dry climate at about 30° north and south latitude.

2. Variation in the marine environment is determined on a global scale by the major ocean currents. These currents redistribute heat over the surface of the earth and greatly affect climates on land. Upwelling currents, caused by winds, ocean basin topography, and variations in water density related to temperature and salinity, bring cold, nutrient-rich water to the surface in some areas.

3. Seasonality in terrestrial environments is caused by the annual progression of the solar equator northward and southward and by the latitudinal movement of associated belts of temperature, wind, and precipitation. At high latitudes, the seasons are expressed primarily as annual cycles of temperature; within the Tropics, seasonality of precipitation is more pronounced.

4. Seasonal warming and cooling profoundly change the characteristics of lakes in the temperate zone. During summer, such lakes are stratified, with a warm surface layer (epilimnion) separated from a cold bottom layer (hypolimnion) by a sharp thermocline. In spring and fall, the profile of temperature with depth becomes more uniform, allowing vertical mixing.

5. Irregular and unpredictable variations in climate, such as El Niño–Southern Oscillation events, may cause major changes in temperature and precipitation and disrupt biological communities on a global scale.

6. Topography and geology superimpose local variation in environmental conditions on more general climatic patterns. Mountains intercept rainfall, creating arid rain shadows in their lees. Conditions at higher altitudes resemble conditions at higher latitudes.

7. The characteristics of soil reflect the influences of the bedrock below and the climate and vegetation above. Weathering of bedrock results in the breakdown of some of its minerals and their re-formation into clay particles, which mix with organic detritus entering the soil from the surface. These vertically graded processes usually result in distinct soil horizons.

8. In acid (podsolized) soils of cool, moist parts of the temperate zone and in deeply weathered (laterized) tropical soils, clay particles break down and the fertility of the soil is much reduced.

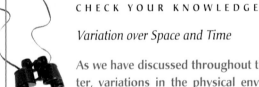

PRACTICING ECOLOGY

CHECK YOUR KNOWLEDGE

Variation over Space and Time

As we have discussed throughout this chapter, variations in the physical environment play an important role in ecological interactions that determine the distributions and abundances of organisms. The way in which individual organisms respond to changes in climate is critical to their success in surviving and reproducing. Therefore, it is important to understand how past and present patterns of climatic variability affect plants and animals so that we can predict the probable effects of future changes in the environment.

In regions subject to cold winters, one of the most critical periods of an individual's life is the initiation of activity in the spring. Insects come out of their resting state (diapause), mammals come out of hibernation, seeds germinate, dormant buds break open and trees begin to leaf out, and birds return from their tropical wintering grounds. These responses are cued by a variety of environmental factors, including day length, which indicates time of year and therefore predicts the average conditions to be expected (see Chapter 9). Organisms are also sensitive to factors that vary between years, such as air temperature, snow depth, soil temperature, and soil water content, and they may adjust the timing of their spring activity accordingly. For example, years with particularly mild winter and spring temperatures usually stimulate earlier plant growth and flowering. Insect herbivores and pollinators usually adjust their activity in parallel by responding to the same cues.

Examples of the relationship between the timing of spring activity and annual variation in climate for a number of European organisms can be found at *http://www.student.wau.nl/~arnold/gpmn.html*. Some of these studies have recorded the timing of activity, or **phenology**, over many decades. The records show variation in phenology over time. A few of the studies show long-term trends in date of flowering, for example, in response to recent warming of the environment.

A recent study in the Rocky Mountains suggests that when different organisms respond to different cues, the phenology of one organism might become unsynchronized with the phenology of others with which it interacts. David Inouye, Billy Barr, Ken Armitage, and Brian Inouye, working at the Rocky Mountain Biological Laboratory (RMBL) at an elevation of about 3,000 m (10,000 feet) in Colorado, examined springtime records of plant emergence, bird migrations from lower elevations, and emergence of marmots from hibernation. They tracked the dates at which these events occurred in relation to date of snowmelt and air temperature from 1975 to 1999. Although temperature, snow cover, and plant phenology varied from year to year, there was no trend in the date of onset of plant growth, in spite of generally increasing temperatures over the period of the study. Increasing winter snowfall, which resulted in later disappearance of the snowpack from mountain meadows, might possibly have offset the effect of increasing temperatures. Indeed, the depth of the snowpack, and the date at which it is fully melted, is tightly linked to the emergence of many plant species.

Temperatures have also been increasing at lower elevations in Colorado; however, there is little snow at lower elevations, and biological activity has shown a trend toward beginning earlier there, just as it has in much of eastern North America and Europe. For organisms that are resident at low elevations, this response to milder winter and spring conditions is appropriate. However, several animals migrate from lowlands to higher elevation in the spring, and if they take their cue from temperatures in their wintering areas, they are likely to show up on their breeding grounds too early. For example, American robins now arrive at RMBL two weeks earlier than they did in 1981, probably cued by warmer conditions at lower elevations, but they are greeted by still heavy snow cover. The changing relationship between spring temperatures and snow cover may also be a problem for hibernating animals, which probably respond

to changes in temperature in their burrows. Inouye and colleagues reported that marmots now emerge from hibernation 38 days earlier than they did in the 1970s, more than a month ahead of the growth of most of their food plants. This disparity between the phenological responses of different populations that take their cues from different signals is bound to affect survival and reproductive success, and may even lead to the decline and disappearance of some populations from the area over time. Climatic change brought about by human activity is bound to magnify problems of this kind and cause changes in the species composition of many ecosystems.

CHECK YOUR KNOWLEDGE

1. What was the purpose of the Inouye, Barr, Armitage, and Inouye study?

2. Read ▌Figure 4.28 and determine by how much winter snowfall has changed over the period from 1975 to 1999.

Understanding phenological responses to variation in the physical environment depends initially on using correlations between biological and climatic data to develop hypotheses about the cues that are important for plant and animal responses. It is important to select the appropriate cue for your question of interest, and to evaluate variation on the right spatial and temporal scales.

MORE ON THE WEB   3. Suppose you were assigned the task of comparing the phenology of plant growth and flowering with climatic variations resulting from El Niño events.

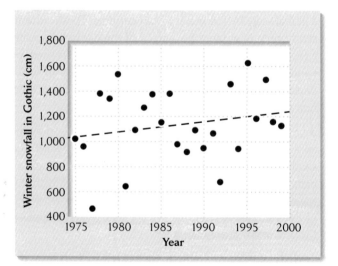

▌**Figure 4.28** Winter snowfall at Rocky Mountain Biological Laboratory (RMBL) 1975–1999. After D. W. Inouye et al., *Proceedings of the National Academy of Sciences USA* 97: 1630–1633 (2000).

A. How would you monitor the phenology of a plant species? Which attribute of plant phenology (e.g., date of the appearance of first leaves, first flower opening date, etc.) would you use to chart the timing of plant development? To find out more about monitoring phenology you might want to visit the Lilac Phenology Network Home Page through *www.whfreeman.com/ricklefs*.

B. What data would you use to monitor El Niño activity? For some hints, see Figure 4.16 or go to the El Niño page of the National Oceanic and Atmospheric Administration through *www.whfreeman.com/ricklefs*.

MORE ON THE WEB **4. You can view the article by Inouye et al. at *http://www.pnas.org/cgi/content/full/97/4/1630* or at *www.whfreeman.com/ricklefs*. Read in the introduction section of the journal article about other studies that suggest climate change is affecting the seasonal timing of animal and plant activity at low altitudes. Why is this important to animals that live at higher altitudes?**

Answers are found at *www.whfreeman.com/ricklefs*.

 Suggested Readings

Barber, R. T., and F. P. Chavez. 1983. Biological consequences of El Niño. *Science* 222:1203–1210.

Cairns, S. C., and G. C. Grigg. 1993. Population dynamics of red kangaroos (*Macropus rufus*) in relation to rainfall in the south Australian pastoral zone. *Journal of Applied Ecology* 30:444–458.

Cane, M. A., G. Eshel, and R. W. Buckland. 1994. Forecasting Zimbabwean maize yield using eastern equatorial Pacific sea surface temperature. *Nature* 370:204–205.

Graedel, T. E., and P. J. Crutzen. 1995. *Atmosphere, Climate, and Change*. Scientific American Library, New York.

Inouye, D. W., B. Barr, K. B. Armitage, and B. D. Inouye. 2000. Climate change is affecting altitudinal migrants and hibernating species. *Proceedings of the National Academy of Sciences USA* 97: 1630–1633.

Jenny, H. 1980. *The Soil Resource: Origin and Behavior*. Springer-Verlag, New York.

McManus, J. F., D. W. Oppo, and J. L. Cullen. 1999. A 0.5-million-year record of millennial-scale climate variability in the North Atlantic. *Science* 283:971–975.

Philander, G. 1989. El Niño and La Niña. *American Scientist* 77:451–459.

Rasmussen, E. M. 1985. El Niño and variations in climate. *American Scientist* 73:168–177.

Shelford, V. E. 1963. *The Ecology of North America*. University of Illinois Press, Urbana.

Sherman, K., L. M. Alexander, and B. D. Gold (eds.). 1990. *Large Marine Ecosystems: Patterns, Processes, and Yields*. American Association for the Advancement of Science, Washington, D.C.

Suplee, C. 1999. El Niño, La Niña. *National Geographic* 195(3): 73–95.

Willis, K. J., M. Braun, P. Sümegi, and A Tóth. 1997. Does soil change cause vegetation change or vice versa? A temporal perspective from Hungary. *Ecology* 78(3):740–750.

Wurtsbaugh, W. A. 1992. Food-web modification by an invertebrate predator in the Great Salt Lake (USA). *Oecologia* 89:168–175.

CHAPTER 5

Biological Communities: The Biome Concept

Climate is the major determinant of plant distribution

Variations in topography and soils influence local distributions of plants

Form and function are adapted to match the environment

Climate defines the boundaries of terrestrial biomes

Walter climate diagrams distinguish the major terrestrial biomes

Temperate climate zones have average annual temperatures between 5°C and 20°C

Boreal and polar climate zones have average temperatures below 5°C

Equatorial and tropical climate zones have average temperatures exceeding 20°C

The biome concept must be modified for aquatic systems

Imagine that you are on safari on an East African savanna and one of your group shouts, "Look over there, a cactus tree!" With your training in botany, you know immediately that this can't be so, because the cactus family (Cactaceae) is restricted to the New World. Yet the plant looks just like cacti you have seen in similar environments in Mexico (❚ Figure 5.1). Closer inspection of the flowers shows that the African plant is a cactus look-alike, a member of the spurge family (Euphorbiaceae).

Your friend was fooled by a common phenomenon in biology, that of convergence. Convergence is the process by which unrelated organisms evolve a resemblance to each other in response to common environmental conditions. The leafless, thick, fleshy branches of the cactus and the cactus-like euphorb evolved as adaptations to reduce water loss in semiarid environments. The two plants look alike because they evolved in response to the same conditions, although they descended from different-looking ancestors. Natural selection and evolution are oblivious to the ancestry of a particular organism as long as it is capable of an adaptive response to a particular condition of the environment.

The principle of convergence explains why we can recognize an association between the forms of organisms and their particular environments anywhere in the world. Tropical rain forest trees have the same general attributes of form no matter where they are found or to which evolutionary lineage they belong. Trees that live in mangrove communities in the Tropics all have thick, leathery leaves, and many have root systems that project above the surface of the waterlogged sediments on which they grow. Several species of mangrove trees share a trait found nowhere else among plants—viviparity, or the germination and growth of seedlings before they are shed from the parent plant (❚ Figure 5.2). Mangrove seedlings disperse by floating in salt water, and their ability to tolerate these conditions and become established on tidal mudflats is greatly enhanced by viviparity. Thus, although different species of

(a)

(b)

mangroves have evolved from many different lineages of terrestrial plants, they share certain traits because of convergent evolution adapting them to the stressful conditions of their environment.

Climate, topography, and soil—and parallel influences in aquatic environments—determine the changing character of plant and animal life over the surface of the earth. Although no two locations harbor exactly the same assemblage of species, we can group biological communities in categories based on their dominant plant forms, which give commu-

nities their overall character. These categories are referred to as **biomes.** The biome concept is a system of classifying biological communities and ecosystems based on similarities in their vegetational characteristics. Thus, biomes provide convenient reference points for comparing ecological processes in various kinds of communities and ecosystems.

The important terrestrial biomes of North America, for example, are tundra, boreal forest, temperate seasonal forest, temperate rain forest, shrubland, grassland, and subtropical desert. To the south in Mexico and Central America, the important biomes are tropical rain forest, tropical deciduous forest, and tropical savanna. As one would

(a)

(b)

▌Figure 5.2 **Some traits are shared among many species in a particular environment.** The seeds of the red mangrove *Rhizophora mangle* germinate while still attached to the tree (a), so that the developing seedling can become established rapidly (b). This pattern of viviparity occurs in many mangrove plants, but is unknown in other environments. Photos by R. E. Ricklefs.

expect, the geographic distributions of these biomes correspond closely to the major climate zones of North America. Although each biome is immediately recognizable by its distinctive vegetation, it is important to realize that different systems of classification make coarser or finer distinctions among biomes, and that the characteristics of one biome usually intergrade gradually into the next. The biome concept is nonetheless a useful tool that enables ecologists to work together to understand the structure and functioning of large ecological systems.

That biomes can be distinguished at all results from the simple fact that no single type of plant can endure the whole range of conditions at the surface of the earth. If plants had such broad tolerance of physical conditions, then the earth would be covered by a single biome. To the contrary, trees, for example, cannot grow under the dry conditions that shrubs and grasses can tolerate, simply because the physical structure, or **growth form,** of trees creates a high demand for water. The grassland biome exists because grasses and nongrass herbs (called **forbs**) can survive the cold winters typical, for example, of the Great Plains of the United States, the steppes of Russia, and the pampas of Argentina.

This matching of growth form and environment allows us to make generalizations about the distributions of life forms and the extents of biomes. If that were the whole of it, however, the study of ecology could simply focus on the biological relationships of individual organisms to their physical environments, and everything else in ecology would emanate from that point. We must remind ourselves that life is not so simple. In addition to physical conditions of the environment, two other kinds of factors influence the distributions of species and growth forms. The first of these is the myriad interactions between species—competition, predation, and mutualism—that determine whether a species or growth form can persist in a particular place. For example, grasses grow perfectly well in eastern North America, as we see along roadsides and on abandoned agricultural lands, but trees are the dominant growth form there, and in the absence of disturbance, they exclude grasses, which cannot grow and reproduce under the deep shade of trees.

The second kind of factor is that of chance and history. The present biomes have developed over long periods, during which the distributions of landmasses, ocean basins, and climate zones have changed continually. Most species fail to occupy some perfectly suitable environments simply because they have not had the opportunity to get to all ends of the earth. This fact is amply illustrated by the successful introduction by humans of such species as European starlings and Monterey pines to parts of the world that had suitable environmental conditions but which were far outside the restricted natural distributions of these species.

In addition, evolution has proceeded along independent lines in different parts of the world, leading in some cases to unique biomes. Australia has been isolated from other continents for the past 40–50 million years, which accounts both for its unusual flora and fauna and for the absence of many of the kinds of plants and animals familiar to northerners. Because of its unique history, areas of Australia that have a climate like California's could support the shrub vegetation—referred to as chaparral—or oak savanna found there, but instead are clothed with tall eucalyptus woodland. Similarities between chaparral and eucalyptus forest include drought and fire resistance, but the predominant plant growth forms in these areas of Australia and California differ, primarily because of historical accident. We shall consider these biological and historical factors later in this book. As we shall see in the present chapter, the physical environment ultimately determines the character and distribution of the major biomes.

Climate is the major determinant of plant distribution

Species ranges are often limited by physical conditions of the environment. In terrestrial environments, temperature and moisture are the most important variables. The distributions of several species of maples in eastern North America show how these factors operate. Sugar maple, a common forest tree in the northeastern United States and southern Canada, is limited by cold winter temperatures to the north, by hot summer temperatures to the south, and by summer drought to the west (▌ Figure 5.3). Attempts to grow sugar maples outside their normal range have shown that they cannot tolerate average monthly summer temperatures above about 24°C or winter temperatures below about −18°C. The western limit of the sugar maple, determined by dryness, coincides with the western limit of forest in eastern North America. Because temperature and rainfall interact to control the availability of moisture, sugar maples require less annual precipitation at the northern edge of their range (about 50 cm) than at the southern edge (about 100 cm). To the east, the range of the sugar maple stops abruptly at the Atlantic Ocean.

The distributions of the sugar maple and other tree-sized maple species—black, red, and silver—reflect differences in their **ecological tolerances,** or the range of conditions within which each species can survive (▌ Figure 5.4). Where their geographic ranges overlap, the maples exhibit distinct preferences for certain local environmental conditions created by differences in soil and topography. Black maple frequently occurs in the same areas as the closely

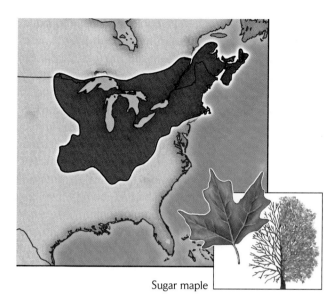

Figure 5.3 The distributions of species are limited by the physical conditions of the environment. The red area shows the range of sugar maple in eastern North America. After H. A. Fowells, *Silvics of Forest Trees of the United States*, U.S. Department of Agriculture, Washington, D.C. (1965).

related sugar maple, but usually on drier, better-drained soils higher in calcium content (and therefore less acidic). Silver maple occurs widely in the eastern United States, but especially on the moist, well-drained soils of the Ohio and Mississippi river basins. Red maple grows best either under wet, swampy conditions or on dry, poorly developed soils—that is, under extreme conditions that limit the growth of the other species.

Variations in topography and soils influence local distributions of plants

The distributions of plants reveal the effects of many factors, which vary over different scales of distance. You saw in Chapter 4 that topography may cause local variation in climate within small areas, and that geology may cause variation in soil characteristics at even finer scales. Characteristics of soil that influence the distributions of plants are referred to as **edaphic** factors. These factors vary most in mountainous regions, and ecologists frequently turn to the varied habitats of mountains to study plant distribution.

Along the coast of northern California, mountains create conditions that support a variety of plant communities, ranging from coastal chaparral to tall needle-leaved (coniferous) forests of Douglas fir and redwood. When localities are ranked on scales of available moisture, the distribution of each plant species among the localities reveals a distinct **optimum**—the type of site in which it does best (**Figure 5.5**). The coast redwood dominates the central portion of the moisture gradient and frequently forms pure stands. Cedar, Douglas fir, and madrone—a broad-leaved evergreen species with small, thick leaves—occur at the drier end of the moisture gradient. Big-leaf maple—a deciduous species—occupies the wetter end, along with the broad-leaved, evergreen California bay tree.

MORE ON THE WEB | *Edaphic specialization.* **Why are some species of plants restricted to particular soil types?**

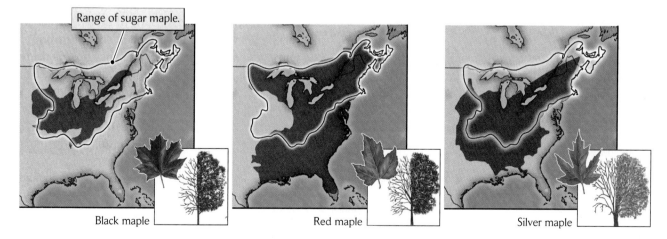

Figure 5.4 Related species may differ in their ecological tolerances. The red areas show the ranges of black, red, and silver maples in eastern North America. The range of the sugar maple is outlined on each map to show the area of overlap. After H. A. Fowells, *Silvics of Forest Trees of the United States*, U.S. Department of Agriculture, Washington, D.C. (1965).

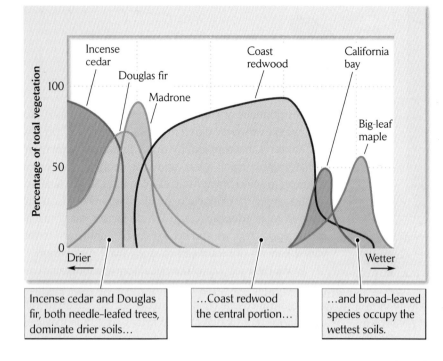

Figure 5.5 The distribution of a plant species reveals its environmental optimum. Distributions of tree species in the northern coastal region of California differ along a gradient of minimum available soil moisture. After R. H. Waring and J. Major, *Ecological Monographs* 34:167–215 (1964).

Incense cedar and Douglas fir, both needle-leafed trees, dominate drier soils...

...Coast redwood the central portion...

...and broad-leaved species occupy the wettest soils.

Form and function are adapted to match the environment

The adaptations of an organism cannot easily be separated from the environment in which it lives. Insect larvae from stagnant aquatic environments in ditches and sloughs can survive longer without oxygen than related species from well-aerated streams and rivers; species of marine snails that occur high in the intertidal zone, where they are frequently exposed to air, tolerate desiccation better than do species from lower levels. These are examples of **specializations**, or adaptations that suit organisms to particular ranges of environmental conditions. Species may be broad or narrow in their range of specialization. Those that have a relatively narrow range of tolerance are called **specialists;** those that have a wide range of tolerance are called **generalists.** Compare the leaves of deciduous forest trees growing in mesic environments with those of desert species. The former are typically broad and thin, providing a large surface area for light absorption. Desert trees have small, finely divided leaves (**Figure 5.6**)—or sometimes none at all. (Cacti rely

1 Mesquite (*Prosopis*) leaves are subdivided into leaflets that facilitate dissipation of heat.

2 The paloverde–Spanish for "green stick"–(*Cercidium*) carries this adaptation even further: its leaflets are tiny, and the thick stems, which contain chlorophyll, are responsible for much of the plant's photosynthesis.

3 Unlike most desert plants, limberbush (*Jatropha*) has broad, succulent leaves, produced for only a few weeks during the summer rainy season.

Figure 5.6 The leaves of desert plants are adapted to hot, dry conditions. These three species from the Sonoran Desert in Arizona all have adaptations that help them cope with desert conditions. Photos by R. E. Ricklefs.

Table 5.1   Characteristics of chaparral and coastal sage vegetation in southern California

Characteristic	Vegetation type	
	Chaparral	Coastal sage
Roots	Deep	Shallow
Leaves	Evergreen	Summer deciduous
Average leaf duration (months)	12	6.0
Average leaf size (cm^2)	12.6	4.5
Leaf weight (g dry weight per dm^2)*	1.8	1.0
Maximum transpiration (g H$_2$O per dm^2 per h)	0.34	0.94
Maximum photosynthetic rate (mg C per dm^2 per h)	3.9	8.3
Relative annual CO$_2$ fixation	49.8	46.8

Source: *A. T. Harrison, E. Small, and H. A. Mooney,* Ecology *52:869–875 (1971); H. A. Mooney and E. L. Dunn,* Am. Nat. *104:447–453 (1970).*

*dm = decimeter (10 cm).

entirely on their stems for photosynthesis; their leaves are modified into thorns for protection.) Leaves heat up in the desert sun. Structures lose heat by convection most rapidly at their edges, where wind currents disrupt insulating boundary layers of still air. The more edges, the cooler the leaf, and the lower the water loss per unit of photosynthetic area. Small leaf size means that a large proportion of each leaf is close to its edge.

Coastal sage and chaparral plants in southern California have divergent forms and lifestyles due to the different levels of water stress in their respective environments (Table 5.1). Chaparral habitat ranges over higher elevations than that of the coastal sage and thus is cooler and moister. Both plant types are exposed to prolonged summer drought, but the soils present greater water stress in the sage habitat. Coastal sage plants typically have shallow roots and small, delicate leaves (❚ Figure 5.7). It is not surprising, therefore, that most coastal sage species shed their leaves during the summer drought period. Chaparral species have deep roots that often extend through tiny cracks and fissures far into the bedrock; their thick evergreen leaves have a waxy outer covering that reduces water loss. Sage and chaparral plants are differently specialized. Black sage is active only during the rainy season of winter and early spring. Its leaves are designed for high rates of photosynthesis and high rates of growth, but they are dropped, and the plant becomes dormant, as soon as water becomes scarce in the soil. Black sage is thus specialized for the transient moisture of the Mediterranean climate winter. Chamise and other chaparral species make use of the more limited water that lies deeper in the soil, but persists longer through the year.

However, they fail to cash in on the winter water bonanza in the upper layers of the soil the way that coastal sage species do.

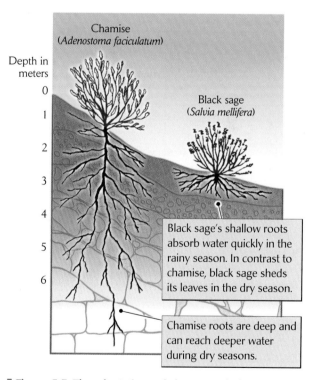

❚ **Figure 5.7 The adaptations of plants match their environments.** Profiles of the root systems of chamise (*Adenostoma fasciculatum*), a chaparral species (left), and black sage (*Salvia mellifera*), a member of the coastal sage community (right), show their differing adaptations to different levels of water stress. After H. Hellmers et al., *Ecology* 36:667–678 (1955).

MORE
ON THE
WEB *Living together on different resources.* Where chaparral and coastal sage species grow together near the overlapping edges of their ranges, they exploit different parts of the environment.

Certain kinds of plants make the environment more favorable for themselves. As we saw in the last chapter, the tough needles of pine, spruce, and fir trees produce abundant organic acids when they decompose. This acid leaches minerals from the soil under podsolizing conditions and makes the soils even poorer than they were. The needle-leaved trees tolerate these conditions better than deciduous trees. Thus, the different effects of these two types of vegetation on soil conditions reinforce the boundary between deciduous and needle-leaved forest.

These examples drive home the important point that growth form is closely related to the physical conditions of the environment. With respect to terrestrial plants, larger growth forms are often competitively superior because they shade smaller growth forms. But they also require moister soils. Thus, we should not be surprised that the availability of water is the predominating factor determining the char-

acter and distribution of terrestrial biomes. Because heat influences moisture stress and moisture availability, temperature and precipitation together are the important determinants of the boundaries of biomes.

Climate defines the boundaries of terrestrial biomes

One of the most widely adopted climate classification schemes is the **climate zone** system developed by the German ecologist Heinrich Walter. This system, which has nine major divisions, is based on the annual course of temperature and precipitation. The important attributes of climate and the characteristic vegetation in each of these zones are presented in ▌Figure 5.8. The values of temperature and precipitation used to define climate zones correspond to conditions of moisture and cold stress that are particularly important determinants of plant form. For example, within the Tropics, the tropical climate zone is distinguished from the equatorial climate zone by the occurrence of water stress

Biome name	Climate zone		Vegetation
Tropical rain forest	**I**	Equatorial: Always moist and lacking temperature seasonality	Evergreen tropical rain forest
Tropical seasonal forest/ savanna	**II**	Tropical: Summer rainy season and "winter" dry season	Seasonal forest, scrub, or savanna
Subtropical desert	**III**	Subtropical (hot deserts): Highly seasonal, arid climate	Desert vegetation with considerable exposed surface
Woodland/shrubland	**IV**	Mediterranean: Winter rainy season and summer drought	Sclerophyllous (drought-adapted), frost-sensitive shrublands and woodlands
Temperate rain forest	**V**	Warm temperate: Occasional frost, often with summer rainfall maximum	Temperate evergreen forest, somewhat frost-sensitive
Temperate seasonal forest	**VI**	Nemoral: Moderate climate with winter freezing	Frost-resistant, deciduous, temperate forest
Temperate grassland/ desert	**VII**	Continental (cold deserts): Arid, with warm or hot summers and cold winters	Grasslands and temperate deserts
Boreal forest	**VIII**	Boreal: Cold temperate with cool summers and long winters	Evergreen, frost-hardy needle-leaved forest (taiga)
Tundra	**IX**	Polar: Very short, cool summers and long, very cold winters	Low, evergreen vegetation, without trees, growing over permanently frozen soils

▌**Figure 5.8** Heinrich Walter classified the climate zones of the world according to the annual course of temperature and precipitation. Biome names for these zones under Whittaker's classification scheme are shown in the left-hand column.

during a pronounced dry season. Subtropical climate zones are perpetually water-stressed. The typical vegetation of these three climate zones is evergreen rain forest, deciduous forest or savanna, and desert scrub, respectively. We will look at Walter's climate zones in more detail below.

Cornell University ecologist Robert H. Whittaker took a somewhat different approach to the relationship between terrestrial biomes and climatic variables. Whittaker defined biomes by their vegetation structure, and then devised a simple climate diagram on which he plotted the approximate boundaries of the major biomes with respect to average temperature and precipitation (▌Figure 5.9). The result is similar to Walter's scheme. Plotted on Whittaker's diagram, most locations on earth fall within a triangular area whose three corners represent warm-moist, warm-dry, and cool-dry climates. (Cold regions with high rainfall are rare because water does not evaporate rapidly at low temperatures and because the atmosphere in cold regions holds little water vapor.)

At tropical and subtropical latitudes, where mean temperatures range between 20°C and 30°C, vegetation ranges from rain forest, which is wet throughout the year and

generally receives more than 250 cm (about 100 inches) of rain annually (Walter's equatorial climate zone), to desert, which generally receives less than 50 cm of rain (Walter's subtropical climate zone). Intermediate climates support seasonal forests (150–250 cm rainfall), in which some or all trees lose their leaves during the dry season, or scrub and savannas (50–150 cm rainfall).

Plant communities at temperate latitudes follow the pattern of tropical communities with respect to rainfall, having four vegetation types: rain forest, seasonal forest, woodland/shrubland, and grassland/desert. At high latitudes, however, precipitation varies so little from one locality to another that vegetation types are poorly differentiated by climate. Where mean annual temperatures are below −5°C, all plant communities may be lumped into one type: tundra.

Toward the drier end of the precipitation spectrum within each temperature range, fire plays a distinct role in shaping plant communities. The influence of fire is greatest where moisture availability is intermediate and highly seasonal. Deserts and moist forests burn infrequently: deserts rarely accumulate enough plant debris to fuel a fire, and moist forests rarely dry out enough to be highly flammable.

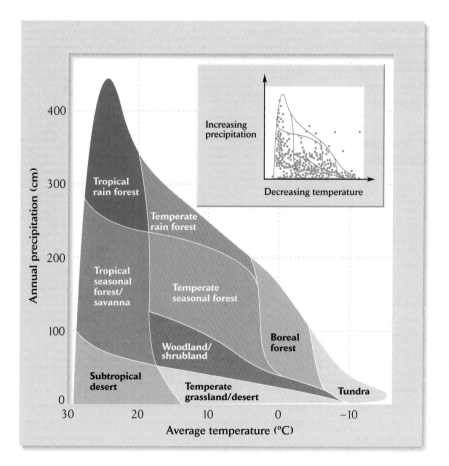

▌Figure 5.9 Whittaker's biomes are delineated according to average temperature and precipitation. *Main diagram*: Whittaker plotted the boundaries of observed vegetation types with respect to average temperature and precipitation. In climates intermediate between those of forest and desert biomes, fire, soil, and climatic seasonality determine whether woodland, grassland, or shrubland develops. *Inset*: Average annual temperature and precipitation for a sample of localities more or less evenly distributed over the land area of the earth. Most of the points fall within a triangular region that includes the full range of climates, excluding those of high mountains. From R. H. Whittaker, *Communities and Ecosystems*, 2d ed., Macmillan, New York (1975).

Grassland and shrub biomes have the combination of abundant fuel and seasonal drought that make fire a frequent visitor. In these biomes, fire is a dominating factor to which all community members must be adapted and, indeed, for which many are specialized: some species require fire for germination of their seeds and growth of their seedlings. In many areas of African savanna and North American prairie, frequent fires kill the seedlings of trees and prevent the encroachment of forests, which could be sustained by the local precipitation if it were not for fire. Burning favors perennial grasses and forbs with extensive root systems that can survive underground. After an area has burned over, grass and forb roots sprout fresh shoots and quickly establish new vegetation above the surface of the soil. In the absence of frequent fires, tree seedlings become established and eventually shade out savanna and prairie vegetation.

As in all classification systems, exceptions appear, and boundaries between biomes are fuzzy. Moreover, not all plant growth forms correspond with climate in the same way; as mentioned earlier, Australian eucalyptus trees form forests under climatic conditions that support only shrubland or grassland on other continents. Finally, plant communities reflect factors other than temperature and rainfall. Topography, soils, fire, seasonal variations in climate, and herbivory all leave their mark. The overview of the major terrestrial biomes at the end of this chapter emphasizes the distinguishing features of the physical environment and how these are reflected in the form of the dominant plants.

MORE ON THE WEB *Biomes and animal forms.* Why are biome definitions based on the predominant life forms of plants, rather than referring to their animal inhabitants?

MORE ON THE WEB *Characterizing climate.* Integrated descriptions of climate emphasize the interaction of temperature and availability of water.

Walter climate diagrams distinguish the major terrestrial biomes

Temperature and precipitation interact to determine conditions and resources available for plant growth. It is not surprising, then, that the distributions of the major biomes of the earth follow patterns of temperature and precipitation. Because of this close relationship, it is important to describe climate in a manner that reflects the availability of water, taking into consideration changes in temperature and precipitation through the year.

Heinrich Walter developed a climate diagram that conveys seasonal periods of water deficit and abundance and

therefore permits ecologically meaningful comparisons of climates between localities (Figure 5.10). The Walter climate diagram portrays average monthly temperature and precipitation throughout the course of a year. The vertical scales of temperature and precipitation are adjusted so that when precipitation is higher than temperature on the diagram, water is plentiful, and plant production is limited primarily by temperature. Conversely, when temperature is higher than precipitation, plant production is limited by availability of water. Walter's scales equate 20 mm of monthly precipitation with 10°C in temperature. Thus, as a rule of thumb, at an average temperature of 20°C, 40 mm of monthly precipitation provides sufficient moisture for plant growth. We'll use Walter climate diagrams to compare the biomes characterized below.

Climate diagrams for locations in each of Walter's climate zones are shown in Figure 5.11. The seasonal distributions of wet and dry periods differ among climate

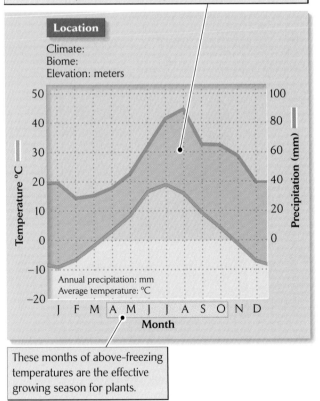

Figure 5.10 **Walter climate diagrams allow ecologically meaningful comparisons between localities.** These diagrams portray the annual progression of monthly mean temperature (left-hand scale) and precipitation (right-hand scale).

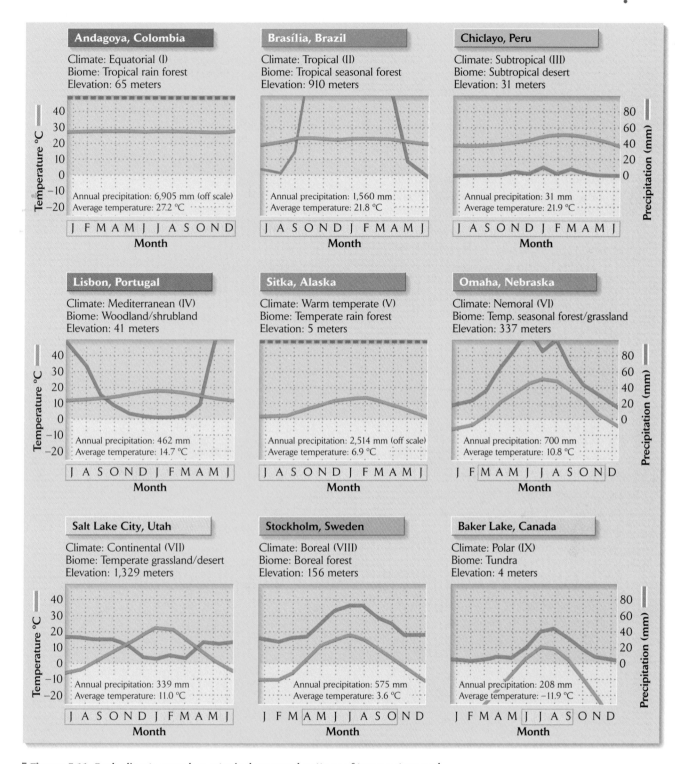

Figure 5.11 **Each climate zone has a typical seasonal pattern of temperature and precipitation.** Climate diagrams for representative locations in each of the nine major terrestrial climate zones are shown. The dashed blue line at the top of the graphs for climate zones I and V indicates monthly precipitation exceeding 100 mm all year long. From H. Walter and S.- W. Breckle, *Ecological Systems of the Geobiosphere, I, Ecological Principles in Global Perspective,* Springer-Verlag, Berlin (1985).

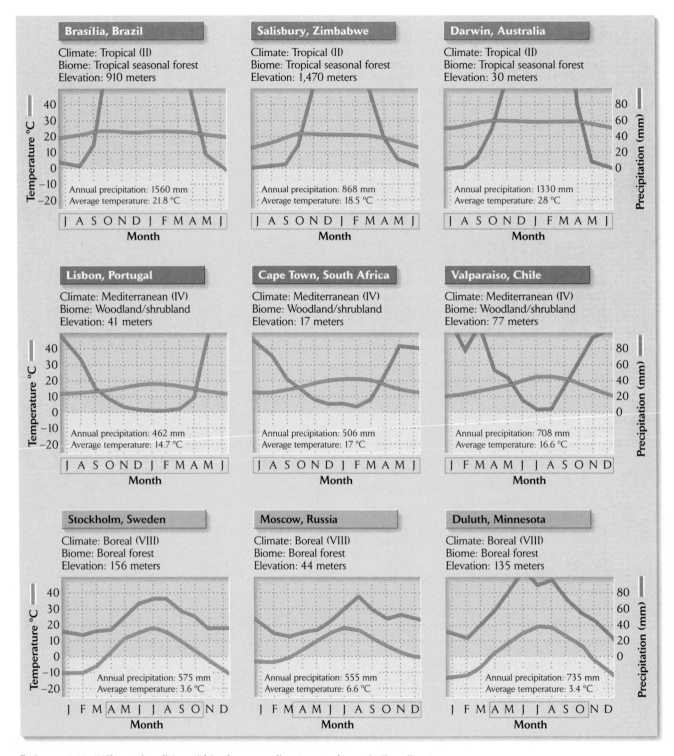

Figure 5.12 Different localities within the same climate zone have similar climate patterns. Climate diagrams of three localities within each of three terrestrial climate zones show consistency within zones. The biomes found at all three localities within each zone are the same. From H. Walter and S.-W. Breckle, *Ecological Systems of the Geobiosphere, I, Ecological Principles in Global Perspective,* Springer-Verlag, Berlin (1985).

zones at lower latitudes. Equatorial climates (climate zone I) like that at Andagoya, Colombia, are aseasonal; that is, they are warm and wet throughout the year. Subtropical climates (III), such as that of Chiclayo, Peru, are warm and dry throughout the year. Tropical climates (II, Brasília, Brazil) are characterized by summer rains, and Mediterranean climates (IV, Lisbon, Portugal) by winter rains. The climate of Sitka, Alaska (warm temperate, V), is wet and mild throughout the year and supports an evergreen forest type of vegetation.

Seasonality of temperature is a major factor in climate zones VI–IX, which occur at middle and high latitudes. Precipitation is typically low, but because of the low temperatures, moisture is generally not limiting during the short summer growing season. Continental climates (VII, Salt Lake City, Utah) are typically dry throughout the year and become warm enough in summer to develop significant water stress. Such areas, which include much of the Great Basin of the western United States, support shrubby desert vegetation.

In striking contrast to the variation in patterns of temperature and precipitation between climate zones, different localities within the same climate zone have similar climates, regardless of their geographic location (▌Figure 5.12). For example, the tropical climates of Brasília (Brazil), Salisbury (Zimbabwe), and Darwin (Australia) all share the even year-round warm temperatures and summer rainfall typical of climate zone II. And each of these areas supports deciduous forest vegetation grading into savanna where precipitation is particularly low. The similar patterns of climate within each climate zone support characteristic vegetation that defines the biome type and makes it easy for us to recognize the general attributes of ecosystems in any region.

The worldwide distribution of biome types follows patterns of temperature and precipitation over the earth (▌Figure 5.13). We shall consider the biomes and general ecological characteristics of each of the major climate zones in the series of vignettes that follow. Because most readers of this book live in the so-called temperate zone, it is a good place to start.

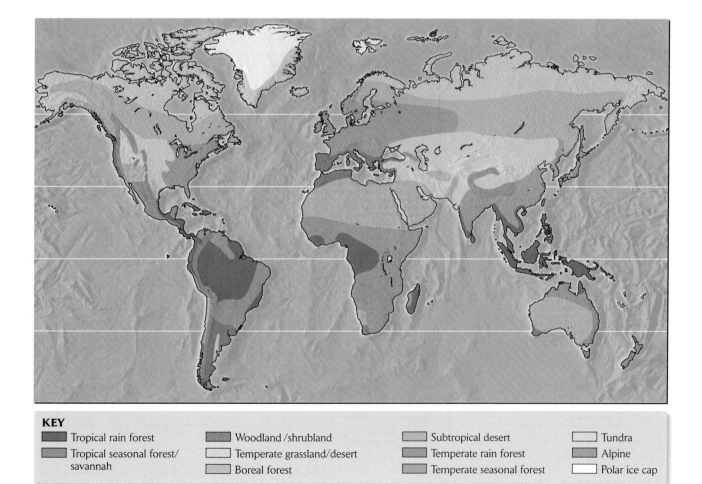

KEY

■ Tropical rain forest	■ Woodland /shrubland	■ Subtropical desert	□ Tundra
■ Tropical seasonal forest/ savannah	□ Temperate grassland/desert	■ Temperate rain forest	■ Alpine
	■ Boreal forest	■ Temperate seasonal forest	□ Polar ice cap

▌**Figure 5.13 Global distribution of the major biomes.**

Temperate climate zones have average annual temperatures between 5°C and 20°C

Temperate climates are characterized by average annual temperatures in the range of 5°–20°C at low elevations. Such climates are distributed approximately between 30°N and 45°N in North America and between 40°N and 60°N in Europe, which is warmed by the Gulf Stream current. Frost is an important factor throughout the temperate zone, perhaps even a defining one. Within the temperate zone, biomes are differentiated primarily by total amounts and seasonal patterns of precipitation, although the length of the frost-free season, which is referred to as the **growing season,** and the severity of frost also are important.

Temperate seasonal forest biome (climate zone VI)

Often referred to as deciduous forest, the temperate seasonal forest biome occurs under moderate conditions with winter freezing. In North America, this biome is found principally in the eastern United States and southern Canada; it is also widely distributed in Europe and eastern Asia (▌Figure 5.14). It is poorly developed in the Southern Hemisphere (New Zealand and southern Chile) because of the milder winter temperatures at moderate latitudes caused by the relatively small continental areas there. The length of the growing season varies from 130 days at higher latitudes to 180 days at lower latitudes. Precipitation usually exceeds evaporation and transpiration; consequently, water tends to move downward through soils and to drain from the landscape as groundwater and as surface streams and rivers. Soils are podsolized, tend to be slightly

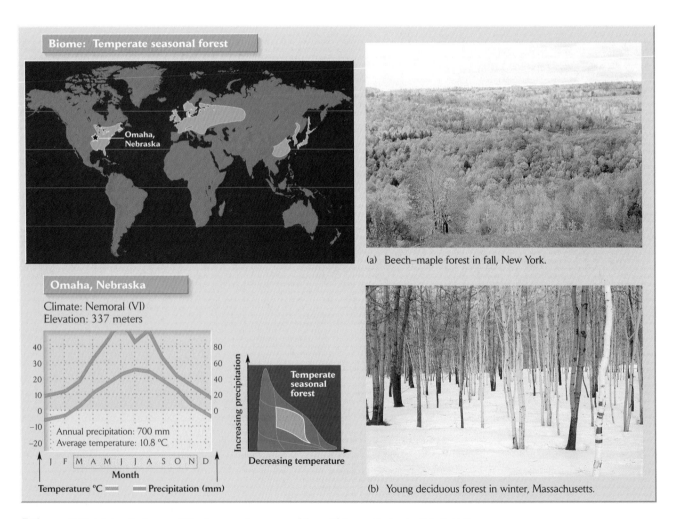

(a) Beech–maple forest in fall, New York.

(b) Young deciduous forest in winter, Massachusetts.

▌Figure 5.14 **Major features of the temperate seasonal forest biome.** Photos by R. E. Ricklefs.

acid and moderately leached, and are brown in color owing to abundant organic humus. Deciduous trees are the dominant plant growth form. The vegetation often includes a layer of smaller trees and shrubs beneath the dominant trees, as well as herbaceous plants on the forest floor, many of which complete their growth and flower early in spring, before the trees have fully leafed out.

Warmer and drier parts of the temperate seasonal forest biome, especially where soils are sandy and nutrient-poor, tend to develop needle-leaved forests, dominated by pines. The most important of these formations in North America are the pine forests of the coastal plains of the Atlantic and Gulf states of the United States; pine forests also occur at higher elevations in the western United States. Because of the warm climate in the southeastern United States, soils are usually lateritic and have low nutrient levels. The low availability of nutrients and water favors evergreen, needle-leaved trees, which resist desiccation and give up nutrients slowly because they retain their needles for several years. Because soils tend to be dry, fires are frequent, and most species are able to resist fire damage.

Temperate rain forest biome (climate zone V)

In warm temperate climates near the coast in the northwestern North America, and also in southern Chile, New Zealand, and Tasmania, mild winters, heavy winter rains, and summer fog create conditions that presently support extremely tall evergreen forests (▮Figure 5.15). In North America these forests are dominated toward the south by coast redwood and toward the north by Douglas fir. Trees are typically 60–70 m high and may grow to over 100 m. It is not well understood why these sites are dominated by needle-leaved trees, but the fossil record shows that these plant formations are very old and that they are remnants

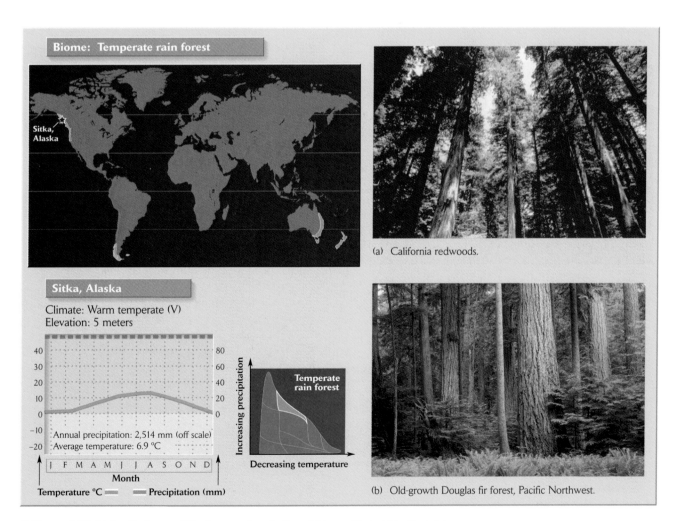

Biome: Temperate rain forest

Sitka, Alaska

Sitka, Alaska

Climate: Warm temperate (V)
Elevation: 5 meters

Annual precipitation: 2,514 mm (off scale)
Average temperature: 6.9 °C

Month

Temperature °C ▬▬ ▬▬ Precipitation (mm)

Increasing precipitation

Temperate rain forest

Decreasing temperature

(a) California redwoods.

(b) Old-growth Douglas fir forest, Pacific Northwest.

▮**Figure 5.15 Major features of the temperate rain forest biome.** Photo (a) by PhotoSphere Images/PictureQuest; photo (b) by Tom & Pat Leeson/Photo Researchers.

of forests that were vastly more extensive during the Mesozoic era, as recently as 70 million years ago. In contrast to that of rain forests in the Tropics, the diversity of temperate rain forests typically is very low.

Temperate grassland/desert biome (climate zone VII)

In North America, grasslands develop within continental climate zones, where the rainfall ranges between 30 and 85 cm per year (▐ Figure 5.16). Summers are hot and wet; winters are cold. The growing season increases from north to south from about 120 to 300 days. These grassland biomes are often called **prairies.** Extensive grasslands are also found in central Asia, where they are called **steppes.** Because precipitation is low, organic detritus does not decompose rapidly, and the soils are rich in organic matter. Because of their low acidity, prairie soils, which belong to the mollisol group, are not heavily leached and tend also to be rich in nutrients. The vegetation is dominated by grasses, which grow to over 2 m in the moister parts of the grassland biome and to less than 0.2 m in more arid regions. There are also abundant nongrass forbs. Fire is a dominant influence in grasslands, particularly where the habitat dries out during the late summer. Most grassland species have fire-resistant underground stems, or **rhizomes,** from which shoots resprout, or they have fire-resistant seeds.

Where precipitation ranges between 25 and 50 cm per year, and winters are cold and summers are hot, grasslands grade into deserts. The temperate desert biome covers most of the Great Basin of the western United States. In the northern part of the region, sagebrush (*Artemisia*) is the dominant plant, whereas toward the south and on somewhat moister soils, widely spaced juniper and piñon trees predominate, forming open woodlands of less than 10 m stature with sparse coverings of grass. In these temperate deserts, the evaporation and transpiration potential of the habitat exceeds precipitation during most of the

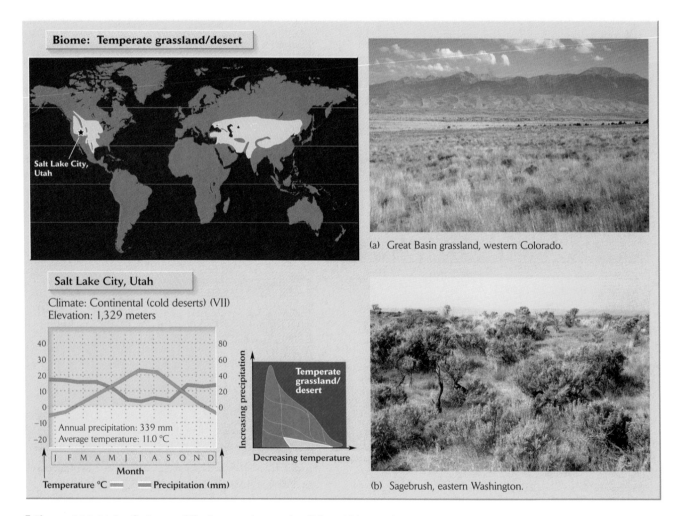

(a) Great Basin grassland, western Colorado.

(b) Sagebrush, eastern Washington.

Biome: Temperate grassland/desert

Salt Lake City, Utah

Salt Lake City, Utah

Climate: Continental (cold deserts) (VII)
Elevation: 1,329 meters

Annual precipitation: 339 mm
Average temperature: 11.0 °C

Temperature °C ▬▬ ▬▬ Precipitation (mm)

Increasing precipitation

Temperate grassland/desert

Decreasing temperature

▐ **Figure 5.16** **Major features of the temperate grassland/desert biome.** Photos by R. E. Ricklefs.

year, so soils are dry and little water percolates through them to form streams and rivers. The soils tend to accumulate, at depths to which water usually penetrates, calcium carbonate leached from the surface layers. Fires occur infrequently in temperate deserts because the habitat produces little fuel. However, because of the low productivity of the plant community, grazing can exert strong pressure on vegetation and may even favor the persistence of shrubs, which are not good forage. Indeed, many dry grasslands in the western United States and elsewhere in the world have been converted to deserts by overgrazing.

Woodland/shrubland biome (climate zone IV)

The Mediterranean climate zone is distributed at 30°–40° north and south of the equator—somewhat higher in Europe—on the western sides of continental landmasses where cold-water currents and winds coming off the con-

tinents dominate the climate. Mediterranean climates are found in southern Europe and southern California in the Northern Hemisphere and in central Chile, the Cape region of South Africa, and southwestern Australia in the Southern Hemisphere. Mediterranean climates are characterized by mild winter temperatures, winter rain, and summer drought. These climates support thick, evergreen, shrubby vegetation 1–3 m in height, with deep roots and drought-resistant foliage (Figure 5.17). The small, durable leaves of typical Mediterranean-climate plants have earned them the label of **sclerophyllous** (hard-leaved) vegetation. Fires are frequent in woodland/shrubland biomes, and most plants have either fire-resistant seeds or root crowns that resprout soon after a fire.

Subtropical desert biome (climate zone III)

What people call deserts varies tremendously. Many people refer to the dry areas of the Great Basin and of central

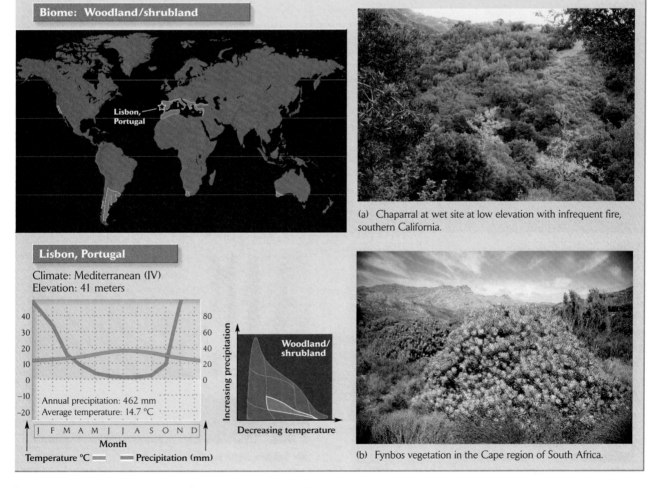

(a) Chaparral at wet site at low elevation with infrequent fire, southern California.

(b) Fynbos vegetation in the Cape region of South Africa.

Biome: Woodland/shrubland

Lisbon, Portugal

Lisbon, Portugal

Climate: Mediterranean (IV)
Elevation: 41 meters

Annual precipitation: 462 mm
Average temperature: 14.7 °C

Temperature °C   Precipitation (mm)

Woodland/shrubland

Increasing precipitation

Decreasing temperature

Figure 5.17 Major features of the woodland/shrubland biome. Photo (a) by Earl Scott/
Photo Researchers; photo (b) by Fletcher & Baylis/Photo Researchers.

Asia as deserts—the Gobi Desert is a name familiar to most of us. But the climates of these "deserts" fall within Walter's continental climate zone, characterized by low precipitation and cold winters. In contrast, subtropical deserts (▌Figure 5.18) develop at latitudes of 20°–30° north and south of the equator in areas having high atmospheric pressure, very sparse rainfall (less than 25 cm), and generally long growing seasons. Because of the low rainfall, the soils of subtropical deserts (aridosols) are shallow, virtually devoid of organic matter, and neutral in pH. Impermeable hardpans of calcium carbonate often develop at the limits of water penetration—depths of a meter or less. Whereas sagebrush dominates Great Basin (continental climate) deserts, creosote bush takes its place in the subtropical deserts of the Americas. Moister sites support a profusion of succulent cacti, shrubs, and small trees, such as mesquite and paloverde. Most subtropical deserts receive summer rainfall, during which many herbaceous plants sprout from dormant seeds, grow quickly, and reproduce

before the soils dry out again. Many of the plants in subtropical deserts are not frost-tolerant. Species diversity is usually much higher than it is in temperate arid lands.

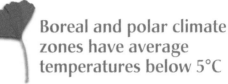

Boreal and polar climate zones have average temperatures below 5°C

Boreal forest biome (climate zone VIII)

Stretching in a broad belt centered at about 50°N in North America and about 60°N in Europe and Asia lies the boreal forest biome, often called **taiga** (▌Figure 5.19). The average annual temperature is below 5°C, and winters are severe. Precipitation is in the range of 40–100 cm, and because evaporation is low, soils are moist throughout most of the growing season. The vegetation consists of seemingly endless dense stands of 10–20 m high ever-

Biome: Subtropical desert

Chiclayo, Peru

Chiclayo, Peru

Climate: Subtropical (hot deserts) (III)
Elevation: 31 meters

Annual precipitation: 31 mm
Average temperature: 21.9 °C

Month

Temperature °C ▬   Precipitation (mm) ▬

Increasing precipitation

Subtropical desert

Decreasing temperature

(a) Cholla cactus in northern Sonora, Mexico.

(b) Sahuaro cactus in southern Arizona.

▌**Figure 5.18 Major features of the subtropical desert biome.** Photos by R. E. Ricklefs.

green needle-leaved trees, mostly spruce and fir. Because of the low temperatures, litter decomposes very slowly and accumulates at the soil surface. The needle-leaf litter produces high levels of organic acids, so the soils are acid, strongly podsolized, and generally of low fertility. Growing seasons are rarely as much as 100 days, and often half that. The vegetation is extremely frost-tolerant, as temperatures may reach −60°C during the winter. Species diversity is very low.

Tundra biome (climate zone IX)

To the north of the boreal forest, in the so-called polar climate zone, lies the arctic tundra, a treeless expanse underlain by permanently frozen soil, or **permafrost** (❙ Figure 5.20). The soils thaw to a depth of 0.5–1 m during the brief summer growing season. Precipitation is generally less and often much less, than 60 cm, but in low-lying areas where drainage is prevented by the permafrost, soils

may remain saturated with water throughout most of the growing season. Soils tend to be acid because of their high organic matter content, and they are very low in nutrients. In this nutrient-poor environment, plants hold their foliage for years. Most plants are dwarf, prostrate, woody shrubs, which grow low to the ground to gain protection under the winter blanket of snow and ice. Anything protruding above the surface of the snow is sheared off by blowing ice crystals. For most of the year, the tundra is an exceedingly harsh environment, but during the 24-hour-long summer days, the rush of activity in the tundra biome testifies to the remarkable adaptability of life.

At high elevations within temperate latitudes, and even within the Tropics, one finds vegetation resembling that of the arctic tundra and even including some of the same species, or their close relatives. These areas above the tree line occur most broadly in the Rocky Mountains of North America and, especially, on the Tibetan Plateau of central Asia. In spite of their similarities, alpine and arctic tundra

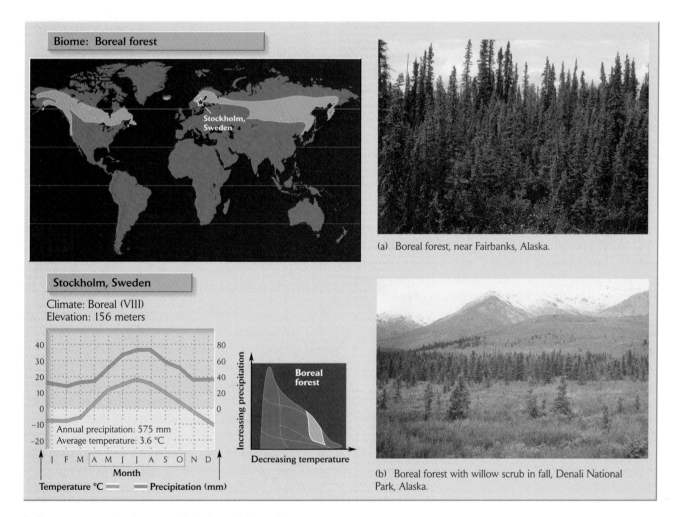

(a) Boreal forest, near Fairbanks, Alaska.

(b) Boreal forest with willow scrub in fall, Denali National Park, Alaska.

❙ **Figure 5.19 Major features of the boreal forest biome.** Photos by R. E. Ricklefs.

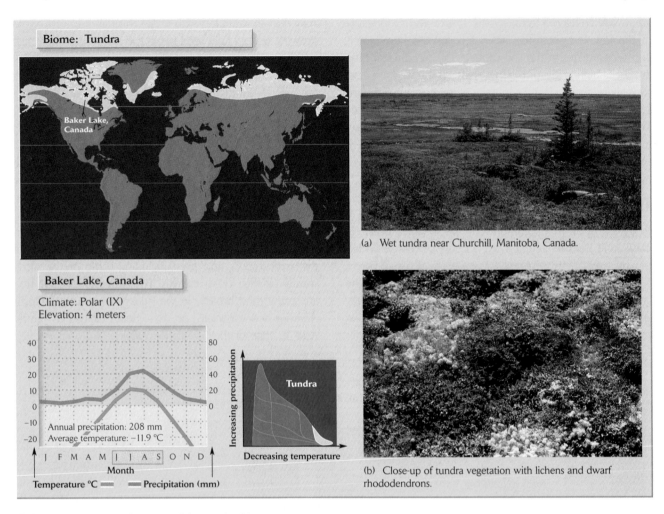

Biome: Tundra

Baker Lake, Canada

Climate: Polar (IX)
Elevation: 4 meters

Annual precipitation: 208 mm
Average temperature: −11.9 °C

J F M A M J J A S O N D
Month

Temperature °C   Precipitation (mm)

Increasing precipitation

Tundra

Decreasing temperature

(a) Wet tundra near Churchill, Manitoba, Canada.

(b) Close-up of tundra vegetation with lichens and dwarf rhododendrons.

▌Figure 5.20 Major features of the tundra biome. Photos by R. E. Ricklefs.

have important points of dissimilarity as well. Areas of alpine tundra generally have warmer and longer growing seasons, higher precipitation, less severe winters, greater productivity, better-drained soils, and higher species diversity than arctic tundra. Still, as in the high-latitude tundra, harsh winter conditions ultimately limit the growth of trees.

Equatorial and tropical climate zones have average temperatures exceeding 20°C

Within 20° of the equator, the temperature varies more throughout the day than average monthly temperatures vary throughout the year. Average temperatures at sea level generally exceed 20°C. Environments within tropical latitudes are distinguished by differences in the seasonal pattern of rainfall. This creates a continuous gradient of vegetation from wet, aseasonal rain forest through sea-

sonal forest, scrub, savanna, and desert. Frost is not a factor in tropical biomes, even at high elevations, and tropical plants and animals generally cannot tolerate freezing.

Tropical rain forest biome (climate zone I)

Climates under which tropical rain forests develop are always warm and receive at least 200 cm of precipitation throughout the year, with not less than 10 cm during any single month. These conditions prevail in three important regions within the Tropics (▌Figure 5.21). First, the Amazon and Orinoco basins of South America, with additional areas in Central America and along the Atlantic coast of Brazil, constitute the American rain forest. Second, the area from southernmost West Africa and extending eastward through the Congo River basin constitutes the African rain forest. Third, the Indo-Malayan rain forest covers parts of Southeast Asia (Vietnam, Thailand, and the Malay Peninsula), the islands between Asia and Australia, includ-

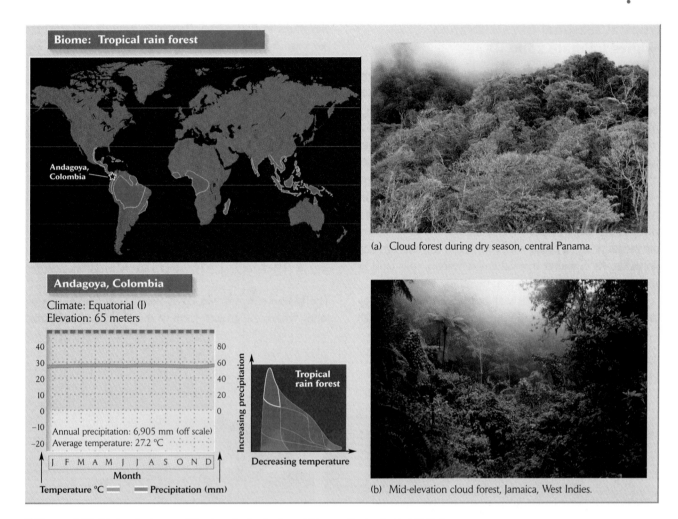

Biome: Tropical rain forest

Andagoya, Colombia

Andagoya, Colombia

Climate: Equatorial (I)
Elevation: 65 meters

Annual precipitation: 6,905 mm (off scale)
Average temperature: 27.2 °C

J F M A M J J A S O N D
Month

Temperature °C ▬   ▬ Precipitation (mm)

Increasing precipitation

**Tropical
rain forest**

Decreasing temperature

(a) Cloud forest during dry season, central Panama.

(b) Mid-elevation cloud forest, Jamaica, West Indies.

▌**Figure 5.21 Major features of the tropical rain forest biome.** Photos by R. E. Ricklefs.

ing the Philippines, Borneo, and New Guinea, and the Queensland coast of Australia.

The tropical rain forest climate often exhibits two peaks of rainfall centered around the equinoxes, corresponding to the periods when the intertropical convergence lies over the equatorial region. Rain forest soils are typically old and deeply weathered oxisols. Because they are relatively devoid of humus and clay, they take on the reddish color of aluminum and iron oxides and have poor ability to retain nutrients. In spite of the low nutrient status of the soils, rain forest vegetation is dominated by a continuous **canopy** of tall evergreen trees rising to 30–40 m, with occasional **emergent trees,** which rise above the canopy to heights of 55 m. Because water stress on emergents is great due to their height and exposure, they are often deciduous, even in an evergreen rain forest. Tropical rain forests have several **understory** layers beneath the canopy, containing smaller trees, shrubs, and herbs, but these are usually quite sparse because so little light pene-

trates the canopy. Climbing **lianas,** or woody vines, and **epiphytes,** plants that grow on the branches of other plants and are not rooted in soil (also called air plants; see Figure 1.7), are prominent in the forest canopy itself. Species diversity is higher than anywhere else on earth.

Per unit of area, the biological productivity of rain forests exceeds that of any other terrestrial biome, and their standing biomass exceeds that of all other biomes except temperate rain forests. Because of the continuously high temperatures and abundant moisture, plant litter decomposes quickly, and vegetation immediately takes up the released nutrients. This rapid nutrient cycling supports the high productivity of the rain forest, but also makes the rain forest ecosystem extremely vulnerable to disturbance. When tropical rain forests are cut, many of the nutrients are carted off in logs or go up in smoke. The vulnerable soils erode rapidly and fill the streams with silt. In many cases, the environment degrades rapidly and the landscape becomes unproductive.

Tropical seasonal forest/savanna biomes (climate zone II)

Within the Tropics, but beyond 10° from the equator, tropical climates often exhibit a pronounced dry season, corresponding to winter at higher latitudes. This is Walter's tropical climate zone. Seasonal forests in the Tropics have a preponderance of deciduous trees that shed their leaves during the season of water stress (▌Figure 5.22). Increasingly longer and more severe dry seasons generally result in vegetation with lower stature and more thorns to protect leaves from grazing. Progressive aridity leads to dry forest, thorn forest, and finally true deserts in the rain shadows of mountain ranges or along coasts with cold ocean currents running alongside. As in wetter tropical environments, soils tend to be strongly laterized and poor in nutrients.

Savannas are grasslands with scattered trees, and they spread over large areas of the dry Tropics, especially at high elevations in East Africa. Rainfall is typically 90–150 cm per year, but the driest three or four months receive less than 5 cm each. Fire and grazing undoubtedly play important roles in maintaining the character of the savanna biome, particularly in wetter regions, as grasses can persist better than other forms of vegetation under both influences. Often when grazing and fire are controlled within a savanna habitat, dry forest begins to develop. Vast areas of African savanna owe their character to the influence of human activity, including burning, over many millennia.

The biome concept must be modified for aquatic systems

The biome concept was developed for terrestrial ecosystems, where the growth form of the dominant vegetation

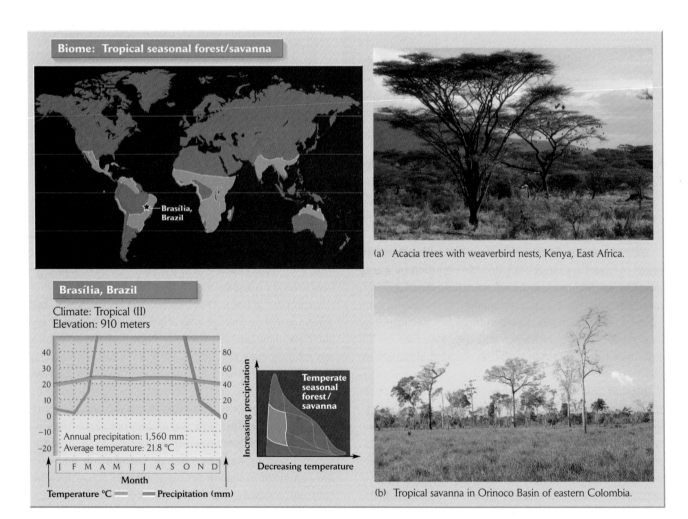

(a) Acacia trees with weaverbird nests, Kenya, East Africa.

Brasília, Brazil

Climate: Tropical (II)
Elevation: 910 meters

Annual precipitation: 1,560 mm
Average temperature: 21.8 °C

Temperature °C ▬▬ Precipitation (mm) ▬▬

Increasing precipitation

Temperate seasonal forest / savanna

Decreasing temperature

(b) Tropical savanna in Orinoco Basin of eastern Colombia.

▌**Figure 5.22 Major features of the tropical seasonal forest/savanna biome.** Photos by R. E. Ricklefs.

reflects climatic conditions. Moreover, terrestrial and aquatic ecologists have generated concepts and descriptive terms for ecological systems independently. As a consequence, aquatic "biomes" do not exist in the sense in which the term is applied to terrestrial systems. Indeed, defining aquatic biomes according to vegetation would be impossible, because the primary producers in many aquatic systems are single-celled algae, which do not form "vegetation" with a characteristic structure. As a result, classifications of aquatic systems have been based primarily on such physical characteristics as salinity, water movement, and depth. The major kinds of aquatic environments are streams, lakes, estuaries, and oceans, and each of these can be subdivided further with respect to many factors. **Streams** form wherever precipitation exceeds evaporation and excess water drains from the surface of the land. Within small streams, ecologists distinguish areas of **riffles,** where water runs rapidly over a rocky substrate, and **pools,** which are deeper stretches of slowly moving water (■ Figure 5.23). Water is well oxygenated in riffles, whereas pools tend to accumulate silt and organic matter. Production in small streams is often dominated by **allochthonous** material—organic material, such as leaves, that enters the aquatic system from the outside. Streams are usually bordered by a **riparian zone** of terrestrial vegetation that is influenced by seasonal flooding and elevated water tables. Streams grow with distance as they join together to form rivers. The larger a river, the more of its production is home-grown, or **autochthonous.**

A **river continuum** concept has grown up around the continuous change in environments and ecosystems between the headwaters and the mouth of a fluvial system. As one moves downstream, water is warmer, more slowly flowing, and richer in nutrients; ecosystems are more complex and generally more productive. Fluvial systems, as rivers are called, are also distinguished by the fact that currents continually move material, including animals and plants, downstream. To maintain a fluvial system in a steady state, this so-called downstream drift must be balanced by active movement of animals upstream, by the productivity of the upstream portions of the system, and by input of allochthonous materials.

Lakes form in any kind of depression. For the most part, such bodies of water are the products of recent glaciation, which leaves behind gouged-out basins and blocks of ice buried in glacial deposits, which eventually melt. Lakes are also formed in geologically active regions, such as the Rift Valley of Africa, where vertical shifting of blocks of the earth's crust creates basins within which water accumulates. Broad river valleys, such as those of the Mississippi and Amazon, have oxbow lakes, which are broad bends of the former river, cut off by shifts in the main channel.

■ **Figure 5.23 Conditions of a stream environment differ in pools and riffles.** Photo by Ed Reschke/Peter Arnold.

An entire lake could be considered a biome, but it is usually subdivided into regions, each of which has its own character. The **littoral zone** is the shallow zone around the edge of a lake or pond within which one finds rooted vegetation, such as water lilies and pickerel weed (■ Figure 5.24). The open water beyond the littoral zone is the **limnetic zone,** where primary production is accomplished by floating single-celled algae, or phytoplankton. Lakes may also be subdivided vertically on the basis of light penetration and the formation of thermally stratified layers of water (see Chapter 4). The sediments at the bottoms of lakes and ponds constitute the **benthic zone,** which provides habitat for burrowing animals and microorganisms.

Estuaries are found at the mouths of rivers, especially where the outflow is partially enclosed by landforms or barrier islands (■ Figure 5.25). Estuaries are unique because of their mix of fresh and salt water. In addition, the nutrients carried by rivers and the rapid exchange between surface waters and sediments contribute to extremely high biological productivity. Because estuaries tend to be shallow areas within which sediments are deposited, they are often edged by extensive tidal marshes. Indeed, the marshes that surround many estuaries are among the most productive habitats on earth, owing to a combination of high

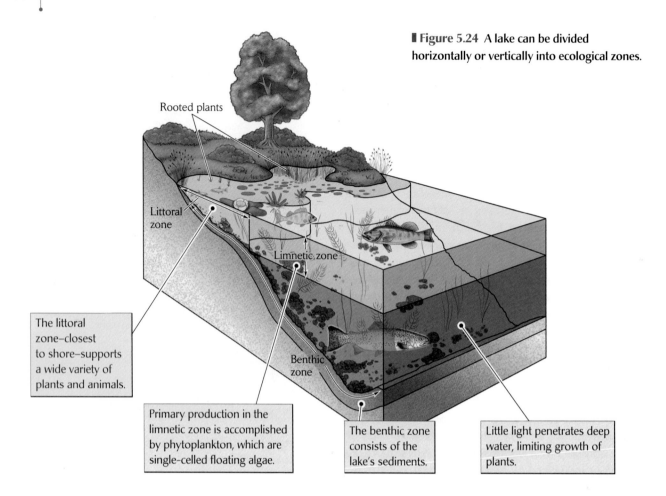

Figure 5.24 A lake can be divided horizontally or vertically into ecological zones.

Rooted plants

Littoral zone

Limnetic zone

Benthic zone

The littoral zone–closest to shore–supports a wide variety of plants and animals.

Primary production in the limnetic zone is accomplished by phytoplankton, which are single-celled floating algae.

The benthic zone consists of the lake's sediments.

Little light penetrates deep water, limiting growth of plants.

Figure 5.25 **Estuaries are extremely productive ecosystems.** They develop at the mouths of rivers and are often bordered by extensive salt marshes, as in this view along the Georgia coast. Photo by S. J. Krasemann/Peter Arnold.

nutrient levels and freedom from water stress. These marshes contribute abundant additional organic matter to the estuarine ecosystem, which in turn supports abundant populations of estuarine and marine species.

Oceans cover the largest portion of the surface of the earth. Beneath the surface of the water lies an immensely complex realm harboring a great variety of ecological conditions and ecosystems (**Figure 5.26**). Variation in marine environments comes from temperature, depth, current, substrate, and, at the edge of the seas, tides.

Many marine ecologists categorize oceanic zones according to depth. The **littoral zone** (also called the intertidal zone) extends between the highest and lowest tidal levels and so, to a varying extent depending on position within the intertidal range, is exposed periodically to air (**Figure 5.27**). The rapid changes in ecological conditions within the littoral zone often create sharp **zonation** of organisms according to their ability to tolerate the stresses of terrestrial conditions. Beyond the range of the lowest tide level, the **neritic** zone extends to depths of about 200 m, which corresponds to the edge of the continental shelf. Often the neritic zone is a region of high productivity because the sunlit surface layers of water are not far removed from the regeneration of nutrients in the sedi-

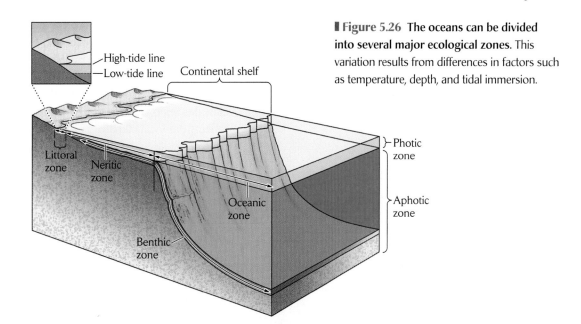

Figure 5.26 The oceans can be divided into several major ecological zones. This variation results from differences in factors such as temperature, depth, and tidal immersion.

ments below. Strong waves can move suspended materials from depths of 100–200 m to the surface. Beyond the neritic zone, the seafloor drops rapidly to the great depths of the **oceanic zone.** Here, production usually is strictly limited by low availability of nutrients. The seafloor beneath the oceanic zone constitutes the **benthic zone.** Both the neritic and the oceanic zones may be subdivided vertically into a superficial **photic zone,** in which there is sufficient light for photosynthesis, and an **aphotic zone** without light, in which organisms depend mostly on organic material raining down from above.

Whereas the open ocean has been compared to a desert, **coral reefs** are like tropical rain forests, both in the richness of their biological production and the diversity of their inhabitants (**Figure 5.28**). Reef-building corals occur in shallow waters of warm oceans, usually where water temperatures remain above 20°C year-round, and often surround volcanic islands. The high production of the reef is fed by nutrients eroding from the rich volcanic soil and by deep-water currents forced upward by the profile of the island. Corals are doubly productive because photosynthetic algae within their tissues generate the carbohydrate energy base for the coral's phenomenal rates of growth.

The unique qualities that characterize each biome or aquatic ecosystem are manifested in most aspects of ecosystem structure and function. The most direct way to evaluate these attributes is to measure the flux of energy through the ecosystem and the cycling of nutrients within the ecosystem. These aspects of ecological structure and function, and how they differ among the terrestrial biomes and aquatic ecosystems, are the subjects of the next part of this book.

Figure 5.27 The littoral zone is exposed to terrestrial conditions twice each day. Nonetheless, it may support prolific growth of algae and a variety of marine animals, as in this area of the New Brunswick coast in Canada. Photos by R. E. Ricklefs.

▌Figure 5.28 Coral reefs are highly productive ecosystems. In contrast to the open ocean, the reef ecosystem provides abundant food for a diverse biological community. This photo was taken in the Red Sea, near Egypt. Photo by Eric Hanauer.

Summary

1. The geographic distributions of plants on continental scales are determined primarily by climate, whereas local distributions within regions may vary according to topography and soils.

2. Climate profoundly affects the evolution of plants and animals, which become specialized to particular conditions of the physical environment. As a consequence, each climatic region has characteristic types of vegetation that differ in growth form, leaf morphology, and seasonality of foliage.

3. Because plant growth form is directly related to climate, we can match major types of vegetation to temperature and precipitation. This relationship emphasizes the way in which temperature and precipitation interact to determine water availability. Major vegetation types can be used to classify ecosystems into categories called biomes. Soil, climatic seasonality, fire, and grazing additionally influence the character of biomes.

4. Two major approaches to the classification of biomes are the climate zone approach of Walter and the vegetation approach exemplified by Whittaker. The first classifies regions on the basis of the climate within which a characteristic type of vegetation normally develops. The second classifies regions according to vegetation, which generally reflects the local climate.

5. Climate zones and biomes are broadly divided into tropical, temperate, boreal, and polar zones according to their latitudes north and south of the equator. Principally temperature, and the adaptations of plants to temperature, distinguish these latitudinal bands. Within these latitudinal zones, the annual level of precipitation and its seasonality further distinguishes biomes.

6. Within temperate climate zones, the major biomes are seasonal forest, rain forest, and grassland/desert. At lower latitudes within temperate regions, are Mediterranean-climate woodlands and shrublands. Subtropical deserts lie between temperate and tropical climate zones.

7. At high latitudes, one encounters boreal forests, usually consisting of needle-leaved trees with persistent foliage and low growth rates on nutrient-poor, acid soils, and tundra, a treeless biome that develops on permanently frozen soils, or permafrost.

8. Tropical climate zones are dominated by evergreen rain forest and seasonal forest, which grades from partly to fully deciduous to thorn forest in drier climates, and sometimes savanna, a grassland with scattered trees that is maintained by fire and grazing pressure.

9. Aquatic systems are not classified as biomes because they lack the equivalent of terrestrial vegetation. One may, however, distinguish streams, lakes, estuaries, and oceans, and each of these systems can be further subdivided on the basis of other factors, such as depth of water.

PRACTICING ECOLOGY

CHECK YOUR KNOWLEDGE

Shifting Biome Boundaries

Climate and soil types dictate the geographic distribution of biomes that we observe today. Various lines of evidence indicate that over long periods of time, biomes have occupied different locations around the globe. Changes in the distribution of biomes over time are caused by large-scale processes including continental drift, the uplift of mountains, and natural climatic variations, such as periods of glaciation. It is important to understand the dynamics of biomes, as subtle changes in their composition may be sensitive indicators of changing ecological conditions.

Where conditions of climate and soil types are changing, the types of plants that occur at that location are likely to change over time. One might expect such changes to be most apparent at the borders or transition zones between biomes (sometimes referred to as **ecotones**). These borders can be caused by a variety of processes. Ecotones can be abrupt zones of vegetation change over space, such as occurs when temperatures decrease as elevation increases up a mountain. In the midwestern United States, the transition between eastern deciduous forest and shortgrass prairie is determined in part by recurring grass fires that kill young tree seedlings. Ecotones can also be created by human activities. Cutting tropical forests to create cattle pastures leaves an abrupt edge along the forest that can dramatically affect temperature, humidity, light, and wind at the forest-pasture boundary.

Biomes are large-scale categories of ecological communities. As such, is important to ask how it is possible to detect a change in a biome's distribution. The vegetation type would determine the ease with which a shifting ecotone might be detected. For example, changes at the boundary between desert and grassland might be easier to see compared to the ecotone between tropical rainforest and tropical thorn scrub. The visibility of changes would also depend in part on the scale of observations. In other words, the right combination of time scale and spatial scale would have to be used. Over time, one might examine and re-examine certain locations to determine whether vegetation composition has changed. Jim Hastings and Raymond Turner, of the U. S. Geological Survey in Arizona, have done this using repeat photography. Hastings and Turner started by finding photographs of various locations in the Southwest taken at different time intervals going back to the 1800s. They then went back to the same locations and took new photographs, which they then compared to the older pictures. The results were striking. In some locations, such as in Saguaro National Monument (now Saguaro National Park), the saguaro forests had dramatically declined in density in just a few decades.

Airborne photographs can also be used to analyze vegetation dynamics over large areas. In 1930s the U.S. Army began a campaign of aerial photography to map the nation. At certain times, such photography has been repeated in order to update or increase the accuracy of maps, thereby creating a database of repeat photography. Craig Allen (of the U.S. Geological Survey) and David Breshears (of Los Alamos National Laboratory) used repeat airborne photography to study the changes in the boundary between

ponderosa pine forest and piñon-juniper woodland in New Mexico. In theory, the distribution of a community type can increase or decrease by expansion or contraction of its borders. A community might appear to migrate (i.e., move over time and space) by a combination of range expansion on one boundary and contraction at the opposite boundary. Indeed, Allen and Breshears found that the ecotone between ponderosa forest and piñon-juniper woodland shifted by over 2 km in less than five years in response to a regional drought. In particular, ponderosa forest died off at its lower elevation boundary and was replaced by piñon-juniper woodland. There was no change at the upper elevation boundary. The rapid change resulted in fragmented (isolated) portions of ponderosa forest and considerably enhanced rates of soil erosion.

CHECK YOUR KNOWLEDGE

1. How are ecotones similar or different from one pair of biomes to another?

2. Read the data in ▌Figure 5.29. Why do you think that tree diameter increase was greatest at high elevations as compared to medium and lower elevations?

MORE
ON THE
WEB
3. Read Allen and Breshear's journal article by linking to *Practicing Ecology on the Web* at *http://www.whfreeman.com/ricklefs*. What do Allen and Breshears conclude regarding the importance of

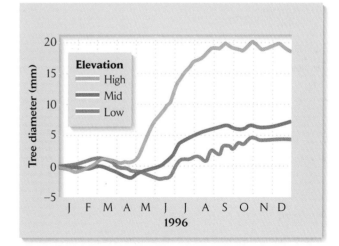

▌Figure 5.29 **Mean changes in stem diameter during calendar year 1996 for ponderosa pine along an elevation/moisture gradient.** Ten trees were measured at each of three sites: 2,010 m elevation ("Low" line) with 41 cm/year precipitation; 2,320 m ("Mid" line) with 51 cm/year; 2,780 m ("High" line) with 89 cm/year.

vegetation mortality when predicting ecotone shifts over time?

4. Why do you think that biomes are classified according to their vegetational characteristics and not differences in animal form?

 Suggested Readings

Allen, C. D., and D. D. Breshears. 1998. Drought-induced shift of a forest-woodland ecotone: Rapid landscape response to climate variation. *Proceedings of the National Academy of Sciences USA.* 95: 14839–14842.

Eyre, S. R. 1968. *Vegetation and Soils: A World Picture.* 2d ed. Aldine, Chicago.

Forman, R. T. T. (ed.). 1979. *Pine Barrens: Ecosystem and Landscape.* Academic Press, New York.

Hastings, J. R., and R. M. Turner. 1965. *The Changing Mile: An Ecological Study of Vegetation Change with Time in the Lower Mile of an Arid and Semiarid Region.* University of Arizona Press, Tucson.

Jaeger, E. C. 1957. *The North American Deserts.* Stanford University Press, Stanford, CA.

Jeffree, E. P., and C. E. Jeffree. 1994. Temperature and biogeographical distributions of species. *Functional Ecology* 8:640–650.

Levinton, J. S. 1982. *Marine Ecology.* Prentice-Hall, Englewood Cliffs, NJ.

Prentice, I. C., et al. 1992. A global biome model based on plant physiology and dominance, soil properties and climate. *Journal of Biogeography* 19:117–134.

Smith, T. M., H. H. Shugart, and F. I. Woodward. 1997. *Plant Functional Types: Their Relevance to Ecosystem Properties and Global Change.* Cambridge University Press, Cambridge.

Teal, J., and M. Teal. 1969. *Life and Death of a Salt Marsh.* Little Brown, Boston.

Terborgh, J. 1992. *Diversity and the Tropical Rain Forest.* Scientific American Library, New York.

Weaver, J. E. 1956. *Grasslands of the Great Plains.* Johnsen, Lincoln, NB.

Whitmore, T. C. 1990. *An Introduction to Tropical Rain Forests.* Oxford University Press, New York.

Whittaker, R. H., and W. A. Niering. 1965. Vegetation of the Santa Catalina Mountains, Arizona: A gradient analysis of the south slope. *Ecology* 46:429–452.

Woodward, F. I. 1987. *Climate and Plant Distribution.* Cambridge University Press, Cambridge.

CHAPTER 6

Energy in the Ecosystem

Alfred J. Lotka developed the first thermodynamic concept of the ecosystem

Primary production is the assimilation of energy and production of organic matter by photosynthesis

Only 5% to 20% of energy passes between trophic levels

Energy moves through ecosystems at different rates

Ecosystem energetics summarizes the movement of energy through the ecosystem

We humans consume a large proportion of the earth's production. The total production of dry plant biomass over the surface of the earth amounts to 224 billion tons per year. Of this, approximately 59% is produced in terrestrial ecosystems. Of that terrestrial production, an astonishing 35% to 40% is used by humans, either directly as food and fiber crops or indirectly as feed for animals. The oceans, traditionally a source of food for people living near the coast, are now becoming a major source of food for much of the world's population. By the early 1980s, the global fish catch amounted to 75 million tons per year, and this has since increased substantially.

How much of the production of algae in the oceans is required to sustain the fisheries on which humans depend? How much of the total algal production is represented in the 75 million tons of fish and other seafood we harvest each year? How much more food can we expect to harvest from the oceans? Two marine ecologists, D. Pauly and V. Christensen, who were working at the International Center for Living Aquatic Resources Management in the Philippines, sought to answer these questions from their understanding of energy flow in natural ecosystems.

Pauly and Christensen assumed that for each step in the chain of feeding relationships that leads from microscopic algae to the fish we eat, about 90% of consumed energy is used to maintain the consumer. This means that only 10% is converted through growth and reproduction to biomass, and thus potential food for other organisms. From studies of the diets of marine organisms, Pauly and Christensen estimated the number of feeding steps leading from algae to fish. These varied from averages of about 1.5 for coastal and reef ecosystems to 3 for the open ocean. Knowing the number of feeding steps and assuming an energy transfer efficiency of 10% per step, they used a simple calculation to convert harvested fish into amounts of algae needed to sustain them. Such calculations showed that for inshore fisheries, which produce most of the fish consumed by humans, the algal growth required to sustain the harvested fish amounted to 24% to 35% of the total production of the ecosystem. Because much of the production in these systems consists of species not eaten by humans, our harvesting may be approaching its upper limit. Only in the open ocean, which is much more difficult for humans to exploit, do we usurp a small fraction (about 2%) of the total production.

During the early part of the twentieth century, several new concepts emerged that led the study of ecology in novel directions. One of these was the realization that feeding relationships link organisms into a single functional entity. Foremost among the proponents of this new ecological viewpoint during the 1920s was the English ecologist Charles Elton. Elton argued that organisms living in the same place not only had similar tolerances of physical factors in the environment, but also interacted with one another, most importantly in a system of feeding relationships that he called a **food web.** Of course, every organism must feed in some manner to gain nourishment, and each may be fed upon by some other organism. However, regarding these feeding relationships as an ecological unit was a novel idea early in the twentieth century.

A decade later, the English plant ecologist A. G. Tansley took Elton's idea an important step further by considering animals and plants, together with the physical factors of their surroundings, as ecological systems. Tansley called this concept the **ecosystem,** and regarded it as the fundamental unit of ecological organization. Tansley envisioned the biological and physical parts of nature together, unified by the dependence of animals and plants on their physical surroundings and by their contributions to maintaining the conditions and composition of the physical world.

Alfred J. Lotka developed the first thermodynamic concept of the ecosystem

Working independently of the ecologists of his day, Alfred J. Lotka, a chemist by training, was the first to consider populations and communities as energy-transforming systems. He suggested that each system can be described in principle by a set of equations that represent exchanges of matter and energy among its components. Such exchanges include the assimilation of carbon dioxide into organic carbon compounds by plants, the consumption of plants by herbivores, and the consumption of animals by carnivores.

Lotka believed that the size of a system and the rates of energy and material transformations within it obeyed certain **thermodynamic principles** that govern all energy transformations. Just as heavy machines and fast machines require more fuel to operate than lighter and slower ones, and inefficient machines require more fuel than efficient ones, the energy transformations of ecosystems grow in direct proportion to their size (roughly, the total mass of their constituent organisms), productivity (rate of transformations), and inefficiency. The earth itself is a giant thermodynamic machine in which the circulation of winds and ocean currents and the evaporation of water are driven by the energy in sunlight. Part of that energy is assimilated by the photosynthesis of plants, and this energy ultimately fuels all biological systems.

Lotka's ideas were not widely appreciated by ecologists of his time. His mathematical representations were difficult and unfamiliar, and he did little to promote his ideas. The concept of the ecosystem as an energy-transforming system was brought to the attention of many ecologists for the first time in 1942 by Raymond Lindeman, a young aquatic ecologist at the University of Minnesota. Lindeman's framework for understanding ecological systems on the basis of thermodynamic principles made a deep impression. He adopted Tansley's notion of the ecosystem as the fundamental unit in ecology and Elton's concept of the food web, including inorganic nutrients at the base, as the most useful expressions of ecosystem structure.

The sequence of feeding relationships by which energy passes through the ecosystem is referred to as a **food chain.** A food chain has many links—plant, herbivore, and carnivore, for example—which Lindeman referred to as **trophic levels.** (The Greek root of the word *trophic* means "food.") Furthermore, Lindeman visualized a **pyramid of energy** within the ecosystem, with less energy reaching each successively higher trophic level (❚ Figure 6.1). Lindeman argued that energy is lost at each level because of the work performed by organisms at that level and because of the inefficiency of biological energy transformations. Thus, plants gather only a portion of the light energy available from the sun. Herbivores harvest even less of that light energy because plants use a portion of the energy they assimilate to maintain themselves, and that energy is not available to herbivores as plant biomass. The same may be said of the consumers of herbivores, and of each successively higher level of the food chain.

By the 1950s, the ecosystem concept had fully pervaded ecological thinking and had spawned a new branch of ecology, called **ecosystem ecology,** in which

Figure 6.1 An ecological pyramid of energy. The breadth of each bar represents the net productivity of a trophic level in the ecosystem. For this particular system, ecological efficiencies are 20%, 15%, and 10% between trophic levels, but these values vary widely in different communities.

the cycling of matter and the associated passage of energy through an ecosystem provided a basis for characterizing that system's structure and function. Energy and the masses of elements, such as carbon, provided a common "currency" that ecologists could use to compare the structure and functioning of different ecosystems in terms of the energy and matter residing in, and transferred among, the plants, animals, microbes, and abiotic components of the ecosystem. Measurements of energy assimilation and energetic efficiencies became the tools for exploring this new thermodynamic concept of the ecosystem.

With this new conceptual framework, ecologists began to measure energy flow and the cycling of nutrients. One of the strongest proponents of this approach has been Eugene P. Odum of the University of Georgia, whose text *Fundamentals of Ecology,* first published in 1953, influenced a generation of ecologists. Odum depicted ecosystems as energy flow diagrams (**Figure 6.2a**). For any one trophic level, such a diagram features a box representing the biomass (or its energy equivalent) of all the organisms making up that trophic level at any given time. For example, a box might represent all the plants or all the herbivores in a particular ecosystem. Superimposed on this box are pathways representing the flow of energy through that trophic level. These diagrams simplified nature, but nonetheless conveyed the important principle that energy passes from one

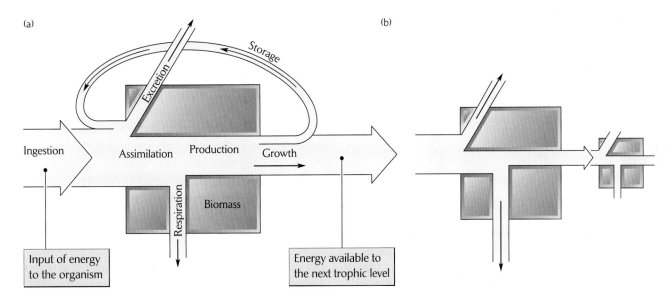

Figure 6.2 E. P. Odum's "universal" model of ecological energy flow. (a) A single trophic level. (b) Representation of a food chain. The net production of one trophic level becomes the ingested energy of the next higher level.

link in the food chain to the next, diminished by respiration and the shunting of unused foodstuffs to detritus-based food chains. Odum depicted feeding relationships as two or more energy flow diagrams linked into food chains, as shown in ▮ Figure 6.2b.

Unlike energy, which enters ecosystems as light and leaves as heat, nutrients are regenerated and retained within the system. Odum extended his model to include this cycling of elements. He showed that matter cycles within the ecosystem, being taken up in inorganic forms by plants and converted to biomass and eventually returned to inorganic forms by the process of decomposition. The most obvious recycling of material in this manner is the production of carbon dioxide by respiration and its uptake by plants during photosynthesis. However, each element is eventually returned to an inorganic form in its cycle through the ecosystem. In ecosystem energetics, studies of the cycling of elements have assumed equal standing with studies of the flow of energy. Amounts of elements and their movement among ecosystem components provide a convenient index to the flow of energy, which is difficult to measure directly. For instance, because light energy is transferred to the chemical energy content of organic molecules during photosynthesis, tracking of the movement of biological forms of carbon can be used to follow the movement of energy through the ecosystem.

A second reason for the prominence of nutrient cycling in ecosystem ecology is the fact that, in many circumstances, the quantities of certain nutrients regulate the production of biomass by plants, which is the material and energetic base of the entire ecosystem. For example, availability of water, rather than sunlight or minerals in the soil, limits the productivity of desert plants. In contrast, the open oceans are deserts by virtue of their scarce nutrients, particularly nitrogen. Understanding how elements cycle among components of the ecosystem is crucial to understanding the regulation of ecosystem structure and function.

Primary production is the assimilation of energy and production of organic matter by photosynthesis

Plants, algae, and some bacteria capture light energy and transform it into the energy of chemical bonds in carbohydrates. This process is referred to as **primary production,** and its rate is quantified as **primary productivity.** As we have seen, photosynthesis chemically unites two common inorganic compounds, carbon dioxide (CO_2) and water (H_2O), to form the sugar glucose ($C_6H_{12}O_6$),

with the release of oxygen (O_2). The overall chemical balance of the photosynthetic reaction is

$$6CO_2 + 6H_2O \rightarrow C_6H_{12}O_6 + 6O_2.$$

Photosynthesis transforms carbon from an oxidized (low-energy) state in CO_2 to a reduced (high-energy) state in carbohydrates. Because work is performed on the carbon atoms to increase their energy level, photosynthesis requires energy. This energy is provided by visible light. In quantitative terms, for each gram of carbon assimilated, a plant transfers 39 kilojoules (kJ) of energy from sunlight to the chemical energy of carbon in carbohydrates.

The pigments that capture the energy of light for photosynthesis actually absorb only a small fraction of the total incident solar radiation. In addition, because of inefficiencies in the many biochemical steps of photosynthesis, plants assimilate no more than a third (and usually much less) of the light energy absorbed by those photosynthetic pigments. The rest is lost as heat.

Photosynthesis supplies the carbohydrates and energy that a plant needs to build tissues and grow. Rearranged and joined together, glucose molecules become fats, starches, oils, and cellulose. Glucose and other organic compounds (starches and oils, for example) may be transported throughout the plant or stored as a source of energy for future needs. Combined with nitrogen, phosphorus, sulfur, and magnesium, simple carbohydrates derived from glucose produce an array of proteins, nucleic acids, and pigments. Plants cannot grow unless they have all these basic building materials. For example, the photosynthetic pigment chlorophyll contains an atom of magnesium, and so even when all other necessary elements are present in abundance, a plant lacking sufficient magnesium cannot produce chlorophyll, and thus cannot engage in photosynthesis.

Plants and other photosynthetic autotrophs form the base of all food chains, and therefore they are referred to as the **primary producers** of the ecosystem. Ecologists are interested in the rate of primary production because this determines the total energy available to the ecosystem. The total energy assimilated by photosynthesis is referred to as **gross primary production.** Plants use some of this energy to support the synthesis of biological compounds and to maintain themselves, and so their biomass contains substantially less energy than the total assimilated (▮ Figure 6.3). The energy accumulated in plants, and which is therefore available to consumers, is referred to as **net primary production.** The difference between gross and net primary production is the energy of respiration, which is the amount used by plants for maintenance and biosynthesis.

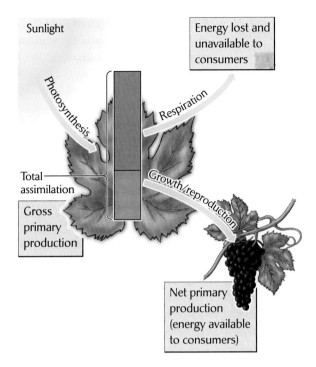

Figure 6.3 Gross primary production can be partitioned into respiration and net primary production.

ECOLOGISTS IN THE FIELD

Primary production can be measured by gas exchange or the growth of plants

Plant production involves fluxes of carbon dioxide, oxygen, minerals, and water and the accumulation of biomass (Figure 6.4). In principle, the rates of any of these flows could provide an index to the rate of primary production. It is worth discussing the measurement of primary production in some detail, as this will provide a better understanding of the processes involved in production and of the difference between gross and net production.

The unit of production is energy per unit of area per unit of time. For comparing production, ecologists often use kilojoules per square meter per year (kJ per m² per yr) or watts per square meter (W per m²). Production need not be measured only in terms of energy, however. Net production can be quantified conveniently as grams of carbon assimilated, dry weight of plant tissues, or their energy equivalents. Ecologists use such indices interchangeably because they are highly correlated. The energy equivalent of an organic compound depends primarily on its carbon content. Organic compounds contain approximately 39 kJ of metabolizable energy per gram of carbon, with some

energy added or subtracted during various biochemical transformations.

In terrestrial ecosystems, ecologists often estimate net production by the amount of plant biomass produced in a year. In areas of seasonal growth, annual production may be estimated by cutting, drying, and weighing plants at the end of the growing season. Clearly, such harvesting methods measure net, rather than gross, production. Root growth is often ignored because roots are difficult to remove from most soils; thus harvesting measures annual aboveground net productivity (AANP), the most common basis for comparing terrestrial communities. Production of small plants or individual leaves is most often quantified directly by carbon dioxide uptake. Because the atmosphere contains so little carbon dioxide (0.03%), plants can measurably reduce its concentration in an enclosed chamber within a short period. This change in CO_2 concentration can provide a direct estimate of photosynthetic rate. A convenient application of this method is to enclose leaves (or whole herbaceous plants or branches of trees) in a clear chamber (light must penetrate for photosynthesis) and measure the change in the concentration of CO_2 in air passed through the chamber. The technology for doing this is now so advanced that rates of carbon dioxide uptake can be measured on a few square centimeters of leaf under natural conditions in a matter of seconds. The change in CO_2 concentration per gram of dry weight or per square centimeter of leaf surface area is then extrapolated to the

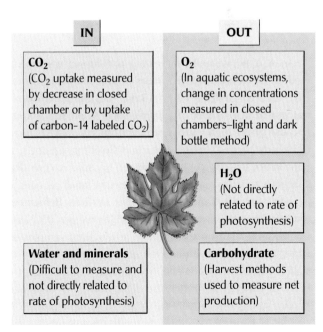

Figure 6.4 The fluxes involved in photosynthesis can be measured and used to estimate primary productivity.

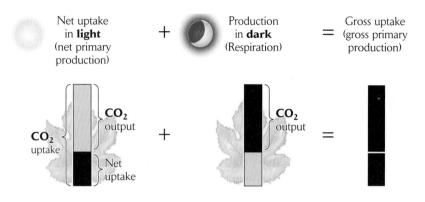

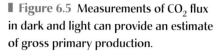

Figure 6.5 Measurements of CO_2 flux in dark and light can provide an estimate of gross primary production.

entire tree or forest. When a plant is exposed to light, carbon dioxide flux includes both assimilation (uptake) and respiration (output), and thus measures net production. Respiration can be measured separately by carbon dioxide production in the absence of light. Gross production can then be estimated by adding respiration to the net production (▌Figure 6.5).

The radioactive isotope carbon-14 (^{14}C) provides a useful variation on this method of measuring productivity. When a known amount of ^{14}C-labeled carbon dioxide is added to an airtight chamber, plants assimilate the radioactive carbon atoms roughly in proportion to their occurrence in the air inside the chamber. Thus, the rate of carbon fixation can be calculated by dividing the amount of ^{14}C in the plant by the proportion of ^{14}C in the chamber at the beginning of the experiment. For example, if a plant takes up 10 mg of ^{14}C in an hour, and ^{14}C constitutes 5% of the carbon in the chamber, we can conclude that the plant assimilates carbon at a rate of about 200 mg per h (10 divided by 0.05).

In aquatic systems, harvesting provides a convenient method for estimating the primary production of large photosynthetic organisms, such as kelps, but this technique is not practical for small organisms, such as phytoplankton. Because of the high concentrations of bicarbonate in most waters, measuring changes in carbon dioxide in aquatic systems is not practical either. However, because oxygen dissolves so poorly in water, one can measure small changes in oxygen concentration in most aquatic systems. Remember that photosynthesis produces molecular oxygen (O_2) as a by-product. To estimate primary production, samples of water containing phytoplankton are suspended in pairs of sealed bottles at desired depths beneath the surface of a body of water. One bottle (the "light bottle") is clear and allows sunlight to enter; the other (the "dark bottle") is opaque (▌Figure 6.6). In the light bottle, photosynthesis and respiration occur together, and part of the oxygen produced by the first process is consumed by the second. In the dark

bottle, respiration consumes oxygen without its being replenished by photosynthesis. Thus, gross production can be estimated by adding the change in oxygen concentration in the dark bottle (respiration alone) to that in the light bottle (photosynthesis and respiration). In unproductive waters, such as those of deep lakes and the open ocean, changes in oxygen concentration are too slow to measure easily. In such situations, the uptake of ^{14}C by plants and algae provides a more sensitive measure of carbon assimilation.

Light and temperature influence rates of photosynthesis

Primary production is sensitive to variations in light and temperature. For plants growing in full sunlight, light levels usually exceed the saturation point of their photosynthetic pigments (see Figure 3.10); therefore the photosynthetic rate of such plants generally is not restricted by light availability. For plants growing in shade or at depth in aquatic systems, however, the rate of photosynthesis often is limited by light. In addition, any particular leaf does not al-

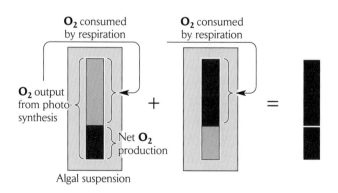

Figure 6.6 Paired light and dark bottles are used to measure photosynthesis in aquatic phytoplankton.

ways operate at its maximum possible photosynthetic rate. Cloud cover, shading by other leaves or plants, and low levels of light early and late in the day reduce the photosynthetic rate below its maximum.

Photosynthetic efficiency is the percentage of the energy in sunlight that is converted to net primary production during the growing season. This measure provides a useful index to rates of primary production under natural conditions. Where water and nutrients do not limit plant production severely, the photosynthetic efficiency of an ecosystem as a whole varies between 1% and 2%. What happens to the remaining 98–99% of the light energy? Leaves and other surfaces reflect anywhere from 25% to 75% of it. Molecules other than photosynthetic pigments absorb most of the remainder, which is converted to heat and either radiated or conducted across the leaf surface or dissipated by the evaporation of water from the leaf (transpiration).

Like the rates of most other physiological processes, the rate of photosynthesis generally increases with temperature, at least up to a point. The optimum temperature for photosynthesis varies with the prevailing temperature of the environment—from about 16°C in many temperate species to as high as 38°C in tropical species. Net production depends on the rate of respiration as well as on the rate of photosynthesis, and respiration generally increases with increasing leaf temperature as well. Thus, net production, and therefore net assimilation of CO_2, may actually decrease with increased temperature.

Water limits primary production in many terrestrial habitats

As we saw in Chapter 3, the tiny openings (stomates) in leaves through which carbon dioxide and oxygen are exchanged with the atmosphere also allow water to leave the leaf by transpiration. When the soil moisture approaches a plant's wilting point, the stomates close to reduce water loss. This prevents uptake of CO_2, and photosynthesis slows to a standstill. Consequently, the rate of photosynthesis depends on the availability of soil moisture, a plant's ability to tolerate water loss, and the influence of air temperature and solar radiation on the rate of transpiration.

Agronomists quantify the drought resistance of crop plants in terms of **transpiration efficiency,** also called **water use efficiency,** which is the number of grams of dry matter produced (net production) per kilogram of water transpired. In most plants, transpiration efficiencies are less than 2 g of production per kilogram of water, but they may be as high as 4 g per kilogram in drought-tolerant crops. Because transpiration efficiency varies little among a wide

variety of plant species, production can be directly related to water availability in the environment, as we saw in the case of maize crops in Zimbabwe (see Figure 4.16). However, much of the precipitation received by an area is never taken up by plants. Ground water, surface water (streams), and evaporation from the soil account for the remainder of the water budget. For example, in perennial grassland in southern Arizona, production varies in direct proportion to precipitation during the summer growing season, but at a rate of only about 200 kilograms of dry matter per hectare for each 10 cm of precipitation. Ten centimeters of rainfall is equivalent to a million kilograms of water per hectare. Thus, the water use efficiency of the grassland biome as a whole is only 0.2 g per kilogram, about one-tenth that based on transpired water. This finding indicates that only about 10% of the precipitation is taken up and transpired by plants in this habitat. Most of the rain comes in extremely heavy thundershowers during the summer months, and most of the water quickly runs off the land.

Nutrients stimulate plant production in both terrestrial and aquatic ecosystems

Fertilizers stimulate plant growth in most environments. When nitrogen and phosphorus fertilizers were applied singly and in combination to chaparral habitat in southern California, most species responded with increased production to additions of nitrogen, but not phosphorus (❚ Figure 6.7). This result suggests that production in most chaparral species is limited by the availability of nitrogen. However, the growth of California lilac bushes (*Ceanothus greggii*), which harbor nitrogen-fixing bacteria in their root systems, responded to the addition of phosphorus, but not to nitrogen. The production of annual plants (forbs and grasses) in the same environment increased when nitrogen was applied, but was depressed somewhat by the application of phosphorus alone. When equal amounts of nitrogen and phosphorus were applied together, however, production soared. Evidently, the annual plants could take advantage of increased phosphorus only in the presence of high levels of nitrogen.

Nutrients limit primary production most strongly in aquatic environments, particularly in the open ocean, where the scarcity of dissolved minerals reduces production far below terrestrial levels. Even in shallow coastal waters, where vertical mixing, upwelling currents, and runoff from the land maintain nutrients at high concentrations, the addition of fertilizers (as often occurs inadvertently through pollution) may greatly enhance aquatic production, upsetting the natural balance of aquatic ecosystems.

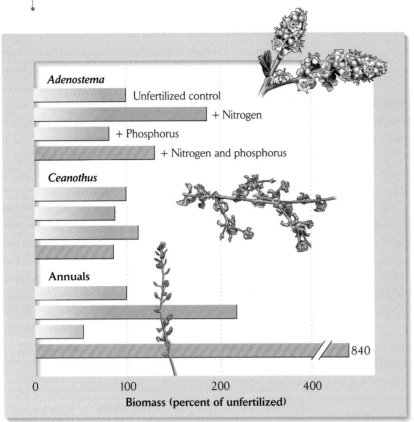

❚ **Figure 6.7 Fertilizing stimulates plant growth in natural habitats.** Response of the chaparral shrubs *Adenostema* (a typical chaparral plant), *Ceanothus* (which harbors nitrogen–fixing bacteria), and annual grasses and forbs to fertilization with nitrogen, phosphorus, or both. *After* G. S. McMaster, W. M. Jow, and J. Kummerow, *J. Ecol.* 70: 745–756 (1982).

Primary production varies among ecosystems

The favorable combination of intense sunlight, warm temperature, abundant rainfall, and ample nutrients in parts of the humid Tropics results in the highest terrestrial productivity on earth. In temperate and arctic ecosystems, low winter temperatures and long winter nights curtail production. Within a particular latitude belt, where light does not vary appreciably from one locality to the next, net production is related directly to temperature and annual precipitation. Above a certain threshold of water availability, net production increases by 0.4 g of dry matter per kilogram of water in hot deserts and by 1.1 g per kilogram in short-grass prairies and cold deserts. Thus, a given amount of water supports almost three times as much plant production in the cooler climates as in the hotter climates within a given latitudinal belt.

Global patterns of net primary production are summarized in ❚ Figure 6.8. The production of terrestrial vegetation is highest in the humid Tropics and lowest in tundra and desert habitats. Swamp and marsh ecosystems, which occupy the interface between terrestrial and aquatic habitats, can produce as much biomass annually as tropical forests because of the continuous availability of water and the rapid regeneration of nutrients in mucky sediments surrounding plant roots.

In the open ocean, scarcity of mineral nutrients limits productivity to a tenth that of temperate forests, or even less. Upwelling zones (where nutrients reach the surface from deeper waters) and continental shelf areas (where bottom sediments in shallow water rapidly exchange nutrients with surface waters) support greater production. In estuaries, coral reefs, and coastal algal beds, production approaches levels observed in terrestrial habitats. Primary production in freshwater environments is considerably higher than that in the open oceans, achieving the highest levels in rivers, shallow lakes, and ponds and the lowest levels in clear streams and deep lakes.

Only 5% to 20% of energy passes between trophic levels

Primary production by plants, algae, and some bacteria forms the base of ecological food chains. Animals, fungi, and most microorganisms obtain their energy and most of their nutrients from plants or animals, or the dead remains of either. These organisms, therefore, have dual roles as food producers and food consumers. These roles give the ecosystem a trophic structure that is determined by food webs through which energy flows and nutrients cycle. The food chain shown in Figure 6.1, from grass to rabbit to fox to hawk, traces one particular path that energy may follow through the trophic structure. As we have seen, biochemical transformations dissipate much of the energy of gross primary production before it can be consumed by organisms feeding at the next higher trophic level. With each step in the food chain, 80–95% of energy is lost. All the grass in Africa piled together would dwarf a mound of all the grasshoppers, gazelles, zebras, wildebeests, and other animals that eat grass. That mound of herbivores, in turn, would overwhelm the pitiful heap of all the lions, hyenas, and other carnivores that feed on them.

As Raymond Lindeman first pointed out in 1942, the amount of energy reaching each trophic level depends on the net primary production at the base of the food chain and on the efficiencies of energy transfers at each higher trophic level. Of the light energy assimilated by photosynthesis, plants use between 15% and 70% for maintenance, thereby making that portion unavailable to consumers. Herbivores and carnivores are more active than plants, and expend correspondingly more of their assimilated energy on maintenance. As a result, the production of each trophic level is typically only 5% to 20% that of the level below it.

Ecologists refer to the percentage of energy transferred from one trophic level to the next as **ecological efficiency** or **food chain efficiency.** To understand why ecological efficiencies are only 5–20%, we must examine how organisms make use of the energy they consume.

Regardless of the source of its food, an organism uses the energy from that food to maintain itself, to fuel its activities, and to grow and reproduce. Once ingested, the energy in food follows a variety of paths through the organism. To begin with, many components of food are not easily digested: hair, feathers, insect exoskeletons, cartilage, and bone in animal foods, and cellulose and lignin in plant foods (▌Figure 6.9). These substances may be defecated or regurgitated, and the energy they contain is referred to as egested energy. What an organism digests and absorbs constitutes its assimilated energy. The portion of this assimilated energy used to meet metabolic needs, most of which escapes the organism as heat, makes up the respired energy. Animals excrete another, usually smaller, portion of the assimilated energy in the form of nitrogen-containing organic wastes (primarily ammonia, urea, or uric acid), produced when the diet contains an excess of nitrogen; this is called excreted energy. Assimilated energy

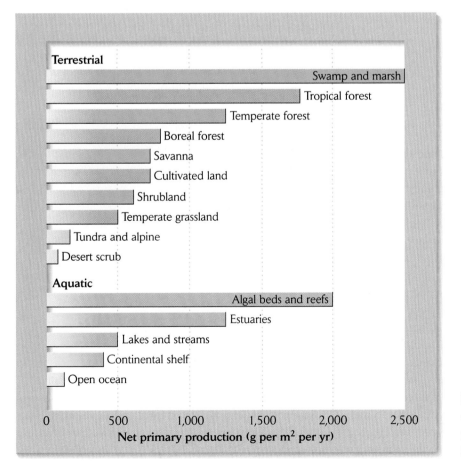

▌ **Figure 6.8 Net primary production varies among ecosystems.** Data from R. H. Whittaker and G. E. Likens, *Human Ecol.* 1:357–369 (1973).

Figure 6.9 Not all components of food can be assimilated. The undigested fibrous plant material in this elephant dung represents egested energy. Photo by R. E. Ricklefs.

retained by the organism becomes available for the synthesis of new biomass (production) through growth and reproduction, which animals feeding at the next higher trophic level may then consume. Thus, the various components of an organism's energy budget are connected to each other in the following relationships:

$$\text{ingested energy} - \text{egested energy} = \text{assimilated energy.}$$

$$\text{assimilated energy} - \text{respiration} - \text{excretion} = \text{production.}$$

Assimilation efficiency depends on the digestibility of the diet

The overall ecological efficiency of the food chain begins with the efficiency with which organisms assimilate the food they consume. **Assimilation efficiency** is the ratio of assimilation to ingestion, usually expressed as a percentage. The energy value of plants to their consumers depends on their food quality—that is, on how much cellulose, lignin, and other indigestible materials they contain. Herbivores assimilate as much as 80% of the energy in seeds and 60–70% of that in young vegetation. Most grazers and browsers (elephants, cattle, grasshoppers) extract 30–40% of the energy in their food. Millipedes, which eat decaying wood composed mostly of cellulose and lignin (and the microorganisms that occur in decaying wood), assimilate only 15% of the energy in their diet.

Food of animal origin is more easily digested than food of plant origin. Assimilation efficiencies of predatory species range from 60% to 90%. Vertebrate prey are digested more efficiently than insect prey because the indigestible

exoskeletons of insects constitute a larger proportion of the body than the hair, feathers, and scales of vertebrates. Assimilation efficiencies of insect eaters vary between 70% and 80%.

The most active animals have the lowest net production efficiencies

Each organism grows and produces offspring. The biomass it adds in this manner represents the organism's production, and is also potentially food for other organisms. The ratio of the energy contained in this production to the total assimilated energy is referred to as **net production efficiency,** and is usually expressed as a percentage. Active, warm-blooded animals exhibit low net production efficiencies: those of birds are less than 1% and those of small mammals with high reproductive rates range up to 6% (**Figure 6.10**). These organisms use most of their assimilated energy to maintain salt balance, circulate blood, produce heat for thermoregulation, and move. In contrast, sedentary, cold-blooded animals, particularly aquatic species, channel as much as 75% of their assimilated energy into growth and reproduction.

Production efficiency can be based on total energy ingested rather than on energy assimilated. In this case, it is referred to as **gross production efficiency,** which is the product of assimilation efficiency and net production efficiency. Thus,

$$\text{gross production efficiency} = (\text{assimilation/ingestion}) \times (\text{production/assimilation}) \times 100 = (\text{production/ingestion}) \times 100.$$

Figure 6.10 High rates of energy metabolism usually result in low production efficiencies. Because this hummingbird expends so much energy in hovering flight, less than 1% of its assimilated energy is converted to growth and egg production over its lifetime. Photo by Gary W. Carter/Corbis.

Gross production efficiency represents the overall energetic efficiency of biomass production within a trophic level. Gross production efficiencies of warm-blooded terrestrial animals rarely exceed 5%, and those of some birds and large mammals fall below 1%. For insects, these efficiencies lie within the range of 5% to 15%, and for some aquatic animals they exceed 30%.

Production efficiency in plants

The concept of production efficiency differs somewhat between plants and animals because plants do not digest and assimilate food. For plants, net production efficiency is defined as the ratio of net production to gross production. Net production efficiency varies between 30% and 85% in plants, depending on environment and growth form. Rapidly growing plants in temperate zones—whether trees, old-field herbs, crop species, or aquatic plants—have uniformly high net production efficiencies, typically between 75% and 85%. Similar types of vegetation in the Tropics exhibit lower net production efficiencies (40–60%). As we might expect because of the higher temperature, respiration increases relative to photosynthesis in tropical latitudes.

Detritus food chains

Terrestrial plants, especially woody species, allocate much of their production to structures that are difficult to ingest, let alone digest. As a result, even though herbivores have specialized adaptations to extract energy from plants, they still tend to have low assimilation efficiencies. Consequently, most of the production of terrestrial plants is consumed as **detritus**—dead remains of plants and undigestible excreta of herbivores—by organisms specialized to attack wood, leaf litter, and fibrous plant egesta. This partitioning between herbivory and detritus feeding establishes two parallel food chains in terrestrial communities (Figure 6.11). The first originates when relatively large animals feed on leafy vegetation, fruits, and seeds; the second originates when relatively small animals and microorganisms consume detritus in the litter and soil. These separate food chains sometimes mingle considerably at higher trophic levels, but the energy of detritus tends to move into the food chain much more slowly than the energy assimilated by herbivores.

The relative importance of herbivore-based and detritivore-based food chains varies greatly among communities. Herbivores predominate in plankton communities,

(a)

(b)

▌Figure 6.11 Plant litter forms the basis of detritus food chains in forest soils. (a) In western Tennessee, decomposition of litter proceeds relatively slowly and soils have a proportion of organic matter. (b) On Barro Colorado Island, Panama, leaves and wood decompose rapidly under the tropical conditions, leaving exposed mineral soil on the forest floor. Photos by R. E. Ricklefs.

detritivores in terrestrial communities. The proportion of net production that enters each of these food chains depends on the relative allocation of plant tissue between structural and supportive functions, on one hand, and growth and photosynthetic functions, on the other. A variety of studies have shown that herbivores consume 1.5–2.5% of the net primary production in temperate deciduous forests, 12% of that in old-field habitats, and 60–99% of that in plankton communities.

Exploitation efficiency

Because most biological production is consumed by one organism or another, little energy accumulates in any one trophic level. Rather, a balance is achieved between the production of biomass at one level and its consumption at another, so that the trophic structure of an ecosystem remains relatively constant. Viewed in this way, the ecological efficiency of a particular link in the food chain is equivalent to gross production efficiency. Under some conditions, however, production and consumption are not balanced, and energy may accumulate in an ecosystem, whether as organic matter in soil or as organic sediments in aquatic ecosystems. In such a case, we may say that

exploitation efficiency—that is, the proportion of production on one trophic level that is consumed by organisms on the next higher level—is less than 100%. In this case, the overall ecological efficiency of the ecosystem is discounted by the exploitation efficiency:

ecological efficiency =
exploitation efficiency × gross production efficiency.

Energy moves through ecosystems at different rates

Ecological efficiencies describe what proportion of the energy assimilated by plants eventually reaches each higher trophic level of an ecosystem. The rate of transfer of energy between trophic levels or, inversely, its **residence time** in each trophic level, provides a second index to the energy dynamics of an ecosystem. For a given rate of production, the residence time of energy and the storage of energy in living biomass and detritus are directly related: the longer the residence time, the greater the accumulation of energy (**Figure 6.12**).

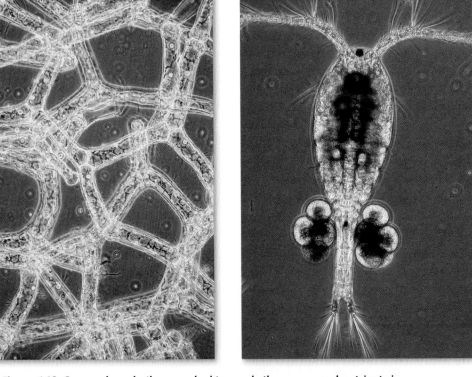

(a) (b)

 Figure 6.12 **Copepods and other zooplankton cycle the energy and nutrients in their algal food very rapidly.** Photo (a) by M. I. Walker/Science Source/Photo Researchers; photo (b) by Roland Birke/Photo Reseachers.

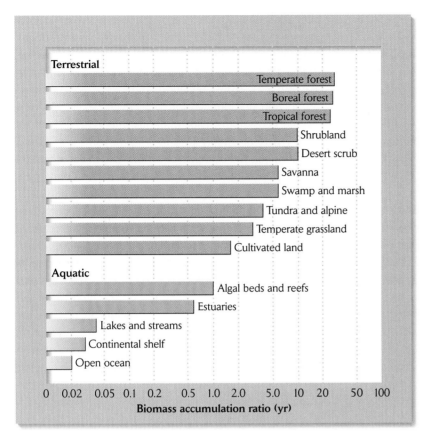

Figure 6.13 Biomass accumulation ratios for primary producers vary among ecosystems. Data from R. H. Whittaker and G. E. Likens, *Human Ecol.* 1:357–369 (1973).

The average residence time of energy at a particular trophic level equals the energy stored divided by the rate at which energy is converted into biomass:

$$\text{residence time (yr)} = \frac{\text{energy stored in biomass (kJ per m}^2)}{\text{net productivity (kJ per m}^2 \text{ per yr)}}.$$

We may also calculate the residence time defined by this equation in terms of mass rather than energy, in which case it expresses the **biomass accumulation ratio.** Accordingly,

$$\text{biomass accumulation ratio (yr)} = \frac{\text{biomass (kg per m}^2)}{\text{rate of biomass production (kg per m}^2 \text{ per yr)}}.$$

Plants in humid tropical forests produce dry matter at an average rate of 1.8 kg per m² per yr and have an average living biomass of 42 kg per m². Inserting these values into the above equation, we obtain 23 years (42/1.8) for the average residence time of biomass in plants. Biomass accumulation ratios for primary producers may average from more than 20 years in forested terrestrial environments to less than 20 days in aquatic phytoplankton-based communities (**Figure 6.13**). In all ecosystems, however,

some energy remains for a long time, and some disappears quickly. For example, leaf eaters and root feeders consume much of the energy assimilated by forest trees during the year of its production, some of it within days of assimilation by the plant. Energy accumulated in the cellulose and lignin in the trunks of trees, on the other hand, may not be recycled for centuries.

Figure 6.13 underestimates the average residence time of energy in energy-containing organic matter because it does not include the accumulation of dead organic matter in leaf litter. The residence time of energy in accumulated litter can be determined by an equation analogous to that for the biomass accumulation ratio:

$$\text{residence time (yr)} = \frac{\text{litter accumulation (g per m}^2)}{\text{rate of litter fall (g per m}^2 \text{ per yr)}}.$$

For forested ecosystems, this value varies from 3 months in the humid Tropics to 1–2 years in dry and montane tropical habitats, 4–16 years in the southeastern United States, and more than 100 years in temperate mountains and boreal regions. Warm temperatures and the abundance of moisture in lowland tropical regions create optimal conditions for rapid decomposition of litter.

Ecosystem energetics summarizes the movement of energy through the ecosystem

The flux of energy and the efficiency of its transfer describe certain aspects of the structure of an ecosystem: number of trophic levels, relative importance of detritus feeding and herbivory, steady-state values for biomass and accumulated detritus, and turnover rates of organic matter. The importance of these measures to understanding ecosystem function was argued by Lindeman, who constructed the first energy budget for an entire biological community—that of Cedar Bog Lake in Minnesota (▌Figure 6.14). The proliferation of energy flow studies during the 1950s and 1960s clearly reflected energy's acceptance as a universal currency, a common denominator to which all populations and their acts of consumption could be reduced.

The overall energy budget of an ecosystem reflects a balance between income and expenditure, just as in a bank account. The ecosystem gains energy through the photosynthetic assimilation of light by autotrophs and through the transport of organic matter into the system from external sources. Organic materials produced outside the system are referred to as **allochthonous** inputs (from the Greek *chthonos*, "of the earth," and *allos*, "other"; (▌Figure 6.15). Photosynthesis that occurs within the system is referred to as **autochthonous** production. In Root Spring,

near Concord, Massachusetts, the herbivores assimilated energy at a rate of 0.31 W per m², but the net productivity of aquatic plants and algae was only 0.09 W per m²; the balance was transported into the spring in the form of leaves from nearby vegetation. In general, autochthonous production predominates in large rivers, lakes, and most marine ecosystems; allochthonous imports make up the largest part of energy flux in small streams and springs under the closed canopies of forests. Life in caves and the abyssal depths of the oceans, to which no light penetrates, subsists entirely on energy transported in from outside.

Lindeman constructed the Cedar Bog Lake energy budget from measurements of the harvestable net production at each of three trophic levels—plants and algae, herbivores, and carnivores—and from laboratory determinations of respiration and assimilation efficiencies. Lindeman's findings were somewhat startling in that the herbivores consumed only 20% of net primary production, and the carnivores consumed only 33% of the net production of the herbivores. These are extremely low exploitation efficiencies. The majority of plant and herbivore biomass that was not consumed ended up as organic sediments at the bottom of the lake.

Even with this sedimentation, the Cedar Bog Lake ecosystem achieved a 12% overall ecological efficiency of energy transfer between trophic levels. After comparing similar analyses of other aquatic communities, ecologist D. G. Kozlovski concluded that (1) assimilation efficiency increases at higher trophic levels; (2) net and gross production efficiencies decrease at higher trophic levels; and (3) ecological efficiency averages about 10%. A simple, and rather surprising, consequence of this 10% rule of thumb for ecological efficiencies is that only 1% of the total energy assimilated by plants and algae ends up as production on the third trophic level. Very little energy is available to support consumers at even higher trophic levels. Thus, as shown in Figure 6.1, the pyramid of energy narrows very quickly as one climbs from one trophic level to the next. For humans, who already command a large proportion of the total primary production of the earth for their own use, this means that food supplies can be increased primarily by eating lower on the food chain—that is, eating more plant products and fewer animal products.

▌**Figure 6.14** Cedar Bog Lake in Minnesota has been the site of important studies on energy flow in aquatic ecosystems. Courtesy of G. David Tilman, University of Minnesota.

Summary

1. An ecosystem is the entire complex of organisms and the physical environments they inhabit. It is also a giant thermodynamic machine that dissipates energy continu-

▌ Figure 6.15 Some energy comes from allochthonous inputs. A small stream running through a forest receives much of its energy from detritus produced by trees. Photo by Mark Gamba/The Stock Market.

ously in the form of heat. This energy initially enters the biological realm of the ecosystem via photosynthesis and plant production, which provide energy for animals and nonphotosynthetic microorganisms.

2. Charles Elton described biological communities in terms of feeding relationships, which he emphasized as a dominant organizing principle in community structure.

3. A. G. Tansley coined the term *ecosystem* to include the organisms and all the abiotic factors in a habitat.

4. Alfred. J. Lotka provided a thermodynamic perspective on ecosystem function, showing that the movements and transformations of mass and energy conform to thermodynamic laws.

5. Raymond Lindeman, in 1942, popularized the idea of the ecosystem as an energy-transforming system.

6. The study of ecosystem energetics dominated ecology during the 1950s and 1960s, due largely to the influence of Eugene P. Odum, who championed energy as a common currency for describing ecosystem structure and function.

7. Gross primary production is the total energy assimilated by photosynthesis. Net primary production is the energy accumulated in plant biomass; hence it is gross primary production minus respiration.

8. Primary production can be measured by one or some combination of methods, such as harvesting, gas exchange (carbon dioxide in terrestrial habitats, oxygen in aquatic habitats), or assimilation of radioactive carbon (^{14}C).

9. Photosynthetic efficiency (gross production divided by total incident light energy) during daylight periods in the growing season is 1–2% in most habitats.

10. Because plants lose water in direct proportion to the amount of carbon dioxide they assimilate, plant production in dry environments is limited by, and varies with, the availability of water. Transpiration efficiency, also called water use efficiency, is the ratio of production (in grams of dry mass) to water transpired (in kilograms). Transpiration efficiency typically ranges between 1 and 2 g per kilogram; it occasionally reaches 4 g per kilogram in drought-adapted species.

11. Production in both terrestrial and aquatic environments can be enhanced by the addition of various nutrients, especially nitrogen and phosphorus, indicating that nutrient availability limits production. Stimulation of production by nutrients is greatest in systems where nutrient inputs are lowest.

12. Primary production of ecosystems is greatest in the humid Tropics. In other terrestrial environments, production is less, due to cold, dark, or drought. Among aquatic ecosystems, estuaries, coral reefs, and coastal algal beds are the most productive.

13. The movement of energy and materials through a food chain can be characterized by assimilation efficiency (the ratio of assimilation to digestion) and net production efficiency (the ratio of production to assimilation). Overall, the ecological efficiency of energy transfer from one trophic level to the next averages 5–20%.

14. Assimilation efficiency depends on the quality of the diet, particularly the amount of digestion-resistant structural material (cellulose, lignin, chitin, keratin) it contains. Assimilation efficiency varies from about 15% to 90%.

15. Material not assimilated is egested and becomes part of the detritus food chain.

16. Net production efficiency is lowest in animals whose costs of maintenance and activity are greatest, especially warm-blooded vertebrates. Typical net production efficiencies of 1–6% for warm-blooded vertebrates contrast with the values of 15–75% that are typical of invertebrates.

17. Gross production efficiency (the ratio of production to ingestion) varies from less than 1% up to 30%.

18. Exploitation efficiency measures the proportion of available food in a trophic level that is consumed by the next higher trophic level.

19. The average residence time of biomass or energy at a trophic level is the ratio of biomass, or of energy stored in biomass, to the rate of net production. Average residence times for primary production vary from 20 years in some forests to 20 days or less in aquatic, plankton-based communities.

PRACTICING ECOLOGY

CHECK YOUR KNOWLEDGE

How Long Are Food Chains?

The generalization that 10% of energy is passed from one trophic level to another is not a fixed law of ecological energetics. Because of a variety of complicating factors, the energy transferred from one level to another can be greater or less than 10%. Indeed, Pauly and Christensen (1995) summarized the energy transfer rate for 40 studies of trophic energy transfer in aquatic or marine communities, and found that the values varied from 2% to 24%. Ecological efficiencies are often lower in terrestrial compared to aquatic ecosystems. From this information, one might be led to ask how long the longest food chain in an ecosystem is.

If we can determine three pieces of information about ecosystem energetics, some simple calculations can shed light on how high a single population of consumers can feed on the energy available within the ecosystem. The required information includes: (1) the net primary production (NPP), which is the basic amount of energy that is available to be transferred to higher trophic levels; (2) the energy needed by a population of consumers (E); and (3) the average ecological efficiency for energy transfer from level to level within a food chain (Eff). A small portion of the energy may actually go through several trophic-level transfers, but it is important to note that these small amounts of energy are not sufficient to support the large predators that usually occupy the top of food chains.

For an ecosystem with some number of trophic levels (n, where plants are at level 1), the amount of energy available to a consumer at level n is denoted E(n). The energy available is determined by multiplying the net primary productivity by the ecological efficiencies leading up to level n:

$$E(n) = (NPP)(Eff^{n-1}).$$

In this equation, *Eff* is the geometric mean of the efficiencies of energy transfer between each trophic level. Rearranging the equation, it is possible to solve for *n*:

$$n = 1 + \frac{\log[E(n)/NPP]}{\log(Eff)}.$$

CHECK YOUR KNOWLEDGE

1. Why is it important to understand the length of food chains?

2. What can food chains tell us about the cycling of energy and materials in an ecosystem?

Table 6.1 Rough estimates of food chain data from various field studies

Community	Net primary production (kcal/m^2/yr)	Consumer ingestion (kcal/m^2/yr)	Ecological efficiency (%)	Number of trophic levels (n)
Open ocean	500	0.1	25	7.1
Coastal marine	8,000	10.0	20	5.1
Temperate grassland	2,000	1.0	10	4.3
Tropical forest	8,000	10.0	5	3.2

Values are approximations based on many studies.

3. From the estimates presented in Table 6.1, what factor contributes most to variations in food chain length among ecosystems?

4. From the material in this and previous chapters, what biological factors account for variation in the factor you chose as your answer to Question 3?

 5. Read the article from *World Watch Magazine* on *Practicing Ecology on the Web* at *http://www. whfreeman.com/ricklefs* about fishing down the food chain. What effects do you think overfishing will have on nutrient cycling and energy balance in the ocean?

 Suggested Readings

Clarkson, D. T., and J. B. Hanson. 1980. The mineral nutrition of higher plants. *Annual Review of Plant Physiology* 31:239–298.

Cook, R. E. 1977. Raymond Lindeman and the trophic-dynamic concept in ecology. *Science* 198:22–26.

Fenchel, T. 1988. Marine plankton food chains. *Annual Review of Ecology and Systematics* 19:19–38.

Golley, F. B. 1994. *A History of the Ecosystem Concept in Ecology.* Yale University Press, New Haven, CT.

Griffin, D. H. 1981. *Fungal Physiology.* Wiley, New York.

Howarth, R. W. 1988. Nutrient limitation of net primary production in marine ecosystems. *Annual Review of Ecology and Systematics* 19:89–110.

Laws, R. M. 1985. The ecology of the Southern Ocean. *American Scientist* 73:26–40.

Lawton, J. H. 1994. What do species do in ecosystems? *Oikos* 71:367–374.

Lindeman, R. 1942. The trophic-dynamic aspect of ecology. *Ecology* 23:399–418.

Odum, E. P. 1968. Energy flow in ecosystems: A historical review. *American Zoologist* 8:11–18.

Pauly, D., and V. Christensen. 1995. Primary production required to sustain global fisheries. *Nature* 374:255–257.

Webb, W., S. Szarek, W. Lauenroth, R. Kinerson, and M. Smith. 1978. Primary productivity and water use in native forest, grassland, and desert ecosystems. *Ecology* 59:1239–1247.

Whittaker, R. H., and G. E. Likens. 1973. Primary production: The biosphere and man. *Human Ecology* 1:357–369.

Wiegert, R. G. 1988. The past, present, and future of ecological energetics. In L. R. Pomeroy and J. J. Albert (eds.), *Concepts of Ecosystem Ecology: A Comparative View*, pp. 29–55. Springer-Verlag, New York.

CHAPTER 7

Pathways of Elements in the Ecosystem

Energy transformation and element cycling are intimately linked

Ecosystems may be modeled as a series of linked compartments

Water provides a physical model of element cycling in the ecosystem

The carbon cycle is closely tied to the flux of energy through the biosphere

Nitrogen assumes many oxidation states in its cycling through ecosystems

The phosphorus cycle is uncomplicated chemically

Sulfur exists in many oxidized and reduced forms

Microorganisms assume diverse roles in element cycles

Should you be worried about the change in carbon dioxide concentration in the earth's atmosphere? Combustion of fossil fuels and burning of forests to clear land for agriculture has increased the atmospheric concentration of CO_2 from 280 to 360 parts per million during the past century. Most of the change has been produced in recent decades, and projections show this trend increasing. The projected increases could bring about dramatic changes in climate through global warming, perhaps on the order of what we experience during extreme El Niño events. Such a scenario may disrupt agriculture and displace some of the human population. But major worries also include a rise in sea level due to melting ice caps and expansion of the surface waters of the oceans as they warm. These changes could flood vast coastal areas, causing economic disaster and rearranging human geography.

Yet the earth has witnessed far greater changes in atmospheric carbon dioxide concentrations in the past. Before the Industrial Revolution, CO_2 concentrations probably were as low as they have ever been in the geologic history of the earth. There is an important difference, however, between the present and the past. CO_2 levels are changing much more rapidly than they ever have before. Carbon and other chemical elements cycle continually through the ecosystem. The routes they take are determined by the particular chemical transformations in which each element participates. Organisms—including humans—move elements through their cycles within the ecosystem whenever they carry out energy transformations. This chapter shows how physical, chemical, and biological processes result in the cycling of elements within ecosystems. We shall see that many aspects of element cycling make sense only when one understands that chemical transformations and energy transformations go hand in hand.

Chemical elements, unlike energy, remain within the ecosystem, where they continually cycle between organisms and the physical environment. Materials used to form biological compounds originate in rocks of the earth's crust or in the earth's atmosphere, but within the ecosystem they are reused over and over by plants, animals, and microbes before being lost in sediments, streams, and groundwater or escaping to the atmosphere as gases. Though all the energy assimilated by green plants is "new" energy received from outside the ecosystem, most nutritive materials taken up by plants have been used before. Ammonia absorbed from the soil by roots might have been leached out of decaying leaves on the forest floor that same day. The carbon dioxide assimilated by a green plant might have been produced recently by animal, plant, or microbial respiration.

Energy transformation and element cycling are intimately linked

Organisms help to move elements through their cycles within the ecosystem whenever they make the chemical transformations needed to carry out their life processes. Transformations that incorporate inorganic forms of elements into the molecules of plants, animals, and microbes are referred to as **assimilatory** processes. One example of an assimilatory transformation of an element is photosynthesis, in which plants use energy to change an inorganic form of carbon (carbon dioxide) into the organic form of carbon found in carbohydrates. In the overall cycling of carbon, photosynthesis is balanced by respiration, a complementary **dissimilatory** process that involves the transformation of organic carbon back to an inorganic form, accompanied by release of energy.

Not all transformations of elements in the ecosystem are biologically mediated, nor do all involve the net assimilation or release of useful quantities of energy. Many chemical reactions take place in the air, soil, and water. Some of these, such as the weathering of bedrock, release certain elements (potassium, phosphorus, and silicon, for example) from compounds in rock and make them available to the ecosystem. Lightning storms produce small amounts of reduced nitrogen (ammonia, NH_3) from molecular nitrogen (N_2) and water vapor (H_2O) in the atmosphere, which plants and microbes can assimilate. Such reactions may have been involved in the origin of life itself. Other physical and chemical processes, such as sedimentation of calcium carbonate in the oceans, remove elements from circulation and incorporate them into rocks in the earth's crust, where they may remain untouched for eons.

Most biological energy transformations are associated with the biochemical oxidation and reduction of carbon, oxygen, nitrogen, and sulfur. As we saw in Chapter 2, an atom is oxidized when it gives up electrons, and it is reduced when it accepts electrons. In a sense, the electrons carry with them a portion of the energy content of an atom. In biological transformations, an energy-releasing oxidation is paired with an energy-requiring reduction, and energy shifts from the reactants in one transformation to the products in the other (■ Figure 7.1). Such coupled transformations are possible only when the oxidation side releases at least as much energy as the reduction side requires. The energy changes associated with various transformations vary widely depending on the compounds involved and the number of electrons exchanged. It is in the nature of the physical world that the energies of the two transformations rarely match. Energy supplied by an oxidation reaction in excess of that required by a coupled reduction reaction cannot be used, and is lost in the form of heat. These imbalances account for the thermodynamic inefficiency of life processes.

A typical coupling of transformations might involve the oxidation of carbon in a carbohydrate (glucose, for example), which releases energy, and the reduction of nitrate-nitrogen to amino-nitrogen (which forms the building blocks of proteins), which requires energy. This, like many biochemical transformations, ties an energy-releasing transformation to the assimilation of an element—nitrogen, in this case—required for growth and reproduction. In animals, such biochemical transformations are also used to maintain the cellular environment and to effect movement. Some of these transformations involve many

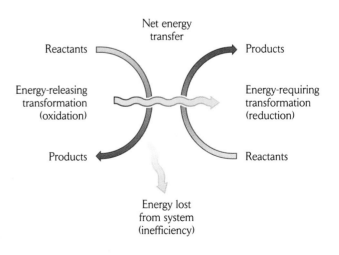

■ Figure 7.1 The coupling of energy-releasing and energy-requiring transformations is the basis of energy flow in ecosystems.

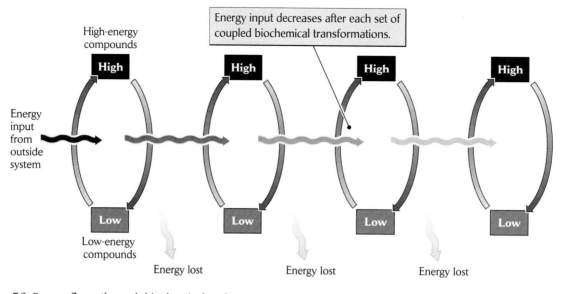

▌Figure 7.2 Energy flows through biochemical pathways. As energy flows through an ecosystem, elements alternate between assimilatory and dissimilatory transformations, thus going through cycles.

intermediate steps of the type shown in Figure 7.1, linked together into a biochemical pathway (▌Figure 7.2). Plants accomplish the initial input of energy into the ecosystem by an assimilatory reduction of carbon in which light, rather than a coupled dissimilatory process, serves as the source of energy. A portion of that energy escapes biological systems with each subsequent transformation. The cycling of elements between the living and nonliving parts of the ecosystem is thus connected to energy flow by the coupling of the dissimilatory part of one cycle to the assimilatory part of another.

Ecosystems may be modeled as a series of linked compartments

With each biochemical transformation, one or more elements are changed from one form to another. Each form of an element within an ecosystem may be thought of as a separate compartment, like a room of a house, into and out of which atoms move as physical and biological processes transform them. The entire ecosystem may be thought of as a set of compartments among which elements are cycled (▌Figure 7.3). For example, photosynthesis moves carbon from the inorganic carbon compartment to that containing organic forms of carbon (assimilation); respiration returns it to the inorganic compartment (dissimilation). Such **compartment models** of ecosystems can be organized hierarchically, having subcompartments within compartments.

The inorganic carbon compartment includes carbon dioxide both in the atmosphere and dissolved in water, carbonate and bicarbonate ions dissolved in water, and calcium carbonate, mostly as a precipitate in the water column and in sediments. The organic carbon compartment also has many subcompartments: autotrophs, animals, microorganisms, and detritus. As organisms feed on others, they move carbon among these subcompartments.

The movement of elements within and between compartments often involves energy. Photosynthesis adds energy to carbon, which we may think of as lifting the element to the second floor of a house. In descending the respiration "staircase," carbon releases this stored chemical energy, which an organism can then use for other purposes.

Elements cycle rapidly among some compartments of ecosystems and much more slowly among others. The movement of an element between living organisms and inorganic forms occurs over periods ranging from a few minutes to the life spans of organisms or their subsequent existence as organic detritus. We saw in the last chapter that some organic matter in some terrestrial environments has an average residence time on the order of centuries. Both organic and inorganic forms of elements occasionally leave rapid circulation within ecosystems for compartments that are not readily accessible to transforming agents. For example, coal, oil, and peat contain vast quantities of organic carbon that has been removed from circulation in ecosystems, often for many millions of years. Inorganic carbon is removed from circulation in aquatic ecosystems by precipitation of calcium carbonate, which forms thick layers of

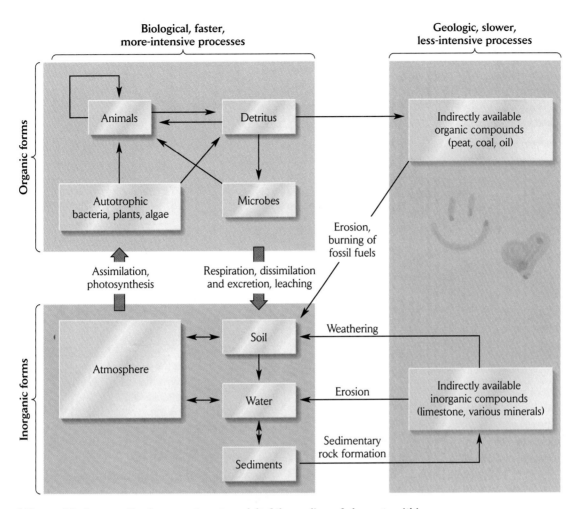

Figure 7.3 A generalized compartment model of the cycling of elements within ecosystems. Within each compartment, we can recognize subcompartments; for example, the compartment that represents available organic forms of nutrients is further subdivided into compartments occupied by autotrophs, animals, detritus, and microbes.

marine sediments that may eventually turn to limestone. These forms of carbon are returned to the rapidly cycling compartments of the ecosystem only by the slow geologic processes of volcanism, uplift, and erosion.

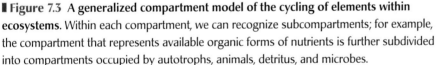

Water provides a physical model of element cycling in the ecosystem

Water is involved chemically in photosynthesis, but it is evaporation, transpiration, and precipitation that drive most movement of water through terrestrial ecosystems (**Figure 7.4**). These physical processes nonetheless couple the movement of water to transformations of energy. Thus, the global hydrologic cycle illustrates many basic features of the cycles of elements.

Light energy absorbed by water performs the work of evaporation. Water vapor has a potential energy, which is the energy of separation of individual water molecules from each other. When atmospheric water vapor condenses to form clouds, water molecules aggregate, and the potential energy in water vapor is released as heat, which eventually escapes the earth as long-wave radiation. From a thermodynamic standpoint, evaporation and condensation resemble photosynthesis and respiration.

Water in the biosphere totals about 1.4 billion cubic kilometers, or $1,400,000 \times 10^{18}$ g. It's hard to get a feeling for such a large number. 10^{18} g of water is a billion times a billion, or a quadrillion, grams. Each cubic meter contains 10^6 g, or 1,000 kg (a metric ton, T) of water, and so 10^{18} g is a trillion (10^{12}) metric tons—that is, a teraton (TT). Numbers on the order of 10^{18} generally are reserved for astronomy and the Federal budget, but we'll use

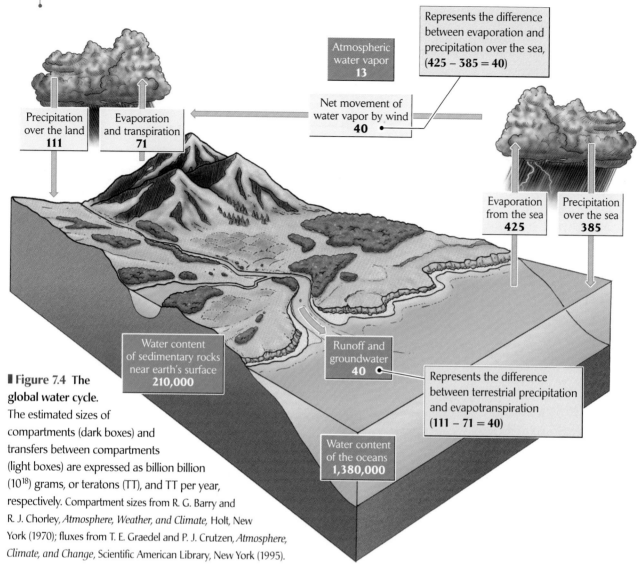

Atmospheric water vapor **13**

Represents the difference between evaporation and precipitation over the sea, **(425 − 385 = 40)**

Precipitation over the land **111**

Evaporation and transpiration **71**

Net movement of water vapor by wind **40**

Evaporation from the sea **425**

Precipitation over the sea **385**

Water content of sedimentary rocks near earth's surface **210,000**

Runoff and groundwater **40**

Represents the difference between terrestrial precipitation and evapotranspiration **(111 − 71 = 40)**

Water content of the oceans **1,380,000**

▌Figure 7.4 The global water cycle. The estimated sizes of compartments (dark boxes) and transfers between compartments (light boxes) are expressed as billion billion (10^{18}) grams, or teratons (TT), and TT per year, respectively. Compartment sizes from R. G. Barry and R. J. Chorley, *Atmosphere, Weather, and Climate,* Holt, New York (1970); fluxes from T. E. Graedel and P. J. Crutzen, *Atmosphere, Climate, and Change,* Scientific American Library, New York (1995).

teratons here as a unit of global water to keep the number of zeros to a minimum.

More than 97% of the water in the biosphere resides in the oceans. Other reservoirs of water include ice caps and glaciers (29,000 TT), underground aquifers (8,000 TT), lakes and rivers (100 TT), soil moisture (100 TT), water vapor in the atmosphere (13 TT), and all the water in living organisms (1 TT). Each of these may be regarded as a separate compartment in a compartment model of water in the biosphere.

Over land surfaces, precipitation (111 TT per yr, which is 22% of the global total) exceeds evaporation and transpiration (71 TT per yr; 16% of the global total). Over the oceans, evaporation exceeds precipitation by a similar amount. Much of the water that evaporates from the surface of the oceans is carried by winds to the continents, where it is captured as precipitation by the land. This net flow of atmospheric water vapor from ocean to land

(40 TT per yr) is balanced by runoff from the land by way of rivers back into ocean basins.

Evaporation determines how fast water moves through the biosphere. The absorption of radiant energy by liquid water to create water vapor couples an energy source to the hydrologic cycle. We can calculate the energy that drives the global hydrologic cycle by multiplying the total weight of water evaporated (456 TT per yr) by the energy required to evaporate 1 g of water (2.24 kJ). The product, approximately 10^{21} kJ per yr (about 32 billion megawatts), represents about one-fourth of the total energy of the sun's radiation striking the earth. Condensation of water vapor to form precipitation releases the same amount of energy as heat. Evaporation and precipitation are closely linked because the atmosphere has a limited capacity to hold water vapor; any increase in the evaporation of water into the atmosphere creates an excess of vapor and causes an equal increase in precipitation.

The amount of water vapor in the atmosphere at any one time corresponds to an average of about 2.5 cm of water spread evenly over the surface of the earth. An average of 65 cm of rain or snow falls each year (the water flux), which is 26 times the average amount of water vapor. Thus the steady-state content of water in the atmosphere—the atmospheric compartment—replaces itself 26 times each year on average. (Conversely, water has an average residence time in the atmosphere of 1/26 of a year, or 2 weeks.) Soils, rivers, lakes, and oceans contain more than 100,000 times as much water as exists in the atmosphere. Fluxes through both compartments are the same, however, because evaporation balances precipitation. Thus the average residence time of water in its liquid form at the earth's surface (about 2,800 years) is about 100,000 times longer than its residence time in the atmosphere.

The carbon cycle is closely tied to the flux of energy through the biosphere

The carbon cycle resembles the hydrologic cycle in that energy from the sun provides its driving force. The carbon cycle is much more complex, however, owing to the various chemical reactions of carbon. Three classes of processes cause carbon to cycle through aquatic and terrestrial ecosystems (Figure 7.5): (1) assimilatory and dissimilatory reactions of carbon, primarily in photosynthesis and respiration, (2) exchange of carbon dioxide between the atmosphere and the oceans, and (3) sedimentation of carbonates.

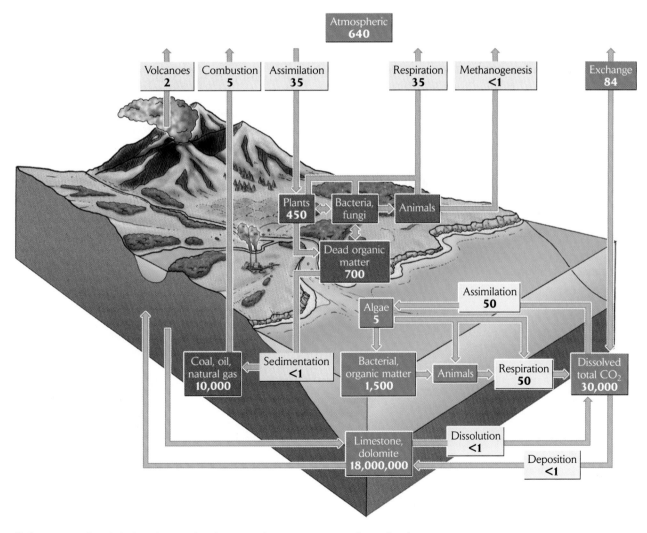

 Figure 7.5 The global carbon cycle. The sizes of compartments and transfers between compartments are in billions of metric tons (10^{15} g), or gigatons (GT), and GT per year. After T. Fenchel and T. H. Blackburn, *Bacteria and Mineral Cycling*, Academic Press, New York (1979); W. D. Grant and P. E. Long, *Environmental Microbiology*, Wiley, New York (1981).

Photosynthesis and respiration

Photosynthesis and respiration are the major energy-transforming reactions of life. Approximately 85 billion metric tons (85×10^{15} g) of carbon enter into such reactions worldwide each year. (We will refer to a billion metric tons as a gigaton, using the symbol GT). During photosynthesis, carbon gains electrons and is reduced (Figure 7.6). This gain of electrons is accompanied by a gain in chemical energy. An equivalent amount of energy is released by respiration, which results in a loss of electrons and a loss of chemical energy.

Although it is difficult to estimate the total carbon in organic matter within the biosphere, it probably adds up to something like 2,650 GT, including both living organisms and organic detritus and sediments. Thus, considering that 85 GT of carbon are assimilated by photosynthesis each year, the average residence time of carbon in biological molecules is approximately 2,650 GT divided by 85 GT per yr, which equals 31 years.

Ocean–atmosphere exchange

The second class of carbon cycling processes involves the physical exchange of carbon dioxide between the atmosphere and oceans, lakes, and streams. Carbon dioxide dissolves readily in water; indeed, the oceans contain about 50 times as much CO_2 as the atmosphere. Exchange across the air–water boundary links the carbon cycles of terrestrial and aquatic ecosystems. In fact, the ocean is an important sink for the carbon dioxide produced by the burning of fossil fuels. As the CO_2 content of the atmosphere increases, the rate of solution of CO_2 in the ocean increases, thereby reducing the rate of increase of CO_2 in the atmosphere below what it would be in the absence of air–water interchange.

Of the total carbon in the atmosphere in the form of carbon dioxide (640 GT), approximately 35 GT is assimilated by land plants, and 84 GT dissolves in the ocean and other surface waters, each year. Respiration and the escape of dissolved carbon dioxide from water to the atmosphere replace these amounts. Overall, the average residence time of carbon in the atmosphere is about 5 years. Because of this short residence time, the amount of carbon dioxide in the atmosphere is very sensitive to the rate of CO_2 production, increasing very nearly in parallel with the burning of fossil fuels. By 1990, combustion of fossil fuels contributed about 6 GT of carbon annually, equivalent to almost 1% of the total atmospheric carbon dioxide and a sixth of the total assimilation of carbon by land plants.

Precipitation of carbonates

The third class of carbon cycling processes occurs only in aquatic systems. It involves the dissolution of carbonate compounds in water and their precipitation (deposition) as sediments, particularly limestone and dolomite. On a global scale, dissolution and precipitation approximately balance each other, although certain conditions favoring precipitation have led to the deposition of extensive layers of calcium carbonate sediments in the past. Dissolution and deposition in aquatic systems occur about 100 times more slowly than assimilation and dissimilation by biolog-

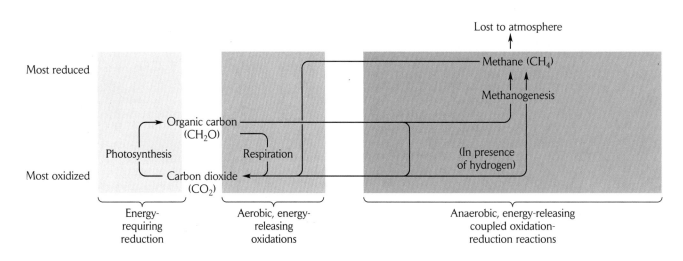

Figure 7.6 Schematic diagram of transformations of compounds in the carbon cycle. Most of the carbon in the biosphere cycles between organic forms and carbon dioxide. Methane is produced only by certain archaebacteria under anaerobic conditions.

■ **Figure 7.7 Most of the earth's carbon is in sedimentary rocks.** These sedimentary deposits of limestone in the mountains of southern Texas represent calcium carbonate precipitated out of solution in the shallow seas that once covered the area. Photo by Gerald & Buff Corsi/Visuals Unlimited.

ical systems. Thus, the exchange between sediments and the water column is relatively unimportant to the short-term cycling of carbon in the ecosystem. Locally and over long periods, however, it can assume much greater importance; in fact, most of the ecosystem's carbon is locked up in sedimentary rocks (■ Figure 7.7).

As we saw in Chapter 2, when carbon dioxide dissolves in water, it forms carbonic acid,

$$CO_2 + H_2O \rightarrow H_2CO_3$$

which readily dissociates into hydrogen, bicarbonate, and carbonate ions:

$$H_2CO_3 \rightarrow H^+ + HCO_3^- \rightarrow 2H^+ + CO_3^{2-}.$$

Calcium, when present, also equilibrates with the carbonate ions to form calcium carbonate:

$$Ca^{2+} + CO_3^{2-} \rightleftharpoons CaCO_3.$$

Calcium carbonate has low solubility under most conditions, and readily precipitates out of the water column to form sediments. This sedimentation effectively removes carbon from aquatic ecosystems, but the rate of removal is less than 1% of the annual cycling of carbon in these ecosystems, and this amount is added back by input from rivers, which are naturally somewhat acid and tend to dissolve limestone (carbonate) sediments.

Dissolution and dissociation may be affected locally by the activities of organisms. In the marine system, under approximately neutral pH conditions, carbonate and bicarbonate are in chemical equilibrium:

$$CaCO_3 \text{ (insoluble)} + H_2O + CO_2 \rightleftharpoons Ca^{2+} + 2HCO_3^- \text{ (soluble)}.$$

Uptake of CO_2 for photosynthesis by aquatic algae and plants shifts the equilibrium to the left, resulting in the formation and precipitation of calcium carbonate. Many algae excrete this calcium carbonate to the surrounding water, but reef-building algae and coralline algae incorporate it into their hard body structures (■ Figure 7.8). In the system as a whole, when photosynthesis exceeds respiration (as it does during algal blooms), calcium tends to precipitate out of the system.

Changes in the carbon cycle over time

Geologists can estimate the amounts of carbon removed from the atmosphere by burial of organic matter and precipitation of carbonates in marine sediments, as well as when these sediments were formed. From this information, they

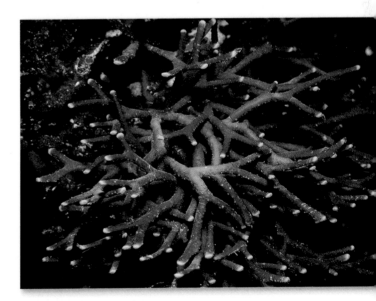

■ **Figure 7.8 The "skeleton" of coralline algae is made of calcium carbonate.** This is precipitated in conjunction with the uptake of dissolved carbon dioxide during photosynthesis. Photo by L. Newman & A. Flowers/Photo Researchers.

can estimate the original concentration of carbon dioxide in the atmosphere and its change over time (▌Figure 7.9).

These estimates indicate that during the early part of the Paleozoic era, roughly 550–400 million years ago (Mya), the atmosphere held 15 to 20 times more carbon dioxide than at present. This amount decreased precipitately early in the Devonian period, nearly 400 Mya, dipping to nearly its present levels by 300 Mya. This decline was initiated by a sharp increase in the rate of weathering of terrestrial environments following the development of forests on land, and by the deposition of the vast accumulations of organic sediments that make up most of the earth's coal beds. Toward the end of the Paleozoic era, at about 250 Mya, CO_2 concentration in the atmosphere again increased, to nearly five times its present level, remained high for approximately 100 million years through the early Mesozoic, and has been declining steadily ever since.

The early Paleozoic and early Mesozoic eras were truly greenhouse times. Average temperatures throughout the earth were hot, and tropical life flourished even at high latitudes. Declining CO_2 in the Devonian set the stage for cooler climates and extensive glaciations at the end of the Mesozoic era, much like those the earth has experienced during the past million years. The current increase in atmospheric CO_2, troubling as it is, will not return the earth to the hothouse conditions of former times, at least not any

time soon. Most of the "geologic" carbon taken from the earth's primitive atmosphere is bound up in limestone sediments. This carbon is returned to the atmosphere very slowly as limestone is subducted below the edges of continental plates, carbonates are turned to carbon dioxide under intense heat and pressure deep in the earth, and carbon dioxide is finally outgassed in volcanic eruptions.

ECOLOGISTS IN THE FIELD

What caused the precipitate decline in atmospheric carbon dioxide during the Devonian?

Why did concentrations of carbon dioxide in the atmosphere decline precipitately over a period of 50 million years during the Devonian period (409–363 Mya)? How can we infer events in the biosphere that occurred so long ago? Were ecological changes in ancient ecosystems involved? Geologist Gregory Retallack of the University of Oregon used several lines of evidence to provide a plausible explanation for this change. Retallack studied fossilized soils (paleosols) formed during the Devonian period in what is now Antarctica. (At that time, the climate there was warm, and vegetation flourished.) By comparing paleosols with modern soils, it is possible to interpret many processes occurring in soils in the past and their consequences for the biosphere.

The beginning of the Devonian was marked by two striking changes (▌Figure 7.10). One was a change in soil chemistry, indicated by a marked increase in clay content. The other was a dramatic increase in the density and depth of plant roots. The middle of the Devonian witnessed a striking increase in the diameter of trunks, stems, roots, and rhizomes of plants, indicating the development of the first forests on earth. These changes were followed at the end of the Devonian and beginning of the Carboniferous by the appearance of thick peat deposits, which later turned into coal.

The level of atmospheric CO_2 itself may be estimated by the ratio of ^{13}C to ^{12}C carbon isotopes in soil carbonates. Carbon dioxide in soil comes both from atmospheric sources directly and from respiration of soil organisms. The $^{13}C/^{12}C$ ratio indicates the relative amounts of atmospheric carbon compared with respired CO_2. With some assumptions about the production of CO_2 in soil, one can estimate the concentration of CO_2 in the atmosphere.

Why did the earth lose its greenhouse atmosphere during the Devonian? Retallack surmised that the increase in terrestrial vegetation, particularly the penetration of soil by fine roots, would have dramatically increased the rate of weathering of soil, thus increasing its clay content. Roots and

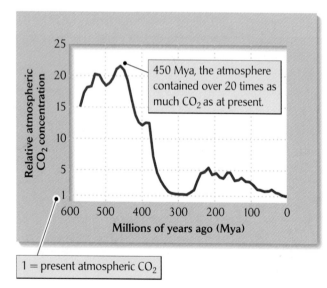

1 = present atmospheric CO_2

▌**Figure 7.9 Concentrations of carbon dioxide in the atmosphere have decreased since the beginning of the Paleozoic era.** Values are expressed in multiples of the concentration (approximately 300 parts per million) at the beginning of the Industrial Revolution. *After R. A. Berner,* Science 276:544–546 (1997).

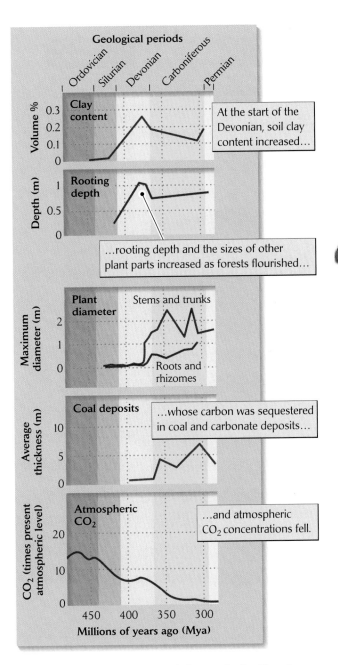

Geological periods

Ordovician · Silurian · Devonian · Carboniferous · Permian

Clay content — At the start of the Devonian, soil clay content increased…

Rooting depth

…rooting depth and the sizes of other plant parts increased as forests flourished…

Plant diameter — Stems and trunks — Roots and rhizomes

Coal deposits — …whose carbon was sequestered in coal and carbonate deposits…

Atmospheric CO_2 — …and atmospheric CO_2 concentrations fell.

Figure 7.10 Fossil soils reveal changes in the biosphere. The features of paleosols during the middle of the Paleozoic era reveal increases in terrestrial vegetation and progressive weathering of soils accompanied by sequestration of atmospheric carbon dioxide in carbonate precipitates. After G. J. Retallack, *Science* 276:583–585 (1997).

their associated microflora secrete organic acids to mobilize soil minerals; decomposition of organic matter also forms acids, which speed the breakdown of minerals. Roots hold clay particles in the soil and therefore enhance the soil's water-holding capacity, which further helps chemical weathering. This enhanced weathering would have caused tremen-

dous amounts of calcium and magnesium to be washed out of soil wherever terrestrial vegetation had developed. As solubilized calcium and magnesium ions entered the oceans, they formed insoluble compounds with the abundant bicarbonate ions there and precipitated out of the water as sediments. As bicarbonate was withdrawn from the oceans, it was replaced by carbon dioxide diffusing in from the atmosphere. Thus, as vegetation promoted weathering, new sedimentary rock was formed, partly from constituents of old continental crust and partly from the missing atmospheric carbon dioxide.

Nitrogen assumes many oxidation states in its cycling through ecosystems

The ultimate source of nitrogen for life is molecular nitrogen (N_2) in the atmosphere. This form of nitrogen dissolves to some extent in water, but is absent from native rock. Lightning discharges convert some molecular nitrogen to forms, such as ammonia, that plants can use, but most enters the biological pathways of the nitrogen cycle (**Figure 7.11**) through its assimilation by certain microorganisms in a process referred to as **nitrogen fixation**. Although this pathway ($N_2 \rightarrow NH_3$) constitutes only a small fraction of the earth's annual nitrogen flux, most biologically cycled nitrogen can be traced back to nitrogen fixation. Once in the biological realm, nitrogen follows pathways more complicated than those of carbon because more oxidized and reduced forms are possible for nitrogen atoms.

Ammonification

Let's begin with the reduced (organic) nitrogen found in proteins. Plants obtain nitrogen from the soil, either as ammonia or as nitrate, which they must then reduce to an organic form. From this point, the first step in the nitrogen cycle is **ammonification**. Ammonification involves the breaking down of proteins into their component amino acids by hydrolysis and the oxidation of the carbon in those amino acids. This results in the production of ammonia (NH_3). Ammonification is carried out by all organisms. Although carbon is oxidized, releasing energy, the nitrogen atom itself is not oxidized, and so its energy potential does not change during ammonification.

Nitrification and denitrification

Nitrification involves the oxidation of nitrogen, first from ammonia to nitrite (NO_2^-), then from nitrite to nitrate

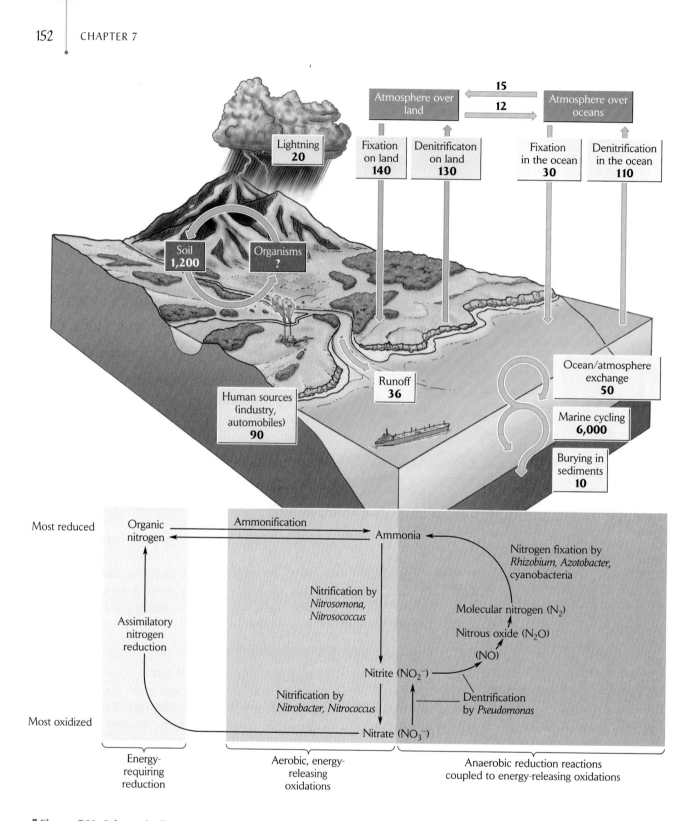

Figure 7.11 Schematic diagram of transformations of compounds in the nitrogen cycle.
The most reduced state of the nitrogen atom has the highest chemical energy potential.

(NO_3^-), during which nitrogen atoms are stripped of six, and then two more, of their electrons. These oxidation steps release much of the potential chemical energy of organic nitrogen. Each step is carried out only by specialized bacteria: $NH_3 \rightarrow NO_2^-$ by *Nitrosomonas* in the soil and by *Nitrosococcus* in marine systems; $NO_2^- \rightarrow NO_3^-$ by *Nitrobacter* in the soil and *Nitrococcus* in the oceans. The overall pathway for nitrification is thus

$$NH_3 \rightarrow NO_2^- \rightarrow NO_3^-.$$

Because both nitrification steps are oxidations, they can occur only in the presence of a powerful oxidizing agent, such as molecular oxygen, that can act as an electron acceptor. However, in waterlogged, anaerobic soils and sediments and in oxygen-depleted bottom waters, nitrate and nitrite are more oxidized than the surrounding environment, and they themselves can act as electron acceptors (oxidizers). Under these conditions, reduction reactions are thermodynamically favorable, and nitrogen may be reduced to nitric oxide (NO):

$$NO_3^- \rightarrow NO_2^- \rightarrow NO.$$

This reaction, called **denitrification**, is accomplished by bacteria such as *Pseudomonas denitrificans*. Denitrification is important for breaking down organic matter in oxygen-depleted soils and sediments, but it also results in the loss of nitrogen from soils because some nitric oxide escapes as a gas. Additional chemical reactions under anaerobic, reducing conditions in soils and water can produce molecular nitrogen,

$$NO \rightarrow N_2O \rightarrow N_2,$$

with the consequent loss of nitrogen from general biological circulation.

Denitrification may be a major cause of low availability of nitrogen in marine systems. When organic remains of plants and animals sink to the depths of the oceans, their oxidation by bacteria in deep waters and bottom sediments often is accomplished anaerobically using nitrate as an oxidizer. This results in the conversion of nitrate and nitrite to the dissolved gases NO and N_2, which cannot be used by algae.

Nitrogen fixation

The loss of readily available nitrogen to ecosystems by denitrification is offset by nitrogen fixation. This assimilatory reduction of nitrogen is accomplished by bacteria such as *Azotobacter*, which is a free-living species; *Rhizobium*, which occurs in symbiotic association with the roots of some legumes (members of the pea family) and other plants (▮ Figure 7.12); and cyanobacteria. The enzyme responsible for nitrogen fixation by these microorganisms—

(a)

(b)

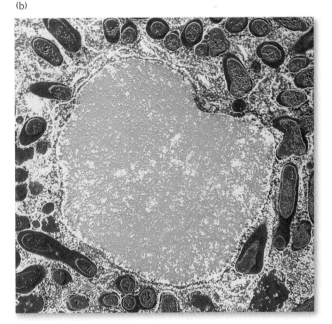

▮ **Figure 7.12 Some bacteria carry out nitrogen fixation.** The nodules on the roots of soybeans (*left*) harbor symbiotic nitrogen–fixing bacteria, as shown in red in the false–color transmission electron micrograph (*right*). Photo (a) courtesy of Thomas R. Sinclair; photo (b) by Dr. Jeremy Burgess/Science Photo Library/Photo Researchers.

nitrogenase—is inactivated by oxygen and works efficiently only under extremely low oxygen concentrations. This explains why *Azotobacter* bacteria, living freely in the soil, exhibit only a small fraction of the nitrogen-fixing capacity of *Rhizobium* bacteria, which are sequestered in the relatively anoxic cores of root nodules. In these nodules, root cells infected by *Rhizobium* form membrane-bounded structures called symbiosomes, within which the bacteria are maintained. Oxygen within a symbiosome is maintained at a very low level so as not to interfere with the activity of nitrogenase. This level of oxygen would limit respiration of plant root cells severely, but is suitable for respiration by *Rhizobium*. Although symbiosomes contain little free oxygen, they do have an abundant supply bound to a special kind of hemoglobin. This so-called leghemoglobin has a high affinity for oxygen and therefore keeps the concentration of free oxygen very low while providing a continuous supply for respiration.

Nitrogen fixation proceeds by reducing nitrogen and therefore requires energy, though no more than the conversion of an equivalent amount of nitrate to ammonia by plants. The reduction of one atom of molecular nitrogen to ammonia requires approximately the amount of energy released by the oxidation of an atom of organic carbon to carbon dioxide. Nitrogen-fixing microorganisms obtain the energy and reducing power they need to reduce N_2 to NH_3 by oxidizing sugars or other organic compounds. Free-living bacteria must obtain these resources by metabolizing organic detritus in the soil, sediments, or water column. More abundant supplies of energy are available to the *Rhizobium* bacteria that enter into symbiotic relationships with plants, which provide them with malate, a four-carbon carbohydrate produced as an end product of glycolysis.

On a global scale, nitrogen fixation approximately balances the production of N_2 by denitrification. These fluxes amount to about 2% of the total cycling of nitrogen through the ecosystem. On a local scale, nitrogen fixation can assume much greater importance, especially in nitrogen-poor habitats. When land is first exposed to colonization by plants—as, for example, are areas left bare by receding glaciers or newly formed lava flows—species with nitrogen-fixing capabilities dominate the colonizing vegetation.

The phosphorus cycle is uncomplicated chemically

Ecologists have studied the role of phosphorus in ecosystems intensively because organisms require this element at a rela-

tively high level (though only about one-tenth that of nitrogen). Phosphorus is a major constituent of nucleic acids, cell membranes, energy transfer systems, bones, and teeth. Phosphorus is thought to limit plant productivity in many aquatic habitats. Influxes of phosphorus into rivers and lakes in the form of sewage and runoff from fertilized agricultural lands can artificially stimulate production in aquatic habitats, which can upset natural ecosystem balances and alter the quality of aquatic habitats. Pollution by phosphorus-containing detergents was a major contributor to this problem until phosphorus-free alternative detergents were developed.

The phosphorus cycle (❚ Figure 7.13) has fewer steps than the nitrogen cycle because, except in a very few microbial transformations, phosphorus does not undergo oxidation–reduction reactions in its cycling through the ecosystem. Plants assimilate phosphorus as phosphate ions (PO_4^{3-}) directly from soil or water and incorporate it directly into various organic compounds. Animals eliminate excess phosphorus in their diets by excreting phosphate salts in urine; phosphatizing bacteria also convert phosphorus in detritus to phosphate ions. Phosphorus does not enter the atmosphere in any form other than dust, so the phosphorus cycle involves only soil and aquatic compartments of the ecosystem.

Acidity greatly affects the availability of phosphorus to plants. In acidic soils, phosphorus binds tightly to clay particles and forms relatively insoluble compounds with iron and aluminum. In basic soils, it forms other insoluble compounds—for example, with calcium. When both calcium and iron or aluminum are present under aerobic conditions, the highest concentration of dissolved phosphate—that is, the greatest availability of phosphorus—occurs at a pH of between 6 and 7.

In well-oxygenated aquatic systems, phosphorus readily forms insoluble compounds with iron or calcium and precipitates out of the water column. Thus, marine and freshwater sediments act as a phosphorus sink, continually removing precipitated phosphorus from rapid circulation in the ecosystem. Phosphorus compounds readily dissolve and enter the water column only in oxygen-depleted aquatic sediments and bottom waters. Under such conditions, iron tends to form soluble sulfides rather than insoluble phosphate compounds.

Sulfur exists in many oxidized and reduced forms

Sulfur is part of two amino acids—cysteine and methionine—and is therefore required by plants and animals. But the

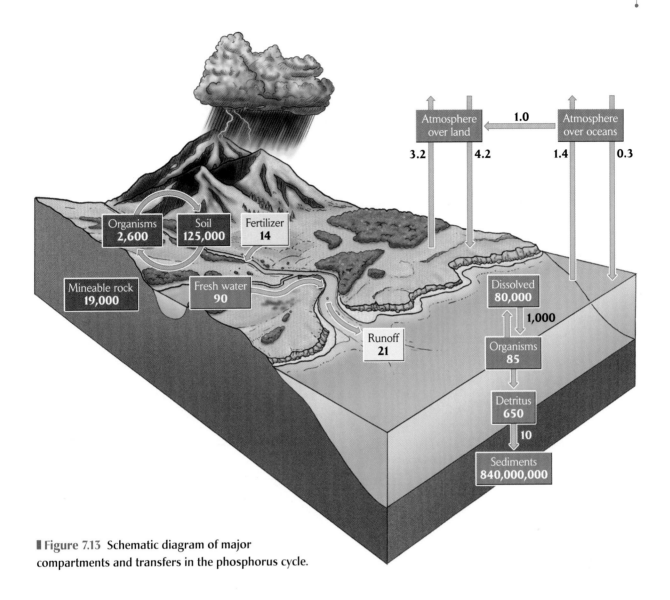

▌Figure 7.13 Schematic diagram of major compartments and transfers in the phosphorus cycle.

importance of sulfur in the ecosystem goes far beyond this role. Like nitrogen, sulfur exists in many reduced and oxidized forms, and so it follows complex chemical pathways and affects the cycling of other elements (▌Figure 7.14).

The most oxidized form of sulfur is sulfate (SO_4^{2-}); the most reduced forms are hydrogen sulfide (H_2S) and organic forms of sulfur, such as those found in amino acids. Under aerobic conditions, energy-requiring assimilatory sulfur reduction by organisms ($SO_4^{2-} \rightarrow$ organic S) balances the oxidation of organic sulfur back to sulfate, which occurs either directly or with sulfite (SO_3^{2-}) as an intermediate step. This oxidation occurs when animals excrete excess dietary organic sulfur and when microorganisms decompose plant and animal detritus.

Under anaerobic conditions, such as those in waterlogged sediments, sulfate, like nitrate, may function as an oxidizer. In such reducing environments, the bacteria *Desul-*

fovibrio and *Desulfomonas* can use energetically favorable sulfate reduction to oxidize organic carbon. The coupling of these reactions makes some energy available to the organisms. The reduced sulfur may then be used by photosynthetic bacteria to assimilate carbon by pathways analogous to photosynthesis in green plants. In these reactions, sulfur takes the place of the oxygen atom in water as an electron donor. As a result, elemental sulfur (S) accumulates unless the sediments are exposed to aeration or oxygenated water, at which point sulfur may be further oxidized by aerobic chemoautotrophic bacteria, such as *Thiobacillus,* to sulfite and sulfate.

The fate of reduced sulfur produced under anaerobic conditions depends on the availability of positive ions. Frequently, hydrogen sulfide (H_2S) forms; it escapes from shallow sediments and mucky soils as a gas having the characteristic smell of rotten eggs. Anaerobic conditions

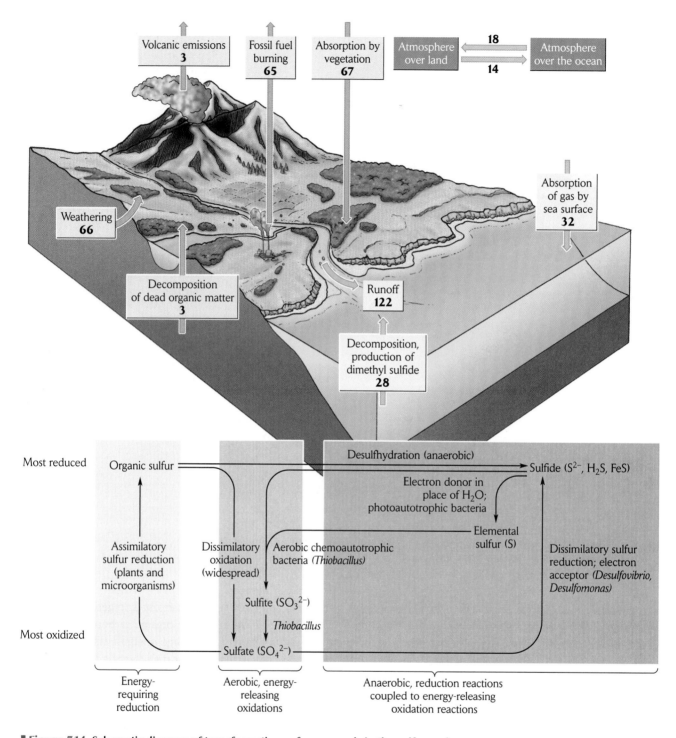

Figure 7.14 Schematic diagram of transformations of compounds in the sulfur cycle.

generally favor the reduction of ferric iron (Fe^{3+}) to ferrous iron (Fe^{2+}), which can combine with sulfide ions to form iron sulfide (FeS). For this reason, sulfides are commonly associated with coal and oil deposits. When these materials are exposed to the atmosphere in mine wastes or burned for energy, the reduced sulfur oxidizes (with the help of *Thiobacillus* bacteria in mine wastes) to sulfate. This oxidized sulfur combines with water to produce sulfuric acid (H_2SO_4), which leads to acid rain and acid mine drainage (**Figure 7.15**).

Figure 7.15 Streams draining from the refuse of coal mines may be extremely acid. Mine spoils in Tioga County, Pennsylvania. Photo by Tim McCabe, courtesy of the U.S. Department of Agriculture, Soil Conservation Service.

Microorganisms assume diverse roles in element cycles

As you may have noticed, many of the transformations discussed in this chapter are accomplished mainly or entirely by bacteria. In fact, were it not for the activities of such microorganisms, many element cycles would be altered drastically and the productivity of the ecosystem much reduced. For example, without the capacity of some microbes to use nitrogen, sulfur, and iron as electron acceptors, little decomposition would occur in anaerobic organic sediments, and their resulting accumulation would reduce the amount of inorganic carbon in the ecosystem. Without nitrogen-fixing bacteria, denitrification under anaerobic conditions would slowly deplete ecosystems of available nitrogen and reduce biological productivity proportionally.

Many of the transformations carried out by microorganisms, such as metabolism of sugars and other organic molecules, are accomplished in similar ways by plants and animals. The bacteria and cyanobacteria are distinguished physiologically by the ability of many species to metabolize substrates under anaerobic conditions and to use substrates other than organic carbon as energy sources.

Every organism needs, above all, a source of carbon for building organic structures and a source of energy to fuel the life processes. As pointed out earlier in this book, we can distinguish organisms in terms of their sources of carbon. Heterotrophs obtain carbon in reduced (organic) form by consuming other organisms or organic detritus. All animals and fungi, and many bacteria, are heterotrophs. Autotrophs assimilate carbon as carbon dioxide and expend energy to reduce it to an organic form. **Photoautotrophs** use sunlight as their source of energy for this reaction (photosynthesis). All green plants and algae are photoautotrophs, as are cyanobacteria, which use H_2O as an electron donor (reducing agent) and are aerobic; and purple and green bacteria, which have light-absorbing pigments different from those of green plants, use H_2S or organic compounds as electron donors, and are anaerobic.

Chemoautotrophs all use CO_2 as a carbon source, but they obtain energy for its reduction by the aerobic oxidation of inorganic substrates: methane (for example, *Methanosomonas* and *Methylomonas*); hydrogen (*Hydrogenomonas* and *Micrococcus*); ammonia (the nitrifying bacteria *Nitrosomonas* and *Nitrosococcus*); nitrite (the nitrifying bacteria *Nitrobacter* and *Nitrococcus*); hydrogen sulfide, sulfur, and sulfite (*Thiobacillus*); or ferrous iron salts (*Ferrobacillus* and *Gallionella*). Chemoautotrophs are almost exclusively bacteria, which apparently are the only organisms that can become so specialized biochemically as to make efficient use of inorganic substrates in this way and efficiently dispose of the waste products of chemoautotrophic metabolism.

The special role of microorganisms in ecosystem function is illustrated nicely by the highly productive communities of marine organisms that develop around deep-sea hydrothermal vents (**Figure 7.16**). Scientists from the Woods Hole Oceanographic Institution in Massachusetts first discovered these miniature ecosystems in deep water off the Galápagos archipelago in 1977. Vent communities have since been found to be widely distributed in the ocean basins of the world. The most conspicuous members of the community are giant white-shelled clams and tube worms (pogonophorans) that grow to 3 meters long, but numerous crustaceans, annelids, mollusks, and fish also cluster at great densities around hydrothermal vents. The high productivity of vent communities contrasts strikingly with the desertlike appearance of the surrounding ocean floor.

How do these communities obtain energy? These deep-sea vents occur well below the level of light penetration, and so there can be no photosynthesis. As you might suspect, the productivity of the vent communities is based on the unique qualities of the water issuing from the vents themselves. This water is hot and loaded with a reduced form of sulfur, hydrogen sulfide (H_2S). Where vent water and seawater mix, conditions are ideal for chemoautotrophic

Figure 7.16 Chemoautotrophic sulfur bacteria form the base of the food chain in hydrothermal vent communities. Other vent organisms, such as these tube worms (*Riftia pachyptila*) at a Pacific hydrothermal vent, rely on these bacteria to produce food. Photo by C. Van Dover, courtesy of OAR/National Undersea Research Program (NURP).

sulfur bacteria. These bacteria use oxygen in seawater to oxidize hydrogen sulfide in vent water. In turn, this oxidation provides a source of energy for the assimilatory reduction of inorganic carbon and nitrogen in seawater. All the other members of the vent community feed on these bacteria, which thus form the base of the local food chain. The pogonophoran worms have gone so far as to house symbiotic colonies of the bacteria within the tissues of a specialized organ, the trophosome, providing a protected place to live in return for a share of the carbohydrate and organic nitrogen produced by the bacteria.

In this chapter we have examined the cycling of several important elements from the standpoint of their chemical and biochemical reactions. Elements are cycled through ecosystems primarily because the metabolic activities of organisms result in chemical transformations of elements. The kinds of transformations that predominate depend on the physical and chemical conditions of the system. Each type of habitat presents a different chemical environment, particularly with respect to the presence or absence of oxygen and possible sources of energy. It stands to reason, therefore, that the patterns of element cycling should differ greatly among habitats and ecosystems. In the next chapter, we shall contrast element cycling in aquatic habitats and in terrestrial habitats by focusing on how some of the unique physical features of each of these environments affect the chemical and biochemical transformations involved in organic production and recycling of elements.

Summary

1. Unlike energy, nutrients are retained within ecosystems, where they are cycled between physical and biotic components. The paths that elements follow through ecosystems depend on chemical and biological transformations, which themselves depend on the chemistry of each element, the physical and chemical conditions of the environment, and the ways in which each element is used by various organisms.

2. Movement of energy through ecosystems parallels the paths of several elements, particularly carbon, whose transformations either require or release energy.

3. Energy transformations in biological systems occur primarily in the course of oxidation–reduction reactions. An oxidizer is a substance that readily accepts electrons; a reducer is one that readily donates electrons. Upon being reduced, an atom gains energy along with the electrons it accepts; upon being oxidized, an atom releases energy along with the electrons it gives up.

4. The cycling of each element may be thought of as movement between compartments of ecosystems. The major compartments are living organisms, organic detritus, immediately available inorganic forms, and unavailable organic and inorganic forms, for the most part locked away in sediments.

5. The water cycle, or hydrologic cycle, provides a physical analogy for element cycling in ecosystems. Energy is required to evaporate water because molecules of water vapor have a higher energy content than molecules of liquid water. This energy is released as heat when water vapor condenses in the atmosphere to produce precipitation.

6. All organisms require organic carbon as the primary substance of life. Organic carbon is also the major source of energy for most animals and microorganisms. Carbon shuttles between organic forms and the inorganic compartments of ecosystems by way of photosynthesis and respiration.

7. The carbon cycle involves nonbiological processes such as the dissolution of carbon dioxide in surface waters. Dissolved carbon dioxide enters into a chemical equilibrium with bicarbonate and carbonate ions, which, in the presence of calcium, tend to precipitate and form sediments. Thick accumulations of these marine sediments can become limestone rock.

8. Nitrogen has many reduced and oxidized forms and consequently follows many pathways through ecosys-

tems. Quantitatively, most nitrogen follows the cycle leading from nitrate through organic nitrogen (following assimilation by plants), ammonia, nitrite (following nitrification by bacteria), and then back to nitrate (following further nitrification). The last two steps are accomplished by certain bacteria in the presence of oxygen.

9. Under anaerobic conditions in soils and sediments, certain bacteria can use nitrate in place of oxygen as an oxidizing agent (denitrification): in this process, nitrate leads to nitrite and (eventually) to molecular nitrogen (N_2). This loss of nitrogen from general biological cycling is balanced by nitrogen fixation by some microorganisms.

10. Plants assimilate phosphorus in the form of phosphate ions (PO_4^{3-}). The availability of phosphorus varies with the acidity and oxidation level of the soil or water. The energy potential of the phosphorus atom does not change during its cycling through ecosystems.

11. Sulfur is an important element in anaerobic habitats, where it may serve as an oxidizer in the form of sulfate ions (SO_4^{2-}) or as a reducing agent (for photoautotrophic bacteria) in the forms of elemental sulfur and sulfide.

12. Many elemental transformations, particularly under anaerobic conditions, are accomplished by biochemically specialized microorganisms. These organisms therefore play important roles in the cycling of elements through the ecosystem.

PRACTICING ECOLOGY

CHECK YOUR KNOWLEDGE

Methanogenesis

Understanding oxidation and reduction reactions can help to clarify many aspects of element cycling in ecosystems. Each reduction reaction is coupled to an oxidation reaction; thus they are referred to together as a reduction-oxidation (redox) reaction. For the chemical reactions that lead to nutrient transformations in ecosystems, oxidation generally involves oxygen because it readily accepts electrons from other atoms. In contrast, the reduction reaction that is coupled to oxidation usually involves organic forms of carbon because carbon atoms are good at donating electrons during redox reactions.

In some kinds of habitats, such as waterlogged sediments in swamps or marshes, oxygen is not available to serve as a terminal electron acceptor for respiration. Certain kinds of bacteria that live in such sediments (Archaebacteria) have evolved the ability to use organic carbon to oxidize organic carbon when oxygen is not available. They utilize organic carbon in the form of methanol or acetate as a substrate. Organic carbon acts as an electron acceptor to produce methane (CH_4) *and* as an electron donor to result in the production of carbon dioxide. The overall reaction is:

$$4CH_3OH \rightarrow 3CH_4 + CO_2 + 2H_2O$$

The resulting methane is released from the surface of the water and results in the phenomenon known as "swamp gas."

The factors that control methane production have received considerable research attention of late. Methane is an important greenhouse gas that is contributing to the increased absorption of infrared radiation in the atmosphere and global warming. In fact, the heating effectiveness of one molecule of methane is about 25 times that of one carbon dioxide molecule. Moreover, methane production is increasing due to growing numbers of cattle and land conversion to rice paddies (both cattle and the sediments of rice paddies contain methanogenic bacteria). Several factors facilitate the production and release of methane from waterlogged sediments. For example, Grünfeld and Brix (1999) showed that methanogenesis and emission from water is affected by differences in the composition of the sediments, the depth of the water level below the sediments, and the presence of plants that emerge above the water (Table 7.1).

Table 7.1 Estimated methane (CH_4) production rate and CH_4 oxidation from treatments that differ in plant cover and water level

	Treatment			
Emergent vegetation	+	+	+	−
Water level	Low	Int	High	High
Methanogenesis (mmol per m^2 per day)	41	55	68	92
CH_4 emission (mean) (mmol per m^2 per day)	22	46	55	85
CH_4 oxidation (% of methanogenesis)	46	20	18	7

CHECK YOUR KNOWLEDGE

1. Of the elements that cycle from the surface of the earth to the atmosphere and back, which is the most important for organisms?

2. From the results presented by Grünfeld and Brix, what is the effect of vegetation cover on methane production, oxidation, and emission?

3. How might a changing climate affect methane production and emission in light of the results presented by Grünfeld and Brix?

MORE ON THE WEB 4. Read the article "Ruminant Livestock and the Global Environment" from the EPA through *Practicing Ecology on the Web* at *http://www.whfreeman.com/ricklefs*. Why is the cow-calf sector of the beef industry the largest emitter of methane within U.S. livestock industries? How can livestock producers reduce the amount of greenhouse gas emissions coming from their cows?

Suggested Readings

Berner, R. A., and A. C. Lasaga. 1989. Modeling the geochemical carbon cycle. *Scientific American* 260:74–81.

Berner, R. A. 1997. The rise of plants and their effect on weathering and atmospheric CO_2. *Science* 276 (25 April 1997): 544–546.

Coleman, D. C., C. P. P. Reid, and C. V. Cole. 1983. Biological strategies of nutrient cycling in soil systems. *Advances in Ecological Research* 13:1–55.

Fenchel, T., and B. J. Finlay. 1995. *Ecology and Evolution in Anoxic Worlds.* Oxford University Press, Oxford.

Grant, W. D., and P. E. Long. 1981. *Environmental Microbiology.* Wiley (Halsted Press), New York.

Grassle, J. F. 1985. Hydrothermal vent animals: Distribution and biology. *Science* 229:713–717.

Grünfeld, S. and H. Brix. 1999. Methanogenesis and methane emissions: Effects of water table, substrate type and presence of *Phragmites australis*. *Aquatic Botany* 64: 63-75.

Howarth, R. W. 1993. Microbial processes in salt-marsh sediments. In T. E. Ford (ed.), *Aquatic Microbiology*, pp. 239–259. Blackwell Scientific Publications, Oxford.

Jannasch, H. W., and M. J. Mottl. 1985. Geomicrobiology of deep-sea hydrothermal vents. *Science* 229:717–725.

Post, W. M., et al. 1990. The global carbon cycle. *American Scientist* 78(4):310–326.

Retallack, G. J. 1997. Early forest soils and their role in Devonian global change. *Science* 276 (25 April 1997):583–585.

Schlesinger, W. H. 1991. *Biogeochemistry: An Analysis of Global Change.* Academic Press, San Diego.

Sprent, J. I. 1987. *The Ecology of the Nitrogen Cycle.* Cambridge University Press, New York.

Stacey, G., R. H. Burris, and H. J. Evans (eds.). 1992. *Biological Nitrogen Fixation.* Chapman & Hall, New York.

CHAPTER 8

Nutrient Regeneration in Terrestrial and Aquatic Ecosystems

Nutrient regeneration in terrestrial ecosystems occurs primarily in the soil

The quality of plant detritus influences the rate of nutrient regeneration

Mycorrhizae are mutualistic associations of fungi and plant roots

Climate affects rates of nutrient regeneration

In aquatic ecosystems, nutrients are regenerated slowly in deep layers of water and sediments

Thermal stratification hinders vertical mixing in aquatic ecosystems

Nutrients frequently limit production in the oceans

Oxygen depletion facilitates regeneration of some nutrients in deep waters

Phosphorus concentration controls the trophic status of lakes

High external and internal nutrient input makes estuaries and marshes highly productive

During the 1960s, scientists in the northeastern United States and central Europe became alarmed by the deaths of many trees in forests on poor soils. These deaths appeared to be correlated with acid rain. Acid precipitation is caused by the products of coal and oil burning spewed into the atmosphere. Among these products are vast quantities of sulfur dioxide (SO_2), which is further oxidized in the atmosphere to sulfate (SO_4^{2-}). This sulfate, in turn, forms sulfuric acid when it comes into contact with water droplets in the atmosphere. Consequently, the pH of rainwater fell to as low as 4.0, and the acidity of streams dropped close to pH 5.

It was clear that acid rain harmed forests. In 1970, the Congress of the United States passed the Clean Air Act to reduce emissions of SO_2 and particulate matter from factories and power plants. The effect of this legislation was striking. Sulfur oxides and, especially, particulate matter in the atmosphere began to decline almost immediately. Scientists were surprised to find, however, that the forests did not show signs of recovery.

Puzzled by this, ecologist Gene Likens and his colleagues began to examine records of soil and water chemistry and tree growth that had been kept since 1963 at the Hubbard Brook Experimental Forest in New Hampshire. By examining inputs and outflows of various ions, they were able to piece together a three-part explanation for the failure of trees to respond to the change in rainwater inputs. First, sulfur and nitrogen emissions began to decline after the Clean Air Act, but sulfur emissions remained relatively high and the acidity of precipitation and streams declined (pH rose) slowly. Second, particulates from factory and power plant emissions dropped precipitously after 1970, but this also removed a major source of calcium for the Hubbard Brook ecosystem, accounting for perhaps half of the pre-1970 inputs.

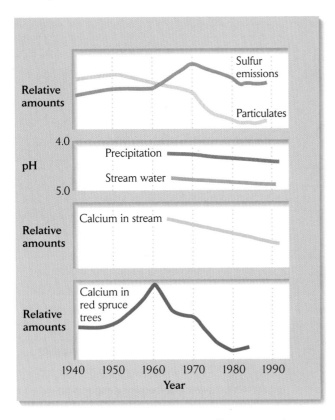

Figure 8.1 Acid rain has long-term effects on nutrient cycling and regeneration. Particulate and sulfur emissions and measures of pH and calcium in the Hubbard Brook watershed are shown for the period before and after passage of the Clean Air Act in 1970. After G. E. Likens, C. T. Driscoll, and D. C. Buso, *Science* 272:244–246 (1996).

Calcium and other positive ions reduce the acidity of water and soil. Without this input of calcium, hydrogen ions could not be removed, and the system remained acid.

The third factor preventing recovery from acid conditions was the long-term leaching of calcium and other positive ions from the soil by the hydrogen ions in the acidified rainwater. Trees require calcium and magnesium for proper growth, but levels of these elements in the soil had decreased dramatically. Acid rain was literally stealing nutrients from the soil. This is shown most clearly by the concentrations of calcium in the trunks of red spruce trees over a 50-year period (Figure 8.1). At one locality, the concentration increased at first as acid rain mobilized calcium in the soil, which made it more readily available to the trees. Eventually, however, soil calcium decreased and began to limit tree growth.

The Hubbard Brook study held several important lessons for forest ecologists. First, trees in regions of acid rain die not because of direct effects of high hydrogen ion concentrations, but because of long-term leaching of nutrients from the soil. Second, the natural recovery of forests growing on nutrient-poor soils will require restoration of soil nutrients through the slow process of soil weathering, which, as we shall see, could take a century or more. Thus, although causes of environmental deterioration may be ameliorated quickly, their overall effects on ecosystem function may remain for many years. More generally, the study demonstrated that an understanding of nutrient cycling and regeneration processes is crucial for understanding how ecosystems function.

Elements cycle through ecosystems along paths charted by their chemical properties, which determine their chemical and biochemical reactions in the biosphere. These reactions are uniquely modified by the physical and chemical conditions created in each type of terrestrial and aquatic ecosystem. Because all organisms rely on the presence of nutrients in forms they can use, the cycling and regeneration of nutrients is an important regulator of ecosystem function. In this chapter, we shall discuss how biochemical processes in soils, water, and sediments influence the productivity of the ecosystem and the cycling of elements within it.

The processes of nutrient regeneration are different in terrestrial and aquatic systems. To be sure, both systems exhibit similar chemical and biochemical transformations: oxidation of carbohydrates, nitrification, and chemoautotrophic oxidation of sulfur, among many others. But terrestrial and aquatic systems differ in the material basis for nutrient regeneration. In terrestrial ecosystems, most ecosystem metabolism is aerobic, and most elements cycle through detritus at the soil surface, where plant roots have ready access to nutrients. In aquatic habitats, sediments are the ultimate source of regenerated nutrients; such sediments at the bottom of lakes and oceans are often far removed from sites of primary production in surface waters.

Nutrient regeneration in terrestrial ecosystems occurs primarily in the soil

A major source of *new* nutrients in terrestrial systems is the formation of soil through the weathering of bedrock and other parent materials. How rapidly does this occur? Weathering normally takes place under deep layers of soil, where it is impossible to measure directly. Soil scientists can estimate the rate of weathering indirectly, however, by measuring the net loss of certain elements from a system. Positive ions, such as calcium (Ca^{2+}), potassium (K^+), sodium (Na^+), and magnesium (Mg^{2+}), are good candidates for such measurements because they dissolve readily in water and leave the soil in groundwater and eventually in streams, where they can be measured easily. When a soil attains equilibrium, as it may in undisturbed areas, the loss of an element from a system equals the weathering input of that element plus any gains from other sources, such as precipitation. Thus, it is possible to estimate weathering input from information about precipitation input and total loss.

A **watershed** is the entire drainage area of a stream or river, from which all surface water and groundwater leave at a single point. Scientists have obtained detailed nutrient budgets for several small watersheds by measuring inputs in rainwater collected at various locations in the watersheds (**∎ Figure 8.2**) and outputs in the streams that drain them (**∎ Figure 8.3**). For soil to be in equilibrium, net loss (precipitation input minus stream output) must equal input to the soil from weathering.

The best-known watershed study comes from the Hubbard Brook Forest of New Hampshire, described at

∎ Figure 8.3 Stream gauges are used to measure nutrient outputs. This stream gauge has been placed at the lower end of a watershed at the Coweeta Hydrological Laboratory, North Carolina. The V-shaped notch is engineered so that the flow of water through the weir can be estimated from the water level in the basin behind the notch. Photo by Barry Near, USDS.

∎ Figure 8.2 Rain gauges are used to measure nutrient inputs. These rain gauges have been installed in a ponderosa pine stand in California to intercept precipitation falling through the canopy of the forest and running down the trunks of trees. Analyses of the nutrient content of water collected in gauges like this one help to determine the overall nutrient budget of the forest and the specific routes of mineral cycles. Courtesy of the U.S. Forest Service.

the opening of this chapter. During the 1960s and 1970s, the annual input of calcium in precipitation at Hubbard Brook averaged 2 kilograms per hectare (kg per ha), while loss of dissolved Ca^{2+} in stream flow was 14 kg per ha. Therefore, the net loss to the system equaled 12 kg per ha. Living and dead plant biomass increased in the watershed during the study period because the forest was recovering from earlier clearing. Net assimilation of calcium in vegetation and detritus brought its overall removal from the mineral soil to 21 kg per ha per yr. Because calcium constitutes about 1.4% of the weight of the bedrock in the area, making up this annual loss would have required the weathering of about 1,500 kg (21/0.014) of bedrock per hectare, or approximately 1 mm of depth, per year. We now know from later analyses of the Hubbard Brook Forest ecosystem that this was a period of high acidity and rapid leaching of ions from soil. Thus, the soil was not in equilibrium, and inputs of calcium through weathering were undoubtedly much lower than previously thought. Nonetheless, this example illustrates how little the slow weathering of bedrock contributes to the annual uptake of nutrients by vegetation. The bulk of nutrients are made available to plants by breakdown of detritus and small organic molecules within the soil profile. Typically, weathering of bedrock provides only 10% of the soil nutrients taken up by vegetation each year.

The quality of plant detritus influences the rate of nutrient regeneration

Plants assimilate elements from soil far more rapidly than weathering generates them from parent material. Ions such as Ca^{2+}, Mg^{2+}, K^+, and Na^+ do not figure prominently in biochemical transformations, although they are required for plant growth. For the most part, plants assimilate these ions with water, which they take up in great quantities. Other important nutrients, such as nitrogen, phosphorus, and sulfur, are typically not abundant in parent material. Igneous rocks—granite and basalt, for example—contain no nitrogen, only 0.3% phosphate, and only 0.1% sulfate by mass. Most sedimentary rocks contain little more. Hence weathering adds little of these nutrients to soil; inputs from precipitation and nitrogen fixation also are small. Plant production therefore depends on rapid regeneration of these nutrients from detritus and their retention within ecosystems.

Organic detritus is everywhere, most conspicuously in terrestrial habitats, where parts of plants not consumed by herbivores accumulate at the soil surface, along with animal excreta and other organic remains (▌Figure 8.4). Ninety percent or more of the plant biomass produced

▌**Figure 8.4 The decay of plant detritus releases nutrients.** Plant detritus accumulating on a forest floor is broken down by soil organisms, and the nutrients it contains are released in forms that can be taken up and used by plants. Photo by R. E. Ricklefs.

in forested habitats passes through this detritus reservoir. The processes of decay break down the detritus, releasing the nutrients it contains in forms that can be reused by plants.

Breakdown of leaf litter on the forest floor occurs in four ways: (1) leaching of soluble minerals and small organic compounds by water; (2) consumption by large detritus-feeding organisms (millipedes, earthworms, wood lice, and other invertebrates); (3) breakdown of the woody components of leaves by fungi; and (4) decomposition of almost everything by bacteria. Between 10% and 30% of the substances in newly fallen leaves dissolve in cold water. Leaching rapidly removes most salts, sugars, and amino acids from the litter, making them available to soil microorganisms and the roots of plants; complex carbohydrates, such as cellulose, and other large organic compounds remain behind. Large detritus feeders typically assimilate only 30% to 45% of the energy available in leaf litter, and even less from wood. They nonetheless speed decay beyond what they themselves extract because they macerate plant detritus in their digestive tracts, and the finer particles in their egested wastes expose new surfaces to feeding by fungi and bacteria.

Leaves of different tree species decompose at different rates depending on their composition. For example, in eastern Tennessee, weight loss of shed leaves during the first year after leaf fall ranged from 64% for mulberry to 39% for oak, 32% for sugar maple, and 21% for beech. Needles of pines and other conifers also decomposed slowly. These differences among species depend to a large extent on the lignin content of the leaves, which determines their toughness. Lignins are long, complex chains of organic molecules. They lend wood many of its structural qualities and are even more difficult to digest than cellulose; in fact, only the so-called "white rot" fungi can break down lignins. The decomposition rate of detritus also depends on its content of nitrogen, phosphorus, and other nutrients required by bacteria and fungi for their own growth. The higher the concentration of these nutrients, the faster microbes can grow, and the more rapidly they decompose plant detritus.

The resistance of some types of litter to degradation highlights the unique role of fungi in regenerating nutrients. Most fungi consist of a network of threadlike structures called hyphae, which can penetrate plant litter and wood where bacteria cannot reach. The familiar mushrooms and shelf fungi are merely fruiting structures produced by the mass of hyphae deep within the litter or wood (▌Figure 8.5; see also Figure 1.9). Like bacteria, fungi secrete enzymes into the substrate and absorb simple sugars and amino acids produced by this extracellular digestion. Fungi differ

Figure 8.5 Shelf fungi speed the decomposition of a fallen log. The visible fruiting structures are produced by the fungal hyphae that grow throughout the interior of the log, slowly destroying its structure. Photo by R. E. Ricklefs.

rhizae are more effective than plant roots alone at extracting certain mineral nutrients, such as phosphorus, from the soil. Mycorrhizae, especially ectomycorrhizal forms, may also protect plant roots from disease by physically excluding pathogens or by producing antibiotics (antibacterial toxins). The main advantage of this association for the fungi appears to be a reliable source of organic carbon in the form of simple sugars transported from the leaves to the roots of their host plants.

Endomycorrhizae are associated with most species of vascular plants. Because their hosts include primitive vascular plants, such as lycopods and ferns, endomycorrhizal associations evidently date back to the initial development of vegetation on land. Endomycorrhizal fungi do not grow freely in the soil, but rather infect plant roots through spores left behind from dead, previously infected roots. (Presumably, these spores blow into virgin soils along with other dust.) These fungi derive their carbon exclusively from plant roots. Their mineral nutrients are derived from soil, into which they send long hyphae.

from bacteria in being able to digest cellulose (which only a few bacteria, protozoa in the guts of termites, and snails also can accomplish) and, especially, lignin.

Mycorrhizae are mutualistic associations of fungi and plant roots

In addition to their role in decomposing detritus, some kinds of fungi grow on the surfaces of or inside the roots of many plants, especially woody species. This association of fungus and root, which is called a **mycorrhiza** (plural, *mycorrhizae;* literally, "fungus root"), enhances a plant's ability to extract mineral nutrients from soil and greatly increases primary production, especially on poor soils. Although many forms of mycorrhizae are recognized, they are classified as **endomycorrhizae** when the fungus penetrates the root tissue and as **ectomychorrhizae** when it forms a sheath over the root surface (Figure 8.6).

Mycorrhizae occur everywhere, but they promote plant growth most strongly in soils that are relatively depleted of nutrients (Figure 8.7). Mycorrhizae increase a plant's uptake of minerals by penetrating a greater volume of soil than the roots could accomplish alone and by increasing the total surface area available for nutrient assimilation. In addition, because the fungi secrete enzymes and acid (hydrogen ions) into surrounding soil, mycor-

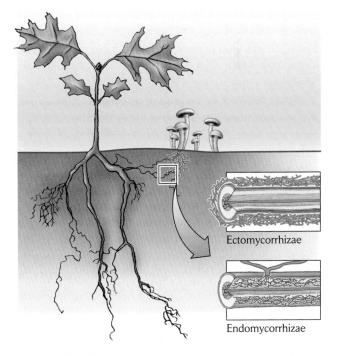

Figure 8.6 Two types of mycorrhizae are recognized. In ectomycorrhizae, the fungus forms a sheath around the root; in endomycorrhizae, the fungus penetrates the root.

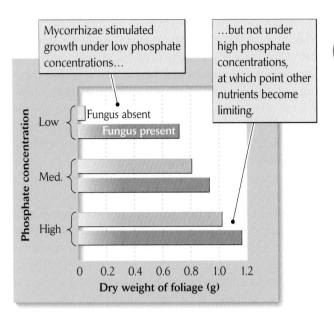

Mycorrhizae stimulated growth under low phosphate concentrations...

...but not under high phosphate concentrations, at which point other nutrients become limiting.

Phosphate concentration

Low { Fungus absent / Fungus present

Med.

High

Dry weight of foliage (g)
0   0.2   0.4   0.6   0.8   1.0   1.2

▌Figure 8.7 **Mycorrhizae promote plant growth most strongly in poor soils.** The effects of different amounts of phosphate fertilizer and inoculation with the mycorrhizal fungus *Enmdogene macrocarpa* on the growth of tomato plants (*Lycopersicon esculentum*) are shown. After J. L. Harley and S. E. Smith, *Mycorrhizal Symbiosis*, Academic Press, London (1983).

Apparently, endomycorrhizal fungi can use forms of phosphorus that plants cannot, such as the highly insoluble "rock phosphorus" $Ca_3(PO_4)_2$, which dissolves only under acid conditions. Over and above their ability to secrete hydrogen ions and organic acids, the fungi may be aided in this task when their hyphae grow in association with phosphate-solubilizing bacteria.

Ectomycorrhizae are widespread, especially among the roots of trees and shrubs in regions of the temperate zone with poor soils. Their hyphae extend far out into the soil and may be responsible for taking up organic forms of nitrogen, especially ammonium, that plants are otherwise incapable of using. Ectomycorrhizal fungi form a tough sheath around roots; the fungus may account for as much as 80% of the mass of a mycorrhizal association (root plus fungus) in heavily infected heath plants. The fungal sheath stores soil-derived nutrients and carbon compounds, which may be one of the advantages that mycorrhizae offer plants. Regardless of how they function, ectomycorrhizal fungi can account for a major part of the energy and nutrient budgets of some environments. For example, the carbon assimilated by ectomycorrhizal fungi in a fir (*Abies*) forest accounted for 15% of the total net primary production. The fungus was undoubtedly responsible for a much greater share of the mineral nutrient uptake.

Climate affects rates of nutrient regeneration

Nutrient cycling differs in tropical and temperate ecosystems because of the effects of different climates on weathering, soil properties, and decomposition of detritus. Tropical soils tend to be deeply weathered and have little clay, which means that they don't retain nutrients well. As a consequence, unless nutrients are taken up rapidly by plants, they wash out of the soil. In spite of this, tropical forests often exhibit extremely high primary production. Their high productivity is supported by (1) the rapid decay of detritus under warm, humid conditions, (2) the rapid uptake of nutrients by plants and other organisms from the uppermost layers of soil, and (3) the efficient retention of nutrients by plants and their mycorrhizal associates. In typical tropical ecosystems, most nutrients are found in living biomass rather than in soil, and elements are regenerated and assimilated rapidly. This pattern has important implications for tropical agriculture and conservation.

Over extensive regions of old, deeply weathered soils in the Tropics, planting crops such as corn on clear-cut land has predictable adverse consequences for soil fertility (▌Figure 8.8). The practice of cutting and burning felled trees releases many mineral nutrients, which may support two or three years of crop growth, but these nutrients are leached quickly out of the soil when natural vegetation is no longer present to assimilate them. Consequently, levels of mineral nutrients in the soil decline rapidly. Furthermore, as exposed tropical soils dry out, upward movement of water draws iron and aluminum oxides toward the surface, where they form a bricklike substance called **laterite.** Surface runoff of water over the impenetrable laterite accelerates erosion, further depleting nutrients and choking streams with sediment. Traditional slash-and-burn agriculture on such soils in the Tropics usually alternates two or three years of crops with fifty to a hundred years of forest regeneration to rebuild soil quality. Where population density no longer allows this practice, soils cannot be replenished naturally, and they deteriorate rapidly unless expensive and environmentally damaging inputs of fertilizers and intense cultivation are used.

A comparison of forested ecosystems cleared for crops in Canada, Brazil, and Venezuela demonstrates the importance of soil organic matter in sustaining soil fertility under intensive agriculture. Carbon contents of undisturbed soils were 8.8 kilograms per square meter (kg per m²) at a prairie site in Canada, 3.4 kg per m² at a semiarid thorn forest site

Figure 8.8 Large areas of tropical forest are cleared for agriculture each year. Trees in the forest of lowland Panama have been cut and burned to make room for planting crops. Soil fertility will decrease dramatically within two or three years. Photo by R. E. Ricklefs.

in Brazil, and 5.1 kg per m² under a Venezuelan rain forest. After 65 years of cultivation, the carbon content of the Canadian soil had been reduced by 51%, which is equivalent to decline at an exponential rate of about 1% per year. In marked contrast, the carbon content of the Brazilian soil had decreased by 40% after six years of cultivation (9% per year), and that of the Venezuelan soil had decreased by 29% after three years of agriculture (11% per year). These results suggest that cultivated temperate soils retain organic matter ten times longer than tropical soils, and thereby provide a more persistent store of mineral nutrients that can be released slowly by decomposition.

Vegetation and soil fertility

Obviously, vegetation is critical to the development and maintenance of soil fertility in many tropical systems. Even in temperate zones, removal of vegetation reveals its important role in the retention of soil nutrients (**Figure 8.9**). In one study, a small watershed in the Hubbard Brook Forest was clear-cut and its nutrient flux compared with that in similar undisturbed forest systems. The clear-cutting increased stream flow severalfold because there were no longer trees present to take up water; losses of nutrients, particularly calcium, increased 3–20 times over losses in

comparable undisturbed systems. The nitrogen budget of the clear-cut watershed sustained the most striking change. Plants assimilate available soil nitrogen so rapidly that undisturbed forest gains nitrogen at the rate of 1–3 kg per hectare per year from precipitation and nitrogen fixation. In the clear-cut watershed, net loss of nitrogen as nitrate soared to 54 kg per ha per year, a value comparable to the annual assimilation of nitrogen by vegetation in undisturbed forest and many times the precipitation input (7 kg per ha per year). As in the undisturbed watershed, organic nitrogen was converted to nitrate by soil microbes. However, in the cleared watershed, trees were not present to take up nitrates and, because nitrate ions do not bind well to particles of clay and humus, they were lost from the soil.

Comparative studies of nutrient dynamics in temperate and tropical forests further illustrate their differences. Litter on the forest floor constitutes an average of about 20% of the total biomass of vegetation (including trunks and branches) and detritus in temperate needle-leaved forests, 5% in temperate hardwood forests, and only 1–2% in tropical rain forests. The ratio of litter to the biomass of living leaves is between 5 and 10 to 1 in temperate forests, but less than 1 to 1 in tropical forests. Of the total organic carbon in the system as a whole, more than 50% occurs in soil and litter in northern forests, but less than 25% in tropical rain forests. The rest is in living biomass. Clearly, litter and other detritus decompose rapidly in the Tropics and do not form as substantial a nutrient reservoir as they do in temperate regions.

Figure 8.9 Clear-cutting experiments demonstrate the role of vegetation in nutrient retention. This clear-cut watershed at the Coweeta Hydrological Laboratory, North Carolina, was employed in studies of evapotranspiration and runoff in forest ecosystems. Courtesy of the U.S. Forest Service.

Table 8.1   Distribution of mineral nutrients in the soil and living biomass of a temperate and a tropical forest ecosystem

Forest (Locality)	Biomass (T per ha)*	Nutrients (kg per ha)		
		Potassium	Phosphorus	Nitrogen
Ash and oak (Belgium)	380			
Living vegetation		624	95	1,260
Soil		767	2,200	14,000
Ratio of soil to biomass		1.2	23.1	11.1
Tropical deciduous (Ghana)	333			
Living vegetation		808	124	1,794
Soil		649	13	4,587
Ratio of soil to biomass		0.8	0.1	2.0

*T = metric tons.

Source: P. Duvigneaud and S. Denayer-de-Smet, in D. E. Reichle (ed.), Analysis of Tropical Forest Ecosystems, Springer-Verlag, New York (1970), pp. 199–225; D. J. Greenland and J. M. Kowal, Plant Soil 12:154–174 (1960); J. D. Ovington, Biol. Rev. 40:295–336 (1965).

Fewer data exist on the relative proportions of nutrient elements in soil and living vegetation. Distributions of potassium, phosphorus, and nitrogen in a temperate and a tropical forest with similar living biomass are compared in Table 8.1. Two main points emerge from this comparison. First, the accumulation of nutrients in vegetation, on a weight-for-weight basis, was somewhat greater in the tropical forest. For example, the total dry weight of living vegetation in the Belgian ash–oak forest exceeded that in the tropical deciduous forest of Ghana by 14%, but accumulation of the three elements per gram of dried vegetation was 32–38% lower in the temperate forest. Second, the soil-to-biomass ratio of each element was much lower in the Tropics. In the temperate forest, 96% of the phosphorus occurred in the soil; in the tropical forest, more than 90% of the much smaller amount of phosphorus was found in the living biomass.

Eutrophic and oligotrophic soils

While recognizing the general nutrient poverty of many tropical soils, we must also distinguish between nutrient-rich and nutrient-poor soils within the Tropics. **Eutrophic,** or "well-nourished," soils develop in geologically active areas where natural erosion is high and soils are relatively young. With bedrock closer to the surface, weathering adds nutrients more rapidly and soils retain nutrients more effectively. In tropical regions of the Western Hemisphere—the Neotropics—such eutrophic soils occur widely in the Andes, in Central America, and in the West Indies.

By contrast, **oligotrophic,** or nutrient-poor, soils develop in old, geologically stable areas, particularly on sandy alluvial deposits (as in much of the Amazon basin), where intense weathering removes clay and reduces the capacity of soils to retain nutrients.

Especially in nutrient-poor areas, nutrient retention by vegetation is crucial to high productivity in tropical ecosystems. In these environments, plants retain nutrients by keeping their leaves for long periods and by withdrawing nutrients from them before they are dropped. They also grow dense mats of roots (and associated fungi) that remain close to the soil surface (where litter decomposes) and even extend up the trunks of trees to intercept nutrients washing down from the forest canopy. Data from Africa reveal that between 68% and 85% of the root biomass of tropical forests is concentrated within the top 25–30 cm of soil. In other tropical areas, radioactively labeled compounds have been used to show that root mats intercept nutrients regenerated by leaching and decomposition of detritus before they can penetrate into mineral soil and be washed out of the system.

ECOLOGISTS IN THE FIELD

Will global warming speed decomposition of organic matter in boreal forest soils?

Temperature has an overwhelming effect on the rate of regeneration of nutrients within soils. At one extreme, in tropical regions, plant litter breaks down rapidly, and the released nutrients are either taken up by plants or leave the forest in ground and surface waters. At the opposite extreme, decomposition is so slow in boreal forests and tundra environments that thick layers of organic matter accumulate in the soil. Decomposition is slow at high latitudes in part because soils are frozen much of the year; below a certain depth, soils may be permanently frozen. By one estimate, permanently and seasonally frozen soils of boreal forests worldwide hold 200–500 gigatons (GT) of carbon, which represents almost 80% as much as exists as carbon dioxide in the atmosphere. If temperatures of boreal soils were to increase because of global warming, soil microorganisms and animals might metabolize a substantial fraction of this soil carbon. Respiration by soil organisms would contribute to the carbon dioxide already in the atmosphere and further accelerate global warming. Ecologists are concerned about this potential, and several studies have sought to understand how decomposition of organic matter in boreal forests responds to temperature.

How does one measure the breathing of a forest? M. L. Goulden led his team of researchers to a spruce forest near Thompson in Manitoba, Canada, in an attempt to answer this question. The spruce forest ecosystem has three main components: soil; aboveground vegetation, mostly spruce trees and moss (❙Figure 8.10); and the atmosphere above the forest with which it exchanges oxygen and carbon dioxide. Carbon flux in and out of the forest was determined by measuring the absorption of a particular wavelength of infrared light by carbon dioxide gas. Carbon dioxide leaving the soil was trapped in containers at the soil surface. Carbon dioxide moving between the forest and the atmosphere was estimated by measuring the concentrations of CO_2 in air at a particular height (29 meters in this case) and the rate of movement of air vertically at a large number of sites. When these values were averaged over rising and falling currents of air for long periods, they provided a measure of net movement of carbon dioxide into or out of the forest as a whole.

Air temperatures in the forest studied were highly seasonal, with minimums averaging −10°C to −25°C in the winter months and 15°C to 25°C in summer. Variation in soil temperature decreased with depth, but followed the seasonal trend in air temperature with a lag of one to two months. Soils thawed in late spring and remained unfrozen late into the year. For the forest as a whole, net uptake of carbon dioxide began in early May and reached a peak of 10–15 kg C per hectare per day in June and early July, when air temperature reached its peak (❙Figure 8.11). Net uptake declined to near zero by August and September as increasing soil temperature stimulated soil respiration. After the end of September, the forest lost carbon, at 6–8 kg per ha

❙Figure 8.10 **The boreal forest may be affected by global warming.** The floor of this boreal forest near Fairbanks, Alaska, has a thick layer of moss on the ground. Photo by R. E. Ricklefs.

per day in October and 2–3 kg per ha per day from December through April. Overall, annual net carbon flux over the three-year study varied from a loss of 0.7 tons (T) per hectare to a gain of 0.1 T per hectare, with a net loss for the whole period. This figure compares with an annual primary production of 8.0 T per ha per yr, and a total carbon content of 95 T in aboveground living and dead biomass and 200 T in the soil.

Thus, observed losses were a small part of the total carbon in the system. However, soil respiration also was clearly sensitive to soil thawing, and even a small increase in the depth of thawing would increase significantly the release of carbon dioxide to the atmosphere. Whether thawing of frozen soils at high latitudes will accelerate global warming substantially is largely a matter of speculation at this point. What Goulden's study shows, however, is that many ecosystems are not in equilibrium and that we can expect major readjustments in the sizes of compartments in the biosphere as global conditions change.

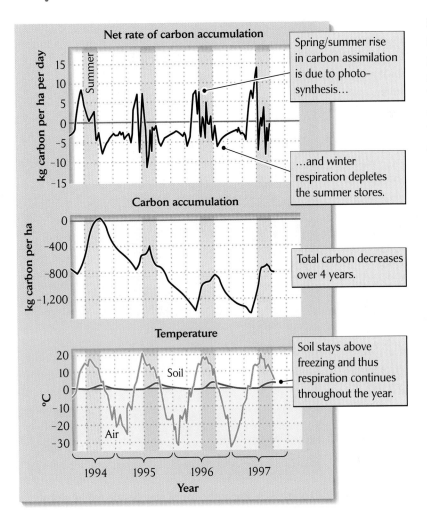

Net rate of carbon accumulation

Spring/summer rise in carbon assimilation is due to photosynthesis…

…and winter respiration depletes the summer stores.

Carbon accumulation

Total carbon decreases over 4 years.

Temperature

Soil stays above freezing and thus respiration continues throughout the year.

▌**Figure 8.11 Carbon flux in a boreal forest is sensitive to temperature.** In spite of a positive carbon balance during the warm summer months when photosynthesis occurs, this forest loses more carbon during the rest of the year as respiration continues in the soil, whose temperature remains above freezing. *After M. L. Goulden, S. C. Wofsy, J. W. Harden, et al., Science 279:214–217 (1998).*

In aquatic ecosystems, nutrients are regenerated slowly in deep layers of water and sediments

Because most cycling of elements takes place in an aqueous medium, the chemical and biochemical processes involved do not differ markedly between terrestrial and aquatic systems. What is distinctive about most rivers, lakes, and oceans is the movement of nutrients into deep layers of water and benthic sediment deposits, from which they are regenerated and returned to zones of productivity relatively slowly.

Sediments in aquatic systems resemble terrestrial soils superficially, but the roles of soils and sediments in ecosystem processes differ in two important ways. First, regeneration of nutrients from terrestrial detritus takes place close to plant roots, where nutrients are assimilated. In contrast, algae and aquatic plants assimilate nutrients from the water column in the uppermost sunlit (photic) zones, often far removed from sediments at the bottom. Second, decomposition of terrestrial detritus occurs, for the most part, aerobically, and hence relatively rapidly. In contrast, aquatic sediments often become depleted of oxygen, which greatly slows most biochemical transformations and changes the way in which some elements are cycled.

Maintenance of high aquatic productivity depends on the proximity of bottom sediments to the photic zone at the surface, or on the presence of some means of bringing nutrients regenerated in the sediments back to the photic zone. A map of primary productivity of the oceans (▌Figure 8.12) reveals high rates of carbon fixation in shallow seas, both in the Tropics (for example, the Coral Sea and the waters surrounding Indonesia) and at high latitudes (the Baltic Sea, the Sea of Japan). Areas having strong upwelling currents, such as the western coasts of Africa and the Americas, also exhibit high primary productivity.

Excretion and microbial decomposition regenerate some nutrients in the photic zone where assimilation and

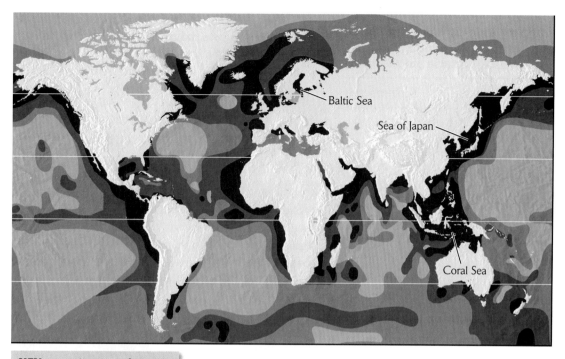

KEY: mg carbon per m² per day

100 150 250 500

▌Figure 8.12 Productivity in aquatic ecosystems is highest where nutrients regenerated in sediments can reach the photic zone. The map shows primary production in the world's oceans, in milligrams of carbon fixed per square meter per day. Productivity is greatest on continental shelves and in regions of upwelling on the west coasts of Africa and South and Central America.

After R. K. Barnes and K. H. Mann, *Fundamentals of Aquatic Ecosystems*, Blackwell, Oxford (1980).

production take place, just as they do in terrestrial soils. In some situations, elements may cycle rapidly within productive surface layers of the water column with little loss to sedimentation. For example, a study conducted in deep water off the western coast of North America showed that phytoplankton assimilate nitrogen about as rapidly as zooplankton excrete it, mostly in the form of ammonium. About half the nitrogen present was taken up by phytoplankton directly as ammonium, and about half was first nitrified ($NH_4^+ \rightarrow NO_3^-$) by bacteria and then taken up by phytoplankton.

Nitrogen budgets for the Bay of Quinte, Lake Ontario, illustrate the relative magnitudes of assimilation and regeneration in a freshwater aquatic ecosystem. These studies were conducted within columns of water enclosed by "limnocorrals," which are triangular or circular in cross section and are formed by sheets of plastic suspended by floats at the surface and entrenched in sediment at the bottom (in this case, 4 meters beneath the surface). Such enclosures make it possible to study the fluxes of elements by adding isotopically labeled compounds. The limnocorrals had large enough cross sections that water at the surface and immediately over sediments at the bottom could mix in parallel with the lake as a whole.

Measurements of nitrogen cycling in the Bay of Quinte are summarized in Table 8.2 for one day in late spring (June 5) and a second day in late summer (September 4). Nitrate accounted for 30–40% of the nitrogen in the water column, and ammonium accounted for the rest. Availability of nitrogen was similar in late spring and late summer. Uptake of isotopically labeled nitrogen revealed that algae assimilated more ammonium than nitrate, by ratios of 4 to 1 early in the growing season (June 5) and 30 to 1 late in the growing season (September 4). Although levels of available nitrogen were similar, uptake in September exceeded that in June by nearly five times, probably because the water was warmer late in the season and perhaps because some other nutrient that limits primary production was more abundant. Remember that vertical mixing of water occurs in temperate lakes in late

Table 8.2 Estimates of release and uptake of nitrogen in a limnocorral in the Bay of Quinte, Lake Ontario, in 1974

Characteristic	June 5	September 4
Concentration (μg N per liter)		
Ammonia (NH_4^-)	102	167
Nitrate (NO_3^-)	8	21
Particulate (>30 μm)	75.6	175
Particulate (<30 μm)	47.4	223
Primary production (μg C per liter per day)		
Gross	135	1,264
Net	−62	1,062
Uptake (μg N per liter per day)		
Ammonia	16	115
Nitrate	2.5	10.4
Nitrogen fixation	0	4.0
Total	18.5	18.5
Release (μg N per liter per day)		
Zooplankton grazing	9.7	26.8
Sedimentation	2.6	35.7
Total	12.3	62.5

Source: *C. F. H. Liao and D. R. S. Lean,* J. Fish. Res. Bd. Can. *35:1102–1108 (1978).*

summer, and that this mixing can bring regenerated nutrients, such as phosphorus, from deep waters to the surface. The ratio of nitrogen concentration to rate of nitrogen uptake is the residence time (see Chapter 6) of nonbiological nitrogen in the system. In June, this was 7.8 days; by late summer, residence time had decreased to 2.0 days.

During both periods, uptake of nitrogen by algae exceeded losses to grazing by zooplankton and by particulate organic matter sinking out of the water column. Thus, total biological nitrogen was increasing, which is generally the case through the summer growing season. Without vertical mixing and return of regenerated sediments to the surface, however, sedimentation would soon remove most of the nitrogen from the water. Nitrogen lost in sinking particulate matter averaged 14% as much as uptake in the water column in June and 28% as much as uptake in September. In September, total nitrogen in the system (NH_4^+, NO_3^-, and particulate, including living organisms)

was 586 μg per liter, and nitrogen loss was 63 μg per liter per day. These figures show that physical removal of nitrogen from the system could have depleted the resource within a few weeks in the absence of vertical mixing.

Thermal stratification hinders vertical mixing in aquatic ecosystems

Vertical mixing of water requires an input of energy to accelerate water masses and keep them moving. Winds supply most of this energy, causing turbulent mixing of shallow water and upwelling currents along some seacoasts, although variations in water density related to temperature and salinity establish vertical currents in other marine ecosystems.

Vertical movement of water can be hindered in aquatic systems when sunlight heats surface water, establishing a thermocline (see Chapter 4), or when fresh water floats over denser salt water. The latter happens in estuaries, at the edges of melting ice, and where precipitation is extremely high. Other processes promote vertical mixing. In marine systems, when evaporation exceeds freshwater input, surface layers of water become more saline, hence denser, and literally fall through the lighter water below. This also occurs when ice forms and salt is excluded from the crystallized water. Temperate zone lakes experience fall overturn when their surface waters cool, become denser, and sink through the warmer, less dense layers below.

Vertical mixing of water affects production in two opposing ways. On one hand, mixing can bring nutrient-rich water from the depths to the photic zone and thereby promote production. On the other hand, mixing can carry phytoplankton below the photic zone and thereby reduce production. Indeed, when vertical mixing extends far below the photic zone, phytoplankton cannot maintain themselves, much less reproduce. Under such conditions, primary production may shut down altogether, resulting in the seeming contradiction of nutrient-rich water without primary production.

A more typical situation in many temperate zone lakes and ponds is one in which thermal stratification during summer prevents vertical mixing; then, as sedimentation removes nutrients from surface layers, production decreases. Nutrients may be regenerated in the deeper layers of a lake, but they cannot reach the surface until stratification breaks down and vertical mixing ensues with cool fall temperatures.

Thermal stratification develops only weakly, if at all, in lakes at high and low latitudes (■ Figure 8.13). In arctic and

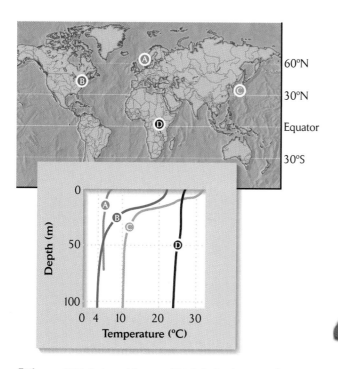

Figure 8.13 Lakes at low and high latitudes experience little thermal stratification. Temperature profiles are shown for four lakes at different latitudes at the height of summer stratification. From G. E. Hutchinson, *A Treatise on Limnology*, Vol. 1, Wiley, New York (1957).

subarctic regions, too little heat enters lakes to establish a thermocline and block turbulent mixing. Thus, the water column tends to warm uniformly, to the extent that water temperature rises at all. In the Tropics, the lack of a pronounced seasonal temperature cycle reduces the sharpness of thermal stratification because the sun and constant high air temperatures warm water uniformly to the deepest parts of the lake.

In marine systems, currents can produce more complex conditions. For example, two very different water masses, one stratified and the other not, may meet at a front, and here intermixing may create excellent conditions for phytoplankton growth. Sometimes at the boundary of a shallow-water system and a deep-water system, mixed (deep) and stratified (shallow) water masses are brought together. On the mixed side, nutrients may be abundant, but phytoplankton may not remain within the photic zone. On the stratified side, nutrients may have been depleted from the surface waters. Where the two systems meet, some of the nutrient-laden mixed water may enter the stratified water mass, creating ideal conditions for production (**Figure 8.14**).

Nutrients frequently limit production in the oceans

On the whole, primary productivity in marine ecosystems is closely related to the supply of nutrients, particularly nitrogen, in surface layers of water. As a result, the highest levels of production occur in shallow seas, where vertical mixing reaches to the bottom, and in areas of strong upwelling. However, some areas of open ocean have nitrogen and phosphorus in abundance, but phytoplankton concentrations and primary production are low. These conditions suggest limitation by other elements, primarily iron and silicon. Iron is an important component in many metabolic pathways. Silicon is the primary material in the silicate shells of diatoms (see Figure 2.14), which are the

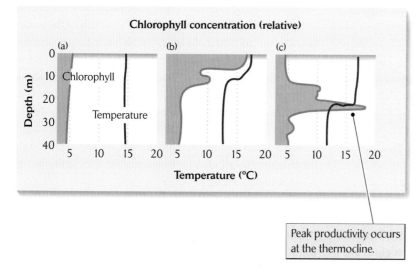

Peak productivity occurs at the thermocline.

Figure 8.14 The meeting of water masses can affect productivity. Chlorophyll concentrations and temperatures are shown as a function of water depth for three locations in the western English Channel in July 1975. (a) A well-mixed water mass. (b) A "front" in the region of mixing between water masses a and c. (c) A stratified water mass. Productivity is greatest at the thermocline because of the presence of regenerated nutrients in the water below this level. Chlorophyll concentrations provide an index to the rate of primary production. After R. K. Barnes and K. H. Mann, *Fundamentals of Aquatic Ecosystems*, Blackwell, Oxford (1980).

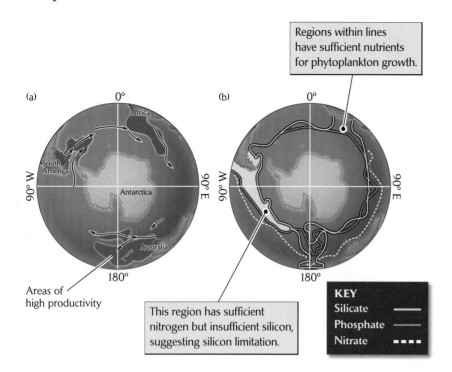

Regions within lines have sufficient nutrients for phytoplankton growth.

(a)

Areas of high productivity

This region has sufficient nitrogen but insufficient silicon, suggesting silicon limitation.

(b)

KEY
Silicate ——————
Phosphate ——————
Nitrate ▪▪▪▪

Figure 8.15 Not all ocean waters containing abundant nutrients are productive. (a) Areas of high phytoplankton concentration in the Southern Ocean, with predominant surface current directions indicated. (b) Regions within which nutrients are sufficient for abundant phytoplankton growth (nitrogen > 10 μM nitrate; phosphorus > 1 μM phosphate; silicon > 5 μM silicate). From C. W. Sullivan et al., *Science* 262:1832–1837 (1993).

predominant kind of phytoplankton in the oceans. Silicon is lost from the photic zone when diatoms die and their dense shells fall to the bottom.

High phytoplankton densities in the Southern Ocean are clearly associated with nearby continental sources of nutrients: strong plankton production is concentrated in waters downcurrent of Australia and New Zealand, South America and the Antarctic Peninsula, and southern Africa, where nutrients are picked up from shallow-water sediments (■ Figure 8.15).

Not all the Southern Ocean, however, is equally productive. Concentrations of nitrogen and phosphorus are high enough to sustain high phytoplankton densities throughout the entire area. However, phytoplankton are sparse in much of the region. This observation suggests limitation by other nutrients. In particular, the area west of southern South America between 40°S and 50°S appears to have too little silicon, probably because it has sedimented out of the water more rapidly than nitrogen and phosphorus across the long stretch of the southern Pacific Ocean.

ECOLOGISTS IN THE FIELD

Can iron limit marine productivity?

Some 20% of the open oceans appear to have abundant nitrogen and phosphorus, but low densities of phytoplankton. These regions are referred to as high-nutrient low-chlorophyll (HNLC) areas, and they have puzzled marine biologists for years. One hypothesis is that phytoplankton populations in these areas are kept low by zooplankton grazers, although it is unclear why this would happen in some regions of the sea and not others.

In the late 1980s, John H. Martin, of the Moss Landing Marine Laboratories in California, proposed that production in these areas is limited by iron. In well-aerated surface waters, the oxidized ferric form of iron (Fe^{3+}) complexes with other elements, including phosphorus, and precipitates out of the system. Whereas inshore areas receive iron from rivers, inputs to remote parts of the oceans come almost exclusively from windblown dust.

In an enormous experiment conducted in 1993 off the Pacific coast of South America, about 5° south of the equator, Martin and his colleagues fertilized a target area by distributing 450 kg of dissolved iron—roughly the amount in an automobile—over 64 km^2 of ocean, which increased the concentration of iron in that area almost a hundredfold. Within a few days, phytoplankton populations inside the fertilized patch, as measured by the concentration of the photosynthetic pigment chlorophyll in surface waters, tripled. This result clearly demonstrated iron limitation in natural surface waters.

The original motivation for Martin's experiment was to determine whether increased photosynthesis by increased phytoplankton populations could quickly remove carbon dioxide from the atmosphere to counteract the increased input of CO_2 from combustion of fossil fuels and clearing and burning of forests. In this respect, however, the experiment was a failure. Apparently, zooplankton populations increased along with the phytoplankton and regenerated much of the assimilated CO_2 by respiration. Nonetheless, the point that marine production might be limited by particular essential nutrients, heterogeneously distributed throughout the oceans, was well made.

Oxygen depletion facilitates regeneration of some nutrients in deep waters

During prolonged periods of stratification in freshwater lakes, bacterial respiration in the hypolimnion (the layer of water below the thermocline) depletes the oxygen supply in that layer (■ Figure 8.16), provided that enough organic matter exists there for bacteria to oxidize. In such anoxic bottom waters, bacterial respiration continues, reducing sulfate rather than molecular oxygen. This results in increasing concentrations of reduced sulfur, primarily in the form of hydrogen sulfide.

In the oxygen-depleted environment of bottom sediments and in the waters immediately over them, there is often insufficient oxygen for bacteria to nitrify (oxidize) ammonium (see Chapter 7). Additionally, such elements as iron and manganese shift from oxidized to reduced

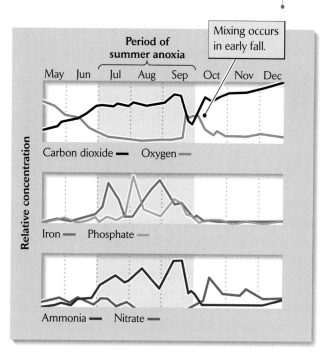

■ Figure 8.17 Oxygen depletion in the hypolimnion changes water chemistry, as shown by the seasonal course of water chemistry in the hypolimnion of Esthwaite Water, England. From G. E. Hutchinson, *A Treatise on Limnology,* Vol. 1, Wiley, New York (1957).

forms, which greatly affects their solubility. In particular, as ferric iron (Fe^{3+}) is reduced to ferrous iron (Fe^{2+}), insoluble iron—phosphate complexes become soluble, and both elements tend to move into the water column.

Changes observed in the water chemistry of the hypolimnion of an English lake, Esthwaite Water, during the course of a single season show the effects of anaerobic conditions (■ Figure 8.17). After stratification becomes established in June, oxygen at the deepest level of the lake decreases gradually, while dissolved carbon dioxide increases. The water becomes depleted of oxygen by early July, and remains so until the end of stratification and the onset of vertical mixing in late September. During the period of oxygen depletion, levels of ferrous iron, phosphate, and ammonia (the reduced forms of iron, phosphorus, and nitrogen, respectively) increase dramatically in sediments. At the sediment—water boundary, these materials become soluble and enter the water column. The return of oxidizing conditions in the fall reverses the chemistry of bottom water, initially because it is replaced by surface water, but ultimately because of the effect of oxygen on several elements. The oxidized forms of these elements produce insoluble compounds, which precipitate out of the water column. Nitrogen is a conspicuous exception:

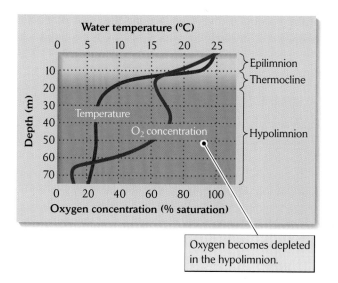

■ Figure 8.16 Stratification in lakes affects oxygen levels in water. These profiles of temperature and oxygen concentration are from Green Lake, Wisconsin, during summer stratification. From G. E. Hutchinson, *A Treatise on Limnology,* Vol. 1, Wiley, New York (1957).

under oxic conditions, nitrifying bacteria convert ammonium to nitrate, which generally remains in solution.

Phosphorus concentration controls the trophic status of lakes

Natural lakes exhibit a wide range of productivity, depending on external inputs of nutrients (rainfall, streams) and internal regeneration of nutrients in the lake. In shallow lakes lacking a hypolimnion, nutrient input occurs continuously through resuspension of bottom sediments. In somewhat deeper lakes where the thermocline develops only weakly, vertical mixing may occur periodically as a result of occasional strong winds or during unusual periods of summer cold. Such mixing returns regenerated nutrients to the surface and stimulates production. In very deep lakes, bottom waters rarely mix with surface waters, and production depends almost entirely on external nutrient sources. As was the case with forest soils discussed earlier in this chapter, aquatic ecologists classify lakes on a continuum ranging from oligotrophic (poorly nourished) to eutrophic (well nourished), depending on their nutrient status and production.

Phosphorus is the most important contributor to the fertility of most lakes, and low levels of phosphorus limit production in these systems. Phosphorus is often scarce in the well-oxygenated surface waters of lakes. In small lakes on the Canadian Shield, productivity increased dramatically in response to the experimental addition of phosphorus, but not nitrogen or carbon (Figure 8.18). Naturally eutrophic lakes have characteristic temporal patterns of production and phosphorus cycling that maintain the system in a well-nourished, dynamic steady state.

Sewage and drainage from fertilized agricultural lands can cause inappropriate nutrient loading and greatly alter natural cycles in lakes. Increased primary production is not bad in and of itself; indeed, many lakes and ponds are artificially fertilized to increase commercial fish production. But overproduction of organic matter within a lake or river (eutrophication) can lead to imbalance when natural regeneration processes cannot handle the increased demands for cycling of organic matter. Heavy organic pollution, which results, for example, from dumping raw sewage into rivers and lakes, creates biological oxygen demand (BOD), resulting from the oxidative breakdown of organic detritus by microorganisms. Inorganic nutrients, including runoff from fertilized agricultural land, stimulate the production of organic matter, adding to the BOD. The problem is heightened in winter, when photosynthesis rates are low and little oxygen is generated within the water column. In its worst manifestations, this type of pollution can deplete oxygen all the way to the surface, causing fish and other obligately aerobic organisms to suffocate.

High external and internal nutrient input makes estuaries and marshes highly productive

Shallow estuaries, which are semi-enclosed coastal regions at the mouths of rivers, are among the most productive ecosystems on earth. Salt marshes, which are areas with

Figure 8.18 Phosphorus is critical to the productivity of freshwater lakes. An experiment in a natural lake on the Canadian Shield demonstrated the crucial role of phosphorus in eutrophication. The near basin, fertilized with carbon (in sucrose) and nitrogen (in nitrates), exhibited no change in organic production. The far basin, separated from the first by a plastic curtain, received phosphorus in addition to carbon and nitrogen, and was covered by a heavy bloom of photosynthetic cyanobacteria within 2 months. Courtesy of D. W. Schindler, from D. W. Schindler, *Science*, 184:897–899 (1974).

Figure 8.19 Salt marshes are highly productive ecosystems. They are a common feature of protected bays along most temperate coasts. Photo by R. E. Ricklefs.

emergent vegetation growing between the highest and lowest tide levels (**Figure 8.19**), are also highly productive. The productivity of these ecosystems results from rapid and local regeneration of nutrients and external inputs in the form of nutrients brought in by rivers and tidal flow.

The effects of high production in estuaries and coastal marshes extend to marine ecosystems in many areas through their net export of organic matter. A Georgia salt marsh exports nearly 10% of its gross primary production and almost half of its net primary production to surround-

ing marine systems in the form of organisms, particulate detritus, and dissolved organic material carried out with the tides (**Figure 8.20**). Because of their high productivity and the hiding places they offer prey organisms, coastal marshes and estuaries are important feeding areas for larvae and immature stages of many fishes and invertebrates that later complete their life cycles in the sea.

 ## Summary

1. Nutrient cycles in terrestrial and aquatic ecosystems result from similar chemical and biochemical reactions expressed in different physical and chemical environments.

2. Nutrient regeneration in terrestrial ecosystems takes place in the soil. Weathering of bedrock and the associated release of new nutrients proceed slowly compared with assimilation of nutrients from soil by plants. Therefore, the productivity of vegetation depends on regeneration of nutrients from plant litter and other organic detritus.

3. Nutrients are regenerated from leaf litter by leaching of soluble substances; consumption by large detritus feeders; fungi that break down cellulose and lignin; and the eventual mineralization of phosphorus, nitrogen, and sulfur, primarily by bacteria.

4. Mycorrhizae are symbiotic associations of certain types of fungi with the roots of plants. The fungi, which may either penetrate root tissue or form a dense sheath around roots, enhance a plant's uptake of soil nutrients.

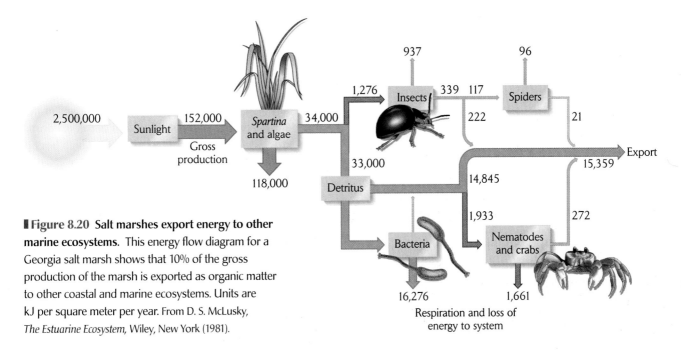

Figure 8.20 Salt marshes export energy to other marine ecosystems. This energy flow diagram for a Georgia salt marsh shows that 10% of the gross production of the marsh is exported as organic matter to other coastal and marine ecosystems. Units are kJ per square meter per year. From D. S. McLusky, *The Estuarine Ecosystem*, Wiley, New York (1981).

They do this primarily by enlarging the volume of soil accessible to roots and by secreting acid into surrounding soil to increase the solubility of such nutrients as phosphorus. In return, they obtain a reliable source of carbon from their host plant.

5. In many tropical climates, deeply weathered soils retain nutrients poorly. In such environments, regeneration and assimilation of nutrients proceed rapidly, and most nutrients, especially phosphorus, are in living vegetation. When such soils are clear-cut for agriculture, they soon lose their fertility because nutrients are removed along with the vegetation, organic matter in soil decomposes rapidly, and nutrients that are released and not assimilated are washed out of the soil.

6. Sediments at the bottoms of lakes and oceans resemble terrestrial soils, but differ from them in two important respects: aquatic sediments are spatially removed from sites of nutrient assimilation by aquatic plants and algae, and aquatic sediments often develop anoxic conditions that retard regeneration of some nutrients.

7. The primary productivity of aquatic ecosystems is maintained by the transport of nutrients from bottom sediments to the surface, as occurs in shallow waters and areas of upwelling, by the recycling of nutrients regenerated within the photic zone, and by the import of nutrients from other systems.

8. Vertical mixing is inhibited by thermal stratification. Stratification enhances aquatic production by retaining phytoplankton within the photic zone, but it diminishes production to the extent that sedimentation of detritus carries nutrients below the depth where light is sufficient for photosynthesis.

9. The primary productivity of marine systems is generally limited by availability of nutrients. Limiting nutrients may be silicon or iron in the open ocean, where both elements tend to leave the water column as sediments: silicon in the shells of diatoms and iron in precipitated complexes with other elements, such as phosphorus. Large-scale fertilization experiments have demonstrated iron limitation of photosynthesis in parts of the ocean.

10. Annual cycles of temperate zone lakes include a period of temperature stratification during summer between spring and fall periods of vertical mixing. Nutrients are brought to the photic zone near the surface during periods of mixing. During summer, nutrients are depleted by sedimentation of organic material.

11. Nutrients are regenerated in aquatic sediments by bacterial decomposition of organic matter. Anaerobic conditions develop beneath the thermocline because bacteria consume oxygen in the process. These conditions result in the chemical reduction of iron, magnesium, and sulfur and the solubilizing of phosphate compounds.

12. Because phosphorus forms insoluble compounds with iron and precipitates readily under the oxic conditions of surface waters, it frequently is in short supply and limits production in fresh waters. Sewage and agricultural runoff add phosphorus and other nutrients to streams and lakes and may greatly alter natural patterns of production and nutrient cycling, upsetting natural balances in freshwater ecosystems.

13. Shallow-water marine communities, particularly estuaries and salt marshes, are extremely productive because of rapid local regeneration of nutrients and the external input of additional nutrients from nearby terrestrial and marine systems. Marshes and estuaries are major exporters of both organic carbon and mineral nutrients to surrounding marine systems.

PRACTICING ECOLOGY
CHECK YOUR KNOWLEDGE

Nutrients and Productivity near Deep Sea Vents

You have seen that organisms use the heterogeneity of the natural environment to obtain necessary nutrients or to maintain suitable internal conditions. Recall how ground squirrels shuttle back and forth between vegetation and underground burrows to regulate body temperature. Productivity in aquatic habitats, as mentioned, is often limited because sunlit, oxygenated layers of water near the surface are physically separated from the dark, oxygen-depleted deep waters where sediments are rich in available nutrients. Any autotrophic organism that could take advantage of the shallow and deep portions of the environment would be able to maintain a high level of productivity. Unfortunately, these two realms are often so far apart that they cannot be traversed by a single organism. However, we have already seen one situation in the deep sea where conditions exist to allow production to occur, namely in the vicinity of hydrothermal vents. How are conditions there appropriate for reducing and oxidizing reactions, and how do organisms take advantage of this unusual circumstance?

The white sulfur bacterium *Thioploca* forms dense, thick mats on the surface of sediments at 40–280 m water depth off the coast of Peru and Chile. The habitat of *Thioploca* occurs within an upwelling zone that supports a rich fishery, but the water chemistry there is unusual in that little oxygen exists in the deep waters. The sediments themselves are oxygen free because heterotrophic bacteria scavenge all the available oxygen. Most cannot use other sources of oxidizing potential, yet primary production of new organic material in this environment proceeds relatively quickly.

The sediments contain relatively high concentrations of sulfate (SO_4^{2-}), a good oxidizing agent that can be used for respiration by the *Thioploca*. But the *Thioploca* cells have not generated energy. To do this, the bacterium, which is chemoautotrophic, will reoxidize the reduced sulfur back to sulfate using nitrate (NO_3^-) which is abundant in the water lying over the sediments. How does the bacterium get at the nitrate, which is restricted to the water column above the sediment? *Thioploca* cells form filaments encased within tubular sheaths, which extend from the water about 5–10 cm into the sediment. The bacterial filaments glide up and down slowly in the sheaths, shuttling between the nitrate-rich water and sediment environments. At the top of the sheath the bacteria take up nitrate, which is stored at high concentrations in a large vacuole inside the center of the bacteria. Armed with a supply of nitrate, *Thioploca* slides down the sheath into the sediment, where it converts H_2S to elemental sulfur, which forms globules of pure sulfur in the bacterial cytoplasm, and then further oxidizes the sulfur to sulfate. When the nitrate supply runs low, the filament slides back up to the water to restock. The energy gained from oxidizing sulfide is used to convert the abundant carbon dioxide in the sediments into organic forms, just as plants do by photosynthesis.

CHECK YOUR KNOWLEDGE

1. What role do deep sea vents play in regenerating nutrients?

2. How is *Thioploca* uniquely adapted to survival in an oxygen free environment?

MORE ON THE WEB 3. The metabolic processes of *Thioploca* and other organisms of deep sea vents have had to adapt to the unusually extreme conditions surrounding the hydrothermal vents. Our knowledge of these conditions and the way that organisms have evolved to cope with deep sea vent environments have been made possible by development of deep sea submersible vehicles and research efforts such as Extreme 2000. Link to Extreme 2000's Web site through *Practicing Ecology on the Web* at *http://www.whfreeman.com/ricklefs*. What is the difference between chemosynthesis and photosynthesis?

4. Why is it important that we understand how organisms survive at depths at which we will never live?

 Suggested Readings

Baskin, Y. 1995. Can iron supplementation make the equatorial Pacific bloom? *BioScience* 45:314–316.

Bertness, M. D. 1992. The ecology of a New England salt marsh. *American Scientist* 80:260–268.

Binkley, D., and D. Richter. 1987. Nutrient cycles and H^+ budgets of forested ecosystems. *Advances in Ecological Research* 16:1–51.

Fossing, H., V. A. Gallardo, B. B. Jørgensen, et al. 1995. Concentration and transport of nitrate by the mat-forming sulphur bacterium *Thioploca*. *Nature* 374 (20 April 1995):713–715.

Gage, J. D., and P. A. Tyler. 1991. *Deep-Sea Biology: A Natural History of Organisms at the Deep-Sea Floor.* Cambridge University Press, Cambridge.

Goulden, M. L., S. C. Wofsy, J. W. Harden, et al. 1998. Sensitivity of boreal forest carbon balance to soil thaw. *Science* 279 (9 January 1998):214–217.

Jordan, C. F. 1982. Amazon rain forests. *American Scientist* 70:394–401.

Libes, S. M. 1992. *An Introduction to Marine Biogeochemistry.* Wiley, New York.

Likens, G. E., C. T. Driscoll, and D. C. Buso. 1996. Long-term effects of acid rain: Response and recovery of a forest ecosystem. *Science* 272 (12 April 1996):244–246.

Mann, K. H., and J. R. N. Lazier. 1991. *Dynamics of Marine Ecosystems: Biological-Physical Interactions in the Oceans.* Blackwell Scientific Publications, Boston.

Martin, J. H., K. H. Coale, K. S. Johnson, et al. 1994. Testing the iron hypothesis in ecosystems of the equatorial Pacific Ocean. *Nature* 371 (8 September 1994):123–129.

McLusky, D. S. 1989. *The Estuarine Ecosystem.* 2d ed. Chapman & Hall, New York.

Richards, B. N. 1987. *The Microbiology of Terrestrial Ecosystems.* Wiley, New York.

Stevenson, F. J. 1986. *Cycles of Soil: Carbon, Nitrogen, Phosphorus, Sulfur, Micronutrients.* Wiley, New York.

Sullivan, C. W., K. R. Arrigo, C. R. McClain, J. C. Comiso, and J. Firestone. 1993. Distributions of phytoplankton blooms in the Southern Ocean. *Science* 262 (17 December 1993):1832–1837.

Tréguer, P., D. M. Nelson, A. J. Van Bennekom, et al. 1995. The silica balance in the world ocean: A reestimate. *Science* 268 (21 April 1995):375–379.

Tunnicliffe, V. 1992. Hydrothermal-vent communities of the deep sea. *American Scientist* 80:336–349.

Van Cleve, K., F. S. Chapin III, C. T. Dyrness, and L. A. Viereck. 1991. Element cycling in taiga forest: State-factor control. *BioScience* 41:78–83.

CHAPTER 9

Adaptation to Life in Varying Environments

Adaptation results from natural selection on traits that affect evolutionary fitness

The phenotype is the expression of the genotype in the form and function of the individual organism

Each type of organism has an activity space defined by conditions of the environment

Organisms can select microhabitats

Acclimation is a reversible change in structure in response to environmental change

Developmental responses are irreversible changes in response to persistent variation in the environment

Migration, storage, and dormancy enable organisms to survive extreme conditions

Animals forage in a manner that maximizes their fitness

The Mojave Desert of southern California has a climate with little rain, searing summer heat, and chilling winter cold. These conditions are so forbidding that, except for a few struggling plants, the desert appears nearly devoid of life for most of the year. But the desert's silence is occasionally broken during the milder days of winter by swarms of insects and other creatures that appear on the surface or fly above it for a few hours, and then disappear as mysteriously as they came. One of the more conspicuous of these creatures is the giant red velvet mite (▌Figure 9.1).

Several decades ago, biologists Lloyd Tevis and Irwin Newell began a study of the behavior of the giant red velvet mite in relation to the physical conditions of its environment. They found that the mites spend most of the year in burrows dug in the sand. The particular conditions that favor the emergence of mites occur infrequently in the Mojave Desert. During four years of observation, adults appeared aboveground only ten times, always during the cooler months of December, January, or February, when they can tolerate the temperatures on the desert's surface. An individual mite appeared only once each year. Tevis and Newell could predict from their observations that an emergence would occur on the first sunny day after a rain of more than 8 millimeters, provided that air temperatures were moderate. On the day of a major emergence, the mites came out of their burrows between 9:00 and 10:00 A.M., and by late morning one could find thousands of mites scurrying across the desert sands in all directions. At midday, between 11:30 and 12:30, the mites dug back into the sand, not to emerge again until the following year.

▌Figure 9.1 The giant red velvet mite lives in a stressful desert environment. An adult mite is shown on the surface of the ground close to its burrow, within which it spends all but a few hours of its life. Photo by P. Ward/Bruce Coleman.

During its 2- to 3-hour stay above ground each year, each mite must perform two important functions: feeding and mating. The mites feed on termites, which appear on the same day the mites emerge, flying in large swarms over the desert sand, their own emergence presumably triggered by the same physical cues that urge the mites to leave their burrows. Because mites cannot fly, they can feed only after the termites have dropped to the ground and shed their wings, but before they have burrowed into the sand to form new colonies. If a mite feeds successfully during this narrow window of opportunity, it soon mates and prepares to re-enter the sand itself.

About midday, after the mites have fed and mated, they congregate in troughs on the windward sides of sand dunes, where surface temperature and the size of the sand particles are "just right" (less than half a millimeter in diameter). Here they re-enter the sand almost simultaneously. The mites continue digging their new burrows until the coolness of the late winter afternoon slows their activity. Burrowing continues on subsequent days when the sand becomes warm enough, until the burrows are completed. During the rest of the year, the adult mite spends its time moving up and down in its burrow to follow the movement of its preferred temperature zone as the surface of the sand heats and cools each day (▌Figure 9.2).

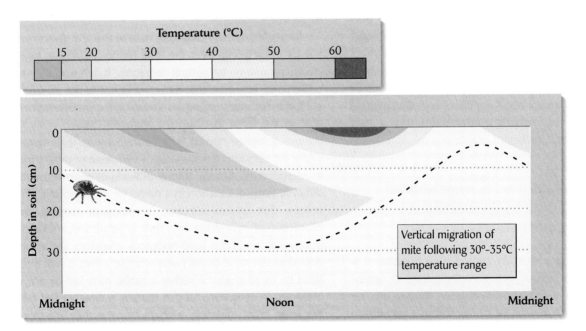

▌Figure 9.2 The giant red velvet mite must adjust to changing environmental conditions. The vertical migration of a mite in its burrow follows changes in soil temperature throughout the course of a typical summer day.

The red velvet mite's world is extremely variable in time and space. Rain comes sporadically. Desert temperatures vary between day and night extremes almost as much as between summer and winter. The mite's world is so forbidding that an individual can be active in only a small portion of its environment, or for only a very limited amount of time. Indeed, all organisms—except, perhaps, for those living at great depths in the seas and in the farthest reaches of caves—must cope with a varied and constantly changing environment. Organisms that can adjust to these changes have the best possible chance of surviving and producing offspring for the next generation.

Each response of an organism to a change in its environment affects the number of descendants it leaves in future populations. Individuals that make the "wrong" response are more likely to die or fail to reproduce than those that respond appropriately. Of course, what is "right" or "wrong" depends on the qualities of the organism and its particular ecological circumstances. For example, whether a sparrow should store fat during times of food abundance depends on whether it is likely to need energy reserves in the near future for a long-distance migration, or as insurance to carry it through a spell of bad weather. In the absence of such need, extra fat is disadvantageous because it reduces speed and maneuverability and increases risk of predation.

In the course of this chapter, we shall learn how different kinds of environmental variation demand different adaptations of individuals. Some kinds of variation occur over space, in which case an organism can make choices about where to live. Other kinds of variation occur over time and are unavoidable; each individual, or its lineage, must be able to survive all the extremes of the environment to persist. Because most of the traits organisms possess have evolved in response to the particular environments in which they live, we shall begin our discussion of adaptation to life in varied and varying environments with a brief explanation of some important aspects of evolution.

Adaptation results from natural selection on traits that affect evolutionary fitness

Each individual in a sexually reproducing population is endowed with a unique genetic constitution, or **genotype,** made up of a combination of genes from its mother and its father. Such genetic variability within a population has many consequences, the most important of which for the study of ecology is evolution by natural selection. The term **evolution** pertains to any change in the genetic makeup of a population. When genetic factors cause differences in fecundity and survival among individuals, evolutionary change comes about through **natural selection.** Individuals whose attributes enable them to achieve higher rates of reproduction leave more descendants, and therefore the genes responsible for these attributes increase in the population. The reproductive success of an individual is referred to as its evolutionary **fitness.**

Consider how these principles apply in the following example of evolutionary change in a California citrus pest. Early in the twentieth century, certain species of scale insects were serious pests in citrus orchards in southern California. An effective means of controlling scale populations was to fumigate orchards with cyanide gas. However, after several years of such treatment, the gas killed fewer of the insects, and before long the scale regained its pest status. Researchers determined that scale insects had evolved a genetically based resistance to cyanide poisoning. Furthermore, when they surveyed orchards in areas that had never been fumigated, they found that small numbers of individuals possessed an innate resistance to cyanide. Thus, despite their initial successes, fumigation programs in the end had favored reproduction by cyanide-resistant individuals, whose progeny then increased to epidemic proportions (Figure 9.3). The citrus scale story illustrates the three main ingredients of evolution by natural selection: (1) variation among individuals, (2) inheritance of that variation (the genetic basis of evolution), and (3) differences in reproductive success, or fitness, related to genetic variation.

Most evolutionary biologists believe that the diversification of living beings over the long history of life has been guided primarily by natural selection. It is important to understand, however, that natural selection is not an external force that urges organisms toward some predetermined goal, in the sense that humans artificially "select" cows to achieve a higher rate of milk production in their herds. Quite the opposite. Selection occurs because of differences in reproductive success among individuals endowed with different form or function in a particular environment. The process that creates selection is ecological—namely, the interaction of individuals with their environment, including its physical conditions, food resources, predators, and so on. A cold winter wind doesn't care whether a bird is well insulated by its plumage. Whether a rabbit runs fast or not is irrelevant to evolution. All that matters is whether fast rabbits leave more offspring, perhaps because they are more likely to escape foxes. One presumes that a fox would prefer to chase slow rabbits, but, alas, by catching slow ones, it ends up favoring reproduction by faster ones.

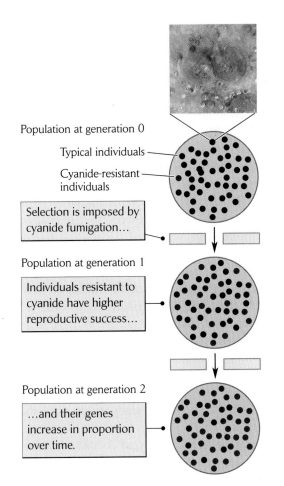

Population at generation 0

Typical individuals —

Cyanide-resistant — individuals

Selection is imposed by cyanide fumigation...

Population at generation 1

Individuals resistant to cyanide have higher reproductive success...

Population at generation 2

...and their genes increase in proportion over time.

▍**Figure 9.3 Evolutionary change in a population may result from a change in the environment.** Genes that confer resistance to cyanide are present at low frequencies in populations of scale insects that have never been exposed to cyanide, simply because of recurrent mutations. In the absence of cyanide, the trait may actually be mildly harmful. When populations are fumigated on a regular basis, however, the gene for cyanide resistance confers high fitness, and its frequency in populations rapidly rises. Photo by Jack Kelly Clark, courtesy of the University of California Statewide IPM Project.

The phenotype is the expression of the genotype in the form and function of the individual organism

Each individual's genotype includes all of its genes. The outward expression of its genotype, called the **phenotype,** is its structure and function. Thus, a genotype is a set of genetic instructions, and a phenotype is the rendering, or expression, of a genotype in the form of an organism. Of course, the environment also influences this rendering. To

put it another way, the genotype is to the phenotype as blueprints are to the structure of a building. In this analogy, the effects of environmental influences are like details in a blueprint that are left up to the discretion of the building contractor, which may hinge, for example, on unpredictable changes in the availability of certain construction materials.

Most genes encode a particular protein, which may be used as part of an organism's structure or may function as an enzyme or hormone. Different forms of a particular gene are referred to as **alleles.** In many cases, alleles create perceptible and measurable differences in an organism's phenotype. For example, blue-eyed and brown-eyed humans have different alleles of a single gene, which controls one of the pigment systems that determines eye color. Many genetic disorders, such as sickle-cell anemia, Tay-Sachs disease, cystic fibrosis, and albinism, as well as tendencies to develop certain cancers and Alzheimer's disease, are caused by defective alleles of individual genes.

Every individual has two copies of each gene, one inherited from its mother and one from its father (exceptions include sex-linked genes and organisms that reproduce without the sexual union of gametes). An individual that has two different alleles of a particular gene is said to be **heterozygous** for that gene. When both copies of a gene are the same, that individual is **homozygous.** When an individual is heterozygous, the two different alleles may produce an intermediate phenotype, or one may mask the expression of the other. In the latter case, one allele is said to be **dominant** and the other **recessive.** When heterozygotes have an intermediate phenotype, the alleles are said to be **codominant.** Most harmful alleles are recessive, and the normal gene product of the dominant allele masks the defective function of their gene products in heterozygous individuals.

While all phenotypic traits have a genetic basis, they are also influenced by variations in the environment, either through the effects of environmental conditions on individuals (as in the effect of food supply on growth and development) or through the responses of individuals to variation in their environments. Such environmentally induced variation in the phenotype is referred to as **phenotypic plasticity.** The capacity of an individual to exhibit different responses to its environment may itself be an evolved trait. That is, the way in which the individual responds to environmental variation is also subject to evolution by natural selection. We shall look at phenotypic plasticity in more detail in the next chapter, but let us keep in mind the difference between these plastic responses by individuals and evolutionary responses by populations as we consider the relationship of organisms to their environments.

Each type of organism has an activity space defined by conditions of the environment

Each organism functions best within a limited range of conditions, which we may refer to as its **activity space.** This concept applies to all aspects of an individual's life, whether it is literally active or not; here we may think of "activity" as synonymous with "performance." For some environmental factors, the activity of individual organisms tends to be highest within a relatively narrow range (■ Figure 9.4). Activity might be measured as rate of photosynthesis, survival, or swimming speed, all of which influence an individual's reproductive success in some way. The environmental factor might be temperature, soil acidity, nutritional quality of food items, or structure of the foraging substrate. Away from the optimum conditions, activity decreases, and consequently so does the individual's probability of surviving and ability to produce offspring. Close to the optimum conditions, reproductive success is high enough to maintain a population. Under marginal conditions, an individual might be able to maintain itself indefinitely, but not replace itself in future populations. Extreme conditions are unsuitable for individual maintenance, and an individual can venture into such conditions only for short periods.

Organisms can select microhabitats

Plants have relatively little choice as to where they live, even though, as we have seen, roots can "forage" for high concentrations of soil minerals, growing shoots can seek out light gaps, and wind, water, and animal dispersers may distribute plant seeds nonrandomly through the environment. Unlike plants, most animals have freedom to move about the environment and choose a habitat in which to live. Nonetheless, even within a habitat, there are distinct differences in temperature, moisture, salinity, and other factors. Parts of the environment that can be distinguished by their conditions are referred to as **microhabitats** or **microenvironments.** In deserts, for example, the shaded ground under a shrub is often cooler and moister than surrounding areas exposed to direct sunlight, although clearly these conditions vary through the course of the daily cycle and with the seasons.

Responses of animals to the changing array of microhabitats in their environments can be illustrated by the diurnal behavioral cycles of lizards. Although lizards do not regulate their body temperatures by generating heat metabolically, they do take advantage of solar radiation and warm surfaces to maintain their temperatures within a suitable range during the day. Thus, it is not surprising that lizards respond to the temperatures of different microhabitats. At night, external sources of heat disappear, and the lizard's body temperature gradually drops to that of the surrounding air.

The desert iguana (*Dipsosaurus dorsalis*) of the southwestern United States lives in a severe environment. Shade temperatures can reach 45°C in summer and plunge below freezing in winter. Desert iguanas have a preferred body temperature range of 39°–43°C. During mid-July, the thermal environment changes rapidly between day and night extremes. Desert iguanas can move about the desert surface in search of food and remain within their preferred range for only about 45 minutes in mid-morning and a similar period in the early evening (■ Figure 9.5). During the remainder of the day, they seek the shade of plants or the coolness of their burrows, where temperatures rarely exceed their preferred range. At night,

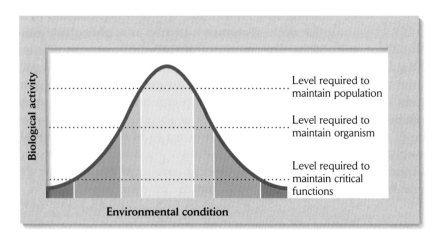

■**Figure 9.4 Biological activity is related to environmental conditions.** For some environmental factors, the activity of individual organisms is sufficient to maintain a population only within a narrow intermediate range. Organisms can maintain themselves for long periods over a broader range of conditions and briefly over a yet broader range.

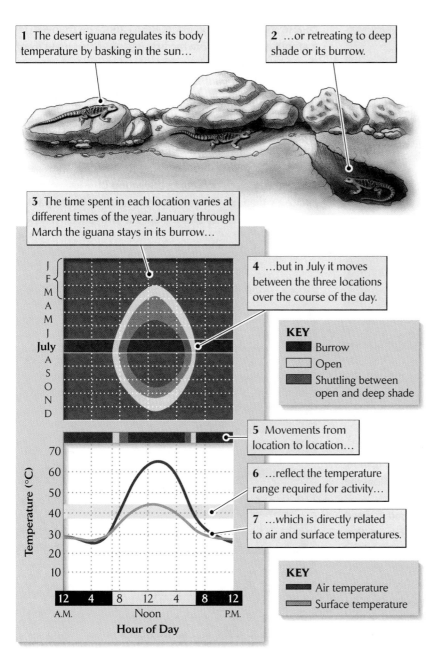

1 The desert iguana regulates its body temperature by basking in the sun...

2 ...or retreating to deep shade or its burrow.

3 The time spent in each location varies at different times of the year. January through March the iguana stays in its burrow...

4 ...but in July it moves between the three locations over the course of the day.

KEY
■ Burrow
□ Open
▨ Shuttling between open and deep shade

5 Movements from location to location...

6 ...reflect the temperature range required for activity...

7 ...which is directly related to air and surface temperatures.

KEY
■ Air temperature
▨ Surface temperature

Figure 9.5 The desert iguana regulates its body temperature by microhabitat selection. The activity space of the desert iguana (*Dipsosaurus dorsalis*) within its habitat in southern California is shown over the entire year and over a summer day (July 15). After W. A. Beckman, J. W. Mitchell, and W. P. Porter, *J. Heat Transfer* (May 1973):257–262.

desert iguanas retreat to the safety of their burrows. If an iguana were to remain aboveground in the cool evening air, its body temperature would drop rapidly and it would become too sluggish to escape predators.

Winter cold restricts *Dipsosaurus* to brief periods of activity in the middle of the day, when air temperatures rise to the point at which individuals can come aboveground and forage. Between early December and the end of February, most days are so cold that desert iguanas cannot even venture from their burrows. Spring offers more favorable temperatures for iguanas. In May, individuals forage actively on the ground surface for a businesslike 9:00 A.M. to 5:00 P.M., only occasionally seeking the cool shade of plants.

ECOLOGISTS IN THE FIELD

Temperature and microhabitat selection by the cactus wren

Unlike the desert iguana, the cactus wren (■ Figure 9.6), an insectivorous bird that lives in deserts of the southwestern United States and northern Mexico, maintains a constant body temperature. However, because the wren has no source of free water, it must avoid gaining too much heat from its environment. Otherwise, it would have to dissipate excess body heat by the cooling effect of evaporation from its respiratory tract (see Chapter 2).

Figure 9.6 The cactus wren (*Campylorhynchus brunneicapillus*) is a conspicuous resident of deserts in the southwestern United States and northern Mexico. Photo by Craig K. Lorenz/Photo Researchers.

course, its position and orientation cannot be changed. For a month and a half, from the first egg until the young fly off, the nest must provide a suitable environment day and night, in hot and cool weather.

During the long breeding period (March through September) in southern Arizona, cactus wrens usually rear several broods of young. Early in spring, they build their nests so that the entrances face away from the direction of the cold winds; during the hot summer months, they orient their nests to face prevailing afternoon breezes, which circulate air through the nest chamber and facilitate heat loss (**Figure 9.8**). This strategy makes a difference! Nests ori-

Thus, the wren's activity space, like that of the desert iguana, reflects changes in environmental conditions throughout the day and season.

Observations made by Robert E. Ricklefs and F. Reed Hainsworth in deserts near Tucson, Arizona, showed that cactus wrens seek favorable microhabitats within which to feed as the thermal environment changes throughout the day. During cool early mornings, wrens forage throughout most of the environment, searching for food among foliage and on the ground. As the day brings warmer temperatures, wrens select cooler parts of their habitat, particularly the shade of small trees and large shrubs, always managing to avoid feeding where the temperature exceeds 35°C (**Figure 9.7**). When the minimum temperature in the environment rises above 35°C, at which point birds must use evaporative cooling to maintain their body temperatures even when inactive, the wrens stop feeding and perch quietly in deep shade.

Although an adult cactus wren can move without restraint to any part of its habitat, its nest is fixed in place: wren chicks cannot move among microhabitats until they are old enough to leave the nest. The microenvironment of the nest must therefore be within the tolerance range of chicks at all times. Cactus wrens appear to achieve this both by choosing particular nest sites and by orienting their nests in particular directions. Cactus wrens build untidy, enclosed nests—bulky, somewhat haphazardly constructed balls of grass—with side entrances. Once a nest is built, of

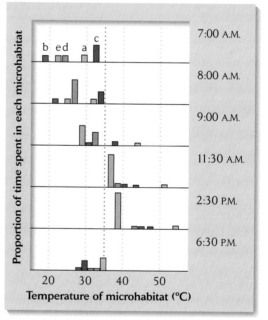

Figure 9.7 Temperature affects microhabitat use by cactus wrens. Microhabitat use is shown over the course of a day in late spring. Microhabitats vary in degree of thermal stress between exposed ground (a) and the deep shade of trees (e). From R. E. Ricklefs and F. R. Hainsworth, *Ecology* 49:227–233 (1968). Photo by R. E. Ricklefs.

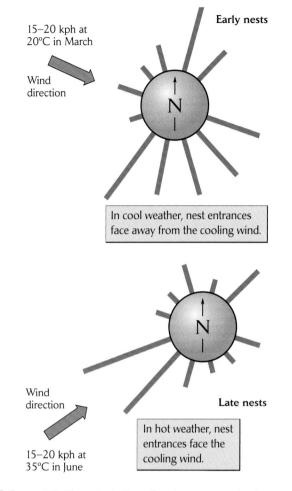

15–20 kph at 20°C in March

Wind direction

Early nests

N

In cool weather, nest entrances face away from the cooling wind.

Wind direction

Late nests

N

In hot weather, nest entrances face the cooling wind.

15–20 kph at 35°C in June

▌**Figure 9.8 The orientation of cactus wren nest entrances changes during the breeding season.** Lengths of bars represent relative numbers of nests with each orientation. After R. E. Ricklefs and F. R. Hainsworth, *Condor* 71:32–37 (1969).

ented properly for the season are consistently more successful (82% produce viable offspring) than nests facing in the wrong direction (only 45% are successful).

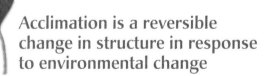

Acclimation is a reversible change in structure in response to environmental change

Growing thicker fur in winter, producing smaller leaves during the dry season, increasing the number of red cells in the blood at high altitude, and producing enzymes with different temperature optima or lipids that remain fluid at different temperatures are all forms of **acclimation.** Acclimation may be thought of as a shift in the range of physiological tolerances of the individual. Because these changes involve modifications of the body's structure and metabolic machinery, they require days to weeks. Thus, acclima-

tion is a strategy restricted to seasonal and other persistent variations in conditions. Acclimation is reversible, and allows organisms to follow the ups and downs of their environments. As long as the environmental change is persistent, it is a good strategy. However, increased tolerance of one extreme often brings reduced tolerance of the other.

By producing enzymes and other molecules having different temperature optima, a cold-blooded (poikilothermic) animal can adjust its activity space in response to prevailing environmental conditions. The relationship between the swimming speed of goldfish and water temperature shows both the advantages and the limitations of acclimation. Goldfish swim most rapidly when acclimated to 25°C and placed in water between 25° and 30°C, conditions that closely resemble those of their natural habitat (▌Figure 9.9). Lowering the acclimation temperature to 5°C increases the swimming speed at 15°C, but reduces it at 25°C.

An organism's capacity for acclimation often reflects the range of conditions experienced in its natural environment. *Larrea divaricata* (creosote bush) inhabits interior deserts in western North America and maintains photosynthetic activity during the cool winters as well as the hot summers. Measurements of the rate of photosynthesis in this plant show a shift in the temperature optimum characteristic of thermal acclimation. Specifically, photosynthetic rate reaches the same level in plants grown at 20°C and 45°C, but plants grown at 20°C do not perform as well at 45°C as plants acclimated to that temperature. The basis for this acclimation seems to be changes in the viscosity of membranes directly related to photosynthetic pathways.

Where the environment normally is relatively constant, we would not expect organisms to have evolved the ability to respond strongly to environmental variation or to tolerate conditions that differ from the norm. Evolution favors

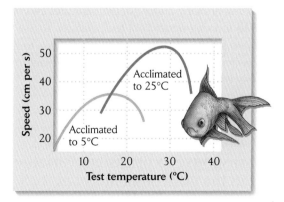

▌**Figure 9.9 Acclimation can shift an organism's activity space in response to environmental conditions.** Swimming speed as a function of temperature is shown for goldfish acclimated to 5°C and to 25°C. After F. E. J. Fry and J. S. Hart. *J. Fish. Res. Bd. Can.* 7:169–174 (1948).

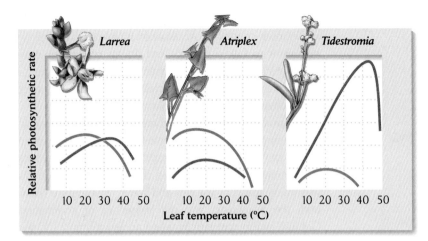

Figure 9.10 The capacity for acclimation may reflect the range of conditions in the environment. Photosynthetic rate as a function of leaf temperature is shown for three species of plants (genera *Larrea*, *Atriplex*, and *Tidestromia*) grown under moderate (blue line) and hot temperatures (red line). From P. W. Hochachka and G. N. Somero, *Biochemical Adaptation*, Princeton University Press, Princeton, NJ (1984); after O. Bjorkman, M. R. Badger, and P. A. Arnold, in N. C. Turner and P. J. Kramer (eds.), *Adaptation of Plants to Water and High Temperature Stress*, Wiley, New York (1980), pp. 231–249.

economical designs, and we presume that the capacity to respond to environmental change imposes a cost for the organism. That these mechanisms have been dispensed with when plants experience only narrow ranges of temperatures is shown by the photosynthetic rates of two other plants from western North America. *Atriplex glabriuscula* is a species of saltbush native to cool coastal regions of California, where temperatures during the growing season rarely exceed 20°C. Unlike *Larrea*, *Atriplex* does not increase its photosynthetic rate at high temperatures when acclimated to 40°C, although it may respond in other ways. However, whatever physiological changes that do occur during acclimation to high temperature cause saltbush plants to perform less well at lower temperatures (**Figure 9.10**). In contrast, the thermophilic (heat-loving) species *Tidestromia oblongifolia* cannot acclimate to cool temperatures. Photosynthesis is reduced uniformly over a wide range of leaf temperatures from 10°C to 40°C when plants are maintained for long periods in cool temperatures. The responses of *Atriplex* to growing under hot temperatures and of *Tidestromia* to growing under cool temperatures appear to be generalized stress responses that allow individuals to survive under extreme conditions, rather than mechanisms that effectively broaden their activity spaces.

Developmental responses are irreversible changes in response to persistent variation in the environment

Light intensity, among many other factors, influences the course of development in plants. Loblolly pine seedlings grown in shade have smaller root systems and more foliage than seedlings grown in full sunlight. Because a shaded environment taxes a plant's water economy less, shade-grown seedlings can allocate more of their production to stem and needles; sun-grown seedlings develop more extensive root systems to obtain sufficient water. The larger proportion of foliage in a shade-grown seedling results in a higher rate of photosynthesis per unit of plant mass under given light conditions, particularly under low light intensities (**Figure 9.11**). These growth responses of pine seedlings show how plants allocate their production in such a

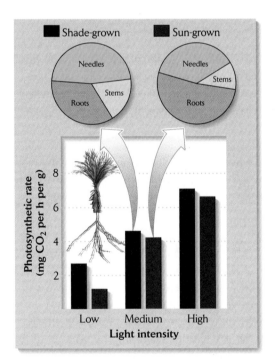

Figure 9.11 Plants show developmental responses to light intensity. Distribution of dry matter and rates of photosynthesis are shown for loblolly pine (*Pinus taeda*) seedlings grown under shade and in full sunlight. After F. H. Bormann, in D. V. Thimann (ed.), *The Physiology of Forest Trees*, Ronald Press, New York (1958), pp. 197–215.

way as to acquire more of the resource that most limits their growth.

Another striking example of a developmental response is the coloration of several species of locusts and grasshoppers. It is important for the color of such insects to match the color of their backgrounds if they are to avoid detection by predators that use sight to locate prey. In tropical habitats with seasonal precipitation, the onset of the wet season stimulates the growth of lush, green vegetation. During the early part of the dry season this vegetation browns and dies, often exposing red-brown earth. As the seasonal drought intensifies, natural fires and those set by humans blacken the ground over vast areas. Consequently, there is a regular seasonal progression of color from green to brown to black and back to green again. Where this happens, many species of grasshoppers match the background coloration of the environment in which they develop (Figure 9.12).

The epidermis of the African grasshopper *Gastrimargus africanus* has a pigment system that permits any given area of skin to be either green or brown; both colors may occur on a single animal, but not in the same area of the body. The green and brown colors represent small biochemical variations on a single pigment molecule. In combination with brown, additional pigments may produce colors ranging from yellow through orange and red to black. Furthermore, black pigment (melanin) may be deposited in the cuticle that covers the epidermis. Between developmental stages a grasshopper sheds its epidermis, discarding its pattern of camouflaging coloration. A new layer of epidermis develops underneath, and thus a young grasshopper can change its color with each molt if the background color of its environment has changed in the meantime. Coloration in *Gastrimargus* responds to environmental conditions that are correlated with the color of its background, particularly quality and intensity of light, which are perceived by the eye and transmitted to the epidermis by hormones produced in the brain.

Developmental responses generally do not reverse themselves; once fixed during development, they remain unchanged for the rest of an individual's life (or particular developmental stage). Because of their long response times and irreversibility, developmental responses cannot accommodate short-term environmental changes. As a rule, therefore, only plants and animals in environments with persistent variation in the conditions experienced by different individuals exhibit developmental responses. Such organisms include plant species, such as loblolly pines, whose seeds may settle in many different kinds of habitats. In such cases, spatial rather than temporal heterogeneity in the environment may create the kind of persistent environmental variation that favors developmental

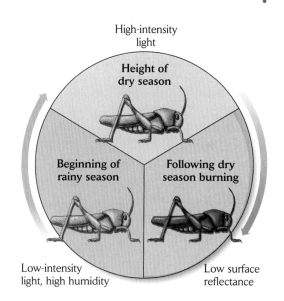

Figure 9.12 Developmental responses can match organisms to environmental conditions. The epidermal coloration of the grasshopper *Gastrimargus africanus* responded to laboratory conditions designed to mimic light and humidity experienced during the wet season (green), the dry season (brown), and following burning (black). After C. H. Fraser Rowell, *Anti-locust Bull.* 47:1–48 (1970).

responses. Observed developmental responses generally match environmental change rather well, because individuals that respond inappropriately do not survive to produce descendants.

MORE ON THE WEB | *Rate of phenotypic response.* **The mechanisms that organisms use to respond to the environment, as shown by the example of wing-length polymorphism in water striders, must match the pattern of environmental change.**

Migration, storage, and dormancy enable organisms to survive extreme conditions

In many parts of the world, extremes of temperature, drought, darkness, and other adverse conditions are so severe that individuals cannot change enough to maintain their normal activities, or if they could, the change would not be worth the cost. Under such conditions, organisms resort to a number of extreme responses. These responses include **migration,** moving to another region where conditions are more suitable; **storage,** relying on resources accumulated under more favorable conditions; and **dormancy,** becoming inactive.

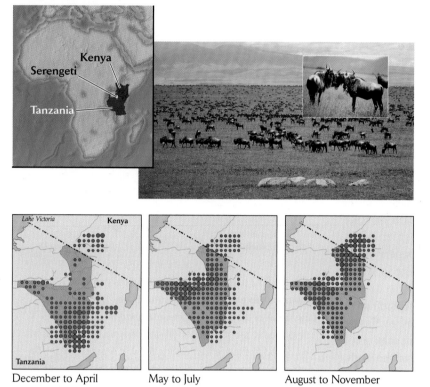

Serengeti National Park, Africa

Kenya

Serengeti

Tanzania

Lake Victoria   Kenya

Tanzania

December to April   May to July   August to November

Figure 9.13 The migration of wildebeests follows their food supply. The distribution of wildebeest populations of the Serengeti ecosystem (shaded area) of northern Tanzania and southern Kenya is shown for three times during the annual cycle during 1969–72. The migrations follow lush growth of grasses following seasonal rains in each area. The size of each dot indicates the relative size of the population in that area. Adapted from L. Pennycuick, in A. R. E. Sinclair and M. Norton-Griffiths (eds.), *Serengeti: Dynamics of an Ecosystem*, University of Chicago Press, Chicago (1979), pp. 65–87. Photos courtesy of A. R. E. Sinclair.

Migration

Many animals, particularly those that fly or swim, undertake extensive migrations. Arctic terns probably hold the record for long-distance migration. Individuals make yearly round trips of 30,000 km between their North Atlantic breeding grounds and Antarctic wintering grounds (where it is the local summer). Each fall hundreds of species of land birds leave temperate and arctic North America, Europe, and Asia for the south in anticipation of cold winter weather and dwindling supplies of their invertebrate food. Populations of monarch butterflies migrate between wintering areas in the southern United States and Mexico to summer breeding areas far to the north into southern Canada. In East Africa, many large ungulates, such as wildebeests, migrate long distances, following the geographic pattern of seasonal rainfall and fresh vegetation (**Figure 9.13**).

Some migratory movements occur in response to occasional failure or depletion of local food supplies, which forces individuals to move out of an area in search of new feeding places. Such movements are perhaps best known from outbreaks of migratory locusts. These migrations occur when locusts leave areas of high local density where food has been depleted. They can reach immense proportions and cause extensive crop damage over wide areas (**Figure 9.14**). Irruptive behavior in locusts is a develop-

mental response to population density. When locusts occur in sparse populations, they become solitary and sedentary as adults. In dense populations, however, frequent contact with other locusts stimulates young individuals to develop gregarious, highly mobile behavior, which can develop into a mass migration.

Storage

Where environmental changes plunge organisms from feast into famine and migration is not a possibility, storage of resources acquired during periods of abundance for use in times of scarcity may be a way to cope. During infrequent rainy periods, desert cacti swell with water stored in their succulent stems. Plants growing on infertile soils absorb, in times of abundance, more nutrients than they require, and use them when soil nutrients are depleted. In habitats that frequently burn—such as the chaparral of southern California—perennial plants store food reserves in fire-resistant root crowns, which sprout and send up new shoots shortly after a fire has passed (**Figure 9.15**).

Many temperate and arctic animals accumulate fat during mild weather in winter as a reserve of energy for periods when snow and ice make food sources inaccessible. The problem with fat reserves is that heavier animals are often slower and less agile, and therefore are more likely to

Figure 9.14 The migration of locusts is a developmental response to high population density. A dense swarm of migratory locusts moves over Somalia, Africa, in 1962. Courtesy of the U.S. Department of Agriculture.

be caught by predators. One way to avoid this problem is to store food before consuming it. Some winter-active mammals (beavers, squirrels, and pikas) and birds (acorn woodpeckers and jays, for example) cache food supplies underground or under the bark of trees for later retrieval. Often these hoards are immense and may sustain individuals for long periods.

Dormancy

Environments sometimes become so cold, dry, or nutrient-depleted that animals and plants can no longer function normally. In such circumstances, some species that are not capable of migration enter physiologically dormant states. Many tropical and subtropical trees shed their leaves during seasonal periods of drought; many temperate and arctic trees shed theirs in the fall before the onset of winter frost and long nights. Many mammals, such as ground squirrels, **hibernate** (spend winter in a dormant state) because they cannot find food in winter, not because they are physiologically unable to cope with the harsh physical environment.

In most species, environmental conditions requiring dormancy are anticipated by a series of physiological changes in the individual (for example, production of antifreezes, dehydration, and fat storage) that prepare it for a partial or complete shutdown of activity. Before winter, some insects enter a resting state known as **diapause,** in which water is chemically bound or reduced in quantity to prevent freezing and metabolism drops so low that it is barely detectable. Drought-resistant insects that enter a summer diapause dehydrate themselves and tolerate the desiccated condition of their bodies, or secrete an impermeable outer covering to prevent drying. Plant seeds and spores of bacteria and fungi exhibit similar dormancy mechanisms. Indeed, there are many cases of seeds stored in burial chambers or recovered in other archeological settings that have sprouted after hundreds of years of dormancy. By whatever mechanism it

Figure 9.15 Chaparral plants store food reserves in fire-resistant root crowns. In chaparral habitat in southern California, chamise (*Adenostoma fasciculatum*) and other species resprout from root crowns following a fire. These photos were taken September 12, 1979, and April 20, 1980, following a fire near Los Angeles, California. Photos by Tom McHugh/Photo Researchers.

occurs, dormancy reduces exchange between organisms and their environments, enabling animals and plants to "ride out" unfavorable conditions.

Stimuli for change

What stimulus indicates to birds wintering in the Tropics that spring is approaching in northern forests? What urges salmon to leave the seas and migrate upstream to their spawning grounds? How do aquatic invertebrates in the Arctic sense that if they delay entering diapause, a quick freeze may catch them unprepared for winter?

In 1938, J. R. Baker made an important distinction between two kinds of cues that trigger these changes. **Proximate factors** are cues, such as day length, by which organisms can assess the state of the environment but that do not directly affect its well-being. **Ultimate factors** are features of the environment, such as food supplies, that bear directly on the well-being of the organism. Virtually all plants and animals sense **photoperiod** (the length of the day) as a proximate factor that indicates season, and many can distinguish periods of lengthening and shortening days. Different populations of a single species may differ strikingly in their responses to photoperiod in different locations, reflecting different relationships of environmental changes to day length. Under controlled cycles of light and dark, southern populations (at 30°N) of side oats grama grass flower in autumn, when day length is 13 hours, whereas more northerly populations (at 47°N) flower in summer, only when the light period exceeds 16 hours each day. In Michigan, at 45°N, populations of small freshwater crustaceans known as water fleas (*Daphnia*) form diapausing broods at photoperiods of 12 hours (mid-September) or less. In Alaska, at 71°N, related species enter diapause when the light period decreases to fewer than 20 hours per day, which happens in mid-August. Warm temperatures and low population densities tend to shorten the day length that triggers diapause (and hence delay the inception of diapause in autumn), suggesting that these factors portend more favorable environmental conditions for *Daphnia*.

Animals forage in a manner that maximizes their fitness

Because animals live in varied and variable environments, they are constantly forced to make decisions about how to behave. Many of these decisions concern food: where to forage, how long to feed in a certain patch of habitat, which types of foods to eat, and so on. Theories of **optimal foraging** seek to explain these decisions in terms of the likely costs and benefits of each possible behavior. Animals are expected to select the behavior that gives the greatest benefit. Cost can be measured in terms of time and energy expended, but the benefit is best judged in terms of evolutionary fitness. However, it is often difficult to measure the consequence of a particular behavioral choice for an individual's survival and reproductive success. Consequently, ecologists usually measure benefit in terms of factors that are likely to be correlated with fitness, such as amount of food gathered per unit of time. We shall examine a number of behavioral decisions from the standpoint of such costs and benefits. Each of these cases features some aspect of variation in time or space.

Central place foraging

When birds feed their offspring in a nest, the chicks are tied to a single location, while the parents are free to search for food at a distance. This situation is referred to as **central place foraging.** The greater the foraging range, the greater the amount of food that is potentially available to the parent. But traveling a longer distance also increases the time, energy costs, and risks of travel. Is there some best distance from the nest at which a parent should forage, and how much food should the parent bring to its brood with each trip? That is, how much time should the parent spend gathering food before it returns to its nest?

Studies on the foraging behavior of European starlings allowed investigators to approach these questions from an economic standpoint. During the summer season, starlings typically forage on lawns or pastures for leatherjackets, which are the larvae of tipulid flies (craneflies). Starlings feed by thrusting their bills into the soft turf and spreading the mandibles to expose prey. When they are gathering food for their young, they hold captured leatherjackets at the base of the bill. You can imagine that the more leatherjackets a starling has in its bill, the more difficult it is to capture the next one. For this reason, the time between captures increases as more prey are caught (**Figure 9.16**). That is, as a predator captures more prey, the rate of capture decreases, and the total number of prey increases less steeply. Indeed, a starling cannot continue to feed efficiently with eight leatherjackets in its bill.

Now, from the standpoint of feeding its offspring, the rate at which a parent delivers food to its young is the number of prey caught divided by the length of the for-

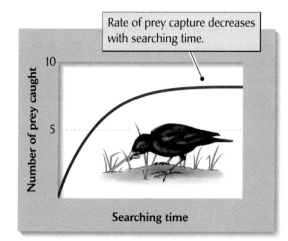

Rate of prey capture decreases with searching time.

■ Figure 9.16 Time between captures increases as more prey are caught. The rate of prey captured by starlings declines as foraging (searching) time increases.

aging trip. The foraging trip includes both the time spent at the foraging site and the time spent traveling between the foraging site and the nest. A starling can maximize

the rate at which it delivers food to its chicks by spending an intermediate amount of time in the feeding area during each trip and bringing back something less than the maximum possible amount of food (**■** Figure 9.17). Imagine yourself in a grocery store where you have to buy as much food as you can in an hour and you have to carry your items by hand. How frequently would you take your items to the cashier? Carrying one item at a time clearly is silly, particularly if there is a long line waiting to be checked out (analogous to a long foraging distance). Trying to carry more than you can handle well, and having to spend time picking up dropped items off the floor and rearranging them, also seem uneconomical. As in the case of the starling with a bill full of leatherjackets, the "law of diminishing returns" sets in. The best strategy is somewhere in the middle. The optimum load varies in direct relation to the traveling time, or more generally, to any fixed cost per trip.

MORE ON THE WEB *Spatially partitioned foraging by oceanic seabirds.* **Albatrosses and other seabirds may intersperse long and short foraging trips to gather food alternately for themselves and their chicks.**

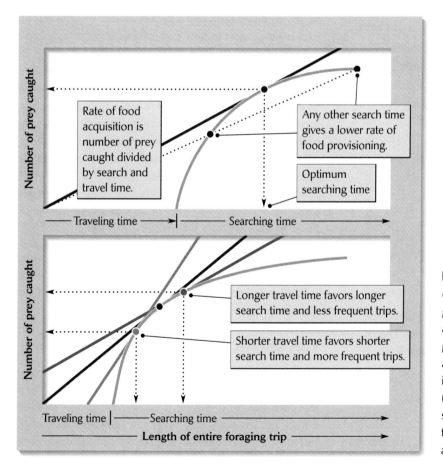

Rate of food acquisition is number of prey caught divided by search and travel time.

Any other search time gives a lower rate of food provisioning.

Optimum searching time

Longer travel time favors longer search time and less frequent trips.

Shorter travel time favors shorter search time and more frequent trips.

■ Figure 9.17 Optimal foraging models can be used to predict behavior. For a given prey accumulation curve (orange line), a line passing through the origin of the graph (the beginning of the foraging trip) and tangent to the prey accumulation curve indicates the maximum rate of capture (number of prey caught per unit of time). As shown in the lower graph, the optimal search time on an individual foraging trip increases as the length of the travel time increases.

ECOLOGISTS IN THE FIELD

Optimal foraging by starlings

To what extent do organisms actually forage optimally? Figure 9.17 is theory. In reality, we have all seen some inefficient shoppers in our local grocery stores. How good are starlings as economists?

This question was addressed in a clever experiment by behavioral ecologist Alex Kacelnik of Oxford University. Instead of letting starlings feed on their natural prey, he trained them to visit feeding tables at which mealworms could be provided through a plastic tube at precisely timed intervals. A starling would arrive at the table, eat the first mealworm, and then wait for the next one to be delivered. Kacelnik adjusted the timing so that each successive mealworm would arrive at a progressively longer interval, mimicking the longer intervals at which a starling would catch leatherjackets as its beak became full. Kacelnik then placed feeding tables at different distances from nests and observed how many mealworms a starling would wait for at different travel times. As expected, starlings increased their load size as travel time increased (∎ Figure 9.18). Kacelnik concluded that starlings are good economists, at least when it comes to gathering food.

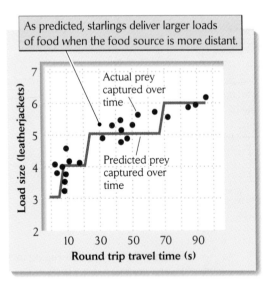

As predicted, starlings deliver larger loads of food when the food source is more distant.

Actual prey captured over time

Predicted prey captured over time

∎ **Figure 9.18 Food loads increase with travel times.** The number of mealworms brought by starlings to their broods increased with the total length of the foraging trip. From A. Kacelnik, *J. Anim. Ecol.* 53:283–299 (1984).

Risk-sensitive foraging

The value of a feeding area depends not only on the rate at which an individual can gather food, but also on the relative safety of the place. Every activity carries a risk of mortality. For many animals, predation is the most significant mortality risk, and the presence of a predator, or even the perceived threat of predation, can reduce the value of an otherwise good foraging place. The extra food is simply not worth the increased risk of becoming food. The predation factor has been incorporated into foraging theory in studies of **risk-sensitive foraging.**

ECOLOGISTS IN THE FIELD

Experiments with risk-sensitive foraging

James F. Gilliam and Douglas F. Fraser demonstrated the principle of risk-sensitive foraging elegantly in a simple experiment with fish. They constructed cages having two compartments and placed these directly in an experimental stream. The subjects were small minnows (juvenile creek chubs), and the predators in the system were adult creek chubs. The minnows were provided with tubifex worms buried in mud placed in small trays in the compartments. A refuge area that permitted passage of the minnows, but not the adult chubs, connected the two compartments.

In the experiment, minnows were presented with a low density of food (0.17 worms per cm^2) and only one predator in one compartment, and a higher density of food but two or three predators in the other compartment. The design of the experiment was to increase the amount of food in the more dangerous part of the cage to determine at what point the minnows would expose themselves to greater risk in order to obtain more food. Chub minnows were very sensitive to predation risk. When the more dangerous side of the cage had two adult chubs, minnows switched to foraging there only after prey density was increased to more than 0.33 worms per cm^2, or twice the level in the less risky side of the cage (∎ Figure 9.19). When there were three predators, the food level had to be more than four times that on the safer side to entice the minnows to switch.

MORE ON THE WEB

Variable food supplies and risk-sensitive foraging. **Would you choose a predictable supply of a lower-quality food or a more variable food supply with a higher average reward?**

Prey choice

Foraging decisions also include choices concerning particular prey items. An actively foraging individual, such as an insectivorous bird flitting among the foliage of trees, encounters in sequence a variety of potential foods of different types. Each type of food has an intrinsic value based on its nutrient and energy content, difficulty of handling, and potential danger from toxins. The cost of selecting a

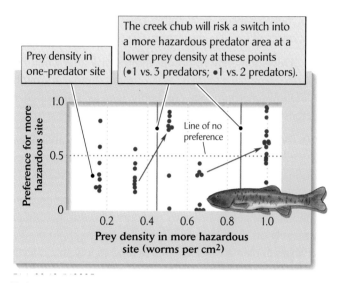

Figure 9.19 Foraging fish are risk-sensitive. Chub minnows switch to a more hazardous feeding site only when the prey density in the hazardous site exceeds a certain critical level. The switch point increases with the relative risk of feeding. After J. F. Gilliam and D. F. Fraser, *Ecology* 68:1856–1862 (1987).

poor food item is that it requires time to handle (capture and ingest), during which the forager cannot be looking for other food items. Thus, taking the poorer item may mean that the forager misses an opportunity for a better item. Low-quality food items may also take more time to digest per unit of nutrient or energy obtained, thus reducing an individual's overall feeding rate.

MORE ON THE WEB *Optimal prey choice in the great tit.* How poor must an alternative type of prey be for a consumer to pass it up?

Diet mixing

One reason that some foragers consume a varied diet is that one or a few food items might not provide all necessary nutrients, but these might be present in other food items. Different food types are **complementary** when each contains a required nutrient missing in the other. Humans, for example, can subsist on a diet of rice and beans, but not on either of these alone, because rice and beans each contain essential amino acids missing in the other. The principle of complementarity also applies when foods contain small amounts of different toxins that individually would be dangerous in large doses, but are relatively harmless in the smaller doses ingested with a mixed diet.

The benefits of diet mixing were demonstrated by Elizabeth Bernays and her colleagues at the University of Arizona using nymphs (immature stages) of the grasshopper *Schistocerca americana*. Grasshopper nymphs grew faster when fed a mixture of kale, cotton, and basil than when they were offered any one of these food plants alone (**Figure 9.20**) The effect was even more pronounced on lower-quality, natural food plants, such as mesquite and mulberry: nymphs with mixed diets grew almost twice as fast as those feeding on either one of these plant species alone. Similar results were obtained on artificial diets that were low in either protein or carbohydrates, both of which are required for proper growth. Grasshoppers on mixed diets grew more rapidly than those provided either of the lower-quality foods alone.

Experiments with birds feeding on fruits in the fall in the Morton Arboretum, Chicago, also demonstrated diet mixing. Fruits of two species of shrubs were presented together on artificial "bushes" against a natural background of either one or the other shrub species—that is, with many

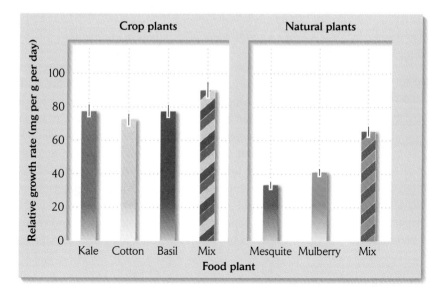

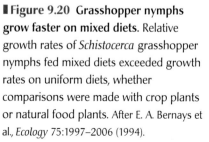

Figure 9.20 Grasshopper nymphs grow faster on mixed diets. Relative growth rates of *Schistocerca* grasshopper nymphs fed mixed diets exceeded growth rates on uniform diets, whether comparisons were made with crop plants or natural food plants. After E. A. Bernays et al., *Ecology* 75:1997–2006 (1994).

shrubs of the background species in the immediate vicinity—and fruit consumption was then recorded. The results support the hypothesis of complementarity in that birds selected the fruit that contrasted with the background fruit supply. For example, rough-leaved dogwood (*Cornus drummondi*) and pokeweed (*Phytolacca americana*) are similarly sized (7.4 and 8.9 mm) fruits with high lipid/low sugar and low lipid/high sugar contents, respectively. In a paired choice test, in an area with abundant natural dogwood, 3.2% of dogwood and 29% of pokeweed fruits were consumed; in an area with abundant natural pokeweed, 97% of dogwood and 71% of pokeweed fruits were consumed. In other words, the birds showed a preference for the less abundant alternative.

 Summary

1. Most of the traits of organisms have evolved in response to their environments, including variation in environmental conditions and resources. For this reason, an understanding of evolution is essential to interpreting adaptations to varying environments.

2. Evolution by natural selection occurs when genetic factors influence survival and reproductive success. The genetic characteristics of those individuals that achieve the highest reproductive success increase in the population with time.

3. The genotype includes all the genetic factors that determine the structure and functioning (which together constitute the phenotype) of an individual. Many genetic factors have unique, measurable effects on the phenotype.

4. Individual organisms can respond to changes in their environments by altering their behavior, physiology, or morphology. Such changes are referred to as phenotypic plasticity.

5. Organisms have characteristic activity spaces defined by the conditions within which they can live and reproduce.

6. The availability of suitable activity space for an individual depends on the range of conditions and resources in the environment at any given time.

7. Animals select microhabitats whose physical conditions fall within their activity space.

8. Acclimation involves reversible changes in structure (for example, fur thickness) or biochemical pathways (changes in the amounts of different enzymes). Such changes require longer periods (usually days or weeks) than behavioral or metabolic changes. Acclimation plays a prominent role in responses of long-lived organisms to seasonal change.

9. Developmental responses express the interaction between an organism and its environment during its growth. Different environmental conditions lead to different characteristic, irreversible structures and appearances.

10. When conditions exceed the range of tolerance, organisms may migrate elsewhere, rely on materials stored during periods of abundance, or enter inactive states.

11. In many cases, the individual must anticipate environmental changes in order to respond successfully. Organisms rely on proximate cues, such as day length, to predict changes in ultimate factors, such as food supply, that directly affect their well-being.

12. Food supplies vary spatially, temporally, and with respect to the quality of prey items. Thus, animals must make choices about when, where, and how to feed that maximize their reproductive success.

13. Central place foragers, which must deliver food to a fixed place, such as a nest with young, balance the costs and risks of travel against the size of the area within which they can forage.

14. The quality of a feeding area is affected by the risk of predation on a foraging individual. Many animals avoid feeding in high-risk areas even though food may be plentiful. This strategy is referred to as risk-sensitive foraging.

15. Some foragers consume a mixed diet to obtain an appropriate balance of required nutrients or to reduce levels of toxic substances in their diets. Diet mixing is especially common among animals that feed on plants.

PRACTICING ECOLOGY

CHECK YOUR KNOWLEDGE

Tolerance of Variable Environments

We started this chapter by discussing how organisms cope with climate variability in the Mojave Desert. Indeed, this desert is hot in the summer. But it also can be cold in the winter, when air temperatures regularly fall below 0°C. Thermal extremes, which can limit the distributions of plants and animals, often occur in brief episodes, such as heat waves and cold snaps, and are not so apparent in annual averages. However, this kind of short-term variability can severely decrease the capacity of organisms to survive and reproduce. Consequently, extreme conditions help explain the geographic distributions of many species. For example, the northern limit of the creosote bush *Larrea tridentata* matches the southern extent of minimum temperatures down to –18°C in the Mojave Desert. Creosote bush, particularly seedlings, cannot tolerate colder winter temperatures.

Besides accommodating temperature extremes, plants will have to adapt to variations in the environment caused by increasing atmospheric carbon dioxide. Burning fossil fuels and cutting forests will alter the global climate, perhaps substantially. The resulting increases in air and soil temperatures and changes in soil water and nutrient content are expected to have dramatic implications for the productivity and distribution of terrestrial vegetation in both natural and managed ecosystems. For example, as winter temperatures increase, the distributions of plants such as creosote bush will likely shift northward and to higher elevations. Plants also respond directly to elevated CO_2. Elevated atmospheric CO_2 increases photosynthetic efficiency, but this effect often decreases upon long-term exposure. In certain ecosystems, exposure to elevated CO_2 can also cause a reduction in leaf water loss. Plants in arid lands, such as the Mojave Desert, may benefit from this water saving more than those in other biomes.

Recent studies by Stan Smith, James Coleman, Robert Nowak, and Jeffrey Seemann at the Nevada Desert FACE Facility have focused on the ability of plants to tolerate stressful conditions. "FACE" is an acronym for "Free Air Carbon Enrichment," a procedure in which large quantities of carbon dioxide gas are released directly to the atmosphere at ground level over small areas. Thus plants can be exposed to elevated CO_2 in natural environments. This method avoids complicating factors caused by exposing plants inside greenhouses, which alter the climate in unrealistic ways. One of the studies from this project has compared the ability of *Yucca brevifolia* (the Joshua Tree), *Yucca schidigera* (the Mojave Yucca), and *Yucca whipplei* to withstand high-temperature extremes under predicted CO_2 concentrations of the future. In this and other studies, elevated CO_2 has been shown to have a direct impact on the responses of plants to environmental variability.

CHECK YOUR KNOWLEDGE

1. Why is it important to understand the responses of desert organisms to their physical environment?

2. Refer to ▌Figure 9.21. Which of the *Yucca* species exhibits the greatest effect due to elevated CO_2? What is the effect of elevated CO_2 on the magnitude of photosynthesis?

3. How do the three different *Yucca* species vary in their responses to elevated CO_2 and high-temperature events?

From this, what general conclusions can you draw about the results of global change experiments?

MORE ON THE WEB 4. Go to the Weather page of the Nevada Desert FACE Facility from *Practicing Ecology on the Web* at *http://www.whfreeman.com/ricklefs* and examine the temperature and precipitation records measured at this site. By how much do they vary from summer to winter? And from year to year? What does this require of plants and animals in terms of evolution of adaptations to climatic variability?

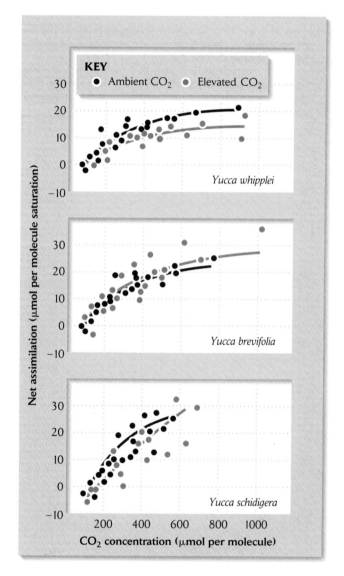

▌Figure 9.21 Curves for three species of *Yucca* exposed to either ambient or elevated CO_2 in a glasshouse, prior to the temperature increase. Red symbols and lines are from ambient plants, while blue symbols dotted lines are from elevated plants. From T. E. Huxman, et al., *Plant, Cell and Environment* 21:1275–1283.

Suggested Readings

Alerstam, T. 1990. *Bird Migration*. Cambridge University Press, Cambridge.

Bernays, E. A., K. L. Bright, N. Gonzalez, and J. Angel. 1994. Dietary mixing in a generalist herbivore: Tests of two hypotheses. *Ecology* 75:1997–2006.

Boyer, B. B., and B. M. Barnes. 1999. Molecular and metabolic aspects of mammalian hibernation. *BioScience* 49:713–724.

French, A. R. 1988. The patterns of mammalian hibernation. *American Scientist* 76:568–575.

Gilliam, J. F., and D. F. Fraser. 1987. Habitat selection under predation hazard: Test of a model with foraging minnows. *Ecology* 68:1856–1862.

Hutchins, M. J., and H. de Kroon. 1994. Foraging in plants: The role of morphological plasticity in resource acquisition. *Advances in Ecological Research* 25:159–238.

Huxman, T. E., Hamerlynck, E. P., Loik, M. E., and Smith, S. D. 1998. Gas exchange and chlorophyll fluorescence responses of three south-western *Yucca* species to elevated CO_2 and high temperature. *Plant, Cell and Environment* 21:1275-1283.

Johnston, I. A., J. D. Fleming, and T. Crockford. 1990. Thermal acclimation and muscle contractile properties in cyprinid fish. *American Journal of Physiology* 259:R231–R236.

Kacelnik, A. 1984. Central place foraging in starlings (*Sturnus vulgaris*). I. Patch residence time. *Journal of Animal Ecology* 53:283–299.

Krebs, J. R., and N. B. Davies (eds.). 1993. *Behavioural Ecology: An Evolutionary Approach* (3d ed.). Blackwell, Oxford.

Lyman, C., et al. 1982. *Hibernation and Torpor in Mammals and Birds*. Academic Press, New York.

Pough, F. H. 1980. The advantages of ectothermy for tetrapods. *American Naturalist* 115:92–112.

Ricklefs, R. E., and F. R. Hainsworth. 1968. Temperature dependent behavior of the cactus wren. *Ecology* 49:227–233.

Tevis, L., Jr., and I. M. Newell. 1962. Studies on the biology and seasonal cycle of the giant red velvet mite, *Dinothrombium pandorae* (Acari, Thrombidiidae). *Ecology* 43:497–505.

Whelan, C. J., K. A. Schmidt, B. B. Steele, W. J. Quinn, and S. Dilger. 1998. Are bird-consumed fruits complementary resources? *Oikos* 83:195–205.

Wilmshurst, J. F., J. M. Fryxell, B. P. Farm, A. R. E. Sinclair, and C. P. Henschel. 1999. Spatial distribution of Serengeti wildebeest in relation to resources. *Canadian Journal of Zoology* 77:1223–1232

Life Histories and Evolutionary Fitness

Trade-offs in the allocation of resources provide a basis for understanding life histories

Phenotypic plasticity allows an individual to adapt to environmental change

Life histories vary along a slow–fast continuum

A life history represents the best resolution of conflicting demands on the organism

Life histories balance trade-offs between current reproduction and future reproduction

Semelparous organisms are those that breed once and then die

Senescence is a decline in physiological function with increasing age

One of the remarkable facts concerning reproductive success is that the end result is always approximately the same. That is, each individual, on average, produces one offspring that lives to reproduce. This must be so, for otherwise populations would either dwindle rapidly to extinction because individuals failed to replace themselves, or they would grow out of all bounds.

Nonetheless, how organisms grow and produce offspring also varies in all imaginable ways. A female sockeye salmon, after swimming up to 5,000 km from her Pacific Ocean feeding ground to the mouth of a coastal river in British Columbia, faces another 1,000-km upriver journey to her spawning ground. There she lays thousands of eggs, and then promptly dies, her body wasted from the exertion. A female African elephant produces a single offspring at intervals of several years, lavishing intense care on her baby until it is old enough and large enough to fend for itself in the world of elephants (▌Figure 10.1). Thrushes start to reproduce when they are 1 year old, and may produce several broods of three or four young each year, but rarely live beyond 3 or 4 years. Storm petrels, which are seabirds the size of thrushes, do not begin to reproduce until they are 4 or 5 years old, and rear at most a single chick each year, but may live for 30 or 40 years.

Such attributes of the schedule of an individual's life—age at maturity, number of offspring, life span—make up what ecologists call the **life history** of the individual. Life histories are complex phenomena influenced by factors in the environment, the general body plan and lifestyle of organisms, and their individual and evolutionary responses to physical conditions, food supply, predators, and other aspects of the environment.

Figure 10.1 Organisms differ in their degree of parental investment. Elephant mothers care intensively for their offspring for many years. Photo by R. E. Ricklefs.

As we have seen, organisms are generally well suited to the conditions of their environments. Form and function vary in parallel with the ranges of temperature, water availability, salinity, oxygen, and other factors encountered by each species. We have seen how organisms choose their activity spaces and make foraging decisions in response to temporal and spatial variation in their environments. Whether modifications of form and function have evolved or result from individual responses to different environments, we presume that they are adaptive and that they increase the reproductive success of individuals. We can see why desert plants have small leaves with thick cuticles to reduce water loss. The close color matching of grass-

hoppers to their backgrounds makes sense when one understands that they are eaten by visually hunting predators. Life histories, too, are shaped by natural selection. An organism's life history represents its solution to the problem of allocating limited time and resources so as to achieve maximum reproductive success.

For decades, ecologists have used observation, mathematical modeling, and experimentation to unravel the causes of variation in life histories among living beings. Songbirds in the Tropics, for example, lay fewer eggs (two or three, on average) than their counterparts at higher latitudes (generally four to ten, depending on the species) (**Figure 10.2**). Professor David Lack of Oxford first placed

Figure 10.2 Organisms differ in the number of offspring they produce. Birds breeding at high latitudes typically lay larger clutches of eggs than tropical species. The five-egg clutch was laid by a snow bunting in Alaska, and the two-egg clutch by a red-headed manakin in Panama. Photos by R. E. Ricklefs.

this life history observation in an evolutionary context. Lack recognized that birds could increase their overall reproductive success by increasing the size of their clutches (the set of eggs laid together in a nest), unless something reduced the survival of offspring in large broods. He hypothesized that the ability of adults to gather food for their young was limited and, accordingly, that chicks in large broods, where there were too many mouths to feed relative to the food supply, would be undernourished and survive poorly. Lack further noted that at temperate and arctic latitudes, birds had longer days in which to gather food during summer, when their offspring are reared. Therefore, it made sense that birds at high latitudes could rear more offspring than birds breeding in the Tropics, where day length remains close to 12 hours year-round.

Lack made three important points. First, he stated that because life history traits, such as number of eggs in a clutch, contribute to reproductive success, they also influence evolutionary fitness. Second, he demonstrated that life histories vary consistently with respect to factors in the environment, such as the length of time available for feeding. This finding suggested the possibility that life history traits are molded by natural selection. Third, he proposed a hypothesis that could be subjected to experimental testing. In the case of clutch size, Lack suggested that the number of offspring that parents can rear is limited by their food supply. To test this idea, one could add eggs to nests to create enlarged clutches and broods. According to Lack's hypothesis, parents should be unable to rear added chicks because they cannot gather the additional food required by a larger brood.

This experiment has been conducted many times over the last several decades, generally with the result predicted by Lack. For example, the Swedish ecologist Gören Hogstedt manipulated the clutch sizes of magpies (relatives of crows) by moving eggs between nests. His results showed that magpies lay a clutch that corresponds to the maximum number of offspring that a pair can rear. Either adding or subtracting eggs results in fewer offspring fledged (▌Figure 10.3), just as Lack had predicted.

Trade-offs in the allocation of resources provide a basis for understanding life histories

Organisms have limited time, energy, and nutrients at their disposal. Adaptive modifications of form and function either increase the resources available to individuals or allow them to use those resources to their best advantage. Many modifications involve **trade-offs,** meaning that limited time, energy, or materials devoted to one structure or

▌Figure 10.3 **Clutch size reflects the maximum number of offspring the parents can rear.** The average European magpie clutch size of seven was manipulated by adding or removing eggs to make up clutches of five to nine eggs. The most productive clutch size was seven. After G. Hogstedt, *Science* 210:1148–1150 (1980).

function cannot be allotted to another. Therefore, each organism is faced with the problem of **allocation.** Given that time and resources are limited, how can the organism best use them to achieve its maximum possible evolutionary fitness?

Practical solutions to the allocation problem depend on how a change in any given structure or function affects fitness. When modification of a trait influences several components of survival and reproduction, as is often the case, the evolution of that trait can be understood only by considering the entire life strategy. For example, an increase in the number of seeds produced by an oak tree may contribute to fitness by increasing the number of offspring. But such a modification may also reduce the survival of seedlings (if seed size is reduced to make more of them), the survival of adult trees (if resources are shifted from root growth to support increased seed production), or subsequent seed production (if seed production in one year reduces tree growth, and therefore size, in subsequent years).

From an evolutionary point of view, an individual's reason for existence is to produce successful progeny—as many as possible. Reproduction involves choosing among many options: when to begin to breed, how many offspring to have at one time, how much care to bestow upon them. The set of rules and choices influencing an individual's survival and reproduction at each age governs its life history. Each life history has many components, the most important of which are **maturity,** or age at first

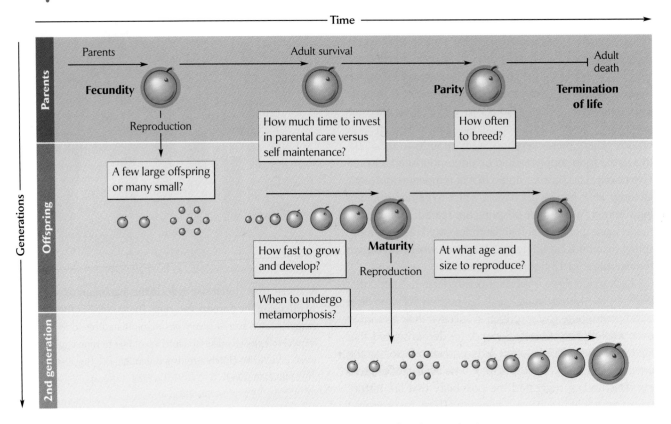

Figure 10.4 A life history is a set of rules and choices influencing survival and reproduction.

reproduction; **parity,** or number of episodes of reproduction; **fecundity,** or number of offspring produced per reproductive episode; and **aging** (■ Figure 10.4).

The optimum life history for an individual depends on the particular circumstances of its environment. As we have seen in the last chapter, when an individual's environment varies, it can respond by altering its behavior, its physiology, or even its development. Individuals also can alter their life history traits. Thus, the capacity to respond to variation in the environment is itself an aspect of the life history that is subject to natural selection. Before we discuss life histories in detail, we shall take some time to distinguish between evolutionary adaptations of populations and responses of individuals to the range of environmental conditions they normally encounter. Both are governed by sets of decision rules concerning allocation of time and resources. Individual responses are nongenetic. However, the ways in which individuals can respond to their environments may be under genetic control and subject to evolutionary change by natural selection. Thus, phenotypic plasticity may be considered a part of the life history strategy.

 Metabolic ceilings. **Can organisms increase overall performance without trading one function off against another?**

Phenotypic plasticity allows an individual to adapt to environmental change

Virtually all attributes of an individual are affected by environmental conditions and by the response of the individual to those conditions. The observed relationship between the phenotype of an individual and the environment is referred to as a **reaction norm** (■ Figure 10.5). Many of the responses discussed in Chapter 9 are examples of reaction norms. The general responsiveness of the phenotype to the surroundings is called **phenotypic plasticity.**

Some reaction norms are a simple consequence of the influence of the physical environment on life processes. Heat energy accelerates most life processes. Therefore, we should not be surprised that caterpillars of the swallowtail butterfly *Papilio canadensis* grow faster at higher temperatures. The relationship between growth rate and temperature for an individual describes the reaction norm of growth rate with respect to temperature for that individual. However, individuals of the same species from Michigan and from Alaska exhibit different relationships between

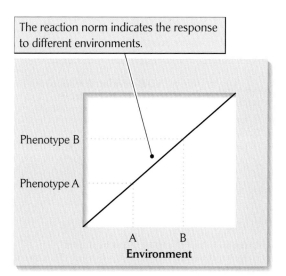

Figure 10.5 A reaction norm is the observed relationship between the phenotype and the environment. The graph shows the reaction norm of a single genotype over a range of environments. Each particular environment (for example, A or B) produces a characteristic phenotype.

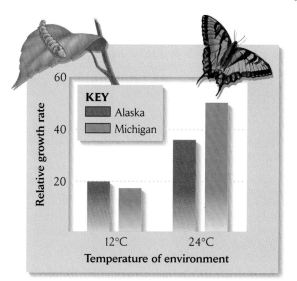

Figure 10.6 The reaction norms of different populations may differ. Fourth-instar larvae of the swallowtail butterfly *Papilio canadensis* were obtained from populations in Alaska and Michigan and reared on balsam poplar in the laboratory at temperatures of 12°C and 24°C. After M. P. Ayres and J. M. Scriber, *Ecol. Monogr.* 64:465–482 (1994).

growth rate and temperature (**Figure 10.6**). In one experiment, larvae from Alaskan populations grew more rapidly at low temperatures and larvae from Michigan grew more rapidly at high temperatures, as one might have predicted from the typically warmer temperatures in Michigan during the growing season. This finding indicates that reaction norms may be modified by evolution to improve performance under the particular conditions experienced by a population, as shown diagrammatically in **Figure 10.7**.

Genotype–environment interaction

As swallowtail growth rates show, the genetic makeup of the individual and the individual's environment interact to determine its performance. When the reaction norms of two genotypes cross, as they do in the case of the swallowtail butterfly, then individuals with each genotype perform better in one environment and worse in the other. Such a relationship is referred to as a **genotype–environment**

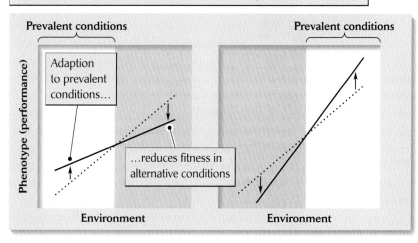

Figure 10.7 Reaction norms may be modified by evolution. Reaction norms may diverge when two populations of the same species exist for long periods under different conditions. Very often an increase in performance under the prevalent conditions is accompanied by a decrease in performance when individuals are exposed to conditions outside the population's normal range.

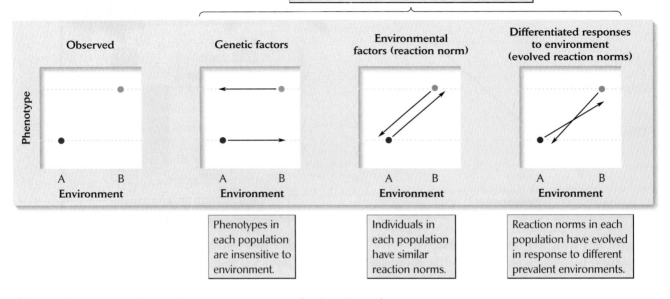

Figure 10.8 Reciprocal transplant experiments are used to investigate the cause of differences between populations. The results of reciprocal transplant experiments show different patterns when phenotypic differences between populations are determined solely by genetic factors, solely by environmental factors (phenotypic plasticity), or by reaction norms that have differentiated between the populations.

interaction because each genotype responds differently to variation in the environment.

Interactions between genetic and environmental factors are the basis for the evolution of **specialization.** Over time, when two populations are exposed to different ranges of environmental conditions, genotype–environment interactions will cause different genotypes to predominate in each population. The populations will therefore become differentiated and will have different reaction norms, each of which enables individual organisms to perform better in their own environment, as illustrated in Figure 10.7.

Reciprocal transplant experiments

Whether differences between populations are due to genetic differences or to phenotypic plasticity often can be revealed by **reciprocal transplant experiments.** Transplant studies compare the observed phenotypes of individuals kept in their native environment with those of individuals transplanted to a different environment (Figure 10.8). Reciprocal transplants involve the switching of individuals between two localities. When phenotypic values of native and transplanted individuals do not vary between the two environments, we may conclude that the traits of interest are genetically determined. That is, the trait values reflect the population from which an individual comes (genotype) rather than where it is living (environment). When trait values reflect where an individual is living (environment) rather than where it comes from (genotype), then the results of the experiment are consistent with phenotypic plasticity. Of course, intermediate results are possible, in which case one might conclude that the reaction norm has been subject to evolutionary modification.

ECOLOGISTS IN THE FIELD

A reciprocal transplant experiment

Peter Niewiarowski and Willem Roosenberg, then at the University of Pennsylvania, transplanted fence lizards (*Sceloporus undulatus*) between nutrient-poor pine barrens in New Jersey and nutrient-rich tall-grass prairies in Nebraska. The effect of the switch on the lizards' growth rates revealed both genetic determination and phenotypic plasticity (Figure 10.9). The growth rates of Nebraska lizards, about twice those of New Jersey lizards in their native environments, decreased by half—to the New Jersey level—when Nebraska individuals were transplanted to New Jersey. In contrast, New Jersey lizards did not grow faster in Nebraska.

A simple interpretation of these results is that resources for growth are consistently scarcer in New Jersey than in

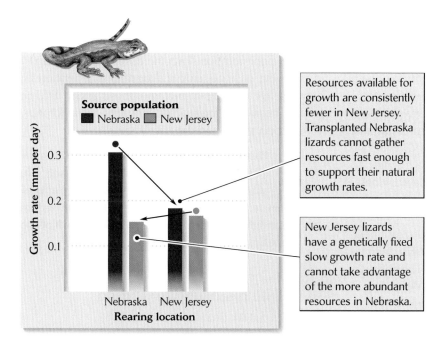

Resources available for growth are consistently fewer in New Jersey. Transplanted Nebraska lizards cannot gather resources fast enough to support their natural growth rates.

New Jersey lizards have a genetically fixed slow growth rate and cannot take advantage of the more abundant resources in Nebraska.

■ **Figure 10.9** **The growth rates of fence lizards reveal both genetic determination and phenotypic plasticity.** Juvenile eastern fence lizards (*Sceloporus undulatus*) from populations in Nebraska and New Jersey were exchanged in a reciprocal transplant experiment. Arrows indicate the growth responses of the transplanted populations. From data in P. H. Niewiarowski and W. Roosenberg, *Ecology* 74:1992–2002 (1993).

Nebraska and that Nebraska lizards transplanted to New Jersey cannot gather resources fast enough to support their natural growth rates. Apparently, New Jersey lizards have a genetically regulated growth rate that is adjusted to a low resource level. That is, they have lost the ability to modify individual growth rates in response to higher resource levels—levels that they probably experience rarely, if ever.

It is fair to ask whether the slower growth of Nebraska lizards under New Jersey conditions is adaptive or merely a consequence of reduced resources and more stressful conditions. If their slower growth was an example of adaptive phenotypic plasticity, it would reduce the negative effects of this environmental change on their fitness. That is, we would expect adaptation to compensate in some beneficial way for a change in environmental conditions. The fence lizard shows that organisms may have little control over their rate of growth under poor conditions. If resources are not available, individuals cannot grow rapidly and achieve large stature. Nonetheless, other aspects of their lives can be modified in response to growth performance.

MORE
ON THE
WEB
Ecotypes and reaction norms. The response of yarrow plants to variation in growing conditions has been studied in a reciprocal transplant experiment.

Food supply and timing of metamorphosis

Many types of organisms undergo dramatic changes during the course of their development. Metamorphosis from larval to adult forms and sexual maturation are the most prominent of these changes. The best time to undergo such transitions depends on the presence of resources and natural enemies in the environment, and their timing is made more complicated by variations in rate of growth due to food supply, temperature, and other environmental factors.

Imagine two growth curves resulting from two levels of food supply (■ Figure 10.10). Let us suppose that under a

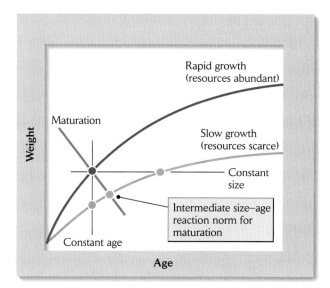

■ **Figure 10.10** **Relationships between age and size at maturation may differ when growth rates differ.** Individuals may switch at a constant age, a constant size, or some intermediate between the two. From S. C. Stearns, *Am. Zool.* 23:65–76 (1983), and S. C. Stearns and J. Koella, *Evolution* 40:893–913 (1986).

good nutritional regime resulting in rapid growth, an individual matures at a given mass and age. Poorly nourished individuals clearly cannot reach the same mass at a given age, and therefore must mature at a different point with respect to size, age, or both. Faced with such environmental variation, an individual can follow one of two pathways, or some intermediate between them. First, the individual may mature when it achieves a certain mass, however long this takes. With poor nourishment, it will take longer to achieve this mass, and maturation will be delayed. Consequently, the individual will be exposed to a longer period of risk prior to reproduction. Alternatively, the individual may mature at a predetermined age. With poor nourishment, this strategy will result in a smaller size at maturity, and perhaps a reduced reproductive rate as an adult. The optimum solution is usually somewhere in between, depending in part on the risk of death as a juvenile (high risk favors earlier maturation at a smaller size) and the slope of the relationship between fecundity and size at maturity (higher values favor delayed maturation at a larger size because the fecundity payoff is greater).

Tadpoles raised under conditions of high and low food availability exhibit different growth rates, as expected. In one experiment, the tadpoles given the poorer diet metamorphosed into adult frogs at a smaller size, but a later age, than those reared with abundant food (❚ Figure 10.11).

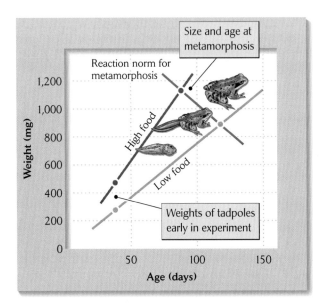

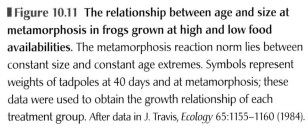

❚ **Figure 10.11 The relationship between age and size at metamorphosis in frogs grown at high and low food availabilities.** The metamorphosis reaction norm lies between constant size and constant age extremes. Symbols represent weights of tadpoles at 40 days and at metamorphosis; these data were used to obtain the growth relationship of each treatment group. After data in J. Travis, *Ecology* 65:1155–1160 (1984).

This finding supports the theoretical conclusion that the timing of metamorphosis should be sensitive to both age and size: poor nutrition slows the developmental program in frogs, but does not stop it altogether. The relationship between age and size at metamorphosis under different feeding regimes is the reaction norm of metamorphosis with respect to age and size.

MORE ON THE WEB | *Phenotypic plasticity and contrasting mechanisms of growth and reproduction in animals and plants.* The modular organization of plants allows them to respond to the challenge of herbivory and changes in light and nutrients by differential growth of root and shoot tips.

Life histories vary along a slow–fast continuum

Life history traits, such as age at maturity, fecundity, and longevity, vary widely among different species and even among different populations of the same species. Two points can be made about this variation. First, life history traits often vary consistently with respect to habitat or conditions in the environment. Seed size, for example, is generally larger among plants of the forest than among plants of grasslands.

Second, variation in one life history trait is often correlated with variation in other traits. For example, the number of independent offspring produced each year is positively correlated with adult annual mortality rate (❚ Figure 10.12). As a result, life history characteristics are generally organized together along a single continuum of values. At one extreme, which we can refer to as the "slow" end of the spectrum, organisms such as elephants, albatrosses, giant tortoises, and oak trees exhibit long life, slow development, delayed maturity, high parental investment, and low reproductive rates. At the fast end of the spectrum are mice, fruit flies, and weedy plants, which exhibit the opposite life history characteristics. The correlation between mortality and fecundity across species must in part reflect the fact that in persistent populations, births and deaths must balance on average. In addition, however, these life history traits may be modified by evolution.

The English plant ecologist J. P. Grime emphasized the relationship between the life history traits of plants and certain conditions of the environment. He envisioned variation in life history traits lying between three extreme apexes, like points of a triangle, and called plants with life histories at these extremes "stress tolerators," "ruderals" (weeds), and "competitors" (Table 10.1). As their name implies, stress tol-

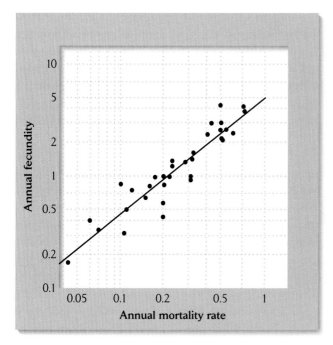

Figure 10.12 Fecundity and mortality rates vary together. Plotted logarithmically, the relationships between annual fecundity (number of offspring produced) and adult mortality rate for 33 species of birds cluster around a straight line. From R. E. Ricklefs, *Condor* 102:9–22.

erators grow under extreme environmental conditions. They grow slowly and conserve resources. Because seedling establishment is difficult in stressful environments, vegetative spread is emphasized. Where conditions for plant growth are more favorable, ruderals and competitors occupy opposite ends of a spectrum of disturbance. Ruderals, which colonize disturbed patches of habitat, exhibit rapid growth, early maturation, high reproductive rates, and easily dispersed seeds. These traits enable them to reproduce quickly and disperse their progeny to other disturbed sites—"growing like weeds"—before being overgrown by superior competitors. Competitors tend to grow to large stature, mature at large sizes, and exhibit long life spans. The competitor life history therefore requires stable conditions for its success.

Larger organisms tend to have longer life spans and lower reproductive rates than smaller organisms. This pattern is partly a function of the physical and physiological consequences of body size and partly results from the different environmental factors affecting large and small organisms.

MORE ON THE WEB *Allometry and the consequences of body size for life histories.* Metabolic rate, fecundity, and mortality tend to decrease with increasing body mass.

Table 10.1 Typical life histories of plants in environments with different selective factors

Competitors	Ruderals	Stress tolerators
Herbs, shrubs, or trees	Herbs, usually annuals	Lichens, herbs, shrubs, or trees; usually evergreen
Large, with a fast potential growth rate	High potential growth rate	Potential growth rate slow
Reproduction at a relatively early age	Reproduction at an early age	Reproduction at a relatively late age
Small proportion of production to seeds	Large proportion of production to seeds	Small proportion of production to seeds
Seed bank sometimes, vegetative spread often important	Seed bank and/or highly vagile seeds	
Vegetative spread important		

Competitors

Increasing disturbance — Increasing stress

Stress tolerators — Ruderals

Increasing resources and stability

Source: *J. P. Grime, Plant Strategies and Vegetation Processes, Wiley, Chichester (1979).*

A life history represents the best resolution of conflicting demands on the organism

Life history traits are, to a greater or lesser extent, under the control of the individual, but not without constraint. For example, electing to produce larger offspring inevitably results in fewer being produced. Watching more carefully for predators takes time away from feeding, and so while this tactic may increase survival, fewer young may be produced. Each of these choices affects other aspects of an individual's life. Because breeding takes time and resources from other activities and entails risks, investment in offspring generally diminishes the survival of parents. In many cases, rearing offspring drains a parent's resources so much that fewer offspring are produced later.

An optimized life history represents a resolution of conflicts between the competing demands of survival and reproduction to the best advantage of the individual in terms of reproductive success. A critical effort in the study of life histories has been to understand the fitness consequences of changing the allocation of limited time and resources to such competing functions.

ECOLOGISTS IN THE FIELD

The cost of parental investment in the European kestrel

Although it is widely believed that trade-offs between functions constrain life histories, demonstrating such trade-offs has proved difficult. One useful approach is to manipulate experimentally individual components of the phenotype. As we saw above, adding and subtracting eggs in the nests of birds has, in many cases, revealed an inverse relationship between the number of chicks in a nest and their survival. Consequently, production of offspring is often greatest from clutches of intermediate size. Sometimes, however, having more mouths to feed stimulates parents to increase their effort to hunt for food for their chicks. In this case, an artificially enlarged brood might result in higher reproductive success, but impose a cost on parents in the form of smaller future broods or decreased survival.

Cor Dijkstra, Serge Daan, and their colleagues at the University of Groningen in The Netherlands have conducted extensive studies of the ecology and reproduction of European kestrels. Kestrels are small hawks that search for voles and shrews in open fields, often by hovering high overhead. Thus, kestrel foraging requires a high rate of energy expenditure, but small mammals are so abundant that individual

kestrels normally can catch enough prey to feed their brood in a few hours each day. Kestrels lay an average of five eggs. When the broods in a sample of nests were about a week old, the investigators either removed two chicks or added two chicks, creating reduced and enlarged broods as well as unchanged controls. Parents that were provided extra chicks worked harder to feed their enlarged broods, increasing their foraging time and energy expenditure.

The fruits of the increased parental investment were an increase in the number of chicks successfully reared per brood. However, in spite of the increased efforts of their parents, chicks in the enlarged broods were somewhat undernourished, and only 81% survived to fledging, compared with 98% in control and reduced broods. Consequently, the extra parental investment netted the harder-working parents only an extra 0.8 chick per nesting attempt, and this gain may have been diminished by subsequent lower survival of the underweight fledglings. A more telling effect of the increased parental effort was seen in the lower survival of adults with enlarged broods to the next breeding season (▮ Figure 10.13). Clearly, at some level of parental investment, the law of diminishing returns sets in, and further parental effort reduces the possibility of future reproductive success more than it increases the success of the present brood.

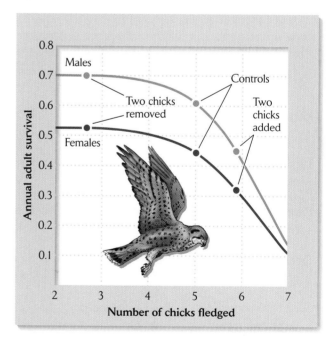

▮ **Figure 10.13 Parental investment affects parental survival.** The annual survival rate of male and female European kestrels (*Falco tinnunculus*) was affected by the number of chicks they reared. From data in C. Dijkstra et al., *J. Anim. Ecol.* 59:269–285 (1990).

Life histories balance trade–offs between current reproduction and future reproduction

Most issues concerning life histories can be phrased in terms of three questions: When should an individual begin to produce offspring? How often should it breed? How many offspring should it attempt to produce in each breeding episode? The variation among species in these life history traits illustrates the different ways of resolving the fundamental trade-off between fecundity and adult growth and survival—that is, between present and future reproduction.

Age at first reproduction

When should an animal or plant begin to breed? Long-lived organisms typically begin to reproduce at an older age than short-lived ones (∎Figure 10.14). Why should this be so? At every age, an individual must choose, whether consciously or not, between breeding and abstaining from breeding. Thus, we may understand age at first reproduction in terms of the benefits and costs of breeding at a particular age. The benefits appear as an increase in fecundity at that age. Costs may appear as reduced survival to older ages or reduced fecundity at older ages, or both.

Consider the following hypothetical example. A type of lizard continues to grow only until it reaches sexual maturity. Its fecundity varies in direct proportion to its body size at maturity. Suppose that the number of eggs laid per year increases by ten for each year that an individual delays reproduction. Thus, individuals that begin to breed in their first year produce ten eggs that year and the same number each year thereafter; individuals first breeding in their second year produce twenty eggs per year; individuals maturing in their third year produce thirty eggs; and so on. Comparing the cumulative egg production of early-maturing and late-maturing individuals (Table 10.2) reveals that the age at maturity that maximizes lifetime reproduction varies in direct proportion to the life span. For example, for a lizard with a life span of 3 years, maturing at 2 years results in the greatest lifetime reproduction. When the life span is 7 years, 4 years is the best age to mature.

For organisms that do not grow after their first year (most birds, for example), the decision whether to breed in a particular year after the onset of sexual maturity may reflect a trade-off between current reproduction and survival (which is, of course, related to future reproduction). Nonbreeding individuals avoid the risks of preparations for reproduction, such as courtship, nest building, and migration to breeding areas. Presumably, life experience gained with age also reduces the risks associated with breeding or increases the number of offspring resulting from a certain level of parental investment, or both, and thereby favors delayed reproduction. Among birds, age at maturity varies directly with the annual survival rate of adults, up to about 10 years in certain long-lived seabirds (see Figure 10.14).

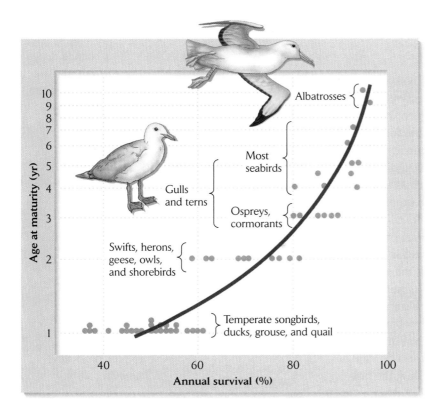

∎ **Figure 10.14 Long-lived organisms mature later than short-lived ones.** Age at maturity is correlated with annual adult survival rate, which is directly proportional to life span, in a variety of birds. From R. E. Ricklefs, in D. S. Farner (ed.), *Breeding Biology of Birds*, National Academy of Sciences, Washington, D.C. (1973), pp. 366–435.

Table 10.2 Total eggs produced by individuals in a hypothetical population as a function of life span and age at first reproduction

Age at first reproduction (years)	Life span (years)							
	1	2	3	4	5	6	7	8
1	**10***	**20**	30	40	50	60	70	80
2	0	**20**	**40**	**60**	80	100	120	140
3	0	0	30	**60**	**90**	**120**	150	180
4	0	0	0	40	80	**120**	**160**	**200**
5	0	0	0	0	50	100	150	**200**
6	0	0	0	0	0	60	120	180

Bold type indicates most productive ages at first reproduction for a given life span.

Tending to offset the advantages of delayed reproduction are many factors that reduce the expectation of future reproduction. These factors include high predation rates, encroaching senescence at old age, and, for organisms with a life span of a single year or less living in seasonal environments, the end of the reproductive season.

MORE ON THE WEB *Annual and perennial life histories. Why should some plants grow, reproduce, and die within one season while others persist from year to year?*

The trade-off between fecundity and survival

When a particular life history attribute affects both current fecundity and future growth or survival, selection would be expected to optimize the trade-off between the two. Intuitively, we would expect high mortality rates for adults to tip the balance in favor of current fecundity. Conversely, when the life span is potentially long, individuals should not increase current fecundity so much as to jeopardize future reproduction. This insight has a simple algebraic proof, which is as follows.

We use a model in which the geometric rate of population growth (λ) is equal to $S_0B + S$, where S is annual adult survival, B is annual fecundity, and S_0 is the survival of offspring to maturity at 1 year of age. Thus, the annual per capita growth rate of the population is the sum of the surviving adults (S) and new recruits to the breeding population (S_0B). We now partition adult survival into two components, one directly related to reproduction (S_R) and the other independent of reproduction (S). Thus, the rate of population growth may be expressed as

$$\lambda = S_0B + SS_R$$

$$\begin{bmatrix} \text{per capita} \\ \text{population} \\ \text{growth} \end{bmatrix} = \begin{bmatrix} \text{recruitment} \\ \text{of offspring} \end{bmatrix} + \begin{bmatrix} \text{adult} \\ \text{survival} \end{bmatrix}$$

Changes in reproductive traits that cause small changes in the values of survival (ΔS_R) and fecundity (ΔB) will influence the rate of population growth ($\Delta\lambda$) as follows:

$$\Delta\lambda = S_0\Delta B + S\Delta S_R$$

$$\begin{bmatrix} \text{increase} \\ \text{or decrease} \\ \text{in fitness} \end{bmatrix} = \begin{bmatrix} \text{change in} \\ \text{fecundity} \\ \text{weighted by} \\ \text{prereproductive} \\ \text{survival} \end{bmatrix} + \begin{bmatrix} \text{change in} \\ \text{reproductive} \\ \text{risk weighted} \\ \text{by annual} \\ \text{survival} \end{bmatrix}$$

When changes that enhance fecundity (make ΔB positive) also reduce survival (make ΔS_R negative), as we have seen in the case of the European kestrel, their effects on $\Delta\lambda$ depend on the relative values of S and S_0. Indeed, by rearranging the last equation, we can show that a change in life history traits will be favored ($\Delta\lambda > 0$) when

$$-\frac{\Delta B}{\Delta S_R} > \frac{S}{S_0}$$

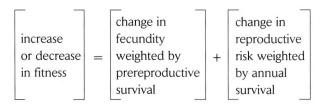

$$\begin{bmatrix} \text{change in} \\ \text{fecundity with} \\ \text{respect to change in} \\ \text{reproductive risk} \end{bmatrix} > \begin{bmatrix} \text{ratio of adult} \\ \text{to prereproductive} \\ \text{survival} \end{bmatrix}$$

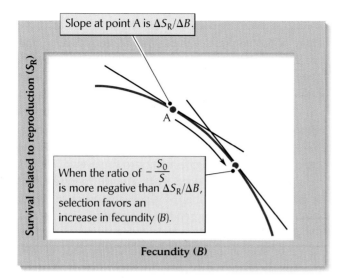

Slope at point A is $\Delta S_R/\Delta B$.

When the ratio of $-\dfrac{S_0}{S}$ is more negative than $\Delta S_R/\Delta B$, selection favors an increase in fecundity (B).

A

Survival related to reproduction (S_R)

Fecundity (B)

▌Figure 10.15 Investment in reproduction is associated with survival rates. When there is a trade-off between survival (S_R) and fecundity (B), indicated by the curved line, the optimum life history occurs where the slope of the tangent to the curve is $\Delta S_R/\Delta B = -S_0/S$.

Thus, we see that the ratio of adult to juvenile survival plays a dominant role in determining the life history. In general, when S is large compared with S_0, increases in fecundity ($\Delta B > 0$) must be very large compared with the resulting decrease in adult survival ($\Delta S < 0$) to be favored by natural selection (▌Figure 10.15). Thus, one would expect to find high parental investment associated with low relative adult survival (low S/S_0), and vice versa. This relationship may partly account for the positive correlation observed

between fecundity and mortality rates. If adults are likely to die before their next breeding opportunity, it is to their advantage to increase their investment in current offspring.

The trade-off between growth and fecundity

Many plants and invertebrates, as well as some fishes, reptiles, and amphibians, do not have a characteristic adult size. They grow, although often at a continually decreasing rate, throughout their adult lives, a condition referred to as **indeterminate growth.** Fecundity is directly related to body size in most species with indeterminate growth. Because egg production and growth draw on the same resources of assimilated energy and nutrients, increased fecundity during one year must be weighed against reduced fecundity in subsequent years. Accordingly, a long life expectancy should favor growth over fecundity during each year. For organisms with less chance of living to reproduce in future years, allocating limited resources to growth rather than eggs wastes potential fecundity at a young age.

Consider two hypothetical fish species, both of which weigh 10 g at sexual maturity, but which allocate resources to growth and reproduction differently. Both gather enough food each year to reproduce their weight in new tissue or eggs. Fish A allocates 20% of its production to growth and 80% to eggs, whereas fish B allocates half of its production to growth and half to eggs. In this model, fecundity is directly proportional to body size. Calculations of growth, fecundity, and accumulated fecundity (Table 10.3) show that for fish living 4 or fewer years, on average, high fecundity and slow

Table 10.3   Numerical comparisons of the strategies of slow growth/high fecundity and rapid growth/low fecundity in two hypothetical fish

Characteristic	Years					
	1	2	3	4	5	6
Slow growth/high fecundity						
Body weight	10	12	14.4	17.3	20.8	25.0
Growth increment	2	2.4	2.9	3.5	4.2	5.0
Weight of eggs	8	9.6	11.5	13.8	16.6	20.0
Cumulative weight of eggs	8	17.6	29.1	42.9	59.5	79.5
Rapid growth/low fecundity						
Body weight	10	15	22.5	33.8	50.7	76.1
Growth increment	5	7.5	11.3	16.9	25.4	38.1
Weight of eggs	5	7.5	11.3	16.9	25.4	38.1
Cumulative weight of eggs	5	12.5	23.8	40.7	66.1	104.2

Note: *All weights in grams. Body weight + growth increment = next year's body weight. Cumulative weight of eggs to last year + weight of eggs = cumulative weight of eggs to this year. Growth increment and weight of eggs in each year are equal to the body weight.*

growth result in greater overall productivity (cumulative weight of eggs), whereas for fish living longer than 4 years, low fecundity and rapid growth are more productive. Adult mortality, therefore, determines the optimal allocation of resources between growth and reproduction.

Semelparous organisms are those that breed once and then die

Unlike most fish, which breed repeatedly, some species of salmon grow rapidly for several years, then undertake a single episode of breeding. During this one burst of reproduction, females convert a large portion of their body tissues into eggs, and then die shortly after spawning. Because salmon make such a great effort to migrate upriver to reach their spawning grounds, it may be to their advantage to make the trip just once. After arriving at their breeding areas, they should then produce as many eggs as possible, even if this supreme reproductive effort results in the wastage of most body tissues and ensures death. This pattern is called **programmed death** because it is a direct consequence of adaptation to maximize reproductive success.

The salmon life history pattern is sometimes called "big-bang" reproduction, but is more properly referred to as **semelparity.** This term comes from the Latin *semel* ("once") and *pario* ("to beget"); the opposite of semelparity is **iteroparity,** from *itero* ("to repeat"). Semelparity is not the same as annual reproduction. For one thing, annuals may have more than one episode of reproduction, or prolonged continuous reproduction, within a season; for another, like perennials, semelparous individuals must survive at least one nonbreeding season—and usually many—before maturing sexually, reproducing, and then dying. Semelparity is rare among long-lived animals and plants.

The best-known cases of semelparous reproduction in plants occur in agaves and bamboos, two distinctly different groups, although this life history pattern has been reported even for some tropical forest trees. Most bamboos are tropical or warm temperate zone plants that form dense stands in disturbed habitats. Reproduction in bamboos does not appear to require substantial preparation or resources, as might be needed to grow a heavy flowering stalk. But there are probably few opportunities for successful seed germination. Once established, a bamboo plant increases by asexual reproduction (vegetative growth), continually sending up new stalks until the habitat in which it germinated is fairly packed with bamboo. Only then, when vegetative growth becomes severely limited, do the plants benefit from producing seeds, which can colonize new disturbed sites. In many species of bamboo, breeding is highly synchronous over large areas, after which the future of the entire population rests with the crop of seeds. Synchronous breeding may facilitate fertilization in this wind-pollinated plant group, or perhaps overwhelm seed predators, which cannot consume such a large crop of seeds.

The environments and habits of agaves occupy the opposite end of the spectrum from those of bamboos. Most species of agaves live in arid climates with sparse and erratic rainfall. Each agave plant grows vegetatively as a rosette of leaves produced from a single meristem over several years (the number of years varies from species to species). The plant then sends up a gigantic flowering stalk. Physiological studies have shown that the growth of the flowering stalk is too rapid to be fully supported by photosynthesis or uptake of water by the roots. As a consequence, the nutrients and water necessary for stalk growth are drawn from the leaves, which die soon after the seeds are produced (▌Figure 10.16).

Agaves frequently live side by side with yuccas, a closely related group of plants that have a similar growth

▌**Figure 10.16 Agaves are semelparous plants.** The Parry agave (*Agave parryi*) of Arizona grows as a rosette of thick, fleshy leaves for many years. Then it rapidly sends up its flowering stalk and sets fruit, after which the rosette dies. Photo by Tom Bean/DRK Photo.

form but which are iteroparous. Yuccas typically are branched and have many terminal rosettes of leaves. The root systems of agaves and yuccas also differ markedly. Yucca roots descend deeply to tap persistent sources of groundwater; agaves have shallow, fibrous roots that catch water percolating through the surface layers of desert soils after rain showers, but are left high and dry during drought periods. Thus, the semelparous agaves may experience greater variation in moisture availability from year to year than the iteroparous yuccas.

Several explanations have been proposed for the occurrence of semelparous and iteroparous reproduction in plants. First, variable environments might favor iteroparity, which would reduce the variation in lifetime reproductive success by spreading reproduction over both good and bad years. This tactic is referred to as **bet hedging**. However, this hypothesis can be rejected because semelparous plants tend to occur in more variable (usually drier) environments than their iteroparous relatives. On the other hand, variable environments might favor semelparity when a plant can time its reproduction to occur during a very favorable year. Storing resources for the big event makes sense, just as not holding back resources for an uncertain episode of future reproduction also makes sense. *Carpe diem:* seize the day. Semelparity is particularly favored when adult survival is relatively low and the interval between good years is long. Finally, attraction of pollinators to massive floral displays might favor plants that put all their effort into one reproductive episode. The few ob-

❙ Figure 10.17 The semelparous plant *Lobelia telekii* is found on the slopes of Mount Kenya. The giant inflorescenses in the foreground are *L. telekii,* while the stalked rosette plants in the background are *L. keniensis.* Courtesy of Truman P. Young.

servations on this point are mildly supportive. For example, in the semelparous rosette plant *Lobelia telekii,* which grows high on the slopes of Mount Kenya in Africa (❙ Figure 10.17), a doubling of inflorescence size was seen to result in a fourfold increase in seed production.

Comparison of *Lobelia telekii* with its iteroparous relative *L. keniensis* (Table 10.4), like the comparison between

Table 10.4   Ecological, life history, demographic, and reproductive traits of *Lobelia telekii* and *Lobelia keniensis* on Mount Kenya

Trait	*Lobelia telekii*	*Lobelia keniensis*
Life history	Semelparous	Iteroparous
Habitat	Dry rocky slopes	Moist valley bottoms
Growth form	Unbranched	Branched
Reproductive output	Larger inflorescences, more seeds	Smaller inflorescences, fewer seeds
Variation in inflorescence size	Highly variable, increases with soil moisture	Relatively invariable, independent of soil moisture
Demography	Virtually no adult survivorship	Populations in drier sites have lower adult survivorship and less frequent reproduction
Variation in number of seeds per pod	Strongly positively correlated with inflorescence size	Independent of inflorescence size, positively correlated with number of rosettes
Effects of pollinators	Increased seed quality, but not seed quantity	Increased seed quality, but not seed quantity

Source: *T. P. Young,* Evol. Ecol. *4:157–171 (1990).*

agaves and yuccas, suggests that semelparity is associated with dry habitats that are highly variable in both space and time. Presumably, infrequent conditions that are highly favorable for the establishment of seedlings trigger the massive flowering episodes in these plants. In summary, semelparity appears to arise either when preparation for reproduction is extremely costly, as it is for species that undertake long migrations to breeding grounds, or when the payoff for reproduction is highly variable but favorable conditions are predictable from environmental cues.

Senescence is a decline in physiological function with increasing age

Although few long-lived organisms exhibit programmed death associated with reproduction, most do experience a gradual increase in mortality and a decline in fecundity resulting from the deterioration of physiological function. This phenomenon is known as **senescence,** and humans are no exception to the general pattern seen in virtually all animals. Most physiological functions in humans decrease in a roughly linear fashion between the ages of 30 and 85 years, by 15–20% for rates of nerve conduction and basal metabolism, 55–60% for volume of blood circulated through the kidneys, and 60–65% for maximum breathing capacity. Birth defects in offspring and infertility generally occur with increasing prevalence in women after 30 years of age, and fertility decreases dramatically in males after 60 years. Reproductive decline and death in old age do not result from abrupt physiological changes. Rather, they follow upon a gradual decrease in physiological function. This decline includes the function of the immune system and other repair mechanisms, and with their decline, the incidence of deaths from tumors and cardiovascular disease rises (Figure 10.18)

Why does senescence exist, when survival and reproduction presumably confer advantages on an individual at any age? Perhaps physiological decline is just a fact of life and evolution can do nothing about it. Senescence may simply reflect the accumulation of molecular defects that fail to be repaired, just as an automobile eventually wears out and has to be junked. Ionizing radiation and highly reactive forms of oxygen break chemical bonds; macromolecules become cross-linked; DNA accumulates mutations. However, this wear and tear cannot be the entire explanation for patterns of aging because maximum longevity varies widely even among species of similar size and physiology. For example, many small insectivorous bats achieve ages in captivity of 10–20 years, whereas

mice of similar size rarely live beyond 3–5 years. In addition, cellular mechanisms for repairing damaged DNA and protein molecules appear to be better developed in long-lived animals than in their short-lived relatives. These observations suggest that, while senescence may be inevitable, rates of senescence are under the influence of natural selection and evolutionary modification. For instance, prolonging the maximum potential life span by postponing senescence may exact costs in terms of reduced reproduction at younger ages. If repair processes require time and resources, and if mortality is so high that an individual has little chance of living to old age, it may be more productive for the individual to allocate resources to reproduction early in life and let the body fall apart with age.

Even in the absence of senescence, accidental causes of death result in fewer and fewer individuals remaining alive at older ages. Consequently, progressive physiological deterioration and loss of reproductive capacity have little effect on average lifetime reproduction because most adults in a population never achieve old age. By this reasoning, selection on changes in survival or fecundity in old age is weak. Thus, selection should tend to favor improvements in reproductive success at young ages over those later in life. Because this effect is greater in populations with low adult survival rates than in populations with high

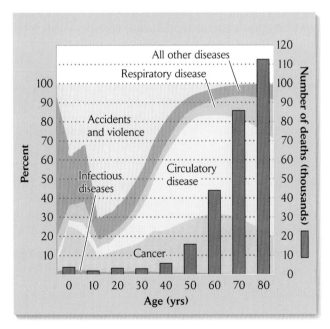

Figure 10.18 The incidence of death from cancer and circulatory disease rises with age. Ages at death and causes of death for males in the English population in the early 1980s are shown. From N. Coni, W. Davison, and S. Webster, *Ageing: The Facts* (2d ed.), Oxford University Press, Oxford (1992).

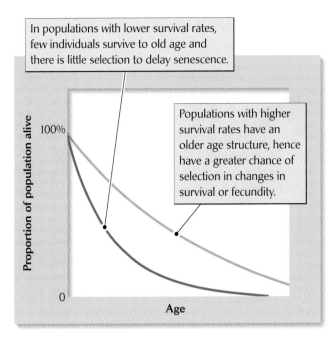

In populations with lower survival rates, few individuals survive to old age and there is little selection to delay senescence.

Populations with higher survival rates have an older age structure, hence have a greater chance of selection in changes in survival or fecundity.

Figure 10.19 The strength of selection varies with mortality rates. The strength of selection on changes in mortality and fecundity at a particular age is related to the proportion of individuals in the population alive at that age, which depends largely on rates of mortality caused by extrinsic factors earlier in life.

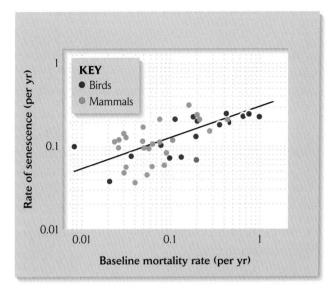

Figure 10.20 Populations with higher extrinsic mortality rates age faster. The relationship between senescence and extrinsic mortality rate is shown for a number of populations of birds and mammals. Senescence is measured as the rate of increase in mortality with age. From R. E. Ricklefs, *Am. Nat.* 152:24–44 (1998).

adult survival rates (**Figure 10.19**), senescence should appear earlier and progress faster in populations with high rates of mortality due to extrinsic causes (accidents, predators, weather). This hypothesis can be tested by comparing the rate of increase in mortality due to senescence with the "baseline" mortality rate caused by extrinsic factors and experienced by young adults in a population prior to the onset of aging. Data for birds and mammals seem to bear out this prediction (**Figure 10.20**).

Studies of aging in a variety of animals demonstrate that senescence is an inevitable consequence of natural wear and tear. It is impossible to build a body that will not wear out eventually. It is also clear, however, that the rate of wear can be modified by a variety of physiological mechanisms that either prevent or repair damage. These mechanisms are under genetic control and therefore, like most life history traits, can be modified by evolution. Mechanisms of prevention and repair involve investments of time, energy, nutrients, and tissues, and therefore the allocation of resources to these mechanisms depends on the expected life span of the individual. In this case, as in so many others we have seen in this chapter, consideration of life history patterns demonstrates the power of evolutionary theory to explain variation in biological systems.

 Summary

1. Life history traits include maturity (age at first reproduction), parity (number of episodes of reproduction), fecundity (number of offspring produced per reproduction episode), and aging. The values of these traits can be interpreted as solutions to the problem of allocating limited time and resources among various structures, physiological functions, and behaviors.

2. Most phenotypic traits of individuals are sensitive to variations in the environment. This response of form and function to the environment is referred to as phenotypic plasticity, and the quantitative relationship between phenotypic values and environmental variables is called the reaction norm.

3. Phenotypic plasticity itself is under genetic control. Differences in the sensitivities of individuals with different genotypes to variation in the environment are referred to as genotype–environment interactions. Such interactions can be revealed by reciprocal transplant experiments, in which individuals with the same genotypes are allowed to develop in each of several different environments.

When the performance of a genotype is superior to others over some range of environmental conditions, genotype–environment interactions can lead to specialization.

4. Life history traits often vary consistently with respect to the environment. Variation in one life history trait is often correlated with variation in others. Delayed reproduction, long life, and low reproductive rates are frequently associated with one another.

5. Plant ecologists have recognized clusters of life history attributes, including relative allocation of resources to reproduction and size of seeds. One such scheme includes three life history strategies associated with ruderal, stressful, and highly competitive environments.

6. Larger organisms tend to have longer life spans and lower reproductive rates.

7. Many theories concerning life history variation among species, including correlations among life history traits, are based on the principle that limited time and resources are allocated among competing functions in such a way as to maximize lifetime reproductive success.

8. Delayed reproduction is favored when life span is relatively long and when immature individuals benefit from increased growth or accumulation of experience by having greater fecundity later in life.

9. High extrinsic adult mortality rates favor increased reproductive effort, or investment in offspring, at the expense of adult survival and future reproduction.

10. When reproduction requires costly preparation, selection may favor a single all-consuming reproductive event followed by death, as in salmon. This pattern of reproduction, called semelparity, is the converse of iteroparity, or repeated reproduction.

11. Senescence, the progressive deterioration of physiological function with age, causes declines in fecundity and probability of survival. Senescence is caused by the wear and tear and the detrimental biochemical changes brought about just by living. Senescence is also subject to evolutionary modification.

12. Owing to accidental deaths, few individuals in most natural populations survive to old age. As a result, the strength of selection diminishes on traits expressed at progressively later ages. Individuals in populations subjected to higher extrinsic mortality rates age faster.

PRACTICING ECOLOGY

CHECK YOUR KNOWLEDGE

Life History Surprises

Experiments are usually designed to test a certain hypothesis, and not all experiments turn out as expected. When the results contradict the hypothesis, one must reject the hypothesis. On the bright side, this can lead to the creation of new hypotheses and further experimentation. David Lack observed that both bird clutch size and day length are greater at high latitudes compared to the equator, leading him to hypothesize that parent birds rear only as many young birds as they can feed during the day. One way to test this hypothesis is to manipulate clutch size by providing more mouths to feed. In many cases such experiments show that parent birds are unable to nourish the larger brood adequately and that adding eggs or chicks to a nest reduces the reproductive success of the parents.

As mentioned earlier in this chapter, Gören Hogstedt conducted brood size manipulation experiments in magpies. He found that the most productive clutch size corresponded to the size of the clutch laid by the female bird. Adding eggs to the nest or removing eggs always reduced the reproductive success, independent of the actual numbers of eggs present. In certain other cases, however, the most productive natural clutch size is larger than the most common clutch size observed within a population (❚ Figure 10.21).

Other studies have failed to demonstrate expected trade-offs between life history traits. David Reznick, of the University of California, Riverside, and colleagues have studied life history traits of guppies. They prevented females from mating with males. If we assume that growth and reproduction compete for resources, the experimental females (the nonmated ones) should have grown larger than control females that were mated with males. In fact, there was little difference in female reproductive tissue mass when mated and nonmated females were compared. Indeed, mated females simply ate more food to gain the energy necessary to produce the mass of eggs.

CHECK YOUR KNOWLEDGE

1. Using Figure 10.21, determine which clutch size has the highest survival rate. What is the approximate survival rate for the most common clutch size?

2. How can you explain the pattern depicted in Figure 10.21 from an evolutionary standpoint?

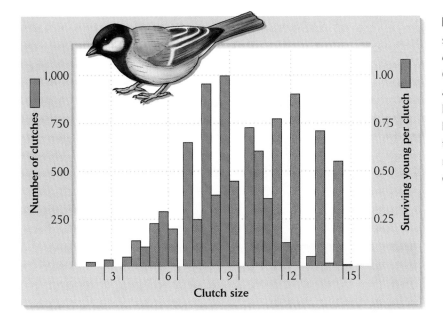

Figure 10.21 The frequencies of clutch sizes (blue bars, left-hand scale) among clutches of the great tit (*Parus major*) near Oxford, England, between 1960 and 1982, and number of young per clutch surviving at least to the next season (orange bars, right-hand scale) as a function of clutch size. Note that the most common clutch size is not the most productive. From M. S. Boyce and C. M. Perrins, *Ecology* 68:142–153.

MORE ON THE WEB 3. Go to the *Wild Wings* Web site through *Practicing Ecology on the Web* at *http://www. whfreeman.com/ricklefs*. Why do so many bird species migrate? What is the selective advantage of migrating in terms of maximizing reproductive success?

4. How is life history different from natural history?

 Suggested Readings

Bazzaz, F. A., N. R. Chiarello, P. D. Coley, and L. F. Pitelka. 1987. Allocating resources to reproduction and defense. *BioScience* 37: 58–67.

Dijkstra, C., A. Bult, S. Bijlsma, S. Daan, T. Meijer, and M. Zijlstra. 1990. Brood size manipulations in the kestrel (*Falco tinnunculus*): Effects on offspring and parental survival. *Journal of Animal Ecology* 59:269–286.

Fleming, I. A., and M. R. Gross. 1989. Evolution of adult female life history and morphology in a Pacific salmon (Coho: *Oncorhynchus kisutch*). *Evolution* 43:141–157.

Grime, J. P. 1979. *Plant Strategies and Vegetation Processes*. Wiley, Chichester.

Gross, M. R. 1996. Alternative reproductive strategies and tactics: Diversity within sexes. *Trends in Ecology and Evolution* 11:92–98.

Janzen, D. H. 1976. Why bamboos wait so long to flower. *Annual Review of Ecology and Systematics* 7:347–391.

Reznick, D. 1985. Costs of reproduction: An evaluation of the empirical evidence. *Oikos* 44:257–267.

Reznick, D. N., H. Bryga, and J. A. Endler. 1990. Experimentally induced life-history evolution in a natural population. *Nature* 346:357–359.

Ricklefs, R. E., and C. E. Finch. 1995. *Aging: A Natural History*. Scientific American Library, New York.

Rose, M. R. 1991. *Evolutionary Biology of Aging*. Oxford University Press, Oxford.

Schlichting, C. D. 1989. Phenotypic integration and environmental change. *BioScience* 39:460–464.

Sibly, R. M., and P. Calow. 1986. *Physiological Ecology of Animals: An Evolutionary Approach*. Blackwell Scientific Publications, Oxford.

Stearns, S. C. 1992. *The Evolution of Life Histories*. Oxford University Press, Oxford.

Strathmann, R. R. 1990. Why life histories evolve differently in the sea. *American Zoologist* 30:197–207.

Williams, G. C. 1966. Natural selection, the costs of reproduction, and a refinement of Lack's principle. *American Naturalist* 100:687–690.

Young, T. P., and C. K. Augspurger. 1991. Ecology and evolution of long-lived semelparous plants. *Trends in Ecology and Evolution* 6:285–289.

CHAPTER 11

Sex and Evolution

Sexual reproduction mixes the genetic material of two individuals

Sexual reproduction is costly

Sex is maintained by the advantages of producing genetically varied offspring

Individuals may have female function, male function, or both

The sex ratio of offspring is modified by evolution to maximize individual fitness

Mating systems describe the pattern of pairing of males and females within a population

Sexual selection can result in male–male combat and elaborate male ornamentation

Nature is full of bizarre creatures, and few are more bizarre-looking than the stalk-eyed flies (*Cyrtodiopsis*) of Malaysia, whose eyes are widely separated at the ends of long projections emanating from the head (■ Figure 11.1). Both males and females have the stalks, but in some species they are up to twice as long in males as they are in females. Stalk-eyed flies aggregate at night to mate, and field observations have shown that the mating success of males increases in direct relation to their eye span. Thus, the sex difference in eye span results from selection by females on the expression of this trait in males. Ecologists refer to this mechanism of evolution as **sexual selection**.

Why does this sexual dimorphism exist? If a wide eye span improved visual detection of food or predators, one would expect similar eye spans to be selected in males and females. Indeed, in some species of *Cyrtodiopsis*, males and females do not differ in this respect. But how can we explain the larger eye span of males in species in which the sexes do differ? Eye span in the dimorphic species seemingly must provide information about some aspect of male quality that is important to females.

Two dimorphic species (*C. dalmanni* and *C. whitei*) have biased sex ratios in nature. Their populations contain only 33–35% males, whereas in most flies and in monomorphic species of *Cyrtodiopsis,* the sex ratio is close to 50% males. Genetic analyses have shown that this biased sex ratio is caused by defective production in some males of sperm that bear the Y chromosome, so that most of their progeny are female (XX) rather than male (XY).

While most males of *C. dalmanni* and *C. whitei* produce few Y-bearing sperm, some males with wider eye spans than average produce practically normal sperm counts, and thus roughly equal

of such mating, the average eye span of males in the experimental population had increased about 1 mm, or 10%. Moreover, the proportions of males in the progeny of these flies had increased to 50% or more. These results show a genetic link between factors in the males that corrected the deficiency in sperm production and factors that increased eye span. So females that choose males with a wider eye span are also choosing males that will father more male offspring.

Why should a female want to produce more male offspring than the population average? The answer is relatively simple: when a population contains a smaller proportion of individuals of one sex, it is to the advantage of an individual to produce more of that sex among its progeny. Each individual receives one set of genes from its mother and one set from its father. Therefore, each generation has equal amounts of genetic material contributed by males and by females. When there are many females in a population, it is to a mother's advantage to produce male offspring, because each of her sons will, on average, contribute more sets of genes (her genes) to future generations than will each of her daughters. Thus, when females choose males that are likely to produce more sons than the population average, they increase their own evolutionary fitness.

Sex and gender are important components of the life histories of all species. The proportions of males and females among offspring, allocation of resources among male and female sexual function, even the presence of sexual reproduction itself, vary greatly from species to species. Understanding this diversity has been a major challenge to ecologists and evolutionary biologists.

Reproduction, as we have seen, is the ultimate goal of an individual's life strategy. Some of the most important and fascinating attributes of life concern sexual function. Among these are gender differences, sex ratios, and the various devices and behaviors used to enhance the success of an individual's gametes. The peacock's glorious tail, whose purpose is to make its bearer more attractive to females, is one of nature's most fantastic productions (❚ Figure 11.2). Indeed, sex underlies much of what we see in nature. In this chapter, we shall consider how sexual function influences the evolutionary modification of organisms and many of their behaviors as individuals. A good place to start is with sex itself.

❚ **Figure 11.1 Sexual dimorphism results from sexual selection.** At right, two large *Cyrtodiopsis whitei* males on a rootlet approach each other to compare eye spans at dusk in peninsular Malaya. Courtesy of G. S. Wilkinson; from G. S. Wilkinson and G. N. Dodson, in J. Choe and B. Crespi (eds.), *The Evolution of Mating Systems in Insects and Arachnids*, Cambridge University Press, Cambridge (1997), pp. 310–328.

numbers of male and female offspring. This was demonstrated dramatically in artificial selection experiments carried out by Gerald Wilkinson, at the University of Maryland, and his colleagues Daven Presgraves and Lili Crymes. They set out to duplicate the effects of sexual selection by mating only male flies with exceptionally wide eye spans. After 22 generations

▌**Figure 11.2 Much of what we observe in nature has evolved to improve an organism's reproductive success.** A male peacock spreads his elaborate tail to attract females. Photo by Norbert Rosing/Animals Animals.

Sexual reproduction mixes the genetic material of two individuals

In most animals and plants, reproduction is accomplished by the production of **gametes.** A male and a female gamete join together in an act of **fertilization** to form a single cell, called a **zygote,** from which a new individual develops. This mixing of genetic material from two parents results in new combinations of genes in the offspring. Because of this mixing, siblings can differ from one another genetically. Thus, in a variable environment, at least some offspring of a sexual union are likely to have a genetic constitution that enables them to survive and reproduce, regardless of the particular conditions. Sexual reproduction may also produce new combinations of genes previously absent from a population. The expression of any one gene can be influenced by other genes, and so new combinations of old genes may provide new variation for natural selection to work on. Indeed, it is the opinion of many biologists that sexual reproduction evolved very early in the history of life as a means of generating the genetic diversity necessary to respond through evolution to a varied and changing environment.

The gametes themselves are formed by a special type of cell division, called **meiosis,** which occurs in germ cells, specialized cells within the primary sex organs, or **gonads.** The cell products of meiosis are **haploid**—that is, they contain only one member of each of the chromosome pairs present in the individual's other, **diploid,** cells. Each of these haploid cells contains a single full set of chromosomes, but whether a particular chromosome has a paternal or maternal origin is, in most instances, random. These haploid cells are the ones that eventually develop into gametes: eggs in females and sperm in males. As a consequence of meiosis, the genetic makeup of each zygote is a unique, random combination of the genetic material of each of the individual's four grandparents.

In contrast to sexual reproduction, progeny produced by **asexual reproduction** are generally identical to one another and to their single parent, and thus none of them is likely to be well adapted to novel conditions. Asexual reproduction occurs in many plants, most of whose cells retain the ability to produce an entire new individual. In various plant species, shoots may sprout from roots or even from the margins of leaves, and then become separated from the parent plant to become new individuals (▌Figure 11.3). A group of such asexual individuals, all descended from the same parent and bearing the same genotype, is called a **clone.** Many simple animals, such as hydras, corals, and their relatives, can form buds in their body walls that develop into new individuals. When these remain attached to the parent individual, a colony develops, as in the case of hydroids, corals, bryozoans, and many other aquatic animals; when buds detach, independent new individuals are formed.

Some animals reproduce asexually by producing eggs with both sets of chromosomes (diploid). We see such modifications in all-female populations of fishes, lizards, and some insects, to name a few. In these animals, germ cells may be transformed directly into egg cells without going through meiosis, in which case all of an individual's eggs are genetically identical. Alternatively, meiosis may proceed through the replication and crossing-over stages and the first meiotic division; suppression of the second meiotic division at this point will result in diploid egg cells that nonetheless differ from one another genetically because of recombina-

Figure 11.3 Many plant species propagate asexually.
The walking fern sprouts a fully formed plant from the tip of
a leaf. After V. A Greulach and J. E. Adams, *Plants: An Introduction
to Modern Botany*, Wiley, New York (1962).

tion. In another variation, meiosis proceeds to completion,
but female gamete-forming cells then fuse to form diploid
eggs. This process is a type of self-fertilization, and its products vary genetically, but not as much as when two parents
are involved in producing them. Finally, individuals that
have both male and female sexual organs may form both
male and female gametes and then fertilize themselves. This
method of reproduction is frequently encountered in
plants. It is sexual in that both types of gametes are made
and fertilization takes place, but it is asexual in that progeny
are produced by a single parent.

MORE ON THE WEB *Environmental sex determination.* In many reptiles, sex is determined by the temperature at
which the embryo grows.

Sexual reproduction is costly

Both sexual and asexual reproduction are viable life strategies. Asexual reproduction is widespread among plants and
is found in all major groups of animals, with the exception
of birds and mammals. Perhaps it is surprising that sex
occurs at all, considering its costs to the organism. Gonads
are expensive organs that confer little direct benefit to the
individual and require resources that could be devoted to
other purposes. Mating itself is a major production for animals and plants, involving floral displays to attract pollinators and elaborate courtship rituals to please mates.

For organisms in which the sexes are separate—that is,
in which individuals are either male or female—sexual
reproduction has a much higher cost. This cost is a consequence of the fact that only half the genetic material of
each individual offspring comes from each parent. Compared with asexually produced offspring, which contain
only the genes of their single parent, the progeny of a sexual union contribute only half as much to the evolutionary
fitness of either parent (**Figure 11.4**). This 50% cost of
sexual reproduction to the individual parent is sometimes
referred to as the **cost of meiosis.**

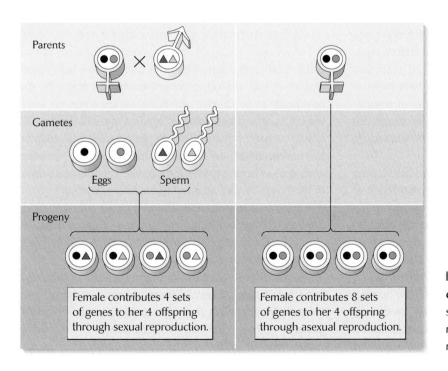

**Figure 11.4 Sexual reproduction is
costly.** A female contributes half as many
sets of her genes to her progeny when she
reproduces sexually as when she
reproduces asexually.

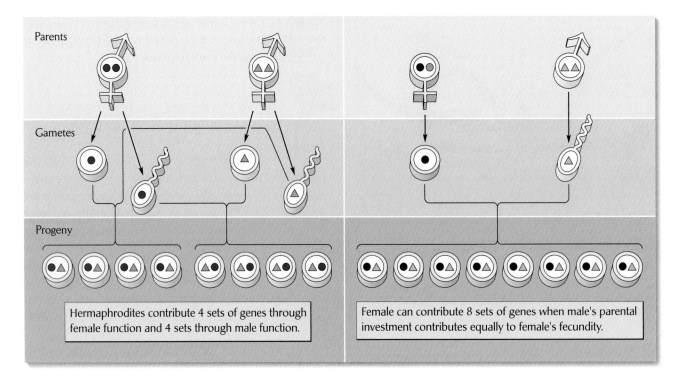

❚ Figure 11.5 Hermaphrodites and male parental care avoid the cost of meiosis.
Hermaphrodites (*left*) can contribute genes to their progeny through both female
and male function. A male's parental investment (*right*) may double his female
partner's fecundity.

Natural selection favors genetically determined traits
that reproduce the greatest number of copies of their genes
in future generations. Genes for asexual reproduction propagate themselves much faster than genes for sexual reproduction whenever the production of varied offspring is not
a consideration. A female can produce only a limited number of eggs. Thus, from the standpoint of an individual
female, asexually produced offspring would have twice as
many copies of her genes as the same number of sexually
produced offspring. Under this scenario, males would not
only be superfluous, but mating with a male would reduce
a female's genetic contribution by 50%.

The cost of meiosis does not apply to individuals having
both male and female sexual function (hermaphrodites)
because each individual contributes one set of its genes to
each of its own offspring produced through female function
and an equivalent number of sets, on average, through male
function (❚ Figure 11.5). The cost of meiosis also does not
apply when the sexes are separate but males contribute, by
means of parental care, as much as females to the number
of offspring produced. When male parental investment
doubles the number of offspring that a female could rear on
her own, the cost of meiosis to the female is canceled.

Sex is maintained by the advantages of producing genetically varied offspring

If sex is so costly, then why does it exist? The high fitness
cost of sexual reproduction presumably is offset by the
advantage of producing genetically varied offspring when
the environment itself varies in time or space (❚ Figure 11.6).
Clearly, a parent that survives to reproduce is well adapted
to the conditions of its environment. Genetic variation
among its offspring makes it more likely that at least some
of them will be well adapted to conditions that differ from
the parental environment. But are these advantages of
producing varied offspring enough to overcome the 50%
disadvantage of sex in species like ourselves, in which the
sexes are separate individuals?

A partial answer to this question comes from the sporadic distribution of asexual reproduction among complex
animals. Most species that reproduce asexually belong to
genera, such as *Ambystoma* (salamanders), *Poeciliopsis*
(fishes), and *Cnemidophorus* (lizards), in which other species

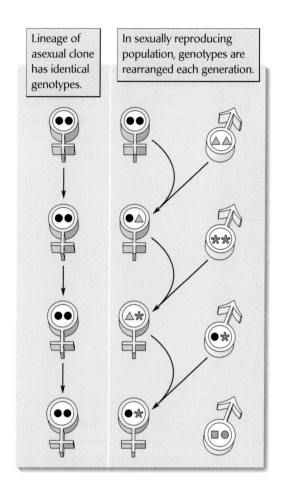

| Lineage of asexual clone has identical genotypes. | In sexually reproducing population, genotypes are rearranged each generation. |

▌ Figure 11.6 Sexual reproduction rearranges genotypes. The advantage of sexual reproduction may be that it produces offspring that differ genetically from the parents, allowing the population to evolve in response to changing conditions.

are sexual. This indicates that, for the most part, the asexual species do not have a long evolutionary history. If they did, one would expect to see larger taxonomic groups of species sharing this derived trait. Thus, it appears that the long-term evolutionary potential of asexual populations is low, possibly because of their greatly reduced genetic variability, and that their lines may die out over time. A combination of infrequent origination from sexual ancestors and a high rate of extinction could limit the occurrence of asexual reproduction in complex animals.

Ecologists believe that it is also important to find a short-term advantage to sexual reproduction. Most theoretical models based on temporal and spatial variation in the physical environment simply do not find a great enough advantage to sex to offset the cost of meiosis. One promising alternative is that sex provides the genetic variation necessary to respond to *biological* change in the environment—

particularly the evolution of virulence by parasites. Parasites that cause disease in their hosts are called **pathogens.** These organisms can evolve very rapidly because their population sizes are large and their generation times short compared with those of their hosts. If host populations could not respond rapidly, they would probably be driven to low numbers and perhaps extinction by increasingly virulent pathogens. Any parent would benefit by producing offspring genetically different from itself, with unique combinations of genes to which the parent's pathogens were not well adapted. In this way, sex and genetic recombination could provide a moving target for the evolution of pathogens and keep them from getting the upper hand. This idea is called the **Red Queen hypothesis,** after the famous passage in Lewis Carroll's *Through the Looking Glass and What Alice Found There,* in which the Red Queen tells Alice, "Now, *here,* you see, it takes all the running *you* can do, to keep in the same place." For this model to work, pathogens must have the potential for severe effects on the fitness of their hosts, and these effects must be strongly dependent on the genotypes of their hosts.

ECOLOGISTS IN THE FIELD

Parasites and sex in freshwater snails

One of the most compelling tests of the Red Queen hypothesis has been conducted by Curt Lively and his coworkers at Indiana University on the freshwater snail *Potamopyrgus antipodarum,* a common inhabitant of lakes and streams in New Zealand. Most of the snail population consists of asexual, all-female clones, but some populations in lakes have about 13% males, which is indicative of a reasonable level of sexual reproduction. A trematode worm of the genus *Microphallus,* which effectively sterilizes its host, commonly infects the snails. *Microphallus* is most abundant in shallow waters of lakes, where ducks, which are a reservoir for the parasite, feed (▌ Figure 11.7). The ducks are the definitive host in the complex life cycle of *Microphallus,* meaning that the sexual stages of the parasites occur in ducks.

A laboratory competition experiment showed that asexual snails reproduce faster than sexual individuals. Asexual clones tend to predominate in areas where *Microphallus* is absent or uncommon, particularly in the deep water of large lakes. Where the prevalence of *Microphallus* infection is high, however, sexual individuals are common. This finding suggests that in spite of their higher reproductive rate, asexual clones cannot persist in the face of high rates of parasitism. According to the Red Queen hypothesis, the problem with asexual clones is that, because they are

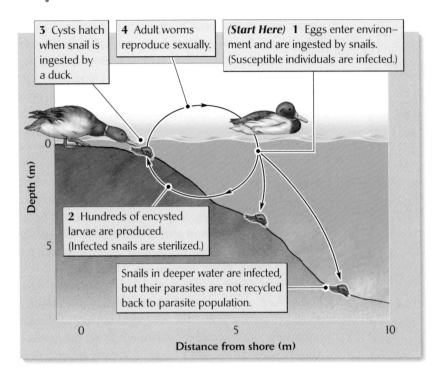

3 Cysts hatch when snail is ingested by a duck.

4 Adult worms reproduce sexually.

(Start Here) 1 Eggs enter environment and are ingested by snails. (Susceptible individuals are infected.)

2 Hundreds of encysted larvae are produced. (Infected snails are sterilized.)

Snails in deeper water are infected, but their parasites are not recycled back to parasite population.

Figure 11.7 The trematode worm *Microphallus* parasitizes snails in freshwater habitats in New Zealand. Adult worms reproduce sexually in ducks and larval stages reproduce asexually in snails, rendering infected snails sterile. After C. M. Lively and J. Jokela, *Proc. R. Soc. Lond.* B 263:891–897 (1996).

genetically uniform, *Microphallus* can evolve to infect them with high efficiency.

Lively and his coworkers were able to test this idea by taking snails from three different depths of Lake Alexandrina, New Zealand, and exposing them to parasites obtained from snails from each of these three depths. If the parasites had evolved to specialize on local (depth-specific) populations of snails, then they should have the greatest success in infecting those same populations. This is in fact what happened, as shown in **Figure 11.8**: snails taken from shallow water were infected most readily by parasites taken from shallow water, and so on. Furthermore, infection rates were relatively low in deep-water snails because the parasites had had little opportunity to specialize on these snail populations. Remember that the definitive hosts—ducks—feed mostly in shallow water, and so only the shallow-water lineages of parasites cycled regularly through snail host populations and had a chance to evolve greater infectivity on them. In Lake Alexandrina, snails from deeper water were infected mostly by parasites from shallow-water populations when ducks roosted in deep water and left parasites behind in fecal matter. Because deep water provided a partial refuge from parasites, sexual lineages of snails did not compete well against asexual lineages there, and the prevalence of male snails was low.

If parasites evolve higher levels of infectivity on more common clones of snails over time, then one would expect rare clones to be parasitized least frequently. When a clone is rare, there should be little selection on the parasites to specialize to infect it with greater efficiency. However, rare clones that experience low rates of parasitism should tend to become more abundant over time because they remain fertile and outcompete highly infected clones. Then, as a clone becomes common, parasites should evolve to specialize on

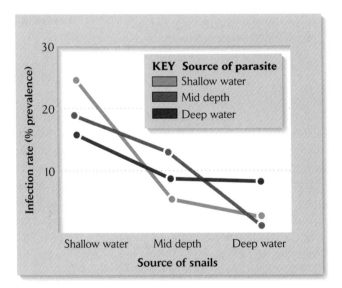

Figure 11.8 Parasites are best at infecting those populations they have evolved with. Populations of snails (*Potamopyrgus*) obtained from each of three different depths in Lake Alexandrina were exposed in the laboratory to parasites (*Microphallus*) obtained from snails at each of three different depths. From data in C. M. Lively and J. Jokela, *Proc. R. Soc. Lond.* B 263:891–897 (1996).

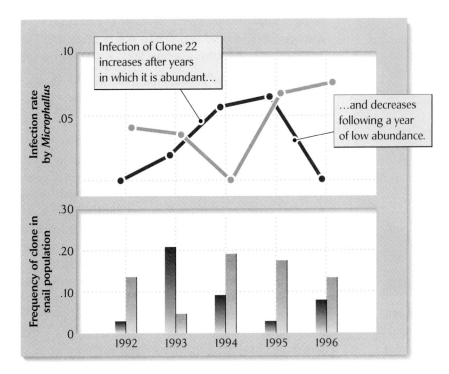

Infection of Clone 22 increases after years in which it is abundant...

...and decreases following a year of low abundance.

Figure 11.9 Cycles of parasite prevalence follow cycles of abundance in asexual clones. Frequencies of two clones in a population of *Potamopyrgus* snails from Lake Poerua and rates of infection of each clone by *Microphallus* are shown over 5 years. From M. F. Dybdahl and C. M. Lively, *Evolution* 52:1057–1066 (1998).

it, and eventually reduce its abundance. This process should lead to a cycling in the relative abundance of clones in an asexual population. Mark Dybdahl and Curt Lively found just such a cycle when they surveyed the relative abundance of clones of *Potamopyrgus* along the shoreline of Lake Poerua, New Zealand, over 5 years, or about fifteen snail generations. The data showed marked variation in the abundance of four common clones and a marked increase in infection by *Microphallus* in years following increases in populations of a clone (**Figure 11.9**). These patterns are consistent with the Red Queen hypothesis.

In spite of the success of research programs such as Curt Lively's, sex remains one of the most challenging questions that face biologists. At this point, however, we shall accept the fact that sex is with us, and we shall turn to exploring some of the consequences of sexual reproduction in the lives of organisms.

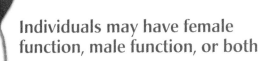

Individuals may have female function, male function, or both

We humans are used to thinking in terms of two sexes, female and male. But female and male sexual functions may be combined in the same individual, or an individual may change its sex during its lifetime. When both functions occur in the same individual, biologists label that individual a **hermaphrodite,** after the mythological Hermaphroditus, son of Hermes and Aphrodite, who while bathing became joined in one body with a not-so-shy nymph. Male and female functions may be **simultaneous,** as in the case of many snails and most worms, or they may be **sequential:** male first in some mollusks, echinoderms, and plants, female first in some fishes.

Plants that exhibit separate sexes in different individuals are called **dioecious,** from the Greek *di-* ("two") and *oikos* ("dwelling," also the root of the word "ecology") (**Figure 11.10**). **Monoecious** plants bear distinct male and female

Figure 11.10 Dioecious plants have two separate sexes. The dioecious tree *Clusia grandiflora* has sexually dimorphic flowers, female (*above*) and male (*below*). Photos by Volker Bittrich.

Figure 11.11 Perfect flowers contain both male and female sex organs. The perfect flower of *Miconia mirabilis* possesses both anthers and carpels. Photo by R. E. Ricklefs.

flowers on the same individual. The most common configuration, however, is seen in plants that bear **perfect flowers** (**Figure 11.11**), flowers that include both male and female parts. Though by one estimate, perfect-flowered hermaphrodites account for 72% of plant species, nearly all the imaginable combinations of sexual patterns are known. Populations of some species have both hermaphrodites and either male or female individuals, or male, female, and monoecious individuals, or hermaphroditic individuals with both perfect flowers and either male or female flowers. Most populations of hermaphrodites are fully **outcrossing,** which means that fertilization takes place between the gametes of different individuals. The rarer case of self-fertilization will be discussed in a subsequent chapter.

Which gender arrangement occurs in a given sexual, outcrossing population depends on the relative fitness costs and benefits to an individual of having either or both sexual functions. One can measure the fitness contributions of male and female sexual function by the number of sets of genes transmitted to offspring through either male or female gametes. When females can achieve the added benefit of a certain amount of male function by giving up a smaller amount of their female function, selection favors individuals that shift some resources to male function. Similarly, males that can add female function and not cut deeply into their male productivity are also favored by selection (**Figure 11.12**). It would seem that both male and female flowers can add the other sexual function with little cost. After all, the basic flower structure and the floral display necessary to attract pollinators are already in

place in single-sexed flowers. Thus, we would expect hermaphroditism to arise frequently, as it has among plants and most simple forms of animal life.

MORE ON THE WEB *Sequential hermaphroditism.* **Some organisms are male first, then switch to female later in their lives, or vice versa.**

Separate sexes provide the best compromise when gains from adding the function of one sex bring about even greater losses in the function of the other (**Figure 11.13**). This may occur when establishing a new sexual function in an individual entails a substantial fixed cost before any gametes can be produced. Sexual function in complex animals requires gonads, ducts, and other structures for transmitting gametes, as well as secondary sexual characteristics for attracting mates and competing with individuals of the same sex. In many animals in which maleness requires specializations for mate attraction and antagonistic interaction with other males, or in which femaleness requires specializations for egg production or brood care, such fixed costs may put hermaphroditism at a disadvantage compared with sexual specialization. In fact, hermaphroditism occurs rarely among animal species that actively seek mates and that engage in brood care. It is much more common among sedentary aquatic animals that simply shed their gametes into the water.

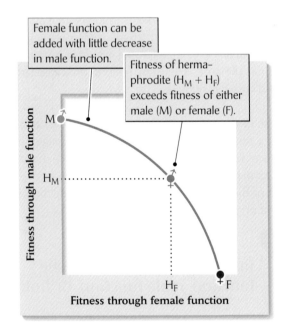

Figure 11.12 Selection sometimes favors hermaphrodites. When male or female function can be added with little depressing effect on the opposite sexual function, hermaphrodites can exclude males and females from a population.

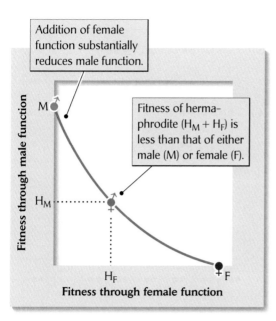

Figure 11.13 box text:
Addition of female function substantially reduces male function.

Fitness of herma-phrodite ($H_M + H_F$) is less than that of either male (M) or female (F).

Fitness through male function

H_M

H_F

Fitness through female function

■ Figure 11.13 Selection sometimes favors separate sexes. When male and female functions interfere with each other in the same individual, hermaphrodites are less successful than males and females and are excluded from a population.

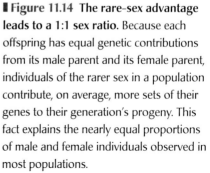

The sex ratio of offspring is modified by evolution to maximize individual fitness

When the sexes are separate, one may define a **sex ratio** among the progeny of an individual, or within a population, as the number of males relative to the number of females. Because females and males occur in the human population in a ratio of approximately 1:1, and because roughly equal numbers of females and males typify populations of most species, we consider the 1:1 sex ratio the usual condition and regard deviations from this ratio as special cases. Yet there are many such deviations.

We can explain the predominant 1:1 sex ratio by the following simple reasoning. Every product of a sexual union has exactly one mother and one father. Consequently, if the sex ratio of a population is not 1:1, individuals of the rarer sex will enjoy greater reproductive success because they will compete for matings with fewer others of the same sex (**■ Figure 11.14**). For example, if a population

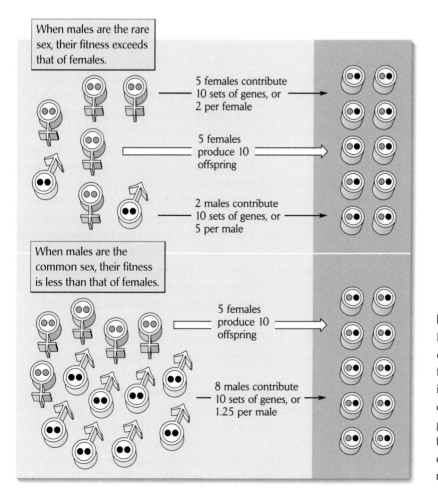

When males are the rare sex, their fitness exceeds that of females.

5 females contribute 10 sets of genes, or 2 per female

5 females produce 10 offspring

2 males contribute 10 sets of genes, or 5 per male

When males are the common sex, their fitness is less than that of females.

5 females produce 10 offspring

8 males contribute 10 sets of genes, or 1.25 per male

■ Figure 11.14 The rare-sex advantage leads to a 1:1 sex ratio. Because each offspring has equal genetic contributions from its male parent and its female parent, individuals of the rarer sex in a population contribute, on average, more sets of their genes to their generation's progeny. This fact explains the nearly equal proportions of male and female individuals observed in most populations.

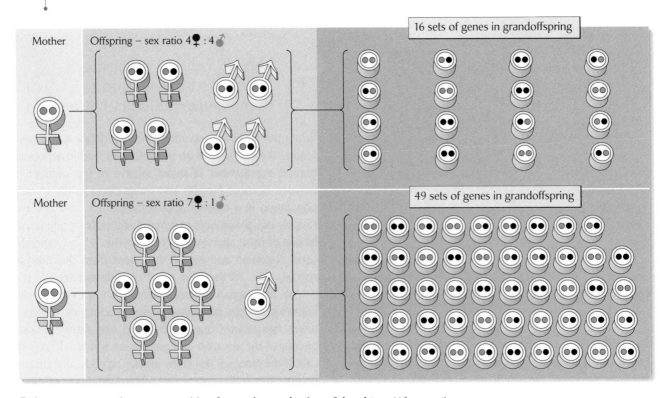

Figure 11.15 Local mate competition favors the production of daughters. When mating opportunities are restricted to siblings, and total fecundity (number of female plus male offspring) is limited, females should produce a high proportion of daughters in their progeny.

of 5 males and 10 females produced 100 offspring, each male would contribute 20 sets of genes, but each female would contribute only 10 sets of genes. Consequently, individuals of the rarer sex would contribute more sets of their genes to subsequent generations. Thus, when a population has more females than males, natural selection will favor any genetic tendency on the part of a parent to produce a larger proportion of male offspring. This will increase the frequency of males in the population to bring the sex ratio closer to 1:1. Similarly, when females are the rarer sex, genotypes that increase the proportion of female progeny will be favored, and the frequency of females will increase. When males and females are equally numerous, individuals of both sexes contribute equally to future generations on average, and different frequencies of males and females among the progeny of one individual are of no consequence to its relative long-term reproductive success. Because the fitness of genes affecting the sex ratio depends on the frequency of males and females in a population, the evolution of the sex ratio is said to be the product of **frequency-dependent selection.**

MORE ON THE WEB *Female condition and offspring sex ratio. Female mammals should produce male offspring only when they are in excellent condition and can raise sons to be good competitors.*

This explanation for the 1:1 sex ratio depends, however, on individuals having the opportunity to mate with unrelated individuals within a large population. When individuals do not disperse far from where they are born, or when they mate prior to dispersal, mating often takes place among close relatives (a situation known as **inbreeding**). In the extreme case, mating may occur among the progeny of an individual parent. In this situation, known as **local mate competition,** competition among males for mates takes place among brothers (**Figure 11.15**). From the standpoint of the parent of these siblings, one son would serve just as well as many to fertilize his sisters and propagate the parent's genes. In the case of brother-sister (sib) mating, the number of copies of her genes that a mother passes on to her grandoffspring depends only on the number of daughters she produces, because each son's genetic contribution to the daughters' offspring will also come from the mother. Thus, females that produce daughters at the expense of sons will have more grandoffspring and greater evolutionary fitness.

Sib mating occurs commonly in certain wasps that parasitize other insects or that lay their eggs and complete their larval development within the fruits of certain plants. For many of these species, hosts are so scarce, and mates are so difficult to find, that females mate where they hatch before they disperse to find new hosts on which to lay their own eggs. These wasps can alter the sex ratio among their progeny, and they do so in a manner predicted by the degree of inbreeding their offspring will experience.

When a single female wasp parasitizes a host, her female offspring will be limited to mating with their brothers. In this circumstance, male offspring make a reduced contribution to the reproductive success of their mothers, as described above, and therefore these wasps skew the sex ratios of their progeny greatly in favor of females—to the point of producing only one male per brood in some species. Males of many of these species lack wings, and in extreme cases, fertilize their sisters as larvae within the host. However, when two or more females lay their eggs in the same host, their male offspring can mate either with their sisters or with the other females' daughters. As the possibility that sons might inseminate the female offspring of another wasp increases, the proportion of males in broods increases, just as one would expect.

How do wasps control the sex of their offspring? Hymenopterans (bees, ants, and wasps) have an unusual sex-determining mechanism by which fertilized eggs produce females and unfertilized eggs produce males. Consequently, females are diploid and males are haploid, a condition known as **haplodiploidy.** Reproductive females can control the sex ratio of their offspring simply by storing sperm when they mate and using it—or not—to fertilize their eggs.

Mating systems describe the pattern of pairing of males and females within a population

The **mating system** of a population is the pattern of matings between males and females—for example, the number of simultaneous or sequential mates each individual has and the permanence of the pair bond between them. Like the sex ratio, the mating system of a population is subject to natural selection and evolutionary modification. Consequently, the mating system of a population usually can be understood in terms of the ecological relationships of individuals.

Mating systems reflect variation in male and female reproductive success

It is a basic asymmetry of life that female and male functions contribute differently to an individual's evolutionary fitness. A female's reproductive success depends on her ability to make eggs and otherwise provide for her offspring. Large female gametes individually require more resources than tiny male gametes, so a female's ability to gather resources to make eggs determines her fecundity. A

male's reproductive success usually depends on the number of eggs he can fertilize.

When a male mates with as many females as he can locate and persuade, and provides his offspring with nothing more than a set of genes, he is said to be promiscuous. **Promiscuity** usually precludes a lasting pair bond. Among animal taxa as a whole, promiscuous mating is by far the commonest system, and it is universal among outcrossing plants. Promiscuity is associated with a high degree of variation in mating success among males compared with females: some individual promiscuous males may obtain dozens of matings while others get none. When eggs and sperm are shed directly into the water, or pollen is shed into wind currents, much of the variation in male mating success is simply random. Whether a particular sperm is the first to find an egg is largely a matter of chance. However, when males attract or compete for mates, reproductive success depends largely on body size and the quality of courtship displays, which are influenced by genetic factors and also by the condition of the male, as we shall see below. Even when fertilization occurs at random, males that produce the most sperm or pollen are bound to procure the most matings on average.

A mating system in which a single individual of one sex forms long-term bonds with more than one individual of the opposite sex is called **polygamy.** Most often, it is the male that mates with more than one female, in which case the system is referred to as **polygyny** (literally, "many females"). Polygyny may involve defense of several females against mating attempts by other males (a harem) (■ Figure 11.16), or defense of territories or nesting sites to which females gravitate to raise their young. Thus, polygyny may arise because a male can prevent access by other males to more than one female, in which case his contribution to his progeny may be primarily genetic, or because he can control or provide resources that females need for reproduction.

Monogamy is the formation of a pair bond between one male and one female that persists through the period that is required for them to rear their offspring, and which may endure even until one of the pair dies. Monogamy is favored primarily when males can contribute substantially to the number and survival of their offspring by providing parental care. Hence it is most common in species with dependent offspring that can be cared for equally well by either sex. Monogamy is not common in mammals because males neither carry the developing embryo nor produce milk. But it is common among birds, especially those in which parents feed their offspring. Male and female birds can incubate eggs and feed young equally well.

Recent genetic surveys of monogamous bird populations have revealed that males other than a female's mate may father some of her offspring, a result of so-called

Figure 11.16 Elephant seals are polygynous. Successful males attract many females to their harems and defend them against the sexual advances of other males. Photo by C. K. Lorenz/Photo Researchers.

extra-pair copulations, or **EPCs.** As many as a third of the broods produced by some species contain one or more offspring sired by a different male. Most EPCs involve males on neighboring territories, indicating considerable opportunism and infidelity in natural bird populations. This behavior surely increases, at relatively little cost, the fitness of neighboring males. It is not known whether EPCs benefit a mated female, but they could do so if the neighboring males have better genotypes than her mate or if her reproductive success is improved by greater genetic variation among her offspring. Regardless of whether females benefit or not, the constant threat of EPCs also has selected strongly for **mate guarding** behaviors on the part of males during their mates' periods of fertility.

The polygyny threshold model: relating mating system to ecology

As long as his territory holds sufficient resources, a male gains by increasing the number of his mates. A female increases her fecundity choosing a territory or a mate of high quality. Thus, polygyny arises when a female can obtain greater reproductive success by sharing a male with one or more other females than she can by forming a monogamous relationship.

Suppose that the quality of two males' territories differs so much that a female could rear as many offspring on the better territory, sharing it with other females and having little or no help from her mate, as she could on the poorer territory with help from a monogamous mate. The difference in territory quality at which polygynous and monogamous females do equally well is referred to as the **polygyny threshold** (Figure 11.17). According to the polygyny threshold model, polygyny should occur only when the quality of male territories varies so much that some females

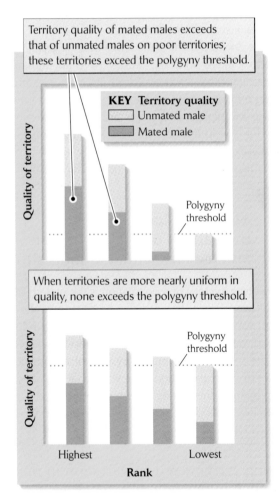

Figure 11.17 The polygyny threshold model predicts the variation in habitat quality at which polygamy will occur. When habitat quality varies enough, females may enjoy greater reproductive success by mating polygynously with males holding the best territories than by choosing an unmated male on a poor-quality territory.

Figure 11.18 The territories of male red-winged blackbirds vary in quality. The male red-winged blackbird has a conspicuous display to attract females and defend his territory against other males. Photo by Richard Day/Animals Animals.

will have higher reproductive success mated to a polygynous male on a high-quality territory than they would mated monogamously to a male on a poor-quality territory.

In cattail marshes throughout North America, male red-winged blackbirds establish territories in early spring (**Figure 11.18**). Marsh habitat is heterogeneous with respect to vegetation cover and water depth, which affect food supply and the safety of nests, and therefore territories vary greatly in their intrinsic quality. Females return to the breeding grounds after the males, by which time the males have established territories. Female blackbirds appear to assess the quality of male territories, and the first individuals to arrive pair monogamously with the best males—that is, those holding the best territories. Latecomers are faced with the choice between pairing monogamously with a low-quality male or pairing polygynously with a high-quality male, but sharing his territory's resources with one or more other females. In contrast to blackbirds, birds of forests live in habitats that are more homogeneous than marshes. Bird territories in forests vary less in quality, and most species of forest birds are primarily monogamous; few territories rise above the polygyny threshold.

MORE ON THE WEB *Alternative male reproductive strategies.* Males in different species take different approaches to winning a female's favor.

Sexual selection can result in male–male combat and elaborate male ornamentation

Regardless of mating system, the initial stages of reproduction involve choosing a mate. In polygynous and promiscuous mating systems, males gain by mating with as many females as they can, and choice of a mate is usually the prerogative of females. How should a female choose among the males that court her attention? If males differ in obvious ways that could affect a female's reproductive success, and if her offspring could inherit those features, she should choose to mate with the male of highest quality. Of course, males should do everything in their power to advertise their quality—that is, they should strut their stuff. This sets the stage for intense competition among males for mates and has resulted in male attributes evolved for use in combat with other males or in attracting females.

A usual result of sexual selection is **sexual dimorphism,** meaning a difference in the outward appearance of male and female individuals of the same species. Sexual selection tends to produce dimorphism especially in body size, ornamentation, coloration, and courtship behavior. Such traits, which distinguish sex over and above the primary sexual organs, are known as **secondary sexual characteristics.** Charles Darwin, in his book *The Descent of Man and Selection in Relation to Sex,* published in 1871, was the first to propose that sexual dimorphism could be explained by selection applied uniquely to one sex.

Sexual dimorphism can arise in three ways. First, the dissimilar sexual functions of males and females emphasize different considerations in the evolution of their life histories and ecological relationships. For example, because females produce large gametes, the number of their offspring often increases in direct relation to body size; this may explain why females are larger than males in many species, particularly when fertilization is internal and producing large numbers of sperm is not a major consideration.

Second, sexual dimorphism may result from contests between males, which may favor the evolution of elaborate weapons for combat, such as the antlers of elk (**Figure 11.19**) and the horns of mountain sheep (see Figure 14.11). Whichever males win such contests are more likely to gain access to females. If large size confers an advantage in these contests, males may be larger than females.

▌Figure 11.19 Sexual selection may favor elaborate weapons. Male elk use their immense antlers during contests to establish control over harems of females. Photo by William Grentell/Visuals Unlimited.

Third, sexual dimorphism may arise through the direct exercise of choice by individuals of the opposite sex. With few exceptions, females do the choosing, and males attempt to influence their choices with magnificent courtship displays. That females choose, and males compete among themselves for the opportunity to mate, is a consequence of the asymmetry of parental investment that defines the male and female conditions. As we saw above, males enhance their fecundity in direct proportion to the number of matings they obtain; females are limited in number of offspring by the number of eggs they can produce, but they stand to improve the quality of their offspring by choosing to mate with males that have superior genotypes.

Female choice

Most males experience female choice at some level. One of the first demonstrations of female choice came from an experimental study of tail length in male long-tailed widowbirds (*Euplectes progne*). This polygynous species inhabits open grasslands of central Africa. The females, which are about the size of a sparrow, are mottled brown, short-tailed, and altogether ordinary in appearance. During the breeding season, the males are jet black, with a red shoulder patch, and they sport a half-meter-long tail that is conspicuously displayed in courtship flights (▌Figure 11.20). Males may attract up to a half dozen females to nest in their territories, but they provide no care for their offspring. The tremendous variation in male reproductive success in this species provides classic conditions for sexual selection.

In a simple yet elegant experiment, researchers cut the tail feathers of some males to shorten them, and glued the clipped feather ends onto the feathers of other males' tails to lengthen them. Length of tail had no effect on a male's ability to maintain a territory, but males with experimentally elongated tails attracted significantly more females

▌Figure 11.20 Sexual selection may favor elaborate courtship displays. The tail of the male long-tailed widowbird (*Euplectes progne*) is a handicap in flight, but is attractive to females. Photo by Gregory G. Dimijian, M. D./Photo Researchers.

than those with shortened or unaltered tails (■ Figure 11.21). This result strongly suggests female choice of mates on the basis of tail length. Many subsequent studies have demonstrated that females choose their mates on the basis of such conspicuous differences among males.

MORE ON THE WEB *The origin of female choice.* **Many issues regarding female choice remain unresolved: Which came first, female choice or male traits that indicate intrinsic quality? How are the various ornaments of males related to fitness attributes? Why don't low-quality males cheat by taking on a high-quality appearance?**

Runaway sexual selection

Regardless of how female choice arises, once it is established in a population, it exaggerates fitness differences among males and may create what is known as **runaway sexual selection.** Whether female widowbirds intrinsically preferred males with longer tails, or tail length indicated fitness and females therefore evolved a preference for long tails, their mating preference would give long-tailed males a fitness advantage. If females choose by comparing among males, rather than by comparing males to some idealized standard of beauty, then mating preferences will continually select for further elaboration of male traits. In other words, if longer tails in males are what

females prefer, then longer tails will evolve. The peacock's tail, as well as the other outlandish (to our eyes) sexual ornaments and behaviors liberally spread throughout the animal kingdom, provide convincing evidence that some sort of runaway process must be at work.

If sexually selected traits indicate—at least initially, before runaway sexual selection takes hold—intrinsic attributes of male quality, we are then faced with a paradox. Presumably, such outlandish traits as the tail of the long-tailed widowbird burden males by making them more conspicuous to predators and by requiring energy and resources to maintain. How, then, can such traits indicate, let alone contribute to, male quality?

The handicap principle

One intriguing possibility, suggested by the Israeli biologist Amotz Zahavi, is that elaborate male secondary sexual characteristics act as handicaps. That a male can survive while bearing such a handicap indicates to a female that he has an otherwise superior genotype. This idea is known as the **handicap principle.** It may sound crazy, but if you wanted to demonstrate your strength to someone, you might make your point by carrying around a large set of weights. A weaker individual couldn't do it, and thus could not falsely advertise strength. Accordingly, the greater the handicap borne, the greater the ability of the individual to offset the handicap by other virtues—and to pass genes for those virtues on to his offspring. One small European songbird, the wheatear, takes the iron-pumping analogy literally and festoons its nesting ledge with up to 2 kilograms of small stones carried from a distance in its beak.

One virtue that males might possess, and which might be demonstrated by producing a showy plumage, is resistance to parasites and other disease-causing organisms. William D. Hamilton and Marlene Zuk first proposed this idea in 1982. They suggested that only individuals having genetic factors to resist parasite infection could produce or maintain a bright and showy plumage. Thus, an elaborate and well-maintained courtship display may provide a convincing demonstration of high male fitness, even when the display itself is an encumbrance (■ Figure 11.22). The importance of parasites to this theory is that they evolve rapidly and thereby continually apply selection for genetic resistance factors. We have already discussed similar reasoning for the evolutionary maintenance of sex itself.

The Hamilton–Zuk hypothesis, along with its subsequent modifications, comes under the general heading of **parasite-mediated sexual selection.** Its general assumptions—that parasites reduce host fitness, that parasites alter male showiness, that parasite resistance is inherited, and that

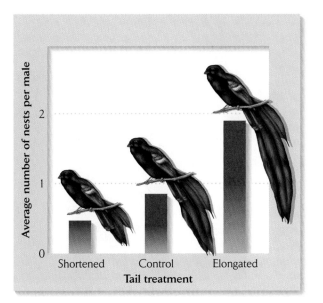

■ **Figure 11.21 The longer the tail, the more attractive the male.** Male long-tailed widowbirds with artificially elongated tails attracted more females to nest in their territories than did control males or males with shortened tails. *After M. Andersson, Nature 299:818–820 (1982).*

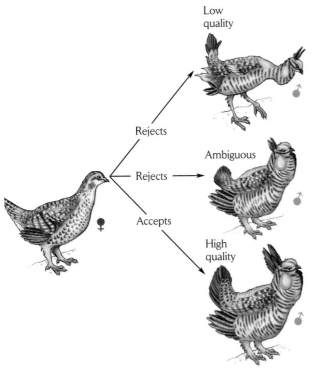

Low
quality

Rejects

Rejects

Ambiguous

Accepts

High
quality

❚ Figure 11.22 Parasites influence mate choice. Given a choice of three males, a female rejects the one with a short tail, which is too inconspicuous to reveal his parasite load. She also rejects the male whose long tail is obviously damaged by parasites, and chooses the male whose well-kept tail reveals that he is parasite-free. After D. H. Clayton, *Parasitology Today* 7:329–334 (1991).

females choose less parasitized males—are generally supported by experiments and field observations. For example, feather lice produce obvious damage by eating the downy portions of feathers and the barbules of feather vanes (❚ Figure 11.23). In feral rock doves, highly infested males had higher metabolic requirements in cold weather because of the reduced insulation provided by their plumage, and they were lighter in body mass. Female rock doves preferred clean to lousy males by a ratio of three to one.

A particularly elegant set of studies on ring-necked pheasants, performed by Torbjorn von Schantz and his colleagues at the University of Lund in Sweden, showed that females prefer males with long spurs (a spikelike projection from the back of a male pheasant's lower leg), and that long spurs are linked genetically to the major histocompatibility (MHC) genes that influence susceptibility to disease. Males with longer spurs had MHC alleles that were linked to longer life spans. Therefore, females that pick long-spurred males as mates should tend to produce offspring with a higher chance of surviving to reproduce as adults.

Sexual selection remains an active area of research, and much has yet to be learned. Studies of sexual displays show quite clearly, however, the power of natural selection to produce evolutionary modification of structures and behaviors, and how these changes can be directed by the asymmetry of sexual function in males and females.

 Summary

1. In most species, reproductive function is divided between two sexes. Sexual reproduction involves the production of male and female gametes with haploid chromosome numbers. Male and female gametes unite to form the zygotes that start a new generation. Haploid gametes are formed by meiosis, in which chromosome number is halved and maternal and paternal sets of genes are mixed.

2. The origin and maintenance of sexual reproduction remain controversial topics. Sex is thought to benefit individuals by increasing genetic variation among their progeny, which increases the probability that at least some may be well suited to varied conditions. Balancing this potential advantage in species with separate sexes is the so-called cost of meiosis: sexual females pass on only half as many genes to their progeny as asexually reproducing individuals.

(a)

(b)

❚ Figure 11.23 Feather lice produce substantial damage. (a) Scanning electron microscopic view of a louse on a host's feather. The louse is about 1 mm long, and is seen from a dorsal view. (b) Average (*center*) and heavy (*right*) damage to abdominal contour feathers by feather lice. A normal feather is at left. Courtesy of D. H. Clayton, from D. H. Clayton, *Am. Zool.* 30:251–262 (1990).

3. An alternative hypothesis for the maintenance of sex is the Red Queen hypothesis, which states that the production of genetically varied offspring slows the evolution of virulence in parasites and pathogens.

4. Most plants and some animals are hermaphrodites, meaning that they have both male and female sexual organs. Separation of the sexual functions between individuals (dioecy) occurs infrequently among plants, but commonly among animals. It is favored when either sexual function imposes large fixed costs or when limited resources must be allocated between the requirements of male and female function.

5. The sex ratio in a population balances the contributions of genes to progeny through male and female function. In general, because the rarer sex is favored, most populations at evolutionary equilibrium have equal numbers of males and females.

6. In some parasitic wasps, males compete with siblings for matings, and the sex ratio is shifted in favor of producing female offspring. In wasps and other hymenopterans, the sex of offspring is determined by whether an egg is fertilized or not, and thus is under direct control of the mother.

7. Mating systems may be monogamous (a lasting bond is formed between one male and one female), polygamous (more than one mate, usually female, per individual), or promiscuous (mating at large within the population, without lasting pair bonds).

8. Promiscuity may arise when males contribute little, other than their genes, to the number or survival of their offspring; this is the common condition in all plants and most animals. Males of promiscuous species often defend territories to which females come to mate.

9. Monogamy usually occurs in species in which males can increase their individual fitness more by caring for offspring than by seeking additional matings. In birds, monogamy is most frequent in species in which offspring are fed by their parents.

10. Polygyny arises when males can monopolize either resources or mates through intrasexual competition. According to the polygyny threshold hypothesis, some females may gain greater fitness by joining an already mated male that holds a superior territory than by joining an unmated male on an inferior territory.

11. When males compete among themselves for mates, females can choose among them. Female choice leads to sexual selection of traits in males that indicate fitness.

12. Sexually selected structures or behaviors may function as "handicaps" that only the more fit males in a population can bear without encumbrance.

13. Because parasites can evolve rapidly, and because they may directly affect the appearance or survival of males with elaborate ornaments or displays, resistance to parasites may be one factor that female preferences judge. This idea is referred to as parasite-mediated sexual selection.

PRACTICING ECOLOGY

CHECK YOUR KNOWLEDGE

Gynodioecy: Thyme over Time

Gynodioecy is the coexistence of both hermaphrodites and females in the same population. It occurs when genes arise among hermaphroditic plants that lead to male sterility, thereby creating plants that are for all intents and purposes females. In many cases, these genes occur in the chloroplast and are transmitted by cytoplasmic inheritance (through the cells of the ovule of the female plant). Such is the case for common thyme (*Thymus vulgaris*), a native of the Mediterranean region that is best known as an herb used to flavor food.

Among flowering plants, cytoplasmic genes are transmitted only through female gametes. Cytoplasmic genes that cause male sterility are strongly selected. If we assume that increased male function decreases female function, it is reasonable to suppose that such genes will be transmitted at a higher rate to future generations. However, male sterility genes are at odds with nuclear genes, which are transmitted equally well through female and male sexual function, and are favored in populations with high proportions of all-female plants. Thus, there is strong selection to reduce male sterility, which is accomplished via nuclear restorer genes that block the action of the cytoplasmic sterility genes. These opposing selective factors set up a constant conflict between nuclear and cytoplasmic genes in determining the sex ratio of thyme populations.

When the number of female thyme plants becomes high there are not enough males to provide pollen for fertilization, and female reproductive success declines. Hermaphrodites avoid the problem of having too few males by being able to self-fertilize. The actual sex ratio of a population depends on the availability of male sterility and restorer genes. Each of these types of genes arises sporadically and spontaneously by mutation. Over time, the sex ratio for thyme is rarely at equilibrium—it is constantly changing back and forth between relatively high and low levels of male sterility. Denis Couvet and his colleagues have studied the temporal variation in male sterility for gynodioecious populations of thyme. In addition, Manicacci et al. (1998) found that the frequency of females was high and varied considerably, as you will see as you answer the following questions.

CHECK YOUR KNOWLEDGE

1. Why is it important that we understand the reproductive patterns of plants?

2. Consult ▌ Figure 11.24, taken from the research of Manicacci et al. What can you conclude regarding the relative fecundity of females? How does the relative male investment differ across species?

MORE ON THE WEB
3. Read the paper by Manicacci et al. on the Web through *Practicing Ecology on the Web* at *http://www.whfreeman.com/ricklefs*. What do the authors conclude regarding the importance of temporal variation in female frequency on the effects of natural selection?

4. Some species of plants (such as *Opuntia bigelovii*, the teddy bear cholla) reproduce exclusively by asexual or vegetative means. Pieces break off, take root, and grow into a new plant. How can such species, lacking sex, evolve to match changes in their environment?

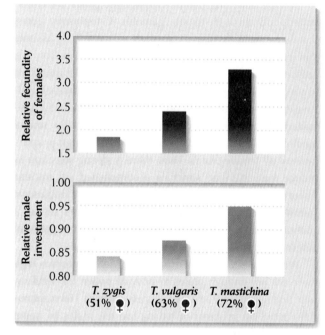

▌**Figure 11.24 The relative fecundity of females** compared with hermaphrodites (*top*) and the relative male investment in hermaphrodites (*bottom*), plotted against female frequency at the species level for *Thymus zygis, Thymus vulgaris,* and *Thymus mastichina.* Manicacci et al., *Int. J. Plant Sci.* 159: 948–957 (1998).

Suggested Readings

Alcock, J. 1980. Natural selection and the mating systems of solitary bees. *American Scientist* 68:146–153.

Andersson, M., and Y. Iwasa. 1996. Sexual selection. *Trends in Ecology and Evolution* 11:53–58.

Barrett, S. C. H., and L. D. Harder. 1996. Ecology and evolution of plant mating. *Trends in Ecology and Evolution* 11:73–79.

Borgia, G. 1995. Why do bowerbirds build bowers? *American Scientist* 83:542–547.

Charnov, E. L. 1982. *The Theory of Sex Allocation.* Princeton University Press, Princeton, NJ.

Clayton, D. H. 1991. The influence of parasites on host sexual selection. *Parasitology Today* 7:329–334.

Clutton-Brock, T. H., S. D. Albon, and F. E. Guinness. 1986. Great expectations: Dominance, breeding success and offspring sex ratios in red deer. *Animal Behaviour* 34:460–471.

Dybdahl, M. F., and C. M. Lively. 1998. Host-parasite coevolution: Evidence for rare advantage and time-lagged selection in a natural population. *Evolution* 52:1057–1066.

Ebert, D., and W. D. Hamilton. 1996. Sex against virulence: The coevolution of parasitic diseases. *Trends in Ecology and Evolution* 11:79–82. (The Red Queen hypothesis: the joint evolution of virulence and sex, in which recombination provides the genetic variation necessary to respond to parasite evolution)

Emlen, S. T., and L. W. Oring. 1977. Ecology, sexual selection, and the evolution of mating systems. *Science* 197:215–223.

Godfray, H. C. J., and J. H. Werren. 1996. Recent developments in sex ratio studies. *Trends in Ecology and Evolution* 11:59–63.

Klinkhamer, G. L., T. J. de Jong, and H. Metz. 1997. Sex and size in cosexual plants. *Trends in Ecology and Evolution* 12(7):260–265. (Sequential hermaphroditism: some plants change sex with size.)

Lively, C. M. 1996. Host–parasite coevolution and sex. *BioScience* 46:107–114.

Lively, C. M., and J. Jokela. 1996. Clinal variation for local adaptation in a host–parasite interaction. *Proceedings of the Royal Society of London B* 263:891–897.

Manicacci, D., A. Atlan, J. A. E. Rosello, and D. Couvet. 1998. Gynodioecy and reproductive trait variation in three *Thymus* species. *International Journal of Plant Sciences* 159:948-957.

Møller, A. P. 1994. *Sexual Selection and the Barn Swallow.* Oxford University Press, Oxford.

Reynolds, J. D. 1996. Animal breeding systems. *Trends in Ecology and Evolution* 11:68–72.

Slater, P. J. B., and T. R. Halliday (eds.). 1994. *Behaviour and Evolution.* Cambridge University Press, Cambridge.

Small, M. F. 1992. Female choice in mating. *American Scientist* 80:142–151.

Soler, M., J. J. Soler, A. P. Møller, J. Moreno, and M. Lindén. 1996. The functional significance of sexual display: Stone carrying in the black wheatear. *Animal Behavior* 51:247–254.

von Schantz, T., H. Wittzell, G. Göransson, M. Grahn, and K. Persson. 1996. MHC genotype and male ornamentation: Genetic evidence for the Hamilton–Zuk model. *Proceedings of the Royal Society of London* B 263:265–271.

Werren, J. H. 1987. Labile sex ratios in wasps and bees. *BioScience* 37:498–506.

Wilkinson, G. S., D. C. Presgraves, and L. Crymes. 1998. Male eye span in stalk-eyed flies indicates genetic quality by meiotic drive suppression. *Nature* 391:276–279.

Family, Society, and Evolution

- Territoriality and dominance hierarchies organize social interaction within populations

- Individuals gain advantages and suffer disadvantages from living in groups

- Natural selection balances the costs and benefits of social behaviors

- Kin selection favors altruistic behaviors toward related individuals

- Cooperation among individuals in extended families implies the operation of kin selection

- Game theory analyses illustrate the difficulties for cooperation among unrelated individuals

- Parents and their offspring may come into conflict over levels of parental investment

- Eusocial insect societies arise out of sibling altruism and parental dominance

Male lizards interact with one another through a variety of social behaviors. They perform displays that show off their size and coloration, presumably in an attempt to intimidate other males, and they also engage in chases and fighting when intimidation fails to achieve the desired result. Barry Sinervo, now at the University of California at Santa Cruz, and Curt Lively, whom we met through his study of sex and parasitism in freshwater snails, analyzed the peculiar social organization of the side-blotched lizard (*Uta stansburiana*) in an area of northern California. In this local population, male side-blotched lizards come in three varieties, or morphs, which are genetically determined. The orange, or O morph, lizards are large, aggressive, dominant over blue (B morph) lizards, and short-lived. B lizards, which are smaller than O lizards, are vigilant and are dominant toward yellow (Y) lizards, which mimic females in coloration and behavior (■ Figure 12.1).

Although all three morphs coexist in the same population, this coexistence is not stable, and the frequencies of the morphs vary over time. Consider how this works. When O are numerous, numbers of B are depressed by their aggression, but Y can sneak onto other males' territories to mate with females because the O males are busy chasing B males, and they don't discriminate Y males from females. Thus, the frequency of Y males increases among the progeny produced in each generation. However, when Y become numerous, the vigilant B males are not fooled by their female appearance, and they chase them out of their territories, Thus, B males increase when Y males are common. When B are numerous, O males can dominate B males, and the proportion of O increases. As you can see, this pattern leads to cycling in the frequencies of the male morphs in the population (■ Figure 12.2).

The whole story is further complicated by female choice. For example, when O males are most common, females should benefit

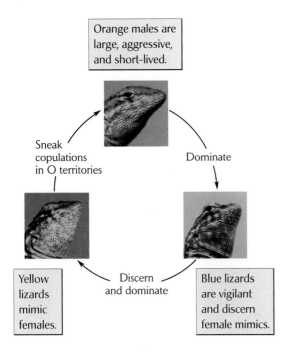

Figure 12.1 The side-blotched lizard has three male morphs. The three morphs of *Uta stansburiana* differ in their behavior as well as in coloration. Photos by R. E. Ricklefs.

by producing a high proportion of Y males among their offspring, and therefore should chose Y males as their mates. Because the frequency of Y increases when O is common, Y males have the higher fitness under this condition.

Sinervo and Lively noted that the fitness relationships among the three male morphs resemble the rock-paper-scissors game many of us played

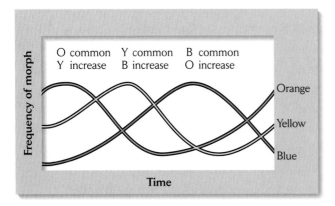

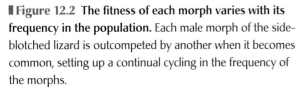

Figure 12.2 The fitness of each morph varies with its frequency in the population. Each male morph of the side-blotched lizard is outcompeted by another when it becomes common, setting up a continual cycling in the frequency of the morphs.

when we were children. Rock can be covered by paper, which can be cut by scissors, which can be broken by rock. Thus, the outcome of any one choice depends on whether the opponent plays rock, paper, or scissors. In the same way, the fitness of each male morph of the side-blotched lizard depends on the frequency of the other morphs in the population. The most common morph is always being replaced by a less common morph, leading to a cycling of frequencies.

Clearly, the social and family environment of an individual, along with its relationship to members of the opposite sex, applies strong selection on behavior and, indirectly, on life histories and ecological relationships.

During the course of its life, each individual interacts with many others of the same species: mates, offspring, other relatives, and unrelated members of its social group. Each interaction requires that the individual perceive the behavior of others and make appropriate responses. When individuals behave in a more friendly, supportive manner toward close relatives than they do toward unrelated individuals, it may be because relatives share genes inherited from a common ancestor and therefore have a common evolutionary interest. Mates, too, must cooperate if they are to raise offspring successfully. Nonetheless, social behaviors emphasize that all interactions between members of the same species delicately balance conflicting tendencies of cooperation and competition, altruism and selfishness.

Humans are the most social of all animals. Our societies are sustained by role specialization among their members, the interdependence attendant upon specialization, and the cooperation that interdependence requires. Yet humans also are competitive, to the point of violence, within this mutually supportive structure. Our social life balances contrasting tendencies toward mutual help and conflict. Some animal populations exhibit much of the complexity of human societies. The social insects—ants, bees, wasps, and termites—are remarkable for their division of labor and behavioral integration within the hive or nest. Similar subtlety of social interaction, including role specialization and altruistic behavior, is being discovered increasingly among other animals, especially mammals and birds.

Social behavior includes all kinds of interactions between individuals, from cooperation to antagonism. In this chapter we shall explore some of the consequences for individuals of interactions within social and family groups, and we shall describe various ways in which social relationships are managed by the behaviors of individuals toward others in the population.

Figure 12.3 Territoriality is often most conspicuous in highly mobile animals. Researchers mark male damselflies with painted numbers to follow their movements and behavioral interactions. Photo by R. E. Ricklefs.

Territoriality and dominance hierarchies organize social interaction within populations

Any area defended by an individual against the intrusion of others may be regarded as a **territory.** Territories may be transient or more or less permanent, depending on the stability of resources and an individual's need for those resources. Territoriality is most conspicuously displayed by birds and other highly mobile animals, which may actively defend areas throughout the year or only during the breeding season (**Figure 12.3**). Many migratory species establish territories on both the breeding and wintering grounds; shorebirds defend feeding areas for a few hours or days at stopover points on their long migrations. Hummingbirds and other nectar feeders defend individual flowering bushes and abandon them when their flowering periods are over (**Figure 12.4**). Male ruffs and grouse defend a few square meters of space on a communal display ground. During the egg-laying period, males of many species stay close to their mates and chase off would-be interlopers. As long as a resource is defensible and the rewards outweigh the cost of defense, animals are likely to maintain territories.

In some situations, the establishment of territories may not be practical because of the pressures of high population density, the transience of critical resources, or the overriding benefits of living in groups. In such circumstances, when conflicts occur, social rank rather than space may be the winner's prize. Once individuals order themselves into a **dominance hierarchy** of social status, subsequent contests between them are resolved quickly in favor of higher-ranking individuals. When a social hierarchy is linearly ordered among individuals in a group, the first-ranked

member dominates all others, the second-ranked dominates all but the first-ranked, and so on down the line to the last-ranked individual, who dominates none.

Occupation of space and social rank are opposite sides of the same coin, and often are directly related. Even within a social group, the position of an individual in a dominance hierarchy is sometimes reflected by its spatial position within the flock or herd. In large foraging flocks of wood pigeons, for example, individuals low in dominance tend to be at the periphery, where they are more vulnerable to predators than the dominant individuals at the flock's center. Peripheral birds appear nervous, and because they spend much of their time looking up from feeding, they are often undernourished. Birds in the center of the flock remain calm and feed more because they are protected from surprise attack by the vigilance of the birds at the periphery.

Whether an individual lives within a territorial system or in a group setting, its social rank is determined by its

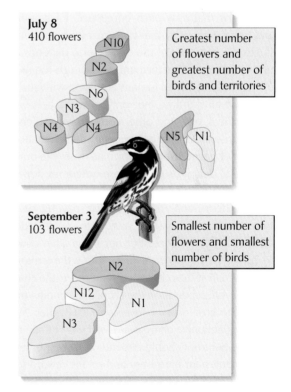

Figure 12.4 Degree of territoriality varies with resource abundance. Territory size in the New Holland honeyeater (*Phylidonyris novaehollandiae*) increases as the abundance of nectar-producing *Banksia* flowers decreases. Numbers within territories refer to individual birds followed through the season. After D. C. Paton and H. A. Ford, in C. E. Jones and R. J. Little (eds.), *Handbook of Experimental Pollination Biology*, Van Nostrand Reinhold, New York (1983), pp. 235–248.

ability to win contests. The outcomes of these contests are all-important to the individual because they determine the quality and amount of space it can defend and its access to food and mates. Each contest between two individuals can be resolved only through behavioral decisions taken by each participant. A spider confronting another over a particularly good place to build a web assesses the situation and decides either to back down or to escalate the contest. Sometimes the projected outcome of a physical contest is obvious, and the smaller individual retreats. When the outcome is more difficult to judge in advance, the two spiders may engage in a series of elaborate displays that help them to weigh each other's fighting abilities, each hoping (though probably not consciously) that the other will be duly impressed and back down. If the match appears to be close and the outcome uncertain, the contest may then escalate to actual fighting, with the risk of serious injury or death to one or both participants.

Optimal behavior in a contest depends on each contestant's assessment of the likely outcome of the contest and the payoffs of winning or losing. What actually happens—that is, how the contest plays out—also depends on the decisions made by each contestant. Each individual should behave in a way that maximizes the net benefit to itself, but the outcome of the contest also depends on the behavior of the other participant, over which the first individual has little control. Humans are faced with such decision making all the time, not only in social behavior but also in business, war, and other competitive and cooperative enterprises. Optimal behaviors in these situations are the subject of **game theory,** which analyzes the outcomes of behavioral decisions when these outcomes depend on the behavior of other players.

Game theory analysis is based on payoffs, or fitness consequences, of behaviors. Consider the spider's decision whether to escalate a contest or not. If the other contestant backs down, the payoff to the first spider is the territory, and the cost is small. If the other contestant meets the challenge, then the payoff depends on the chance of winning the contest, and the cost—win or lose—is much higher. Without making a quantitative analysis, it is still easy to see that an individual's behavior should depend on its best estimate of the other contestant's response and on the reward for winning. When the first spider is much larger than the second, it is likely that the second will back down from any confrontation, so escalation carries little risk of harmful conflict—an easy win, so to speak. When the two are evenly matched, both the response of the second spider and the outcome of the conflict are more difficult to predict, and the probability of getting hurt is higher. Under such circumstances, both escalation and meeting the challenge are likely

to occur only when the potential rewards for winning a contest are large. It is no surprise, then, that spiders are observed to fight only over the best web sites, and only when the two contestants are similar in size.

MORE ON THE WEB *Ritualized antagonistic behavior reduces the incidence of fighting.* **Certain appearances or behaviors signal high social status and discourage aggression by subordinate individuals.**

Individuals gain advantages and suffer disadvantages from living in groups

Animals get together for a variety of reasons. Sometimes they are independently attracted to the same habitat or resources and form aggregations, such as those of vultures around a carcass or dung flies on a cowpat. Within such groups, individuals may interact, usually to compete for space, resources, or mates. In other cases, offspring remain with their parents to form family groups, and aggregation results from their failure to disperse. True social groups, however, arise through the attraction of unrelated individuals to one another—that is, through a purposeful joining together.

Animals form groups to increase their chances of surviving, feeding, or finding mates. When they are in groups, individuals tend to spend more time feeding and less time looking out for predators. Consider the data presented in ▮ Figure 12.5 for the European goldfinch (*Carduelis carduelis*), which feeds on seed heads of plants in open fields and hedgerows. Two factors control optimal flock size in these birds. As flock size increases, each individual spends less time looking out for predators. If you watch closely as birds feed, you will notice that they raise their heads and look around from time to time. In a larger group, an individual goldfinch can spend more time going about the business of eating and can gather and husk seeds more rapidly, because the total vigilance of the flock is higher. Balancing this advantage of reduced individual vigilance time, a larger flock depresses a local food supply faster, and individuals are forced to fly farther between suitable foraging patches, using valuable feeding time and energy and perhaps increasing their vulnerability to predators. Thus, joining a flock is a good choice for an individual as long as the flock is not too large.

MORE ON THE WEB *Social groups as information centers.* **Watching your neighbors can provide valuable information about food resources and habitat quality.**

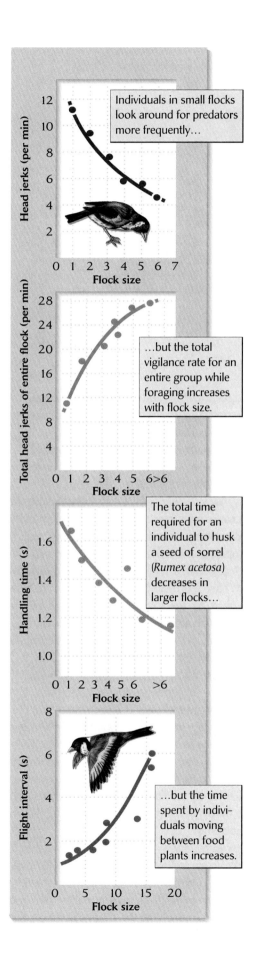

Figure 12.5 Individuals in groups spend less time in vigilance and more time feeding. From top to bottom, these four graphs show how flock size in the European goldfinch (*Carduelis carduelis*) affects: mean rates of individuals looking up from foraging; total vigilance rate for the entire group; time required to husk each seed of sorrel (*Rumex acetosa*); and time spent by individuals moving from one plant to the next. After E. Gluck, *Ethology* 74:65–79 (1987).

Callout (top graph): Individuals in small flocks look around for predators more frequently…

Callout (second graph): …but the total vigilance rate for an entire group while foraging increases with flock size.

Callout (third graph): The total time required for an individual to husk a seed of sorrel (*Rumex acetosa*) decreases in larger flocks…

Callout (bottom graph): …but the time spent by individuals moving between food plants increases.

Natural selection balances the costs and benefits of social behaviors

Any social interaction other than mutual display can be dissected into a series of behavioral acts by one individual, the **donor** of the behavior, directed toward another, the **recipient.** One individual delivers food, the other receives it; one threatens, the other is threatened. When one individual attacks another, the attacker may be thought of as the donor of a behavior. The attacked individual (the recipient in this case) may respond by standing its ground or by fleeing; in either case, it thereby becomes the donor of a subsequent behavior. The donor–recipient distinction is useful because each act has the potential to affect the reproductive success of both the donor and the recipient of the behavior. These increments of reproductive success, or fitness, may be positive or negative, depending on the interaction.

Four combinations of cost and benefit to donor and recipient can be used to organize social interactions into four categories (**Figure 12.6**). **Cooperation** and **selfishness**

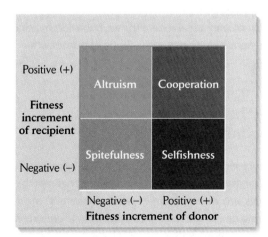

Figure 12.6 Social interactions can be organized into four categories. Classifications are according to the effects of the actions on the fitness of donors and recipients.

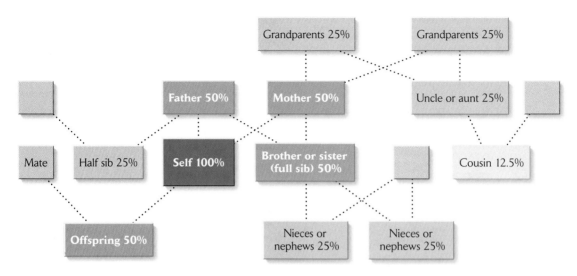

❚ Figure 12.7 Kin selection is based on degrees of genetic relationship among relatives. Identity by descent, which is the probability of occurrence in a relative of a copy of a gene carried by oneself, varies with the coefficient of relationship.

both benefit the donor of the behavior, and therefore should be favored by natural selection. **Spitefulness**—behavior that reduces the fitness of both donor and recipient—cannot be favored by natural selection under any circumstance, and presumably does not occur in natural populations. The fourth type of behavior, **altruism,** benefits the recipient at a cost to the donor.

Altruism presents a difficult problem because it requires the evolution of behaviors that reduce the fitness of the individuals performing them. We would expect selfish behaviors to prevail to the exclusion of altruism because selfishness increases the fitness of the donor. However, altruism appears to have arisen in colonies of social insects, in which workers forgo personal reproduction to rear the offspring of the queen, their mother. We humans also like to think that we are not only capable of altruistic behavior, but that such interactions hold together the fabric of our society.

Kin selection favors altruistic behaviors toward related individuals

The evolutionary dilemma posed by the apparent altruism of social insects is resolved when one recognizes that their colonies are discrete family units, containing mostly the offspring of a single female (the queen). Therefore, behavioral interactions within an ant colony or beehive occur between close relatives—in this case, between siblings. When an individual directs a behavior toward a sibling or other close relative, it influences the fitness of an individual with which it shares more genes than it does with an

individual drawn at random from the population. This special outcome of social behavior among close relatives is referred to as **kin selection.**

Close relatives have a certain probability of inheriting copies of the same gene from a particular ancestor. The likelihood that two individuals share copies of any particular gene is the probability of **identity by descent,** the value of which varies with their degree of relationship (❚ Figure 12.7). For example, two siblings have a 50% probability of inheriting copies of the same gene from one parent. This probability is also called their **coefficient of relationship.** Two cousins have a probability of one in eight (12.5%) of inheriting copies of the same gene from one of their grandparents, which are their closest shared ancestors.

When an individual behaves in a particular way toward a close relative, that act influences not only its own personal fitness, but also the fitness of an individual that shares a portion of its genes. Suppose that an act of altruism is directed toward a sibling. The probability that the recipient of the behavior (the sibling) will also have a copy of any particular one of the donor's genes is 50%. Therefore, if a tendency toward the behavior is inherited, the fitness of a gene influencing that behavior will be determined both by its influence on the fitness of the donor and by its influence on the fitness of the recipient, weighted by their coefficient of relationship.

Biologists refer to the total fitness of a gene responsible for a particular behavior as its **inclusive fitness.** Inclusive fitness is the contribution of the gene to the fitness of the donor, resulting from its own behavior plus the product arrived at by multiplying the change in fitness of the recipient times the probability that the recipient carries a copy of the same gene (■ Figure 12.8). Therefore, the inclusive fitness of a gene for altruistic behavior would exceed that of its selfish alternative as long as the cost to the altruist was less than the benefit to the recipient multiplied by the average genetic relationship of donor and recipient. Thus, a gene promoting altruistic behavior will have a positive inclusive fitness and will increase in the population when the cost (C) of a single altruistic act is less than the benefit (B) to the recipient times their coefficient of relationship (r); that is, when $C < Br$. This equation can be rearranged to show that the condition for the evolution of altruism is $C/B < r$; that is, the cost–benefit ratio, which is a measure of how altruistic the behavior is, must be less than the average coefficient of relationship between donor and recipient.

While inclusive fitness makes possible the evolution of altruism among close relatives, it also constrains the evolution of selfish behavior toward relatives. For a selfish behavior, B represents the benefit to the donor and C the cost to the recipient. Accordingly, selfish behavior among close relatives can evolve only when $B > Cr$, or $C/B < 1/r$. The cost–benefit ratio (C/B) is, in this case, a measure of the selfishness of the behavior. A higher coefficient of relationship (r) between donor and recipient reduces the level of selfishness that can evolve (■ Figure 12.9).

The maintenance of altruistic behavior by kin selection requires that such behaviors have a low cost to the donor and be restricted to close relatives. Individuals of many

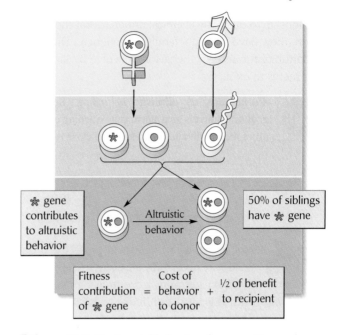

■ Figure 12.8 **Most social behaviors have positive and negative consequences for inclusive fitness.** The inclusive fitness of a gene controlling behavior toward relatives is the cost of the behavior to the donor plus the benefit to the recipient multiplied by their coefficient of relationship.

species tend to associate in family groups, and limited dispersal often keeps close relatives together. Moreover, individuals of many species can sense their degree of

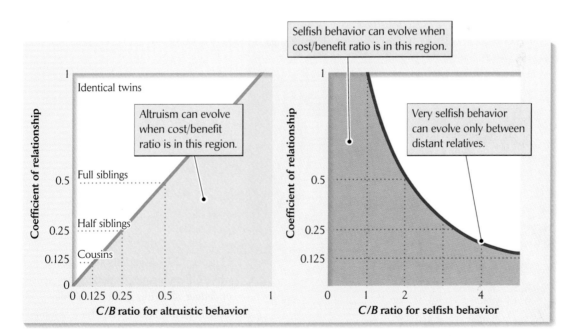

■ Figure 12.9 **Inclusive fitness constrains the evolution of behaviors between close relatives.** The potential level of altruism that can evolve in response to kin selection increases with the degree of genetic relationship between the interacting individuals. The level of selfish behavior is increasingly constrained as genetic relationship increases.

relationship to others by chemical or behavioral cues, even when they have had no family experience. Thus, the opportunity for altruistic behavior to evolve by kin selection seems to exist.

MORE ON THE WEB *Alarm calls as altruistic behaviors.* Belding's ground squirrels give alarm calls warning of predators more often in the presence of close relatives.

ECOLOGISTS IN THE FIELD

Are cooperative acts always acts of altruism?

Not all behaviors that help protect a social group are necessarily altruistic. In the meerkat (*Suricata suricatta*), a group-living mongoose of southern Africa (Figure 12.10), individuals take up positions on raised structures such as mounds or dead trees and stand guard while others in the group forage. Timothy Clutton-Brock, of the University of Cambridge, and his coworkers have spent thousands of hours observing guarding behavior in natural groups of this species in the field in Kalahari Gemsbok Park, South Africa. The major issue they have addressed is whether guarding behavior is altruistic or not. That is, does a guarding individual suffer a decrease in personal fitness to increase the fitness of other members of its group?

In this situation, the answer appears to be no. An individual assumes a guard position only after it has filled its own stomach by foraging for small invertebrates dug out of the soil. Thus, guarding does not detract from foraging. Furthermore, a guarding individual is free to keep watch from a safe site close to a burrow. Guards are usually the first to see approaching predators and, after emitting warning calls, they are the first to reach safety underground. So the cost of guarding is likely to be low. The larger a group of meerkats, the more likely it is to be guarded at any particular time (Figure 12.11). In a larger group, more individuals are potentially available to guard, and individuals that forage fill their stomachs more quickly because they spend less time in vigilant behavior.

The role of satisfying food requirements in guarding behavior was demonstrated by a simple experiment in which ten individuals were fed 25 grams of hard-boiled egg and their guarding behavior compared with that on previous days when they had not received food supplements. When fed, individuals stood guard more often for longer periods and were more likely to go on guard before foraging in the morning. That meerkats bother to assume sentinel positions at all may reflect the fact that their groups are extended families with a single, dominant breeding female and therefore most of the individuals are close relatives. Regardless of the benefits of guarding to its relatives, the cost to the donor evidently is small.

Figure 12.10 Meerkats stand guard while others forage. Alerted by their sense of danger, all of this group in Kalahari Gemsbok Park, South Africa, have assumed a stance typical of sentinel individuals that guard foraging group members and warn them of approaching enemies. Photo by J & B Photographers/Animals Animals.

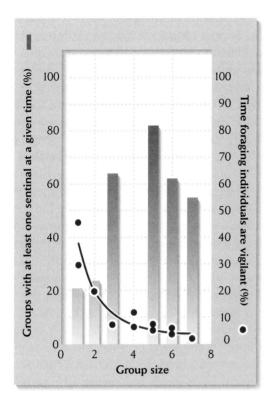

Figure 12.11 Larger groups of meerkats are more likely to be protected by sentinels. Nonetheless, foraging individuals in larger groups spend more time foraging and less time being vigilant, and therefore satisfy their food needs in less time and are available to stand as sentinels. From T. Clutton-Brock et al., *Science* 284:1640-1644 (1999).

Cooperation among individuals in extended families implies the operation of kin selection

Human extended families include the nuclear family of a mated pair and their young progeny as well as, to varying degrees, grandparents, uncles and aunts, cousins, nephews and nieces, and sometimes individuals of uncertain relationship to the rest. These families are complex social units within which occur a tremendous variety of social interactions, most of them cooperative, but many competitive enough to stress the bonds that hold a family unit together. Rarely do human extended families include more than one child-producing pair, and at least a portion of the behavior of non-nuclear members of the family is directed toward supporting the well-being and upbringing of the children.

Studies of the white-fronted bee-eater (*Merops bullockoides*) in East Africa by Stephen Emlen, Peter Wrege, and Natalia Demong, of Cornell University, have revealed complex extended families in this species (Figure 12.12). These families are typically multigenerational groups of three to seventeen individuals, often including two or three mated pairs plus assorted single birds—unpaired young and widowed older individuals. Careful observations of individually marked birds over several years have shown that these social groups are truly extended families, made up of related individuals and their mates, which normally come from other families. Although relationships within extended families tend to be cooperative, bee-eater family groups are hardly models of harmonious behavior; one sees the usual squabbling over food, nest sites, and mates. Remarkably, however, selfless and selfish acts appear to be directed toward other individuals very much in accordance with degree of relationship: brothers and sisters are

Figure 12.12 A family of white-fronted bee-eaters.
Courtesy of Natalia Demong.

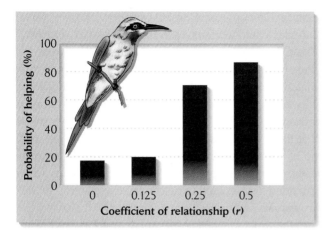

❙ Figure 12.13 **The frequency of altruistic behavior varies with degree of relationship.** White-fronted bee-eaters engaged in more helping behaviors toward close relatives than toward more distant relatives. From S. T. Emlen, P. H. Wrege, and N. J. Demong, *Am. Sci.* 83:148–157 (1995).

treated better than half-siblings and uncles, for example, and cousins fare almost as badly as nonrelatives outside the family group (❙ Figure 12.13).

Through their behavior, bee-eaters tell us that individuals know who their relatives are and can distinguish subtle differences in degree of relationship. We can also conclude from the distribution of helpful and harmful behaviors in this species that inclusive fitness is the appropriate measure of selection on social behavior. In other words, altruistic behaviors can indeed evolve among close relatives by kin selection.

We will return to interactions between family members later in this chapter. Before doing so, however, we shall consider whether or not cooperative or altruistic behavior can evolve among nonrelatives. Clearly, social groups can form owing to the self-interest of group members seeking protection from predators, or perhaps to some efficiency gained by foraging or hunting with other individuals. Whether groups of unrelated individuals can take the next step toward true cooperation is a fundamental issue in the evolution of social behavior. We shall address this issue by means of a simple game theory analysis.

Game theory analyses illustrate the difficulties for cooperation among unrelated individuals

Self-interest rules behavior among nonrelated individuals. A paradox of social behavior is that conflict can reduce the

fitness of selfish individuals below that of cooperative individuals. Because natural selection favors increased fitness, it should be possible for cooperation to evolve within societies at large. The problem with this reasoning is that when most of a society consists of cooperative members, a selfish individual can greatly increase its personal fitness by "cheating." Thus, selfish behavior will always be favored by natural selection, which will prevent groups from crossing the threshold of cooperative behavior to become true societies.

The logic of this somewhat pessimistic argument can be shown by a simple game theory analysis. The approach we will use is called the **hawk–dove game** (it is also known in a different context as the prisoner's dilemma). Let's assume that one type of individual will always behave selfishly in conflict situations, always being willing to fight over contested resources and taking all the reward when it wins: this is hawk behavior. In contrast, doves never compete over a resource, but share it evenly with other doves. Each contest between two individuals has a potential reward, or benefit (B), and it may have a cost (C) if a contest results in physical conflict. The payoff to an individual—either hawk or dove—depends on the behavior of the second contestant—that is, whether it is a hawk or a dove. For example, two hawks always fight and, on average, get half the reward, so the payoff for hawk behavior toward another hawk is $\frac{1}{2}B - C$, which is one-half of the average benefit minus the cost of fighting. When a hawk confronts a dove, the hawk gains the entire uncontested reward at no cost; thus, the payoff for the hawk is B, and the payoff for the dove is zero. When two doves come together at a resource, they share it, and incur no cost of conflict, so the payoff is $\frac{1}{2}B$.

The average payoff (fitness) to hawks and doves depends on the relative proportions of the two kinds of individuals in a population. Let p be the proportion of hawks and ($1 - p$) the proportion of doves. The payoffs are now as follows: hawks receive $p(\frac{1}{2}B - C) + (1 - p)B$, and doves receive $\frac{1}{2}(1 - p)B$. A population consisting only of hawks ($p = 1$) has an average payoff of $\frac{1}{2}B - C$, which is less than the average payoff of $\frac{1}{2}B$ in a population consisting only of doves ($p = 0$). Clearly, the dove strategy would be the best all around from a social point of view, because resources are distributed evenly without the cost of fighting.

The problem is that dove behavior is not an **evolutionarily stable strategy**: that is, it cannot resist evolutionary invasion by an alternative strategy—namely, hawkish behavior. A single hawk in a population of doves (p close to 0) receives twice the average payoff that doves do (B versus $\frac{1}{2}B$) because it never encounters another hawk, and con-

4. reproductive dominance by one or a few individuals, including the presence of sterile **castes.**

Thus defined, eusociality is limited among the insects to the termites (Isoptera) and the ants, bees, and wasps (Hymenoptera). Elements of eusociality are present in at least one mammal, the naked mole-rat of Africa.

From its distribution across taxonomic groups, it is clear that eusociality has evolved independently many times in bees, wasps, and ants. The evolutionary steps that lead to eusociality are less clear. The most widely accepted sequence of evolutionary events includes a lengthened period of parental care for the developing brood, with parents either guarding their nests or continuously provisioning their larvae in a manner similar to birds feeding their young. If parents lived and continued to produce eggs after their first progeny emerged as adults, then their offspring would be in a position to help raise subsequent broods consisting of their younger siblings. Once progeny remain with their mother after they attain adulthood, the way is open to relinquishing their own reproductive function solely to support hers.

Organization of insect societies

Insect societies are dominated by one or a few egg-laying females, which are referred to as **queens.** The queens in colonies of ants, bees, and wasps mate only once during their lives and store enough sperm to produce all their offspring, up to a million or more over 10–15 years in some army ants. Nonreproductive progeny of a queen gather food and care for developing brothers and sisters, some of which become sexually mature, leave the colony to mate, and establish new colonies.

Bee societies are organized simply: the offspring of a queen are divided among a sterile worker caste, which is all genetically female, and a reproductive caste, consisting of both males and females, that is produced seasonally. Whether an individual becomes a sterile worker or a fertile reproductive is controlled by the quality of nutrition it receives as a developing larva. In general, the differentiation of sterile castes is stimulated by environmental (usually nutritional) factors. Substances produced by a queen and fed to her larvae can inhibit the development of reproductive forms. In bees, the worker caste represents an arrested stage in the development of reproductive females, stopped short of sexual maturity.

Unlike ant, bee, and wasp societies, termite colonies are headed by a mated pair—the king and queen—which produce all the workers by sexual reproduction. Termite workers are both male and female, but neither sex matures sexually unless either the king or the queen dies.

Coefficients of genetic relationship in hymenopteran societies

Hymenopterans have a haplodiploid sex-determining mechanism (see Chapter 11). The workers are all females produced from fertilized eggs (▌Figure 12.18). Males, which develop from unfertilized eggs, appear in colonies only as reproductives (drones) that leave to seek mates. The haplodiploid sex determination system creates strong asymmetries in coefficients of genetic relationship within insect societies (Table 12.1). In particular, a female worker's coefficient of relationship to a female sibling is 0.75, whereas to a male sibling it is 0.25. The queen herself has the same genetic relatedness to sons and to daughters (0.50), so she can be relatively ambivalent about the sex of her offspring, especially when the sex ratio among reproductive individuals in the population as a whole is near equality. The skewed genetic relatedness among siblings means that cooperation is likely to be greater among all-female castes than among male castes or, especially, among mixed castes. This may explain why workers in hymenopteran societies are all female, and why broods of reproductive individuals usually favor females, by about 3:1 on a weight basis. Furthermore, when a female worker can help to rear more

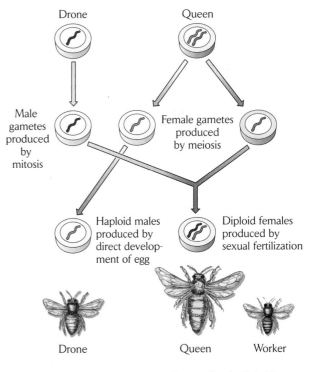

▌**Figure 12.18 The Hymenoptera have a haplodiploid sex determination system.** A queen can determine the sex of her offspring by using stored sperm to fertilize eggs, producing diploid females (workers), or by not fertilizing them, producing haploid males (drones).

Table 12.1	Probabilities of identity by descent between male and female individuals and their relatives in eusocial hymenopterans		
Probability of identity by descent with		Male	Female
Mother		0.50	0.50
Father		0.00	0.50
Brother		0.50	0.25
Sister		0.25	0.75
Son		0.00	0.50
Daughter		1.00	0.50

female than male reproductive individuals, her own inclusive fitness may actually be higher than it would be if she raised a brood of her own consisting of an equal number of males and females. Under these bizarre circumstances, it is not surprising that sterile castes might have evolved.

Behavioral relationships among the social insects represent one extreme along a continuum of social organization from animals that live alone except to breed to those that aggregate in large groups organized by complex behavior. Regardless of their complexity, all behaviors balance costs and benefits to the individual and to close relatives affected by its behavior. Like morphology and physiology, behavior is strongly influenced by genetic factors and thus is subject to evolutionary modification by natural selection. The evolution of behavior becomes complicated when individuals interact within a social setting and the interests of individuals within a population may either coincide or conflict. Understanding the evolutionary resolution of social conflict in animal societies continues to be one of the most challenging and important concerns of biology.

Summary

1. Selection imposed by interactions with family members and with unrelated individuals within a population provides the basis for evolutionary modification of social behavior.

2. Territoriality is the defense of an area or resource from intrusion by other individuals. Animals are more likely to maintain territories when the resources they gain by doing so are rewarding and defensible.

3. Dominance hierarchies order individuals within social groups by rank, which is often established by direct confrontation in social contests. Because rank is generally respected, dominance relationships may reduce conflict within the group.

4. Living in large social groups may benefit individuals by enabling them to better detect and defend against predators or to obtain food more efficiently. Groups form to the extent that such benefits outweigh the costs of competition among group members.

5. Isolated acts of social behavior involve a donor and a recipient. When both benefit from their interaction, the behavior is termed cooperation; when the donor benefits at a cost to the recipient, the behavior is selfish; when the recipient benefits at a cost to the donor, the behavior is altruistic.

6. The presence of altruistic behavior in populations has been explained in terms of kin selection. When an individual interacts with a relative, it affects the fitness of that portion of its own genotype that is also inherited by the relative directly from a common ancestor.

7. Inclusive fitness expresses the benefit (or cost) of a behavior to the donor plus the benefit or cost to the recipient adjusted by the coefficient of their relationship. In the case of interactions between siblings, which have a coefficient of relationship of 0.50, selection will favor any altruistic behavior whose cost to the donor is less than one-half the benefit it confers on the recipient.

8. In general, the distribution of cooperation and altruism within social groups is sensitive to degree of genetic relatedness between individuals.

9. Game theory analyses, such as the hawk–dove game, indicate that cooperative behavior cannot evolve among nonrelatives even though the average benefit to individuals in a purely cooperative social group exceeds that gained by individuals through confrontation and conflict. The reason is that cooperative behavior is not an evolutionarily stable strategy, but can be invaded by selfish cheaters.

10. Conflict may arise between parents and offspring over the optimal level of parental investment. All siblings are genetically equal in the eyes of their parents, but siblings are genetically related to one another by only ½. Therefore, individual offspring should prefer unequal parental investment in themselves, even when parental fitness is reduced as a result.

11. Social insects (termites, ants, wasps, and bees) live in extended family groups in which most offspring are retained in a colony as sterile workers, increasing their mother's fitness by rearing reproductive siblings.

12. The haplodiploid sex determining mechanism of the Hymenoptera results in females having a ¾ coefficient of relationship to sisters, but only ¼ to brothers. This skew probably has contributed to the workers in ant, bee, and wasp colonies all being female and to the production of more female reproductives than males.

PRACTICING ECOLOGY

CHECK YOUR KNOWLEDGE

Angry Ants

In many species of social insects, colonies have more than one queen, and often they are not close relatives. In such situations, the degree of genetic relationship among workers is, on average, greatly reduced. Moreover, kin selection on altruistic traits may not be strong enough to maintain peace within the colony. Indeed, the fact that such multi-queen social colonies exist at all suggests that the queens exercise control over the development of eggs into various caste types. Otherwise, female progeny could increase their inclusive fitness by maturing sexually and reproducing.

The fire ant (*Solenopsis invicta*) is an introduced noxious species in the southern United States. It has two kinds of colonies. One colony type has a single queen while the other type can have hundreds of queens, each of which lays relatively few eggs. These colonies produce smaller workers and a number of sexually mature females, or new queens. New colonies are established by unrelated queens, and surprisingly for the Hymenoptera (that is, the ants, bees, and wasps), new queens may be adopted from outside the colony. Both types of colonies produce reproductive adults, some of which mate outside their natal colonies or may move on and join other colonies.

Recent research on the Argentine ant, another species of invasive ant, shows how relatedness can impact aggression and invasiveness. This species was introduced to the United States from South America about 100 years ago and has displaced several native species of ants. It appears that the introduced population experienced a genetic bottleneck that reduced the species' genetic variability. Tsutsui et al. (2000) measured genetic variability at seven gene loci and found it to be about half as great as in ant populations back home in Argentina. Ants usually act aggressively toward nonrelated individuals, but tolerate related individuals. Because Argentine ants in the United States are relatively uniform genetically, individuals from different colonies don't recognize colony differences and cooperate with each other. Thus, Argentine ants may form networks of connected colonies. This may help explain why they are so effective at working together to displace native species.

CHECK YOUR KNOWLEDGE

1. What can an understanding of ant genetics and behavior provide for us humans?

MORE ON THE WEB 2. Go to *Practicing Ecology on the Web* at *http://www.whfreeman.com/ricklefs* and read the paper by Tsutsui et al. What was the goal of studying ants in multiple geographic locations in the United States, Chile, Bermuda, and Argentina?

3. Consider ❚ Figure 12.19, taken from the paper by Tsutsui et al. How was aggression measured? What is the relationship between aggression and genetic relatedness?

4. What do the authors conclude about intraspecific aggression in populations of Argentine ants here in the United States?

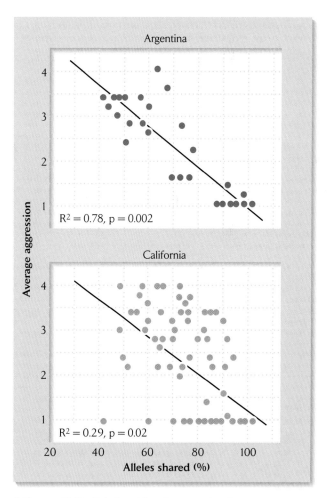

❚ **Figure 12.19 Relationship of aggression between nests and their genetic similarity.** Aggression between nests was measured by using a behavioral assay ranging from 1 (no aggression) to 4 (high aggression). After N. D. Tsutsui, A. V. Suarez, D. A. Holway, and T. J. Case, *Proc. Natl. Acad. Sci. USA* 97:5948–5953 (2000).

Suggested Readings

Clutton-Brock, T. H., M. J. O'Riain, P. N. M. Brotherton, D. Gaynor, R. Kansky, A. S. Griffin, and M. Manser. 1999. Selfish sentinels in cooperative mammals. *Science* 284:1640–1644.

Dugatkin, L. A., and H. K. Reeve (eds.). 1998. *Game Theory and Animal Behavior.* Oxford University Press, New York.

Elgar, M. A. 1989. Predator vigilance and group size in mammals and birds: A critical review of the empirical evidence. *Biol. Rev.* 64:13–33.

Emlen, S. T., P. H. Wrege, and N. J. Demong. 1995. Making decisions in the family: An evolutionary perspective. *American Scientist* 83:148–157.

Gordon, D. M. 1995. The development of organization in an ant colony. *American Scientist* 83:50–57.

Hammerstein, P., and S. E. Riechert. 1988. Payoffs and strategies in territorial contests: ESS analyses of two ecotypes of the spider *Agelenopsis aperta. Evolutionary Ecology* 2:115–138.

Heinrich, B., and J. Marzluff. 1995. Why ravens share. *American Scientist* 83:342–349.

Honeycutt, R. L. 1992. Naked mole-rats. *American Scientist* 80:43–53.

Ketterson, E. D., V. Nolan, Jr., M. J. Cawthorn, P. G. Parker, and C. Ziegenfus. 1996. Phenotypic engineering: Using hormones to explore the mechanistic and functional bases of phenotypic variation in nature. *Ibis* 138:70–86.

Krause, J. 1994. Differential fitness returns in relation to spatial position in groups. *Biological Reviews* 69:187–206.

Lima, S. L. 1995. Back to the basics of anti-predatory vigilance: The group-size effect. *Animal Behavior* 49:11–20.

Mock, D. W. and G. A. Parker. 1997. *The Evolution of Sibling Rivalry.* Oxford University Press, Oxford.

Queller, David C., and J. E. Strassmann. 1998. Kin selection and social insects. *BioScience* 48(3):165–175.

Seeley, T. D. 1989. The honey bee colony as a superorganism. *American Scientist* 77:546–553.

Sherman, P. W., J. U. M. Jarvis, and S. H. Braude. 1992. Naked mole-rats. *Scientific American* 267:72–78.

Sinervo, B., and C. M. Lively. 1996. The rock-paper-scissors game and the evolution of alternative male strategies. *Nature* 380:240–243.

Trivers, R. L. 1974. Parent-offspring conflict. *American Zoologist* 14:249–264.

Trivers, R. L. 1985. *Social Evolution.* Benjamin/Cummings, Menlo Park, CA.

Tsutsui, N. D., A. V. Suarez, D. A. Holway, and T. J. Case. 2000. Reduced genetic variation and the success of an invasive species. *Proceedings of the National Academy of Sciences USA* 97:5948–5953.

Wilkinson, G. S. 1988. Social organization and behavior. In A. M. Greenhall and U. Schmidt (eds.), *Natural History of Vampire Bats,* pp. 85–97. CRC Press, Boca Raton, FL.

Wilson, E. O. 1975. *Sociobiology.* Harvard University Press, Cambridge, MA.

Winston, M. L., and K. N. Slesser. 1992. The essence of royalty: Honey bee queen pheromone. *American Scientist* 80:374–385.

CHAPTER 13

Population Structures

The geographic distributions of populations are determined by ecologically suitable habitats

The dispersion of individuals within populations reflects habitat heterogeneity and social interactions

Populations exist in heterogeneous landscapes

Population size may be estimated by several techniques

Movement of individuals maintains the spatial coherence of populations

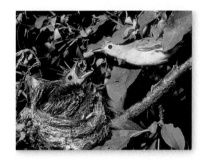

One of the primary threats to the stability of a population is the fragmentation of its habitat into small patches. This is happening all over the world as forests are cleared, roads are built, and rivers are channeled. The consequences of habitat fragmentation emphasize the effects of individual movements and habitat structure on population dynamics.

In recent years, a new branch of ecology, called landscape ecology, has begun to address how the size and arrangement of habitat patches influence the activities of individuals, the growth and regulation of populations, and interactions between species. Plants and animals can use a particular habitat patch only if they can gain access to it by moving through less favorable surrounding habitats. However, changes in human land use patterns have reduced such access and driven home the lessons of landscape ecology in many settings. Small, isolated subpopulations in habitat fragments are vulnerable to extinction because of loss of genetic diversity and random environmental perturbations.

Habitat fragmentation also means that any particular location is closer to a habitat edge, which may have a number of consequences. For example, in the Amazon basin, the increased exposure of trees within 100 meters of the edge of a clear-cut results in the drying of vegetation and excessive wind damage that would otherwise be prevented by surrounding vegetation. Such edge effects have caused losses of up to 15 tons of tree biomass per hectare annually (∎ Figure 13.1).

The importance of landscape structure comes very close to home in many parts of the world. Throughout much of the eastern and midwestern United States, forest fragmentation has brought populations of forest birds into contact with the parasitic brown-headed cowbird, which lays its eggs in the nests of other birds, reducing the reproductive success of its hosts. Cowbirds prefer open farms and fields, but they

Figure 13.1 Habitat fragmentation places organisms closer to the edges of suitable habitat. Patches of rain forest near Manaus, Brazil, were created when forests were converted to pastures. Mortality of trees at the edges of the patches was several times higher than that of trees in the middle of intact forest, owing to drying and wind damage. Courtesy of Eduardo M. Venticinque, National Institute for Research in the Amazon.

enter the edges of woodlots to seek out host nests. Rates of nest parasitism on Kentucky warblers are as high as 60% within 300 meters of the edge of forests in southern Illinois, and this edge effect is still discernible more than a kilometer into the forest (**Figure 13.2**). Nest predators, including many small rodents, that normally hunt in fields typically do not venture far into the forest; however, in a highly fragmented landscape, they can prey upon a larger proportion of eggs and young. Consequently, populations of some species of forest birds have declined precipitately in parts of eastern North America.

There are many threats to population stability, and habitat fragmentation is just one of them. Yet overall, populations of most species have persisted over many thousands or even millions of years. That so many are now threatened with extinction is cause for great concern. To understand why this has happened and what we can do about it, we must understand how changes in the environment influence the structure and dynamics of populations.

A **population** is made up of the individuals of a species within a given area. Each population lives primarily within patches of suitable habitat. The number of individuals in

the population may vary with food supplies, predation rates, nest site availability, and other ecological factors within that habitat.

Habitats naturally exist as a mosaic of different habitat patches, for example, areas of woodland within savannas, or mountains with shrubby vegetation on dry south-facing slopes next to forest on moister north-facing slopes. The patchy distribution of suitable habitat means that many populations are divided into smaller **subpopulations** between which individuals move less frequently than they would if the habitat were homogeneous.

Population structure refers to the density and spacing of individuals within suitable habitat and the proportions of individuals in each age class. Mating systems and genetic variation, which are also part of a population's structure, will be considered in a later chapter. Together, these measures provide us with a snapshot of a population at an instant in time.

Populations exhibit **dynamic behavior,** continuously changing over time because of births, deaths, and movements of individuals. These processes are influenced by the interactions of individuals with their environment and with one another. An understanding of population dynamics also illuminates community structure and ecosystem function, allowing us to address questions such as, Will a population persist within a habitat? How does it affect the flux of energy and contribute to the cycling of elements within an ecosystem? Hence much of ecology focuses on processes at the population level.

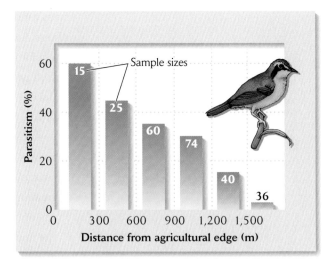

Figure 13.2 Habitat fragmentation can affect population dynamics. The rate of parasitism on nests of the Kentucky warbler by brown-headed cowbirds varies with distance from the edges of forest plots adjacent to agricultural lands. From S. F. Morse and S. K. Robinson, *Conserv. Biol.* 13:327–337 (1999).

The geographic distributions of populations are determined by ecologically suitable habitats

The **distribution** of a population is its **geographic range.** The presence or absence of suitable habitat often determines the extent of a population's distribution, although other factors, such as competitors, disease organisms, and barriers to dispersal, also have an influence. As discussed in Chapter 5, the natural range of the sugar maple in the United States and Canada corresponds primarily to the limits of tolerable physical conditions. Of course, sugar maples can occur no farther east than the edge of the Atlantic Ocean. But their distribution is limited more gradually to the west by aridity, to the north by cold winters, and to the south by hot summers (see Figure 5.3). Much suitable habitat for the sugar maple exists in other parts of the world, especially in Europe and Asia, where close relatives live and the sugar maple itself has been transplanted successfully. But this particular species evolved in North America and did not have the opportunity to colonize these areas on its own.

That barriers to long-distance dispersal may limit geographic ranges is often shown dramatically when introduced species expand successfully into new regions. For example, in 1890 and 1891, 160 European starlings were released in the vicinity of New York City by someone who wished to introduce all the birds mentioned in Shakespeare's works to the New World. Within 60 years, the population had expanded to cover more than 3 million square miles and stretched from coast to coast. Such introductions are as old as human movements. Aboriginal peoples brought dogs to Australia. Polynesians distributed pigs and rats throughout the small islands of the Pacific. In more recent times, foresters have transplanted fast-growing eucalyptus trees and pines all over the earth for timber and fuelwood. Other species have hitched rides on human conveyances, hidden among cargo, in ballast, and on the hulls of ships, and have done quite well on their own in new lands (and waters). These introduced species may become far more widespread and numerous in their new homes than the natural populations from which they were derived. We'll have more to say about such invasive organisms later in the book, but their success in many places emphasizes the important role of barriers to dispersal in limiting geographic distributions.

Within the geographic range of a population, individuals are not equally numerous in all regions. Individuals generally live only in suitable habitat. For example, sugar maples are absent from marshes, serpentine barrens, newly formed sand dunes, recently burned areas, and a variety of other habitats that simply lie outside their range of ecological tolerance. Hence, the geographic range of the sugar maple is a patchwork of occupied and unoccupied areas.

Climate, topography, soil chemistry, and soil texture similarly exert progressively finer influences on the geographic distribution of the perennial shrub *Clematis fremontii* (▌Figure 13.3). Climate and perhaps interactions with

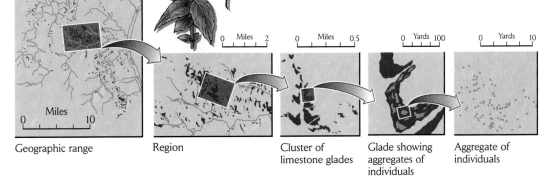

Missouri

Miles 0 — 10

Miles 0 — 2

Miles 0 — 0.5

Yards 0 — 100

Yards 0 — 10

Geographic range   Region   Cluster of limestone glades   Glade showing aggregates of individuals   Aggregate of individuals

▌**Figure 13.3 Within a population's geographic range, only suitable habitats are occupied.** Different scales of mapping reveal a hierarchy of patterns in the geographic distribution of *Clematis fremontii* var. *riehlii* in east central Missouri. After R. O. Erickson, *Ann. Mo. Bot. Gard.* 32:416–460 (1945).

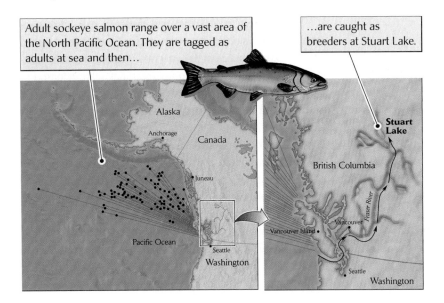

Adult sockeye salmon range over a vast area of the North Pacific Ocean. They are tagged as adults at sea and then...

...are caught as breeders at Stuart Lake.

▌Figure 13.4 Migration expands the geographic distribution of a population. Individuals of the sockeye salmon (*Onchorynchus nerka*) range as adults over a vast area of the Gulf of Alaska, then migrate up rivers to their birthplaces to breed. Spawning areas include Stuart Lake, more than 1,000 km upriver from the mouth of the Fraser River. Adapted from C. Groot and T. P. Quinn, *Fishery Bull.* 85 (1967).

ecologically similar plants restrict this species of *Clematis* to a small part of the midwestern United States. The distinctive variety of *Clematis fremontii* named *riehlii* occurs only in Jefferson County, Missouri. Within its geographic range, *Clematis fremontii* var. *riehlii* is restricted to dry, rocky soils on outcroppings of limestone. Small variations in relief and soil quality further confine the distribution of this plant within each limestone glade to sites with suitable conditions of moisture, nutrients, and soil structure. Local aggregations occurring on each of these sites consist of many more or less evenly distributed individuals.

It is important to remember that the geographic range of a population includes all of the areas its members occupy during their life cycle. Thus the distribution of salmon includes not only the rivers that are their spawning grounds, but also vast areas of the sea where individuals grow to maturity before making the long migration back to their birthplace (▌Figure 13.4).

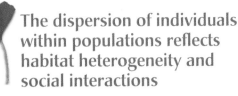

The dispersion of individuals within populations reflects habitat heterogeneity and social interactions

The **dispersion** of individuals within a population describes their spacing with respect to one another. (Keep in mind the distinction between *dispersion* and *dispersal,* which refers to the movement of individuals, which we shall discuss in detail below.) Patterns of spacing range from **clumped** distributions, in which individuals are found in discrete groups, to evenly **spaced** distributions, in which each individual maintains a minimum distance between

itself and its neighbors (▌Figure 13.5). Between these extremes one finds **random** distributions, in which individuals are distributed throughout a homogeneous area without regard to the presence of others.

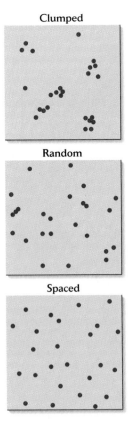

▌Figure 13.5 Dispersion patterns describe the spacing of individuals. Populations may have clumped, random, or evenly spaced distributions.

Figure 13.6 Evenly spaced distributions result from interactions between individuals. The even spacing of these desert shrubs in Sonora, Mexico, results from competition for water in the soil. Photo by R. E. Ricklefs.

Spaced and clumped distribution patterns derive from different processes. Even spacing most commonly arises from direct interactions between individuals. Maintenance of a minimum distance between oneself and one's nearest neighbor results in even spacing. For example, plants situated too close to larger neighbors often suffer from shading and root competition; as these individuals die, the spacing of individuals becomes more even (▌Figure 13.6).

Clumped distributions may result from social predisposition to form groups, clumped distributions of resources, and tendencies of progeny to remain with their parents. Birds often travel in large flocks to find safety in numbers. Salamanders that live under logs exhibit clumped distributions corresponding to the pattern of fallen deadwood. Trees form clumps of individuals by vegetative reproduction or when their seeds disperse poorly (▌Figure 13.7).

Finally, in the absence of social antagonism and mutual attraction, individuals may distribute themselves at random, without regard to the positions of other individuals in the population.

MORE ON THE WEB | *A statistical test for nonrandom dispersion.* The three general patterns of dispersion—clumped, random, and spaced—may be distinguished by comparison with a Poisson distribution.

Populations exist in heterogeneous landscapes

The natural world is extremely varied. Uniform, homogeneous habitats extending over vast areas simply do not exist. Instead, we may regard the natural world as a mosaic of habitat patches. For any particular species, some of these patches are suitable to live in and others are not. Consequently, most populations are divided into subpopulations of individuals living within homogeneous patches of suitable habitat, separated from other subpopulations by areas of unfavorable habitat.

Figure 13.7 Vegetative reproduction gives rise to clumped distributions. This photograph, taken in Coconino National Forest, Arizona, shows many different clones of aspen trees, which are distinguished by timing of leaf fall. Within each clone, each individual stem ("tree") has grown from a common root system that developed from a single seedling. Photo by Tom Bean/DRK Photo.

Depending on the distances between subpopulations, the nature of the intervening environment, and the mobility of the species, areas of unfavorable habitat may or may not be substantial barriers to the movement of individuals. As shown in ▌Figure 13.8, adult snail kites move quite readily among the areas of wetland habitat that they use for feeding and breeding in southern Florida. Thus, although their habitat is patchy, the mobility of these relatives of hawks ties all the snail kites in southern Florida into a single population.

In contrast, the small arboreal gecko *Oedura reticulata* is restricted to small habitat patches of natural eucalypt woodland in the wheat belt of southwestern Australia (▌Figure 13.9). Unlike the snail kite, it is incapable of traveling through the 1 to 20 kilometers of farmland separating patches of suitable habitat, and thus exists in largely unconnected subpopulations. Habitat patches in this landscape are 0.4 to 5 hectares, contain 33 to 570 individual eucalyptus trees, and sustain populations of 12 to 430 geckos. It is difficult to determine by direct observation that individual geckos do not move between subpopulations. However, differences in genetic composition between the subpopulations indicate that many of them have lost genetic variation. This finding suggests that there has been little movement of individuals between habitat patches, as we will see in more detail in a later chapter.

The patchiness of the natural world has given rise to three models of populations. One of these is the **metapopulation model,** which views a population as a set of subpopulations occupying patches of a particular habitat type, between which individuals move occasionally. The intervening habitat, which is referred to as the **habitat**

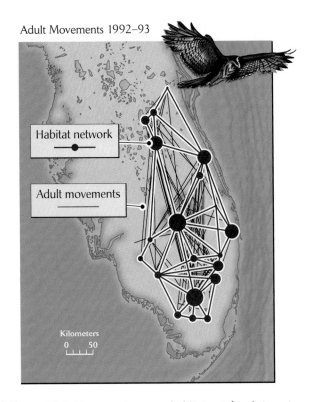

Adult Movements 1992–93

Habitat network

Adult movements

Kilometers
0   50

▌**Figure 13.8 Movement among habitat patches integrates a population.** The black lines represent movements of individual radio-tagged snail kites among wetlands in southern Florida during 1992–1993. The red lines and dots, whose sizes are proportional to areas of suitable habitat, provide a summary of the connections between populations of snail kites in wetland areas based on the recorded movement patterns of adults. From R. E. Bennetts and W. M. Kitchens, U.S. Geol. Surv./Biol. Res. Div., Tech. Rep. No. 56 (1997).

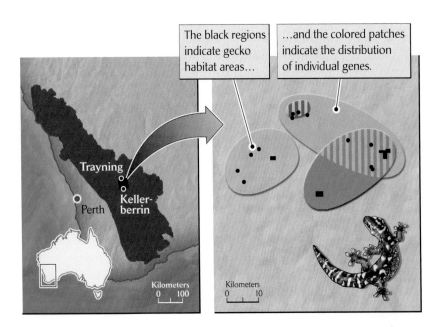

The black regions indicate gecko habitat areas…

…and the colored patches indicate the distribution of individual genes.

Trayning

Kellerberrin

Perth

Kilometers
0   100

Kilometers
0   10

▌**Figure 13.9 Genetic differences between populations in small, isolated habitat patches indicate lack of movement between them.** Over the century since the wheat belt of southwestern Australia was cleared for agriculture, subpopulations of the gecko *Oedura reticulata* have become isolated in small habitat patches. The genetic differences among the subpopulations show that there is little movement of individuals between them. After S. Sarre, G. T. Smith, and J. A. Meyers, *Biol. Conserv.* 71:25–33 (1995); S. Sarre, *Mol. Ecol.* 4:395–405 (1995).

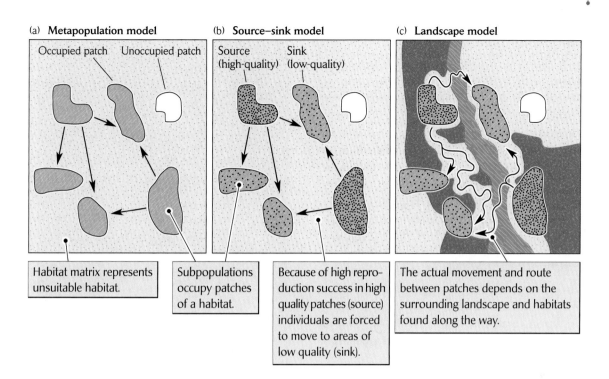

Figure 13.10 Models of population structure differ with respect to variation in habitat patch quality and the intervening matrix of environments. (a) A metapopulation with six patches of suitable habitat, five of which are occupied. Arrows represent interpatch movements. According to the metapopulation concept, habitat patches occur against a featureless background of unsuitable habitat. Thus, the quality of the background matrix has no effect on interpatch movements, but the arrangement of patches and distance between patches may. (b) Populations in high-quality habitat patches (source populations) produce excess offspring, which disperse to less suitable habitat patches, where immigration maintains less productive sink populations. (c) The same metapopulation overlaid on a landscape composed of a mosaic of many different habitat types, which, as depicted by the arrows, affect the way that individuals move between patches. After J. A. Wiens, in I. A. Hanski and M. E. Gilpin (eds.), *Metapopulation Biology: Ecology, Genetics, and Evolution,* Academic Press, San Diego (1997), pp. 43–62.

matrix, is viewed only as a barrier to the movement of individuals between subpopulations (**Figure 13.10a**).

A second model, the **source–sink model,** assumes that there are differences in the quality of patches of the suitable habitat type. Where resources are abundant, individuals produce more offspring than required to replace themselves, and the surplus offspring disperse to other patches. Such populations serve as source populations. In patches of poor habitat, populations are maintained by immigration of individuals from elsewhere, because too few offspring are produced locally to replace losses to mortality. These populations are known as sink populations (**Figure 13.10b**).

The **landscape model** goes a step beyond the metapopulation model by considering the effects of differences in habitat quality within the habitat matrix (**Figure 13.10c**). Accordingly, the quality of one habitat patch can be altered by the nature of nearby habitats. For example, habitat qual-ity may be improved when other patches in the landscape provide resources such as safe roosting sites, nesting materials, pollinators, or water. Other kinds of neighboring habitat patches may be a serious drawback if they harbor predators and disease organisms. The habitat matrix also influences the movement of individuals from one subpopulation to another. Clearly some travel routes are more attractive than others because of the habitat types encountered along the way.

ECOLOGISTS IN THE FIELD

The scale of variation in coral abundance and recruitment on the Great Barrier Reef

As the examples of the snail kite and gecko show, the effects of habitat variation on population structure depend on the scale of that variation relative to the mobility of individuals. However, measuring spatial variation in populations has proved difficult. A study by Terry Hughes and his colleagues at James Cook University in Australia shows one approach to this problem.

The investigators were interested in variation in abundance of corals along the Great Barrier Reef, which stretches 2,000 kilometers along the tropical Pacific coast of Australia. They established sampling locations within different sectors of the reef (at intervals of 250–500 km), at different patches of reef within sectors (at distances of 10–15 km), at different sites within reefs (0.5–3 km), and at replicated sampling points within sites (1–5 meters). The spatial plan of sampling is shown on the map in ▌Figure 13.11. Adult corals were counted on 10-meter-long sampling transects at each site. New recruits to the population (newly settled larvae) were sampled by bolting more than one thousand 11×11 cm unglazed tiles onto the reef (▌Figure 13.12) and counting the number of larval corals that settled on the tiles over an 8-week period. The study was so large that 41 student volunteers contributed to the fieldwork!

Hughes and his colleagues divided the species of corals into two categories based on their method of reproduction and larval dispersal. Spawners release sperm and eggs directly into the water, and their larvae remain in the water for a week or more before they settle and begin to grow as corals. Brooders release fully grown larvae, which settle much more quickly. The larvae of spawners disperse greater distances, and the investigators postulated that this difference might result in a difference in the scale of spatial variation in abun-

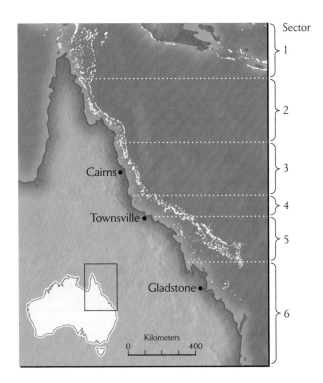

▌Figure 13.11 Spatial patterns of variation can be measured at different scales. Sampling stations were distributed among sectors of the Great Barrier Reef to measure the spatial pattern of variation in corals. Within these sectors, the sampling stations were distributed among different reef patches, and at a still smaller scale, among different sites within patches. From T. P. Hughes, A. H. Baird, E. A. Dinsdale, N. A. Moltschaniwskyj, M. S. Pratchett, J. E. Tanner, and B. L. Willis, *Nature* 397:59–63 (1999).

▌Figure 13.12 Ecologists use a variety of sampling methods to measure spatial variation in populations. Views of the Great Barrier Reef show the location of sampling transects (*left*) and a settling plate (*right*). Courtesy of T. P. Hughes.

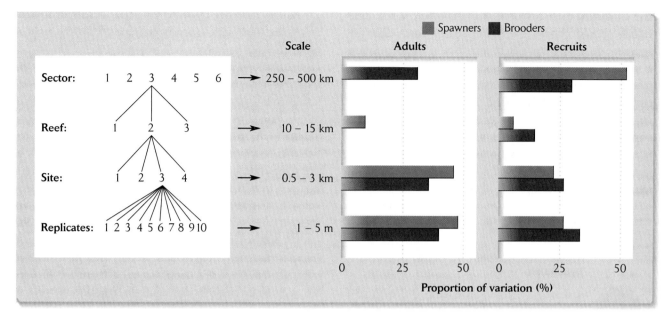

▍Figure 13.13 The abundances of adult corals and recruits vary at different spatial scales. Differences in density of adult corals are apparent primarily at small spatial scales, whereas differences in newly settled recruits also occur on the scale of sectors, at distances of 250–500 km. From data in T. P. Hughes, A. H. Baird, E. A. Dinsdale, N. A. Moltschaniwskyj, M. S. Pratchett, J. E. Tanner, and B. L. Willis, *Nature* 397:59–63 (1999).

dance between the two coral types. The spatial variation was estimated by the amount of variation in numbers of adults and recruits that was attributable to differences between sectors (thus, at a scale of hundreds of kilometers), differences between reefs within sectors (about 10 km), and differences between sites within reefs (about 1 km). Variation within a site—that is, over distances of 1–5 meters—provided an indication of the variability inherent in the sampling methods themselves and of the natural heterogeneity of the habitat.

For adult corals, the data revealed variation in abundance among sites within a reef (0.5–3 km), reflecting small differences in habitat quality and the history of disturbance at a particular site, but no variation among reefs within a sector or, for spawners, among sectors (▍Figure 13.13). These findings suggest that dispersal during the larval stage is adequate to make populations of corals uniform over dimensions of tens to hundreds of kilometers, but that other processes affect the abundances of corals at smaller scales. The significant variation for brooder adults among sectors reflects the fact that brooders were much more abundant in the southern three sectors (6–11 individuals per 10-meter transect) than in the northern, more tropical three sectors (1–5 individuals per 10-meter transect).

The pattern of variation in larval settlement was strikingly different from that of adult abundances for spawner corals, with most of the variation occurring among sectors (see Figure 13.13). Especially high numbers of recruits settled

during the study period in sectors 2 and 3 (38–41 per tile) compared with the other sectors (fewer than 15 per tile). The differences in spatial variation between recruits and adults of spawners probably reflect factors that affect the survival of corals after recruitment. The investigators concluded that production and dispersal of larvae, even at low densities of recruits, is sufficient to distribute corals to all parts of the Great Barrier Reef. Local ecological processes subsequently even out the average abundances of adult corals over the entire extent of the Great Barrier Reef itself.

Ideal free distributions match population density to habitat quality

Many organisms are capable of making behavioral decisions about where to live and where not to live. Individuals can base these decisions on a combination of intrinsic habitat quality and the density of other individuals. A patch becomes less attractive to newcomers as the number of potential competitors in the patch increases, just as a male bird's territory becomes less attractive to a female if he already has a mate (see Chapter 11). Occupied patches are likely to have fewer remaining resources than undiscovered patches; furthermore, competing individuals may precipitate costly behavioral conflicts.

Each individual should choose among patches so as to maximize its own access to resources. Imagine two patches, one containing more resources than the other. At first, individuals should choose the higher-quality patch. But the apparent quality of that patch decreases as the population in the patch builds up. As resources in the patch are depleted and antagonistic interactions increase, the second patch becomes an equally good choice. At this point, individuals should choose the second patch or the first patch impartially (■ Figure 13.14). Thus, each individual in the population exploits a patch of the same apparent quality, regardless of intrinsic patch quality. This outcome is called an **ideal free distribution.**

ECOLOGISTS IN THE FIELD

An ideal free distribution in a laboratory population

Tendencies toward an ideal free distribution have been investigated in several laboratory studies, in which the quality of patches can be controlled. Manfred Milinski conducted a series of such experiments with stickleback fish (genus *Gasterosteus*). The fish were provided with food (water fleas) at different rates at opposite ends of an aquarium. Each half of the tank constituted a patch. The experimental system had the following conditions conducive to establishing an ideal free distribution: (1) the two patches differed in quality, (2) quality decreased as the number of fish using a patch increased, and (3) fish were free to move between patches.

Hungry fish were placed in the aquarium about 3 hours before the start of the experiment. During trials, the number of fish in each half of the tank was recorded at the end of each 20-second interval. Before any food was added to the tank, the fish were distributed equally between the two halves. In one experiment, water fleas were added at a rate of 30 per minute to one end of the tank and at a rate of 6 per minute to the other—a ratio of 5 to 1. Within 5 minutes, the fish had distributed themselves between the two halves in the same ratio as predicted for an ideal free distribution—that is, 5 to 1. In a second experiment, water fleas were provided at rates of 30 and 15 per minute, a 2-to-1 ratio. Again, the distribution of the fish followed suit. When the better and poorer patches in the tank were reversed, the fish reversed their distribution within about 5 minutes. How they achieved such an ideal free distribution was not determined, but cues for such behavioral choices could incorporate the rate of encountering food items and the number of competitors within a patch. Such experiments demonstrate the sensitivity of organisms to the conditions of their environments, as well as their behavioral flexibility in making choices.

Individuals move from source populations in productive habitats to sink populations in marginal habitats

Under an ideal free distribution, the reproductive success of each individual would be the same regardless of the intrinsic quality of the patch it occupied. However, there are many reasons why this ideal may rarely be attained in nature. Among the most important reasons are that individuals do not have perfect knowledge of patch quality and that territorial behavior by dominant individuals reduces free choice in subordinates. In almost every case in which reproductive success has been measured in the field, it varies between habitats. This variation establishes a situation in which some habitats have growing, and others declining, populations. As a result, there is often a net movement of individuals from growing (source) populations to shrinking (sink) populations (see Figure 13.10b).

In southern Europe, a small songbird called the blue tit (*Parus caeruleus*) breeds in two kinds of habitats, one dominated by the deciduous downy oak (*Quercus pubescens*) and the other by the evergreen holm oak (*Quercus ilex*). Comparisons of population densities and reproductive success in the two habitats suggest that deciduous oak habitat is superior for tits (Table 13.1). Indeed, tit populations in deciduous oak habitats produce so many young that they would grow at almost 10% annually if individuals did not disperse. The net rate of population decline in evergreen oak habitats would be about 13% per year in the absence of immigration. Even though densities of blue

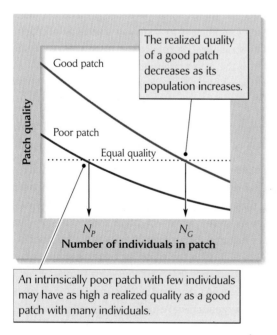

The realized quality of a good patch decreases as its population increases.

An intrinsically poor patch with few individuals may have as high a realized quality as a good patch with many individuals.

■ **Figure 13.14 In an ideal free distribution, each individual exploits a patch of the same apparent quality.** Habitat quality decreases as the number of individuals in a patch increases.

Table 13.1   Comparisons of the density and productive of blue tit (*Parus caeruleus*) populations from deciduous and evergreen oak woodlands

	Habitat	
	Deciduous oak	Evergreen oak
Breeding density (pairs per 100 ha)	90	14
Laying date (average)	10 April	21 April
Clutch size	9.8	8.5
Survival to fledging	0.60	0.43
Fledglings per parent	2.9	1.8
Probable number of recruits per parent	0.59	0.37
Likelihood of death of parent	0.50	0.50
Net productivity per parent	+0.09	−0.13
Type of population	Source	Sink

Source: *J. Blondel, P. C. Dias, M. Maistre, and P. Perret,* Auk *110:511–520 (1993).*

tits in deciduous oak habitats are six times higher than densities in evergreen oak habitats, tits do not achieve an ideal free distribution because the reproductive success of pairs in deciduous oak habitats also is higher. Genetic analyses have been used to estimate the movement of blue tits between local populations in southern France separated by an average of 10 kilometers. The number of young birds moving from deciduous oak forest to breed in evergreen oak forest (estimated at 2,000 individuals per year in the region) was found to be about one hundred times higher than the number moving between equally distant patches of the same kind of habitat.

Population size may be estimated by several techniques

The ultimate measure of a population is the number of individuals it contains. From a management and conservation standpoint, it is important to understand the factors that cause population size to change and the processes that regulate the size of a population. This understanding must begin with an empirical knowledge of the number of individuals in the population.

Total population size has two components, density and area. **Density** is defined as the number of individuals per unit of area. Multiplied by the area occupied, density times area equals total population size.

How do we measure density? When populations are very small, especially when individuals can be distinctively marked by unique tags—or by radios, as in the case of the

snail kite—it may be possible to count all the individuals. Such techniques are often applied to endangered species, particularly large mammals and birds (Figure 13.15). However, most populations contain too many individuals, or are distributed over too large an area, to make a complete count of the population practical. When individuals are not mobile—plants and sessile marine invertebrates, for example—their densities may be estimated by counting individuals within plots of known area to obtain local density and then multiplying by area to extrapolate those counts to

Figure 13.15 Small populations can be counted by marking individuals. Here, a female shoveler duck is fitted with an individually identifiable beak tag and a small radio transmitter. Photo by R. E. Ricklefs.

the entire population. When individuals can move between plots faster than an investigator can count the number within a plot, other methods must be employed.

Estimating population size by the mark–recapture method

One class of methods used to estimate the sizes of animal populations involves capturing and marking a sample of individuals from the population and then releasing them. After enough time has passed to allow the marked individuals to recover from the trauma of capture and marking and to mix thoroughly with the rest of the population, a second sample is taken from the population, and the ratio of marked to unmarked individuals is recorded. If one assumes that the ratio of marked to unmarked individuals in the second sample is representative of the ratio for the entire population, then an estimate of the population size can be calculated. This method of estimating population size is one of the simplest of a class of methods called **mark–recapture** methods.

Suppose we capture 20 fish from a small pond and mark them with colored fin tags. The pond contains a population of N fish, which is what we wish to estimate. After we release the 20 marked fish (M) into the pond, the ratio of marked to unmarked fish in the entire population is M/N. After the marked fish have dispersed evenly throughout the population, any further sample of n fish should include marked individuals in the ratio M/N. Suppose that several days later, we capture another sample of 50 fish, and 6 of those 50 fish have our colored fin tags. The number of recaptures, denoted x, in the second sample is the product of the size of that sample (n) times the proportion of individuals in the population that are marked (M/N), or

$$x = \frac{nM}{N}.$$

The only variable in this simple relationship that we do not know is N, the size of the population. We can rearrange the equation to obtain

$$N = \frac{nM}{x},$$

which provides an estimate of the population size. In this example, we can estimate the population of fish in our small pond to be $N = 50(20)/6 = 167$.

Variation in populations over space and time

Because densities of populations change over time and space, no population has a static structure; one's perception of a population depends on where and when one looks at it. Long-term records of the population of the chinch bug

(*Blissus leucopterus*) in Illinois illustrate this point (▌Figure 13.16). Because they damage cereal crops, the Illinois State Entomologist's Office and later the State Natural History Survey Division realized the importance of monitoring chinch bug populations, which they estimated from county reports of crop damage. These estimates were calibrated by local studies of the relationship between population size (determined by direct counts on small plots) and crop damage.

Consider the numbers involved. During 1873, when the bugs infested crops over most of the state, ballpark estimates of the population indicated an average density of 1,000 chinch bugs per square meter over an area of 300,000 square kilometers, or a total of 3×10^{14} pests (300 trillion, more or less). By contrast, farmers reported

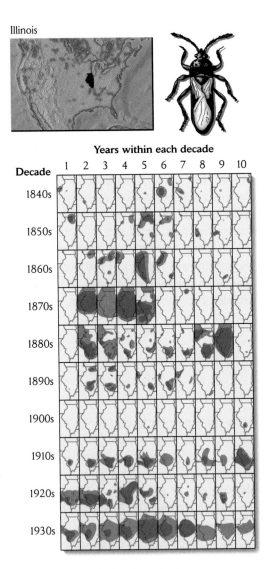

▌**Figure 13.16 Population density changes over time and space.** The distribution of crop damage caused by chinch bugs (*Blissus leucopterus*) in Illinois varied dramatically over the period between 1840 and 1939. Yellow indicates low densities of chinch bugs and blue, high densities. From V. E. Shelford and W. P. Flint, *Ecology* 24:435–455 (1943).

little damage in 1870 and 1875. The population flux reveals itself in Figure 13.16 in the waxing and waning of infestations.

Movement of individuals maintains the spatial coherence of populations

Population biologists refer to movements within populations as **dispersal.** They refer to movements between subpopulations—for example, between sources and sinks—as **emigration** (leaving) and **immigration** (entering) or, more generally, **migration.** When individuals disperse widely through a population, they link together the dynamics of different subpopulations, and make the whole population function and evolve as a single structure. When dispersal is limited, different parts of a population behave independently of one another.

Measuring dispersal, particularly over long distances, requires marking and recapturing individuals. The field methods are laborious because it is difficult to predict how far and in what direction individuals will move. Therefore, large areas must be covered to ensure an accurate sampling of movements. Population biologists often resort to ingenious techniques to measure dispersal. Indeed, one of the first attempts to measure dispersal in natural populations involved measuring movements away from a release point by fruit flies (*Drosophila*) that could be distinguished by a visible mutation.

A convenient measure of movement within a population is the average lifetime dispersal distance, which indicates how far an individual moves from its birthplace to where it settles to mature and reproduce. A circle having a radius equal to the lifetime dispersal distance is the lifetime dispersal area. Such a circle encompasses all the other individuals in a population that one individual potentially could interact with or mate with in its lifetime. The number of these individuals defines the **neighborhood size** of a population. Thus, the lifetime dispersal area times the population density equals the neighborhood size.

MORE ON THE WEB | *A mathematical description of dispersal.* The lifetime dispersal distance can be estimated from a model of random movement.

Small songbirds of eight species marked with leg bands as nestlings and recaptured as breeding adults exhibited lifetime dispersal distances averaging between 344 and 1,681 meters, densities between 16 and 480 individuals per square kilometer, and neighborhood sizes between 151 and 7,679 individuals. In three populations of the land snail *Cepaea nemoralis*, which holds no speed records, dispersal distances varied between 5.5 and 10 meters after 1 year, but because populations are dense, neighborhood sizes were similar to those of the birds: 1,800 to 7,600 individuals. Mark–recapture data on a rusty lizard (*Sceloporus olivaceous*) population near Austin, Texas, revealed average lifetime dispersal distance to be 89 meters and neighborhood size to be between 225 and 270 individuals. Thus, for a variety of different animal populations, neighborhood sizes were rather more similar than one would expect from knowledge of either dispersal distance or population density alone.

In most studies of dispersal, the principal source of information is movements of individuals away from a point of release within an established population. Other observations are also pertinent, particularly the spread of introduced populations, which can occur only by movements of individuals beyond a population's borders. The European starling population spread almost 4,000 kilometers across the United States in 60 years, at an average rate of about 67 kilometers, or 67,000 meters, per year (▌Figure 13.17). This figure greatly exceeds the estimates

1918   1926   1949

▌**Figure 13.17 Some populations can spread rapidly outside their established range.** Western expansion of the range of the European starling (*Sturnus vulgaris*) in the United States has occurred through long-distance dispersal. The shaded areas represent the breeding range; dots indicate records of birds in preceding winters. The population now inhabits the entire country. After B. Kessel, *Condor* 55:49–67 (1953).

reported above for dispersal distances within populations of small songbirds. Adult starlings tend to nest in the same area year after year, so most long-distance dispersal is accomplished by young birds; established populations act as sources for population extension. Furthermore, the westward spread of the starling was characterized by frequent sightings outside the breeding season before breeders were settled in an area. It is unlikely that such long-distance movements of unmarked juveniles would be detected within an established population.

 ## Summary

1. The distribution of a population is its geographic range, which is generally limited by the extent of suitable habitat and by barriers to dispersal. Within the limits of its distribution, the density of a population may vary according to differences in habitat quality.

2. Dispersion describes the spacing of individuals with respect to others in a population. Evenly spaced distributions may result from competitive interactions between individuals. Clumped distributions may result from independent aggregation of individuals in suitable habitats, from spatial proximity of parents and offspring, or from tendencies to form social groups.

3. The heterogeneity of the natural world causes most populations to be divided into subpopulations with varying amounts of migration between them. Such populations are referred to as metapopulations.

4. Populations in which reproduction exceeds mortality are called source populations. Individuals disperse from source populations to sink populations, where local reproduction cannot maintain a population without immigration.

5. The types of habitat surrounding suitable habitat patches make up the landscape, which influences the quality of suitable habitat patches and the movement of individuals between them.

6. Faced with variation in habitat quality and complete freedom to choose where to live, organisms should tend to distribute themselves in proportion to available resources in what is known as an ideal free distribution. Poorer habitats are eventually settled because dense populations reduce the quality of intrinsically superior habitats. Ideal free distributions are rarely realized, and reproductive success is often higher in some habitats than in others.

7. Population density, which is the number of individuals per unit of area, reflects the ecological relationship of individuals to their environment. The total size of a population is the average density multiplied by the area occupied by the population. Because population size is so central to ecological studies, many techniques have been developed to estimate numbers of individuals. One of these is the mark–recapture method.

8. Ecologists characterize movement within populations by the average lifetime dispersal distances of individuals from their birthplaces. Neighborhood size is the number of individuals within a circle whose radius is equal to the average lifetime dispersal distance. Neighborhood size provides an index to the number of individuals with which an individual can potentially interact or mate. For several species, neighborhood sizes have been estimated on the order of 10^2 to 10^4 individuals.

 ## PRACTICING ECOLOGY

CHECK YOUR KNOWLEDGE

Propagules and Currents

As you are surely aware, plants cannot get up and move around. This means that plants have had to evolve ways to get their propagules (seeds and other reproductive structures, such as bulbils) to sites that are favorable for their successful establishment. Many structures and strategies have evolved to meet this challenge. For example, the fruit in which seeds are often embedded attracts animals to feed on them; the seeds are subsequently dispersed in the animals' feces. The seeds of many parasitic mistletoe species are embedded in a sticky pulp. When birds eat the fruit, some seeds invariably stick to the bird's beak. When the bird wipes its beak on a branch, the seed and sticky pulp are transferred to the bark, which is the optimal establishment site for a new parasitic mistletoe plant. The fruits of other species possess barbs or other structures that help the seeds hitch rides on the fur or feathers of animals.

Many plant species do not rely on animals and have evolved structures to make use of currents for dispersal. The "wings" of a maple key provide a certain degree of lift to help the seeds ride the wind for some distance. Dandelion seeds are smaller than maple seeds, and dandelion dispersal is assisted by a pappus, which is a parachute-like structure that helps dandelions to float on the wind. Coconuts are big seeds that are able to float for long periods on seawater, which can be important for long-distance dispersal. The prickly pear cactus, distributed across wide portions of Canada and the United States, reproduces primarily by vegetative methods. Pieces of stems break off and stick to animal fur, roll downhill, or in some cases float downstream when rivers are swollen with snowmelt.

It is relatively easy to determine neighborhood size for slow-moving organisms or organisms that don't move far throughout their lives. However, for organisms in aquatic environments, where currents can transport floating life stages over great distances, estimating dispersal distance can be quite tricky indeed. Mary Ruckelshaus, now of the National Marine Fisheries Service in Seattle, Washington, measured the neighborhood size for the eelgrass *Zostera marina*. Eelgrass is a perennial, marine, flowering plant. It occurs from the intertidal zone to depths of up to 15 meters where the bottom is sandy. It reproduces sexually in the same manner as other flowering plants. The pollen travels from one flower to another by floating on or just under the surface. Seeds are released from fruits, sink to the substrate, and are then dispersed by tides and currents. Ruckelshaus measured pollen dispersal by specially designed traps placed at different distances and directions from a pollen release point. She conducted this experiment in an area without flowering eelgrass plants nearby and released pollen collected elsewhere. She measured seed dispersal by counting seeds and seedlings at increasing distances from edges of patches of eelgrass.

CHECK YOUR KNOWLEDGE

1. Tides and ocean currents presumably can carry pollen and seeds over considerable distance, suggesting that

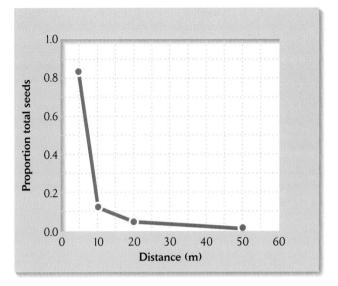

■ **Figure 13.19 Distribution of dispersal distances of eelgrass seeds in False Bay, Wisconsin.** From M. H. Ruckelshaus, *Evolution* 50:856–864 (1996).

neighborhood sizes for these organisms should be large. What aspect of marine organisms might reduce the dispersal of pollen and seeds, thereby reducing neighborhood size?

2. Examine ■ Figure 13.18 and ■ Figure 13.19, which are taken from Ruckelshaus (1996). Compare the dispersal distances of pollen and seeds. What aspects of the biology of eelgrass might explain these differences?

3. Ruckelshaus found that neighborhood size for eelgrass was very large. What consequences does neighborhood size have for the genetic structure of populations and the relative importance of random genetic drift compared to natural selection?

MORE ON THE WEB 4. Read more about eelgrass at the Port Townsend Marine Science Center Web page through *Practicing Ecology on the Web* at *http://www.whfreeman.com/ricklefs*. What important functions does eelgrass provide in the marine habitat?

Suggested Readings

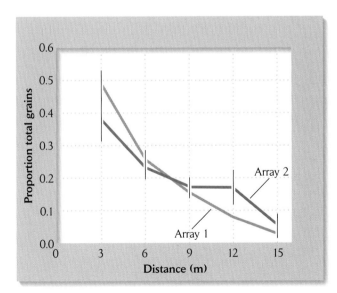

■ **Figure 13.18 Distribution of dispersal distances of eelgrass pollen in experimental arrays.** From M. H. Ruckelshaus, *Evolution* 50:856–864 (1996).

Begon, M., and M. Mortimer. 1986. *Population Ecology.* 2d ed. Blackwell Scientific Publications, Oxford.

Caughley, G. 1977. *Analysis of Vertebrate Populations.* Wiley, New York and London.

Cook, R. E. 1983. Clonal plant populations. *American Scientist* 71:244–253.

Dunning, J. B., B. J. Danielson, and H. R. Pulliam. 1992. Ecological processes that affect populations in complex landscapes. *Oikos* 65:169–175.

Krebs, C. J. 1989. *Ecological Methodology.* Harper and Row, New York.

Laurance, W. F., S. G. Laurance, L. V. Ferreira, J. M. Rankindemerona, C. Gascon, and T. E. Lovejoy. 1997. Biomass collapse in Amazonian forest fragments. *Science* 278:1117–1118.

Marquet, P. A., S. A. Naverrete, and J. C. Castilla. 1995. Body size, population density, and the Energetic Equivalency Rule. *Journal of Animal Ecology* 64:325–332.

Milinski, M. 1979. An evolutionarily stable feeding strategy in sticklebacks. *Zeitschrift für Tierpsychologie* 51:36–40.

Morse, S. F., and S. K. Robinson. 1999. Nesting success of a Neotropical migrant in a multiple-use, forested landscape. *Conservation Biology* 13:327–337.

Ruckelshaus, M. H. 1996. Estimation of genetic neighborhood parameters from pollen and seed dispersal in the marine angiosperm *Zostera marina* L. *Evolution.* 50: 856–864.

Sarre, S. 1995. Mitochondrial DNA variation among populations of *Oedura reticulata* (Gekkonidae) in remnant vegetation: Implications for metapopulation structure and population decline. *Molecular Ecology* 4:395–405.

Sarre, S., G. T. Smith, and J. A. Meyers. 1995. Persistence of two species of gecko (*Oedura reticulata* and *Gehyra variegata*) in remnant habitat. *Biological Conservation* 71:25–33.

Silva, M., and J. A. Downing. 1995. The allometric scaling of density and body mass: A nonlinear relationship for terrestrial mammals. *American Naturalist* 145:704–727.

Wiens, J. A., N. C. Stenseth, B. Van Horne, and R. A. Ims. 1993. Ecological mechanisms and landscape ecology. *Oikos* 66:369–380.

CHAPTER 14

Population Growth and Regulation

Populations grow by multiplication rather than addition

How fast a population grows depends on its age structure

A life table summarizes age-specific schedules of survival and fecundity

The intrinsic rate of increase can be estimated from the life table

Population size is regulated by density-dependent factors

The size of the human population passed the 6 billion mark on October 12, 1999, as close as United Nations demographers could tell. Many fewer humans existed when you were born, and only one-fifth as many were present in 1850. The growth of the human population during the last ten thousand years, since the advent of agriculture, has been one of the most significant ecological developments in the history of the earth. It ranks with the massive displacements caused by glaciation during the last million years and the wholesale extinctions caused by a comet impact 65 million years ago. One of the most remarkable aspects of human population growth is that its rate continued to increase even as the population was becoming more crowded. Estimates of the size of the human population in ancient times are understandably crude, but it is likely that there might have been a million individuals in the population of our ancestors as much as a million years ago. Population increase was very slow until the development of agriculture, by which time the human population might have been 3–5 million. The abundant food produced by agriculture removed a limiting factor on the human population, which then increased a hundredfold by the beginning of the eighteenth century, even with occasional setbacks such as those caused by the bubonic plague (Figure 14.1). A hundredfold increase over nearly 10,000 years is equivalent to an average exponential growth rate of about 2% per century. The Industrial Revolution, which began about 1700, provided another impetus to human population growth, particularly with improvements in public health and medicine and increasing personal material wealth. Over 300 years of industrialization, humans have increased in number from perhaps 300 million to 6 billion, roughly a twentyfold increase, or an average exponential

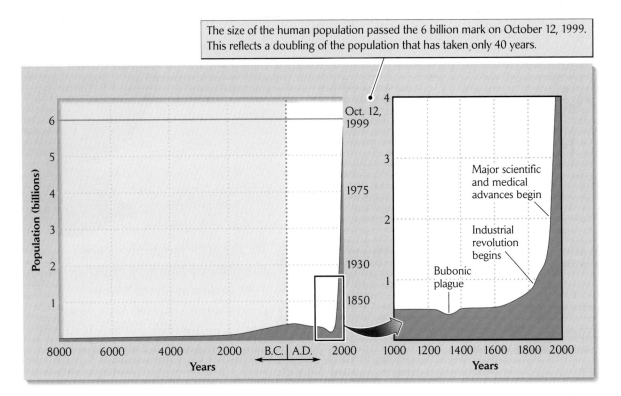

The size of the human population passed the 6 billion mark on October 12, 1999. This reflects a doubling of the population that has taken only 40 years.

Figure 14.1 The human population increased rapidly with the development of technology. After W. K. Purves, G. H. Orians, H. C. Heller, and D. Sadava, *Life: The Science of Biology* (5th ed.), Sinauer Associates, Sunderland, MA (1998).

growth rate of close to 100% per century (1% per year). The most recent doubling of the human population, from 3 billion to 6 billion, has taken only 40 years (1.7% per year).

The earth is becoming a very crowded place (**Figure 14.2**). Many believe that the human population has long since exceeded the ability of the earth to support it and that we are depleting the earth's resources rapidly. How this will be compensated for in the future is uncertain. What is certain is that continued population growth will further stress the biosphere and will lead to the further degradation of many environments.

When, and at what level, will human population growth cease? Predicting the future is difficult because there are so many unknown factors, including changes in technology, emergence of epidemic diseases of humans or their crops and livestock, and changes in material wealth, education, and culture. At present, the growth rate of the human population is decreasing, and current United Nations estimates indicate a population plateau at about 9 billion.

Figure 14.2 Human populations have become extremely crowded in both rich and poor countries. Photo by Will & Deni McIntyre/Photo Researchers.

Ever since humankind began to understand the rapid increase in its own numbers, human population growth has been cause for concern. This concern led to the development of mathematical techniques to predict the growth of populations—the discipline of **demography,** or the study of populations—and to intensive study of natural and laboratory populations to learn about mechanisms of population regulation. Therefore, we now have a general understanding of the causes of fluctuations in natural populations and of the effects of crowding on birth and death rates.

In this chapter, we shall explore the nature of population growth and examine the factors that limit population size, showing how their effects increase with increasing population density in such a way as to bring growth under control.

Populations grow by multiplication rather than addition

A population increases in proportion to its size, just as a bank account earns interest on its principal. Increase in its numbers depends on reproduction by individuals. Therefore, a population growing at a constant rate gains individuals ever faster as the number of individuals increases. For example, a 10% annual rate of increase adds 10 individuals in 1 year to a population of 100, but the same rate of increase adds 100 individuals to a population of 1,000. Allowed to increase at this rate, the population would rapidly climb toward infinity. As Charles Darwin wrote, in *On the Origin of Species,* "There is no exception to the rule that every organic being naturally increases at so high a rate, that, if not destroyed, the earth would soon be covered by the progeny of a single pair." To make his case as forcefully as possible, Darwin offered a conservative example:

> The elephant is reckoned the slowest breeder of all
> known animals, and I have taken some pains to
> estimate its probable minimum rate of natural increase;
> it will be safest to assume that it begins breeding when
> thirty years old, and goes on breeding till ninety years
> old, bringing forth six young in the interval, and
> surviving till one hundred years old; if this be so, after a
> period of from 740 to 750 years there would be nearly
> nineteen million elephants alive, descended from the
> first pair.

Because baby elephants grow up, mature, and themselves have babies, the elephant population grows by multiplication.

Two mathematical expressions describe two kinds of population growth: exponential growth and geometric growth. **Exponential growth** pertains when young individuals are added to the population continuously and a plot of the population increase as a function of time forms a smooth curve. Such a population increases according to the equation

$$N(t) = N(0)e^{rt},$$

where $N(t)$ is the number of individuals in a population after t units of time, $N(0)$ is the initial population size ($t = 0$), and r is the exponential growth rate. The constant e is the base of the natural logarithms; it has a value of approximately 2.72. Exponential growth results in a continuously accelerating curve of increase (or decelerating curve of decrease) whose slope varies directly with the size of the population (❚ Figure 14.3).

HELP ON THE WEB | *Go to Living Graphs at http://www.whfreeman. com/ricklefs.* **Use the interactive tutorials on exponential and geometric growth to understand better the concepts behind population growth.**

The rate at which individuals are added to a population undergoing exponential growth is the derivative of the exponential equation, which is the slope of the curve illustrated in Figure 14.3,

$$\frac{dN}{dt} = rN.$$

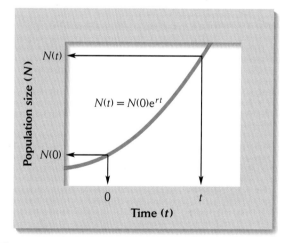

❚ **Figure 14.3 Exponential growth results in a continuously accelerating curve of increase.** The curve of exponential growth is shown for a population growing at rate r between time 0 and time t. During this period, the number of individuals increases from $N(0)$ to $N(t)$. Notice that the slope becomes steeper as the population increases.

This equation encompasses two principles. First, the exponential growth rate (r) expresses population increase (or decrease) on a "per individual basis." Second, the rate of increase (dN/dt) varies in direct proportion to the size of the population (N). In words, this equation could read

$$\left[\begin{array}{c} \text{the rate of} \\ \text{change in} \\ \text{population} \\ \text{size} \end{array}\right] = \left[\begin{array}{c} \text{the contribution} \\ \text{of each individual} \\ \text{to population} \\ \text{growth} \end{array}\right] \times \left[\begin{array}{c} \text{the number} \\ \text{of individuals} \\ \text{in the} \\ \text{population} \end{array}\right].$$

The human population grows continuously, and therefore exponentially, because babies are born and added to the population at all seasons of the year. This situation is unusual in natural populations, where reproduction typically is restricted to a particular time of year when resources are most abundant. Accordingly, populations grow during the breeding season, then decline between one breeding season and the next. In California quail populations, for example, the number of individuals doubles or triples each summer as adults produce their broods of chicks, but then dwindles by nearly the same amount during autumn, winter, and spring (∎Figure 14.4). Within each year, the population growth rate varies tremendously due to seasonal changes in the balance of birth and death processes. If we wished to measure the long-term population growth rate, it would be pointless to compare numbers in August, recently augmented by the chicks born that year, with numbers in May, after winter had taken its toll. One must count individuals at the same time each year, so that all counts are separated by the same cycle of birth and death processes. Such an increase (or decrease) over discrete intervals is referred to as **geometric growth.**

The rate of geometric growth is most conveniently expressed as the ratio of a population size in one year to that in the preceding year (or other time interval). Demographers have assigned the symbol λ, the lowercase Greek letter lambda, to this ratio; hence $\lambda = N(t+1)/N(t)$, where t is an arbitrary time. This definition of λ can be rearranged to provide a formula for projecting the size of the population through a single time interval:

$$N(t+1) = N(t)\lambda.$$

To calculate the growth of a population over many time intervals, we multiply the original population size by the geometric growth rate, once for each interval of time passed. Hence $N(1) = N(0)\lambda$, $N(2) = N(0)\lambda^2$, $N(3) = N(0)\lambda^3$, and

$$N(t) = N(0)\lambda^t.$$

For example, if a population increased at a geometric rate of 50% per year ($\lambda = 1.50$), an initial population of 100 individuals would grow to $N(0)\lambda = 150$ at the end of one year, $N(0)\lambda^2 = 225$ at the end of two years, and $N(0)\lambda^{10} = 5,767$ at the end of 10 years.

Note that the equation for geometric growth is identical to the equation for exponential growth except that λ takes the place of e^r, which equals the amount of exponential growth accomplished in one time period. Thus geometric and exponential growth are related by

$$\lambda = e^r$$

and

$$\log_e\lambda = r.$$

Because of this relationship, the two models of growth can describe the same data equally well (∎Figure 14.5), and there is a direct correspondence between the values of λ

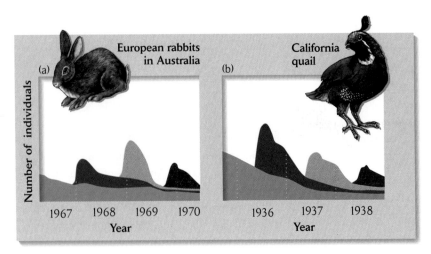

∎ **Figure 14.4 Populations with discrete reproductive periods increase by geometric growth.** (a) Rabbits in a subalpine population in New South Wales, Australia. (b) California quail. Each cohort of individuals born or hatched in a particular year is colored differently. Each population increases during the reproductive period, then declines. After K. Myers, in P. J. den Boer and G. R. Gradwell (eds.), *Dynamics of Populations*, Centre Agric. Publ. Doc., Wageningen, The Netherlands (1970), pp. 478–506; J. T. Emlen, Jr., *J. Wildl. Mgmt.* 4: 2–99 (1940).

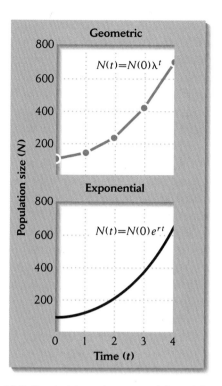

Figure 14.5 Geometric and exponential growth curves can be superimposed. The curves in these diagrams have equivalent rates (geometric: $\lambda = 1.6$; exponential, $r = 0.47$). $N(0) = 100$ individuals.

and r. For example, when a population's size remains constant, $r = 0$ and $\lambda = 1$ [$\log_e(1) = 0$]. Decreasing populations have negative exponential growth rates and geometric growth rates of less than 1 (but greater than 0—a real population cannot have a negative number of individuals). Increasing populations have positive exponential growth rates and geometric growth rates greater than 1 (**Figure 14.6**).

The individual, or **per capita,** contribution to population growth (r in the case of exponential growth) is the difference between birth rate (b) and death rate (d) calculated on a per capita basis. Thus, $r = b - d$. In the case of geometric growth, the per capita annual growth rate λ is the difference between the annual rates of birth (B) and death (D). Thus, $\lambda = B - D$.

Rates of birth and death, whether exponential or geometric, are abstractions having little meaning for the individual. An elephant dies only once, so it cannot have a personal rate of death. Babies are produced in discrete litters separated by intervals needed for gestation and parental care, not at constant rates, such as 0.05 young per day. But when births and deaths are averaged over a population as a whole, they take on meaning as rates of demographic events in the population. If $N = 1,000$ individuals produced $N \times B = 10,000$ progeny in a year, the

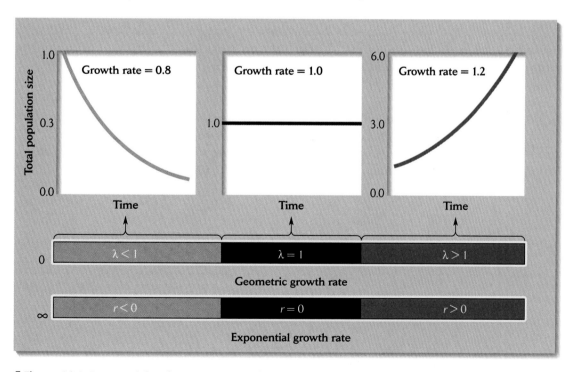

Figure 14.6 Exponential and geometric growth can describe increasing or decreasing populations. This diagram shows population growth curves resulting from values of $\lambda < 1$ ($r < 0$), $\lambda = 1$ ($r = 0$), and $\lambda > 1$ ($r > 0$).

average per capita birth rate would be $B = 10$ per year, and we could assume that a population of 1 million would produce 10 million progeny—still 10 per capita—under the same conditions. Conversely, suppose that only half the groundhogs alive on their big day in one particular year survive to February 2 the next year, a death rate of 50% per year. Although some of the groundhogs have died "completely" and others haven't died at all, 50% represents the probability that any one groundhog will have died during the period of one year.

How fast a population grows depends on its age structure

When birth rates and death rates have the same values for all members of a population, we can estimate future population size from the total population size (N) at the present time. But when birth and death rates vary with the ages of individuals, the contributions of younger and older individuals to population growth must be calculated separately. Two populations having identical birth and death rates at corresponding ages, but different **age structures** (proportions of individuals in each age class), will grow at different rates, at least for a while. A population composed wholly of prereproductive adolescents and postreproductive oldsters, for example, cannot increase until the young individuals reach reproductive age. This represents an extreme case, but smaller variations in age distribution can also profoundly influence population growth rates.

A little pencil-and-paper (or spreadsheet) figuring with a hypothetical population will demonstrate these effects. Let's start with a population of 100 individuals having the age-specific survival and fecundity rates and the age distribution shown in Table 14.1. In this population, all the adult individuals reproduce in a single breeding period each year, and then individuals suffer mortality between that breeding period and the next. All 3-year-olds die after they reproduce (survival at age 3, $s_3 = 0.0$), and so there are no 4-year-olds in the population. Newborns have a fecundity of zero (fecundity at age 0, $b_0 = 0$), as seems biologically reasonable. We count our population just after the breeding period and so our count includes newborn individuals (n_0).

Now, we can use the schedule of births and deaths to project the population into the future (Table 14.2). The first step is to calculate the number of individuals surviving from one year to the next in each age group. You can see that the 20 newborn individuals have a survival rate of 0.50 and become 10 one-year-olds in the following year; the 10 one-year-olds have a survival rate of 0.8 and become 8 two-year-olds in the following year; and so on. After calculating the number of individuals surviving to breed in the following year, we can calculate the total number of newborn individuals as the sum of the number of breeding adults in each age class times their fecundity. In the example in Table 14.2, at year 1 ($t = 1$), the total number of newborn individuals is 10×1 (fecundity of one-year-olds) $+ 8 \times 3$ (fecundity of two-year-olds) $+ 20 \times 2$ (fecundity of three-year-olds) $= 74$. The same exercise is repeated each year into the future. The results of the population projection are shown in Table 14.3, along with the total number of individuals in the population after each breeding season. The geometric rate of population growth is the ratio of the population size after one year to that at the beginning of the year, thus $\lambda = N(t + 1)/N(t)$.

Our hypothetical population at first grows erratically, with λ fluctuating between 1.05 and 1.69. However, provided that age-specific birth and survival rates remain unchanged, the population eventually assumes a **stable age distribution.** Under such conditions, each age class in a population grows or declines at the same rate, and there-

---

Table 14.1   Life table for a hypothetical population of 100 individuals

Age (x)	Survival (s_x)	Fecundity (b_x)	Number of individuals (n_x)
0	0.5	0	20
1	0.8	1	10
2	0.5	3	40
3	0.0	2	30

Table 14.2 Steps in the projection of a population through one time period of survival and reproduction. Each age class becomes one time unit older from one breeding season to the next.

Age	(1) Census of population just after reproduction ($t = 0$)	(2) Survival rate of individuals (s_x) (Table 14.1)	(3) Number surviving to next breeding season (1) × (2)	(4) Number of offspring per adult (b_x) (Table 14.1)	(5) Total number of offspring produced (3) × (4)	(6) Census of population just after reproduction ($t = 1$) (3) and sum of (5)
0	20	0.5		0		74
1	10	0.8	10	1	10	10
2	40	0.5	8	3	24	8
3	30	0.0	20	2	40	20
4	0		0			0
Total	100		38		74	112

fore, so does the total size of the population. As you can see in Table 14.3, λ eventually settles down to a constant value of 1.49, at which point the population has achieved a stable age distribution. Accordingly, the percentages of individuals in each age class also remain constant (∎ Figure 14.7).

The stable age distribution and growth rate achieved by a particular population depend on the birth and survival rates of its individuals. Any change in survival or fecundity rates alters the stable age distribution and results in a new rate of population growth. The effect of population growth rate on age structure stands out in comparisons of stable and growing human populations (∎ Figure 14.8). The population of Sweden in 1997, for example, had been stable for many years, and the age structure of the population primarily reflects the survival of individuals from infancy through old age. Declining birth rates during

Table 14.3 Projection of age distribution and total size through time for the hypothetical population described in Table 14.1

	0	1	2	3	4	5	6	7	8	Percent
					Time Interval					
n_0	20	74	69	132	175	274	399	599	889	63.4
n_1	10	10	37	34	61	87	137	199	299	21.3
n_2	40	8	8	30	28	53	70	110	160	11.4
n_3	30	20	4	4	15	14	26	35	55	3.9
N	100	112	118	200	279	428	632	943	1,403	100
λ		1.12	1.05	1.69	1.40	1.53	1.48	1.49	1.49	

Note: *The population was projected by multiplying the number of individuals in each age class by the survival to obtain the number in the next older age class in the next time period: $n_x(t) = n_{x-1}(t-1)s_x$. Then the number of individuals in each age class was multiplied by its fecundity to obtain the number of newborns: $n_0(t) = \Sigma n_x(t)b_x$.*

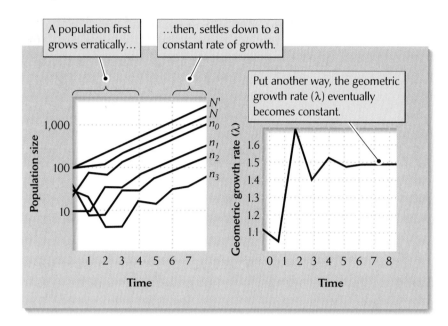

Figure 14.7 In a stable age distribution, each age class grows at the same rate. Notice that the growth rate of the population as a whole also stabilizes eventually. The data used to create this graph are taken from Table 14.3. N' represents the growth of a population of 100 individuals having a stable age distribution at time 0.

the Depression and the baby boom that followed World War II were responsible for irregularities in the age structure. In contrast, the rapid population growth of Costa Rica has resulted in a bottom-heavy age structure with large proportions of young individuals.

MORE ON THE WEB *The effect of birth and death rates on age structure and population growth rate.* Reducing survival and fecundity rates in Table 14.1 creates an older age structure and a slower, or even negative, rate of geometric growth.

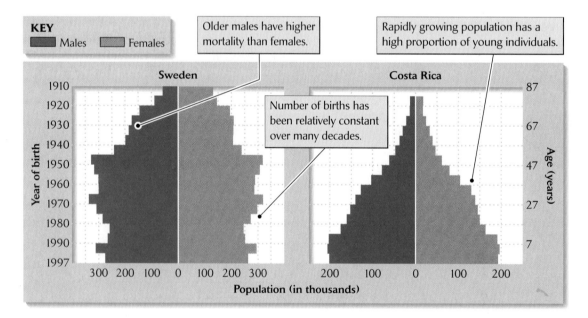

Figure 14.8 Population age structures of Sweden and Costa Rica in 1997 reflect the history of birth and survival rates. Because Sweden's population has grown slowly, its population is weighted toward older age classes. Declining birth rates during the Depression (1930s) and the baby boom that followed World War II (1945–1955) were responsible for irregularities in the age structure. Costa Rica's rapid population growth, caused by a high birth rate, resulted in a bottom-heavy age structure. However, birth rates have decreased over the last two decades.

A life table summarizes age-specific schedules of survival and fecundity

Life tables, such as the hypothetical table of birth and survival rates in Table 14.1, can be used to model the addition and removal of individuals in a local population (in the absence of immigration and emigration). Because it is hard to ascertain paternity in many species, life tables are usually based on females. For some populations with highly skewed sex ratios or unusual mating systems, this can pose difficulties, but in most cases a female-based life table provides a workable population model.

As we have seen in Table 14.1, age is designated in a life table by the symbol x, and the subscript x indicates age-specific variables. When reproduction occurs during a brief breeding season each year, each age class is composed of a discrete group of individuals born at approximately the same time. When reproduction is continuous, as it is in the human population, each age class x is designated arbitrarily as comprising individuals between ages $x - \frac{1}{2}$ and $x + \frac{1}{2}$.

The fecundity of females, which is often expressed in terms of female offspring produced per breeding season or age interval, is designated by b_x (think of b for "births"). Life tables portray the statistics of mortality in several ways. The fundamental measure is probability of survival (s_x) between ages x and $x + 1$. Probabilities of survival over many age intervals are summarized by survivorship to age x, designated by l_x (think of l for "living"), which is the probability that a newborn individual will be alive at age x. Because by definition all newborn individuals are alive at age 0, $l_0 = 1$. The proportion of newborns alive at age 1 is the probability of surviving from age 0 to age 1, hence $l_1 = s_0$. Similarly, $l_2 = s_0 s_1$ and, by extension, $l_x = s_0 s_1 s_2 \ldots s_{x-1}$.

The variables used in a life table are illustrated in Table 14.4 by data for the annual meadow grass Poa annua. This life table follows the survival and fecundity of a planting of the grass under experimental conditions over 2 years, by which time the last individual had died. Age is tabulated in intervals of 3 months. Because Poa annua is hermaphroditic, sexes are not distinguished. Of 843 plants alive at time 0 (germination), 722, or 85.7%, were alive at 3 months; hence s_0 and $l_{3\,mo}$ both equal 0.857. The life table shows that probability of dying increased with age. Fecundity rose to a peak (620 seeds per 3-month period) at 6 months of age and then declined.

The life table for Poa annua (Table 14.4) is a cohort life table in that it follows the fate of a group of individuals

Table 14.4 Life table of the grass *Poa annua*

Age $(x)^*$	Number alive	Survivorship (l_x)	Mortality rate (m_x)	Survival rate (s_x)	Fecundity (b_x)
0	843	1.000	0.143	0.857	0
1	722	0.857	0.271	0.729	300
2	527	0.625	0.400	0.600	620
3	316	0.375	0.544	0.456	430
4	144	0.171	0.626	0.374	210
5	54	0.064	0.722	0.278	60
6	15	0.018	0.800	0.200	30
7	3	0.004	1.000	0.000	31
8	0	0.000			

Number of 3-month periods; in other words, 3 = 9 months.

Source: *M. Begon and M. Mortimer, Population Ecology, 2d ed., Blackwell Scientific Publications, Oxford (1986). After data of R. Law.*

Summary of life-table variables

l_x Survival of newborn individuals to age x
b_x Fecundity at age x

m_x Proportion of individuals of age x dying by age $x + 1$
s_x Proportion of individuals of age x surviving to age $x + 1$

born at the same time from birth to the death of the last individual. This method is readily applied to populations of plants and sessile animals in which marked individuals can be continually resampled over the course of their life spans. Herein lies one of its disadvantages, however: it can take a long time to collect the data (particularly if the subjects are redwood trees!). This method is also difficult to apply to highly mobile animals.

Another method, employing a **static life table,** sidesteps the time problem by considering the survival of individuals of known age during a single time interval. The investigator estimates each age-specific survival value independently for each age class of a population during the same time period. Of course, to apply this technique, it is necessary to know the ages of individuals (which may be estimated by growth rings, tooth wear, or some other reliable index). By combining estimates of survivorship curves with age-specific fecundity into a single life table for a population, it is possible to determine whether, and how rapidly, a population is increasing or decreasing.

ECOLOGISTS IN THE FIELD

Building life tables for natural populations

Dedicated field ecologists have followed marked individuals of many species of animals and plants in their natural environments to understand the dynamics of their populations. One of the finest examples of such work is that of Peter and Rosemary Grant, of Princeton University, who have studied several species of ground finches of the genus *Geospiza* on Daphne Island in the Galápagos archipelago. Because Daphne is isolated (█ Figure 14.9), the populations of birds there have been relatively undisturbed. Daphne's small size (40 hectares) enabled the Grants to capture and mark with uniquely colored plastic leg bands all the birds on the island.

The Grants followed finch populations on Daphne for over 15 years, during which time it was possible to construct cohort life tables for many of the birds born early in the study. For example, the fates of 210 chicks of *Geospiza scandens* fledged in 1978 are shown in █ Figure 14.10. The survival rate was low during the first year of life, and then was higher, but also quite variable, for the next dozen age classes. The variation in survival from year to year reflects swings in precipitation on the island related to El Niño and La Niña climate patterns (see Chapter 4). El Niño years are extremely wet, and vegetation grows luxuriantly, producing abundant food for the finches and resulting in high survival. La Niña years are periods of drought and food scarcity.

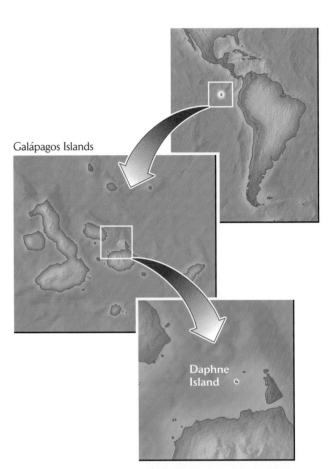

Galápagos Islands

Daphne Island

█ **Figure 14.9 Peter and Rosemary Grant have studied the population of the cactus finch *Geospiza scandens* on Daphne Island, Galápagos, for several decades.** The island is the crater of an extinct volcano. The cactus finches live mostly in the dense vegetation on the steep slopes of the crater, which afford little space for a field camp, but a good view nonetheless. Photo by R. E. Ricklefs.

One disadvantage of a cohort life table can be seen in the data for *Geospiza scandens:* variation in survival with age may be obscured by variation in the environment. The

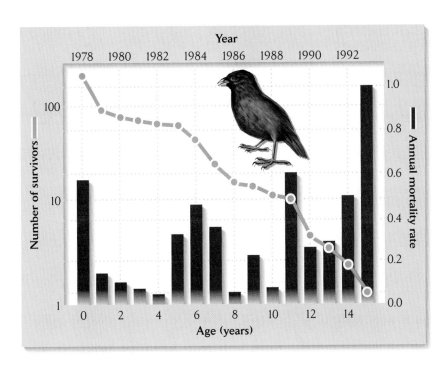

Figure 14.10 Survival rates may vary from year to year. Survival data for the cohort of Daphne Island cactus finches fledged in 1978 show high variation in survival rates. Mortality of immature birds during the first year is high. Adult mortality rates generally increase with age, but also vary with climatic patterns. From data in P. R. Grant and B. R. Grant, *Ecology* 73:766–784 (1992).

use of a static life table can avoid this problem. The mammalogist Olaus Murie used the distribution of ages at death to construct a static life table for Dall mountain sheep (*Ovis dalli*) in Mount McKinley (now Denali) National Park, Alaska, during the 1930s. The size of the horns, which grow continuously during the lifetime of an individual sheep (**Figure 14.11**), provided an estimate of age at death. Of 608 skeletal remains that Murie recovered, he judged 121 to have been less than 1 year old at death, 7 between 1 and 2 years old, 8 between 2 and 3 years old, and so on, as shown in Table 14.5.

Murie constructed the life table by using the following reasoning. All 608 dead sheep must have been alive at birth; all but the 121 that died during the first year must have been alive at the age of 1 year (608 − 121 = 487), all but 128 (the 121 dying during the first year and the 7 dying during the second) must have been alive at the end of the second year (608 − 128 = 480), and so on, until the oldest sheep died during their fourteenth year. Survivorship (l_x; the rightmost column in Table 14.5) was calculated by converting the number of sheep alive at the beginning of each age interval to a decimal fraction of those alive at birth. Thus, for example, the 390 sheep alive at the beginning of the seventh year is decimal fraction 0.640 (or 64.0%) of 608. By proceeding in this way, Murie was able to construct a life table for a population of long-lived organisms during a relatively short time interval. Such static life tables allow investigators to determine how changes in the environment affect population processes.

Figure 14.11 Static life tables require a means of estimating the age of organisms. The size of the horns of the Dall mountain sheep increases with age. Photo by R. E. Ricklefs.

Table 14.5 Life table for Dall mountain sheep constructed from the age at death of 608 sheep in Denali National Park

Age interval (years)	Number dying during age interval	Number surviving at beginning of age interval	Number surviving as a fraction of newborns (l_x)
0–1	121	608	1.000
1–2	7	487	0.801
2–3	8	480	0.789
3–4	7	472	0.776
4–5	18	465	0.764
5–6	28	447	0.734
6–7	29	419	0.688
7–8	42	390	0.640
8–9	80	348	0.571
9–10	114	268	0.439
10–11	95	154	0.252
11–12	55	59	0.096
12–13	2	4	0.006
13–14	2	2	0.003
14–15	0	0	0.000

Source: *Based on data of O. Murie,* The Wolves of Mt. McKinley, *U.S. Department of the Interior, National Park Service, Fauna Series No. 5, Washington, D.C. (1944); quoted by E. S. Deevey, Jr.,* Quarterly Review of Biology *22:283–314 (1947).*

The intrinsic rate of increase can be estimated from the life table

The **intrinsic rate of increase** of a population, indicated by r_m (often called the Malthusian parameter after Thomas Malthus, the eighteenth-century English economist), is the exponential rate of increase (r) assumed by a population with a stable age distribution. In practice, populations rarely achieve stable age distributions and therefore rarely grow at their intrinsic rates of increase. When changes in environmental conditions alter life-table values, the age structure of a population continuously readjusts to the new schedule of birth and death rates. Thus, the actual growth performance of a population depends as much on past conditions, which determine its age structure, as on current life-table values. Therefore, the intrinsic rate of increase shows how a population would grow if environmental conditions remained constant. In a varying world, r_m cannot project long-term population growth accurately.

Each life table determined under a particular set of environmental conditions has a single intrinsic rate of increase. To find the exact value of r_m, one must solve a complicated equation. However, r_m may be approximated by r_a (the a indicating an approximate estimate), which is

obtained by a simple formula based on life-table values. Before we can calculate r_a, however, we have to add a few things to our hypothetical life table in Table 14.1. One of these is a column with the product of l_x and b_x, which represents the expected number of offspring that a newborn individual will produce at age x (Table 14.6). We then must add another column that is the product of x, l_x, and b_x, the expected number of births multiplied by age, which will be used to estimate the average age at which a female gives birth to her offspring. Finally, we add the figures in the $l_x b_x$ and $x l_x b_x$ columns to get the column totals.

HELP ON THE WEB *Go to Living Graphs at http://www.whfreeman. com/ricklefs. Use the interactive tutorials on life tables to understand better how they can be used to estimate the growth rate and age structure of a population.*

The sum of the $l_x b_x$ terms has a name, which is the **net reproductive rate** (R_0) of individuals in the population. One may think of R_0 as the expected total number of offspring of an individual over the course of his or her life span. Thus, in the life table in Table 14.6, newborn individuals produce 2.1 offspring on average during the course of their life span. This is greater than the replacement rate of 1 offspring per individual, and so the population should grow rapidly as long as the life table does not change.

Table 14.6 Estimation of the exponential rate of increase for the hypothetical population described in Table 14.1

x	s_x	l_x	b_x	$l_x b_x$	$x l_x b_x$
0	0.5	1.0	0	0.0	0.0
1	0.8	0.5	1	0.5	0.5
2	0.5	0.4	3	1.2	2.4
3	0.0	0.2	2	0.4	1.2
Net reproductive rate (R_0)				2.1	
Expected number of births weighted by age					4.1

Note: *The sums of the $l_x b_x$ column (net reproductive rate) and the $x l_x b_x$ column are used to estimate r_a according to the equation given in the text. In this case, we calculate r_a to be 0.38; this is equivalent to $\lambda = 1.46$, close to the observed value of about 1.49 after the population achieved a stable age distribution.*

The average age at which an individual gives birth to its offspring (T) is now calculated as

$$T = \frac{\Sigma x l_x b_x}{\Sigma l_x b_x}.$$

T is sometimes referred to as the **generation time.** In our hypothetical population, T is 4.1/2.1, or approximately 2.0 time units.

Now we are prepared to estimate the exponential growth rate of the population, according to the equation

$$r_a = \frac{\log_e R_0}{T}.$$

In this case we can calculate r_a to be $\log_e(2.1)/2.0$, or 0.38. This is equivalent to $\lambda = e^{0.38} = 1.46$, which is close to the observed value of about 1.49 after the population achieved a stable age distribution (see Table 14.3).

You can see that the intrinsic rate of growth of a population depends on both the net reproductive rate (R_0) and the generation time (estimated by T). Clearly, a population grows when R_0 exceeds 1, which is the replacement level of reproduction for a population. The rate of population growth, whether positive or negative, increases as young are born to their mothers at younger ages, that is, as T, or generation time, decreases.

Most populations have a great biological growth potential

We can best appreciate the capacity of a population for growth by following the rapid increase of organisms introduced into a new region with a suitable environment. In 1937, 2 male and 6 female ring-necked pheasants were released on Protection Island, Washington. They increased to 1,325 adults within 5 years. This 166-fold increase represents a 178% annual rate of increase ($r = 1.02$, $\lambda = 2.78$). In other words, the population almost tripled, on average, each year. When domestic sheep were introduced to Tasmania, a large island off the coast of Australia, the population increased from fewer than 200,000 in 1820 to more than 2 million in 1850 (see Figure 15.3). This tenfold increase in 30 years is equivalent to an annual rate of increase of 8% ($r = 0.077$, $\lambda = 1.08$).

Even such an unlikely creature as the elephant seal, whose population along the western coast of North America had been all but obliterated by hunting during the nineteenth century, increased from 20 individuals in 1890 to 30,000 in 1970 ($r = 0.091$, $\lambda = 1.096$). If you are unimpressed, consider that another century of unrestrained growth would find 27 million elephant seals crowding surfers and sunbathers off southern California beaches (■ Figure 14.12). Before the end of the following century, the shorelines of the Western Hemisphere would give lodging to a trillion of the beasts.

Elephant seal populations do not hold any records for growth potential, however—quite the contrary. Life tables of populations maintained under optimal conditions in laboratories have exhibited potential annual growth rates (λ) as great as 24 for the field vole, a small mouselike mammal, 10 billion (10^{10}) for flour beetles, and 10^{30} for the water flea *Daphnia*.

Another way of expressing growth rate is the **doubling time** (t_2) of the population. Doubling time can be calculated by the expression

$$t_2 = \frac{\log_e 2}{\log_e \lambda}.$$

▌**Figure 14.12 Populations have the potential to increase rapidly.** The northern elephant seal, which was nearly extirpated in the nineteenth century, has rebounded from near extinction. These females and young males have gathered on a beach for several weeks while they are molting their fur. Photo by François Gohier/Photo Researchers.

The value of $\log_e 2$ is 0.69. Hence, for the field vole ($\lambda = 24$), $t_2 = 0.69/\log_e 24$, which is 0.22 years, or 79 days. Population doubling times are 246 days for the ring-necked pheasant, 11 days for the flour beetle, and only 3.6 days for the water flea. The potential growth rates of populations of bacteria and viruses under ideal conditions are almost unimaginable.

Environmental conditions influence intrinsic rates of increase

The intrinsic rate of increase of a population depends on how individuals perform in that population's environment. This performance is reflected in the life table as probabilities of survival and rates of fecundity, and therefore the life table (and hence the intrinsic rate of increase) responds to changes in conditions in the environment. These conditions vary spatially and temporally, creating differences in population dynamics from place to place and leading to changing population dynamics over time.

Such effects of environmental conditions can be seen in experimental studies in which groups of individuals from the same population are subjected to different conditions. This method eliminates any effect of different genotypes and shows only the reaction norm of a particular population. In one such study, temperature and moisture were found to influence the intrinsic rates of increase of two species of grain beetles differently (▌Figure 14.13). Neither *Rhizopertha dominica* nor *Calandra oryzae* performed well at low temperatures and humidities. Moreover, the optimal conditions for growth differed between the two species: *Rhizopertha* populations grew most rapidly at somewhat warmer temperatures. Of the two, *Rhizopertha* has the more tropical distribution in nature.

MORE ON THE WEB *Key-factor analysis*. **This method has been used to identify the factors most responsible for population change, as shown in a study of the black-headed budworm.**

A population's intrinsic rate of increase is balanced by extrinsic factors

Experimental laboratory populations are generally reared under controlled conditions, usually with abundant food. It is not surprising that under such circumstances a population can increase under a broad range of physical conditions, as illustrated by the flour beetles in Figure 14.13. Continuous exponential growth leads, with time, to inconceivable numbers. Even the most slowly reproducing species would cover the earth in a short time if its population growth were unrestrained. Yet most populations we observe in nature remain at relatively stable levels.

Nearly two centuries ago, Thomas Malthus understood that this fact "implies a strong and constantly operating check on population from the difficulty of subsistence." In *An Essay on the Principle of Population* (1798), he wrote:

> Through the animal and vegetable kingdoms, nature has scattered the seeds of life abroad with the most profuse and liberal hand. She has been comparatively sparing in the room and the nourishment necessary to rear them. The germs of existence contained in this spot of earth, with ample food, and ample room to expand in, would fill

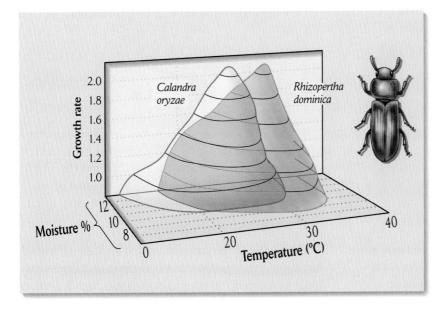

Figure 14.13 Environmental conditions influence intrinsic rates of increase. The geometric growth rates of the grain beetles *Calandra oryzae* and *Rhizopertha dominica* living in wheat varied with temperature and moisture. Rates of increase are indicated by contour lines that describe conditions with identical values of λ. After L. C. Birch, *Ecology* 34:698–711 (1953).

millions of worlds in the course of a few thousand years. Necessity, that imperious all pervading law of nature, restrains them within the prescribed bounds. The race of plants, and the race of animals shrink under this great restrictive law.

Darwin echoed this view in *On the Origin of Species:*

As more individuals are produced than can possibly survive, there must in every case be a struggle of existence, either one individual with another of the same species, or with the individuals of distinct species, or with the physical conditions of life. It is the doctrine of Malthus applied with manifold force to the whole animal and vegetable kingdoms; for in this case there can be no artificial increase of food, and no prudential restraint from marriage. Although some species may be now increasing, more or less rapidly, in numbers, all cannot do so, for the world would not hold them.

This essentially modern view of the regulation of populations grew out of an awareness of the immense capacity of populations for increase. In a sense, a population's growth potential and the relative constancy of its numbers cannot be logically reconciled otherwise.

The growth potential of populations is driven home by the rabbit population in Australia. In 1859, 12 pairs were released on a ranch in Victoria to provide sport for hunters. Within 6 years, the population had increased so rapidly that 20,000 rabbits were killed in a single hunting drive. Even by conservative estimates, the population must have increased by a factor of at least 10,000 in 6 years,

an exponential growth rate (r) of about 1.5 per year (a doubling time of about 5.5 months). Yet the present-day population neither increases nor decreases, on average, from year to year. How can the initial rapid growth rate be reconciled with the eventual stabilization of the rabbit population?

Either birth rate decreased, death rate increased, or both changed when the rabbit population became more numerous. When there are more rabbits, there is less food for each. Fewer resources mean that fewer offspring can be nourished, and those offspring survive less well. Crowded populations also aggravate social strife, promote the spread of disease, and attract the attention of predators, as we shall see in a later chapter. Many such factors may act together to slow, and finally halt, population growth. An important step in understanding how populations are regulated was the development of mathematical descriptions of population growth processes early in the twentieth century.

The logistic equation

In 1920, Raymond Pearl and L. J. Reed, at the Institute for Biological Research of Johns Hopkins University, published a paper in the *Proceedings of the National Academy of Sciences* titled "On the Rate of Growth of the Population of the United States since 1790 and Its Mathematical Representation." Thorough and accurate population data had been gathered even in colonial times. Indeed, the phenomenal population growth of the American colonies had impressed upon Malthus how rapidly humans could multiply; this was not as evident in the more crowded European countries of his time.

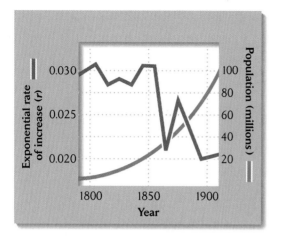

Figure 14.14 Population growth rates become lower as a population grows larger. The population growth rate in the United States slowed as the population increased between 1790 and 1910. The rising curve shows the size of the population. The exponential rate of increase (*r*) for each 10-year period shows a downward trend, indicating a decreasing rate of population growth. From data in R. Pearl and L. J. Reed, *Proc. Natl. Acad. Sci.* 6: 275–288 (1920).

Pearl and Reed wished to project the future growth of the population of the United States, which they supposed must eventually reach a limit. Data to 1910, the latest census then available, had revealed a decline in the exponential rate of growth (**Figure 14.14**). Pearl and Reed reasoned that if this decline followed a regular pattern that could be described mathematically, it would be possible to predict the future course of the population, as long as the decline in the exponential growth rate continued. They also reasoned that changes in the exponential rate of growth must be related to the size of a population rather than to time, because time scales are arbitrary. And so, in place of a constant value of *r* in the differential equation for unrestrained population growth ($dN/dt = rN$), Pearl and Reed suggested that *r* decreases as *N* increases, according to the relation

$$r = r_0 \left(1 - \frac{N}{K} \right).$$

In this expression, r_0 represents the intrinsic exponential growth rate of a population when its size is very small (that is, close to 0), and *K*—the **carrying capacity** of the environment—represents the number of individuals that the environment can support. The differential equation describing restricted population growth now became

$$\frac{dN}{dt} = r_0 N \left(1 - \frac{N}{K} \right).$$

In words, this equation may be expressed as

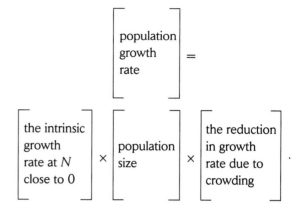

According to this equation, which is called the **logistic equation,** the exponential rate of increase decreases as a linear function of the size of a population. Such a decrease reasonably approximated the data for the population of the United States (**Figure 14.15**).

HELP ON THE WEB *Go to Living Graphs at http://www.whfreeman.com/ ricklefs.* **Use the interactive tutorial to see how the logistic equation represents a population that always tends toward an equilibrium carrying capacity.**

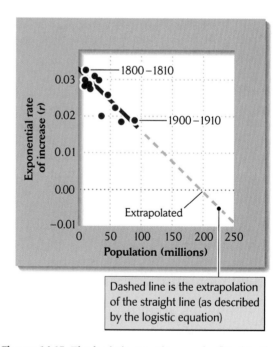

Dashed line is the extrapolation of the straight line (as described by the logistic equation)

Figure 14.15 The logistic equation can be fitted to the population growth pattern of the United States. The exponential rate of increase during each decade between 1790 and 1910 is plotted as a function of population size during that decade (the geometric mean of the beginning and ending numbers) using the data shown in Figure 14.14. The best fit to the data is a straight line, which can be extrapolated. This extrapolation suggested that the population would level off (*r* = 0) at just under 200 million individuals.

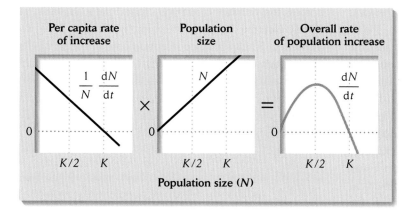

Figure 14.16 The logistic curve of population growth incorporates the influences of increasing population size and decreasing per capita growth rate. The overall rate of population growth (dN/dt) is the product of the per capita exponential growth rate ($r = 1/N \, dN/dt$) and population size (N). The value of r declines as a linear function of population size (N), from r_0 at $N = 0$ to 0 when $N = K$. The overall growth rate of a population reaches a maximum at the inflection point, when population size is one-half the carrying capacity (K).

So long as population size N does not exceed the carrying capacity K—that is, N/K is less than 1—a population continues to increase, albeit at a slowing rate. When N exceeds K, the ratio N/K exceeds 1, the term in parentheses $(1 - N/K)$ becomes negative, and the population decreases. Because populations below K increase and those above K decrease, K is the eventual steady-state, or equilibrium, size of a population growing according to the logistic equation.

The influence of population size (N) on the per capita rate of growth (r) and the overall rate of growth (dN/dt) is shown in **Figure 14.16**. The overall growth rate is small when the population is small just because there are so few individuals to produce babies. The growth rate falls off as the population approaches the carrying capacity because declining resources limit reproduction. The curve for dN/dt has a maximum at intermediate population size, specifically

when $N = K/2$, or half the carrying capacity. This **inflection point** (i) separates the early accelerating phase of population growth from the later decelerating phase.

You can visualize this falling off of growth rate by using a form of the logistic equation that relates number of individuals to time,

$$N(t) = \frac{K}{1 + e^{-r_0(t-i)}}.$$

This equation describes a sigmoid, or S-shaped, curve (**Figure 14.17**). The population grows slowly at first, then more rapidly as the number of individuals increases, and finally more slowly again, gradually approaching the equilibrium number K.

The sigmoid curve is applied to the growth of the U.S. population from 1790 to 1910 in **Figure 14.18**. Pearl and

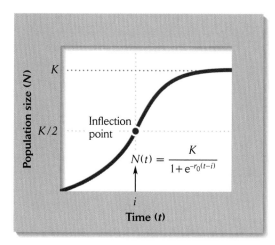

Figure 14.17 Logistic growth follows an S-shaped curve. The curve is symmetrical about the inflection point ($K/2$); that is, accelerating and decelerating phases of population growth have the same shape.

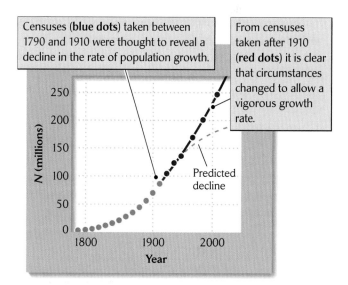

Censuses (**blue dots**) taken between 1790 and 1910 were thought to reveal a decline in the rate of population growth.

From censuses taken after 1910 (**red dots**) it is clear that circumstances changed to allow a vigorous growth rate.

Figure 14.18 Population projections may prove incorrect if life-table values change. A logistic curve fitted to population data for the United States between 1790 and 1910 predicts a leveling off at about 200 million. Subsequent censuses, however, have shown numbers above the projected population curve.

Reed obtained the best fit of their equation to the U.S. population data when the value of K was 197,273,000 and that of r_0 was 0.03134, which is equivalent to about a 3% increase in the population per year and a population doubling time of 22 years. Thus, although the population in 1910 was only 91,972,000, Pearl and Reed could extrapolate its future growth to twice the 1910 level from its earlier growth performance. Projections often prove incorrect, however, when circumstances change. The U.S. population reached 197 million between 1960 and 1970, when it was still growing vigorously. Improved public health and medical treatment raised survival rates substantially, particularly for infants and children, between the 1920s and 1970s. A leveling off in the mid-200 millions can now be predicted on the basis of a much reduced birth rate in recent years, but all this could easily change.

Population size is regulated by density–dependent factors

The logistic equation has been found to describe the growth of populations in the laboratory and in natural habitats. The fact that it does so suggests that factors limiting growth exert stronger effects on mortality and fecundity as a population grows (see Figure 14.16). That is why the intrinsic growth rate of a population decreases as population size increases. But what are these effects, and how do they operate?

Many things influence rates of population growth, but only **density-dependent factors,** whose effect increases with crowding, can bring a population under control. Of prime importance among these factors are supplies of food and places to live, which are relatively fixed in amount and number. Additionally, the effects of predators, parasites, and diseases are felt more strongly in crowded populations than in sparse ones. Other factors, such as temperature, precipitation, and catastrophic events, alter birth and death rates largely without regard to the numbers of individuals in a population. Thus, such **density-independent factors** may influence the growth rate of a population, but they do not regulate its size.

Density dependence in animals

Numerous experimental studies have revealed a variety of mechanisms of density dependence. For example, when a single pair of fruit flies is confined to a bottle with a fixed supply of food, its descendants increase in number rapidly at first, but soon reach a limit. When different numbers of pairs of flies are introduced into otherwise identical culture bottles, the number of progeny raised per pair varies inversely with the density of flies in the bottle (■ Figure 14.19). This effect results from competition among the larvae for food, which causes high mortality in dense cultures. Adult life span also declines, but only at densities well above the levels at which the survival of larvae drops off. It is often the case that juvenile stages suffer the adverse effects of density-dependent factors more than adults do.

MORE ON THE WEB | *Density dependence in laboratory cultures of water fleas.* **How does density influence life-table variables to determine population growth rate?**

Most studies of density dependence have focused on laboratory populations, in which the factors that affect population growth can be controlled experimentally. The simplicity of such systems leaves some doubt about the relevance of laboratory findings to populations in more complex natural surroundings, where physical conditions change continually and the experimenter does not control food supply or predation. For example, winter weather and other factors have caused the population of song sparrows (*Melospiza melodia*) on Mandarte Island—a 6-hectare speck of land off the coast of British Columbia—to fluctuate widely. This population has fluctuated between 4 and 72 breeding females and between 9 and 100 breeding males during recent years. In a sense, this population provides a natural experiment on the effects of environmentally induced variation in population size. But the response of the population to this variation clearly shows that density-

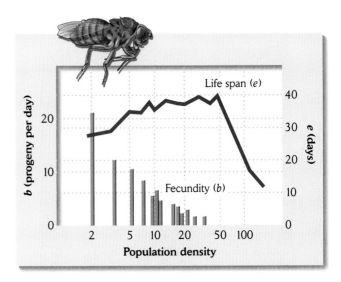

■ **Figure 14.19 Density–dependent factors can regulate population growth.** Fecundity and life span decrease as population density increases in laboratory populations of the fruit fly *Drosophila melanogaster*. After R. Pearl, *Q. Rev. Biol.* 2:532–548 (1927).

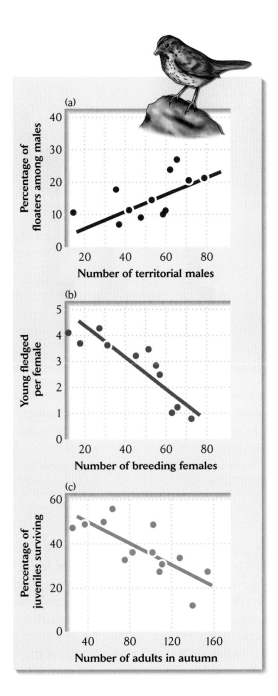

(a)

(b)

(c)

Figure 14.20 Density-dependent factors can control the size of natural populations. The responses of song sparrows on Mandarte Island to population fluctuations caused by environmental variation show that density-dependent factors are at work even in these circumstances. With increased crowding on the small island, the proportion of males prevented from acquiring territories ("floaters") increases (a) and the number of fledglings produced per female (b), as well as the survival of those offspring through the autumn and winter (c), decreases. After P. Arcese and J. N. M. Smith, *J. Anim. Ecol.* 57:119–136 (1988), and J. N. M. Smith, P. Arcese, and W. M. Hochachka, in C. M. Perrins, J.-D. Lebreton, and G. J. M. Hirons (eds.), *Bird Population Studies: Relevance to Conservation and Management*, Oxford University Press, Oxford (1991), pp. 148–167.

iment in nature as in the laboratory—that is, to alter the density of individuals in the population while keeping everything else constant. Such an experiment is difficult to carry out on a large scale. But populations of game animals are sometimes increased or decreased by management practices, and ecologists have taken advantage of such situations to study population processes. A survey of harvested white-tailed deer (*Odocoileus virginianus*) in New York State in the 1940s provides an example of this method.

Deer depend on high-quality food for reproduction and survival. Deer browse leaves, and they require large quantities of new growth with high nutritional content to grow rapidly and reproduce successfully. Studies on white-tailed deer by game biologists in New York State showed that the proportion of females pregnant and the average number of embryos per pregnant female were directly related to range conditions (**Figure 14.21**). The number of *corpora lutea* in each ovary indicates the number of eggs ovulated, and hence the reproductive potential of a female. When the number of *corpora lutea* exceeds the number of embryos, the difference reflects the death and resorption of embryos, which often results from poor nutrition of pregnant females on range of low quality. In the central Adirondack area, where habitat for deer was very poor, even ovulation was greatly reduced.

Selective hunting to thin dense populations can often reverse range deterioration caused by overgrazing. When DeBar Mountain, an area of very poor range, was opened to hunting, the population of white-tailed deer decreased, range quality recovered, and reproduction improved dramatically (**Figure 14.22**).

dependent factors potentially could regulate the size of the population. During years of high density, territorial behavior restricts the number of breeding males, decreases in food supply reduce the average number of offspring reared by each female, and the survival of juveniles in autumn and winter is reduced (**Figure 14.20**).

ECOLOGISTS IN THE FIELD

Density dependence in white-tailed deer populations

Although natural variation in population size provides a method for visualizing density dependence, ideally we would like to conduct the same exper-

MORE ON THE WEB *Positive density dependence.* **In some circumstances, population growth rate increases with increasing density.**

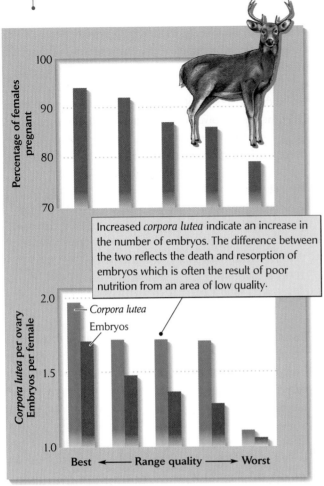

Increased *corpora lutea* indicate an increase in the number of embryos. The difference between the two reflects the death and resorption of embryos which is often the result of poor nutrition from an area of low quality.

■ Figure 14.21 Game management practices may provide natural experiments in population dynamics. Reproductive parameters of white-tailed deer (*Odocoileus virginianus*) harvested in 1939–1949 reflect range quality in five regions of New York State. After E. L. Chaetum and C. W. Severinghaus, *Trans. N. Am. Wildl. Conf.* 15:170–189 (1950).

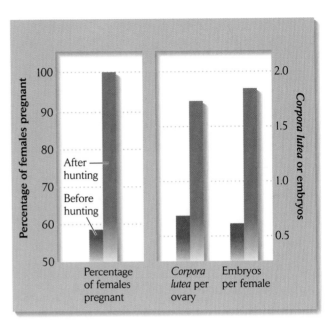

■ Figure 14.22 Reproductive parameters may be density-dependent. Measures of fecundity of white-tailed deer in the DeBar Mountain area of the Adirondack Mountains of New York State increased after hunting had reduced the population. After E. L. Chaetum and C. W. Severinghaus, *Trans. N. Am. Wildl. Conf.* 15:170–189 (1950).

Density dependence in plants

Plants experience increased mortality and reduced fecundity at high densities, just as animals do. A common response of plants to intense competition for resources is slowed growth, which has consequences for fecundity and, to a lesser extent, survival. The sizes of flax (*Linum*) plants grown to maturity at different densities reveal this flexibility (**■ Figure 14.23**). When seeds were sown sparsely at a density of 60 per square meter, the modal dry weight of individuals was between 0.5 and 1 g, and many plants attained weights exceeding 1.5 g by the end of the experiment. When seeds were sown at densities of 1,440 and 3,600 per square meter, most of the individuals weighed less than 0.5 g, and few grew to large size. The variation in average plant size for the three treatments is a reaction norm with respect to resource availability. Variation in the

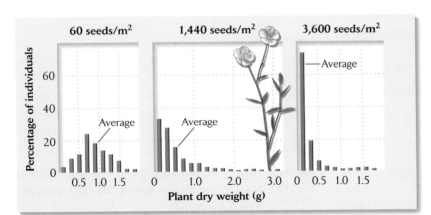
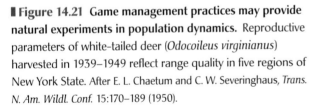

■ Figure 14.23 Plants may respond to competition for resources by slowing their growth. The average sizes of individual flax plants were smaller when seeds were sown at higher density. After J. L. Harper, *J. Ecol.* 55:247–270 (1967).

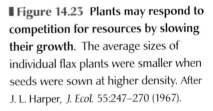

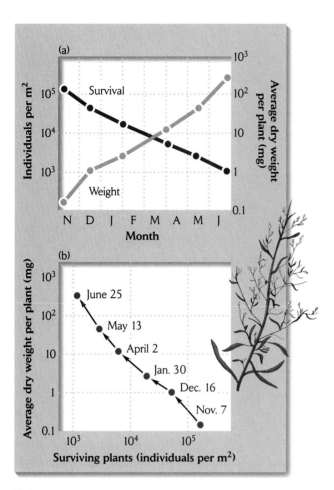

▌Figure 14.24 Plant populations increase in biomass even as numbers of individuals decrease. (a) Progressive change in plant weight and population density in an experimental planting of horseweed (*Erigeron canadensis*) sown at a density of 100,000 seeds per square meter. (b) Relationship between plant density and plant weight as the season progressed. After J. L. Harper, *J. Ecol.* 55:247–270 (1967).

growing season, a thousandfold increase in the average weight of each plant more than balanced the hundredfold decrease in population density.

When the logarithm of average plant weight is plotted as a function of the logarithm of density, the data points recorded during the growing season fall on a line with a slope of approximately $-\frac{3}{2}$ (▌Figure 14.25). Plant ecologists call this relationship between average plant weight and density a **self-thinning curve.** Such is the regularity of this relationship that many have referred to it as the $-\frac{3}{2}$ **power law.**

Density-dependent factors tend to bring populations under control and maintain their size close to the carrying capacity set by the availability of resources and conditions of the environment. Changes in these conditions and resources continually establish new equilibrium values toward which populations grow or decline. Furthermore, density-independent changes in the environment brought about by events such as a sudden freeze, a violent storm, or a shift in an ocean current often reduce populations far below their carrying capacities and initiate periods of rapid population growth. Thus, although density-dependent factors regulate all populations, variations in the environment also cause populations to fluctuate about their equilibrium sizes. We shall explore population variation in more detail in the next chapter.

size of individuals within each planting results from chance factors early in the seedling stage, particularly date of germination and quality of the site on which a seedling grows. Early germination in a favorable spot gives a plant an initial growth advantage over others, which increases as larger plants grow and crowd their smaller neighbors.

The flexibility of plant growth does not preclude mortality in crowded situations. When horseweed (*Erigeron canadensis*) seed was sown at a density of 100,000 per square meter (equivalent to about 10 seeds in the area of your thumbnail), young plants competed vigorously. As the seedlings grew, many died, decreasing the population density for the surviving seedlings (▌Figure 14.24). At the same time, however, the growth rates of the surviving individual plants exceeded the rate of decline of the population, and the total weight of the planting increased. Over the entire

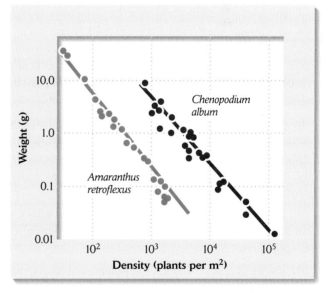

▌Figure 14.25 Plant populations at high densities undergo self-thinning. Changes in plant density and mean plant weight with time illustrate the $-\frac{3}{2}$ power law. The data shown are for plantings of *Amaranthus retroflexus* and *Chenopodium album*. After J. L. Harper, *Population Biology of Plants*, Academic Press, New York (1977).

Summary

1. Population growth can be described by the exponential rate of increase (r) in the expression $N(t) = N(0)e^{rt}$. Exponential growth applies to those populations that grow continuously.

2. The factor by which a population increases in one unit of time (e^r) is also the geometric growth rate of the population (λ). λ and e^r are interchangeable in population equations. Populations with discrete breeding seasons increase geometrically by periodic increments according to the relation $N(t + 1) = N(t)\lambda$.

3. Exponential growth rate is the difference between birth (b) and death (d) rates averaged over individuals (that is, per capita rates) in a population $(r = b - d)$.

4. The instantaneous rate of increase of an exponentially growing population is $dN/dt = rN$, which says that the growth rate of a population depends on both its size and its per capita birth and death rates.

5. When birth rates and death rates vary according to the age of individuals, one must know the proportion of individuals in each age class to calculate future population growth.

6. The life table of a population displays fecundities (b_x) and probabilities of survival (s_x) of individuals by age class (x). These are the principal variables in models of population dynamics.

7. The fates of individuals born at the same time and followed through their lives may be used to produce a cohort life table. Survival of individuals of known age during a single period, the age structure of a population at a particular time, or the age distribution of deaths may be used to produce a static life table.

8. A population in which life-table values do not change assumes a stable age distribution, in which numbers in each age class, as well as in the population as a whole, increase at the same exponential or geometric rate, known as the intrinsic rate of increase (r_m).

9. The life-table values of a population vary with the conditions of the environment and with the density of the population.

10. The incongruity between the potential of all populations for rapid growth and the relative constancy of populations observed over long periods led naturally to the idea that population growth can be slowed if birth and survival rates decrease as populations grow. Dwindling supplies of food and increasing pressure from predators and disease exert their effects on population processes in such a density-dependent manner.

11. Density-dependent population growth is described by the logistic equation, which describes a sigmoid curve

12. Laboratory studies of animal and plant populations under controlled conditions have shown how density-dependent factors are expressed in population processes. Similar studies of plants and of animals in natural habitats have also revealed density-dependent effects.

PRACTICING ECOLOGY

CHECK YOUR KNOWLEDGE

Negative Density Dependence

Increasing population density often has a negative effect on survival and birth rates, and therefore this type of population response is referred to as negative density dependence. Factors causing such a response constitute part of a negative feedback system on population growth. That is, when populations are high—above the capacity of the environment to support them—processes go into action to reverse population growth and restore the population to a level that is in balance with the resources and conditions of the environment.

However, there are situations, especially at low population densities, when population growth rate actually increases with increasing population density. As the population grows, its intrinsic rate of increase (r) goes up, instead of decreasing as it would if it obeyed the logistic growth equation. Because of the positive relationship between growth rate and density, this is referred to as inverse or positive density dependence. At least three processes may be important for positive density dependence. The first is the so-called Allee effect (named for the pioneering population biologist W. C. Allee of the University of Chicago), in which the ability of individuals to find mates is enhanced as population density increases. The second is the fact that increasing population size can result in greater genetic diversity and lowered incidence of deleterious mutations arising from inbreeding. The third is related to the ability of populations to manage populations of their prey species.

Many species of plants avoid inbreeding by means of genetic instructions that block fertilization when a pollen grain carrying a certain self-incompatible allele lands on a flower that carries the same allele. In this manner, self-fertilization and inbreeding between close relatives is prevented. What are the potential implications of this system for small populations of endangered plants? Marc Kéry and colleagues at the University of Zurich attempted to answer this question by studying *Primula veris* and *Gentiana lutea*. These are two small plant species that live on nutrient-poor grasslands in Europe and whose numbers are declining. Kéry and colleagues conducted an experiment to test for the effects of

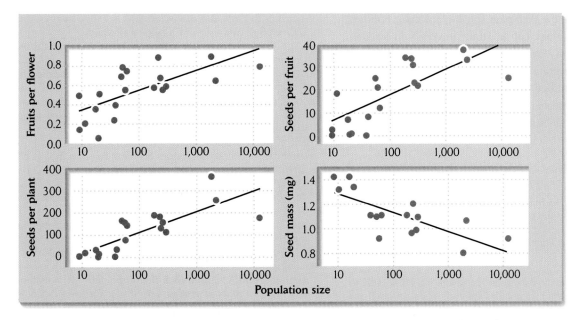

▌Figure 14.26 Relationship between reproduction and population size in *Primula veris*. Seed mass is mean individual seed mass. After M. Kéry et al., *J. Ecol.* 88:17–30 (2000).

small population size on the fitness of offspring. To test the ability of plant populations of different sizes to cope with environmental heterogeneity, they measured fitness in response to a competition and fertilization treatment.

CHECK YOUR KNOWLEDGE

1. **Under what conditions might positive density dependence serve as a useful feature for endangered plants?**

2. **Compare ▌Figure 14.26 and ▌Figure 14.27, both taken from the paper by Kéry et al. What factors did the authors measure to act as indicators of fitness? What do you notice regarding the effect of population density on fitness components for *Primula veris* in comparison with *Gentiana lutea*?**

3. **Consult the figures again. Do the results indicate negative or positive density dependence? Explain.**

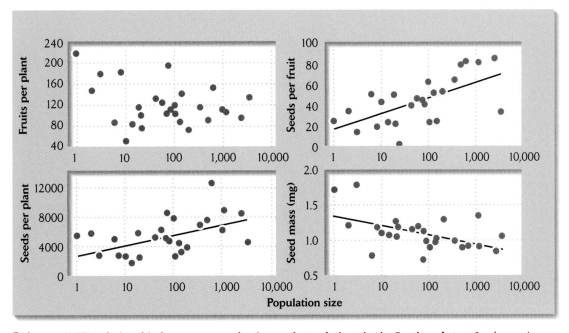

▌Figure 14.27 Relationship between reproduction and population size in *Gentiana lutea*. Seed mass is mean individual seed mass. After M. Kéry et al., *J. Ecol.* 88:17–30 (2000).

MORE ON THE WEB 4. Visit the Web page for the Molecular Population Biology Lab at Lund University in Sweden through *Practicing Ecology on the Web* at *http://www.whfreeman.com/ricklefs*. Read the feature "From the Lab Bench" entitled "More Partners Give Healthier Offspring." By what mechanism does increased female copulation result in higher-fitness offspring?

Suggested Readings

Clutton-Brock, T. H., M. Major, and F. E. Guinness. 1985. Population regulation in male and female red deer. *Journal of Animal Ecology* 54:831–846.

Gotelli, N. J. 1995. *A Primer of Ecology*. Sinauer Associates, Sunderland, MA.

Harrison, S. and N. Cappuccino. 1995. Using density-manipulation experiments to study population regulation. In N. Cappuccino and P. W. Price (eds.), *Population Dynamics: New Approaches and Synthesis*. Academic Press, New York.

Kéry, M., D. Matthies, and H.-H. Spillmann. 2000. Reduced fecundity and offspring performance in small populations of the declining grassland plants *Primula veris* and *Gentiana lutea*. *Journal of Ecology* 88:17–30.

Kingsland, S. E. 1985. *Modeling Nature: Episodes in the History of Population Ecology*. University of Chicago Press, Chicago.

Lack, D. 1954. *The Natural Regulation of Animal Numbers*. Oxford University Press, London.

Murdoch, W. W. 1994. Population regulation in theory and practice. *Ecology* 75:271–287.

Myers, R. A., N. J. Barrowman, J. A. Hutchings, and A. A. Rosenberg. 1995. Population dynamics of exploited fish stocks at low population levels. *Science* 269:1106–1108.

Pollard, E., K. H. Lakhani, and P. Rothery. 1987. The detection of density-dependence from a series of annual censuses. *Ecology* 68: 2046–2055.

Skogland, T. 1985. The effects of density-dependent resource limitations on the demography of wild reindeer. *Journal of Animal Ecology* 54:359–374.

Turchin, P. 1995. Population regulation: Old arguments and new synthesis. In N. Cappuccino and P. W. Price (eds.), *Population Dynamics: New Approaches and Synthesis*. Academic Press, New York.

Weiner, J. 1988. Variation in the performance of individuals in plant populations. In A. J. Davy, M. J. Hutchings, and A. R. Watkinson (eds.), *Plant Population Ecology*, pp. 59–81. Blackwell Scientific Publications, Oxford.

Weller, D. E. 1987. A reevaluation of the $-3/2$ power rule of plant self-thinning. *Ecological Monographs* 57:23–43.

Wolff, J. O. 1997. Population regulation in mammals. *Journal of Animal Ecology* 66:1–13.

CHAPTER 15

Temporal and Spatial Dynamics of Populations

Fluctuation is the rule for natural populations

Temporal variation affects the age structure of populations

Population cycles result from time delays in the response of populations to their own densities

Metapopulations are discrete subpopulations linked by movements of individuals

Chance events may cause small populations to go extinct

An important part of the lore of many countries at high northern latitudes is the dramatic fluctuation of local mammal populations. Lemmings, whose numbers alternate between scarcity and plague proportions, are the most famous of these populations. Lemming plagues occur approximately every four years, at which times the extremely high population densities set off mass dispersal movements reminiscent of locust migrations. That some of these movements might terminate at the edge of the sea, with the forward ranks being pushed into the water by those behind, gave rise to the idea that lemmings commit suicide to reduce the density of their populations. There is no evidence to suggest that such behavior is either suicidal or altruistic.

Many populations fluctuate, varying between high and low numbers more or less periodically. The regularity of some population cycles was first made known to the scientific community in 1924 by Charles Elton, the same ecologist who first proposed the idea of food webs (see Chapter 6). Such cycles had been evident to naturalists, trappers, and others who had kept careful records of economically important species for centuries, but they have intrigued biologists only since Elton brought the matter to their attention.

Elton also called attention to the close parallel between fluctuations of populations of predators and those of their prey. The best example of such a predator–prey interaction comes from records of fur trapping in the boreal regions of Canada. The Hudson's Bay Company paid trappers for the pelts of many mammal species, including the snowshoe hare and one of its most important predators, the lynx. The numbers of pelts brought in by trappers varied by up to a thousandfold between good and bad years. It is hard to believe that the prices of fur or the motivation of trappers drove this variation. If anything, during poor years the higher prices of pelts should have stimulated greater trapping effort. We shall return to the dynamics of such predator–prey interactions in Chapter 19. Here let us consider an additional example of how the economic value of a species led to recording the cycling of a population over many years.

From the late seventeenth century through the end of the eighteenth century, Danish royalty transported gyrfalcons from Danish territories in Iceland to Copenhagen and presented them as diplomatic gifts to the courts of Europe. This was a time when falconry was a popular pastime among European royalty, and the gyrfalcon, the largest of the falcons and one of the most beautiful, was especially prized (▌Figure 15.1). Birds were usually caught by trapping them at their nests, which typically were placed on ledges on high cliffs. Records kept between 1731 and 1793 of the number of gyrfalcons exported from Denmark show 10-year cycles of striking regularity (▌Figure 15.2). After 1770, the numbers declined, and the fluctuations were not so apparent, primarily because falconry had become less popular and the demand for birds could be satisfied even during years of population lows.

▌**Figure 15.1 The gyrfalcon has been a favorite with falconers for centuries.** Photo by J. Krimmel/Cornell Laboratory of Ornithology.

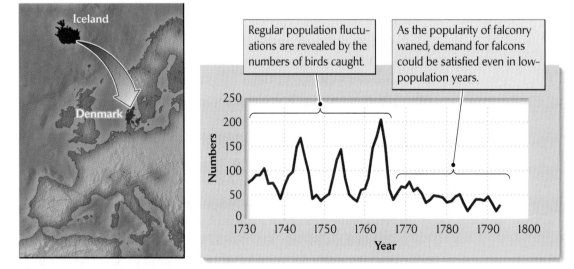

Regular population fluctuations are revealed by the numbers of birds caught.

As the popularity of falconry waned, demand for falcons could be satisfied even in low-population years.

▌**Figure 15.2 The number of gyrfalcons exported from Iceland to Denmark between 1731 and 1770 reflected cyclical changes in the population.** From Ó. K. Nielsen and G. Pétursson, *Wildlife Biol.* 1:65–71 (1995).

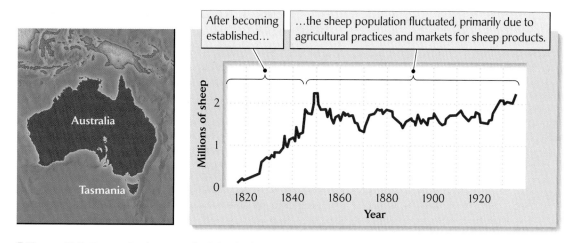

Figure 15.3 Domestic sheep on the island of Tasmania maintain a relatively stable population. Sheep were introduced to Tasmania in the early 1800s. After increasing rapidly to their carrying capacity in less than 30 years, their numbers varied by no more than a factor of 2. After J. Davidson, *Trans. R. Soc. S. Aust.* 62:342–346.

Under the influence of density-dependent factors, populations tend to increase or decrease toward equilibrium numbers determined by the carrying capacities of their environments. However, we also know that populations vary over time. This variation is caused by two different kinds of factors. First, populations respond to changes in environmental conditions, such as temperature, moisture, salinity, acidity, and other similar factors. These responses may either be direct, reflecting the effects of conditions on the performance of individuals, or they may be indirect, resulting from the effects of environmental conditions on the food supply. Second, variation in population size also may result from the intrinsic dynamics of population responses. The dynamics of some biological systems are inherently unstable and result in oscillations in population sizes.

Ecological conditions, moreover, vary from place to place, causing the dynamics of individual subpopulations to differ. Usually, distance isolates subpopulations from one another, and they behave at least partly independently. Changes in an entire population are the sum of changes in all of its subpopulations, but because the dynamics of large and small populations differ, subdivided populations possess unique properties.

In this chapter, we shall discuss the causes of variation in population size, explore how this variation affects small and large populations, and examine the consequences of dispersal among subpopulations. The dynamics of small populations have become increasingly relevant as many species dwindle toward extinction and human activities modify landscapes by fragmenting habitats into smaller, more isolated parcels.

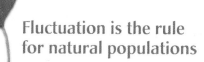

Fluctuation is the rule for natural populations

Variation in the density of a population depends on the two factors just mentioned: the magnitude of fluctuation in the environment and the inherent stability of the population. Some populations tend to remain relatively stable over long periods of time. After domestic sheep became established on the island of Tasmania, their population varied irregularly between 1,230,000 and 2,250,000—less than a factor of 2—over nearly a century (Figure 15.3). We know this because sheep were, and still are, very important to the economy of Tasmania, and their numbers were carefully recorded. Much of the variation in their numbers was related to changes in environmental factors such as grazing practices, markets for wool and meat, and pasture management.

In sharp contrast, populations of small, short-lived organisms may fluctuate wildly over many orders of magnitude within short periods. Populations of the green algae and diatoms that make up the phytoplankton may soar and crash over periods of a few days or weeks (Figure 15.4). These rapid fluctuations overlay changes with longer periods that occur, for example, on a seasonal basis.

Sheep and algae differ in their sensitivity to environmental change and in the response times of their populations. Because sheep are larger, they have a greater capacity for homeostasis and better resist the physiological effects of

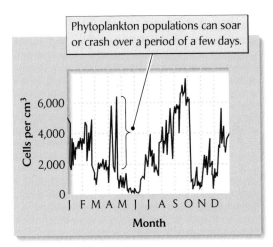

Phytoplankton populations can soar or crash over a period of a few days.

Figure 15.4 Population sizes of phytoplankton fluctuate widely over short periods. The density of phytoplankton in samples of water taken from Lake Erie during 1962 varied over the course of the year. After C. C. Davis, *Limnol. Oceanogr.* 9:275–283 (1964).

environmental change. Furthermore, because sheep live for several years, the population at any one time includes individuals born over a long period; this tends to even out the effects of short-term fluctuations in birth rate. Thus, sheep populations possess a high intrinsic stability. The lives of single-celled algal cells span only a few days, so these intrinsically unstable populations turn over rapidly and bear the full impact of a capricious environment.

Populations of similar species living in the same place often respond to different environmental factors. For example, the densities of four species of moths, whose larvae all feed on pine needles, were found to fluctuate more or less independently in a pine forest in Germany (**Figure 15.5**). The populations varied over three to five orders of magnitude (a thousandfold to a hundred thousandfold) with irregular periods of a few years. Furthermore, the highs and lows of the four populations did not coincide closely, suggesting that, even though each species fed on the same trees, their populations were governed independently by different factors, possibly specialized parasites or pathogens.

Some populations show **periodic cycles**—that is, the period between successive highs or lows is remarkably regular. Among the most striking population phenomena in nature are the periodic cycles of abundance of certain mammals and birds at high latitudes, such as the lemmings described at the beginning of this chapter. Regular trapping of small mammals over a 26-year period in northern Finland revealed six peaks of abundance separated by periods of 4 or 5 years (**Figure 15.6**). As in the 10-year cycle of gyrfalcon populations in Iceland, this regularity is distinctly nonrandom, and the peaks and troughs exhibited by the most abundant species, the vole *Clethrionomys rufocanus*, are roughly paralleled by the less common species. The cause of these cycles has been one of the most interesting and persistent problems in ecology, as we shall see below.

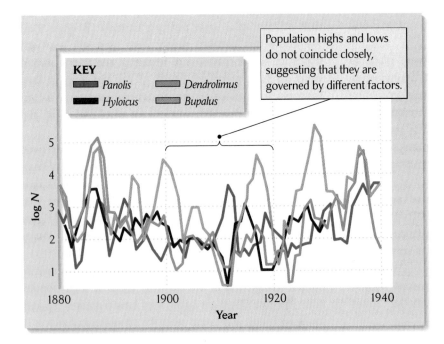

KEY
■ Panolis   ■ Dendrolimus
■ Hyloicus   ■ Bupalus

Population highs and lows do not coincide closely, suggesting that they are governed by different factors.

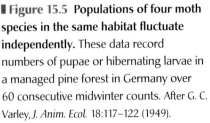

Figure 15.5 Populations of four moth species in the same habitat fluctuate independently. These data record numbers of pupae or hibernating larvae in a managed pine forest in Germany over 60 consecutive midwinter counts. After G. C. Varley, *J. Anim. Ecol.* 18:117–122 (1949).

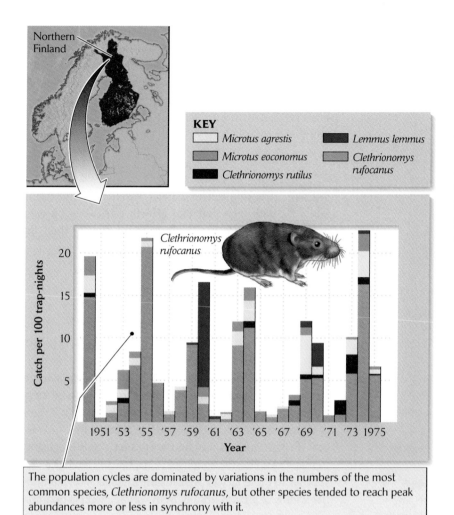

KEY

- ☐ *Microtus agrestis*
- ▨ *Microtus eoconomus*
- ■ *Clethrionomys rutilus*
- ■ *Lemmus lemmus*
- ▨ *Clethrionomys rufocanus*

Clethrionomys rufocanus

The population cycles are dominated by variations in the numbers of the most common species, *Clethrionomys rufocanus*, but other species tended to reach peak abundances more or less in synchrony with it.

❙ Figure 15.6 Population cycles of small mammals in northern Finland are synchronized. Cycles in the numbers of small mammals trapped in an area of northern Finland from 1950 to 1975 are dominated by variation in the most common species, *Clethrionomys rufocanus*, but other species tend to reach peak abundances more or less in synchrony with it. From R. Brewer, *The Science of Ecology*, 2d ed., Saunders, New York (1994), after S. Lahti, J. Tast, and H. Uotila, *Luonnon Tutkija* 80:97–107 (1976).

Temporal variation affects the age structure of populations

Temporal variation in population dynamics often leaves its mark on the age structure of a population—that is, on the relative frequencies of individuals of each age. As we have seen in Chapter 14, a changing age structure can affect the rate of population growth. The sizes of age classes also provide a history of population change in the past. For example, the age composition of samples from the Lake Erie commercial whitefish catch for the years 1945–1951 shows that during 1947, 1948, and 1949, most of the individuals caught belonged to the 1944 year class (❙ Figure 15.7). Biologists estimated the ages of fish from growth rings on their scales. Their data show that 1944 was an excellent year for spawning and recruitment. Thus, the 1944 cohort consti-

tuted an atypically large proportion of the total population over the next several years.

Temporal variation in the annual recruitment of individuals into a population is also evident in the age structure of stands of trees. The ages of trees in seasonal environments may be estimated by counting the growth rings in the woody tissue of the trunk—one ring being added each year under normal circumstances. The age structure of a virgin stand of timber surveyed near Hearts Content, Pennsylvania, in 1928 shows that individuals of most species were recruited sporadically over the nearly 400-year span of the record (❙ Figure 15.8). Many oaks and white pines became established between 1620 and 1710, undoubtedly following a major disturbance, possibly associated with the serious drought and fire year of 1644. Fire can open a forest enough to allow the establishment of white pine seedlings, which do not tolerate deep shade. In contrast, beech—a species whose seedlings can grow under the canopy of a closed forest—exhibited a relatively even age distribution.

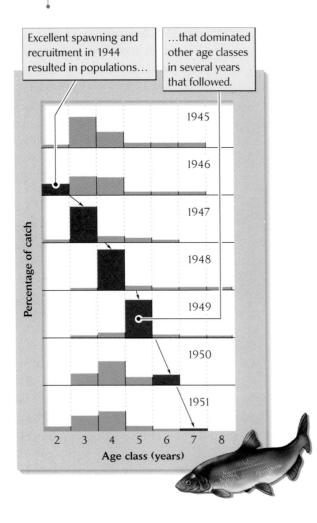

Excellent spawning and recruitment in 1944 resulted in populations...

...that dominated other age classes in several years that followed.

Figure 15.7 Temporal variation in recruitment may be evident in the age structure of a population. Samples from the commercial whitefish catch in Lake Erie between 1945 and 1951 show that the cohort of fish spawned in 1944 dominated the age structure of the population for several years afterward. From G. H. Lawler, *J. Fish. Res. Bd. Can.* 22:1197–1227 (1965).

MORE ON THE WEB *Tracking environmental variation.* Populations with high potential growth rates keep up with environmental change more closely than do those with low potential growth rates.

Population cycles result from time delays in the response of populations to their own densities

Except for factors associated with daily, lunar (tidal), and seasonal cycles, environmental fluctuations tend to be irregular rather than periodic. Historical records reveal that years of abundant rain or drought, extreme heat or cold, and natural disasters such as fires and hurricanes occur irregularly, perhaps even at random (see Chapter 4). Biological responses to these factors are similarly irregular. For example, the widths of growth rings of trees vary in direct proportion

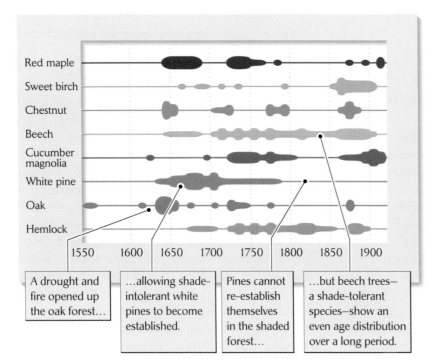

A drought and fire opened up the oak forest...

...allowing shade-intolerant white pines to become established.

Pines cannot re-establish themselves in the shaded forest...

...but beech trees—a shade-tolerant species—show an even age distribution over a long period.

KEY

Percentage of each species

0–5
5–15
15–40

Figure 15.8 Age distributions of forest trees show the effects of disturbances on seedling establishment. These data were collected near Hearts Content, Pennsylvania, in 1928. After A. F. Hough and R. D. Forbes, *Ecol. Monogr.* 13:299–320 (1943).

to temperature and rainfall. The frequency distribution of the number of years between peaks in ring width is often the same as that from a random series (▮ Figure 15.9).

The sizes of many populations do, however, change with periodic frequency (see, for example, Figure 15.6). The distribution of the intervals between peaks in a red fox population was decidedly nonrandom (▮ Figure 15.10). Intervals of 4 and 5 years were more common than would be expected by chance, and intervals of 2 and 3 years were less common. For many years, ecologists believed that such cycles must be caused by extrinsic environmental factors that exhibit similar periodic variation. A regular 11-year cycle in sunspot numbers was frequently mentioned in relation to population cycles of snowshoe hares, but the sunspot cycle never matched the hare cycles well, and no one could see a direct connection between the two.

A mechanism for population cycles was discovered through mathematical models of populations developed in the 1920s and 1930s. These models showed clearly that because of the inherent dynamic properties associated with density-dependent regulation of population size, some populations subjected to even minor, random environmental fluctuations could be caused to oscillate. Populations have an intrinsic periodicity, just as a pendulum has an inherent frequency of swinging depending on its length. Momentum imparted to a pendulum by the acceleration of gravity carries it past the equilibrium point and causes it to swing back and forth periodically. "Momentum" im-

parted to a population by high birth rates at low densities causes the population to grow rapidly and overshoot its carrying capacity. Then, low survival rates at high densities cause the population to overcompensate and decrease below the carrying capacity. This cycling can result from **time delays,** or delays in the response of birth and death rates to changes in the environment.

Time delays and oscillations in discrete-time models

Population models based on the logistic equation (see Chapter 14) have been used to study population cycles. Time delays that cause populations to oscillate are inherent in models of populations with discrete generations. In such discrete-time models, population growth occurs at intervals associated with breeding episodes. Thus, the response of a population to conditions at one point in time is not expressed as a change in population size until the next interval. Because the population responds by discrete increments from one time to the next, it cannot continuously readjust its growth rate as its size approaches the carrying capacity. This can cause the population to overshoot its equilibrium, first in one direction and then in the other, as its size draws closer to the carrying capacity. Whether and how much the population will oscillate depends on whether the growth increment exceeds the difference between the size of the population and the carrying capacity.

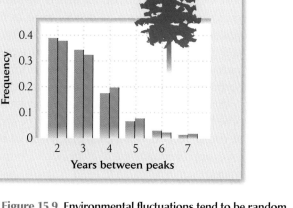

▮ **Figure 15.9 Environmental fluctuations tend to be random.** The frequency distribution of intervals between peaks in the widths of growth rings of Douglas fir resembles that for a series of random numbers. After L. C. Cole, *J. Wildl. Mgmt.* 15:233–252 (1951).

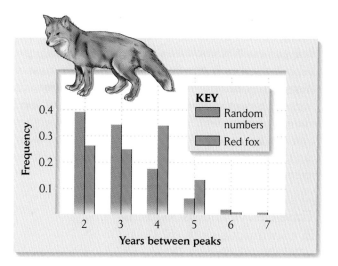

▮ **Figure 15.10 Fluctuations in population numbers tend not to be random.** The frequency distribution of intervals between peaks in a red fox population differs significantly from that for a series of random numbers. After L. C. Cole, *J. Wildl. Mgmt.* 15:233–252 (1951).

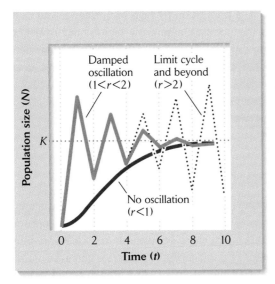

Figure 15.11 A population may adopt one of three oscillation patterns. In a discrete-time model based on the logistic equation, the oscillation pattern depends on the per capita growth rate (r). When r < 1, the population returns without fluctuation to the carrying capacity. When 1 < r < 2, the population exhibits damped oscillations, which become limit cycles when r > 2.

Accordingly, a population may adopt one of three patterns. When the per capita growth rate is small (r_0), the population will approach the carrying capacity (K) directly, without oscillation, as shown in ❚ Figure 15.11. When r exceeds 1 but is less than 2, a population will tend to overshoot its equilibrium, but will nonetheless end up closer to the equilibrium than it was before. Thus, the population will **oscillate** back and forth across the equilibrium value, getting closer with each generation. This behavior is called **damped oscillation.** When r exceeds about 2, the population can end up farther from the equilibrium each generation, and oscillations tend to increase. The population may nonetheless settle into stable oscillations called **limit cycles,** in which numbers bounce back and forth between high and low values. With increasing r, these oscillations can take on very complex, eventually unpredictable forms referred to as **chaos.**

Time delays and oscillations in continuous-time models

Discrete-time models have a built-in delay of one time unit in the response of a population to its environment. Continuous-time models have no built-in time delays. Instead, time delays result from the developmental period that separates reproductive episodes between generations. These time delays can also create cyclic population behavior.

Oscillations in population size can occur when a population responds to its density at some time in the past, rather than its density at the present. The length of such a time delay is referred to by the symbol τ, which is the lowercase Greek letter tau.

In continuous-time models based on the logistic equation, the behavior of the population depends on the product of the intrinsic rate of increase, r, and the time delay, or rτ. High intrinsic rates of increase and long time delays increase oscillations. Oscillations in population size (N) are damped as long as the product rτ is less than π/2 (about 1.6). Below rτ = e^{-1} (0.37), the population increases or decreases without oscillation to the carrying capacity (K). At rτ greater than π/2, the oscillations increase to form limit cycles, whose amplitudes grow as the time lag lengthens. For example, at rτ = 2, the maximum population size is nearly 3 times K and at rτ = 2.5 it is nearly 5 times K. The periods of these limit cycles, measured from peak to peak, increase from about 4τ to more than 5τ with increasing rτ. Thus, a population cycle with a period of 10 years would imply a time delay of about 2 years.

Cycles in laboratory populations

Population cycles have been observed in many laboratory cultures of single species. In one such study, populations of the water flea *Daphnia magna* exhibited marked oscillations when cultured at 25°C, but these oscillations disappeared at 18°C (❚ Figure 15.12). The period of the cycle at 25°C appeared to be just over 60 days, suggesting a time delay in the density-dependent response of about 12–15 days. This is about the average age at which water fleas give birth at 25°C.

The time delay arose in the following manner. As population density increased, fecundity decreased, falling nearly to zero when the population exceeded 50 individuals. Survival was less sensitive to density even at the highest densities, and adults lived at least 10 days. Thus, crowding at the peak of the cycle prevented births; then, when the population fell to densities low enough to permit reproduction, it contained mostly senescent, nonreproducing individuals, and thus the population continued to decline. The beginning of a new cycle awaited the accumulation of young, fecund individuals. The length of the time delay was approximately the average adult life span at high densities.

At the lower temperature, reproductive rate fell quickly with increasing density, and life span was longer that seen at 25°C at all densities. Populations at the colder temperature apparently lacked a time delay because deaths were more evenly distributed over all ages, and some individuals gave birth even at high population densities. Consequently,

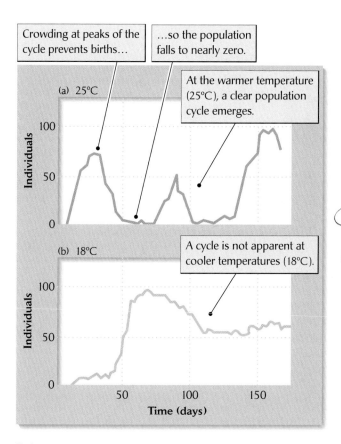

Figure 15.12 Warm temperatures lead to population cycles in *Daphnia magna*. Populations were maintained at (a) 25°C and (b) 18°C. After D. M. Pratt, *Biol. Bull.* 85:116–140 (1944).

generations overlapped more broadly. At the higher temperature, water fleas behaved according to a discrete-generation model with its built-in time delay of one generation. At the lower temperature, they behaved according to a continuous-generation model with little or no time delay.

Other studies of water fleas reveal a connection between life history adaptations and population dynamics. The storage of food reserves by some species of water fleas reduces the sensitivity of mortality to crowding and therefore introduces a time delay into population processes. *Daphnia galeata*, a large species, stores energy in the form of lipid droplets during periods of high food abundance (that is, low population density). It can then live on these stored reserves when food supplies dwindle as a result of overgrazing at high population densities. Females also transfer lipids to each offspring through oil droplets in their eggs, thereby increasing the survival of young, prereproductive water fleas under poor feeding conditions. The smaller *Bosmina longirostris* stores a smaller amount of lipids, so mortality increases quickly in response to increases in population density. The consequences of this difference for population growth are predictable. In one study, *Daphnia*

exhibited pronounced limit cycles with a period of 15 to 20 days, whereas *Bosmina* populations grew quickly to an equilibrium size, with perhaps a single strongly damped overshoot. The r value of *Daphnia* populations was about 0.3 per day. With a cycle period of 15 to 20 days, τ must have been about 4 to 5 days, and therefore $r\tau$ was about 1.2–1.5. Because the value of $r\tau$ was somewhat less than $\pi/2$, the cycles in the *Daphnia* population should have damped out eventually.

ECOLOGISTS IN THE FIELD

Time delays and oscillations in blowfly populations

Slight differences in culture conditions or in the life histories of species can tip the balance between a population that does not oscillate and one that sustains a limit cycle. The Australian entomologist A. J. Nicholson's pioneering experimental manipulations of time delays in laboratory cultures of the sheep blowfly (*Lucilia cuprina*) provide a dramatic demonstration of the relationship of time delays to population cycles.

Under one set of culture conditions, Nicholson provided blowfly larvae with 50 grams of ground liver per day, while giving adults unlimited food. The number of adults in the population cycled from a maximum of about 4,000 to a minimum of 0 (at which point all the individuals were either eggs or larvae) with a period of between 30 and 40 days (**Figure 15.13**). These regular fluctuations were caused by a time delay in the responses of fecundity and mortality to the density of adults in the cages. At high population densities, adults laid many eggs, resulting in strong larval competition for the limited food supply. None of the larvae that hatched from eggs that were laid during adult population peaks survived, primarily because they did not grow large enough to pupate. Therefore, large adult populations gave rise to few adult progeny, and because adults lived less than 4 weeks, the population soon began to decline. Eventually, so few eggs were laid on any particular day that most of the larvae survived, and the size of the adult population began to increase again.

The behavior of this population suggests a time-delayed logistic process, which provides a good fit to the observed oscillations with $r\tau = 2.1$. This value predicts the ratio of the maximum population to the carrying capacity ($N/K = e^{r\tau}$) to be 8.2 and the cycle period to be 4.5τ. The experiment clearly reveals that density-dependent factors did not affect the mortality rates of adults immediately as the population increased, but were felt a week or so later when their progeny were larvae. Larval mortality did not express itself in the size of the adult population until those larvae emerged as

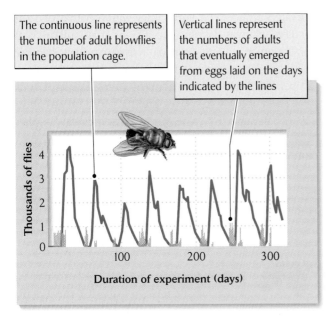

The continuous line represents the number of adult blowflies in the population cage.

Vertical lines represent the numbers of adults that eventually emerged from eggs laid on the days indicated by the lines

Duration of experiment (days)

Figure 15.13 Introduction of time delays results in regular population cycles. Limiting the food supply available to larvae in a laboratory population of the sheep blowfly *Lucilia cuprina* caused a time delay in density-dependent effects and resulted in regular population cycles. Larvae were provided with 50 grams of liver per day; adults were given unlimited supplies of liver and water. After A. J. Nicholson, *Cold Spring Harbor Symp. Quant. Biol.* 22:153–173 (1958).

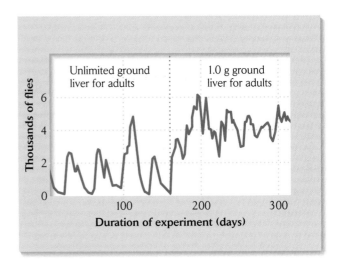

Figure 15.14 Elimination of time delays results in the elimination of population cycles. Limiting the food supply available to adults removed the time delay and eliminated fluctuations in a sheep blowfly population. This experiment was similar in all other respects to that depicted in Figure 15.13. After A. J. Nicholson, *Cold Spring Harbor Symp. Quant. Biol.* 22:153–173 (1958).

adults about 2 weeks after eggs were laid. As in the *Daphnia* population maintained at a high temperature, crowding in the blowfly population created discrete, nonoverlapping generations with an inherent time delay (τ) equal to the larval development period, about 10 days.

The hypothesis that time delays in density-dependent responses cause population cycles can be tested directly by eliminating time delays—that is, by making the deleterious effects of resource depletion at high densities felt immediately. Nicholson did this by adjusting the amount of food provided to his flies so that food availability limited adults as severely as it did larvae. Adult flies require protein to produce eggs. By restricting the amount of liver available to adults to 1 gram per day, Nicholson curtailed egg production to a level determined by the availability of liver. Under these conditions, recruitment of new individuals into the population was determined at the egg-laying stage, and most of the larvae survived. Consequently, fluctuations subsided in the population (**Figure 15.14**).

We have seen that responses of populations to density can be delayed by development time and by storage of nutrients, both of which put off deaths to a later point in the

life cycle or to a later time. Density-dependent effects on fecundity can act with little delay when adults produce eggs quickly from resources accumulated over a short period. Populations controlled primarily by such factors should not exhibit marked oscillations.

Regardless of any time delay in its density-dependent response, a population at its equilibrium point will remain there until perturbed by some outside influence, whether a change in the carrying capacity (K) or a catastrophic change in population size (N). Once displaced from their equilibrium, some populations will move toward stable limit cycles, depending on the nature of the time delay and the response time. Others will return to the equilibrium directly or through damped oscillations. Cycles may be reinforced through interactions with other species—prey, predators, parasites, perhaps even competitors—with similar rates of response to population change, as we shall see in later chapters.

Metapopulations are discrete subpopulations linked by movements of individuals

Areas of habitat with the necessary resources and conditions for a population to persist are called **habitat patches,** or simply **patches.** The individuals of a species that live in a habitat patch constitute a **subpopulation.** Areas of unsuitable habitat often separate suitable habitat patches; individuals may move through these areas on occasion, but cannot persist in them. A set of subpopulations interconnected by occasional movement of individuals between them is referred to as a **metapopulation** (▌Figure 15.15). The metapopulation concept has recently become one of ecology's most important tools for understanding the dynamics of species living in fragmented habitats. Thus, as forest clearing, road building, and other human activities create patchworks of different types of habitats, metapopulation models help us manage and conserve populations that cannot move readily through a fragmented landscape.

Two types of processes contribute to the dynamics of metapopulations. The first includes the growth and regulation of subpopulations within patches—processes that we have already discussed in detail. The second comprises colonization to form a new subpopulation by migration of individuals to an empty patch and extinction of established subpopulations. Because subpopulations are typically much smaller than the metapopulation as a whole, local catastrophes and chance fluctuations in numbers of individuals have greater effects on their population dynamics. Indeed, the smaller the subpopulation, the higher its probability of extinction during a particular time interval.

When individuals move frequently between subpopulations, such fluctuations are damped out, and changes in subpopulation size mirror those of the larger population. Thus, a high rate of migration transforms metapopulation dynamics into the dynamics of a single large population. At the other extreme, when no individuals move between patches,

the subpopulations in each patch behave independently. When these subpopulations are small, they have high probabilities of extinction, as we shall see below, and so the total population gradually goes extinct. Intermediate levels of migration result in the colonization of some patches left unoccupied by subpopulation extinction. Under such circumstances, the entire metapopulation exists as a shifting mosaic of occupied and unoccupied patches. This mosaic has its own dynamics and equilibrium properties, which can be understood in terms of a simple model.

The basic model of metapopulation dynamics

Consider a population divided into discrete subpopulations. We assume that within a given time interval, each subpopulation has a probability of going extinct, which we shall refer to as e. Therefore, if p is the fraction of suitable habitat patches occupied by subpopulations, then subpopulations go extinct at the rate ep. The rate of colonization of empty patches depends on the fraction of patches that are empty $(1 - p)$ and the fraction of patches sending out potential colonists (p). Thus, we may express the rate of colonization as a single rate constant c times the product $p(1 - p)$. From these simple considerations, one can show that a metapopulation attains equilibrium when the proportion of occupied patches is

$$\hat{p} = 1 - \frac{e}{c}.$$

Numbers are estimated carrying capacities of each patch.

Little is known of the movement of owls between patches.

Santa Cruz
Monterey
100
60
6
32
24
80 24 20 24
36 30 40
20 190 266
Santa Barbara
Los Angeles 24 40
20
60 32
10
30 30

▌**Figure 15.15 A metapopulation is a set of discrete subpopulations having partially independent dynamics.** At any one time, some patches of suitable habitat may be occupied while others are not. Southern California spotted owls are distributed as a metapopulation over patches of suitable old-growth forest habitat in the mountains of southern California. The dark areas on the map represent habitat patches, and the numbers next to them represent the estimated carrying capacities of the patches. Little is known about the movement of owls between patches. From W. S. Lahaye, R. J. Gutiérrez, and H. R. Akçakaya, *J. Anim. Ecol.* 63:775–785 (1994).

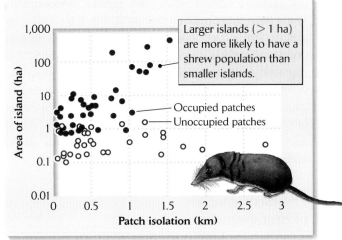

Larger islands (> 1 ha) are more likely to have a shrew population than smaller islands.

Occupied patches
Unoccupied patches

Figure 15.16 Larger patches are more likely to be occupied. In a metapopulation of the shrew *Sorex araneus* on islands in two lakes in Finland, islands larger than 1 ha were more likely to be occupied than smaller islands. The solid red dots represent patches that are occupied; the open white dots, patches that are unoccupied. Patch isolation appears to be relatively unimportant in this metapopulation. From I. Hanski, *Biol. J. Linn. Soc.* 42:17–38 (1991).

The equilibrium proportion of occupied patches is indicated by the little hat (^) over the *p*. The equilibrium is stable because when *p* is below the equilibrium point, colonization exceeds extinction, and vice versa.

This simple model shows the critical importance of the relative rates of extinction and colonization (*e/c*). When $e = 0$, $\hat{p} = 1$, and all patches are occupied. (This does not mean that the patches cease to have independent dynamics, only that they are large enough or otherwise stable enough not to suffer extinction.) When $e = c$, $\hat{p} = 0$, and the metapopulation heads toward extinction. Intermediate values of *e*, that is, greater than 0 but less than the colonization rate, result in a shifting mosaic of occupied and unoccupied patches. Thus, when the rate of colonization exceeds the rate of extinction, the fraction of occupied patches reaches equilibrium between 0 and 1. When extinction exceeds colonization, the fraction of occupied patches declines to zero, and the entire metapopulation goes extinct. This pattern makes clear the importance of keeping habitat patches from becoming too isolated or, alternatively, of maintaining migration corridors between patches in a managed landscape.

The model outlined above assumes that all patches are equal and that rates of extinction and colonization for each patch are the same. More realistically, patches vary in size, habitat quality, and degree of isolation from other patches. Larger patches can support larger subpopulations, which have lower probabilities of extinction than the subpopulations supported by smaller patches. This effect of patch size was shown dramatically by the distribution of the shrew *Sorex araneus* on islands in two lakes in Finland. These islands varied in size from about 0.1 to 1,000 ha and in distance from other islands or the shore of the lake from less than 0.1 to more than 2 km. As you can see from **Figure 15.16**, islands greater than 1 ha in area had a greater proba-

bility of having a shrew population at any one time than did smaller islands. Patch isolation did not exert a marked effect, and therefore we can conclude that shrews can reach more distant islands as readily as they can reach close ones.

ECOLOGISTS IN THE FIELD

Metapopulations in grassland patches along the Rhine River

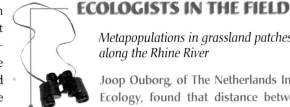

Joop Ouborg, of The Netherlands Institute of Ecology, found that distance between subpopulations was an important factor in the occupancy of patches by herbaceous plants. His study used surveys of patches of dry, nutrient-poor grassland habitat distributed along the IJssel and Rhine rivers. In 1956, 143 such patches of grassland were surveyed and the abundances of plant species were recorded. Data from the same sites were surveyed for presence and absence of the same species again in 1988.

Ouborg divided the 1956 subpopulations of each species into those that were present in 1988 and those that were absent in 1988. For each of these groups, he calculated the average population size in 1956 and the distance to the nearest patch occupied by a subpopulation of the same species. Depending on the species of plant, between 30% and 85% of the subpopulations present in 1956 were absent in 1988; over the same period, between 2% and 44% of patches empty in 1956 had acquired new subpopulations by colonization from other patches. The importance of initial population size was apparent in the fact that nonextinct subpopulations tended to contain more individuals in 1956 than extinct subpopulations (**Figure 15.17a**). That colonization was important was shown by the fact that nonextinct populations tended to be closer to other occupied patches than did extinct populations (**Figure 15.17b**).

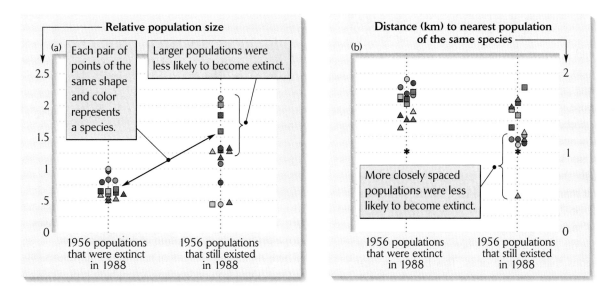

Figure 15.17 Patch occupancy increases with patch size and decreases with patch isolation. (a) The averages of living versus extinct populations of plants with respect to population size in 1956 in isolated patches of habitat along the Rhine River system in The Netherlands. Each point represents one species. (b) The averages of living versus extinct populations of plants with respect to distance to the nearest other subpopulation in 1956. From N. J. Ouborg, *Oikos* 66:298–308 (1993).

The rescue effect

Immigration from large, productive subpopulations can keep declining subpopulations from dwindling to small numbers and eventual extinction. This phenomenon is known as the **rescue effect,** and it can be incorporated into metapopulation models by making the rate of extinction (e) decrease as the fraction of occupied patches (p) increases (that is, with more numerous sources of migrants, or rescuers). In one version of a metapopulation model with the rescue effect built in, $\hat{p}$ either increases to 1 or decreases to 0, depending on the relative values of the parameters for colonization and extinction. The rescue effect can produce positive (inverse) density dependence among subpopulations, in which the survival of subpopulations increases in the presence of more numerous neighboring subpopulations and consequent higher colonization rates.

An important consideration in evaluating data on metapopulations is the length of the sampling interval relative to the expected time to extinction or recolonization. If the sampling period is long, a population may have disappeared and then been restored by colonization, in which case it would have been recorded as persistent. This is one of the reasons that one often finds an inverse relationship between rates of extinction and colonization, as Ouborg did for grassland patches.

Chance events may cause small populations to go extinct

Most population models assume large population sizes and no variation in the average values of birth and death rates due to chance. Such models, whose outcomes we can predict with certainty, are called **deterministic models.** In the real world, however, random variations can influence the course of population growth.

Ecologists recognize three types of randomness that affect populations in the natural world. First, populations may be subjected to unpredictable **catastrophes** that strongly affect all individuals in the population, causing reproductive failure or a high proportion of deaths. The second type of randomness consists of variations in physical conditions and other environmental factors that continually influence rates of population growth and the carrying capacity of the environment in all years. The third type of randomness is due to random sampling, or **stochastic,** processes, which can result in variation in populations even in a constant environment. The death of an individual, for example, is a chance event that has a certain probability of occurring during a particular interval. The number of deaths within a population, however, has a probability distribution. The average value of this distribution is the number of individuals in the population times the probability of death. But the actual number of deaths observed in a particular population will vary above or below this value just by chance.

Coin tossing provides a useful analogy to a stochastic process. Suppose you repeatedly toss a set of 10 pennies. Although the probability of a head turning up on each toss is one-half, any one set of trials might turn up 6 heads, or it might turn up 3 heads. When the test is repeated frequently enough, the average of the outcomes settles down to 5 heads, but many trials turn up 4 or 6 heads, somewhat fewer yield 3 or 7 heads, and runs with all heads occur once in 1,024 trials, on average.

Chance events exert their influence more forcefully in small populations than in large ones. This becomes clear when we consider that the probability of obtaining 5 tails in a row with 5 successive tosses of a coin is 1 in 32, compared with the smaller chance of 1 in 1,024 of obtaining 10 tails in a row. If we visualize each individual in the population as a coin, and equate turning up tails with death, it is clear that a population of 5 individuals has a higher probability of extinction, just by chance, than a population of 10.

Turning to population models, let us suppose that adults successfully rear offspring with a probability of 0.5 per year. We would therefore expect a population of 10 individuals to produce 5 offspring on average, but the actual number is likely to vary from that value. What effect will this have on population growth? Consider a simple birth process (no deaths) in which a population grows exponentially according to $N(t) = N(0)e^{bt}$. Now suppose that the product bt equals 0.5, that is, an average of one-half offspring per adult per time interval. A deterministic model predicts a population at time t equal to 1.65 times the initial population ($e^{0.5}$). Thus, if the initial population were 500 individuals, the number after interval t would be 824. If the initial population were only 5 individuals, however, the number of individuals at time t would average 8.24, but it could vary from as few as 5 (no births occurred) to as many as 20, just by chance (▌Figure 15.18).

Stochastic extinction of small populations

Birth rates and death rates depend on a variety of ecological factors, but whether a particular individual dies or successfully rears one or more progeny during an interval depends largely on chance. With an annual probability of death of one-half, for example, some individuals live and, on average, an equal number die. There exists a finite probability, however, that all the individuals in a population will die, just as 10 out of 10 coin tosses will come up tails with a small but finite probability.

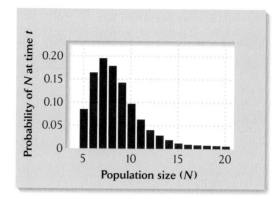

▌**Figure 15.18** **Chance variations in population processes produce a probability distribution of population size.** Plotted here are the probabilities of numbers of individuals (N) at time t in a population undergoing a pure birth process with initial size $N(0) = 5$ and $bt = 0.5$. After E. C. Pielou, *Mathematical Ecology*, Wiley, New York (1977).

Theorists have devoted considerable attention to the probability of extinction of populations. They have derived mathematical expressions for the probability of extinction at time t, which is written $p_0(t)$. p_0 represents the probability that the population will have 0 individuals. Probability of extinction is sensitive to birth rate b, death rate d, and population size N. The simplest model for the probability of stochastic extinction is the case in which b and d are equal—that is, births balance deaths, and the average change in population size is zero—for which

$$p_0(t) = \left[\frac{bt}{1 + bt} \right]^N.$$

Because the term within the brackets is always less than 1, the probability of extinction decreases with increasing population size and increases with larger b (and d), indicating more rapid population turnover. Note that the probability of extinction also increases with time (t). The relationship of the probability of extinction within time period t to population size N is shown in ▌Figure 15.19 for a population in which $b = d = 0.5$. These are reasonable values for adult death and recruitment in a population of terrestrial vertebrates. We see, for example, that for a population with 10 individuals, the probability of extinction is 0.16 within 10 years, 0.82 within 100 years, and virtually certain (0.98) within 1,000 years, provided that b and d do not vary with population size. For a population with an initial size of 1,000, the probability of extinction is about 0.13 within a millennium; it becomes virtually certain (0.999, not shown) within a million years.

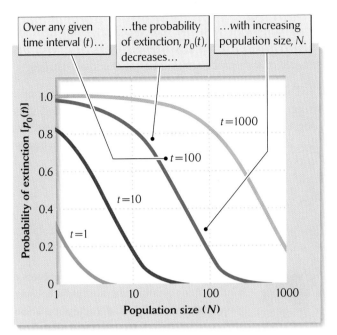

Over any given time interval (t)...

...the probability of extinction, $p_0(t)$, decreases...

...with increasing population size, N.

Figure 15.19 Probability of extinction increases over time (t) but decreases as a function of initial population size (N). In this example, birth rate and death rate equal 0.5.

Stochastic extinction with density dependence

Most stochastic extinction models do not include density-dependent changes in birth and death rates. In those that do, extinction becomes exceedingly rare except in the smallest populations because, as a population drops below its carrying capacity, birth rates typically increase and death rates decrease. This greatly improves the probability that a small population will increase rather than decrease further. Accordingly, we should consider whether density-independent stochastic models are relevant to natural populations. The answer is that they are, for several reasons.

First, land use patterns and habitat fragmentation are such that many species now exist as collections of exceedingly small subpopulations, often so isolated that their eventual demise cannot be prevented by immigration from other populations. Second, changing environmental conditions are likely to reduce the fecundity of populations trapped in isolated habitat patches and bring them closer to the abyss of extinction. Third, when endangered species compete for resources with other species, the advantages that they would gain because of their low density (perhaps more food per individual) may be usurped by their competitors. In this case, high population densities of one species may depress small populations of other species. Finally,

small populations sometimes exhibit positive rather than negative density dependence owing to inbreeding effects and problems of locating mates, and so their numbers may decline all the more rapidly.

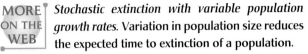 *Stochastic extinction with variable population growth rates.* Variation in population size reduces the expected time to extinction of a population.

Size and extinction of natural populations

When populations dwindle to a small size, they become more susceptible to extinction, particularly on small islands, where populations are restricted geographically and are rarely augmented by immigration. In fact, extinction occurs so often in some island groups that we can determine its probability from historical records. These data confirm the theoretical predictions of models of stochastic extinction. For example, species lists compiled in 1917 and in 1968 for birds on the Channel Islands off the coast of southern California revealed extinctions of island populations during the 51-year interval between censuses. Seven of 10 species disappeared from Santa Barbara Island (3 km² in area), but only 6 of 36 species disappeared from the larger Santa Cruz Island (249 km²). (New colonists of different species replaced some of the species that went extinct on each island.) On an annual basis, these extinction figures can be expressed as 0.1% and 1.7% of the avifauna per year, respectively, with extinction rate and island size inversely related. Proportions of island subpopulations disappearing over this interval were also inversely related to population size:

Number of breeding pairs per population	4–10	11–30	31–60	over 60
Proportion of populations extinct	0.30	0.09	0.15	0.02

Disappearances of populations from isolated islands dramatize the role of extinction of subpopulations in the dynamics of metapopulations. The extinction rate, which influences the equilibrium number of patches occupied, depends on the number of individuals in a subpopulation and hence on the size of the patch it occupies. These considerations emphasize the interaction of spatial and temporal dynamics in population processes and remind us that we must understand the spatial structure of populations if we are to manage them intelligently.

Summary

1. Most populations fluctuate, either in response to variations in the environment or because they have oscillatory properties intrinsic to their dynamics. Owing to their well-developed mechanisms for homeostasis, species with larger body sizes and longer life spans tend to respond less rapidly to changes in their environments. A conspicuous exception occurs in the regular cycling of populations of small mammals in northern latitudes.

2. The age structure of a population often indicates temporal heterogeneity in the recruitment of individuals. For example, seedlings of certain species tend to become established in forests primarily following a major disturbance such as fire, drought, or storm. Thus, population processes may be sporadic rather than uniform over time.

3. Discrete-time models of populations with density dependence show that populations tend to oscillate when perturbed. For r between 0 and 1, the size of a population approaches equilibrium (K) without oscillation. For r between 1 and 2, population size undergoes damped oscillations and eventually settles down to K. When r exceeds 2, oscillations in population size increase in amplitude until either a stable limit cycle is achieved or the population fluctuates irregularly (chaos).

4. Continuous-time models can produce cyclic population changes when density-dependent responses are time-delayed. Defining the time delay as τ, we find that such populations exhibit no oscillation when the product $r\tau$ lies between 0 and e^{-1} (0.37), damped oscillations when $r\tau$ lies between e^{-1} and $\pi/2$ (1.6), and limit cycles with a period of 4τ or more when $r\tau$ exceeds $\pi/2$.

5. Many laboratory populations exhibit oscillations that arise from time delays in the responses of individuals to density. These time delays are related to the period of development from egg to adult and may be enhanced by storage of nutrients. In laboratory populations of sheep blowflies, A. J. Nicholson experimentally circumvented a time delay and was thereby able to eliminate population cycles.

6. Populations that are subdivided into discrete subpopulations occupying patches of suitable habitat are referred to as metapopulations. The dynamics of metapopulations depend not only on birth and death processes within patches, but also on migration of individuals between patches. When the rate of extinction of subpopulations is small compared with the rate of colonization of unoccupied patches, a metapopulation exists as a changing mosaic of an equilibrium number of occupied patches.

7. The dynamics of small populations, such as a subpopulation in an individual habitat patch, depend to a large degree on chance events. Stochastic models demonstrate that the probability of extinction due to random fluctuation in population size is greater in smaller populations.

PRACTICING ECOLOGY

CHECK YOUR KNOWLEDGE

Populations and Nature Reserve Design

We have seen how migration of individuals is important for maintaining movement between subpopulations within a metapopulation. Such movement allows individuals to maintain connections between source and sink populations, and it promotes gene flow. How migration, gene flow, and source–sink relations are altered by anthropogenic factors such as habitat destruction and climate change is an important consideration for the design of nature reserves. If you were in charge of designing a nature reserve, could you predict changes in the geographic distributions of populations over time? Could you be sure that the boundaries of a reserve set aside will include a species' habitat requirements in the future?

Reed Noss, currently of Conservation Science, Inc., coined the term "node" to refer to a geographic area of high biodiversity and the term "corridor" to denote a strip of wild land that allows organisms to move between nodes. According to the classic model of metapopulations, corridors need to be only wide enough to allow for migration. Noss's contrasting view is that organisms sometimes have to use corridors as temporary habitat until more nodes become available. Moreover, the physical requirements for corridor design vary according to the types of species that use them. Some corridors can be as simple as culverts under roads that allow amphibians to cross without risking their lives. Indeed, automatic cameras have shown that these "wildlife tunnels" are used by a wide variety of species. Birds may need only discontinuously distributed stop-over sites between subpopulations, but most mammals require a contiguous strip of habitat for migration. Because of the different ways in which organisms use habitat space and their differing sensitivity to habitat edges and disturbance, some animals need wider corridors than others. Bobcats require corridors more than 2.5 km wide, while wolves in Alaska require corridors as much as 22 km wide.

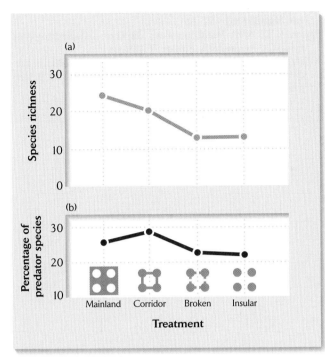

Figure 15.20 Effect of connectivity treatments on diversity in experimental moss islands. (a) Local species richness of individual 10-cm islands. (b) Percentage of the species richness that consists of predators. After F. Gilbert, A. Gonzalez, and I. Evans-Freke, *Proc. R. Soc. London, Series B* 265:577–582.

Experimental studies of corridors and their impacts on metapopulations can be difficult because of the size of landscape patches required, differences in site factors (microclimate, soils, slope, and so on), and influences of human activities. Therefore, it is often useful to employ modeling techniques to study the effect of corridors on metapopulation dynamics. This approach becomes especially powerful if the models can be combined with field experiments. Gilbert et al. (1998) used patches of moss as model ecosystems and sampled microarthropod populations in patches with different degrees of connection to other patches to test the hypothesis that corridors can help to maintain species richness.

CHECK YOUR KNOWLEDGE

1. What are two possible negative consequences of corridors to connect subpopulations?

MORE ON THE WEB **2. Consult the paper by Gilbert et al. through *Practicing Ecology on the Web* at *http://www. whfreeman.com/ricklefs*. What were the benefits** of using moss patches as model ecosystems? How did the authors manipulate connectivity between patches of moss?

3. Refer to Figure 15.20, taken from the paper by Gilbert et al. (1998). What percentage of species were lost from "moss islands" with corridors in comparison to isolated moss islands? What can you conclude about the impact of fragmentation on predators versus nonpredators?

MORE ON THE WEB **4. Visit the Web page for Conservation Science, Inc., through *Practicing Ecology on the Web* at *http://www.whfreeman.com/ricklefs*. Examine the list of clients and the reports they have commissioned. Why would these agencies or companies hire a company like Conservation Science, Inc.?**

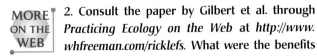 Suggested Readings

Belovsky, G. 1987. Extinction models and mammalian persistence. In M. Soulé (ed.), *Viable Populations for Conservation*, pp. 35–57. Cambridge University Press, Cambridge and New York.

Berryman, A. A. 1996. What causes population cycles of forest Lepidoptera? *Trends in Ecology and Evolution* 11:28–32.

Daniel, C. J., and J. H. Myers. 1995. Climate and outbreaks of the forest tent caterpillar. *Ecography* 18:353–362.

Elliott, J. K. 1985. Population regulation for different life-stages of migratory trout *Salmo trutta* in a Lake District stream, 1966–1983. *Journal of Animal Ecology* 54:617–638.

Eriksson, O. 1996. Regional dynamics of plants: A review of evidence for remnant, source-sink and metapopulations. *Oikos* 77:248–258.

Gilbert, F., A. Gonzalez, and I. Evans-Freke. 1998. Corridors maintain species richness in the fragmented landscapes of a microecosystem. *Proceedings of the Royal Society of London, Series B: Biological Sciences* 265:577–582.

Gilpin, M. E., and I. Hanski. 1991. *Metapopulation Dynamics: Empirical and Theoretical Investigations*. Cambridge University Press, Cambridge and New York.

Gotelli, N. J. 1995. *A Primer of Ecology*. Sinauer Associates, Sunderland, MA.

Goulden, C. E., and L. L. Hornig. 1980. Population oscillations and energy reserves in planktonic Cladocera and their consequences to competition. *Proceedings of the National Academy of Sciences USA* 77:1716–1720.

Hassell, M. P., J. H. Lawton, and R. M. May. 1976. Patterns of dynamical behaviour in single-species populations. *Journal of Animal Ecology* 45:471–486.

Husband, B. C. and S. C. H. Barrett. 1996. A metapopulation perspective in plant population biology. *Journal of Ecology* 84:461–469.

Keith, L. B. 1990. Dynamics of snowshoe hare populations. *Current Mammalogy* 2:119–195.

Kendall, B. E., C. J. Briggs, W. W. Murdoch, P. Turchin, S. P. Ellner, E. McCauley, R. M. Nisbet, and S. N. Wood. 1999. Why do populations cycle? A synthesis of statistical and mechanistic modeling approaches. *Ecology* 80:1789–1805.

Lindström, J., E. Ranta, V. Kaitala, and H. Lindén. 1995. The clockwork of Finnish tetraonid population dynamics. *Oikos* 74:185–194.

Myers, J. H. 1993. Population outbreaks in forest Lepidoptera. *American Scientist* 81:240–251.

Nielsen, Ó. K., and G. Pétursson. 1995. Population fluctuations of gyrfalcon and rock ptarmigan: Analysis of export figures from Iceland. *Wildlife Biology* 1:65–71.

Ouborg, J. 1993. Isolation, population size and extinction: The classical and metapopulation approaches applied to vascular plants along the Dutch Rhine-system. *Oikos* 66:298–308.

Pimm, S. L., H. L. Jones, and J. M. Diamond. 1988. On the risk of extinction. *American Naturalist* 132:757–785.

Pulliam, H. R. 1988. Sources, sinks, and population regulation. *American Naturalist* 132:652–661.

Taylor, A. D. 1990. Metapopulations, dispersal, and predator–prey dynamics: An overview. *Ecology* 71:429–433.

Villard, M.-A., G. Merriam, and B. A. Maurer. 1995. Dynamics in subdivided populations of Neotropical migratory birds in a fragmented temperate forest. *Ecology* 76:27–40.

CHAPTER 16

Population Genetics and Evolution

- The source of genetic variation is mutation and recombination

- The genotypes of all individuals make up the gene pool of a population

- The Hardy–Weinberg law governs the frequencies of alleles and genotypes in large populations at equilibrium

- Most natural populations deviate from Hardy–Weinberg equilibrium

- Natural selection may be stabilizing, directional, or disruptive

- Evolutionary changes in allele frequencies have been documented in natural populations

- Ecologists can draw useful conclusions from population genetics studies

The Galápagos archipelago, lying 1,000 km off the Pacific coast of Ecuador, was a source of inspiration to Charles Darwin 175 years ago on his famous globe-circling voyage on the H.M.S. *Beagle*. Darwin noticed the different forms of several of the archipelago's inhabitants on different islands. He surmised that these differences must have arisen by independent modification of the descendants of the original colonists, which came from the mainland of South America. This idea helped Darwin to develop his theory of evolution by natural selection. Ever since Darwin's time, the Galápagos archipelago has had a special fascination for evolutionary biologists, and many have returned there to pursue evolutionary studies. The most famous of these has been the work of Peter and Rosemary Grant, of Princeton University, who have followed small populations of Darwin's finches on Daphne Island for many years. They showed how the reproductive success and survival of individuals with different-sized beaks differed between El Niño and La Niña years, causing dramatic evolutionary changes in the finch populations.

The Galápagos Islands normally come under the influence of the cold Peru Current and are relatively dry. During El Niño years, however, prolonged warm sea surface temperatures greatly increase rainfall, and the vegetation flourishes (see Figure 4.17). La Niña conditions can bring periods of unremitting drought.

The medium ground finch (*Geospiza fortis*), one of a group known as Darwin's finches, subsists primarily on seeds, which it cracks open with its beak. During one La Niña drought period on tiny Daphne Major Island (see Figure 14.9), the abundance of seeds dropped precipitately as vegetation shriveled and died.

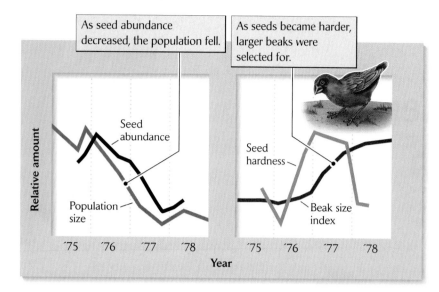

As seed abundance decreased, the population fell.

As seeds became harder, larger beaks were selected for.

Seed abundance

Population size

Seed hardness

Beak size index

Relative amount

Year

'75 '76 '77 '78   '75 '76 '77 '78

Figure 16.1 Darwin's finches show evolutionary responses to climate changes. (*Left*) Decline in the abundance of seeds and population size of the medium ground finch (*Geospiza fortis*) on Daphne Major Island in the Galápagos archipelago during the drought period 1975–1978. (*Right*) Increase in relative seed hardness and the average beak size in the ground finch population during this same period. From P. R. Grant, *Ecology and Evolution of Darwin's Finches*, Princeton University Press, Princeton, NJ (1986).

The seeds that did remain were generally the hardest to crack, and thus avoided by the finches during times of plenty. However, when seeds were scarce, the finches had no choice other than to try to crack the hard ones. Consequently, numbers of the medium ground finch dropped from about 1,400 in 1975 to about 200 at the end of 1977. Mortality did not occur at random, however. Because the average hardness of seeds increased as the drought intensified and soft seeds were consumed, birds with larger beaks that could generate the forces needed to crack the larger seeds survived better than individuals with smaller beaks. Thus, the average beak size of surviving individuals, and their progeny, increased dramatically between 1976 and 1978 (**Figure 16.1**). Here was a case of evolution in action! During the exceedingly wet El Niño year of 1983, small seeds were produced in abundance. Birds with smaller beaks handled the smaller seeds more efficiently, and so they survived better and produced more offspring than individuals with larger beaks. Consequently, the average beak size of the population returned to a lower value.

Although these evolutionary responses were small, they nonetheless illustrate the capacity of a population, because of genetic variation among individuals, to respond to changes in the environment. The field of population genetics studies the ways in which populations respond to such selective pressures with changes in allele frequencies.

Regardless of the form natural selection takes, it cannot cause evolutionary change unless there is genetic variation within a population. We learn in introductory biology courses that genetic information is contained in the molecule **deoxyribonucleic acid,** or **DNA** for short, and that genetic variation is caused by changes in the DNA molecule. The structure of DNA is an unbranched chain of subunits held together by sugar–phosphate links. DNA has four kinds of subunits, which are called **nucleotides:** adenine, thymine, cytosine, and guanine. Each of these nucleotides is known by an acronym, which is its first letter—that is, A, T, C, or G. Thus, a stretch of DNA may be represented by a string of letters showing the order of its subunits; for example, ATGGCATTAACGT. Genetic information is encoded in the particular order of the different nucleotides, just as the order of letters in a word conveys information.

A DNA strand serves as a template from which a cell manufactures proteins and other nucleic acids. Proteins themselves also are unbranched chains of subunits, composed of up to 20 different amino acids. Each amino acid is encoded by one or more unique sequences of three nucleotides. These coding triplets are referred to as **codons.** For example, the DNA sequence AAA (adenine-adenine-adenine) specifies the amino acid phenylalanine; GAG specifies leucine, CTT specifies glutamic acid, and so on.

Because four different subunits taken three at a time yield 64 (4^3) different sequences, the **genetic code** contains considerable redundancy. In other words, because DNA needs to encode only 20 amino acids, and 64 different codons are possible, several codons can specify the same amino acid. The most extreme example is leucine, which is encoded by the sequences AAT, AAC, GAA, GAG, GAT, and GAC.

The source of genetic variation is mutation and recombination

Mutations

Various kinds of errors in the sequence of nucleotides in DNA can occur. The most common of these errors are nucleotide substitutions, but deletions, additions, and rearrangements of nucleotides also occur. These mistakes can be caused by random copying errors when the genetic material replicates during cell division, by certain highly reactive chemical agents, or by ionizing radiation.

A substitution of one of the nucleotides in a DNA codon may change the amino acid that it specifies. Consider this sequence of nucleotides and the corresponding amino acids:

Position:	1	4	7	10	13	16
DNA:	GAA	TGG	CGA	GAA	ATA	GGG
Amino acid:	Leucine	Serine	Alanine	Leucine	Tyrosine	Proline

If the guanine occupying the eighth position were changed to thymine, the third codon would be altered from CGA to CTA, and it would now encode the amino acid aspartine instead of alanine:

Position:	1	4	7	10	13	16
DNA:	GAA	TGG	CTA	GAA	ATA	GGG
Amino acid:	Leucine	Serine	Aspartine	Leucine	Tyrosine	Proline

Such changes do occur, and geneticists call them **mutations.**

Because of the redundancy in the genetic code, some changes in nucleotides do not change the amino acid specified by a codon, and therefore are not expressed in the phenotype. Such changes are often referred to as silent mutations, because they are not "heard from," or neutral mutations, because they have no consequence for fitness. Other mutations may result in the substitution of one amino acid for another in a protein chain. The new protein produced by a mutant gene may or may not have properties different from those of the original protein.

Any altered properties of proteins resulting from mutations may be beneficial. It is more likely, however, that they will be harmful to an individual. The reason for this is simple: Over millions of generations of evolution, natural selection weeds out most deleterious genes, leaving behind only those genes that suit organisms to their environments. Any new variant is more likely to disrupt the well-tuned interaction between an organism and its surroundings than to improve this interaction.

Mutations of hemoglobin, which is the principal oxygen-binding molecule in the bloodstream of vertebrates, including ourselves (see Chapter 3), are a case in point. Most mutations in the genes for hemoglobin impair the ability of the molecule to bind oxygen, and the individual suffers from anemia, or lack of oxygen. For example, the disease sickle-cell anemia is caused by a mutation that changes the sixth amino acid in the beta chain of the hemoglobin molecule from glutamic acid to valine, with drastic consequences. This substitution alters the structure of hemoglobin molecules such that when they release oxygen from the red cells in the bloodstream, they become stacked close together in long helices, which impairs their function and causes severe and debilitating anemia. The aberrant hemoglobin molecules give red blood cells a peculiar sicklelike shape—hence the name of the disease (█ Figure 16.2).

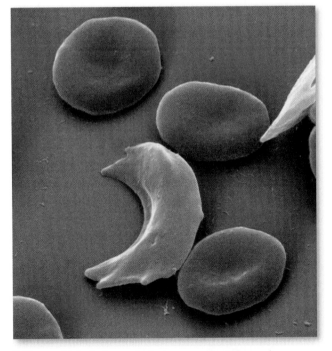

█ **Figure 16.2 A change in just one nucleotide can have phenotypic effects.** A single-nucleotide mutation in the gene for human hemoglobin changes the structure of the hemoglobin molecule and causes red blood cells to assume a sickle shape. This results in severe anemia in individuals who have two copies of mutant gene. Photo by Meckes/Ottawa/Photo Researchers, in A. J. Griffiths et al., *Introduction to Genetic Analysis,* 7th ed., W. H. Freeman and Company, New York (2000).

Mutations are more likely to be beneficial when the relationship between the organism and its environment changes. For example, selection of beneficial mutations in hemoglobin's oxygen-binding properties has accompanied the adaptation of animals to environments with different oxygen concentrations and to more or less active lifestyles. However, it is important to realize that the processes that cause mutations are blind to the selective pressures on a population. That is, mutation is a random force in evolution that produces genetic variation independently of its fitness consequences for the individual that bears the mutation.

Mutation rates

For any particular nucleotide in a DNA sequence, the rate of mutation is extremely low, on the order of 1 in 100 million per generation. However, this low rate multiplied by the hundreds or thousands of nucleotides in a gene, and by the trillion or so nucleotides in such complex organisms as vertebrates, means that each individual is likely to sustain one or more mutations in some part of its genome. Nonetheless, measured rates of expressed gene mutation average about 1 in 100,000 to 1 in 1 million per gene per generation, with rates for particular genes being much higher. For example, mutations having readily apparent visible effects, such as the color of kernels on an ear of corn (∥ Figure 16.3), or defects in the wing structure of fruit flies, occur at rates of 1 in 10,000 to 1 in 100,000 gametes per generation. Plants with chlorophyll deficiencies, which are lethal because plants cannot make sugars without photosynthesis, appear by mutation at rates of almost 1 in 100 to 1 in 10,000. However, these rates are high because they are summed over many genes; any of a number of genes involved in producing chlorophyll may be responsible for such "phenotypic" mutations. In laboratory strains of the fruit fly *Drosophila*, mutations occurring anywhere in the genome and having lethal effects arise in about 2% of individuals each generation, and mutations with mildly detrimental effects arise at a rate of about one per fly. There is no question that mutation is actively churning out genetic variation.

Recombination

Variation is also produced during meiosis when parts of the genetic material inherited from the mother and from the father recombine with each other. **Recombination** is the exchange of homologous sections of maternal and paternal chromosomes, which creates new combinations of genes. Recombination produces new genetic variation rapidly, whereas mutation does so very slowly. Thus, while mutation is the ultimate cause of genetic variation in

∥ **Figure 16.3 Mutations are a source of genetic variation.** Readily visible mutations, such as those in the color of kernels of corn, have been used extensively for analyses of inheritance. Photo by Gregory G. Dimijian/Photo Researchers.

a population, recombination multiplies the level of variation. The new combinations of genetic traits resulting from sexual reproduction and recombination provide natural selection with abundant variation to work on.

MORE ON THE WEB *Evolution of body size in Galápagos marine iguanas.* **Natural and sexual selection have opposing influences on the size of males.**

The genotypes of all individuals make up the gene pool of a population

Each individual has a unique genetic makeup. All the genes in all the individuals in a population make up the **gene pool,** which represents all the genetic variation of the population.

When all individuals mate at random within a population, then all combinations of different alleles are possible, although many of these combinations may not be present in the population at a given time. For example, suppose that alleles A_2 and B_2 are rare mutants of genes A and B, respectively. Most of the individuals in the population have genotypes A_1B_1. A few individuals may have genotype A_2B_1, and a few others may have genotype A_1B_2. If A_2 and B_2 are sufficiently rare, A_2B_2 individuals may not exist in the population. However, they may be produced in future generations by matings between A_1B_2 and A_2B_1 individuals. Thus, the presence of the A_2 and B_2 alleles in the gene pool represents latent variation in the kinds of genotypes that are possible in the population. If the combination A_2B_2 confers particularly high fitness—substantially higher than the A_2 or

B_2 allele alone—then A_2B_2 individuals will produce more offspring than average, and the overall frequency of these alleles will increase in the population.

The Hardy–Weinberg law governs the frequencies of alleles and genotypes in large populations at equilibrium

It is an important principle of genetics that the frequencies of alleles and genotypes remain constant from generation to generation in a population with (1) a large number of individuals, (2) random mating, (3) no selection, (4) no mutation, and (5) no migration between populations. In other words, no evolutionary change occurs through the process of sexual reproduction itself. This principle is called the **Hardy–Weinberg law** after the two geneticists who independently described it in 1908. Hardy's and Weinberg's important discovery showed that changes in allele and genotype frequencies can result only from the action of additional forces on the gene pool of a population. Understanding the nature of these forces has occupied evolutionary biologists ever since.

When a population exists in **Hardy–Weinberg equilibrium**, the proportions of homozygotes and heterozygotes take on equilibrium values, which we can calculate from the proportions of each allele in a population. Suppose that a particular gene A has two alleles, A_1 and A_2, which occur in proportions p and q ($p + q = 1$, and therefore $q = 1 - p$). At Hardy–Weinberg equilibrium, the three genotypes that can result will occur in the following proportions:

Genotype	A_1A_1	A_1A_2	A_2A_2
Frequency	p^2	$2pq$	q^2

Notice that $p^2 + 2pq + q^2 = 1$: the proportions of all the genotypes in the population add up to 1.

These proportions come from the probabilities that each type of zygote will be formed from random combination of any two gametes (**Figure 16.4**). To form an A_1A_1 homozygote, both gametes must have an A_1 allele. When the probability of drawing one A_1 allele in one random pick is p, the probability of drawing two A_1 alleles in two picks is simply the probability of each multiplied by the other, or p^2. The proportion of heterozygotes is $2pq$ because an A_1A_2 heterozygote will result from an A_1 egg and an A_2 sperm, with probability pq, and from an A_2 egg and an A_1 sperm, also with probability pq.

We can calculate the numerical values of proportions of genotypes under Hardy–Weinberg equilibrium as in the following example. Suppose one allele (A_1) occurs in a population with a frequency of 0.7, and the other (A_2) with a frequency of 0.3 ($0.7 + 0.3 = 1$). Accordingly, 49% ($0.7^2 = 0.49$) of the genotypes in the population will be A_1 homozygotes, 42% ($2 \times 0.7 \times 0.3 = 0.42$) will be heterozygotes, and 9% ($0.3^2 = 0.09$) will be A_2 homozygotes. Notice that $0.49 + 0.42 + 0.09 = 1$. Any

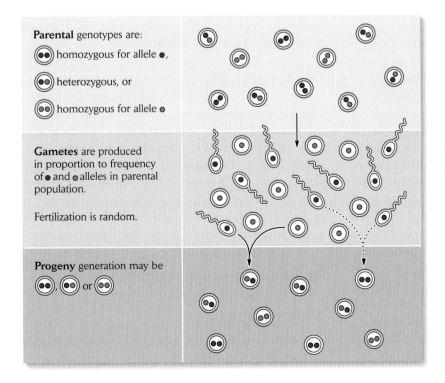

Parental genotypes are:

(●●) homozygous for allele ●,

(●○) heterozygous, or

(○○) homozygous for allele ○

Gametes are produced in proportion to frequency of ● and ○ alleles in parental population.

Fertilization is random.

Progeny generation may be

(●●), (●○) or (○○)

Figure 16.4 The proportions of genotypes in a sexually reproducing population are a result of the random combination of gametes. This diagram is particular to an organism, such as a clam or mussel, that sheds its gametes into the water, in which the gametes fuse at random to form zygotes. When fertilization is internal, each combination of genotypes in a mating pair will produce a unique combination of, or unique proportions of, offspring genotypes. Hardy-Weinberg frequencies nonetheless pertain.

deviations from these genotype frequencies indicate the presence of selection, nonrandom mating, or other factors that influence the genetic makeup of the population.

HELP ON THE WEB *Go to Living Graphs at http://www.whfreeman. com/ricklefs* and use the interactive tutorial to better understand the relationships behind the probabilities in the Hardy-Weinberg Equation.

Most natural populations deviate from Hardy–Weinberg equilibrium

Mutation, small population size, nonrandom mating, migration, and natural selection all cause populations to deviate from Hardy–Weinberg equilibrium. Mutation is a weak force compared with selection because its rate is so low; we will not consider it further here. We shall consider the effects of small population size, nonrandom (assortative) mating, and migration in more detail.

Small populations can lose variation by genetic drift and founder events

Genetic drift is a change in allele frequencies due to random variations in fecundity and mortality in a population. Genetic drift has its greatest effects in small populations, in which the same kinds of stochastic variation in birth and death rates that can cause variation in population size (see Chapter 15) can cause substantial changes in allele frequencies. For example, suppose a population contains 95 individuals with genotype A_1A_1 and 5 individuals with genotype A_1A_2. If each individual had only a 50% chance of surviving to reproduce, then the probability would be 1 in 32 (about 3%) that all 5 of the A_1A_2 individuals will fail to reproduce and the A_2 allele will disappear from the population (at least until new mutants appear). If the A_2 allele is lost in this manner, the A_1 allele is said to become **fixed** in the population. The rate of fixation of alleles is inversely related to the size of a population. Thus, genetic variation decreases more rapidly over time in small populations than in large ones.

A single episode of small population size, as might occur during the colonization of an island or a new habitat by a few individuals from a large parent population, can reduce genetic variation in the colonizing population. Such episodes are known as **founder events.** When founding populations consist of ten or fewer individuals, they typically contain a substantially reduced sample of the total genetic variation of the parent population (▌Figure 16.5).

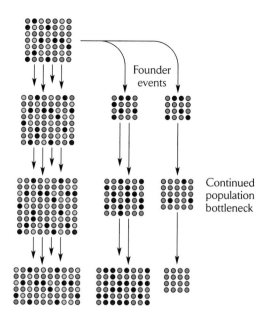

▌**Figure 16.5 Founder events often result in reduced genetic diversity.** When a small number of individuals colonize a new habitat, the colonists take with them a random sample of the alleles in the parent population. Continued low population size following colonization or any other form of population decline (a population bottleneck) results in further loss of genetic diversity through random sampling of each generation's gene pool in the following generation.

Continued existence at low population size results in further loss of genetic variation due to genetic drift. This situation is often referred to as a **population bottleneck.** Such a condition appears to have occurred in the recent past in the population of cheetahs in East Africa (▌Figure 16.6), which exhibit practically no genetic variation.

The significance of founder events and bottlenecks for natural populations is that the fragmentation of populations into small subpopulations may eventually restrict the evolutionary responsiveness of those subpopulations to the selective pressures of changing environments, making them more vulnerable to extinction. Furthermore, loss of genetic variation in the major histocompatibility (*MHC*) genes, which are involved in immune function, may reduce the natural resistance of individuals to disease. As yet, there is no agreement as to whether the cheetah's genetic uniformity poses a serious threat to its future.

Assortative mating changes the frequencies of genotypes within a population

Assortative mating occurs when individuals choose mates nonrandomly with respect to their own genotypes. Like

Figure 16.6 Population bottlenecks often result in reduced genetic diversity. Cheetahs in East Africa exhibit virtually no genetic variation, suggesting that the population recently passed through a severe population bottleneck. Photo by R. E. Ricklefs.

mating with like is referred to as **positive assortative mating.** A tendency for mates to differ genetically is referred to as **negative assortative mating.** Assortative mating does not directly change the frequencies of alleles within populations, only the frequencies of genotypes. Negative assortment—for example, an A_1A_1 homozygote mating with an A_2A_2 homozygote, producing only A_1A_2 offspring—increases the proportion of heterozygotes in a population at the expense of homozygotes. Positive assortment, including **inbreeding,** or mating with close relatives, has the opposite effect of reducing the proportion of heterozygotes, even though the frequencies of the A_1 and A_2 alleles do not change in the population as a whole (▌**Figure 16.7**).

Assortative mating can lead secondarily to changes in allele frequencies due to natural selection if some allele combinations have greater fitness than others do. For example, because positive assortment produces an overabundance of homozygotes among progeny, it increases the expression of recessive alleles, including rare harmful genes that are usually masked by dominant alleles in heterozygous form.

Most species employ mechanisms—including dispersal of progeny, recognition of close relatives, and negative assortative mating—to reduce the occurrence of inbreeding. Mammals, including humans, can distinguish differences in the major histocompatibility genes of potential mates by smell. Unrelated individuals are less likely to share MHC genes than are close relatives. Hermaphroditic species of plants, in which individuals bear both male and female sexual organs, have many mechanisms to prevent self-pollination, including genetic self-incompatibility, temporal separation of male and female function, and elaborate flower structures designed to make selfing difficult.

Although inbreeding generally creates genetic problems, it may confer important benefits in certain situations. In particular, plants that can self-pollinate guarantee fertilization of their flowers in habitats that lack suitable pollinators

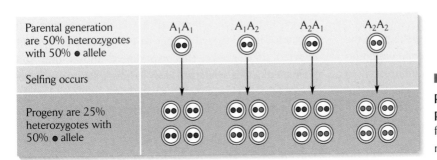

Parental generation are 50% heterozygotes with 50% ● allele	A_1A_1	A_1A_2	A_2A_1	A_2A_2
Selfing occurs				
Progeny are 25% heterozygotes with 50% ● allele				

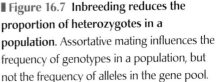

Figure 16.7 Inbreeding reduces the proportion of heterozygotes in a population. Assortative mating influences the frequency of genotypes in a population, but not the frequency of alleles in the gene pool.

or where individuals are widely spaced. Many weedy species that colonize isolated patches of disturbed habitat (for example, dandelions) often self-pollinate. It is assumed that most deleterious variation is weeded out of such populations a little at a time as it is exposed in homozygous individuals during the transition between outcrossing and self-pollination.

In heterogeneous environments, adaptation to particular habitat patches enhances fitness, and mating with distant individuals who are adapted to different habitat conditions may reduce the fitness of progeny. Several studies have reported an **optimal outcrossing distance** in populations of plants. Nearby individuals are likely to be close relatives. Distant individuals are likely to be adapted to different conditions. Optimal outcrossing distance should be somewhere in between. In a study conducted in central Colorado, Mary Price and Nicolas Wasser fertilized flowers of the larkspur *Delphinium nelsoni* with pollen obtained from the same individual and from individuals located at distances of 1, 10, 100, and 1,000 meters. The number of seeds set per flower was greatest when the pollen came from individuals 10 meters away, and smallest for selfed pollen and for pollen obtained from individuals 1,000 meters distant. Furthermore, when these seeds were planted, survival to 1 and 2 years was greatest among the offspring of matings across the optimal outcrossing distance of 10 meters (▌Figure 16.8).

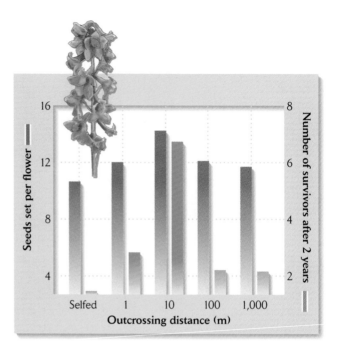

▌**Figure 16.8 The optimal outcrossing distance balances the risks of inbreeding and of mating with individuals adapted to different conditions.** The graph shows the number of seeds set per flower and the number of seeds (out of 98) that produced plants surviving after 2 years as a function of the distance from which pollen was obtained. The data show an intermediate optimal outcrossing distance of about 10 meters, which is well beyond the average distance of pollen transfer within the study population. After M. V. Price and N. M. Wasser, *Nature* 277:294–297 (1979).

ECOLOGISTS IN THE FIELD

Inbreeding depression and selective abortion in plants

Banksia spinulosa is an Australian shrub that is pollinated by small nectar-feeding birds. *Banksia* plants can self-pollinate, but they normally outcross. Each inflorescence has about 800 flowers, but produces fewer than 50 fruits. Many developing seeds in each inflorescence are normally aborted. In other words, the pollinated flowers compete for the resources needed to develop into a fruit. Fruit production appears to be resource-limited rather than pollen-limited, because removal of one-third of the flowers from either the base or the top of an inflorescence does not significantly depress fruit set.

To determine whether these plants can distinguish the source of pollen, the Australian botanists Glenda Vaughton and Susan Carthew hand-pollinated *Banksia* inflorescences with pollen obtained either from the same plant (selfed pollen) or from neighboring plants (outcrossed pollen). In some of the plants, they pollinated half the inflorescence

with selfed pollen and half with outcrossed pollen (mixed pollination). After fruits had developed, the numbers of fruits and seeds (no more than one seed per fruit) were counted on each side of each inflorescence.

Compared with cross-pollination, self-pollination reduced the number of seeds produced by 38% (24 versus 39 seeds per half-inflorescence); fruits with aborted seeds increased from 8% to 16% (▌Figure 16.9). These results clearly indicate **inbreeding depression**, or reduction of fitness caused by inbreeding. When one half of an inflorescence was cross-pollinated and the other half was self-pollinated, the number of seeds produced per self-pollinated half dropped further to 14, and 28% of the fruits aborted their seeds. This experiment shows that self-pollinated ovules, which are likely to have inferior genotypes, do not fare well in competition with cross-pollinated ovules. Thus, plants are capable of making distinctions among developing embryos on the basis of their genotypes.

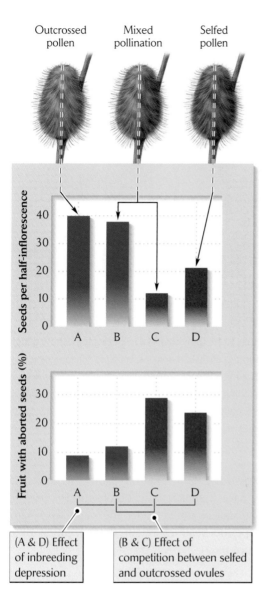

Outcrossed pollen   Mixed pollination   Selfed pollen

(A & D) Effect of inbreeding depression

(B & C) Effect of competition between selfed and outcrossed ovules

❚ Figure 16.9 Inbreeding usually has negative effects on fitness. Results of a pollination experiment with the Australian shrub *Banksia spinulosa*, in which the two halves of individual inflorescences were fertilized with pollen either from the same plant or from other plants. The results show the negative effects of inbreeding (D compared with A) and the further discrimination against selfed ovules in competition with outcrossed ovules growing in the same inflorescence (C compared with B). Data from G. Vaughton and S. M. Carthew, *Biol. J. Linn. Soc.* 50:35–46 (1993).

Migration can result in deviations from Hardy–Weinberg equilibrium

When two populations have different allele frequencies—which may happen if they experience different selective forces—any mixture of the two populations resulting from

migration of individuals between them will generally reduce the proportion of heterozygotes. This phenomenon is called the **Wahlund effect.** Consider an extreme example: Suppose we mix equal numbers of individuals from two populations that have different fixed alleles—that is, one population has only A_1A_1 individuals and the other, only A_2A_2 individuals. The resulting population has 50% A_1 and 50% A_2 alleles, but no heterozygotes. Clearly, it is not in Hardy–Weinberg equilibrium. If these individuals mate, successive generations of random mating should increase the frequency of heterozygotes, but continued mixing of the populations would prevent the achievement of Hardy–Weinberg equilibrium.

This is an extreme example, but in general, a deficiency of heterozygotes in a population strongly indicates either positive assortative mating or a Wahlund effect due to population mixing. A closer look at the biology of the population is likely to help us distinguish between these possibilities. For example, one would not expect mussels, which shed their gametes into the ocean, to exercise strong assortative mating, but it would not be surprising to find that ocean currents had caused some mixing between populations. Such movement of genes between populations is referred to as **gene flow.**

The genotypes of individuals within a population often vary geographically

Differences in selective factors or random changes (genetic drift, founder events) in different parts of a population's range can cause geographic variation in allele frequencies among subpopulations. Such variation is often found between subpopulations that are divided by a natural barrier, such as a river or a mountain range. However, populations do not have to be subdivided for genetic differences to arise within them. If the difference in selective pressures between two localities is strong relative to the rate of gene flow between them, then differences in allele frequencies can be maintained by differential natural selection. This phenomenon often results in a gradual change in allele frequencies, or some phenotypic character under genetic influence, over distance.

Botanists have long recognized that individuals of a species growing in different habitats may exhibit varied forms corresponding to local conditions. In many cases, these differences result from developmental responses (see Chapter 9). However, experiments on some species have revealed genetic adaptations to local conditions. Early in the twentieth century, the Swedish botanist Göte Turesson collected seeds from several species of plants that lived in a variety of habitats and grew them together in his garden. This is referred to as a **common garden experiment.** He found that even when grown under identical conditions, many of the plants exhibited different forms depending on their habitat of origin. Turesson called these forms **ecotypes,** a name that persists to the present. He suggested that ecotypes represent genetically differentiated lineages of a population, each restricted to a specific habitat. Because Turesson grew his plants under identical conditions, he realized that the differences between ecotypes must have had a genetic basis, and that they must have resulted from evolutionary differentiation within the species according to habitat.

Experiments on yarrow (*Achillea millefolium*) also revealed ecotypic variation. This plant grows in many habitats ranging from sea level to more than 3,000 meters in elevation. Plants raised from seed collected at various elevations but grown together at sea level at Stanford, California, retained the distinctive sizes and levels of seed production typical of the populations from which they came (▌Figure 16.10). Similar differentiation has been found over distances of only a few meters where contrasting selective pressures are strong enough to overcome the migration of individuals, seeds, or pollen between habitat patches. Such situations frequently arise on soils that develop on mine tailings, which sometimes exert strong selective pressure for tolerance to toxic metals (for example, copper, lead, zinc, and arsenic).

A trait may exhibit a pattern of gradual change, or **cline,** in response to variation in the environment over distance. In the Japanese field cricket *Teleogryllus,* for example, body size and duration of nymphal (larval) development both

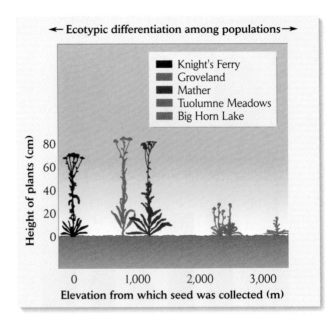

▌**Figure 16.10 Individuals of a species may show geographic variation in allele frequencies.** Ecotypic differentiation in populations of the yarrow, *Achillea millefolium,* was demonstrated by collecting seeds from different elevations and growing them under identical conditions in a common garden. After J. Clausen, D. D. Keck, and W. M. Hiesey, *Carnegie Inst. Wash. Publ.* 58:1–129 (1948).

increase clinally from north to south. We know that these clines have a genetic basis because individuals from different localities raised under the same temperature and photoperiod regimes retain their regional differences (▌Figure 16.11).

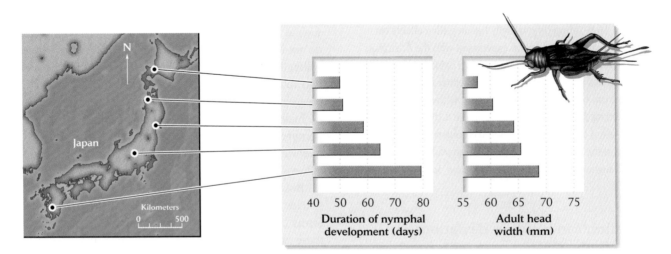

▌**Figure 16.11 Clinal variation may occur over distance with variations in the environment.** The duration of nymphal development and the width of the adult head in males in the field cricket *Teleogryllus emma* show a cline from north to south. When nymphs from five local populations were all raised together at 28°C on a cycle of 16 hours of light and 8 hours of dark, this variation still appeared, demonstrating that it has a genetic basis. Data from S. Masaki, *Evolution* 21:725–741 (1967).

In contrast, the color and banding patterns on the shells of the snail *Bradybaena* vary geographically, but their variations bear no relationship to habitat or locality (Figure 16.12). This snail was introduced to Japan, probably many times over the last 200 years, with the widespread cultivation of sugarcane. Such variation in phenotypic frequencies among subpopulations could occur if each local population was established by a small group of colonists that contained a random, but not necessarily representative, sample of the genetic variation of the parent population.

A final example shows how a geographic barrier that isolates subpopulations from one another can influence genetic variation. The saddleback tamarin (*Saguinus fuscicollis*) is distributed along both sides of the Rio Juruá in the western Amazon basin of Brazil. Toward the lower part of the river, where it is wide, tamarin populations on either side are genetically distinct. Near the headwaters, where the river is narrower and where its changing course may shift areas from one side of the river to the other, genotypes typical of populations on the right bank of the river also appear on the left side (Figure 16.13).

We now have seen that genetic variation is common in populations, and that organisms have adaptations that effectively manage that variation to reduce its potentially

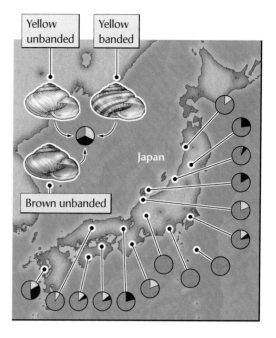

 Figure 16.12 Some traits vary over distance independently of variations in the environment. The frequencies of phenotypes of the land snail *Bradybaena similaris* do not show clinal variation. Data from T. Komai and S. Emura, *Evolution* 9:400–418 (1955).

Rio Juruá, Brazil

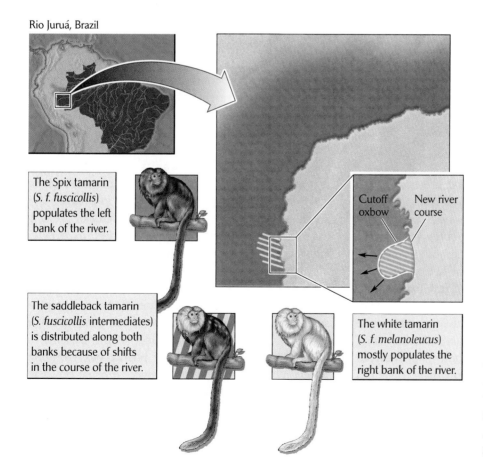

The Spix tamarin (*S. f. fuscicollis*) populates the left bank of the river.

The saddleback tamarin (*S. fuscicollis* intermediates) is distributed along both banks because of shifts in the course of the river.

Cutoff oxbow   New river course

The white tamarin (*S. f. melanoleucus*) mostly populates the right bank of the river.

 Figure 16.13 Geographic isolation may cause two populations to be genetically distinct. The map shows the distribution of different subspecies of saddleback tamarins (*Saguinus fuscicollis*) in the upper reaches of the Rio Juruá in the western Amazon basin of Brazil. (The subspecies are distinguished by coat color patterns and mitochondrial genotypes.) The lower broad reaches of the river are an effective barrier to dispersal of the tamarins, and the populations on either side are distinct. Close to the headwaters, however, the cutting off of oxbows has left some of the right-bank populations on the left bank of the river, where the two subspecies have hybridized. After C. A. Peres, J. L. Patton, and M. N. F. da Silva, *Folia Primatol.* 67:113–124 (1996).

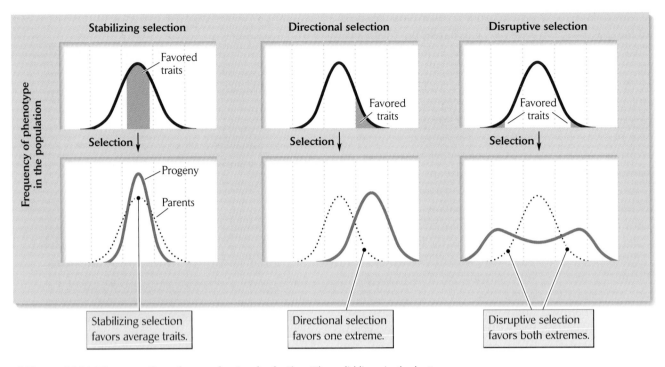

Figure 16.14 There are three forms of natural selection. The solid lines in the bottom panels show the distribution of phenotypes after selection (progeny generation) compared to the pre-selection (parental) generation (dotted line).

negative consequences for individual fitness. Genetic variation also provides the raw material for evolutionary change, which proceeds as natural selection increases the frequency of more fit genotypes within a population.

Natural selection may be stabilizing, directional, or disruptive

As we have seen, evolution by natural selection occurs when genetic factors influence survival and fecundity. Those individuals that achieve the highest reproductive rate are said to be selected, and their proportion in the population increases with time. Selection can take any of three different forms, depending on the heterogeneity of, and rate of change in, the environment. **Stabilizing selection** occurs when individuals with intermediate, or average, phenotypes have higher reproductive success than do those with extreme phenotypes (**Figure 16.14**). Stabilizing selection tends to draw the distribution of phenotypes within a population toward an intermediate, optimum point, and counteracts the tendency of phenotypic variation to increase through mutation and gene flow between populations. Stabilizing selection performs housekeeping

for a population, sweeping away harmful genetic variation. When the environment of a population is relatively unchanging, stabilizing selection is the dominant mode, and little evolutionary change takes place.

Directional selection occurs when the fittest individuals have a more extreme phenotype than the average of the population. In this case, individuals whose phenotypes are to one side of the population average produce the most progeny, and the distribution of phenotypes in succeeding generations shifts toward a new optimum. When that new optimum is reached, selection becomes stabilizing. Runaway sexual selection, discussed in Chapter 11, is a case of directional selection because females prefer to mate with males having extreme phenotypes—exceedingly long tails or bright coloration, for example.

Individuals with a phenotype at either extreme may occasionally have higher fitness than individuals with intermediate phenotypes. This situation leads to **disruptive selection,** which tends to increase genetic and phenotypic variation within a population and, in the extreme, to create a bimodal distribution of phenotypes. Disruptive selection is thought to be uncommon. It might occur, for example, when individuals can specialize on one of a small number of food sources that differ according to size or some other attribute (**Figure 16.15**). Disruptive selec-

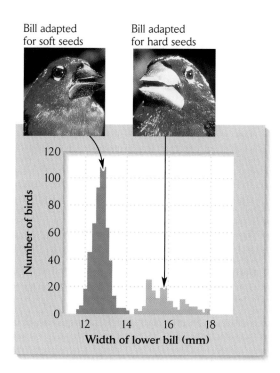

Bill adapted for soft seeds   Bill adapted for hard seeds

❙ Figure 16.15 Disruptive selection increases phenotypic variation in a population. The graph shows dimorphism in beak size in a Cameroon population of the African estrildid finch *Pyrenestes ostrinus*. The large-beaked form can crack the hard seeds of one species of sedge (*Scleria*) more rapidly than the small-beaked form. Although both forms can crack open the soft seeds of another species of sedge more rapidly, the small-beaked form has an advantage. After T. B. Smith, *Nature* 329:717–719 (1987). Photos courtesy of T. B. Smith.

tion might also ensue when interactions among individuals create alternatives to the prevalent life history; for example, territorial behavior by large males might favor small males that sneak copulations with females (see Figure 12.1). Strong competition among individuals for a preferred resource might also increase the fitness of individuals that specialize on one of several alternative resources. Few such cases have come to light, however.

Directional selection changes allele frequencies in the gene pool

The composition of the gene pool changes in response to selection. The rate of this change depends on the relative fitness, or reproductive success, of individuals having different genotypes. Population geneticists have modeled the effects of natural selection on the genetic makeup of a population to determine the rate at which evolution can occur.

 Modeling selection against a deleterious recessive gene. The rate of change in allele frequency depends on how the allele is expressed in the phenotype and the strength of selection.

Population genetic models tell us several important things about the course of evolution. Let us take as an example a population containing an allele A_1 and a deleterious, recessive allele A_2. First, selection against the A_2A_2 genotype always causes a decrease in the frequency of the A_2 allele. Second, the rate of change in the frequency of the A_2 allele varies in direct relationship to the selective pressure on the population and the frequency of the A_2 allele in the population. For example, the change in the frequency of A_2 is fastest when the A_2 allele is relatively common because a larger proportion of the A_2 alleles are then exposed in homozygous form. Third, evolution stops only when either the A_1 or the A_2 allele is fixed in the population and there is no longer any genetic variation for selection to act upon. Thus, directional selection continuously removes genetic variation from populations.

Mutation, spatial and temporal variation, and heterozygote superiority maintain genetic variation in populations

Natural selection cannot produce evolutionary change without genetic variation. Yet one consequence of both stabilizing and directional selection is the reduction of genetic variation by elimination of less fit alleles from the gene pool. How does evolution continue in these circumstances? Does the availability of genetic variation ever limit the rate of evolutionary change?

We have seen that every population is supplied with new genetic variation by mutation and migration. Spatial and temporal variation in the environment tends to maintain this genetic variation by favoring different alleles at different times and places. Often such environmental variation results in heterozygotes having greater fitness than either homozygote. Suppose, for example, that the environment varies in such a way that individuals having the A_1 allele are favored in some years, and individuals with the A_2 allele are favored in others. Because heterozygotes have both alleles, their phenotype is often superior in such a varying environment (❙ Figure 16.16).

When heterozygotes have higher fitness than either homozygote, the relative fitness of each allele depends on its frequency in the population. When an allele is rare (p is small), it occurs most often in high-fitness heterozygous genotypes ($2pq$ is large compared with p^2). When the allele is common (p is large), it occurs most often in low-fitness

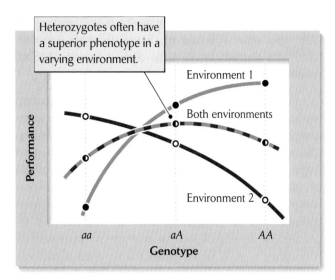

Heterozygotes often have a superior phenotype in a varying environment.

Figure 16.16 In environments that favor different alleles in different years, the heterozygote may be more fit than either of the homozygotes. Such environmental variation maintains genetic variation within a population.

homozygotes (p^2 is large compared with $2pq$). Thus, the fitness of an allele is dependent on its frequency. Such **frequency-dependent selection** maintains both alleles in the population (**Figure 16.17**).

One of the most remarkable cases of heterozygote superiority, or **heterosis,** involves the sickle-cell allele (*S*), which is a mutation of the gene for the beta chain of the hemoglobin molecule in humans (see above). When homo-

zygous (*SS*), this gene produces a debilitating anemia that often leads to an early death. Yet in some areas of the Mediterranean region and in tropical Africa, the frequency of the *S* allele may reach 20% or more of the gene pool. The reason for this is that in the heterozygous state (*AS*), the sickle-cell mutation confers protection against malaria. Where malaria is prevalent and virulent, the fitness of heterozygous individuals may be 25% greater than the fitness of individuals who are homozygous for the normal (*A*) allele.

It is perhaps remarkable that populations contain as much genetic variation as they do. About a third of the genes that encode enzymes involved in cellular metabolism show variation in most species surveyed, and fully 10% of these genes may be heterozygous in any given individual. At any particular time, most of this genetic variation either has no consequences for individuals—that is, it is neutral—or has negative effects when expressed. Thus, most genetic variation either produces no variation in fitness among individuals, and therefore no evolutionary change, or creates stabilizing selection that weeds out harmful genetic variation. In the event that the environment changes, some of this genetic variation may take on positive survival value, and it then fuels the fires of evolution. But this is purely a hit-or-miss consequence of the randomness of mutation. Nonetheless, there seems to be enough genetic variation in most populations so that evolutionary change is a constant presence.

Evolutionary changes in allele frequencies have been documented in natural populations

The story of cyanide resistance in scale insects (see Chapter 9) was one of the first documented cases of a population that responded genetically to a change in selective factors in its environment. Similar cases of pesticide and herbicide resistance among agricultural pests and disease vectors, as well as the increasing resistance of bacteria to antibiotics, are further examples of how rapidly the gene pools of populations can respond to changes wrought by humankind in their environments. In each case, genetic variation in the gene pool allowed the population to respond to the changed conditions.

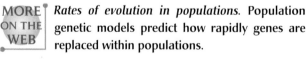

Rates of evolution in populations. **Population genetic models predict how rapidly genes are replaced within populations.**

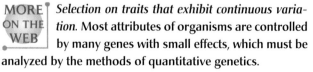

Selection on traits that exhibit continuous variation. **Most attributes of organisms are controlled by many genes with small effects, which must be analyzed by the methods of quantitative genetics.**

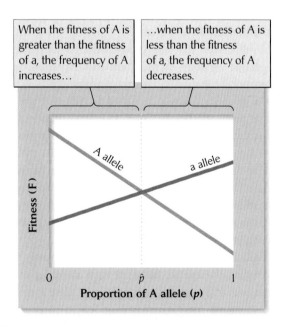

When the fitness of A is greater than the fitness of a, the frequency of A increases…

…when the fitness of A is less than the fitness of a, the frequency of A decreases.

Figure 16.17 Circumstances favoring heterozygotes may give rise to frequency–dependent selection. In these circumstances, the fitness of an allele depends on its frequency in the population.

ECOLOGISTS IN THE FIELD

Selection and change in the frequency of melanistic moths

One of the most striking cases of evolution in action is that of industrial melanism in the peppered moth. Early in the nineteenth century, occasional dark, or **melanistic**, specimens of the common peppered moth (*Biston betularia*) were collected in England. Over the next hundred years, this dark form became increasingly common in forests near heavily industrialized regions, which is why the phenomenon is often referred to as **industrial melanism**. In the absence of factories and other heavy industry, the light, or typical, form of the moth still prevailed. This phenomenon aroused considerable interest among geneticists, who showed by cross-mating light and dark forms that melanism is an inherited trait that is determined by a single dominant gene.

Because the melanistic trait is an inherited characteristic, its spread must reflect genetic changes (evolution) in the population. It seemed reasonable to suppose that natural selection had led to the replacement of typical light individuals with melanistic individuals. That is, where melanistic individuals had become common, the environment must somehow have been altered so as to give dark forms a survival advantage over light forms.

To test this hypothesis, the English biologist H. B. D. Kettlewell first measured the relative fitnesses of the two forms, independently of the fact that the frequency of one had increased over that of the other. To determine whether the melanistic form had greater fitness than the typical peppered moth in areas where melanism occurred, Kettlewell chose the mark-recapture method. He marked adult moths of both forms with a dot of cellulose paint and then released them. The mark was placed on the underside of the wing so that it would not attract the attention of predators to a moth resting on a tree trunk. Kettlewell recaptured moths by attracting them to a mercury vapor lamp in the center of the woods or to caged virgin females at the edge of the woods. (Only males could be used in the study because females are attracted neither to lights nor to virgin females.)

In one experiment, Kettlewell marked and released 201 typicals and 601 melanics in a wooded area near industrial Birmingham. The results were as follows:

	Typicals	Melanics
Number of moths released	201	601
Number of moths recaptured	34	205
Percentage recaptured	16	34

These figures indicated that more of the dark form had survived over the course of the experiment. A similar experiment in a nonindustrial area revealed higher survival by the typical salt-and-pepper form of the moth.

The specific agent of selection was easily identified. Peppered moths inhabit dense woods, where they rest on tree trunks during the day. Kettlewell reasoned that in industrial areas, pollution had darkened the trunks of trees so much that typical moths stood out against them and were readily found by predators (■ Figure 16.18). Any aberrant dark forms were better camouflaged against darkened tree trunks, and their coloration conferred survival value. Eventually, differential survival of dark and light forms would lead to changes in their relative frequencies in a population.

To test this idea, Kettlewell placed equal numbers of light and dark moths on tree trunks in polluted and unpolluted woods and watched them carefully at a distance from behind a blind. (A "blind" is a tentlike structure intended to conceal observers from their subjects; it is more often called a "hide" in England.) He quickly discovered that several species of birds regularly searched tree trunks for moths and other insects, and that these birds more readily found a moth that contrasted with its background than one that resembled the bark it clung to. Kettlewell tabulated the following instances of predation:

	Individuals taken by birds	
	Typicals	Melanics
Unpolluted woods	26	164
Polluted woods	43	15

■ **Figure 16.18 Industrial melanism in the peppered moth demonstrates genetic change in response to selective factors in the environment.** Melanistic (*left*) and typical forms of the peppered moth at rest on a soot-darkened tree trunk. Photo by Stephen Dalton/Photo Researchers.

These data were consistent with the results of the mark-recapture experiments. Together they clearly demonstrated the operation of natural selection, which, over a long period, had resulted in genetic changes in populations of the peppered moth in polluted areas.

One of the most gratifying aspects of the peppered moth story is that, with the advent of smoke control programs and the return of forests to a cleaner state, frequencies of melanistic moths have decreased, as evolutionary theory predicted they should have. In the area around the industrial center of Kirby in northwestern England, for example, the melanistic form decreased from more than 90% of the population to about 30% over a period of 20 years (▌ Figure 16.19).

Ecologists can draw useful conclusions from population genetics studies

Population genetics has a number of important messages for ecologists. First, every population harbors some genetic variation that influences fitness. This means that evolution is potentially a continuing process in all populations. It also means that individual organisms should be expected to have adaptations that help them to reduce the harmful effects of deleterious alleles on themselves and their offspring. Adaptations to ensure outcrossing are one kind of mechanism by which organisms manage the ubiquitous genetic variation in populations.

Second, changes in selective factors in the environment will almost always be met by evolutionary responses that lead to shifts in the frequencies of genotypes within the population. The magnitude of the shift is not always predictable and depends on the particular genetic variation present in the population at a given time. Most continuously variable traits, such as size, have enough genetic variation to respond to selection, but the range and extent of a response may be limited by correlated responses of other traits that have negative fitness consequences. Given enough time, populations may reach some sort of evolutionary optimum and become stabilized, but we have little idea of how much time is required.

Third, rapid environmental changes brought about by human-caused changes in the environment, the introduction of predatory or disease organisms, or the appearance of genetic novelties in those enemies will often exceed the capacity of a population to respond by evolution. In these circumstances, the decline of a population toward extinction is a distinct possibility.

Although much remains to be learned, it is clear that populations have evolutionarily dynamic relationships with their environments, particularly with the biological components of their environments (competitors, predators, and pathogens) that are also evolving in response to other kinds of organisms. The interactions among different species can exert powerful effects on the demography and

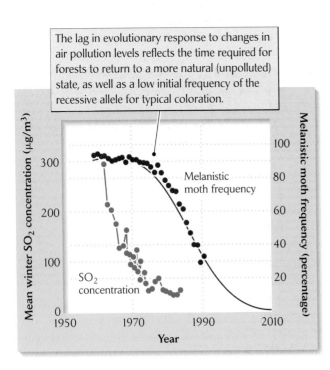

The lag in evolutionary response to changes in air pollution levels reflects the time required for forests to return to a more natural (unpolluted) state, as well as a low initial frequency of the recessive allele for typical coloration.

▌ **Figure 16.19 A change in the environment can result in a change in the frequency of a phenotype.** The frequency of the melanistic form of the peppered moth has decreased since the beginning of pollution control programs in England in the 1950s. The index to pollution is the winter level of sulfur dioxide, which directly affects lichens growing on tree trunks, against which moths rest by day. After C. A. Clarke, G. S. Mani, and G. Wynne, *Biol. J. Linn. Soc.* 26:189–199 (1985); G. S. Mani and M. E. N. Majerus, *Biol. J. Linn. Soc.* 48:157–165 (1993).

genetics of populations, influencing the evolutionary and population dynamics of species and determining whether species can coexist with one another. These interactions are the subject of the next section of this book.

 ## Summary

1. Mutations result from changes in the nucleotide subunits that make up DNA molecules. These changes occur at a very low rate, but they are the ultimate source of all genetic variability. Most mutations are detrimental to the well-being of their carriers.

2. Recombination of the genes inherited from each parent can result in novel combinations of genes for selection to act on.

3. The frequencies of homozygous and heterozygous genotypes in a large population in the absence of selection, mutation, migration, and nonrandom mating can be estimated by the Hardy–Weinberg law. This law states that alleles with frequencies p and q will form homozygous genotypes with frequencies p^2 and q^2, and heterozygotes with frequency $2pq$.

4. Deviations from Hardy–Weinberg equilibrium may be caused by mutation, migration, nonrandom mating, small population size, and selection.

5. In small populations, random variations in reproductive success result in changes in gene frequencies and occasional losses of alleles. This causes a decrease in genetic diversity by a process called genetic drift.

6. Assortative mating changes the frequencies of genotypes within a population. To reduce the consequences of inbreeding on the fitness of offspring, most populations exhibit negative assortative mating, in which mates are genetically unlike each other. Negative assortative mating tends to increase the frequency of heterozygotes in a population.

7. Migration of individuals between populations with different allele frequencies can result in deviations from Hardy–Weinberg equilibrium. The most common effect is fewer heterozygotes in a population than one would expect by chance alone. This is known as the Wahlund effect.

8. Different selective pressures on, or random changes in, different subpopulations result in variation in gene frequencies within a geographic range. This variation is accentuated by population subdivision and barriers to dispersal.

9. Selection may be either stabilizing, in which case intermediate phenotypes in a population are most fit; directional,

in which case one or another extreme phenotype is favored over the most common ones; or disruptive, in which case several extreme phenotypes are favored simultaneously.

10. Even though selection tends to remove genetic variation from a population, variation is maintained by mutation and gene flow from other populations and by varying selective pressures within populations.

11. Under strong selection, allele frequencies may change within a small number of generations. This has occurred in many cases in which humans have drastically altered the environments of populations. The example of melanism in the peppered moth is such a case of evolution in action.

 ## PRACTICING ECOLOGY

CHECK YOUR KNOWLEDGE

Population Genetics and Rates of Evolution

 The mechanics of selection and genetic responses of organisms are part of the branch of ecology known as population genetics. A primary goal of population genetics since the 1920s has been the development of quantitative methods for predicting changes in gene frequencies in response to selection. Such changes are the essence of evolution. Predictions are possible through the use of models for selection on a single gene with one allele dominant over the other. Such equations that make predictions resulting from one generation of selection can be used to show how a population evolves over many generations of continued selection. They can also be used to predict how a population can respond genetically to a change in its environment.

Recall the changes in *Biston betularia* phenotypic frequencies in England that occurred following the onset of the Industrial Revolution. Replacement of the typical peppered moth by the melanistic form in polluted forests is known to have taken about a century. From the results of Kettlewell's experiments, we can estimate that the fitness of the recessive homozygous genotype for typical coloration was 47% of that for the melanistic genotype. Therefore, the fitness differential, or selection strength against the typical form, was 53%. Because melanism is caused by a dominant allele, it is exposed to selection even at low frequency, and the initial frequency of the allele has little effect on the rate of evolutionary change. Using model simulations, the transition to a population that consists mostly of melanistic forms takes about 50 generations. Not surprisingly, when selection is weaker, the transition takes much longer.

Recall that selection on the peppered moth was based on bird predation. Presumably, the birds were better able to see the moths because soot caused the death of lichens that grow on the surface of the trees. This in turn allowed the dark-colored versions of the moth to avoid bird predation. But this reasoning is based on human visual perception. What do the moths look like through a bird's visual perception? It turns out that birds can see ultraviolet (UV) light, so it is important to reassess the story of *Biston betularia* evolution in this context. Majerus et al. (2000) compared the UV reflection of both typical and melanistic forms of *Biston betularia* on different lichens to add new details to this story.

CHECK YOUR KNOWLEDGE

1. Why is it important to continually challenge paradigms of ecology and evolution?

2. Can you think of any other populations that have had to adapt to changes in their environment caused by humans?

3. Why do scientists need to know the rate of evolution of a species?

MORE ON THE WEB
4. Visit the Web site *Lichens as Bioindicators* through *Practicing Ecology on the Web* at *http://www.whfreeman.com*. What role have lichens historically played in air pollution studies?

 Suggested Readings

Berry, R. J. 1990. Industrial melanism and peppered moths (*Biston betularia* L.). *Biological Journal of the Linnean Society* 39:301–322.

Charlesworth, D., and B. Charlesworth. 1987. Inbreeding depression and its evolutionary consequences. *Annual Review of Ecology and Systematics* 18:237–268.

Cook, L. M., C. S. Mani, and M. E. Varley. 1986. Postindustrial melanism in the peppered moth. *Science* 231:611–613.

Endler, J. A. 1986. *Natural Selection in the Wild.* Princeton University Press, Princeton, NJ.

Falconer, D. S. 1989. *Introduction to Quantitative Genetics.* 3d ed. Longman, Harlow, England.

Ford, E. B. 1975. *Ecological Genetics.* 4th ed. Chapman and Hall, London; Wiley, New York.

Gould, F. 1991. The evolutionary potential of crop pests. *American Scientist* 79:496–507.

Grant, P. R. 1991. Natural selection and Darwin's finches. *Scientific American* 265:82–87.

Hartl, D. L. 1988. *A Primer of Population Genetics.* 2d ed. Sinauer Associates, Sunderland, MA.

Kettlewell, H. B. D. 1959. Darwin's missing evidence. *Scientific American* 200:48–53.

Majerus, M. E. N., C. F. A. Brunton, and J. Stalker. 2000. A bird's eye view of the peppered moth. *Journal of Evolutionary Biology* 13:155–159.

Maynard Smith, J. 1989. *Evolutionary Genetics.* Oxford University Press, Oxford.

Merola, M. 1994. A reassessment of homozygosity and the case for inbreeding depression in the cheetah, *Acinonyx jubatus:* Implications for conservation. *Conservation Biology* 8:961–971.

O'Brien, S. J., M. E. Roelke, L. Marker, A. Newman, C. A. Winkler, D. Meltzer, L. Colly, J. F. Evermann, M. Bush, and D. E. Wildt. 1985. Genetic basis for species vulnerability in the cheetah. *Science* 227:1428–1434.

Peres, C. A., J. L. Patton, and M. N. F. da Silva. 1996. Riverine barriers and gene flow in Amazonian saddle-back tamarins. *Folia Primatologia* 67:113–124.

Price, M. V., and N. M. Wasser. 1979. Pollen dispersal and optimal outcrossing in *Delphinium nelsoni. Nature* 277:294–297.

Ralls, K., J. D. Ballou, and A. Templeton. 1988. Estimates of the cost of inbreeding in mammals. *Conservation Biology* 2:185–193.

Schemske, D. W. 1984. Population structure and local selection in *Impatiens pallida* (Balsaminaceae), a selfing annual. *Evolution* 38:817–832.

Stephenson, A. G. 1981. Flower and fruit abortion: Proximate causes and ultimate functions. *Annual Review of Ecology and Systematics* 12:253–279.

Predation and Herbivory

Predators have adaptations for exploiting their prey

Prey have adaptations for escaping their predators

Parasites have adaptations to ensure their dispersal between hosts

Parasite–host systems feature adaptations for virulence and resistance

Plants have structural and chemical adaptations for defense against herbivores

Herbivores effectively control some plant populations

When prickly pear cactus (*Opuntia*) was introduced into Australia as an ornamental and as living fences for pastures, it spread rapidly over the island continent, covering thousands of acres of valuable pasture and rangeland. After several unsuccessful attempts to eradicate the plant, the cactus moth (*Cactoblastis cactorum*) was introduced from South America. The caterpillar of the moth feeds on growing shoots of the prickly pear and quickly destroys the plant—literally nipping it in the bud, and inoculating it with various pathogens and rot-causing organisms. After they became established in Australia, cactus moths exerted such effective control that within a few years, prickly pear had become a pest of the past (❙ Figure 17.1).

The cactus moth has not eradicated the prickly pear, however, because the cactus manages to disperse to predator-free areas, thereby keeping one jump ahead of the moth and maintaining a low-level equilibrium in a continually shifting mosaic of isolated patches. Indeed, one would probably not guess that the cactus moth keeps the prickly pear at its present low population levels; the moths are scarce in the remaining stands of cactus in Australia today. (The same moth probably controls prickly pear populations in some areas of its native range in Central and South America, but its decisive role might have gone unnoticed if the appropriate experiment had not been performed in Australia.)

The cactus–cactus moth example shows the potentially strong influence of consumers on prey populations. The balance between the two depends in part on the adaptations evolved by both predator and prey over generations of interaction. In this chapter, we shall explore the many kinds of predator–prey, parasite–host, and herbivore–plant interactions in nature, focusing in particular on the adaptations of each to gain advantage in their relationships.

(a)

(b)

Figure 17.1 The prickly pear cactus population is controlled by its predator, the cactus moth. Photographs of a pasture in Queensland, Australia, (a) 2 months before and (b) 3 years after the introduction of the cactus moth to control the prickly pear cactus. Main photos from A. P. Dodd, in A. Keast, R. L. Crocker, and C. S. Christian (eds.), *Biogeography and Ecology in Australia*, W. Junk, The Hague (1959), courtesy of W. H. Haseler, Department of Lands, Queensland, Australia. Inset photos by D. Habeck and F. Bennet, University of Florida.

All life forms are both **consumers** and victims of consumers. Predation, herbivory, parasitism, and other kinds of consumption are the most fundamental interactions in nature because everything must eat, and most organisms risk being eaten. Predator–prey, herbivore–plant, and parasite–host relationships are all examples of **consumer-resource interactions,** which organize biological communities into series of **consumer chains.** We have seen these in Chapter 6 as food chains. It is typical of consumer-resource interactions that consumers benefit, and their population sizes may increase, while resource populations are decreased. Thus, although energy and nutrients move up a consumer chain, populations are controlled both from below by resources and from above by consumers. Similarly, natural selection exerts its influence from both directions.

Consumers go by many names. The most familiar of these are predator, parasite, parasitoid, herbivore, and detritivore. From the standpoint of population interactions, some of these distinctions are useful, whereas others are confusing. Let's start with **predator.** The images of an owl eating a mouse and a spider eating a fly capture the essentials of predation. Predators catch individuals and consume them, thereby removing them from the prey population.

In contrast, a **parasite** consumes parts of a living prey organism, or **host.** Some parasites attach themselves to, or invade, the bodies of their hosts and feed on their tissues, blood, or on partially digested food in their intestines. Although parasitism may increase the probability of a host's dying from other causes or reduce its fecundity, a parasite does not by itself remove an individual from a resource population.

Parasitoid is the term applied to species of wasps and flies whose larvae consume the tissues of living hosts—usually the eggs, larvae, and pupae of other insects. This strategy inevitably leads to the host's death, but not until the parasitoid larvae have pupated (**Figure 17.2**). Parasitoids resemble both parasites, because they reside within and eat the tissues of a living host, and predators, because they inevitably kill their hosts.

Herbivores eat whole plants or parts of plants. From the standpoint of consumer–resource relations, herbivores function as predators when they consume whole plants, and as parasites when they consume living tissues but do not kill their victims. Depending on what parts of a plant they eat, herbivores act as either predators or parasites. Thus, a deer browsing on a few leaves and stems functions as a parasite. A sheep that consumes an entire plant, pulling it up by the roots and macerating it into lifeless shreds, behaves as a predator. Consumption of a portion of a plant's tissues is referred to as **grazing** (generally applied to grasses and other herbaceous vegetation, and to algae) or **browsing** (applied to woody vegetation).

Detritivores consume dead organic material—leaf litter, feces, carcasses—and have no direct effect on the populations that produce those resources. Detritivores live off

Figure 17.2 Parasitoid wasps develop inside the larvae or pupae of other insects. Photo by Scott Bauer.

the waste of other species. As a consequence, they do not affect the abundance of their food supplies, and their activities consequently do not influence the evolution of the living sources of their food. Detritivores are important in the recycling of nutrients within an ecosystem. However, because detritivore populations are not dynamically coupled to their resource populations, they will not be considered further in this chapter.

Predators have adaptations for exploiting their prey

When we think of predator and prey, we usually think of lynx and hare, or bird and beetle—predators that pursue, capture, and eat individual prey. Though smaller than their predators, prey like hares and beetles are large enough to be worth pursuing individually. Other organisms consume minute prey in vast numbers, and they are also predators. Blue whales weigh many tons, but live on tiny shrimplike krill, fish fry, and the like, which they filter from the seawater.

As the size of prey increases in relation to that of the predator, prey become more difficult to capture, and predators become specialized for pursuing and subduing them (■ Figure 17.3). Beyond a certain size ratio, however, predators lack sufficient strength and swiftness to capture potential prey items. An individual lion will attack an animal its own size or a little larger, but it is no match for a fully grown elephant. A few species, including lions, wolves, hyenas, and army ants, hunt cooperatively and in this way can run down and subdue prey that are substantially larger than themselves.

A predator's form and function are closely tied to its diet. Seemingly simple differences in tooth structure, for example, indicate important ecological differences (■ Figure 17.4). Herbivores tend to have teeth with large grinding surfaces to macerate tough plant tissues. The upper and lower incisors of horses are strongly opposed so that they can cut the fibrous stems of grasses. Other ungulates, such

Figure 17.3 African lions are specialized for pursuing large prey. With their powerful legs and jaws, lions can subdue prey somewhat larger than themselves. But because they cannot maintain speed over long distances, successful hunting relies on stealth and surprise. Photo by Michael Fairchild/Peter Arnold.

(a) Horse   (b) Deer   (c) Wolf

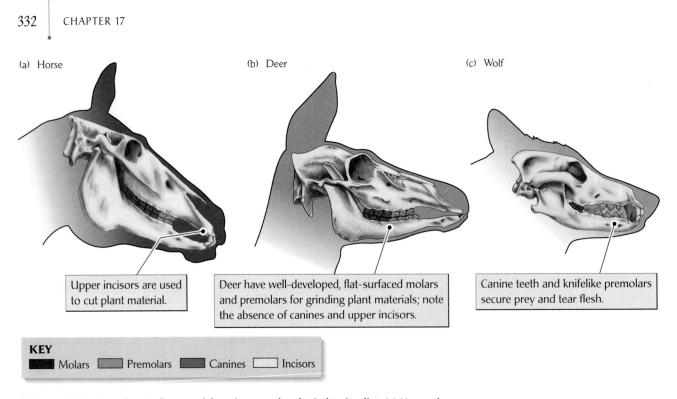

Upper incisors are used to cut plant material.

Deer have well-developed, flat-surfaced molars and premolars for grinding plant materials; note the absence of canines and upper incisors.

Canine teeth and knifelike premolars secure prey and tear flesh.

KEY
■ Molars   ■ Premolars   ■ Canines   □ Incisors

▌**Figure 17.4 A predator's form and function are closely tied to its diet.** (a) Horses have grinding molars but, unlike deer, retain their upper incisors, which are used to cut plant material. (b) Deer have well-developed, flat-surfaced molars and premolars for grinding plant materials; note the absence of canines and upper incisors. The lower incisors are used to secure vegetation against the upper jaw; the deer then rips the leaves from the plant. (c) The wolf has daggerlike canine teeth and knifelike premolars for securing prey and tearing flesh.

as cows, sheep, and deer, lack upper incisors; their lower teeth press against the upper jaw at an angle for gripping and pulling plant material. The teeth of **carnivores** (meat-eating predators) have cutting and biting surfaces that both immobilize prey in the mouth and cut food items into pieces small enough to swallow and digest.

Many predators use their forelegs to help them tear their food into small morsels. Birds such as hawks, eagles, owls, and parrots use their powerful sharp-clawed feet and hooked beaks for this purpose. Diving birds often eat large fish, but they must swallow them whole because their hind legs are specialized for swimming and diving rather than for grasping and dismantling prey. Some species of snakes compensate for their lack of grasping appendages with distensible jaws that enable them to swallow large prey whole (▌Figure 17.5).

The quality of the diet influences the adaptations of a predator's digestive and excretory systems as well. Plants contain long, fibrous molecules, such as cellulose and lignin, that form supportive structures in stems and leaves. These components make vegetation more difficult to digest than the high-protein diets of carnivores. Consequently, the

digestive tracts of many herbivorous animals, such as rabbits and cows, are often greatly elongated. Many herbivores have enlarged, saclike regions of the foregut or hindgut that function like fermentation vats, housing bacteria and pro-

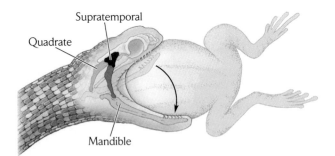

Supratemporal

Quadrate

Mandible

▌**Figure 17.5 The jaws of snakes are adapted for grasping and swallowing large prey.** Some species of snakes have enlarged their gape by as much as 20% through an adaptation that shifts the articulation of the jaw with the skull from the quadrate bone to the supratemporal. The figure shows the position of the jaw elements when the mouth is closed and open. After C. Gans, *Biomechanics*, Lippincott, Philadelphia (1974).

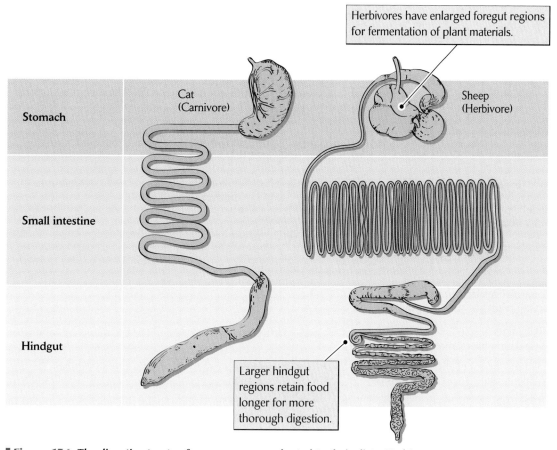

Herbivores have enlarged foregut regions for fermentation of plant materials.

Larger hindgut regions retain food longer for more thorough digestion.

▌Figure 17.6 The digestive tracts of consumers are adapted to their diets. Herbivores tend to have much larger digestive organs than carnivores because of the resistance of plant foods to digestion. After C. E. Stevens, *Comparative Physiology of the Vertebrate Digestive System*, Cambridge University Press, Cambridge (1988).

tozoans that aid digestion (▌Figure 17.6). With these adaptations, an herbivore can keep meals in the digestive tract longer and digest them more thoroughly. However, such herbivores must carry quantities of undigested food in their bellies, adding weight and reducing their mobility. This is one reason that few species of birds make use of fermentative digestion.

ECOLOGISTS IN THE FIELD

The relative sizes of mammalian predators and their prey

Chris Carbone, of the Zoological Society of London, and several colleagues noticed that among mammalian predators, the size of the prey relative to that of the predator depended very strongly on the body size of the predator. Predators weighing less than about 20 kilograms

tend to eat prey less than half their body mass, and many species subsist on very tiny prey, specializing, for example, on termites or millipedes. Above 20 kg, most species of predators consume prey—mostly other mammals—of close to their own body mass. Why the switch? The investigators suspected that larger predators have to pursue large prey because they cannot sustain themselves on small prey items. While small prey tend to be more abundant, they don't provide as much energy per prey item. Also, large predators need to consume proportionately more food than small predators, and there must seemingly be a limit to how rapidly a large predator can gather small prey items.

Carbone's research group used a modeling approach to test this idea. Starting with information on the prey capture rates of small predators, they calculated how rapidly hunting speed, food consumption, and food requirements increased with body mass, and estimated the hunting time per day required to satisfy food requirements as a function of predator

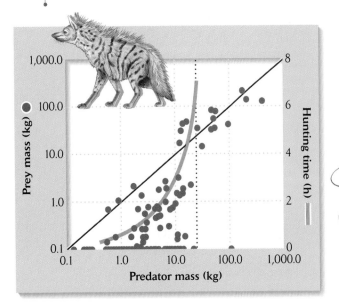

Figure 17.7 Large carnivores tend to pursue large prey. The sizes of prey consumed by a variety of mammalian carnivores are shown in relation to carnivore body size. Superimposed on these data is the estimated hunting time required for a carnivore feeding on small prey items to satisfy its food requirements. Above a predator mass of 10–20 kilograms, time–energy constraints permit only hunting of large prey. After C. Carbone, G. M. Mace, S. C. Roberts, and D. W. Macdonald, *Nature* 402:286–288 (1999).

body mass (**Figure 17.7). This simple exercise indicated that when predator body mass exceeded about 10 to 20 kilograms, daily feeding time exceeded the time available; larger predators could not sustain themselves feeding on small prey. Of course, other factors may play a role in prey choice, but we can see that time and energy considerations may constrain prey choice and force larger predators to pursue rarer, but more profitable, large prey items.

Prey have adaptations for escaping their predators

The ways in which prey organisms avoid being eaten are as diverse as the hunting tactics of their predators. Hiding, escape, and active defense can all be effective, depending on the particular circumstances of a predator–prey relationship. Grasslands offer few hiding places for deer, antelope, and other grazers, so their escape depends on early detection of predators and swift movement. Plants cannot flee like animals, but many produce thorns and defensive

chemicals that dissuade herbivores. Where animals are able to hide or seek refuge in safer microhabitats, they are often very sensitive to the presence of predators and adjust their behavior accordingly. For example, small fish living in ponds with larger predatory fish will avoid the best feeding areas in open water and spend most of their time lurking in safer weed beds close to the water's edge.

ECOLOGISTS IN THE FIELD

Predator avoidance and growth performance in frog larvae

Staying out of harm's way can depress growth rates when predator avoidance limits prey individuals to poor feeding areas. The effects of predation risk on the growth of tadpoles have been demonstrated in laboratory and field experiments on bullfrogs (*Rana catesbiana*) by Rick Relyea and Earl Werner at the University of Michigan. They conducted laboratory experiments in which newly hatched tadpoles were placed in aquaria with caged dragonfly larvae or fish. The tadpoles reduced their activity in the presence of the predators, especially the dragonfly larvae, and also avoided the side of the aquarium where the caged predator was located (**Figure 17.8). Similar experiments within enclosures set in a natural pond further demonstrated that the presence of caged dragonfly larvae reduced growth rates significantly in some species of frogs. Other studies (see Figure 9.19) have shown that the perception of predation risk is widespread in the animal world and has a strong effect on behavior and habitat selection as well as the demography of prey organisms.

Where prey cannot hide, they often adopt protective defenses. These rarely involve physical combat because few types of prey can match their predators, and predators carefully avoid those that can. Instead, many seemingly defenseless organisms produce foul-smelling or stinging chemical secretions to dissuade predators. Whip scorpions and bombardier beetles direct sprays of noxious liquids at threatening animals (**Figure 17.9). Many plants and animals contain chemical substances that make them inedible or poisonous. Slow-moving animals, such as porcupines and armadillos, protect themselves with spines or armored body coverings.

Crypsis and warning coloration

The camouflaged appearances and resting positions by which some prey organisms avoid detection represent

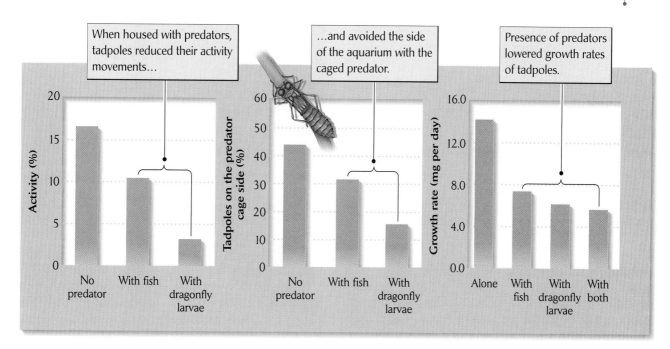

After R. A. Relyea and E. E. Werner, *Ecology* 80:2117–2124 (1999).

Figure 17.8 Avoiding predators may result in reduced growth rates. The graphs show the activity levels of bullfrog tadpoles (*left*), the number of tadpoles found on the side of the aquarium where the predator cage was located (*center*), and the growth rate of tadpoles (*right*) in relation to the presence of two types of predators: fish and dragonfly larvae. The dragonfly larvae, in particular, led to reduced activity, avoidance of areas near predators, and reduced growth rates.

Figure 17.9 Many organisms have evolved chemical defenses to ward off predators. A bombardier beetle sprays a noxious liquid at the temperature of boiling water toward a predator. Courtesy of Thomas Eisner, Cornell University.

another type of defense against predators. Such adaptations testify to the force and pervasiveness of predators as a force of natural selection: those prey individuals that do not hide effectively are discovered and eaten.

Many organisms achieve **crypsis,** or blending in with their backgrounds, by matching the color and pattern of bark, twigs, or leaves (**Figure 17.10**). Various animals resemble sticks, leaves, flower parts, or even bird droppings. These organisms are not so much concealed as they are mistaken for inedible objects and passed over. The stick-mimicking phasmids (stick insects) and leaf-mimicking katydids often conceal their legs in resting positions either by folding them back upon their bodies or by protruding them in an apparently stiff, unnatural fashion. The behavior of cryptic organisms must correspond to their appearance. A leaf-mimicking insect resting on bark, or a stick insect moving rapidly along a branch, would not be likely to fool many predators.

Crypsis is a strategy of palatable, or edible, animals. Others take a bolder approach to antipredator defense: they produce noxious chemicals or accumulate them from food

(a)

(b)

(c)

▮ **Figure 17.10 Many palatable organisms have evolved cryptic appearances to avoid detection by predators.** (a) A mantid; (b) a stick insect; (c) a lantern fly. Photos by R. E. Ricklefs.

plants, and they advertise the fact with conspicuous color patterns in the form of **warning coloration,** or **aposematism** (▮Figure 17.11). Predators learn quickly to avoid markings such as the black and orange stripes of the monarch butterfly, which tastes so bitter that a single experience with this prey is remembered for a long time. It is not a coincidence that many noxious forms adopt similar patterns: black and either red or yellow stripes adorn such diverse animals as yellow-jacket wasps and coral snakes. These color combinations so consistently advertise noxiousness that some predators have evolved innate aversions to such patterns and need not learn to avoid such prey by experience.

Why aren't all potential prey species noxious or unpalatable? Part of the answer is that chemical defenses use a large portion of an individual's energy or nutrients that might otherwise be allocated to growth or reproduction. Furthermore, many noxious organisms rely on their food plants to supply toxic organic compounds that they cannot manufacture themselves, and not all food plants have such compounds. When they do, the consumers must themselves avoid the toxic effects of the chemicals in order to use them effectively against their potential predators.

Batesian mimicry

Unpalatable animals and plants that display warning coloration often serve as models for mimicry by palatable ones, which evolve to resemble the noxious organisms. These relationships are collectively referred to as **Batesian mimicry,** which was named after its discoverer, the nineteenth-century English naturalist Henry Bates. In his journeys to the Amazon region of South America, Bates found numerous cases of palatable insects that had forsaken the cryptic patterns of their close relatives and had come to resemble brightly colored, unpalatable species (▮Figure 17.12).

Experimental studies have demonstrated convincingly that mimicry does confer an advantage on mimics. For example, toads that were fed live bees, and were stung on the tongue, thereafter avoided palatable drone flies, which mimic bees. But when naive toads were fed only dead bees from which the stings had been removed, they relished the drone fly mimics (as well as the now harmless bees). Thus, toads learned to associate the conspicuous and distinctive color patterns of live bees with an unpleasant experience.

(a)

(b)

(c)

Figure 17.11 Many unpalatable organisms have evolved warning coloration. Some unpalatable insects also aggregate to enhance the warning signal. Photo (a) by J. Burgett, photos (b) and (c) by Carl C. Hansen, courtesy of the Smithsonian Tropical Research Institute.

Müllerian mimicry

Another type of mimicry, called **Müllerian mimicry** after its discoverer, the nineteenth-century German zoologist Fritz Müller, occurs among unpalatable species that come to resemble one another. Many species form Müllerian mimicry complexes in which each participant is both model and mimic. When a single pattern of warning coloration is adopted by several unpalatable species, avoidance learning by predators is made more efficient because a predator's bad experience with one species confers protection on all the other members of the

Figure 17.12 Batesian mimics are palatable prey organisms that resemble noxious ones. Here, a harmless, palatable mantid (*center*) and moth (*right*) have both evolved to resemble a wasp (*left*). Photos by Larry Jon Friesen/Saturdaze.

Highly unpalatable

Moderately unpalatable

❙ Figure 17.13 Müllerian mimics are unpalatable organisms that share a pattern of warning coloration. Several groups of Costa Rican butterflies and moths form Müllerian mimicry complexes. Each of these "moderately unpalatable" butterflies is a Müllerian mimic of the "highly unpalatable" butterfly above. These unpalatable insects are also models for palatable Batesian mimics.

mimicry complex. For example, most of the bumblebees and wasps that co-occur in Rocky Mountain meadows share a pattern of black and yellow stripes. In the Tropics, dozens of species of unpalatable butterflies, many of them distantly related, share patterns of black and orange "tiger stripes" or black, red, and yellow coloration patterns (❙ Figure 17.13).

Parasites have adaptations to ensure their dispersal between hosts

Parasites are usually much smaller than their prey, or **hosts,** and live either on their surfaces (ticks, lice, and mites) or inside their bodies (viruses, bacteria, protozoans, various roundworms, flukes, tapeworms, and arthropods). Both types demonstrate characteristic adaptations to their way of life. Parasites that live inside, or in close association with, a larger organism enjoy a benign physical environment regulated by their host. Tapeworms, for example, are bathed in a predigested food supply and retain for themselves little more than a highly developed capacity to produce eggs.

In spite of the advantages of a comfortable environment and a ready supply of nourishment close at hand, the life of a parasite is not easy. Host organisms have a variety of mechanisms to recognize invaders and destroy them. Moreover, parasites must disperse through a hostile environment to get from one host to another. Many ac-

complish this via complicated life cycles, one or more stages of which can cope with the external environment.

The life cycle of the protozoan parasite *Plasmodium,* which causes malaria, shows how complex such cycles can be. It involves two hosts, one a mosquito and the other a human or some other mammal, bird, or reptile. The sexual phase of the *Plasmodium* life cycle takes place within the mosquito (❙ Figure 17.14). When an infected mosquito bites a human, mobile cells called sporozoites are injected into the bloodstream with the mosquito's saliva. The sporozoites at first proliferate by mitosis in liver cells, after which they enter red blood cells (erythrocytes) as merozoites, where they feed on hemoglobin and grow. When a merozoite becomes large enough, it undergoes a series of divisions (asexual reproduction), and the daughter merozoites break out of the red blood cell. Each merozoite can enter a new red blood cell, grow, and repeat the cycle, which takes about 48 hours. (When the infection has built up to a high level, the emergence of daughter cells corresponds to periods of high fever.) After several of these cycles, some of the merozoites that enter red blood cells change into sexual forms. If these are swallowed by a mosquito along with a meal of blood, the sexual cells are transformed into eggs and sperm, and fertilization (sexual reproduction) takes place. Fertilized eggs penetrate the mosquito's gut wall and then undergo a series of divisions to produce sporozoites. These work their way into the salivary glands of the mosquito, from which they may enter a new vertebrate host.

Parasite-host systems feature adaptations for virulence and resistance

The complex life histories of parasites involve a variety of interactions with hosts, and different sets of factors affect each stage of the parasite life cycle. The balance between parasite and host populations is influenced by the virulence of the parasite and by the immune response and other defenses of the host.

The immune systems of vertebrates react to the presence of invaders by inflammation or by producing antibodies. The antibodies recognize and bind to foreign proteins, such as those on the outer surfaces of bacteria and protozoans, targeting them for attack by macrophage cells. The disabled parasites are then transported to the spleen and cleared from the body.

An immune response takes time to develop, however, which gives the parasite a chance to develop and multiply within a host. Parasites also have ways of circumventing the host's immune mechanisms. Some parasites produce

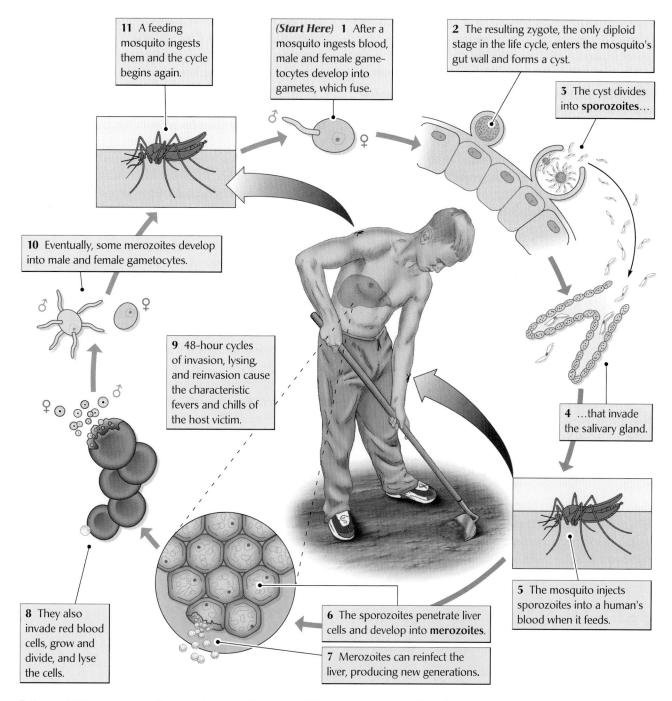

11 A feeding mosquito ingests them and the cycle begins again.

(Start Here) 1 After a mosquito ingests blood, male and female gametocytes develop into gametes, which fuse.

2 The resulting zygote, the only diploid stage in the life cycle, enters the mosquito's gut wall and forms a cyst.

3 The cyst divides into **sporozoites**...

10 Eventually, some merozoites develop into male and female gametocytes.

9 48-hour cycles of invasion, lysing, and reinvasion cause the characteristic fevers and chills of the host victim.

4 ...that invade the salivary gland.

8 They also invade red blood cells, grow and divide, and lyse the cells.

5 The mosquito injects sporozoites into a human's blood when it feeds.

6 The sporozoites penetrate liver cells and develop into **merozoites**.

7 Merozoites can reinfect the liver, producing new generations.

Figure 17.14 Many parasites have complex life cycles. Different stages in the life cycle of the malarial parasite *Plasmodium* are adapted to life in two different hosts and to dispersal between hosts. After R. Buchsbaum, *Animals without Backbones*, 2d ed., University of Chicago Press, Chicago (1948); M. Sleigh, *The Biology of Protozoa*, American Elsevier, New York (1973).

chemical factors that suppress the immune systems of their hosts; this is the most troublesome feature of the AIDS virus. Others have surface proteins that mimic the host's own proteins and thus escape notice by the immune system. Trypanosomes, which are flagellate protozoans that cause sleeping sickness in humans, escape the immune sys-

tem by continually coating their surfaces with novel proteins produced by gene rearrangements.

Some schistosomes (a trematode worm, or blood fluke) are known to excite an immune response when they enter the host, but do not succumb to antibody attack because they coat themselves with proteins of the host before antibodies become numerous. As a consequence, other schistosomes that subsequently infect that host face a barrage of antibodies stimulated by the earlier entrance of the now entrenched individuals. When this response affects closely related species of parasites, it is known as **cross-resistance.** For example, humans in many tropical regions are infected by *Schistosoma*. One extremely virulent species of *Schistosoma* is found only in humans, and causes a debilitating disease called schistosomiasis or bilharziasis. But when a person has been infected previously by other schistosome organisms from wild game or domestic livestock, some of which have little effect on humans, the effect of the parasite is moderated considerably.

Antibodies can persist long after an infection has been controlled, thereby reducing the probability of subsequent infection. On a population level, a serious outbreak of a viral or bacterial disease is often followed by a period during which most of the individuals in the host population have achieved some degree of immunity to reinfection. Until this immunity is lost, or until susceptible individuals are recruited into the population, parasites may be unable to spread (see Figure 1.15).

Plants have structural and chemical adaptations for defense against herbivores

The conflict between herbivore and plant resembles that between parasite and host in that both are waged primarily on biochemical battlegrounds. Plant defenses against herbivores include the inherently low nutritional value of most plant tissues as well as toxic compounds that plants produce and sequester for defense. Structural defenses, such as spines, hairs, tough seed coats, and sticky gums and resins, are important as well (Figure 17.15).

The nutritional quality and digestibility of algal and plant foods is critical to herbivores. Because young animals require a lot of protein for growth, the reproductive success of grazing and browsing mammals depends on the protein content of their food. Herbivores usually select plant food according to its nutrient content. Young leaves and flowers are often preferred over mature leaves because of their low cellulose content; fruits and seeds are particularly nutritious compared with leaves, stems, and buds because of their higher nitrogen, fat, and sugar contents. Many plants use chemicals to reduce the availability of their proteins to herbivores. Oaks and other plants sequester **tannins** in vacuoles in their leaves. These compounds bind to proteins and inhibit their digestion. As a consequence, tannins considerably slow the growth of caterpillars and other herbivores. With the buildup of tannins in oak leaves over the summer, fewer leaves are attacked by herbivores. Insects that feed on tannin-rich plants can reduce the inhibitory effects of tannins by producing detergent-like surfactants in their gut fluids, which tend to disperse tannin–protein complexes.

Whereas tannins exhibit a generalized reaction with proteins of all types, many so-called **secondary compounds** of plants (that is, compounds used not for metabolism, but for other purposes, chiefly defense) interfere with specific metabolic pathways or physiological processes of herbivores. Herbivores may counter their toxic effects by modifying their own physiology and biochemistry.

(a)

(b)

 Figure 17.15 Spines protect the stems and leaves of many plants from herbivores. (a) Cholla cactus and (b) prickly pear cactus (both in the genus *Opuntia*) from Arizona. Photos by R. E. Ricklefs.

Consider the chemical give-and-take between larvae of bruchid beetles and the seeds of legumes (pea family) that they consume. Adult bruchids lay their eggs on developing seed pods. The larvae then hatch and burrow into the seeds, which they consume as they grow. Most legume seeds contain substances that inhibit the proteolytic enzymes produced in an herbivore's digestive organs. Although these toxins provide an effective biochemical defense against most insects, many bruchid beetles have metabolic pathways that either bypass them or are insensitive to them. Among legume species, however, soybeans stand out as being resistant to attack even by most bruchid species. When bruchids lay their eggs on soybeans, the larvae die soon after burrowing beneath the seed coat. Chemicals isolated from soybeans have been shown to inhibit the development of bruchid larvae in experimental situations.

Seeds of the tree-sized tropical legume *Dioclea megacarpa* contain a nonprotein amino acid called L-canavanine that is toxic to most insects. It interferes with the incorporation into proteins of the amino acid arginine, which it closely resembles. One species of bruchid, *Caryedes brasiliensis,* possesses enzymes that discriminate between L-canavanine and arginine during protein formation, as well as enzymes that degrade L-canavanine to forms that can be used as a source of nitrogen. Thus, it seems that for every defense, a new counterattack can be devised. Most plants produce toxic defensive compounds. Many of these, such as pyrethrin, are important sources of pesticides (which is how plants use them, of course). Others, like digitalis, have found use as drugs (some of their pharmacological effects are beneficial in small doses). Secondary plant compounds can be divided into three major classes based on their chemical structure: nitrogen compounds ultimately derived from amino acids, terpenoids, and phenolics. Among the nitrogen compounds are lignin; alkaloids such as morphine (derived from poppies), atropine, and nicotine (from various members of the tomato family); nonprotein amino acids such as L-canavanine; and cyanogenic glycosides, which produce cyanide (HCN). **Terpenoids** include essential oils, latex, and plant resins. Among the **phenolics,** many simple phenols have antimicrobial properties.

Some types of defensive chemicals are maintained at high levels in plant tissues at all times; these are called **constitutive defenses.** Other plant defenses may be **induced** by herbivore damage in a manner analogous to the way that foreign proteins induce an immune response in vertebrate animals. Toxins increase dramatically in many plants following defoliation by herbivores (or the clipping of leaves by investigators). Wounding may cause the production of toxic, noxious, or nutrition-reducing compounds—in the area of a wound or systemically throughout the plant—that reduce subsequent herbivory. In some cases, these responses may take only minutes or hours; in others, they require a new season of growth. When shoots of aspen, poplar, birch, and alder are heavily browsed by snowshoe hares, shoots produced during the following year have exceptionally high concentrations of terpenes and phenolic resins, which are extremely unpalatable to hares. When researchers applied resins in varying concentrations to shoots of unbrowsed trees, hares avoided shoots containing 80 milligrams or more of resin per gram of dry weight, regardless of the amount of other food available.

Many studies have shown that plant responses to herbivory can substantially reduce subsequent herbivory (▌Figure 17.16). This inducibility suggests that some chemical defenses are too costly to maintain under light grazing pressure. Several studies have shown trade-offs between the production of defensive chemicals and plant growth. In addition, where soils are low in the nutrients required

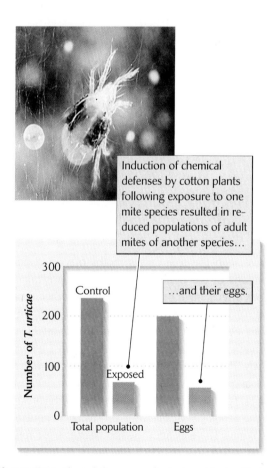

Induction of chemical defenses by cotton plants following exposure to one mite species resulted in reduced populations of adult mites of another species...

...and their eggs.

Control

Exposed

Number of *T. urticae*

300

200

100

0

Total population

Eggs

▌**Figure 17.16 Plant defenses can be induced by herbivory.** Mean numbers of the mite *Tetranychus urticae* were lower on cotton plants that had been previously exposed to a closely related mite species, *T. turkestani,* than on control plants with no previous mite exposure. From R. Karban and J. R. Carey, *Science* 225:53–54 (1984). Photo by J. K. Clark.

(a)

(b)

❚ Figure 17.17 Herbivores can control plant populations.
(a) Klamath weed, or St. John's wort (*Hypericum perforatum*),
became a widespread pest following introduction to the western
United States until it was brought under control by introduced
beetles of the genus *Chrysolina* (b). Klamath weed contains high
concentrations of the alkaloid hypericin, which has therapeutic
effects in small quantities, but is dangerous to cattle and sheep.
Photo (a) by David Sieren/Visuals Unlimited; photo (b) courtesy of
Verein für Naturwissenschaftliche Heimatforschung zu Hamburg.

for the production of defensive chemicals, the costs of
defense are relatively higher. Undoubtedly, the offensive
biochemical tactics of herbivores are also expensive.

Herbivores effectively control some plant populations

We have seen the role cactus moths play in controlling
populations of the prickly pear cactus in Australia. Herbivorous insects have been used in many other situations
to control introduced weeds. Consider the example of
Klamath weed, a European species toxic to livestock, which
accidentally became established in northern California in
the early 1900s (❚ Figure 17.17). By 1944 the weed had
spread over 2 million acres of rangeland in 30 counties.
Biological control specialists borrowed an herbivorous beetle of the genus *Chrysolina* from an Australian control program. Within 10 years after the first beetles were released,
Klamath weed was all but obliterated as a range pest.
Range biologists estimated its abundance to have been reduced by more than 99%.

In grasslands, native herbivores (mostly insects and grazing mammals) typically consume 30–60% of the aboveground vegetation. Their influence on plant production is
revealed by exclosure experiments (❚ Figure 17.18). In one

**❚ Figure 17.18 Herbivory has dramatic effects on plant
production.** The area at the left, on the slope of Mauna Loa in
Hawaii, is protected from cattle grazing by a barbed-wire
fence. Photo by R. E. Ricklefs.

study in California, wire fences were constructed to keep voles out of small areas of grassland. At the end of the 2-year study, the food plants of voles (mostly annual grasses) had grown more and produced more seeds within the fenced plots than outside the exclosures, where voles continued to graze. Perennial grasses and herbs not included in the voles' diet were not directly affected by the exclosures (■ Figure 17.19).

Although herbivores rarely consume more than 10% of forest vegetation, occasional outbreaks of tent caterpillars, gypsy moths, and other insects can completely defoliate or otherwise eradicate entire forests (■ Figure 17.20). In addition, long-term studies of growth and survival of trees after defoliation by insects demonstrate that there may be a considerable lag between an infestation and the expression of its effects.

Grazing and browsing resemble infection by parasites and pathogens in that, like an infected host, the individual grazed-upon plant usually is not killed. Thus plants have time to respond to grazing and browsing by producing defensive chemical compounds. Depending on how long a plant takes to produce these defenses, time lags may be incorporated into grazer–plant systems. As we shall see in the next chapter, time lags may lead to complex dynamics between consumer and resource populations, in some cases even producing regular oscillations in population size.

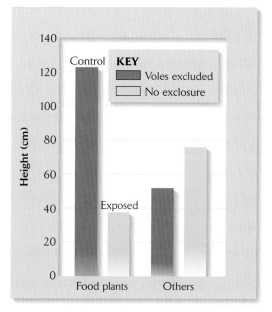

■ **Figure 17.19 The effects of herbivores on plant production can be measured by exclosure experiments.** Relative biomass (summed height per 100 cm²) of food plants and other plants in grassland plots fenced to exclude voles and in unfenced control plots after 2 years. The food plants of voles are mostly annual grasses; nonfood plants include perennial grasses and herbs. After G. O. Batzli and F. A. Pitelka, *Ecology* 51:1027–1039 (1970).

■ **Figure 17.20 Outbreaks of herbivorous insects can defoliate forests.** This forest in Algonquin Provincial Park, Ontario, Canada, has been devastated by an outbreak of the spruce budworm. Photos by R. E. Ricklefs.

Summary

1. Ecologists distinguish three basic kinds of consumers: predators, which remove prey individuals from prey populations as they consume them; parasitoids (mostly small flies and wasps), which kill their hosts, but not immediately; and parasites, which consume portions of a living animal or plant, but do not kill it.

2. Depending on the relative sizes of predators and their prey, predatory behavior may range from the active pursuit of individual prey to the consumption of large numbers of tiny organisms by filtering.

3. Predators are well adapted to pursue, capture, and eat particular types of prey. Carnivorous and herbivorous mammals, for example, differ in the conformation of their teeth and in the size and design of their digestive systems.

4. Organisms avoid predation by avoiding detection; by means of chemical, structural, and behavioral defenses; by hiding in safe refuges; and by escaping. Crypsis and warning coloration are defenses used by palatable and unpalatable organisms respectively.

5. In Batesian mimicry, palatable organisms evolve to resemble unpalatable ones that are rejected by predators, thus gaining some protection by deceiving potential attackers. Müllerian mimicry complexes comprise noxious species that use similar displays to advertise that they are unpalatable.

6. Parasites are characterized by complex life cycles that include stages specialized to make the difficult journey from one host to another.

7. Parasite–host interactions often evolve a delicate balance between the immune response of the host and the parasite's resistance.

8. Plants have evolved numerous structural and chemical defenses to deter herbivores. These defenses include factors that influence the nutritional quality and digestibility of plant parts as well as specialized chemicals—secondary compounds—that have toxic effects on herbivores. Most of these chemicals are nitrogen-containing compounds, terpenoids, or phenolics.

9. Many herbivores have evolved ways to detoxify secondary compounds of plants, enabling them to specialize on plant hosts that are poisonous to most other species.

10. In spite of the defenses of plants, herbivores are able to control the populations of some plants at levels far below the population sizes achieved in the absence of specialized consumers.

PRACTICING ECOLOGY

CHECK YOUR KNOWLEDGE

The Cost of Defense

We have read how consumers and their prey are involved in a constant conflict to eat or avoid being eaten. As a result, traits evolve within consumer populations to capture resources and within prey populations to avoid being eaten. Such traits provide a clear benefit, but they also presumably involve costs in terms of resources or time that could be spent on other behaviors, morphological structures, or biochemical pathways. Although it seems intuitive that limited energy and resources used for one adaptive trait must be taken away from another, it has been difficult to assess experimentally the extent to which this actually occurs.

Ian Baldwin of the Max Planck Institute of Chemical Ecology and his colleagues are engaged in experiments that they hope will provide a more critical examination and quantification of trade-offs. In one experiment, they sought to determine whether nitrogen-based trade-offs among nicotine production, growth, and seed production could be detected in the wild tobacco plant, *Nicotiana attenuate*.

Wild tobacco plants release methyl jasomnate (MeJA) within their leaves in response to herbivory by tobacco hornworms, *Manduca sexta*. The release of this chemical stimulates the production of nicotine which can be a potent anti-herbivory compound. Recall that inducible defenses are produced only in response to herbivore damage, as opposed to constitutive defenses, which are present at all times as a defense chemical in a plant. The beauty of this inducible defense as a system for experimentation is that MeJA can be applied externally to leaves and cause the production of nicotine. Thus, one can study the costs of induced plant defenses in the absence of herbivores.

To determine the cost of producing nicotine to defend these plants in terms of energy or resources, Baldwin and his colleagues used the stable isotope nitrogen-15 (^{15}N) to track nitrogen movement within defense-induced plants. Stable isotopes do not decay as do their radioactive counterparts. They exist in small quantities in comparison to the most common isotopes (in the case of nitrogen, ^{14}N), so they are extremely useful for following the fates of molecules in various pathways, reactions, and even ecosystems. Baldwin and colleagues used this method to determine the differences in the way that nitrogen is allocated to nicotine production, growth, and seed production. Their results quantified the use of nitrogen in the presence and absence of defense mechanisms. However, the results did not indicate whether the use of this defense mechanism lowered the life fitness of the plant.

Figure 18.1 Fur trapping records reveal population cycles. Canadian fur trappers sold pelts to the Hudson's Bay Company in Manitoba, whose records of purchases provided data showing pronounced population cycles of fur-bearing mammals. Courtesy of Hudson's Bay Company Archives, Provincial Archives of Manitoba.

The basic question of population biology is this: What factors influence the size and stability of populations? In Chapter 15, we saw how density-dependent factors and time delays modify the responses of birth and death rates to population density. We shall see in the next chapter that competition for resources from other species can depress the growth rate of a population. Most species are both consumers and resources for other consumers, however, and so it is also important to ask whether populations are limited primarily by what they eat or by what eats them. Studies of predator–prey interactions attempt to answer at least two fundamental questions: First, do predators reduce the size of prey populations substantially below the carrying

capacities set by resources for the prey? Second, do the dynamics of predator–prey interactions cause populations to oscillate? The first question is of great practical concern to those interested in the management of crop pests, game populations, and endangered species. It also has far-reaching implications for our understanding of the interactions among species that share resources and, therefore, for our understanding of the structure of biological communities. The second question is motivated by observations of predator–prey cycles in nature and directly addresses the issue of stability in natural systems. Ecologists have tried to answer these questions with a combination of observation, theory, and experimentation.

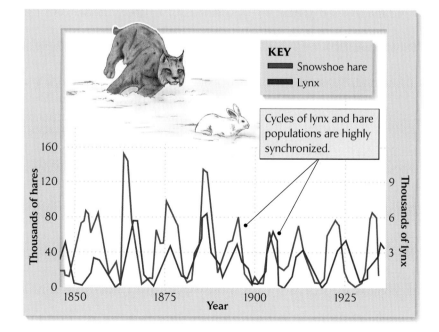

Cycles of lynx and hare populations are highly synchronized.

Figure 18.2 Population cycles of predators and their prey may be highly synchronized. According to the records of the Hudson's Bay Company, population cycles of the lynx and the snowshoe hare track each other closely. After D. A. MacLulich, *University of Toronto Studies,* Biol. Ser. No. 43 (1937).

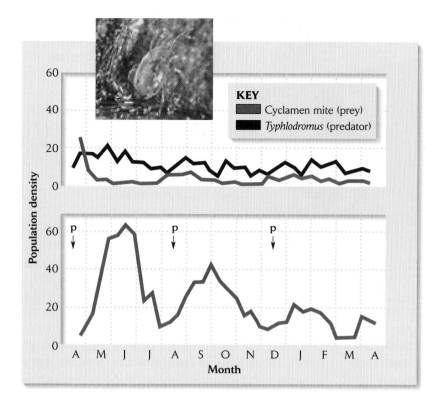

Population density

Figure 18.3 Predators can control prey populations. Infestation of strawberry plots by cyclamen mites (*Tarsonemus pallidus*) in the presence of the predatory mite *Typhlodromus* (*above*) and in its absence (*below*). Prey populations are expressed as numbers of mites per leaf; predator levels are the numbers of leaflets out of 36 on which one or more *Typhlodromus* were found. Parathion treatments are indicated by "**p.**" After C. B. Huffaker and C. E. Kennett, *Hilgardia* 26:191–222 (1956). Photo of *Typhlodromus* courtesy of IPM Program, Cornell University.

Consumers can limit resource populations

The cyclamen mite is a pest of strawberry crops in California. Populations of these mites are usually kept under control by a species of predatory mite of the genus *Typhlodromus*. Cyclamen mites typically invade a strawberry crop shortly after it is planted, but their populations usually do not reach damaging levels until the second year. *Typhlodromus* mites usually invade fields during the second year. Because they are such efficient predators, these mites rapidly reduce cyclamen mite populations and prevent further outbreaks.

Greenhouse experiments have demonstrated the role of predation in keeping cyclamen mites in check. One group of strawberry plants was stocked with both predator and prey mites; a second group was kept predator-free by regular application of parathion, an insecticide that kills the predatory species but does not affect the cyclamen mite. Throughout the study, populations of cyclamen mites remained low in plots they shared with *Typhlodromus*, but their infestation attained damaging proportions on predator-free plants (**Figure 18.3**). In field plantings of strawberries, cyclamen mites also reached damaging levels where predators were eliminated by parathion (a good example of an insecticide having an undesired effect), but

they were effectively controlled in untreated plots. When a cyclamen mite population began to increase in an untreated planting, the predator population quickly mushroomed and reduced the outbreak. On average, cyclamen mites were about 25 times more abundant in the absence of predators than in their presence.

Typhlodromus owes its effectiveness as a predator to several factors in addition to its voracious appetite. Most important, its population can increase as rapidly as that of its prey. Cyclamen mites lay three eggs per day over the 4 or 5 days of their reproductive life span; *Typhlodromus* lay two or three eggs per day for 8–10 days. During winter, when cyclamen mite populations dwindle to a few individuals hidden in crevices and folds of leaves in the crowns of strawberry plants, the predatory mites subsist on honeydew produced by aphids and whiteflies. Whenever predators appear to control prey populations—and *Typhlodromus* is no exception—the predators usually exhibit a high reproductive capacity compared with that of their prey, combined with strong dispersal powers and an ability to switch to alternative food resources when their primary prey are unavailable.

Consumer control is not unique to terrestrial ecosystems—quite the contrary. Experiments exploring the effects of sea urchins on populations of algae have demonstrated consumer control in rocky shore communities. The simplest experiments consisted of removing sea urchins and following the subsequent growth of their algal prey. When urchins are kept out of tide pools and off subtidal rock surfaces, the biomass of algae quickly increases, indicating that herbivory by urchins reduces algal populations below the level that the environment can support. Different kinds of algae also appear after herbivore removal. Large brown algae flourish and begin to replace both coralline algae (whose hard, shell-like structure deters grazers; see Figure 7.8) and small green algae (whose short life cycles and high reproductive rates enable their population growth and re-establishment to keep ahead of grazing pressure by sea urchins). In subtidal plots kept free of predators, brown kelps become established in thick stands that shade out most small species.

Predator and prey populations often increase and decrease in regular cycles

The periods of population cycles vary from species to species, and even within a species. In Canada, most cycles have periods of either 9–10 years (large herbivores, such as snowshoe hares, muskrat, ruffed grouse, and ptarmigan) or 4 years (small herbivores, such as voles and lemmings). Predators that feed on short-cycle herbivores (arctic foxes, rough-legged hawks, snowy owls) themselves have short population cycles; predators of larger herbivores (red foxes, lynx, marten, mink, goshawks, horned owls) have longer cycles. The length of the cycle also appears to be related to habitat: longer cycles are observed in forest-dwelling species and shorter ones in tundra-dwelling species.

The closely synchronized population cycles of some predators and prey suggest that these oscillations could result from the way in which predator and prey populations interact with each other. In simple terms, predators eat prey and reduce their numbers. Consequently, preda-

tors go hungry, and their numbers drop as well. With fewer predators around, the remaining prey survive better, and their populations begin to increase. Of course, with increasing numbers of prey, the predators also begin to increase again, thereby completing the cycle. This sequence results in oscillations in population numbers.

We saw in Chapter 15 that delays in the responses of birth and death rates to changes in the environment can cause population cycles. Most predator–prey interactions also have response lags because of the time required to produce offspring. Population dynamic models predict that the period of a population cycle should be about four to five times the response lag. Thus, the 4-year and 9–10-year population cycles of mammals inhabiting boreal forest and tundra are consistent with time delays of 1 or 2 years, respectively. Such time delays could result from the typical lengths of time between birth and sexual maturity in these mammals. In other words, the influence of conditions in a particular year may not be felt in a predator population until young born in that year are themselves old enough to reproduce.

Time delays in the relationship between pathogens and their hosts can result from the development of immune responses, which remove host individuals from the susceptible population and slow the spread of disease. Measles, which is a highly contagious viral disease that stimulates lifelong immunity, typically produced epidemics at 2-year intervals in pre-vaccine human populations (∎ Figure 18.4). Two years were required for a high enough density of newly susceptible infants to accumulate in a population to sustain a measles outbreak.

Host population density also affects the rate of disease transmission. Pathogens infect individuals more readily in crowded than in sparse populations because the chances of contacting a new host are greater in a dense population.

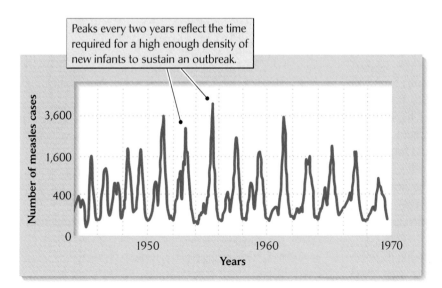

Peaks every two years reflect the time required for a high enough density of new infants to sustain an outbreak.

∎ Figure 18.4 **The development of host immunity influences pathogen population cycles.** Cases of measles reported in London, England, between 1944 and 1968, prior to the availability of a measles vaccine, peaked about every 2 years. After P. Rohani, D. J. D. Earn, and B. T. Grenfell, *Science* 286:968–971 (1999).

Under these conditions, disease rates can be so high that the host population declines to low levels. This breaks the chain of contagion within the population, fewer host individuals become sick, and the host population begins to grow again. This pattern is evident in populations of the forest tent caterpillar, which periodically increase to such high densities that they can defoliate stands of trees over thousands of square kilometers. These infestations are usually brought under control by a virulent disease organism, the nuclear polyhedrosis virus, which causes high mortality of tent caterpillars at high population densities. In many regions, tent caterpillar infestations last about 2 years before the virus brings its host population under control. In other regions, however, infestations may last up to 9 years. Jens Roland, at the University of Alberta, discovered that forest fragmentation, which creates abundant forest edge, tends to prolong outbreaks of the tent caterpillar (▌Figure 18.5). Apparently, increased forest edge exposes caterpillars to more intense sunlight, which inactivates the nuclear polyhedrosis virus. Here is a clear case in which habitat manipulation can have important secondary effects by influencing population cycles.

Long-term observations have confirmed that population fluctuations continue more or less unchanged over many cycles, so this dynamic behavior appears to represent a stable interaction between predators and prey. One of the earliest goals of population biologists was to establish such cycles in experimental populations, for which one could work out the dynamics of the relationship and study potential causes of the cycles.

In laboratory cultures, extremely efficient predators often eat their prey populations to extinction and then become extinct themselves. This hopeless situation can be stabilized, however, if some of the prey can find refuges in which to escape. G. F. Gause demonstrated this principle in one of the earliest experimental studies on predator–prey systems. In one experiment, Gause introduced *Paramecium* as the prey and another ciliated protozoan, *Didinium*, as the predator into a nutritive medium in a plain test tube. By creating such a simple environment, Gause had "stacked the deck" against the prey; the predators readily found all of them, and when the last *Paramecium* had been consumed, the predators starved. In a second experiment, Gause added some structure to the environment by placing glass wool, in which the *Paramecium* could find refuge from the predators, at the bottom of the test tube. The tables thus having been turned, the *Didinium* population starved after consuming all readily available prey, but the *Paramecium* population was restored by individuals concealed in the glass wool.

Gause finally achieved recurring oscillations in predator and prey populations by periodically adding small numbers of predators—restocking the pond, so to speak. Repeated addition of individuals to an experimental culture corresponds to natural repopulation by colonists from other areas in a locality where extinction of either predator or prey has occurred. This pattern is reminiscent of the interaction between the cactus moth and the prickly pear, in which the cactus escapes complete annihilation by dispersing to predator-free areas.

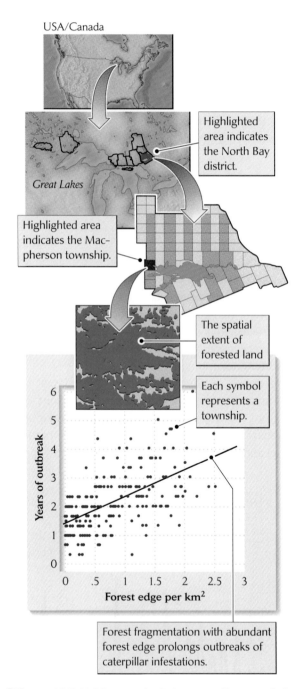

▌Figure 18.5 Habitat manipulation can affect population cycles. The length of infestations of forest tent caterpillars in eastern Canada increases as a function of forest fragmentation. After J. Roland, *Oecologia* 93:25–30 (1993).

ECOLOGISTS IN THE FIELD

Huffaker's experiments on mite populations

C. B. Huffaker, a biologist at the University of California at Berkeley who pioneered the biological control of crop pests, attempted to produce an environment in the laboratory that would allow predator and prey to persist without the restocking of either population. In this study, the six-spotted mite, *Eotetranychus sexmaculatus*, was the prey; another mite, *Typhlodromus occidentalis*, was the predator. Oranges provided the prey's food. Huffaker established experimental populations on trays within which he could vary the number, exposed surface area, and dispersion of the oranges (❚ Figure 18.6).

Each tray had 40 positions arranged in 4 rows of 10 each. Where Huffaker did not place oranges, he substituted rubber balls of about the same size. The exposed surface area of the oranges was varied by covering them with different amounts of paper, the edges of which were sealed in wax to keep mites from crawling underneath. In most experiments, Huffaker first established a prey population with 20 females per tray, then introduced 2 female predators 11 days later. (Both species reproduce parthenogenetically, so males were not required.)

When six-spotted mites were introduced to the trays alone, their populations leveled off at between 5,500 and 8,000 mites per tray. When predators were added, their numbers increased rapidly, and they soon wiped out the prey population. Their own extinction followed shortly. Although the predators always eliminated the prey, the positions of the exposed areas of the oranges influenced the course of extinction. When the exposed areas were in adjacent positions, minimizing dispersal distance between food sources, prey populations reached maxima of only 113–650 individuals and were driven to extinction within 23–32 days of the beginning of the experiment. The same area of exposed oranges dispersed randomly throughout the 40-position tray supported prey populations that reached maxima of 2,000 to 4,000 individuals and persisted for 36 days. Thus, Huffaker could prolong the survival of the prey population by providing remote areas of suitable habitat to which predators dispersed slowly.

Huffaker reasoned that if predator dispersal could be further retarded, the two species might coexist. To accomplish this, he increased the spatial complexity of the environment and introduced barriers to dispersal. The number of possible orange positions was increased to 120, and a feeding area equivalent to six oranges was dispersed over all 120 positions. A mazelike pattern of Vaseline barriers was placed among the oranges to slow dispersal of the predators. *Typhlodromus* must walk to get where it is going, but six-spotted mites spin a silk line that they can use like a parachute to float on wind currents. To take advantage of this behavior, Huffaker placed vertical wooden pegs

(a)

(b)

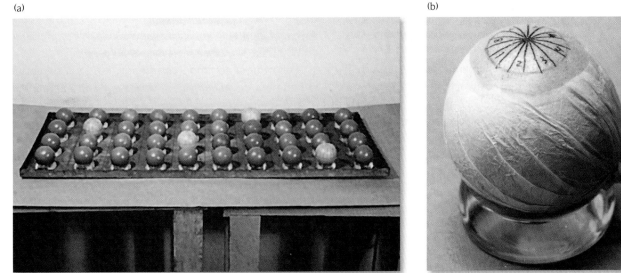

❚ **Figure 18.6 C. B. Huffaker's classic experiment tested the parameters of predator–prey coexistence.** (a) In each experimental tray, four oranges, half exposed, are distributed at random among the 40 positions in the tray. Other positions are occupied by rubber balls. (b) Each orange is wrapped with paper and its edges sealed with wax. The exposed area has been divided into numbered sections to facilitate counting the mites.
Courtesy of C. B. Huffaker, from C. B. Huffaker, *Hilgardia* 27:343–383 (1958).

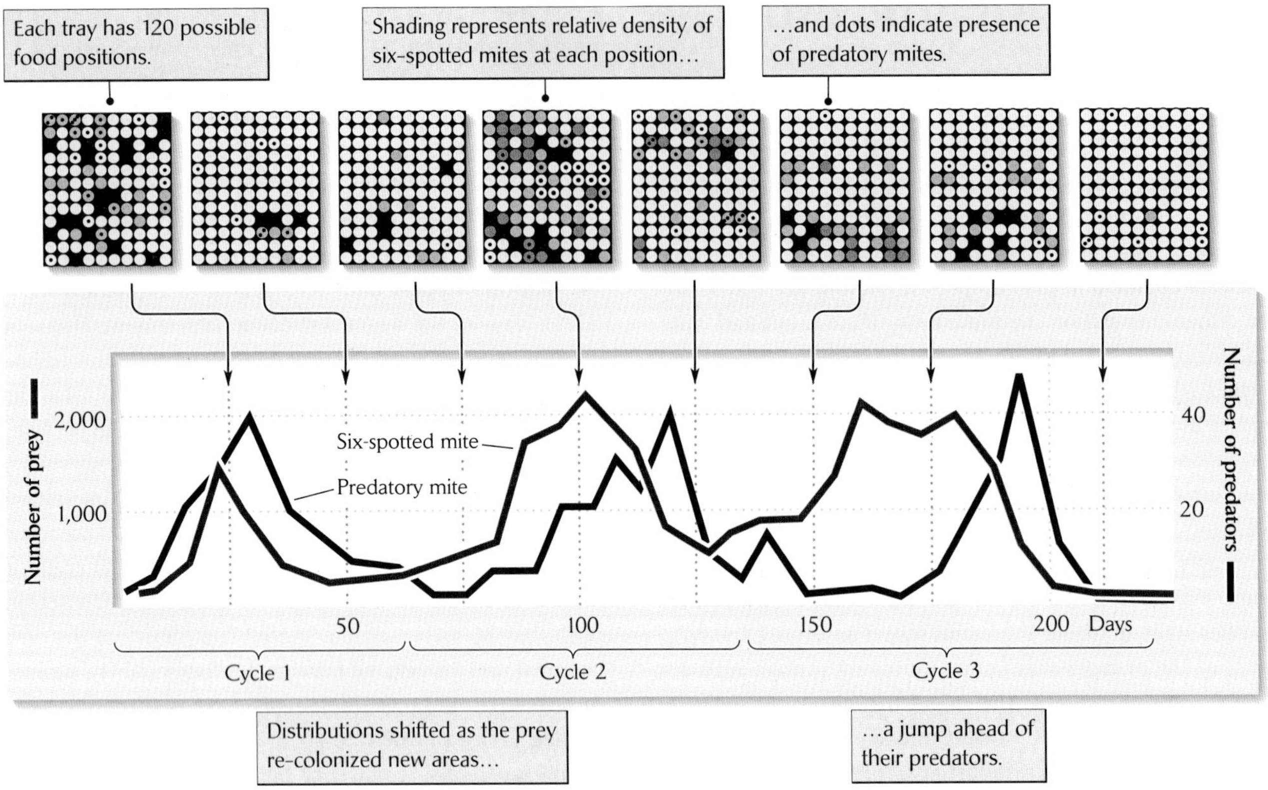

Each tray has 120 possible food positions.

Shading represents relative density of six-spotted mites at each position...

...and dots indicate presence of predatory mites.

Six-spotted mite

Predatory mite

Cycle 1   Cycle 2   Cycle 3

Distributions shifted as the prey re-colonized new areas...

...a jump ahead of their predators.

Figure 18.7 A spatial mosaic of habitats allows predators and prey to coexist. The shaded boxes show the positions and relative densities of Huffaker's mites in the trays at the eight times indicated. From C. B. Huffaker, *Hilgardia* 27:343–383 (1958).

throughout the trays, which the mites used as jumping-off points in their wanderings. This arrangement finally produced a series of three population cycles over the 8 months of the experiment (▌Figure 18.7). The distribution of predators and prey throughout the trays continually shifted as the prey, on the way to extermination in one feeding area, recolonized the next a jump ahead of their predators.

Despite the tenuousness of the predator–prey cycle that was achieved, Huffaker's experiment demonstrated that a spatial mosaic of suitable habitats enables predator and prey populations to coexist through time. But, as we saw in Gause's experiments with protozoans, predator and prey may also coexist locally if some prey can take refuge in hiding places. When the environment is so complex that predators cannot easily find scarce prey, stability can be achieved.

Predator–prey interactions can be modeled by simple equations that exhibit cyclic dynamics

In an attempt to understand the origin of population cycles, Alfred J. Lotka and the Italian biologist Vito Volterra

independently developed the first mathematical descriptions of predator–prey interactions during the 1920s. The **Lotka–Volterra model** predicts oscillations in the abundance of predator and prey populations, with predator numbers lagging behind those of their prey.

Following a common convention, we shall designate the number of predator individuals by P and the number of prey individuals by R (think of R for "resource"). The growth rate of the prey population (dR/dt) has two components: (1) unrestricted exponential growth in the absence of predators, rR, where r is the exponential growth rate (the difference between the per capita birth and death rates); and (2) removal of prey by predators, over and above other causes of death. The Lotka–Volterra model assumes that predation varies in direct proportion to the product of the prey and predator populations, RP, and therefore in proportion to the probability of a random encounter between predator and prey. Accordingly, the rate of increase of the prey population is given by

$$\frac{dR}{dt} = rR - cRP.$$

where c is a coefficient expressing the efficiency of predation (think of c for "capture efficiency"). In words,

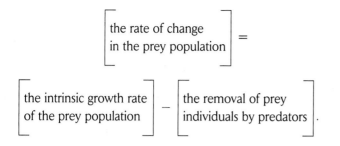

$$\begin{bmatrix} \text{the rate of change} \\ \text{in the prey population} \end{bmatrix} =$$

$$\begin{bmatrix} \text{the intrinsic growth rate} \\ \text{of the prey population} \end{bmatrix} - \begin{bmatrix} \text{the removal of prey} \\ \text{individuals by predators} \end{bmatrix}.$$

The growth rate of the predator population also has two components: (1) the birth rate, which depends on number of prey captured; and (2) a death rate imposed from outside the system:

$$\frac{dP}{dt} = acRP - dP.$$

The birth term is the number of prey captured (cRP) times a coefficient (a) for the efficiency with which food is converted to population growth. The death rate is a constant (d) times the number of predator individuals. The expressions for the growth rate of the prey and predator populations are referred to as differential equations because they describe the change in numbers (dR or dP) with respect to a change in time (dt).

When both predator and prey populations achieve equilibrium ($dR/dt = 0$ and $dP/dt = 0$), $rR = cRP$ and $acRP = dP$. We can rearrange these equations to give

$$\hat{P} = \frac{r}{c} \quad \text{and} \quad \hat{R} = \frac{d}{ac},$$

where $\hat{P}$ and $\hat{R}$ are the equilibrium sizes of the predator and prey populations, respectively. Note that both $\hat{P}$ and $\hat{R}$ are constant values, each independent of the abundance of the other population. When these values are graphed, the point at which the lines representing $\hat{P}$ and $\hat{R}$ cross is called the **joint equilibrium,** and is the only combination of population sizes P and R that is stable.

HELP ON THE WEB | *Go to Living Graphs at http://www.whfreeman. com/ricklefs and use the interactive tutorial to better understand the derivation and properties* of the Lotka-Volterra model.

According to the Lotka–Volterra model, when the populations stray from their joint equilibrium, rather than returning to the equilibrium point, they oscillate around it in a continuous cycle. The period of that oscillation (T) is approximately $2\pi/\sqrt{rd}$. Hence the higher the population growth potential of the prey or the death rate of the predator—that is, the higher the rate of population turnover—the shorter is T, and the faster the system oscillates.

The relationship between predator and prey can be portrayed as a graph with axes representing the sizes

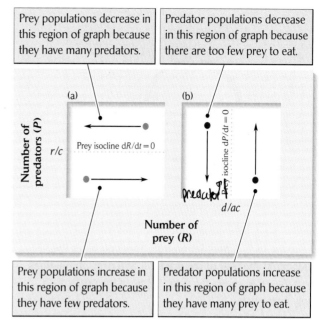

Prey populations decrease in this region of graph because they have many predators.

Predator populations decrease in this region of graph because there are too few prey to eat.

Prey populations increase in this region of graph because they have few predators.

Predator populations increase in this region of graph because they have many prey to eat.

▮ **Figure 18.8 The equilibrium isoclines of predator and prey delineate regions of population increase and decrease.** (a) The prey isocline ($dR/dt = 0$ when $P = r/c$) separates regions of prey population increase (low predator numbers) and decrease (high predator numbers). (b) The predator isocline ($dP/dt = 0$ when $R = d/ac$) separates regions of predator population increase (high prey numbers) and decrease (low prey numbers). The two graphs can be superimposed, as in Figure 18.9, to show the pattern of simultaneous change in both populations.

of the populations, as shown in ▮ Figure 18.8. By convention, predator numbers increase along the vertical axis and prey numbers along the horizontal axis. The horizontal line $\hat{P} = r/c$ in Figure 18.8a represents the condition $dR/dt = 0$ and is called the **equilibrium isocline** (or **zero growth isocline**) of the prey. For any combination of predator and prey numbers that lies in the region below this line, the population of prey increases because there are relatively few predators. In the region above the equilibrium isocline, the prey population decreases because of overwhelming predator pressure.

The population of predators can increase only when the abundance of prey exceeds a vertical line representing $\hat{R} = d/ac$, the equilibrium isocline of the predator (Figure 18.8b). To the right of this line, prey are abundant enough to sustain the population growth of the predator. Left of the line, predator populations decrease because prey are scarce. Thus, the equilibrium population values of predator ($\hat{P}$) and prey ($\hat{R}$) partition the graph into four regions.

The change in predator and prey populations together follows a closed cycle that combines the individual changes

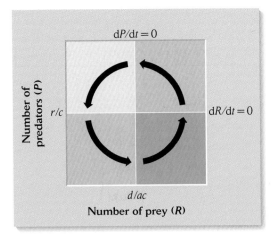

Figure 18.9 A population trajectory combines the individual changes of the predator and prey populations. This trajectory shows the cyclic nature of the predator–prey interaction.

of the predator and prey populations (■ Figure 18.9). This cycle, called a **population trajectory,** can be traced through the four sections of the graph. In the lower right-hand region, for example, both predators and prey increase, and their joint population trajectory moves up and right.

The trajectories in the four regions together define a counterclockwise cycling of predator and prey populations

one-quarter cycle out of phase, with the prey population increasing and decreasing just ahead of predator population (■ Figure 18.10). Referring back to Figures 18.2 and 18.7, you can see that prey populations tend to peak in each cycle ahead of the predator populations.

The equilibrium isocline of the predator ($dP/dt = 0$) is the minimum level of prey ($\hat{R} = d/ac$) that can sustain the growth of the predator population. The equilibrium isocline of the prey ($dR/dt = 0$) is the greatest number of predators ($\hat{P} = r/c$) that the prey population can sustain. If the reproductive rate of the prey (r) were to increase, or the capture efficiency of the predators (c) were to decrease, or both, the prey isocline (r/c) would increase. That is, the prey population would be able to bear the burden of a larger predator population, and the predator population would increase. If the death rate of the predators (d) increased and either the predation efficiency (c) or the reproductive efficiency (a) of predators decreased, the predator isocline (d/ac) would move to the right, and more prey would be required to support the predator population. By itself, increased predation efficiency (c) would simultaneously reduce both isoclines: fewer prey would be needed to sustain a given capture rate (the predator isocline would decrease), and the prey population would be less able to support the more efficient predators (the prey isocline would decrease).

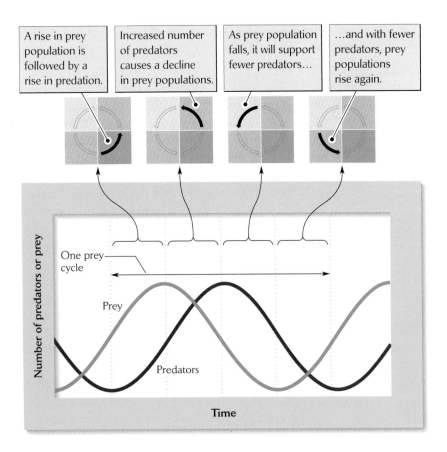

Figure 18.10 The Lotka–Volterra model defines a regular cycling of predator and prey populations. These curves of predator and prey population size show that the two continually cycle out of phase with each other.

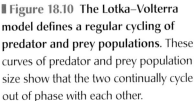

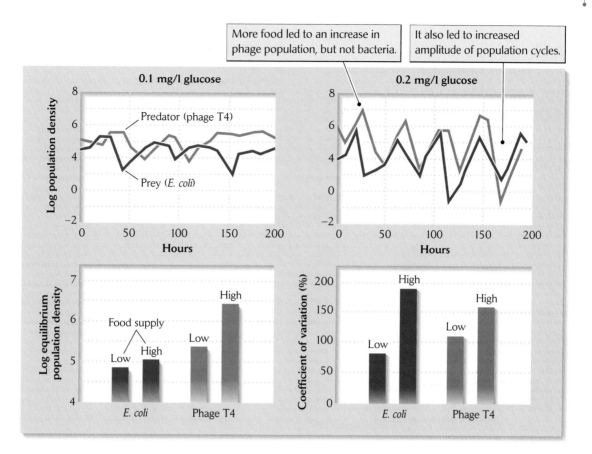

More food led to an increase in phage population, but not bacteria.

It also led to increased amplitude of population cycles.

Figure 18.11 An increase in the birth rate of prey increases the predator population, but not the prey population. This prediction of the Lotka–Volterra model was tested by increasing the rate of resource (glucose) provisioning to cultures of *E. coli* bacteria in chemostats containing the bacteria and their predators, T4 bacteriophage. As shown in the lower left panel, the phage (predator) population was more responsive than the bacteria population (prey) to an increase in the food supply of the prey. After B. J. M. Bohannan and R. E. Lenski, *Ecology* 78:2303–2315 (1997).

MORE ON THE WEB *Simulation models of predator–prey interactions.* Try changing the variables *r, c, a,* and *d* of the Lotka–Volterra model to see the effect on the period and amplitude, or ratio of peak to trough populations, of the predator–prey cycle.

ECOLOGISTS IN THE FIELD

Testing a prediction of the Lotka–Volterra model

One of the more surprising predictions of the Lotka–Volterra model is that an increase in the birth rate of the prey (*r*) should lead to an increase in the population of predators ($\hat{P}$), but not prey ($\hat{R}$). It is as if the benefit to the prey of some improvement in their environment—a better supply of food, for example—is passed on directly to their predators.

This prediction was tested by Brendan Bohannan and Richard Lenski, of Michigan State University, in a simple microcosm experiment. The prey in their system was the bacterium *Escherichia coli*, and the predator was the bacteriophage T4. Populations of bacteria and phage were maintained in a chemostat, a device in which the culture medium is continually replaced by a fresh supply as old medium is removed. In these experiments, the reproductive rate of *E. coli* was limited by the availability of glucose, which was supplied in concentrations of either 0.1 or 0.5 mg per liter of medium. Because a constant influx of new medium was balanced by removal of old medium, the bacteria and phage populations soon reached equilibrium levels. Consistent with the predictions of the Lotka–Voltera model, a higher rate of food provisioning to the bacteria led to an increase in the population of the phage, but not the bacteria themselves. An increased rate of food provisioning also led to increased amplitudes of population cycles (**Figure 18.11**). These results in a very simple predator–prey system seem to verify that the Lotka–Volterra model captures the essence of the interaction.

Modifications of the Lotka–Volterra model incorporate more complex relationships of predators and prey

The Lotka–Volterra model has enabled ecologists to explain the occurrence of population cycles in nature, but the model is so simple that it fails to represent nature in some important ways. For example, according to the Lotka–Volterra model, when either the predator or the prey population is displaced from its equilibrium, the system will oscillate in a closed cycle. Any further perturbation of the system will give these population oscillations a new amplitude and duration until some other outside influence acts upon them. This state of oscillation is said to be a **neutral equilibrium** because no internal forces act to restore the populations to the joint equilibrium. Therefore, random perturbations will eventually increase the oscillations to the point at which the trajectory reaches one of the axes of the predator–prey graph (R or $P = 0$), and one or both populations will die out. This property in itself suggests that the Lotka–Volterra equations greatly oversimplify nature.

Other concerns about the adequacy of the Lotka–Volterra model focus on the predation term (cRP). For a given density of predators (P), the rate of prey capture (cRP) increases in direct proportion to the density of prey (R). Accordingly, predators cannot be satiated; they just keep on eating, no matter how many prey they catch. Clearly this aspect of the model is unrealistic. How would adding a bit more realism here affect the behavior of the model?

The functional response

The relationship of an individual predator's rate of food consumption to the density of its prey has been labeled the **functional response** by entomologist C. S. Holling. As we have seen, the total rate of prey consumption by predators according to the Lotka–Volterra model is cRP, where P is the number of predators, R is the number of prey, and c is the capture efficiency. Dividing this term by the number of predators, P, we see that the rate of consumption per predator is simply cR. Thus, prey are consumed in direct proportion to their abundance or density at rate c. This relationship, called a **type I functional response,** is illustrated in ▌Figure 18.12. This means that the fecundity of individual predators, which in the model is proportional to the number of prey consumed, increases without limit in direct proportion to prey availability. Thus, when there are few predators but many prey, the birth rate of the predators is very high, the predator population grows rapidly, and the prey population can be brought under control. This particular circumstance causes the neutral equilibrium that is characteristic of the Lotka–Volterra model.

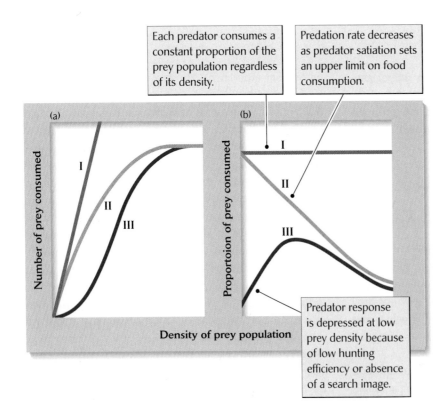

Each predator consumes a constant proportion of the prey population regardless of its density.

Predation rate decreases as predator satiation sets an upper limit on food consumption.

Predator response is depressed at low prey density because of low hunting efficiency or absence of a search image.

▌**Figure 18.12 Predators exhibit three types of functional responses to increasing prey density.** These functional responses are shown in terms of (a) the number of prey consumed and (b) the proportion of prey consumed.

An obvious modification of the type I functional response is seen in the **type II functional response,** in which the number of prey consumed per predator initially rises quickly as the density of prey increases, but then levels off with further increases in prey density. A **type III functional response** resembles the type II response in having an upper limit to prey consumption, but it differs in that the response of predators to prey is depressed at low prey densities.

Two factors dictate that a functional response should reach a plateau. First, predators may become satiated—constantly full—at which point their rate of feeding is limited by the rate at which they can digest and assimilate food. Second, as a predator captures more prey, the time it spends handling and eating prey cuts into its searching time. Eventually, these two factors reach a balance, and the prey capture rate levels off.

At high prey densities, the type II and type III functional response curves differ little: they are both inversely density-dependent. In other words, as the density of prey increases, the proportion of prey consumed decreases. Over the lower range of prey densities, however, type III responses differ from type II responses in that the proportion of the prey consumed is decreased at lower prey densities.

What circumstances might cause a type III functional response? First, a heterogeneous habitat may afford a limited number of safe hiding places, which protect a larger proportion of the prey at lower densities than at higher densities. Second, lack of reinforcement of learned searching behavior owing to a low rate of prey encounter may reduce capture efficiency at low prey densities. Third, **switching** to alternative sources of food when prey are scarce may reduce pressure on the prey. Switching produces a type III response because prey consumption at low prey population densities is reduced as predators switch to more abundant alternative prey.

When the predatory water bug *Notonecta glauca* was presented with two types of prey—isopods and mayfly larvae—in the laboratory, it consumed the more abundant prey species, whichever it was, in a proportion greater than its percentage of occurrence (▌Figure 18.13). The predators' switching depended on variation in the success of attacks on prey as a function of their relative density. When water bugs encountered mayfly larvae infrequently, fewer than 10% of attacks were successful. At higher densities, and therefore higher encounter rates, attack success rose to almost 30%.

Variation in prey availability does not always lead to switching, however. In southern Yukon Territory, Canada, one long-term study through a population cycle of snowshoe hares found that coyotes and lynx, the principal predators of the hares, always fed on hares in greater propor-

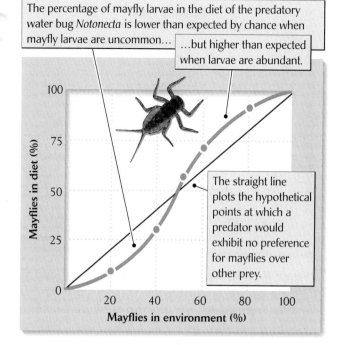

▌**Figure 18.13 Predators may switch to different prey in response to fluctuations in prey density.** From M. Begon and M. Mortimer, *Population Ecology,* 2d ed., Blackwell Scientific Publications, Oxford (1981); after J. H. Lawton, J. R. Beddington, and R. Bonser, in M. B. Usher and M. H. Williamson (eds.), *Ecological Stability,* pp. 141–158, Chapman & Hall, London (1974).

tion to their relative abundance than on red squirrels and other small mammals (▌Figure 18.14). Only at the lowest densities of snowshoe hares did either predator include a substantial proportion of other prey in its diet.

The numerical response

Individual predators can increase their consumption of prey only to the point of satiation. Continued predator response to increasing prey density above this level can be achieved only through an increase in the number of predators, either by immigration or by population growth, which together constitute a **numerical response.** Populations of most predators grow slowly, especially when the reproductive potential of a predator is much lower than that of its prey and the predator's life span is longer.

Immigration from surrounding areas contributes to the numerical responses of mobile predators, which may opportunistically congregate where resources become abundant. The bay-breasted warbler, a small insectivorous bird of eastern North America, exhibits such behavior during periodic outbreaks of the spruce budworm. During years of outbreak in a particular area, the density of warblers

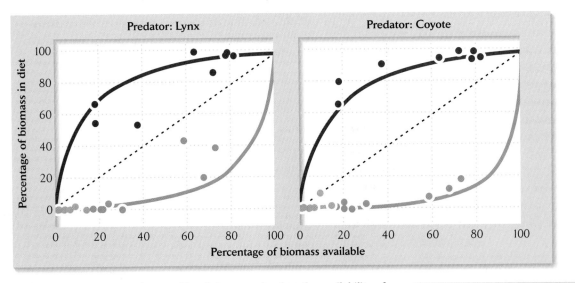

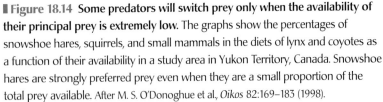

Figure 18.14 **Some predators will switch prey only when the availability of their principal prey is extremely low.** The graphs show the percentages of snowshoe hares, squirrels, and small mammals in the diets of lynx and coyotes as a function of their availability in a study area in Yukon Territory, Canada. Snowshoe hares are strongly preferred prey even when they are a small proportion of the total prey available. After M. S. O'Donoghue et al., *Oikos* 82:169–183 (1998).

KEY
■ Snowshoe hares   ■ Small mammals
■ Red squirrels

may reach 300 breeding pairs per km², compared with about 25 pairs per km² during non-outbreak years. This behavior shows how a predator can take advantage of a shifting mosaic of prey abundance.

In the study area in southern Yukon Territory mentioned above, numbers of coyotes and lynx increased 6-fold and 7.5-fold in response to increasing snowshoe hare populations (**Figure 18.15a**). Because of the great geographic extent of hare population cycles, most of the increase in their predators was due to local population growth rather than immigration from elsewhere. During the phase of hare population increase, the coyote and lynx fed almost exclusively on hares. However, after the number of hares began to decline, snowshoe hares eventually became extremely scarce, and both predators began to capture larger numbers of alternative prey, particularly red squirrels and other small mammals (**Figure 18.15b**). The smaller mammals, whose populations were stable or increasing during the decline phase of the cycle, evidently could not sustain coyote or lynx populations. The food return per unit of hunting time was too low, and populations of both predators began to decline in response.

The numerical response of the predator tends to lag behind that of the prey, whether the prey population is increasing or decreasing. Consequently, the relationship between predation and prey population density differs between the increasing and declining phases of a population cycle (**Figure 18.16**). When prey are increasing, predators

tend to be scarce; when they are decreasing, predators tend to be relatively abundant.

Several factors tend to reduce oscillations in predator–prey models

Stability in predator–prey systems may be thought of as the tendency of populations of predators and prey to achieve nonvarying equilibrium sizes. This is a special, restrictive meaning of the word *stability* because, as we have seen, predator and prey populations may also oscillate in stable cycles. Nonetheless, the presence of cycles indicates the influence of destabilizing factors.

Five factors tend to reduce the amplitude of predator–prey cycles and thus promote stability of predator and prey populations:

1. Predator inefficiency (or enhanced prey escape or defense strategies)

2. Density-dependent limitation of either the predator or the prey by factors external to their relationship

3. Alternative food sources for the predator

4. Refuges from predation at low prey densities

5. Reduced time delays in predator response to changes in prey abundance.

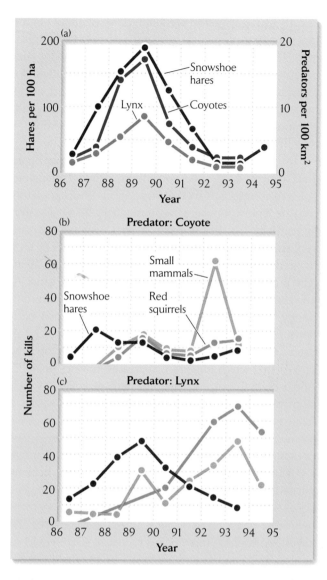

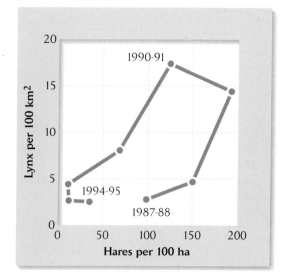

Figure 18.16 The numerical response of a predator population to its prey exhibits cycling out of phase. The lynx population shown in Figure 18.15 responded to the hare population in a manner resembling Lotka–Volterra predator–prey dynamics. After data in M. S. O'Donoghue et al., *Oikos* 82:169–183 (1998).

Figure 18.15 Predator populations exhibit a numerical response to changes in prey density. (a) Population changes in snowshoe hares and in their predators, the lynx and coyote, are shown through a hare population cycle. (b, c) Red squirrels and other small mammals were eaten in large numbers only after the densities of hares fell to a low level. After M. S. O'Donoghue et al., *Oikos* 82:169–183 (1998).

dation allow prey populations to maintain themselves at higher levels in the face of intense predation, thereby facilitating the recovery phase of the population cycle. Indeed, so many factors tend to stabilize predator–prey relationships that the periodic cyclic behavior of some systems seems to require special explanation.

Most probably, population cycles reflect a balance between the stabilizing effects of density dependence, alternative prey, and refuges and the destabilizing effects of time delays in predator–prey interactions. Time delays are ubiquitous in nature, arising from the developmental periods of animals and plants, the time required for numerical responses by predator populations, and the time course of immune responses by animals and of induced defenses in plants. In some circumstances—perhaps in simple ecological systems—these factors outweigh stabilizing influences and result in population cycles.

Several of these factors deserve special comment. Predator inefficiency (lower c in the Lotka–Volterra model) results in higher equilibrium levels for both prey and predator populations (more predators can be supported by the larger prey populations) and in lower turnover rates for both at equilibrium. Both these consequences would seem to enhance the stability of a predator–prey system. Alternative food sources stabilize predator populations because individuals can switch between food types in response to changing prey abundance. Similarly, safe refuges from pre-

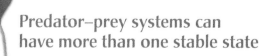

Predator–prey systems can have more than one stable state

The size of any population is influenced by the abundance of its resources and of its consumers. Extremely efficient predators may depress a prey population to levels far below its carrying capacity. Alternatively, a prey population may be limited primarily by its own food supply while

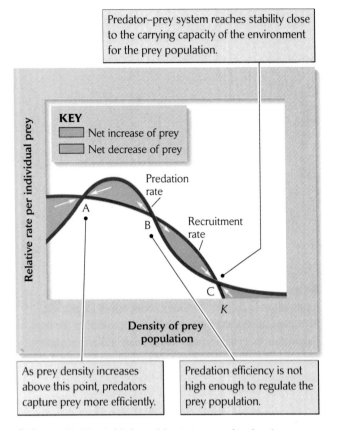

Predator–prey system reaches stability close to the carrying capacity of the environment for the prey population.

KEY
▢ Net increase of prey
▢ Net decrease of prey

Relative rate per individual prey

Predation rate

A

B

Recruitment rate

C

K

Density of prey population

As prey density increases above this point, predators capture prey more efficiently.

Predation efficiency is not high enough to regulate the prey population.

❚ **Figure 18.17 Multiple stable states can develop in some predator–prey systems.** In this model, the predation curve is based on a type III functional response. When predation (blue line) exceeds recruitment (red line), prey populations decrease, and vice versa (as shown by the arrows). Points A and C are stable equilibria for the prey population. The lower point (A—lower in terms of prey density) represents population control by predators; the higher point (C) represents population control by food and other resources.

predators remove an inconsequential number of prey individuals. As we have already seen, equilibrium population size often reflects a balance between the limiting influences of food supply and predators. Under some circumstances, however, a population may have two or more stable equilibrium points, only one of which may be occupied at a given time. This situation is referred to as **multiple stable states.**

A simple model illustrates how multiple stable states can develop (❚ Figure 18.17). This model describes changes in recruitment and predation rates in prey populations as a function of increasing prey density. The recruitment curve represents the net contribution of births and deaths to the prey population in the absence of predators. Hence per capita recruitment is high when the prey population is small and decreases to zero as the population approaches

its carrying capacity. The predation curve is the sum of the functional and numerical responses of predator populations. The predation rate may be low at low prey densities because of switching to alternative prey or difficulty in locating scarce prey; it also tails off at high prey densities because of predator satiation and extrinsic limits on predator populations. The predation curve thus resembles a type III functional response (see Figure 18.12).

The recruitment and predation curves shown in Figure 18.17 can produce three equilibrium points. Equilibrium point A corresponds to the situation in which predators regulate a prey population at substantially below its carrying capacity (*K*). Below point A—that is, when prey are very scarce—either they are difficult for predators to find, or predators develop preferences for other, more abundant prey. As the density of prey increases above point A, however, predators are able to capture them more efficiently, and they tend to eat the prey population back down to point A. Above point B and below point C, predation efficiency is not high enough to regulate the prey population, which therefore increases toward point C. Above point C, however, the prey become limited by their own resources, and the predator–prey system reaches a stable equilibrium close to the carrying capacity of the environment for the prey population. This transition occurs because, as prey become more abundant, predators capture a smaller proportion of them. Individual predators become satiated more quickly, and predator populations become limited by factors other than their food supply—perhaps by disease or social interactions. Thus, the highest (C) and lowest (A) of these three points represent stable equilibria around which populations are regulated; the middle equilibrium (B) is unstable. Point B represents a changeover from strong predator control to strong resource control of a prey population. Below point B, prey populations are further reduced to stable point A; above point B, the prey escape control and increase to stable point C.

The implications of Figure 18.17 for practical concerns such as the control of crop pests are clear. Predators maintain a shaky hold on prey populations at point A. If a heavy frost or an introduced disease were to reduce the predator population long enough to allow the prey population to slip above point B, the prey would continue to increase until they reached the higher (in terms of prey density) stable equilibrium point C, regardless of how quickly the predator population recovered. To the farmer, this means that the population of a crop pest that is normally controlled at harmless levels by predators and parasites suddenly becomes a menacing outbreak. After such a change, predators exert little control over the pest population until some quirk of the environment brings its numbers below point B, back within the realm of predator control.

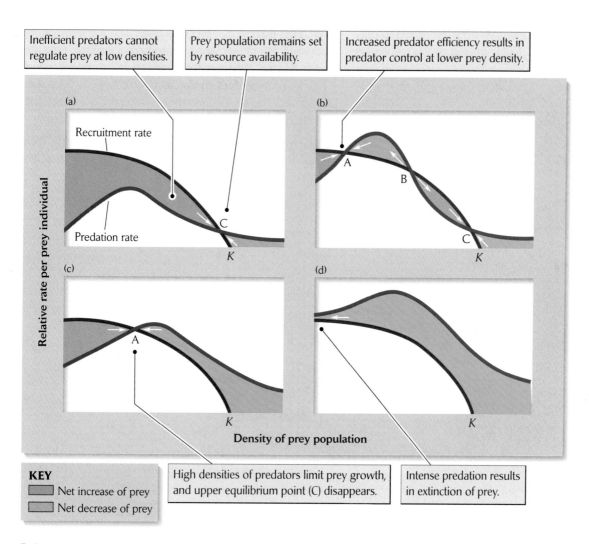

Inefficient predators cannot regulate prey at low densities.

Prey population remains set by resource availability.

Increased predator efficiency results in predator control at lower prey density.

High densities of predators limit prey growth, and upper equilibrium point (C) disappears.

Intense predation results in extinction of prey.

KEY
■ Net increase of prey
■ Net decrease of prey

❚ Figure 18.18 Intensity of predation relative to prey recruitment determines the number of stable predator–prey equilibrium points.

Outbreaks of winter moths in forests of eastern North America can be brought under control by introduced parasitoids of the caterpillars. When the winter moth population is reduced to a low level (point A), it can be kept low for some time by small mammals that prey on the pupae in the leaf litter on the forest floor. However, if the winter moth escapes predator control, which might happen when climate or disease limits its predators, the population may increase back toward point C until parasitoid populations increase enough to regain control of their host population.

MORE ON THE WEB *Predator–prey dynamics in a metapopulation of the cinnabar moth.* The stability of this herbivore–plant interaction depends on isolated refuges for the plant population.

Using the predation–recruitment model in Figure 18.17, we can examine the consequences of different lev-

els of predation for control of a prey population (❚ Figure 18.18). Inefficient predators cannot keep prey populations at low densities; they depress prey numbers slightly, but the prey population remains near the equilibrium level set by its resources (Figure 18.18a, point C). Increased predation efficiency at low prey density, however, can result in predator control at point A, as in the case of the winter moth (Figure 18.18b). When functional and numerical responses are sufficient to maintain high densities of predators, predation may effectively limit prey population growth under all circumstances, and equilibrium point C disappears (Figure 18.18c). Finally, predation may be so intense at all prey densities that the prey are eaten to extinction (Figure 18.18d, no equilibrium point.)

As a general rule, predators are able to drive prey populations to extinction only in simple laboratory systems or when predator populations are maintained at high densities by the availability of some alternative but less preferred prey (hence, there is no switching). Indeed, many ecologists have advocated providing parasites and predators of pest species with alternative prey to enhance biological control. At the very least, the curves in Figure 18.17 suggest that the position of a predator–prey equilibrium, whether at very low prey numbers or close to their carrying capacity, may shift between extremes with small changes in closely matched predation and recruitment curves. Such considerations would appear to make equilibria at intermediate prey densities very unlikely.

MORE ON THE WEB *Maximum sustainable yield.* What is the maximum sustainable level of predation on a prey population? Do predators limit their own populations to achieve the maximum sustainable yield from their prey?

MORE ON THE WEB *Three-level consumer systems.* When predators themselves have predators, their prey may benefit. Birds and wasps reduce the number of herbivorous insects on trees, and trees benefit from the reduced damage by maintaining faster growth and achieving larger size.

Summary

1. Experimental studies of pest species and their natural predators have demonstrated that, in many cases, consumers can reduce resource populations far below their carrying capacities.

2. Predator and prey populations in natural systems often increase and decrease in regular cycles. In systems involving mammalian predators and their prey, the length of the population cycles is typically about 4 years or 9–10 years.

3. Predator and prey populations can be made to oscillate in the laboratory. Maintenance of population cycles usually requires a complex environment in which prey populations can establish themselves in refuges.

4. Alfred J. Lotka and Vito Volterra, in the 1920s, devised a simple mathematical model of predator and prey dynamics that predicted population cycles. The Lotka–Volterra model uses differential equations in which the rate of prey removal is directly proportional to the product of the predator and prey populations.

5. A surprising prediction of the Lotka–Volterra model is that increased productivity of the prey should increase the size of the predator population but not the prey population. This prediction has been verified in several experimental studies.

6. The functional response describes the relationship between the rate at which a predator consumes prey and prey density. Whereas the Lotka–Volterra model, which employs a type I functional response curve, is inherently unstable, type III functional response curves can result in stable regulation of prey populations at low densities.

7. Type III functional response curves can result from predators switching from their preferred prey when it is scarce to a more abundant alternative prey.

8. The numerical response describes the response of a predator population to increasing prey density by population growth and immigration.

9. Stability in predator–prey interactions is promoted by density dependence in either predator or prey, by refuges or hiding places in which prey can escape predation, by low predator efficiency, and under some circumstances, by the availability of alternative prey. Stable population cycles in nature apparently represent a balance between these stabilizing factors and the destabilizing influence of time delays in population responses.

10. Models of consumer–resource systems suggest that such systems can have two equilibrium points (multiple stable states) between which populations may shift, depending on environmental conditions. The lower equilibrium is determined by predation pressure; the upper equilibrium lies close to the carrying capacity of the prey. Sudden climatic or biotic stresses may shift a system from one to the other of these points, resulting in successive controlled and outbreak conditions.

PRACTICING ECOLOGY
CHECK YOUR KNOWLEDGE

Applying Predator–Prey Models to Wildlife Management

We have shown in an earlier *Practicing Ecology* how the use of models can help increase our understanding of complex ecological relationships. One of the best ways to explore the effect of various factors on predator–prey dynamics is to simulate the behavior of the system using computer models. This can be done using various types of software and programming. In fact, simple predator–prey models can be created using Microsoft Excel, in which the programmer can alter the parameters that are input to describe the linkage between predator and prey populations. One can show how predation efficiency and birth and death rates of the predator and prey species alter the relationship between the two species. For example, simple models with discrete time intervals often result in the collapse of either the predator or prey population. Lotka noted, as you read earlier in this chapter, that adding mathematical terms that simulate density dependence tend to increase the stability of the populations and can lead to an equilibrium number of predators and prey.

The use of computer modeling techniques has been especially useful for understanding the dynamics of populations of fish and other aquatic organisms that are important food sources. Such techniques are quite useful when you consider that marking and recapturing fish can be quite expensive and time consuming. Also, some species of economically important fish, such as salmon, die after reproducing and so mark-and-recapture techniques may not be effective for answering the question of interest. The use of models helps in the development and management of scientific research programs that generate the best knowledge for understanding and managing species.

Not surprisingly, the populations of many marine prey species and their predators are the subject of research in Alaska, which has the nation's largest fishery. Patricia Livingston of the National Marine Fisheries Service and her colleagues have been studying fish populations using computer modeling methods. The Bering Sea near Alaska is home to large populations of several bottomfish populations, particularly walleye pollock. Walleye pollock are important prey species in this region. Livingston and Jurado-Molina (2000) used a multispecies population analysis model to simulate conditions during the period 1979–1995. The model is based on diet data from numerous predator–prey combinations and includes walleye pollock, Pacific cod, Greenland turbot, yellowfin sole, arrowtooth flounder, and northern fur seal as predators. The results indicate that large numbers of walleye pollock are cannibalized by adult pollock! Therefore, predation represents an important source of mortality for young fish, which has clear implications for walleye pollock population changes as well as fisheries management.

CHECK YOUR KNOWLEDGE

1. Why did Livingston and her colleagues rely on computer models for understanding the populations of many marine prey species and their predators?

2. How can knowledge of the ecology of predators and their prey affect our ability to feed people?

3. Consult the paper by Livingston and Jurado-Molina through *Practicing Ecology on the Web* at *http://www. whfreeman.com/ricklefs*. What data were used to determine the pattern and quantity of consumption of prey by predators?

MORE ON THE WEB 4. Visit the Web page for the Alaska office of the National Marine Fisheries Service at *http://www. fakr.noaa.gov* and navigate to the research page (*http://www.afsc.noaa.gov/research.htm* or find the link at *www.whfreeman.com/ricklefs*.) What is the current status of the northern rockfish, *Sebastes polyspinis*? How do individuals of this species differ in the Aleutian Islands and the Gulf of Alaska? What might account for such differences?

 Suggested Readings

Anderson, R. M., and R. M. May. 1980. Infectious diseases and population cycles of forest insects. *Science* 210:658–661.

Bohannan, B. J. M., and R. E. Lenski. 1997. Effect of resource enrichment on a chemostat community of bacteria and bacteriophage. *Ecology* 78:2303–2315.

Brooks, J. L., and S. I. Dodson. 1965. Predation, body size and composition of the plankton. *Science* 150:28–35.

Crawley, M. J. 1983. *Herbivory: The Dynamics of Animal–Plant Interactions*. University of California Press, Berkeley.

Crawley, M. J. 1997. Plant–herbivore dynamics. In M. J. Crawley (ed.), *Plant Ecology*, 2d ed., pp. 401–474. Blackwell Scientific, Oxford.

DeBach, P., and D. Rosen. 1991. *Biological Control by Natural Enemies*. 2d ed. Cambridge University Press, New York.

Dobson, A. 1995. The ecology and epidemiology of rinderpest virus in Serengeti and Ngorongoro conservation areas. In A. R. E. Sinclair and P. Arcese (eds.), *Serengeti II: Dynamics, Management, and Conservation of an Ecosystem*, pp. 485–505. University of Chicago Press, Chicago.

Errington, P. L. 1963. The phenomenon of predation. *American Scientist* 51:180–192.

Jefferies, R. L., D. R. Klein, and G. R. Shaver. 1994. Vertebrate herbivores and northern plant communities: Reciprocal influences and responses. *Oikos* 71:193–206.

Livingston, P. A., and J. Jurado-Molina. 2000. A multispecies virtual population analysis of the eastern Bering Sea. *ICES Journal of Marine Science* 57 : 294-299.

May, R. M. 1983. Parasite infections as regulators of animal populations. *American Scientist* 71:36–45.

May, R. M., J. R. Beddington, C. W. Clark, S. J. Holt, and R. M. Laws. 1979. Management of multispecies fisheries. *Science* 205:267–277.

Myers, J. H. 1993. Population outbreaks in forest lepidoptera. *American Scientist* 81:240–281.

O'Donoghue, M., S. Boutin, C. J. Krebs, D. L. Murray, and E. J. Hofer. 1998. Behavioral responses of coyotes and lynx to the snowshoe hare cycle. *Oikos* 82:169–183.

Pech, R. P., A. R. E. Sinclair, A. E. Newsome, and P. C. Catling. 1992. Limits to predator regulation of rabbits in Australia: Evidence from predator removal experiments. *Oecologia* 89:102–112.

Petraitis, P. S., and S. R. Dudgeon. 1999. Experimental evidence for the origin of alternative communities on rocky intertidal shores. *Oikos* 84:239–245.

Ranta, E., V. Kaitala, and P. Lundberg. 1997. The spatial dimension in population fluctuations. *Science* 278:1621–1623.

Roland, J. 1993. Large-scale forest fragmentation increases the duration of tent caterpillar outbreak. *Oecologia* 93:25–30.

Shrag, S. J., and P. Wiener. 1995. Emerging infectious disease: What are the relative roles of ecology and evolution? *Trends in Ecology and Evolution* 10:319–324.

CHAPTER 19

Competition

Consumers compete for resources

Failure of species to coexist in laboratory cultures led to the competitive exclusion principle

The theory of competition and coexistence is an extension of logistic growth models

Field studies demonstrate the pervasiveness of competition in nature

Plant competition differs between nutrient-rich and nutrient-poor habitats

Competition may occur through exploitation of shared resources or direct interference

The outcome of competition can be influenced by predators

The British botanist A. G. Tansley (1917) was the first ecologist to demonstrate experimentally competition between closely related species. Tansley prefaced his report with the observation that closely related plant species living in the same region often grow in different habitats or on different types of soil. The observation was not new, nor was the suggestion that such segregation might have resulted from the species competing for resources, leading to the exclusion of one species or the other. But no one had experimentally investigated the correctness of that hypothesis, or of its alternative—namely, that the two species had such different ecological requirements that each could not grow where the other flourished.

Tansley selected two species of bedstraw (genus *Galium*), which are small, perennial, herbaceous plants. One species, *G. saxatile*, normally lives on acid, peaty soils; the other, *G. sylvestre*, inhabits limestone hills and pastures. These he planted as seeds, both singly and together, in soils taken from areas in which each species grows. Because the seeds were planted together in a common garden (see Chapter 18), soil type and presence or absence of the other species were the only experimental treatments (▌Figure 19.1).

Like many more recent ecological studies, Tansley's experiments were plagued by such technical problems as poor germination and lapses in watering. His results were nonetheless quite clear. When planted alone, each of the species grew and maintained itself on both types of soil, although germination and growth were most vigorous on the soil on which the species normally grows in nature. When grown together on calcareous (limestone) soils, *G. sylvestre* plants overgrew and shaded *G. saxatile*. The reverse occurred on the more acid, peaty soil that is typical of *G. saxatile* habitat.

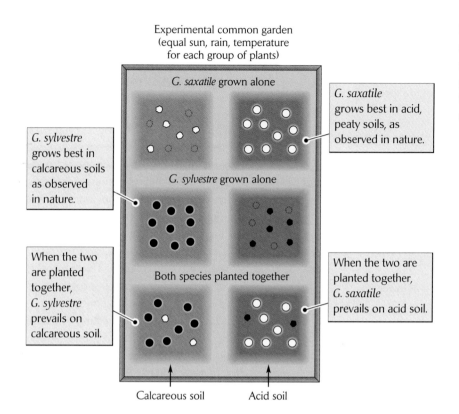

Experimental common garden
(equal sun, rain, temperature
for each group of plants)

G. saxatile grown alone

G. saxatile grows best in acid, peaty soils, as observed in nature.

G. sylvestre grows best in calcareous soils as observed in nature.

G. sylvestre grown alone

When the two are planted together, G. sylvestre prevails on calcareous soil.

Both species planted together

When the two are planted together, G. saxatile prevails on acid soil.

Calcareous soil   Acid soil

Figure 19.1 **Tansley's basic experimental design is still used in most modern studies of competition.** Tansley grew two species of *Galium*, alone and together, on two different soil types in a common garden.

Tansley concluded that *G. saxatile* is at a disadvantage on calcareous soils, and is thus unable to compete effectively with *G. sylvestre* on that soil type. Similarly, *G. sylvestre* is at a disadvantage when growing on acid peat, and is therefore an inferior competitor to *G. saxatile* on that soil type. Both species, however, were able to establish themselves on either soil type. These results suggested to Tansley an explanation for the observed distributions of other similar pairs of species: that they are restricted to certain soil types where the other member of the pair is present, but broadly distributed over soil types where it is not. Where, however, the disadvantage is very great—as in the case of *G. saxatile* on calcareous soils—it is unlikely that the species would survive competition even with other, unrelated vegetation in the community, and thus it would be absent on that soil type even where the congeneric competitor is not present.

Thus, in this brief paper, Tansley put on record (1) that the presence or absence of a species could be determined by competition with other species; (2) that the conditions of the environment affected the

outcome of competition; (3) that competition might be felt very broadly (that is, from other vegetation) throughout the community; and (4) that the present ecological segregation of species might have resulted from competition in the past. Although ecologists did not take up studies of competition again for more than 15 years, Tansley's approach, or some modification of it, is used in most modern studies of competition between species.

Competition is any use or defense of a resource by one individual that reduces the availability of that resource to other individuals. Competition is one of the most important ways in which the activities of individuals affect the well-being of others, whether they belong to the same species, in which case the interaction is **intraspecific competition,** or to different species, in which case it is **interspecific competition.**

As we saw in Chapter 14, competition within populations reduces resource levels in a density-dependent manner and thereby decreases fecundity and survival. The more crowded a population, the stronger is competition between individuals. Thus, intraspecific competition underlies the regulation of population size. Furthermore, when genetic factors cause individuals to differ in the efficiency with

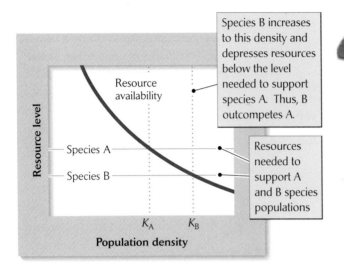

Species B increases to this density and depresses resources below the level needed to support species A. Thus, B outcompetes A.

Resources needed to support A and B species populations

Population density

❚ Figure 19.2 Superior competitors can persist at lower resource levels.

which they exploit resources, more efficient individuals may leave more descendants than less efficient ones, and the proportion of their genes may increase in populations over time. In this way, intraspecific competition is intimately related to evolutionary change.

Competition between individuals of different species causes a mutually depressing effect on the populations of both; each species contributes to the regulation of the other as well as its own population. Under some conditions, particularly when interspecific competition is intense, it may lead to the elimination of one species by the other. Because of this potential, competition is an important factor in determining which species can coexist within a habitat (❚ Figure 19.2).

The outcome of competition between two species depends on the relative efficiencies with which individuals exploit shared resources. Individuals in every population consume resources. When resources are scarce relative to demand for them, each act of consumption by one individual makes a resource less available to others, as well as to itself. As consumption continues, resources decline to levels that no longer support the further growth of the consumer population, and the population may reach an equilibrium size. When one population can continue to grow at a resource level that curtails the growth of a second population, the first will eventually replace the second. Thus, competition and its various outcomes depend on the relationship of consumers to their resources.

In this chapter, we shall consider some of the general principles of competition between species, illustrate the potential effects of competition by examining the results of laboratory experiments, and demonstrate the importance of competition in natural systems.

Consumers compete for resources

The ecologist David Tilman of the University of Minnesota defines a **resource** as any substance or factor that is consumed by an organism and that supports increased population growth rates as its availability in the environment increases. Three things are key to this definition. First, a resource is consumed, and its amount is thus reduced. Second, a consumer uses a resource for its own maintenance and growth. Thus, food is always a resource, and water is a resource for terrestrial plants and animals. Third, when resource availability is reduced, biological processes are affected in such a way as to reduce population growth.

Consumption includes more than just eating. For sessile animals, space (open, available sites) is a resource. Among barnacles growing on rocks within the intertidal zone, individuals require space to grow, and larvae require space to settle and take up adult life (❚ Figure 19.3). Crowding increases adult mortality and reduces fecundity by limiting the growth of adults and the recruitment (settling) of larvae. Open space fosters reproduction and recruitment, and individuals "consume" open sites as they colonize and grow on them.

Hiding places and other safe sites constitute another kind of resource. Each area of habitat has a limited number of holes, crevices, or patches of dense cover in which an organism may escape predation or seek refuge from inclement weather. As some individuals occupy the best sites, others must settle for less favorable places; consequently, they may suffer higher mortality.

What factors are not resources? Temperature is not a resource. Higher temperatures may raise reproductive rates, but individuals do not consume temperature. One individual does not change the temperature of the environment for another. Temperature and other nonconsumable physical and biological factors are important, of course, but they must be considered differently from resources.

Renewable and nonrenewable resources

Resources can be classified according to how their consumers affect them. **Nonrenewable resources,** such as space, are not regenerated. Once occupied, space becomes unavailable; it is "replenished" only when the consumer leaves. In contrast, **renewable resources** are constantly regenerated, or renewed. Births in a prey population continually supply food items for predators. The continuous decomposition of organic detritus in the soil provides a fresh supply of nitrate for plant roots.

(a)

(b)

∥ Figure 19.3 For sessile animals, space is an important resource. (a) When barnacles on the Maine coast are living above their optimal range in the intertidal zone, their density is low, and young can settle in the bare patches. (b) Lower in the intertidal zone, dense crowding of barnacles precludes further population growth; young barnacles can settle only on an older individual. Courtesy of the American Museum of Natural History.

Among renewable resources, ecologists recognize three types. The first type includes resources that have a source external to the system, beyond the influence of consumers. Sunlight strikes the surface of the earth regardless of whether plants "consume" it, and local precipitation is largely independent of the consumption of water by plants.

Renewable resources of the second type are generated within the ecosystem, and their abundances are directly depressed by their consumers. Most predator–prey, plant–herbivore, and parasite–host interactions involve renewable resources because the supply of prey, plants, and hosts is constantly regenerated. Yet, by reducing populations of their prey, predators also reduce the rate of renewal of their food supply. This linkage can have special consequences for the dynamic relationships between consumer and resource populations.

Renewable resources of the third type are regenerated within the ecosystem, but resource and consumer are linked *indirectly,* either through other resource–consumer steps or through abiotic processes. In the nitrogen cycle of a forest, for example, plants assimilate nitrate from the soil. Herbivores and detritivores consume plants, returning large quantities of organic nitrogen compounds to the soil. These compounds are consumed by detritivores and further broken down by microorganisms, which release the nitrogen in a form the plants can use. The uptake of nitrate by plants has little direct effect on its renewal by detritivores. Similarly, consumption of detritus cannot immediately influence plant production. Clearly, however, detritivores and microorganisms do influence plant production

indirectly through the rate at which they release nutrients into the soil.

Limiting resources

Consumption reduces the availability of both renewable and nonrenewable resources. What is used by one organism cannot be used by another. By diminishing their resources, consumers limit their own population growth. As a population grows, its overall resource requirement grows as well. When the requirement increases so much that the decreasing supply of resources can no longer fulfill the need, population size levels off or even begins to decrease. However, whereas all resources are, by definition, reduced by their consumers, not all resources limit consumer populations in this way. All terrestrial animals require oxygen, for example, but they do not depress its level in the atmosphere even noticeably before some other resource, such as food supply, limits population growth.

The potential of a resource to limit population growth depends on its availability relative to demand. At one time, ecologists believed that populations were limited by the single resource that was most scarce. This principle has been called **Liebig's law of the minimum,** after Justus von Liebig, a German chemist who set forth the idea in 1840. According to this law, each population increases until the supply of some resource, the **limiting resource,** no longer satisfies the population's need for it.

The growth of a population under a given set of conditions responds uniquely to the level of each of its resources. David Tilman discovered that when the diatom *Cyclotella meneghiniana* is grown under silicate and phosphate limitation in a laboratory culture, population growth and resource depletion cease when phosphate levels are reduced to 0.2 mM or silicate levels are reduced to 0.6 mM. According to Liebig's law of the minimum, whichever of these resources is reduced to this limiting value first regulates the growth of the *Cyclotella* population.

Liebig's law applies strictly only to resources having an independent influence on the consumer. In many cases, two or more resources interact to determine the growth rate of a consumer population; that is, the growth rate of a population at a particular level of one resource depends on the level of one or more other resources.

ECOLOGISTS IN THE FIELD

Limitation by more than one resource

When two resources together enhance the growth of a consumer population more than the sum of both individually, the resources are said to be **synergistic** (from the classical Greek roots *syn*, "together," and *ergon*, "work"). This principle was illustrated in a study by the British ecologists W. J. H. Peace and P. J. Grubb of the small herbaceous plant *Impatiens parviflora*, which is common in woodlands of England.

In one experiment, Peace and Grubb exposed fertilized (with nitrate and phosphate) and nonfertilized (control) *Impatiens* to different levels of light from the time of seed germination until the end of the experiment at 5 weeks. Added light enhanced the growth of fertilized plants more than that of controls (▌Figure 19.4). Thus, the ability of *Impatiens* to use

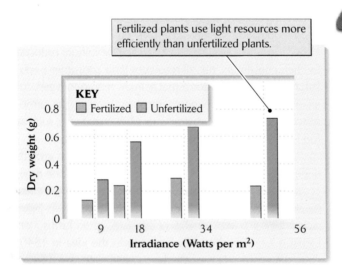

Fertilized plants use light resources more efficiently than unfertilized plants.

KEY
■ Fertilized ■ Unfertilized

▌**Figure 19.4 Resources often act together to influence growth.** The joint influence of light levels and fertilizer on the growth of *Impatiens*. After W. J. H. Peace and P. J. Grubb, *New Phytol.* 90:127–150 (1982).

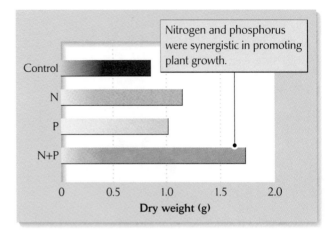

Nitrogen and phosphorus were synergistic in promoting plant growth.

▌**Figure 19.5 Nitrogen and phosphorus are synergistic in promoting plant growth.** The influence of nitrogen (N) and phosphorus (P) fertilization on growth of *Impatiens* to 5 weeks of age is shown. Neither nutrient alone enhanced growth as much as the addition of both. After W. J. H. Peace and P. J. Grubb, *New Phytol.* 90:127–150 (1982).

light depends on the presence of other resources. Plant growth requires both the carbon assimilated by photosynthesis, as a source of energy and for structural carbohydrates, and nitrogen and phosphorus, which are needed for the synthesis of proteins and amino acids.

At the highest light intensities used in the experiment, nitrogen and phosphorus were also shown to interact synergistically in their effect on plant growth (▌Figure 19.5), demonstrating that both are required for normal growth.

Failure of species to coexist in laboratory cultures led to the competitive exclusion principle

Many of the early studies of population dynamics were designed to determine the effects of one species on the population growth of another. In these experiments, two species were first grown separately under controlled conditions and resource levels to determine their carrying capacities in the absence of interspecific competition. The two species were then grown together under the same conditions to determine the effect of each on the other. The difference between the population growth of one species in the presence and in the absence of the other was taken as a measure of competition between them.

Experiments of this kind by the Russian biologist G. F. Gause on protozoans were among the first and most influential on subsequent work in population biology. When the protozoans *Paramecium aurelia* and *P. caudatum* were

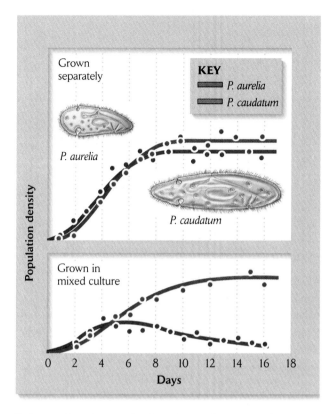

Figure 19.6 Gause's landmark experiments on the coexistence of species in laboratory cultures led to the competitive exclusion principle. Two species of *Paramecium* were grown in separate cultures (*above*) and grown together (*below*). Although both species thrived when grown separately, *P. caudatum* did not persist in the presence of *P. aurelia*. After G. F. Gause, *The Struggle for Existence*, Williams & Wilkins, Baltimore (1934).

prompt us to ask how much difference in resource requirements is sufficient to allow coexistence. Although this question has been very difficult to answer, theoretical analyses of competition suggest some of the general conditions under which species may coexist.

The theory of competition and coexistence is an extension of logistic growth models

Most competition theory springs from mathematical models developed by A. J. Lotka and G. F. Gause, who used the logistic equation for population growth as their starting point. Remember that according to the logistic equation (Chapter 14), the rate of increase of a population is expressed by

$$\frac{1}{N}\frac{dN}{dt} = r\left[\frac{K-N}{K}\right]$$

where r is the exponential rate of increase in the absence of competition and K is the number of individuals that the environment can support (the carrying capacity). Note that this equation is presented in a slightly different form than in Chapter 14 to make the incorporation of interspecific competition more straightforward. In this equation, intraspecific competition appears as the term $(K-N)/K$; as N approaches K (that is, as population size approaches the carrying capacity), $K-N$ approaches 0. As we have seen before, a stable equilibrium is reached when $N = K$ and population size has reached the carrying capacity (**Figure 19.7**).

established separately on the same type of nutritive medium, both populations grew rapidly to limits imposed by resources. When the two species were grown together, however, only *P. aurelia* persisted (**Figure 19.6**). Similar experiments with fruit flies, mice, flour beetles, and annual plants usually produced the same result: one species persisted and the other died out, usually after 30 to 70 generations.

The accumulating results of laboratory experiments on competition were eventually summarized by the **competitive exclusion principle,** which states that two species cannot coexist indefinitely on the same limiting resource. The qualification *limiting* is required in the definition of the principle because competitive exclusion expresses itself only when consumption depresses resources to the degree that it limits population growth. Similar species do, of course, coexist in nature. But as we shall see in later chapters, observations often reveal subtle differences between them in habitat or diet preference. These observations

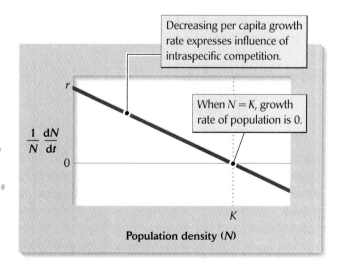

Decreasing per capita growth rate expresses influence of intraspecific competition.

When $N = K$, growth rate of population is 0.

Figure 19.7 Intraspecific competition depresses the per capita population growth rate. The population continues to increase, however, until population size (*N*) is equal to carrying capacity (*K*).

The Italian ecologist Vito Volterra incorporated inter-specific competition into the logistic equation. When modeling the interaction between two species, it is necessary to distinguish each of the populations. We will do this by using a subscript 1 for species 1 and a subscript 2 for species 2. Volterra modeled the effect of interspecific competition from species 2 on the population growth of species 1 by adding the term $-a_{1,2}N_2/K_1$ to the quantity within the brackets. Hence

$$\frac{1}{N_1}\frac{dN_1}{dt} = r_1\left[\frac{K_1 - N_1 - a_{1,2}N_2}{K_1}\right],$$

where N_2 is the number of individuals of species 2, and $a_{1,2}$ is the coefficient of competition—that is, the effect of an individual of species 2 on the exponential growth rate of the population of species 1. We may think of the competition coefficient $a_{1,2}$ as the degree to which each individual of species 2 uses the resources of individuals of species 1. How much individuals of species 2 usurp the resources of species 1 (K_1) determines the effect of population 2 on population 1's rate of growth and on the equilibrium size of population 1 under interspecific competition (Figure 19.8).

Because each species of a pair exerts an effect on the other, the mutual relationship between them requires two equations. One, presented above, incorporates the effect of species 2 on species 1. The second equation, for the effect of species 1 on species 2, is similar to the first, but with the subscripts 1 and 2 reversed.

If two species are to coexist, the populations of both must reach a stable size greater than zero. That is, both dN_1/N_1dt and dN_2/N_2dt must equal zero at some combination of positive values of N_1 and N_2. From the previous equation, we see that $dN_1/N_1dt = 0$ when

$$\hat{N}_1 = K_1 - a_{1,2}N_2$$

and $dN_2/N_2dt = 0$ when

$$\hat{N}_2 = K_2 - a_{2,1}N_1.$$

The little "hat" ($\hat{\ }$) over the Ns indicates that they are equilibrium values. In the absence of interspecific competition ($a_{1,2} = 0$) the equilibrium population size $\hat{N}_1$ is equal to K_1, a measure of the resources available to species 1. Interspecific competition reduces the effective carrying capacity of the environment for species 1 by the amount $a_{1,2}\hat{N}_2$—that is, in proportion to the population size and coefficient of competition of the second species (see Figure 19.8).

In general, these equations tell us that coexistence is most likely—that is, both $\hat{N}_1$ and $\hat{N}_2 > 0$—when coeffi-cients of interspecific competition ($a_{1,2}$ and $a_{2,1}$) are relatively weak, in particular less than 1. Any number of species may coexist as long as this criterion is met for all pairs of them. The simplest way to achieve competition coeffi-cients less than 1 is for competitors to share resources incompletely by partitioning resources among themselves.

HELP ON THE WEB | *Go to Living Graphs at http://www.whfreeman. com/ricklefs and use the interactive tutorial on competition better to understand the dynamics between species.*

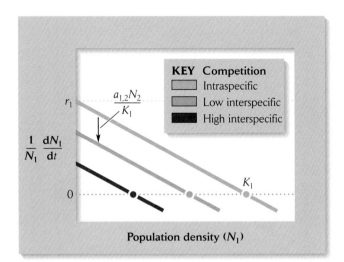

Figure 19.8 Interspecific competition reduces the equilibrium level of a population below its carrying capacity. The per capita population growth rate of species 1 as a function of its population density (N_1) is shown under purely intraspecific competition and under low and high levels of interspecific competition.

Field studies demonstrate the pervasiveness of competition in nature

We can observe the process of competitive exclusion in the laboratory when we mix populations and follow the course of their interaction. In nature, however, the evidence of competitive exclusion is lost when a poor competitor disappears. The closest natural analogy to a laboratory experiment is the accidental or intended introduction of species by humans. For example, when several species of parasites are introduced into an area simultaneously to control a weed or insect pest, the control species are brought together in the same locality to exploit the same resource. It is not surprising that competitive exclusion might occur under these conditions.

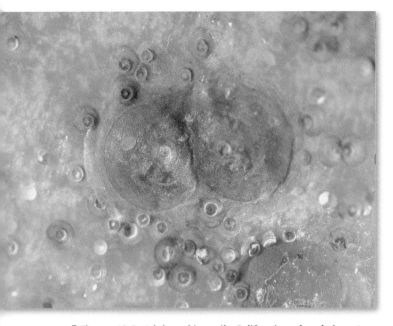

Figure 19.9 Adult and juvenile California red scale insects produce a waxy protective coating. Photo by Jack Kelly Clark, courtesy of the University of California Statewide IPM Project.

A case involving wasps that parasitize scale insects (**Figure 19.9**) has been thoroughly documented in southern California. Scale insects are pests of citrus groves and can cause extensive damage to trees. As the evolution of resistance reduced the effectiveness of chemical pesticides (see Chapter 9), agricultural biologists turned to the importation of parasitoids and predators to control these pests.

Of the many species introduced in an effort to control citrus scale, tiny parasitoid wasps of the genus *Aphytis* (from the Greek *aphysso,* "to suck") have been the most successful. One species, *A. chrysomphali,* was accidentally introduced from the Mediterranean region and became established by 1900 (**Figure 19.10**). Despite its tremendous population growth potential, *A. chrysomphali* did not effectively control scale insects, particularly not in the dry interior valleys. In 1948, a close relative from southern China, *A. lingnanensis,* was introduced as a control agent. This species increased rapidly and widely replaced *A. chrysomphali* within a decade (**Figure 19.10**). When both species were grown in the laboratory, *A. lingnanensis* was found to have the higher net reproductive rate, whether the two species were placed separately or together in population cages.

Although *A. lingnanensis* had excluded *A. chrysomphali* throughout most of southern California, it still did not provide effective biological control of scale insects in the interior valleys because cold winter temperatures greatly reduced its populations. Its larval development slows to a standstill at temperatures below 16°C (60°F), and adults cannot tolerate temperatures below 10°C (50°F). In 1957, a third species of wasp, *A. melinus,* was introduced from areas in northern India and Pakistan where temperatures range from below freezing in winter to above 40°C in summer. As was hoped, *A. melinus* spread rapidly throughout the interior valleys of southern California, where temperatures resemble those of the wasp's native habitat, but it did not become established in the milder coastal areas.

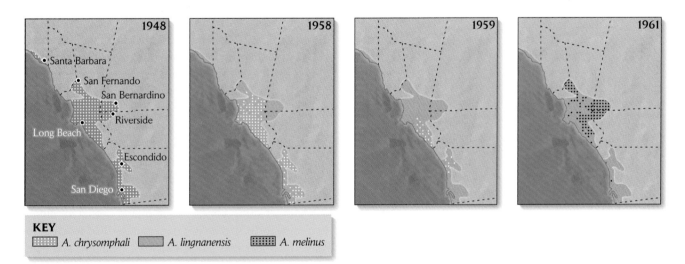

KEY

A. chrysomphali   A. lingnanensis   A. melinus

Figure 19.10 Competitive exclusion may occur when new competitors are introduced. The successive changes in the distributions of three species of parasitoids (genus *Aphytis*) on citrus scale between 1948 and 1961 suggest competitive exclusion. After P. Debach and R. A. Sundby, *Hilgardia* 34:105–166 (1963).

ECOLOGISTS IN THE FIELD

An experimental study of competition in a forest herb

The depressing effect of interspecific competition on the growth of plants has been demonstrated in many experimental studies. One such study, by plant ecologist W. G. Smith, used two species of *Desmodium*, small herbaceous legumes (members of the pea family) common in oak woodlands in the midwestern United States. Small individuals of each of two species, *D. glutinosum* and *D. nudiflorum*, were planted 10 cm from a large individual of the same species (intraspecific test), 10 cm from a large individual of the other species (interspecific test), or at least 3 m from any other *Desmodium* plant (control). (All the plants were, however, surrounded by unrelated plants that occur in their habitat.) The total increase in length of all leaves, both old and new, served as an index to growth.

The results of the experiment (Figure 19.11) showed that both species grew best in the absence of individuals of either species. It was also clear, however, that the growth of *D. nudiflorum* was depressed more by interspecific than by

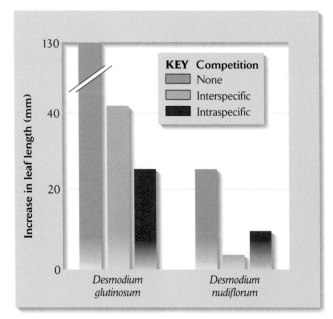

Figure 19.11 Interspecific competition may depress the growth of plants. Growth responses of two species of *Desmodium* are shown when planted near individuals of the same species, near individuals of the other species, and at a distance from individuals of either species. In this case, *D. glutinosum* is the superior competitor. After W. G. Smith, *Am. Midl. Nat.* 94:99–107 (1975).

intraspecific competition, and that interspecific competition was asymmetrical, with *D. glutinosum* exerting the stronger effect and *D. nudiflorum* the weaker..

Plant competition differs between nutrient-rich and nutrient-poor habitats

Several ecologists have suggested that the outcome of plant competition depends on nutrient levels in soils. According to one hypothesis promoted by plant ecologists P. J. Grubb and David Tilman, the intensity of competition is greater where resources are less abundant. Where nutrient levels are high, they are less likely to be limiting to plant populations, and therefore interspecific competition should be weaker. The opposite point of view, espoused by J. P. Grime and Paul Keddy, suggests that lack of water and mineral nutrients should limit plant populations so much that individual plants will be spaced far apart and there will be relatively little competition for light. The difference between the Grubb–Tilman and Grime–Keddy hypotheses lies in the relative importance placed on belowground and aboveground competition for resources—that is, nutrients and light, respectively.

One way to distinguish between these two hypotheses is to conduct competition experiments in high-productivity and low-productivity environments. The results often depend on the particular species used and the way in which the experiments are designed. For example, a competition experiment in Israel with the desert annual *Stipa capensis* showed that competition intensity increased with increasing resource availability, as measured by aboveground biomass (Figure 19.12). Variation in productivity was related to variation in soil water, which differed between the two years of the study and which was additionally manipulated during each year. With little water, plants achieved small stature and did not interfere with each other; apparently, most of the competition in this system occurs above ground. Furthermore, the experiment involved an annual plant, which grows up from seed each year and then dies. Low water availability could prevent these plants from growing large enough root systems to compete for limited water or nutrients in the soil.

Another study, conducted in prairie habitat in Minnesota using three species of prairie grasses, had different results. The researchers established low-, medium-, and high-productivity plots by adding ammonium nitrate fertilizer. Aboveground biomass varied among the plots by factors of only 2 to 3, not by several orders of magnitude as in the study in Israel. Competition intensity did not vary signifi-

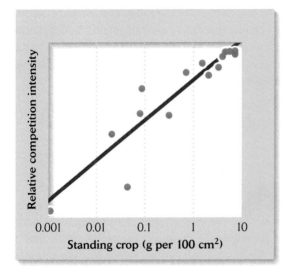

Figure 19.12 The intensity of competition may increase with resource availability. Interspecific competition between the desert annual plant *Stipa capensis* and several other species increases as a function of the productivity of the habitat. Differences in productivity were measured by standing crop and were produced by varying the water availability. After R. Kadmon, *J. Ecol.* 83:253–262 (1995).

cantly over the nutrient gradient. The investigators suggested that competition was intense below ground on low-nutrient plots and above ground on high-nutrient plots, resulting in strong competition across the gradient.

What are we to make of such studies? Competition appears to be pervasive, but the manifestation of competition depends very strongly on the characteristics of the species and habitats in which competitive effects are investigated.

Competition may occur through exploitation of shared resources or direct interference

In the examples that we have seen so far, individuals compete indirectly through their mutual effects on shared resources. This kind of competition is called **exploitation competition** (Figure 19.13). Less frequently, when consumers can profitably defend resources, competitors may interact directly through various antagonistic behaviors. This behavior is referred to as **interference competition.** Hummingbirds chase other hummingbirds, not to mention bees and moths, from flowering bushes. Encrusting sponges use poisonous chemicals to overcome other species of sponges as they expand to fill open space on rock surfaces. Many shrubs release toxic chemicals into the soil that depress the growth of competitors. Even bacteria wage chemical warfare with one another to tip the balance of their competitive interactions.

Allelopathy

Chemical competition, or **allelopathy,** has been reported most frequently in terrestrial plants, in which such interactions may take on a variety of forms. Most frequently, this type of interference competition involves the direct effect of a toxic substance that causes injury (*-pathy*) to other (*allelo-*) individuals. It has also been suggested that the abundant oils in the eucalyptus trees of Australia promote frequent fires in the leaf litter, which kill the seedlings of competitors (Figure 19.14).

In shrub habitats in southern California, several species of sage (genus *Salvia*) use chemicals to inhibit the growth of other vegetation. Clumps of *Salvia* are usually surrounded by bare areas separating the sage from neighboring grassy areas (Figure 19.15). When observed over long periods, *Salvia* can be seen to expand into the grassy areas. But because sage roots extend only to the edge of the bare strip and not beyond, it is unlikely that a toxin is extruded

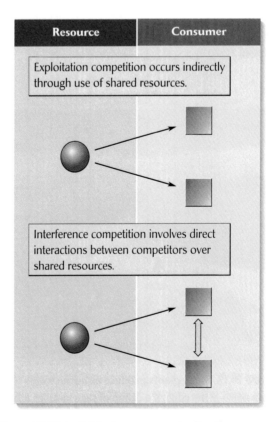

Figure 19.13 Individuals can compete directly or indirectly for resources.

Figure 19.14 Some plants compete by chemical means. The litter and bark of eucalyptus trees in Australia have a high oil content, which promotes fires that kill seedlings of potentially competing species, but leave the eucalyptus trees unharmed. Photo by R. E. Ricklefs.

into the soil directly by the roots. The leaves of *Salvia* produce volatile terpenes (a class of organic compounds that includes camphor and gives foods spiced with sage part of their distinctive taste), which apparently harm nearby plants directly through the atmosphere.

ECOLOGISTS IN THE FIELD

Competition for space among barnacles

The animals that most closely resemble plants in competing for space as a resource are sessile invertebrates of rocky shores. Among the most prominent of these are barnacles, which may form dense, continuous populations. Barnacles gather food in the form of plankton from the water that washes over them. Food is not a limiting resource for barnacles because their feeding cannot reduce the vast numbers of plankton in coastal waters. Rather, populations of barnacles in many areas are limited by space for settling and growth.

Joseph Connell, now at the University of California at Santa Barbara, performed a series of classic experiments on two species of barnacles within the intertidal zone of the rocky coast of Scotland. Adults of *Chthamalus stellatus* normally occur higher in the intertidal zone than those of *Balanus balanoides*, the more northerly of the two species. Although the vertical distributions of newly settled larvae of the two species overlap broadly within the intertidal

(a)

(b)

Figure 19.15 Some plants produce airborne toxins to compete with nearby plants. (a) Bare patch at the edge of a clump of sage includes a 2-m-wide strip with no plants (A to B) and a wider area of inhibited grassland (B to C) lacking wild oat and bromegrass, which are found with other species to the right of C in unaffected grassland. (b) An aerial view shows sage and California sagebrush invading annual grassland in the Santa Ynez Valley of California. Courtesy of C. H. Muller, from C. H. Muller, *Bull. Torrey Bot. Club* 93:332–351 (1966).

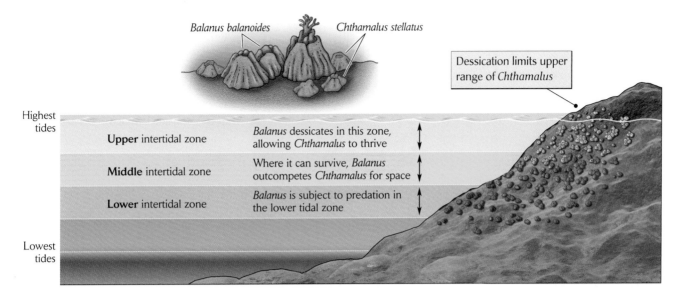

Balanus balanoides *Chthamalus stellatus*

Dessication limits upper range of *Chthamalus*

Highest tides

Upper intertidal zone	*Balanus* dessicates in this zone, allowing *Chthamalus* to thrive
Middle intertidal zone	Where it can survive, *Balanus* outcompetes *Chthamalus* for space
Lower intertidal zone	*Balanus* is subject to predation in the lower tidal zone

Lowest tides

▌**Figure 19.16 Some sessile organisms compete for space.** The distributions of the barnacle species *Balanus balanoides* and *Chthamalus stellatus* within the intertidal zone on rocky shores in Scotland differ because of interference competition as well as physical factors. After J. H. Connell, *Ecology* 42:710–723 (1961) and *Ecol. Monogr.* 31:61–104 (1961).

zone, the line between the vertical distributions of adults is sharply drawn (▌Figure 19.16).

Connell demonstrated that adult *Chthamalus* live only in the portion of the intertidal zone above *Balanus* not because of physiological tolerance limits, but because of interspecific competition. When Connell removed *Balanus* from rock surfaces, *Chthamalus* thrived in the lower portions of the intertidal zone where they normally did not occur. The two barnacle species compete directly for space. *Balanus* have heavier shells and grow more rapidly than *Chthamalus*; as individuals expand, the shells of *Balanus* edge underneath those of *Chthamalus* and literally pry them off the rock! *Chthamalus* can live in the upper parts of the intertidal zone because they are more resistant to desiccation than *Balanus*; even when surfaces in the upper levels are kept free of *Chthamalus*, *Balanus* do not invade.

MORE ON THE WEB *Asymmetry in competition.* The asymmetry between *Balanus* and *Chthamalus* reflects a basic trade-off between competitive ability and tolerance of stressful conditions. Such trade-offs are common in natural systems.

Competition among terrestrial animals

Competition between barnacles results from physical interference rather than from depression of shared food or other resources. Mobile animals may exhibit similar interference competition through occasional aggressive encounters. For example, two species of voles (small mouse-like rodents of the genus *Microtus*) co-occur in some areas of the Rocky Mountain states. In western Montana, the meadow vole (*M. pennsylvanicus*) normally lives in wet habitats surrounding ponds and watercourses, whereas the mountain vole (*M. montanus*) is restricted to dry habitats. When meadow voles were experimentally trapped and removed from an area of wet habitat, mountain voles began to move in from surrounding dry habitats. And when mountain voles were trapped and removed from a dry habitat that they occupied exclusively, meadow voles began to show up there. Each species excludes the other from its preferred habitat by aggressive behavior.

Although territorial defense and social aggression occur frequently within species, they are not as common between species. Interspecific competition occurs more regularly through exploitation of resources. Because exploitative competition expresses its effects indirectly, through differential survival and reproduction of individuals of different species, it may be difficult to detect.

In Big Bend National Park, Texas, the canyon lizard *Sceloporus merriami* and the tree lizard *Urosaurus ornatus* appear to compete for a shared food resource. Both species search for insect prey on exposed surfaces of large rocks. When *Sceloporus* were removed from experimental areas, the numbers of *Urosaurus* increased over those in control

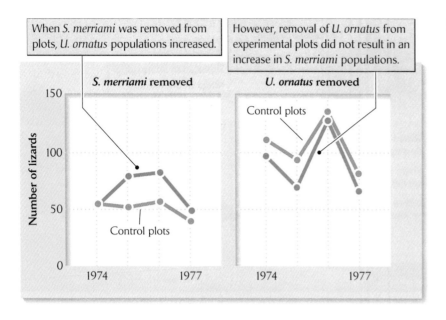

When *S. merriami* was removed from plots, *U. ornatus* populations increased.

However, removal of *U. ornatus* from experimental plots did not result in an increase in *S. merriami* populations.

Figure 19.17 Exploitation competition may affect two species unequally. Populations of two lizard species (individuals per hectare) are shown for plots from which the other species was removed and for control plots where both species remained. From A. E. Dunham, *Ecol. Monogr.* 50:309–330 (1980).

areas during 2 years of a 4-year study (Figure 19.17). In contrast, removal of *Urosaurus* did not result in an increase in *Sceloporus* populations. Fluctuations in both control and experimental populations of *Sceloporus* were closely related to rainfall (1975 and 1977 were dry years), suggesting that physical factors may limit *Sceloporus* more severely, whereas competition influences *Urosaurus* more strongly, a situation recalling the relative ecological positions of *Balanus* and Chthamalus in the intertidal zone.

Competition between distantly related species

Charles Darwin emphasized that competition should be most intense between closely related species or organisms. In *On the Origin of Species,* he remarked, "As species of the same genus have usually, though by no means invariably, some similarity in habits and constitution, and always in structure, the struggle will generally be more severe between species of the same genus, when they come into competition with each other, than between species of distinct genera." Darwin reasoned that similar structure indicates similar ecology, especially with respect to resources consumed.

Although this must generally be the case, many of the same resources are used by distantly related organisms. Barnacles and mussels, as well as algae, sponges, bryozoans, tunicates, and others, occupy space in the intertidal zone and actively compete by preemption and overgrowth. Both fish and aquatic birds prey on aquatic invertebrates. Krill (*Euphausia superba*), shrimplike crustaceans that abound in subantarctic waters (Figure 19.18), are fed upon by virtu-

ally every type of large marine animal, including fish, squid, diving birds, seals, and whales. Recent increases in seal and penguin populations in the Southern Ocean have been related to decreased competition from whales, whose populations have been decimated by commercial exploitation. In terrestrial habitats, spiders, ground beetles, salamanders, and birds consume invertebrates living in forest litter. In desert ecosystems, birds and lizards eat many of the same insect species, and ants, rodents, and birds consume the

Figure 19.18 Common food sources bring distantly related organisms into competition. In the Antarctic Ocean, krill (*Euphausia superba*) supply food for a wide range of marine animal species. Courtesy of Dr. Uwe Kils.

seeds of many of the same plants. These examples illustrate the strong potential for competition between very distantly related as well as unrelated organisms.

The outcome of competition can be influenced by predators

Darwin noted that grazing can maintain a high diversity of plants in grasslands. In the absence of grazers, dominant competitors grow rapidly and exclude others. Similar results have been obtained from experiments on marine algal communities under the pressure of grazing by limpets, snails, and urchins. These studies indicate that predation has a strong hand in shaping the structure of biological communities by influencing the outcome of competitive interactions between prey species.

University of Washington ecologist Robert Paine was one of the first investigators to demonstrate this point experimentally. On the exposed rocky coast of the state of Washington, the intertidal zone harbors several species of barnacles, gooseneck barnacles, mussels, limpets, and chitons (a kind of grazing mollusk). All of these animals are preyed upon by the sea star *Pisaster* (▌ Figure 19.19). Paine removed sea stars from a study area 8 meters in length and 2 meters in vertical extent; an adjacent area of similar size was left undisturbed.

Following the removal of the sea stars, the number of prey species in the experimental plot decreased rapidly, from fifteen at the beginning of the study to eight at the end. Diversity declined in the experimental area because populations of barnacles and mussels increased and crowded out many of the other species. Paine concluded that sea stars maintain the diversity of the area by limiting populations of barnacles and mussels, which are superior competitors for space in the absence of predators.

ECOLOGISTS IN THE FIELD

Predation and competition in anuran communities

Studies conducted in artificial ponds have shown that predators can reverse the outcome of competition among anuran (frog and toad) tadpoles. In one experiment conducted by Peter Morin of Rutgers University, ponds were supplied with 200 hatchlings of the spadefoot toad (*Scaphiopus holbrooki*), 300 of the spring peeper (*Hyla crucifer*), and 300 of the southern toad (*Bufo terrestris*). Each of the ponds, which were identical in all other respects, also received 0, 2, 4, or 8 individuals of the predatory broken-striped newt (*Notophthalmus viridescens*).

In the absence of newt predation, *Scaphiopus* tadpoles grew rapidly, survived well, and dominated the ponds along with smaller numbers of *Bufo; Hyla* tadpoles were all but eliminated by competition (▌ Figure 19.20). The newts, however, apparently preferred toad tadpoles, and at higher numbers of predators, survival of both *Scaphiopus* and *Bufo* decreased markedly. With fewer toads per pond,

(a)　　(b)　　(c)

▌ **Figure 19.19 In the absence of predators, some competitors can dominate others.**
(a) A congregation of sea stars (*Pisaster*) at low tide on the coast of the Olympic Peninsula, Washington. The sea star (b) is an important predator on mussels (c). Photo (a) by Ken Lucas/Visuals Unlimited; photo (b) by Daniel W. Gotshall/Visuals Unlimited; photo (c) by Francis & Donna Caldwell/Visuals Unlimited.

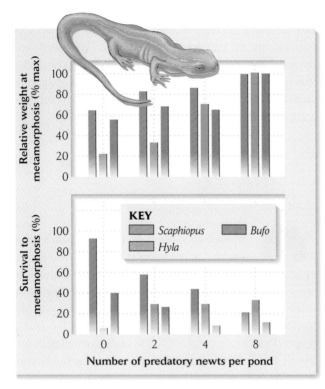

Figure 19.20 Predators can influence the outcome of competition between prey species. Predators strongly affected the growth (measured by weight at metamorphosis) and survival of three species of anurans (frogs and toads) raised in artificial ponds. From P. J. Morin, *Science* 212:1284–1286 (1981).

supplies of food increased, and survival and growth of *Hyla* tadpoles improved immensely, as did the growth of surviving *Scaphiopus* and *Bufo* tadpoles.

Summary

1. Competition is the use or contesting of a resource by two or more individual consumers. When the individuals belong to the same species, their interaction is called intraspecific competition; when they belong to different species, it is called interspecific competition.

2. A resource may be defined as any factor that is consumed and whose increase promotes population growth. Thus, light, food, water, mineral nutrients, and space are resources. Temperature, salinity, and other such conditions are not.

3. Resources may be classified as nonrenewable (space) or renewable (light and food), and the latter may be further distinguished according to the influence of the consumer on the provisioning of the resource: no influence, direct influence, or indirect influence through other consumers.

4. Of all the resources consumed, only one or a few limit the population growth of the consumer. These limiting resources are normally those whose supply relative to demand is least. This principle is known as Liebig's law of the minimum.

5. Competition may be inferred by a change in the population size of one species following the addition or removal of another. When two species compete strongly, the population of the first is sensitive to changes in numbers of the second, and vice versa.

6. Theoretical investigations and laboratory studies have led to the generalization that no two species of competitors can coexist on the same limiting resource. This has come to be known as the competitive exclusion principle.

7. Some mathematical treatments of competition are based on the logistic equation of population growth, to which a term is added for the effect of interspecific competition. The strength of this effect in the model is specified by the coefficient of competition.

8. The equilibrium population sizes of two competing species can be described by an equation including the carrying capacities and competition coefficients for each of the species. In the most general terms, coexistence requires that the product of the competition coefficients of the first and second species be less than 1.

9. Laboratory experiments present clear evidence of competition among species, but natural populations are limited by physical conditions and consumers as well as by shared resources. Ecologists have also conducted numerous field studies designed to reveal the influence of competition on the sizes of natural populations.

10. Transplant experiments with plants under varying conditions of intraspecific and interspecific competition illustrate differences in the mechanisms of competition between habitats of high and low productivity.

11. When individuals compete through their effects on shared resources, competition is exploitative, or indirect; when they defend resources against one another, their behavior is known as direct, or interference, competition.

12. Many plants compete directly by producing volatile substances that impair the growth and survival of individuals of other species. This mechanism is known as allelopathy.

13. Removal experiments involving intertidal invertebrates have demonstrated strong competition among

space-filling animals such as barnacles, mussels, and encrusting sponges. Competitive exclusion is accomplished by direct physical interaction.

14. Rapid invasion of a habitat by one mobile animal species following the removal of another demonstrates direct interference competition through aggressive behavior.

15. Exploitation competition is most convincingly demonstrated in studies that show appropriate changes in resource levels accompanying the demographic response of one species after removal of a competitor.

16. Predators can alter the outcome of competitive interactions if they selectively prey upon superior competitors. This behavior increases the diversity of the prey trophic level by permitting more prey species to coexist.

PRACTICING ECOLOGY

CHECK YOUR KNOWLEDGE

Asymmetric Competition

Experiments conducted in field situations have indicated that many species respond to the addition or removal of a second species, but the second does not respond to manipulation of the first. This is referred to as asymmetric competition, because its effect is not felt equally by both species. According to one survey of interactions between 98 pairs of species, no response was found for either species in 44 cases, reciprocal effects were found in 21 cases, and a response by only one species was found in 33 of the interactions. The large proportion of cases in the last category indicates that asymmetry is more common than reciprocal negative competitive effects.

Asymmetry in competition must occur because of an imbalance in the fundamental ecological relationships between two species. The superior competitor (that is, the one that "wins") is almost always more strongly limited by some other factor, such as environmental stress or predators. In the case of the barnacles mentioned earlier in this chapter, *Balanus* could effectively outcompete *Chthamalus* in the lower portions of the intertidal zone of the rocky coast of Scotland but it lacked the physiological tolerance of desiccation that allowed *Chthamalus* to occur higher up on the shoreline. Thus, when *Balanus* was removed from the lower portion of the shoreline, *Chthamalus* could establish itself, indicating a strong effect of *Balanus* on *Chthamalus*. However, when *Chthamalus* was removed from rocks at the higher and drier levels of the shoreline, *Balanus* did not colonize.

Parasitism may be another factor that can create asymmetric competition. D. S. Maksimowich and A. Mathis, ecologists at Southwest Missouri State University in Springfield, recently examined the effects of parasitism on the ability of salamanders to compete for territories and food resources. In particular, they studied the impacts of an ectoparasitic mite (*Hannemania eltoni*) on aggressive and foraging behavior in the Ozark zigzag salamander, *Plethodon angusticlavius*. In a first experiment, territoriality of male salamanders with high parasite loads was compared with that of males having low parasite loads. In another experiment, the effects on the foraging behavior of female salamanders with and without parasites were tested in response to familiar prey (*Drosophila* flies) and novel prey (termites). Parasitized female salamanders took significantly longer to capture both familiar and novel prey. Also, the nonparasitized salamanders consumed more of both prey types. The results suggest that parasite-mediated effects on competitive interactions may have important influences on the fitness of individuals in natural populations. When two species have different parasites, the depressing effect of a particularly virulent parasite on the competitive ability of one of the species could establish a competitive asymmetry between the species.

CHECK YOUR KNOWLEDGE

1. How could knowledge of the balance of competition affect our ability to manage non-native species, which are invading the nation at an unprecedented rate?

2. Examine Table 19.1; what is the difference in time that it took for the nonparasitized and parasitized females to eat familiar prey? The time it took for the nonparasitized and parasitized females to eat novel prey? How might you explain the differences in the time required for parasitized vs. nonparasitized females to capture a *Drosophila* fruit fly?

Table 19.1 Response of parasitized and nonparasitized females to fruit flies (*Drosophila*) and termites (*Reticulitermes*) (responses are time to capture measured in seconds)

	Parasitized	Nonparasitized
Fruitflies (*Drosophila*)	180	100
Termites (*Reticulitermes*)	115	85

MORE ON THE WEB **3.** Visit the *World of Parasites* Web page from McGill University through *Practicing Ecology on the Web* at *www.whfreeman.com/ricklefs*. Click on the map to see a list of many of the parasitic species of North America. What parasite species is the most prevalent? What disease does it cause?

Suggested Readings

Connell, J. H. 1961. The influence of interspecific competition and other factors on the distribution of the barnacle *Chthamalus stellatus*. *Ecology* 42:710–723.

Connor, E. F., and D. Simberloff. 1986. Competition, scientific method, and null models in ecology. *American Scientist* 74:155–162.

Goldberg, D. E., and A. M. Barton. 1992. Patterns and consequences of interspecific competition in natural communities: A review of field experiments with plants. *American Naturalist* 139:771–801.

Grace, J. B., and D. Tilman (eds.). 1990. *Perspectives on Plant Competition*. Academic Press, San Diego, CA.

Hardin, G. 1960. The competitive exclusion principle. *Science* 131:1292–1297.

Inchausti, P. 1995. Competition between perennial grasses in a Neotropical savanna: The effects of fire and of hydric-nutritional stress. *Journal of Ecology* 83:231–243.

Kadmon, R. 1995. Plant competition along soil moisture gradients: A field experiment with the desert annual *Stipa capensis*. *Journal of Ecology* 83:253–262.

Keddy, P. 1989. *Competition*. Chapman and Hall, London.

Maksimowich, D. S., and A. Mathis. 2000. Parasitized salamanders are inferior competitors for territories and food resources. *Ethology* 106:319–329.

Paine, R. T. 1974. Intertidal community structure: Experimental studies on the relationship between a dominant competitor and its principal predator. *Oecologia* 15:93–120.

Schoener, T. W. 1983. Field experiments on interspecific competition. *American Naturalist* 122:240–285.

Tilman, D. 1982. *Resource Competition and Community Structure*. Princeton University Press, Princeton, NJ.

Wilson, S. D., and D. Tilman. 1993. Plant competition and resource availability in response to disturbance and fertilization. *Ecology* 74:599–611.

CHAPTER 20

Coevolution and Mutualism

- Antagonists evolve in response to each other

- Coevolution in plant–pathogen systems reveals genotype–genotype interactions

- Consumers and resources can achieve an evolutionary equilibrium

- Competitive ability exhibits genetic variation and responds to selection

- Traits of competing populations may diverge through character displacement

- Mutualists have complementary functions

- Coevolution involves mutual evolutionary responses by interacting populations

Shortly after the release of a few pairs of rabbits on a ranch in Victoria in 1859, European rabbits became a major pest in Australia. Rabbit populations increased so quickly that within a few years, local ranchers were erecting rabbit fences and organizing rabbit brigades—shooting parties— in vain attempts to keep their numbers under control. Eventually, hundreds of millions of rabbits ranged throughout most of the continent, where they destroyed pasturelands and threatened wool production. The Australian government tried poisons, predators, and other control measures, all without success.

After much investigation, the answer to the rabbit problem seemed to be a myxoma virus (a relative of smallpox) discovered in populations of a related South American rabbit. The myxoma virus produced a small, localized fibroma (a fibrous cancer of the skin). Its effect on South American rabbits was not severe, but European rabbits infected by the virus died quickly of myxomatosis.

In 1950, the myxoma virus was introduced locally in Victoria. An epidemic of myxomatosis broke out among the rabbits and spread rapidly. The virus was transmitted primarily by mosquitoes, which bite infected areas of the skin and carry the virus on their mouthparts. The first epidemic killed 99.8% of the infected rabbits, reducing their populations to very low levels. But during the following myxomatosis season (which coincides with the mosquito season), only 90% of the remaining rabbit population was killed. During the third outbreak, only 40–60% of infected rabbits succumbed, and their population began to grow again.

The decline in the lethality of the myxoma virus resulted from evolutionary responses in both the rabbit and the virus populations. Before the introduction of the virus, some rabbits had genetic factors that conferred resistance to the disease. Although nothing had spurred an increase in those factors before, they were strongly selected by the

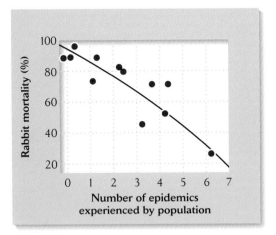

Figure 20.1 Interacting populations evolve in response to each other. The susceptibility of European rabbits in Australia to the introduced myxoma virus declined after the first epidemic. From F. Fenner and F. N. Ratcliffe, *Myxomatosis*, Cambridge University Press, London (1981).

myxoma epidemic, until most of the surviving rabbit population consisted of resistant animals (■ Figure 20.1). At the same time, less virulent virus strains became more prevalent because they did not kill their hosts as quickly and therefore were more readily dispersed to new hosts (mosquitoes bite only living rabbits).

Left on its own, the Australian rabbit–virus system would probably evolve to an equilibrial state of benign, endemic disease, as it had in the population of South American rabbits from which the myxoma virus was isolated. However, pest management specialists keep the system out of equilibrium and maintain the effectiveness of the myxoma virus as a control agent by finding new strains to which the rabbits have yet to evolve immunity.

The less virulent strains of myxoma have a higher rate of growth in rabbit populations as a whole, if not within individual rabbits. This pattern is unlike that of highly contagious diseases that are spread directly through the atmosphere or water. These pathogens, which do not depend on their hosts for dispersal, often exhibit high levels of virulence, with debilitating or even fatal consequences for their hosts. Similarly, most predators do not rely on a third party to find prey, and rather than evolving toward a benign equilibrium of restraint and tolerance, predator and prey tend to become locked in an evolutionary battle of persistent intensity. The outcome of the battle depends on which population gets the evolutionary upper hand.

When populations of two or more species interact, each may evolve in response to those characteristics of the other that affect its own evolutionary fitness. This process is referred to as **coevolution.** Populations of consumers, resources, and competitors select traits in the others that tend to alter their interactions. Plants and animals use a variety of structures and behaviors to obtain food and to avoid being eaten or parasitized. Much of this diversity is the result of natural selection acting on the ways in which plants and animals procure resources and escape predation. Wing markings that blend artfully into resting backgrounds by day enable moths to escape the notice of most predators. Flowers, by their insistent colors and fragrances, call attention to themselves and attract the notice of insects and birds that carry pollen from one flower to the next.

The agents whose influence has shaped such adaptations are biological—they are characteristics of living organisms. Their effects differ from those of physical factors in the environment in two ways. First, biological factors stimulate mutual evolutionary responses in the traits of interacting populations. Through natural selection and evolution, predators shape their prey's adaptations for escape, but their own adaptations for pursuit and capture also respond to attributes of their prey. For example, by capturing those prey individuals that it finds most readily or can catch most easily, a predator leaves behind more cryptic or swifter prey. These prey individuals reproduce and pass on to future generations their protective coloration and speed. Consequently, descendants of today's predators should exploit descendants of today's surviving prey with diminished efficiency over evolutionary time, all other things being equal. At the same time, the predators are evolving to capture the prey more easily. In contrast, adaptations of organisms in response to changes in the physical environment have no effect on that environment. Second, biological agents foster diversity of adaptations rather than promoting similarity. In response to biological factors, organisms tend to specialize, pursuing unique assortments of prey, striving to avoid unique combinations of predators and disease organisms, and engaging in mutually beneficial arrangements with other species. In response to similar physical stresses in the environment, however, many kinds of organisms evolve similar adaptations. We have seen this phenomenon, which is called **convergence,** in the reduced or finely divided leaves that minimize heat stress and water loss in many desert plants (see Chapter 5).

In its broadest sense, the term *coevolution* recognizes that each species influences the evolution of all other species with which it interacts. Often, however, *coevolution* is restricted more narrowly to the situation in which one species evolves an adaptation specifically in response to an adaptation in another species that also evolved in response to their interaction. In this sense, the term *coevolution* is restricted to reciprocal evolution between two interacting populations. Ecologists have some difficulty identifying unambiguous cases of such coevolution, as the following examples illustrate.

Hyenas have jaws and associated muscles that are strong enough to crack the bones of their prey. These modifications clearly are adaptations for eating their prey. However, the hyena's powerful jaws cannot be considered an example of coevolution because the properties of the bones of gazelles did not evolve to resist being eaten by hyenas, or any other predator. By the time a hyena has reached that part of its meal, bone structure has no consequence for the gazelle's survival. In contrast, when an herbivore evolves the ability to detoxify substances produced by a plant specifically to deter that herbivore, the requirements of the strict definition of coevolution are more likely to be met.

In this chapter, we shall explore some of the consequences of evolutionary responses to interactions between predators and their prey, between competitors, and within mutualistic associations. When the coevolutionary relationship between two species is antagonistic, as it is between predator and prey or between parasite and host, the species can become fixed in an evolutionary struggle to increase their own fitnesses, each at the other's expense. Such a struggle may lead to an evolutionary stalemate in which both antagonists continually evolve in response to each other, and the net outcome of their interaction may be a steady state. Alternatively, when one of the antagonists cannot evolve fast enough, it may be driven to extinction. In contrast, coevolution between mutualists may lead to stable arrangements of complementary adaptations that promote their interaction.

Antagonists evolve in response to each other

The term *coevolution* was coined by Charles Mode in a paper published in *Evolution* in 1958. Mode was concerned with the relationship between agricultural crops and their fungal pathogens, especially rusts, which cause millions of dollars worth of crop losses each year. He developed a model of continual evolution of host and pathogen in response to evolutionary changes in each other. Mode's model assumed that pathogen virulence and host resistance each were controlled by a single dominant gene (V and R, respectively), and that both virulence and resistance were, by themselves, costly to the organism. Thus, the fitnesses of both host and pathogen were contingent on the genotype of the other. In these circumstances, frequencies of virulence and resistance genes should tend to oscillate over time

in much the same manner as a predator–prey population cycle (see Figure 18.2).

Thus Mode's model would work as follows: When the host is susceptible (genotype *rr*), selection favors virulent pathogens (genotype *VV* or *Vv*). Virulent pathogens cause selection for host resistance (genotype *RR* or *Rr*). When the host is resistant, selection favors avirulent pathogens (genotype *vv*), because virulence is costly. When the pathogen is avirulent, the host is selected to become susceptible (genotype *rr*), because resistance is costly. These responses of each organism to the other result in a pattern of continual cycling.

In 1964, Paul Ehrlich and Peter Raven, who were assistant professors at Stanford University, published an article, also in *Evolution*, that placed coevolution in a more ecological context and greatly popularized the term. Ehrlich and Raven noted that closely related groups of butterflies tended to feed on closely related species of host plants. For example, species of butterflies in the tropical genus *Heliconius* feed exclusively on passionflower vines of the genus *Passiflora* (■ Figure 20.2). Such tight relationships suggested that butterflies and their host plants had had a long evolutionary

(a)

(b)

■ **Figure 20.2 The taxonomic specificity of some predator–prey relationships suggests a long evolutionary history.** Larvae of *Heliconius* butterflies (a) feed only on passionflower vines (b, *Passiflora*). Photo (a) courtesy of Andy McGregor; photo (b) by Ray Coleman/Photo Researchers.

history together, undoubtedly involving the ability of the butterflies to tolerate the particular defenses of their hosts.

Thus, the study of coevolution initially went in two directions. On one hand, Mode used modeling to address the evolutionary mechanisms of the relationships between populations. On the other hand, Ehrlich and Raven observed empirical patterns of relationships and interpreted them as outcomes of coevolutionary interactions. Most recently, these two approaches have found a common ground in analyses of the evolutionary history of traits directly involved in the relationships between species. Early studies on evolution in laboratory populations, however, were already demonstrating the powerful role of an evolutionary response by one species to another.

ECOLOGISTS IN THE FIELD

Studies on evolution in parasitoid–host systems

In a series of experiments conducted during the 1960s, David Pimentel and his colleagues at Cornell University explored the evolution of host–parasitoid relationships. They used the pupal stage of the housefly as their host and a wasp, *Nasonia vitripennis* (∎ Figure 20.3), as their parasitoid. In one population cage (shown on the left in ∎ Figure 20.4), *Nasonia* was allowed to parasitize a fly population that was kept at a constant number by replenishment from a stock that had not been exposed to the wasp. Any flies that escaped attack by wasp parasitoids were removed from the population cage, so the wasps were provided only with evolutionarily "naive" hosts. In a second population cage

∎ **Figure 20.3 Pimentel's study of coevolution used a parasitoid–host system.** The wasp *Nasonia*, a parasitoid of the housefly, is shown here laying eggs in a fly pupa. Courtesy of D. Pimentel, from D. Pimentel, *Science* 159:1432–1437 (1968).

(shown on the right in Figure 20.4), fly hosts were kept at the same constant number, but because emerging flies were allowed to remain, the population could evolve resistance to the wasps. The population cages were maintained for about 3 years, long enough for evolutionary change to occur.

Over the course of the experiment, the reproductive rate of wasps in the cage that permitted evolution dropped from 135 to 39 progeny per female, and longevity decreased from 7 to 4 days. The average parasitoid population also decreased (1,900 adult wasps versus 3,700 in the nonevolving system), and population size was more nearly constant than in the nonevolving cage. These results suggest that the flies evolved resistance to the parasitoids when subjected to intense parasitism.

Experiments were then established in new population cages in which the numbers of flies were allowed to vary freely. One cage started with flies and wasps that had had no previous contact with each other, and a second was established with animals

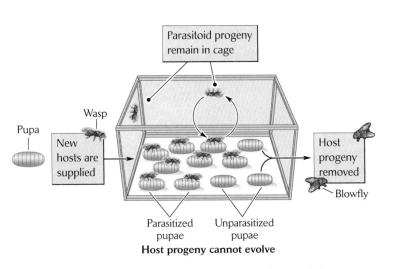
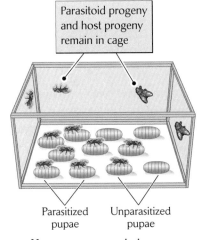

∎ **Figure 20.4 Pimentel's classic experiment tested for a host evolutionary response to a parasitoid.** The difference in population sizes at the end of the experiment between the system in which the host could not evolve (*left*) and the system in which it could evolve (*right*) indicated the effectiveness of the host's evolutionary response.

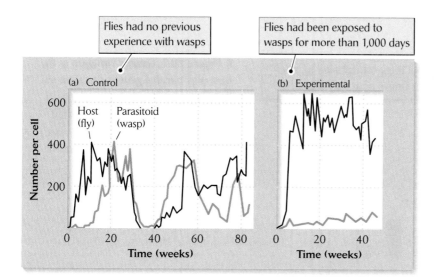

Flies had no previous experience with wasps

Flies had been exposed to wasps for more than 1,000 days

(a) Control

(b) Experimental

Figure 20.5 Population changes in Pimentel's host–parasitoid system reinforced the conclusion that populations evolve in response to each other. Houseflies and parasitoid wasps were placed together in 30-cell population cages. In one cage (a), the fly population had no previous experience with the wasp; in the other (b), the fly population had been previously exposed to wasp parasitism. Numbers of flies and wasps per cell, as well as the pattern of population cycling, differed between the two cages. After D. Pimentel, *Science* 159:1432-1437 (1968).

from the evolving population described above. In the first cage, the wasps were efficient parasitoids, and the system underwent dramatic oscillations. In the second cage, however, the wasp population remained low, and the flies attained a high and relatively constant population level (**Figure 20.5**). This result strongly reinforced the conclusion, drawn from the earlier experiments, that the flies had evolved resistance to the wasp parasitoids.

Coevolution in plant–pathogen systems reveals genotype–genotype interactions

The suggestion that consumer and resource populations evolve in response to each other presupposes that each contains genetic variation for traits that influence their interactions. In the case of the wasp–fly interaction, it was clear that evolution had occurred, but the genetic basis of the

evolutionary change could not be determined. This has been less of a problem in studies on diseases of plants. In these systems, the difference between virulence and avirulence may depend on a single gene, as Mode had assumed, and thus is amenable to simple Mendelian genetic analysis.

Plant geneticists have developed strains of crops, such as flax and wheat, that are resistant to particular genetic strains of pathogens, such as rusts (teliomycetid fungi). These crop strains differ from one another by being either susceptible or resistant to infection by particular strains of rust. Over the course of crop improvement programs, when new strains of rust appear, crop geneticists select new resistant strains of the crop by exposing experimental populations to the pathogen. However, new strains of the pathogen continue to appear in an area, either by migration or by mutation, creating continual evolutionary flux in the system.

A survey of wheat rust (*Puccinia graminis*) in Canada revealed that new virulence genes appear from time to time and sweep through a population (**Figure 20.6**). Genetic

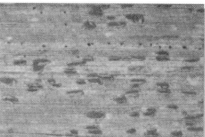

Different virulence strains of wheat rust appear from time to time and sweep through a population.

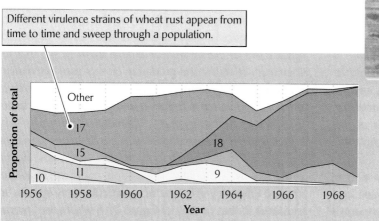

Figure 20.6 Coevolution involves an interaction between the genetic fitness of a host and that of its parasite or pathogen. The relative proportions of different virulence genes in the rust *Puccinia graminis* infecting Canadian wheat (*right*) have changed over time. From G. J. Green, *Can. J. Bot.* 53:1377–1386 (1975). Photo courtesy of Gary Munkvold.

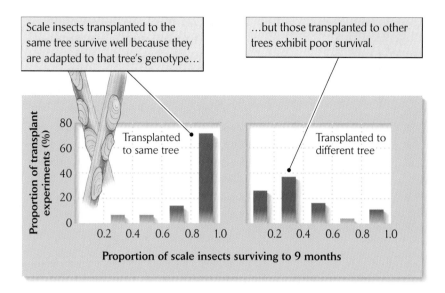

Scale insects transplanted to the same tree survive well because they are adapted to that tree's genotype…

…but those transplanted to other trees exhibit poor survival.

Transplanted to same tree

Transplanted to different tree

Proportion of transplant experiments (%)

Proportion of scale insects surviving to 9 months

Figure 20.7 Genetic variation in the host may parallel genetic variation in the pathogen. The survival rate of scale insects decreases markedly when they are transplanted to different trees. From G. F. Edmunds and D. N. Alstad, *Science* 199:941–945 (1978).

races of wheat rust are distinguished both by their physiological characteristics and by their virulence when tested on strains of wheat containing different resistance alleles. Most of the virulence strains within a single physiological race of rust differ by only one gene. The rust–wheat system contains the essential element of coevolution envisioned by Mode: an interaction between the fitnesses of the genotypes of the host and the genotypes of the pathogen. The system is kept in flux by the introduction of new virulence genes in the rust—and perhaps by new resistance genes in the wheat, although the latter are pretty much controlled by plant geneticists nowadays.

Such genotype–genotype interactions have been found in several natural systems, and they may turn out to be the rule in populations of plants and herbivores and of hosts and pathogens. The genetics of most plant defenses are difficult to work out in as much detail as has been done for wheat resistance genes, but genetic effects can nonetheless be detected. D. N. Alstad and G. F. Edmunds, Jr., at the University of Minnesota, have shown that variation among individuals in the defenses of ponderosa pines is paralleled by variation in the genotypes of the scale insects that infest them (Figure 20.7). Scale insects (see Figure 19.10) are extremely sedentary, exhibiting so little migration from tree to tree that local populations on individual trees have apparently evolved independently. Alstad and Edmunds inferred this from the differing success of scale insects experimentally transferred between trees and between branches on the same tree. The survival rate of scale insects transplanted between trees was much lower than that of controls transferred between branches. It is reasonable to assume that differences between individual trees and between local populations of scales are genetic, so this finding represents a case of genotype–genotype interaction. It could also repre-

sent a case of strict coevolution if the trees respond genetically to infestations of herbivores and pathogens.

Consumers and resources can achieve an evolutionary equilibrium

A simple graphic model that relates the rates of evolution of a consumer and a resource population to the efficiency with which the consumer exploits the resource can depict the evolutionary responses of the two populations (Figure 20.8). In the case of a prey population, for example, the rate at which new adaptations useful in escaping or avoiding predators evolve should vary in direct proportion to the rate at which prey are exploited. In the absence of predation, there can be no selection of adaptations for predator avoidance. But as predation increases, so do selection and the evolutionary response of the prey, at least up to limits set by the availability of genetic variation.

The selection of new adaptations useful to predators in exploiting their prey should vary in the opposite fashion. When a particular prey species is not heavily exploited, adaptations of predators that enable them to use that resource are selected, and predation on that prey population increases. As exploitation of the prey increases, however, the prey population is reduced, and this reduces the selective value of further increases in predation efficiency. Very high rates of predation conceivably could favor individuals that shifted their diets toward other prey species. Hence evolution by a predator population could result in decreased efficiency in its use of a particular prey species, as indicated in that portion of Figure 20.8 where the "predator" (red) curve sinks below the horizontal axis.

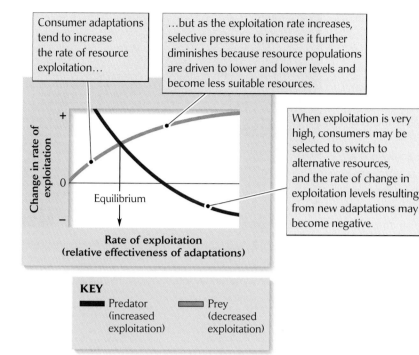

Consumer adaptations tend to increase the rate of resource exploitation…

…but as the exploitation rate increases, selective pressure to increase it further diminishes because resource populations are driven to lower and lower levels and become less suitable resources.

When exploitation is very high, consumers may be selected to switch to alternative resources, and the rate of change in exploitation levels resulting from new adaptations may become negative.

KEY

⬛ Predator (increased exploitation)   ⬛ Prey (decreased exploitation)

Figure 20.8 The rate of exploitation influences the rate of evolution in consumer and resource populations. Selective pressure on resource populations increases as the exploitation rate increases; adaptations by resource populations tend to decrease the rate of exploitation. Exploitation is brought into equilibrium when the population consequences of consumer and resource adaptations balance.

In this simple model, the balancing influences of predator and prey adaptation can achieve an evolutionary steady state where the two curves cross. When predator adaptations are relatively effective and the prey are exploited at a high rate, selection on the prey population tends to improve its escape mechanisms faster than selection on the predator population improves its ability to exploit the prey. Conversely, when the exploitation rate is low, prey populations evolve more slowly than predator populations. A balance between these influences should result in a relatively constant rate of exploitation regardless of the specific predator and prey adaptations. As in any steady state, both antagonists continually evolve to maintain this balance, just as nations continually develop new weapons and defenses to maintain a stalemate in an arms race. This is the Red Queen hypothesis, which we discussed in the context of the evolutionary maintenance of sexual reproduction in populations (see Chapter 11).

Pimentel's experiments on host–parasitoid interactions, discussed above, illustrate the dynamics of this consumer–resource equilibrium. The housefly (host) and the wasp *Nasonia* (parasitoid) undoubtedly had achieved an evolutionary equilibrium in their natural habitat. When brought into a simple laboratory habitat, *Nasonia* wasps were able to exploit housefly populations at a greatly increased rate because they required little time to search out hosts. (Setting up these experimental conditions was equivalent to shifting the exploitation rate of *Nasonia* on houseflies far above the equilibrium level in Figure 20.8.) This shift increased the selective pressure on the housefly to escape parasitism much more than

the selective pressure on the predator to further increase its exploitation rate. Consequently, the ability of houseflies to escape parasitoids increased, and the level of exploitation by *Nasonia* decreased toward a new steady state.

Competitive ability exhibits genetic variation and responds to selection

Competitors, like predators and prey, exert selective pressures on each other's competitive abilities. Sometimes genes that influence competitive ability express themselves in the phenotype so subtly that we cannot detect them by directly examining the traits of individuals. Instead, they must be inferred from the outcome of competition. In this sense, competitive ability summarizes the interaction of a phenotype with its environment. The experiments that follow demonstrate genetic variation in competitive ability. Clearly, such variation makes it possible for competitive ability to evolve.

Laboratory populations of the fruit flies *Drosophila serrata* and *D. nebulosa* were established in population cages by population geneticist Francisco Ayala. The populations quickly achieved a pattern of stable coexistence, with 20–30% *D. serrata* and 70–80% *D. nebulosa* in each cage. In one cage, however, the frequency of *D. serrata* began to increase after the 20th week and attained about 80% by the 30th week, reversing the initial predominance of *D. nebulosa*.

When individuals of both species were removed from the competing populations after the 30th week and tested against stocks of flies that had been maintained in single-species cultures, the competitive ability of each species was found to have increased after exposure to the other in the first competition experiment. When the competitive ability of *D. serrata* individuals from the one cage in which that species predominated was tested against that of unselected stocks of *D. nebulosa, D. serrata* again showed superior competitive ability.

These experiments show that competitive ability has a genetic basis and can evolve in laboratory populations. The particular adaptations responsible for changes in competitive ability were not determined in these experiments. They could conceivably include an increase in the efficiency of using a food resource, in the number of offspring produced per unit of food consumed, or in survival at any stage of the life cycle.

One of the few generalizations to come from this body of work is that sparse populations can evolve interspecific competitive ability more rapidly than dense populations. Why? Possibly because if different, somewhat conflicting adaptations determine the outcomes of intraspecific and interspecific competition, then selection for increased interspecific competitive ability will be stronger on the rarer of two competitors. We return to the work of David Pimentel for laboratory evidence that a rare competitor can evolve a competitive advantage (judged by relative population density) over a formerly superior adversary.

ECOLOGISTS IN THE FIELD

Studies on the evolution of competitive advantage

Pimentel and his colleagues conducted laboratory experiments with flies to determine whether species show frequency-dependent evolutionary changes in their competitive ability. In other words, can one species, as it is being excluded by the second and becoming rare, evolve increased interspecific competitive ability rapidly enough to regain the upper hand? The housefly (*Musca domestica*) and the blowfly (*Phaenicia sericata*), which have similar ecological requirements and comparable life cycles (about 2 weeks), were chosen for the experiments (Figure 20.9). Both species feed on dung and carrion in nature, and they are often found together on the same food resources. The flies were raised in small population cages, with a mixture of agar and liver provided as food for the larvae and sugar for the adults.

The outcomes of four initial competition experiments using individuals from wild populations of houseflies and blowflies were split, with each species winning twice. The mean extinction time for the blowfly, when the housefly won,

Figure 20.9 **The blowfly (shown) and housefly were used as competitors in Pimentel's study.** The two species are often found on the same food resources in nature. Courtesy of L. Higley, University of Nebraska, Lincoln.

was 92 days; it was 86 days for the housefly when the blowfly won. These results showed that the two species were close competitors, but the small cages used did not allow enough time for evolutionary change before one of the species was excluded.

To prolong the housefly–blowfly interaction, Pimentel started a population in a sixteen-cell cage, which consisted of single cages in four rows of four with connections between them (Figure 20.10). Under these conditions, populations of houseflies and blowflies coexisted for almost 70 weeks. The

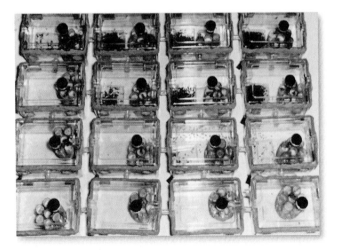

Figure 20.10 **Pimentel used a sixteen-cell cage to study competition between populations of flies.** Note the vials with larval food in each cage and the passageways connecting the cells. The dark objects concentrated in the upper right-hand cells are fly pupae. Courtesy of D. Pimentel, from D. Pimentel et al., *Am. Nat.* 99:97–109 (1965).

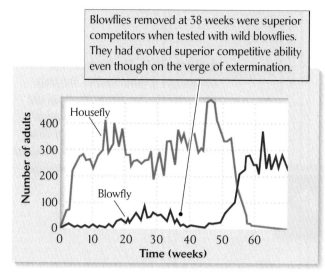

Blowflies removed at 38 weeks were superior competitors when tested with wild blowflies. They had evolved superior competitive ability even though on the verge of extermination.

Figure 20.11 A rare competitor can evolve superior competitive ability. The proportions of houseflies and blowflies in a sixteen-cell cage shifted dramatically at about 50 weeks, after the blowflies had become rare. After D. Pimentel et al., *Am. Nat.* 99:97–109 (1965).

houseflies were more numerous initially, but the two species showed a striking reversal of numbers at about 50 weeks, and the blowflies had excluded the houseflies by the end of the experiment (■ Figure 20.11).

After 38 weeks, when the blowfly population was still low, and just a few weeks prior to its sudden increase, individuals of both species were removed from the population cage and tested in competition with each other and with wild strains of the housefly and blowfly. Captured wild blowflies turned out to be inferior competitors to wild and experimental strains of the housefly. But blowflies that had been removed from the population cage at 38 weeks consistently outcompeted both wild and experimental populations of the housefly. Apparently, the experimental blowfly population had evolved superior competitive ability while it was rare and on the verge of extermination.

Traits of competing populations may diverge through character displacement

Theory suggests that if resources are sufficiently varied, competitors should diverge and specialize on different resources. Laboratory experiments such as those described above have shown that competitive ability may have a genetic component and therefore may be under the influence of selection. However, if competition exerts a potent evolutionary force in nature, we should find evidence that competitors have partly molded each other's adaptations.

Although related species that live together differ in the way they use the environment (using different food re-

sources, for example), we cannot assume that these differences have evolved as a result of their interaction. An alternative explanation for such differences is that each of the species became adapted to different resources in different places, and when their populations subsequently overlapped as a result of range extensions, these ecological differences remained.

We may get around this objection by comparing the ecology of a species in an area where it co-occurs with a competitor with its ecology in another area where the competitor is absent. Where two species coexist within the same geographic area, they are said to be **sympatric**; where their distributions do not overlap, they are said to be **allopatric**. Suppose that species 1 occurs in areas A and B, and species 2 occurs in areas B and C (■ Figure 20.12). The populations of the two species in area B are sympatric, and the population of species 1 in area A is allopatric with the population of species 2 in area C. If areas A, B, and C all have similar environmental conditions and habitats, and if competition causes divergence, we would expect the sympatric populations of species 1 and 2 in area B to differ more from each other than the allopatric populations of those species in areas A and C. This phenomenon is called **character displacement.**

Ecologists disagree on the prevalence of character displacement in nature. Some examples do seem to fit the pattern very well. One of these involves the ground finches (*Geospiza*) of the Galápagos archipelago (see Chapter 16).

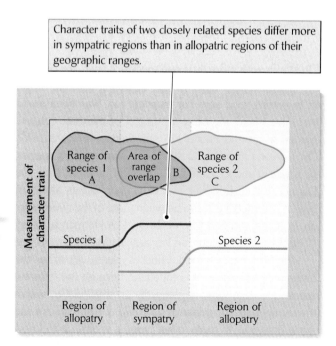

Character traits of two closely related species differ more in sympatric regions than in allopatric regions of their geographic ranges.

Figure 20.12 Character displacement is the divergence of competing populations over time.

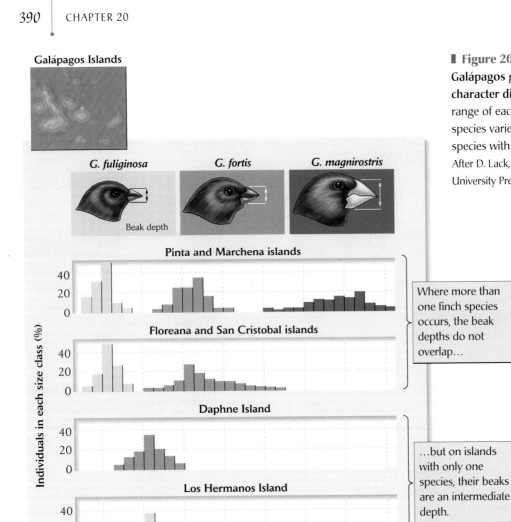

Galápagos Islands

G. fuliginosa **G. fortis** **G. magnirostris**

Beak depth

Pinta and Marchena islands

Floreana and San Cristobal islands

Daphne Island

Los Hermanos Island

Individuals in each size class (%)

Beak depth (mm)

Where more than one finch species occurs, the beak depths do not overlap…

…but on islands with only one species, their beaks are an intermediate depth.

▌ **Figure 20.13 The beaks of the Galápagos ground finches illustrate character displacement.** The beak size range of each ground finch (*Geospiza*) species varies with the number of other species with which it coexists on an island. After D. Lack, *Darwin's Finches,* Cambridge University Press, Cambridge (1947).

On islands with more than one species of finch, the beaks of the species usually differ in size, indicating different ranges of preferred food size. For example, on Marchena Island and Pinta Island, the beak size ranges of the three resident species of ground finches do not overlap (▌Figure 20.13). On Floreana and San Cristobal, the two resident species, *G. fuliginosa* and *G. fortis,* have beaks of different sizes. On Daphne Island, however, where *G. fortis* occurs alone, its beak is intermediate in size between those of the two species on Floreana and San Cristobal. On Los Hermanos Island, *G. fuliginosa* occurs alone, and its beak is intermediate in size.

The Galápagos ground finches clearly illustrate the diversifying influence of competition because the different species are distributed haphazardly on small islands within the archipelago: some islands have two or three species and some only one. In many other cases, it is difficult to know whether differences between two species arose because of competition between them or evolved in response to selec-

tion by other environmental factors in different places, then were retained when the populations re-established contact. In most cases, the genetic differences that lead to the formation of new species evolve in allopatry. So why not differences that allow two species to avoid strong competition? In either case, coexistence depends on some degree of ecological difference between species, whether it is achieved in allopatry or as an evolutionary consequence of competition in sympatry.

Mutualists have complementary functions

It is not only antagonistic relationships between species that can lead to coevolution. There is a wide range of interactions between species that benefit both participants.

Flowers provide bees with nectar, and bees carry pollen between plants and effect fertilization. Mycorrhizal fungi extract inorganic nutrients from the soil and make them available to plants, and plants supply their fungus partners with carbohydrates. These kinds of interactions are known as **mutualisms**. In most cases, each party to a mutualism is specialized to perform a complementary function for the other. In lichens, photosynthetic algae team up with fungi that can obtain nutrients from difficult substrates, such as bark and rock surfaces (see Figure 1.11). Such intimate associations, in which the members together form a distinctive entity, are referred to as **symbioses**—literally, "living together." In very general terms, mutualisms fall into three categories: trophic, defensive, and dispersive.

Trophic mutualisms usually involve partners specialized in complementary ways to obtain energy and nutrients; hence the term *trophic*, which pertains to feeding relationships. We have seen trophic mutualisms in lichens, in mycorrhizae (see Chapter 8), and in the *Rhizobium* bacteria and plant roots that form nitrogen-fixing root nodules (see Chapter 7). In these cases, each of the partners supplies a limiting nutrient or energy source that the other cannot obtain by itself. *Rhizobium* can assimilate molecular nitrogen (N_2) from the soil—a useful feature in nitrogen-poor soil—but requires carbohydrates supplied by a plant for the energy it needs to do this. Bacteria in the rumens of cows and other ungulates can digest the cellulose in plant fibers, which a cow's own digestive enzymes cannot do. The cows benefit because they assimilate some of the by-products of bacterial digestion and metabolism for their own use (they also digest some of the bacteria). The bacteria benefit by having a steady supply of food in a warm, chemically regulated environment that is optimal for their own growth.

Defensive mutualisms involve species that receive food or shelter from their partners in return for defending those partners against herbivores, predators, or parasites. For example, in some marine ecosystems, specialized fishes and shrimps clean parasites from the skin and gills of other fish species (❚ Figure 20.14). These cleaners benefit from the food value of parasites they remove, and the groomed fish are unburdened of some of their parasites. Such relationships, often referred to as cleaning symbioses, are most highly developed in clear, warm tropical waters, where many cleaners display their striking colors at locations, called cleaning stations, to which other fish come to be groomed. As might be expected, a few species of predatory fish mimic the cleaners: when other fish come and expose their gills to be groomed, they get a bite taken out of them instead.

 Ant–acacia mutualisms. **Some ants protect acacia plants from herbivores and are rewarded with food and nesting sites.**

❚ **Figure 20.14 Cleaning symbioses benefit both participants.** The prawn *Lysmata amboiensis* is removing parasites from a moray eel. From this association, the prawn gets food, and the eel gets its parasites removed. Photo by Doug Perrine/DRK Photo.

Dispersive mutualisms generally involve animals that transport pollen between flowers in return for rewards such as nectar, or that eat nutritional fruits and disperse the seeds they contain to suitable habitats. Dispersive mutualisms rarely involve close living arrangements between members of the association. Seed dispersal mutualisms are not usually highly specialized: for example, a single bird species may eat many kinds of fruit, and each kind of fruit may be eaten by many kinds of birds. Plant–pollinator relationships tend to be more restrictive because it is in a plant's interest that a flower visitor carry its pollen to another plant of the same species.

MORE ON THE WEB *Seed dispersal.* **The seeds of many species of plants are widely distributed by animals, often to habitats favorable for germination and growth.**

MORE ON THE WEB *Pollination.* **Plants have many ways of manipulating their pollinators so as to increase the efficiency of pollen transfer between individuals.**

Coevolution involves mutual evolutionary responses by interacting populations

Reciprocal evolutionary responses between pairs of populations are referred to as coevolution. But the term has also been used more broadly to describe the close associations of certain species and groups of species in biological communities. The contrast between these two perspectives raises two questions: First, do pairs of populations undergo reciprocal evolution, or do "coevolved" traits arise from the responses of populations to selective pressures exerted by a variety of species and physical factors, followed by an ecological sorting out of subsets of species with compatible features? Second, are species organized into interacting sets based on their evolved adaptations, whether "coevolved" or not?

Complementary adaptations among pairs of species or small groups of species have often been attributed to coevolution without evidence having been presented for the evolutionary history of the relationship itself. A close match between the adaptations of different species does not prove coevolution.

Coevolution in ants and aphids?

The difficulty we have in recognizing true cases of coevolution is illustrated by a mutualism in which ants protect aphids and leafhoppers from predators and, in return, harvest the nutritious honeydew that those insects excrete. In one such system, aphids, which are small, sedentary homopteran insects, form dense colonies on inflorescences of ironweed (*Vernonia*) in New York State. Also occurring on ironweed is the larger membracid bug (leafhopper) *Publilia*, another homopteran that sucks plant juices from leaves. These insects are tended by three species of ants. One, in the genus *Tapinoma*, is tiny (2–3 mm) but abundant. The other two (both species of *Myrmica*) are larger (4-6 mm) and more aggressive, but less common. The two genera of ants rarely co-occur on the same plant. *Tapinoma* ants greatly enhance the survival of aphid colonies, but have less effect on the survival of leafhoppers. The larger *Myrmica* ants offer substantial protection to leafhoppers but are less effective in warding off predators of aphids.

The ant–aphid–leafhopper mutualism has all the elements expected of coevolution, but how can we be sure that the adaptations of the ant and homopteran participants evolved in response to each other? Most insects that suck plant juices produce large volumes of excreta from which they either do not or cannot extract all the nutrients. Therefore, honeydew production may simply reflect diet rather than having evolved to encourage protection by ants. For their part, many ants are voracious generalists that are likely to attack any insect they encounter; they may need no special adaptations to deter predators of the aphids and leafhoppers upon whose excreta they also feed. The fact that the different genera of ants more effectively protect different honeydew sources may simply reflect their different sizes and levels of aggression, which may have evolved in response to unrelated environmental factors.

Why don't the ants eat the aphids and leafhoppers they tend? Perhaps this restraint is an evolved trait of ants that facilitates the ant–homopteran mutualism. It may even have arisen as an extension of the common ant behavior of defending plant structures that produce nectar, such as flowers or specialized nectaries.

Although the ant–homopteran system has apparent specificity of interactions, this specificity is not sufficient to prove coevolution. The best evidence for coevolution comes from reconstructions of the evolutionary history of traits in coevolving groups of organisms.

ECOLOGISTS IN THE FIELD

Herbivores and the chemical defenses of plants

University of Illinois biologist May Berenbaum has used comparative methods to place elements of the relationship between certain insects and their umbelliferous host plants in the context of coevolution. Umbellifers (parsley family) produce many defensive chemicals, among the most prominent of which are the furanocoumarins. The biosynthetic pathway of these chemicals leads from para-coumaric acid (which, being a precursor of lignin, is found in virtually all plants) to hydroxycoumarins such as umbelliferone, and finally to the furanocoumarins, which occur in both linear and angular chemical forms (Figure 20.15). As one proceeds down this biosynthetic pathway, toxicity increases. Hydroxycoumarins have some biocidal properties; linear furanocoumarins (LFCs) interfere with DNA replication in the presence of ultraviolet light; and angular furanocoumarins (AFCs) interfere with herbivore growth and reproduction quite generally.

The most toxic of these chemicals occur among the fewest plant families. Para-coumaric acid is widespread among plants, occurring in at least a hundred families, while only thirty-one families possess hydroxycoumarins. LFCs are

(a)

(b)

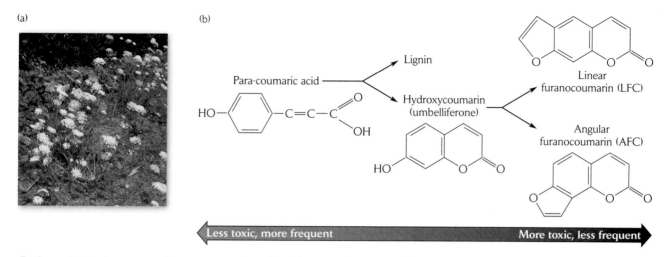

Less toxic, more frequent ← → **More toxic, less frequent**

▌ **Figure 20.15 Some secondary compounds and herbivore resistance may have coevolved.** The taxonomic relationships among certain umbellifers, which produce defensive chemicals called furanocoumarins, and among insects that can feed on these plants suggest that these plants and herbivores have coevolved. (a) Queen Anne's lace (*Daucus carota*) is a familiar umbellifer. (b) Biosynthetic pathways in the synthesis of furanocoumarins. Photo by Alfred Brousseau, courtesy of Saint Mary's College of California.

restricted to eight plant families and are widely distributed in only two: Umbelliferae and Rutaceae (the citrus family). AFCs are known only from two genera of Leguminosae (pea family) and ten genera of Umbelliferae.

Among the species of herbaceous umbellifers in New York State, some (especially those growing in woodland sites with low levels of ultraviolet light) lack furanocoumarins, others contain LFCs only, and some contain both LFCs and AFCs. Berenbaum's surveys of herbivorous insects collected from these plant species showed that (1) host plants containing both AFCs and LFCs were, somewhat surprisingly, attacked by more species of insects than were plants with only LFCs or with no furanocoumarins; (2) insect herbivores found on AFC plants tended to be extreme diet specialists, most being found on no more than three genera of plants; and (3) these specialists tended to be abundant compared with the few generalists found on AFC plants and compared with all herbivores found either on LFC plants or on umbellifers lacking furanocoumarins.

Although LFCs and (especially) AFCs are extremely effective deterrents to most herbivorous insects, some genera that have evolved to tolerate these chemicals have obviously become successful specialists. One can make a strong case for coevolution here. The taxonomic distribution of hydroxycoumarins, LFCs, and AFCs across host plants suggests that plants containing LFCs are a subset of those containing hydroxycoumarins, and that those containing AFCs are an even smaller subset of those containing LFCs. This

pattern is consistent with an evolutionary sequence of plant defenses progressing from hydroxycoumarins to LFCs and AFCs. Furthermore, insects that specialize on plants containing LFCs belong to groups that characteristically feed on plants containing hydroxycoumarins, and those that specialize on plants containing AFCs have close relatives that feed on plants containing LFCs. These patterns of taxonomic distribution are consistent with coevolution within the system.

The story of the evolution of chemical defenses in plants and the evolution of resistance to those defenses by certain groups of insects is conjecture, based on the logic of the evolutionary relationships of the taxa involved. We have no means of directly watching such evolutionary interactions unfold; evolution occurs too slowly in natural systems. Berenbaum's inferences about evolution build on the idea that evolutionarily older and younger characters (such as absence and presence of AFCs) should be found among close relatives if they are linked by evolution. This logic has been elaborated into a branch of evolutionary biology known as phylogenetic reconstruction, which uses similarities and differences among species to determine their evolutionary relationships.

 MORE ON THE WEB   *Inferring phylogenetic history.* How can one reconstruct evolutionary relationships among species from their traits?

The yucca moth and the yucca

The application of phylogenetic reconstruction to the problem of coevolution can be illustrated using the curious pollination relationship between yucca plants (*Yucca*, a member of the lily family) and moths of the genus *Tegeticula* (Figure 20.16). This mutually beneficial and obligatory relationship was first described nearly a century ago, but its details have been worked out only during the past few years, largely due to studies by Olle Pellmyr of Vanderbilt University.

Adult female yucca moths carry balls of pollen between yucca flowers by means of specialized mouthparts. During the act of pollination, a female moth enters a yucca flower, makes cuts in the ovary with her ovipositor, and deposits one to fifteen eggs. After each egg is laid, the moth crawls to the top of the pistil of the flower and deposits a bit of pollen on the stigma. This guarantees that the flower is fertilized and that the moth's offspring will have developing seeds to feed on. After the moth has laid her eggs, she may scrape some pollen off the anthers and add it to the ball she carries in her mouthparts before flying to another flower. Male moths also come to the flowers to mate with the females, but only the females carry pollen.

The relationship between the moth and the yucca is an obligate mutualism. *Tegeticula* larvae can grow nowhere else; *Yucca* has no other pollinator. In return for the pollination of its flowers, the yucca seemingly tolerates the moth larvae feeding on its seeds, but the extent of this loss of potential reproduction is small, rarely exceeding 30% of a seed crop. The moth's restraint concerning the number of eggs laid per flower is a more puzzling aspect of the yucca–moth relationship. Over the short term, it would seem that moths laying larger numbers of eggs per flower might have higher individual reproductive success and evolutionary fitness, even though such behavior over the long term might lead to extinction of the yucca. In fact, it is the yucca that regulates the number of eggs laid per flower. When too many eggs are laid in the ovary of a particular flower—too many being enough to eat a majority of the developing seeds—the flower is aborted and the moth larvae die. While this would also seem to reduce the seed production of the yucca, resources that would have supported the production of seeds in the now aborted flower are diverted to other flowers. Selective abortion of insect-damaged fruit occurs widely among plants, and yuccas use this mechanism to keep their moth pollinators in line.

(a)

(b)

 Figure 20.16 The relationship between yucca and yucca moth is an obligatory mutualism. The mohave yucca (a, *Yucca shidigera*) is pollinated only by a yucca moth of the genus *Tegeticula* (b). The moth larvae develop only on yucca plants. After J. A. Powell and R. A. Mackie, *Univ. Calif. Publ. Entomol.* 42:1–59 (1966). Photo (a) by Alfred Brousseau, courtesy of Saint Mary's College of California; photo (b) by Larry Jon Friesen/Saturdaze.

The moth and the yucca plant have many adaptations that support their mutualistic interaction. On the yucca's part, its pollen is sticky and can easily be formed into a ball that the moth can carry; the stigma is specially modified as a receptacle to receive pollen. On the moth's part, individuals visit flowers of only one species of yucca, mate within the flowers, lay their eggs in the ovary within the flower, exhibit restraint in the number of eggs laid per flower, and have specially modified mouthparts and behaviors to obtain and carry pollen. Because the mutualism of *Tegeticula* and *Yucca* is so tight, one might expect all these characteristics to have evolved as a result of coevolution between the two.

In fact, however, many aspects of this mutualism are present in the larger lineage of nonmutualistic moths (Prodoxidae) within which *Tegeticula* evolved. A diagram of the evolutionary relationships among species, known as a phylogenetic tree, can reveal such patterns. Examination of a phylogenetic tree of the Prodoxidae (▌Figure 20.17) shows that several of the highly specialized characters of *Tegeticula* are found in other members of the family. Indeed, host specialization and mating on the host plant are old features of the family—features found in all its members. The trait of ovipositing in flowers has evolved independently at least three times in the family and has reversed (reverted to the ancestral state) at least twice, in *Parategeticula* and *Agavenema*. Of the species that lay eggs in flowers, only *Tegeticula* and one species of *Greya* actually function as pollinators; the others are strictly parasites of the plants in which their larvae grow. It should be mentioned that *Greya politella* pollinates *Lithophragma parviflorum* in the saxifrage family, which is not even closely related to the yuccas. We can see from this phylogenetic tree that many of the adaptations that occur in the yucca–yucca moth mutualism appear to have been present in the moth lineage before the establishment of the mutualism itself. Such traits are often referred to as **preadaptations.**

Where does this leave us with regard to coevolution? The consensus among ecologists is that all species interactions strongly affect evolution and shape the adaptations of consumer and resource populations alike. This phenomenon may be thought of as coevolution in a broad sense—sometimes called *diffuse coevolution*—in that populations simultaneously respond to an array of complex interactions with many other species. Coevolution in the narrow sense, in which changes in one evolving lineage stimulate evolutionary responses in the other, and vice versa, may be limited to very tight mutualisms in which strong interactions are limited to a pair of species. Even in such cases as that of the yucca and its moth pollinator, what appear to be coevolved traits may have been preadaptations that were critical to the establish-

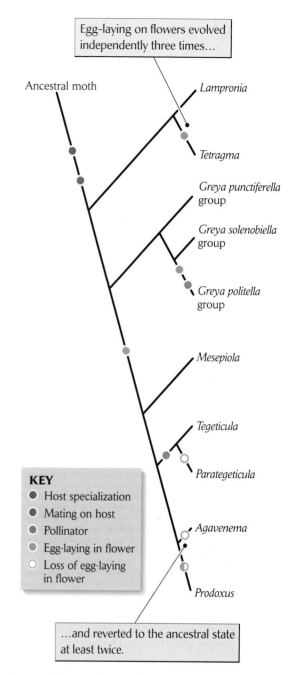

KEY
- Host specialization
- Mating on host
- Pollinator
- Egg-laying in flower
- Loss of egg-laying in flower

Egg-laying on flowers evolved independently three times...

...and reverted to the ancestral state at least twice.

Ancestral moth

Lampronia
Tetragma
Greya punctiferella group
Greya solenobiella group
Greya politella group
Mesepiola
Tegeticula
Parategeticula
Agavenema
Prodoxus

▌ **Figure 20.17 Phylogenetic trees can reveal preadaptations.** The phylogenetic tree of the moth family Prodoxidae shows the sites of the evolution of traits critical to the yucca moth–yucca mutualism in moths of the genus *Tegeticula*. From O. Pellmyr and J. N. Thompson, *Proc. Natl. Acad. Sci. USA* 89:2927–2929 (1992).

ment of the obligate mutualism in the first place. However, no subtlety of definition can detract from the reality that interactions among species are major sources of selection and evolutionary response.

Summary

1. Coevolution is the interdependent evolution of species that interact ecologically. The interactions may be antagonistic (consumer–resource, competition) or cooperative (mutualism). Because each species in a coevolved pair is an important component of the environment of the other, changes in one select adaptive responses in the other, and vice versa.

2. Evidence of evolutionary changes in consumer–resource systems has been obtained in laboratory studies on host–parasitoid interactions. After periods of co-occurrence, rates of parasitoid attack decreased and host populations increased, apparently following selection for improved host defenses against parasitoids.

3. Studies on pathogens of plant crops have revealed a simple genetic basis for virulence and resistance, which determine the outcomes of parasite–host interactions.

4. Because selection for prey defenses increases in proportion to predation rate, and selection for predator efficiency decreases as predation rate increases, predator and prey can achieve an evolutionary steady state at some intermediate level of predation.

5. Experiments on competition between species of flies have revealed reversals of competitive ability over the course of tens of generations. By testing populations against unselected controls, investigators have confirmed genetic changes in competing populations.

6. One may test whether competition can result in evolutionary divergence in nature by comparing ecological (or related morphological) traits of a population in the presence and in the absence of a competitor. When the two differ, the pattern is referred to as character displacement.

7. Mutualisms are relationships that benefit both parties. They may be classified as trophic, defensive, or dispersive. In trophic mutualisms, each partner is specialized to provide a different limiting nutrient. In defensive mutualisms, one partner provides protection or removes herbivores or parasites, usually in return for food.

8. Dispersive mutualisms are plant–animal interactions in which the animal disperses pollen or seeds in the course of harvesting or processing food that the plant supplies. The structures of flowers and fruits limit the variety of animals that perform this function for a particular species of plant, thereby increasing the efficiency of pollen transfer and the likelihood that seeds will reach suitable sites for germination and growth.

9. Analysis of the biosynthetic pathways of secondary compounds has shown that plants may evolve increasingly toxic chemical defenses in response to herbivore pressure. When variations in these pathways and in insect resistance to the chemicals are overlaid upon taxonomic relationships within each group, one can infer the evolutionary history of a plant–insect interaction.

10. The interaction between yucca moths and yuccas is an obligate mutualism in which the moth pollinates the plant, but its larvae consume developing seeds. Both the moth and the yucca have specializations that promote this relationship, but phylogenetic analysis shows that some of the adaptations of the moth are present in close relatives that are not mutualists of yuccas. Such traits are called preadaptations.

PRACTICING ECOLOGY

CHECK YOUR KNOWLEDGE

Ants and Plants

Mutualistic relationships between ants and plants have evolved independently many times. One of the best-studied systems is that of the ant *Pseudomyrmex* and the bull's horn acacia, *Acacia cornigera*, in Central America, which is described in an essay in this chapter's *More on the Web*, reached through *http://www.whfreeman.com/ricklefs*. The relationship is mutualistic because both species benefit from the interaction. Strong mutualistic dependencies like this one are thought to have evolved in many cases from antagonistic interactions. For example, pollination may have evolved from a situation where insects fed on pollen. Mutualisms can also break down when the benefit gained by one of the species outweighs that gained by the partner species.

In East Africa, the whistling thorn acacia *Acacia drepanolobium* occurs in dense, single-species stands. Ants in the genera *Crematogaster* and *Tetraponera* live in swollen thorns on this acacia and protect the plant from herbivores. In fact, four species of ants from these two genera compete for nest sites and extrafloral nectaries on *Acacia drepanolobium*. Colonies that come into contact will fight until one or the other is destroyed, and a single acacia is never occupied by more than one ant colony.

Maureen Stanton and her colleagues at the University of California at Davis have studied how competition among these ants affects the ant–plant mutualism. At their study site

Table 20.1 Percentage of initially empty trees colonized by ants and percentage of initially occupied trees abandoned by ants (for the two smallest classes of trees)

| Species | Percentage of empty trees colonized | | Percentage of occupied trees abandoned | |
| | Tree height | | Tree height | |
	0.0–0.49 m	0.5–0.99 m	0.0–0.49 m	0.5–0.99 m
C. sjostedti	2.5	2.5	26	10
C. mimosae	1.3	5.0	40	11
C. nigriceps	2.5	11.0	19	1
T. penzigi	0.0	2.5	20	5

in Kenya, all but the smallest trees are occupied by one of the four species of ants. In one recent study, these researchers monitored tree size and ant occupancy during a drought period. The results showed that small trees (less than 1 m tall) occupied by aggressive ant species were frequently abandoned because they did not produce enough resources to support a colony of aggressive ants. These abandoned trees grew more slowly than those that were not abandoned. Once abandoned, however, a small tree might be colonized by less aggressive ants. If conditions improved and the trees began to grow again (and presumably contain greater food resources for the ants), less aggressive ants were replaced by the more aggressive species.

CHECK YOUR KNOWLEDGE

1. Examine Table 20.1, which shows data for the two smallest tree sizes. What can you conclude from the data regarding the desirability of trees based on size? What does the relatively low percentage of abandonment of small trees by *C. nigriceps* and *T. penzigi* tell you about where those species are on the dominance hierarchy?

2. Name two reasons why plants inhabited by ants might grow faster than plants without ants.

MORE ON THE WEB 3. Visit the Web page "A Natural History of Extrafloral Nectar-Collecting Ants in the Sonoran Desert" through *Practicing Ecology on the Web* at *http://www.whfreeman.com/ricklefs*. Why do cacti secrete nectar from extrafloral nectaries? What service do ants provide in return for the nectar?

4. Why is it important to understand the dynamics between mutualistic species such as the ants and the acacia trees they inhabit?

Suggested Readings

Armbruster, W. S. 1992. Phylogeny and the evolution of plant–animal interactions. *BioScience* 42:12–20.

Berenbaum, M. R. 1983. Coumarins and caterpillars: A case for coevolution. *Evolution* 37:163–179.

Bogler, D. J., J. L. Neff, and B. B. Simpson. 1995. Multiple origins of the yucca–yucca moth association. *Proceedings of the National Academy of Sciences USA* 92:6864–6867.

Boucher, D. H. (ed.). 1985. *The Biology of Mutualism*. Croom Helm, London.

Brooks, D. R., and D. A. McLennan. 1991. *Phylogeny, Ecology, and Behavior*. University of Chicago Press, Chicago.

Davidson, D. W., and D. McKey. 1993. The evolutionary ecology of symbiotic ant–plant relationships. *Journal of Hymenopteran Research* 2:13–83.

Davies, N. B., and M. Brooke. 1991. Coevolution of the cuckoo and its hosts. *Scientific American* 264:92–98.

Ehrlich, P. R., and P. H. Raven. 1964. Butterflies and plants: A study in coevolution. *Evolution* 18:586–608.

Ewald, P. W. 1994. *Evolution of Infectious Disease*. Oxford University Press, Oxford.

Feldman, R., D. F. Tomback, and J. Koehler. 1999. Cost of mutualism: Competition, tree morphology, and pollen production in limber pine clusters. *Ecology* 80:324–329.

Futuyma, D. J., and M. Slatkin (eds.). 1983. *Coevolution*. Sinauer Associates, Sunderland, MA.

Handel, S. N., and A. J. Beattie. 1990. Seed dispersal by ants. *Scientific American* 263:76–83.

Jackson, R. R., and R. S. Wilcox. 1990. Aggressive mimicry, prey-specific predatory behaviour and predator–recognition in the predator–prey interactions of *Portia fimbriata* and *Euryattus sp.*, jumping spiders from Queensland. *Behavioral Ecology and Sociobiology* 26:111–119.

Janzen, D. H. 1966. Coevolution of mutualism between ants and acacias in Central America. *Evolution* 20:249–275.

Janzen, D. H. 1985. The natural history of mutualisms. In D. H. Boucher (ed.), *The Biology of Mutualism*. Croom Helm, London, pp. 40–99.

Mode, C. J. 1958. A mathematical model for the co-evolution of obligate parasites and their hosts. *Evolution* 12:158–165.

Nitecki, M. H. (ed.). 1983. *Coevolution*. University of Chicago Press, Chicago.

Palmer, T. M., T. P. Young, M. L. Stanton, and E. Wenk. 2000. Short-term dynamics of an acacia ant community in Laikipia, Kenya. *Oecologia* 123:425–435.

Pellmyr, O., and C. J. Huth. 1994. Evolutionary stability of mutualism between yuccas and yucca moths. *Nature* 372:257–260.

Pellmyr, O., J. Leebens-Mack, and C. J. Huth. 1996. Non-mutualistic yucca moths and their evolutionary consequences. *Nature* 380:155–156.

Pellmyr, O., J. N. Thompson, J. M. Brown, and R. G. Harrison. 1996. Evolution of pollination and mutualism in the yucca moth lineage. *American Naturalist* 148:827–847.

Price, P. W. 1977. General concepts on the evolutionary biology of parasites. *Evolution* 31:405–420.

Real, L. (ed.). 1983. *Pollination Biology.* Academic Press, Orlando, FL.

Stanton, M. L., T. M. Palmer, T. P. Young, A. Evans, and M. L. Turner. 1999. Sterilization and canopy modification of a swollen thorn acacia tree by a plant-ant. *Nature* 401:578–581.

Thompson, J. N. 1994. *The Coevolutionary Process.* University of Chicago Press, Chicago.

Tomback, D. F., and Y. B. Linhart. 1990. The evolution of bird-dispersed pines. *Evolutionary Ecology* 4:185–219.

Weiner, J. 1994. *The Beak of the Finch.* Knopf, New York.

Community Structure

Ecologists hold diverse concepts of communities

Ecologists use several measures of community structure

The term "community" has been given many meanings

Is the community a natural unit of ecological organization?

Feeding relationships organize communities in food webs

Trophic levels are influenced from above by predation and from below by production

Species in biological communities vary in relative abundance

Number of species increases with area sampled

Diversity indices weight species richness by relative abundance

Ecologists have puzzled for almost a century over how to define a biological community. To most ecologists, the term *community* means an assemblage of species that occur together in the same place. Ecologists also agree that the species that coexist within a community can interact strongly through consumer–resource and competitive interactions. Yet there is also much disagreement about what a community is. Some ecologists assert that it is a unit of organization having recognizable boundaries and whose structure and functioning are regulated by interactions among species. Some have further suggested that communities are organized in such a way as to increase their efficiency and productivity. Other ecologists regard the community as a loose assemblage of those species that can tolerate the conditions of a particular place or habitat, but which form no distinct boundary where one type of community meets another.

Ecologists who describe communities as organized ecological units think of communities as *superorganisms* in which the functions of various species are connected like those of parts of the body and have evolved so as to enhance their interdependent functioning. This viewpoint requires that communities be discrete entities that can be distinguished from one another in the sense that we distinguish individuals within populations or different species within a community. Certainly the most influential advocate of the organismal viewpoint was the American plant ecologist Frederic E. Clements, who, early in the twentieth century, perceived the community as a discrete unit with sharp boundaries and a unique organization. Clements's idea of the community was closely tied to vegetation types. He pointed out that a forest of ponderosa pine differs from the fir forests that grow in moister habitats and from the shrubs and grasses typical of drier sites. He

■ Figure 21.1 The boundaries of some communities are clearly defined. Hillsides in southern California have chaparral vegetation at higher elevation, grassland on the lower, hotter slopes, and live oaks in the moister valleys between ridges. Photo by Christi Carter/Grant Heilman Photography.

thought that the boundary between such communities could be crossed within a few meters along a gradient of environmental conditions. Indeed, some community boundaries, such as those between deciduous forest and prairie in the midwestern United States and between broad-leaved forest and needle-leaved forest in southern Canada, are clearly defined and are respected by most species of plants and animals (■ Figure 21.1).

An opposite view of community organization was put forward at about the same time by the botanist H. A. Gleason. Gleason suggested that a community, far from being a distinct unit like an organism, is merely a fortuitous association of species whose adaptations and requirements enable them to live together under the particular physical and biological conditions that characterize a particular place. A plant association, he said, is "not an organism, scarcely even a vegetational unit, but merely a coincidence."

Debate over the nature of the community continues today. It is an important issue because the properties of assemblages of species that co-occur in the same place integrate all the interactions between them. Thus, we cannot have a full understanding of ecology until we can understand the nature of the community.

Every place on earth—each meadow, each pond, each rock at the edge of the sea—is shared by many coexisting organisms. These plants, animals, and microbes are linked to one another by their feeding relationships and other interactions, forming a complex whole often referred to as a **biological community.** Interrelationships within communities govern the flow of energy and the cycling of elements within the ecosystem. They also influence population processes, and in doing so determine the relative abundances of species.

The members of a community must be compatible in the sense that the outcomes of all their interactions allow their survival and reproduction. Although the theory of species interactions, as we have seen in the last section of this book, tells us when predator and prey or two competitors can coexist, this theory cannot be applied easily to large numbers of interacting species. Thus, ecologists still debate the factors that determine the number of species that can coexist and why these vary from place to place. Moreover, it is also important to understand how species interactions influence the structure and dynamics of communities. Species assume different functional roles in communities, and their relative abundances reflect how they fit into the entire web of interactions within the community. Assemblages of species also change over time, whether in response to a disturbance or following some intrinsic dynamic process.

Ecologists hold diverse concepts of communities

The view that a community is a superorganism whose functioning and organization can be appreciated only when it is considered as an entire entity—the **holistic concept**—makes sense. We cannot ponder the significance of a kidney's functioning apart from the organism to which it belongs, and many ecologists argue that we similarly cannot consider soil bacteria without reference to the detritus they feed on, their predators, and the plants nourished by their wastes. Accordingly, they argue, one can understand each species only in terms of its contribution to the dynamics of the whole system. Most importantly, ecological and evolutionary relationships among species enhance community properties such as the stability of energy flow and nutrient cycling, making a community much more than the sum of its individual parts.

The view that community structure and functioning simply express the interactions of individual species that make up local associations, and do not reflect any organization, purposeful or otherwise, above the species level, is referred to as the **individualistic concept.** According to this view, because natural selection acts on the reproductive output of individuals, each population in a community evolves so as to maximize the reproductive success of its own members, not to benefit the community as a whole.

An intermediate, or mixed, point of view accepts the individualistic premises that most species interactions are antagonistic and that communities may be assembled haphazardly, but also admits to the holistic premise that some attributes of community structure and function arise only from interactions among species. Furthermore, it assumes that these interactions are often reinforced by coevolution (see Chapter 20), reflecting the strong reciprocal forces of selection that occur among interacting species.

Ecologists use several measures of community structure

Regardless of the debate over the nature of the community, ecologists often wish to characterize its structure and function. Community structure is difficult to define and measure. One of the simplest and most revealing measures of a community's structure is the number of species it includes. This measure is often referred to as **species richness.**

Naturalists have known for centuries that more species live in tropical regions than in temperate and boreal zones. For example, Barro Colorado Island, a 16 km² island in Gatun Lake, Panama (■ Figure 21.2), supports 211 tree species, more than are found in all of Canada. Plots of 1 hectare in some parts of Amazonian Peru and Ecuador contain more than 300 species; every other individual tree in such a plot belongs to a different species! With the exception of taxa especially adapted to harsher conditions unique to higher latitudes, most types of organisms exhibit their highest diversity in the Tropics.

Biologists have hardly catalogued all the species of plants and animals, let alone microbes. About a million and a half species have been described and named worldwide; estimates of the total run upward into the tens of millions. Because many of these species are becoming rare or extinct before they become known to science, ecologists feel an urgent need to understand why some communities are more biologically diverse than others and to find ways to preserve as much of this natural heritage as possible.

Even the simplest biological communities contain overwhelming numbers of species. To manage this complexity, ecologists often partition diversity into numbers of species

■ **Figure 21.2 Tropical forests harbor the greatest diversity of any communities.** The number of different tree species on Barro Colorado Island, Panama, is obvious even in this aerial photograph. Photo by Carl C. Hansen, courtesy of the Smithsonian Tropical Research Institute.

at each trophic level: primary producers, herbivores, and carnivores. Within trophic levels, method or location of foraging distinguishes different **guilds** of species: herbivores, for example, include leaf eaters, stem borers, root chewers, nectar sippers, and bud nippers.

Whenever ecologists attempt to tabulate the diversity of a community or part of a community by identifying all individuals encountered within a given area, they find that a few species are abundant and many more are rare. These patterns of relative abundance are another way in which ecologists have quantified the structure of communities, as we shall see below.

Regular patterns of community structure do not argue for or against a holistic interpretation of the community because organization can result from the independent activities of, and interactions between, a system's components. The holistic point of view argues that community structure reflects attributes of species selected to enhance the functioning of the community as a whole. The individualistic concept sees the structure of a community as a collective property of its individual components, each of which endeavors to function in its own right within the community. Ecologists have come to understand that it is not tenable to hold one or the other extreme viewpoint, and they now strive to determine the extent of community integration and its biological mechanisms.

The term "community" has been given many meanings

Throughout the development of ecology as a science, the term *community* has often denoted assemblages of plants and animals occurring in a particular locality and dominated by one or more prominent species or by some physical characteristic. We speak of an oak community, a sagebrush community, and a pond community, meaning all the plants and animals found in a particular place dominated by the community's namesake (█ Figure 21.3). Used in this way, the term is unambiguous: a community is spatially defined and includes all the populations within its boundaries. Ecologists also have a community concept that encompasses interactions among coexisting populations. This implies a functional rather than a descriptive use of the term.

When populations extend beyond arbitrary spatial boundaries, both the concept and the reality of the community become more difficult to pin down. Migrations of birds between temperate and tropical regions link assemblages of species in each area; within some tropical localities, as many as half the birds present during the northern winter are migrants. Salamanders, which complete their larval development in streams and ponds but pursue their adult existence in the surrounding woods, tie together aquatic and terrestrial assemblages, just as trees do when they shed their leaves into streams and thereby support aquatic detritus-based food chains.

Community structure and function blend a complex array of interactions, directly or indirectly tying together all members of a community into an intricate web. The influence of each population extends to ecologically distant parts of the community. Insectivorous birds, for example, do not eat trees, but they do prey on many of the insects that feed on foliage or pollinate flowers. The ecological and evolutionary effects of a population extend in all directions throughout the trophic structure of a community by way of its influence on predators, competitors, and prey.

(a)

(b)

(c)

█ **Figure 21.3 Communities are often named after their most conspicuous members.**
(a) A ponderosa pine community in the Santa Catalina Mountains of Arizona; (b) a riparian forest community borders a stream coursing through dry mountains in southern Arizona; (c) a deciduous forest community in the Great Smoky Mountains of Tennessee. Photos by R. E. Ricklefs.

Is the community a natural unit of ecological organization?

The holistic and individualistic concepts of community organization predict different patterns of species distribution over ecological and geographic gradients. From a holistic viewpoint, the species belonging to a community are closely associated with one another, which implies that the ecological limits of distribution of each species will coincide with the distribution of the community as a whole. Ecologists call this concept of community organization a **closed community.** From an individualistic viewpoint, each species is distributed independently of others that coexist with it in a particular association. Such an **open community** has no natural boundaries; therefore, its limits are arbitrary with respect to the geographic and ecological distributions of its member species, which may extend their ranges independently into other associations.

The structures of closed and open communities are shown schematically in ❙ Figure 21.4, in which the distribution of each species is plotted on a gradient of environmental conditions (for example, from dry to moist). In the diagram depicting closed communities, the distributions of the species in each community coincide closely. Thus, closed communities are discrete ecological units with distinct boundaries. The boundaries of such communities, called **ecotones,** are regions of rapid replacement of species along the gradient.

In the diagram depicting open communities, species are distributed at random with respect to one another, and so no discrete boundary of species composition is formed between communities. We could arbitrarily define the species occurring at a particular point—perhaps in a dry forest near the left-hand end of a moisture gradient—as a "community." At the same time, we would have to recognize that some of the species included would be more typical of even drier portions of the gradient, and that others would reach their greatest abundance in wetter sites.

The concepts of open and closed communities both have validity in nature (Table 21.1). We observe distinct boundaries between associations under two circumstances. The first of these occurs when the physical environment changes abruptly—for example, at the transition between aquatic and terrestrial communities, between distinct soil types, and between north-facing and south-facing slopes of mountains. The second circumstance occurs when one species or life form so dominates the environment that the edge of its range determines the distribution limits of many other species.

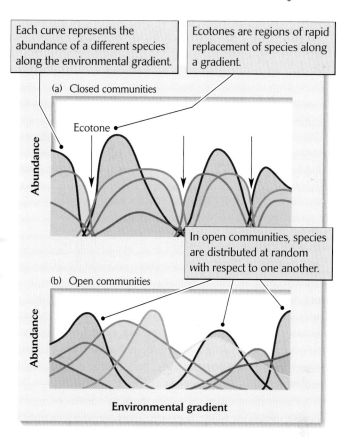

❙ **Figure 21.4 Closed community structure is distinguished from open community structure by the presence of ecotones.** Hypothetical distributions of species along an environmental gradient when the species are organized into distinct assemblages (closed communities, a) and when they are distributed at random along the gradient (open communities, b). Arrows indicate ecotones between closed communities. Each curve represents the abundance of a different species along the environmental gradient.

Ecotones

Ecotones are places where many species reach the edges of their distributions. Thus, ecotones represent boundaries between different closed communities. Ecotones are especially prominent where sharp physical differences separate distinct communities. Such differences occur at the interface between most terrestrial and aquatic (especially marine) environments (❙ Figure 21.5) and where underlying geologic formations cause the mineral content of soils to change abruptly.

An ecotone between plant associations on serpentine and nonserpentine soils in southwestern Oregon is represented in ❙ Figure 21.6. Levels of nickel, chromium, iron, and magnesium increase across the boundary into the serpentine soil; copper and calcium levels of the soil drop off. The edge of the serpentine soil marks the boundaries of

Table 21.1   Characteristics of open and closed communities

	Open community	Closed community
Early proponent	H. A. Gleason	F. E. Clements
Organization	Individualistic	Holistic
Boundaries	Diffuse	Distinct (ecotones)
Species ranges	Independent	Coincident
Coevolution	Uncommon and diffuse	Prominent

many species that either cannot tolerate these soils, such as black oak, or are restricted to them, such as buckbrush and fireweed. A few species, such as collomia and rag-wort, exist only within the narrow zone of transition; others, such as hawkweed and fescue, which are seemingly unresponsive to variations in these soil minerals, extend across the ecotone. Thus, the transition between serpentine and nonserpentine soils only partly conforms to the concept of closed communities; the ecotone is recognized by many, but not all, species.

Plants themselves may contribute to conditions maintaining ecotones (see Chapter 5). Such ecotones occur at the transition between broad-leaved and coniferous needle-leaved forests. The decomposition of conifer needles produces organic acids more abundantly than the breakdown of leaves of flowering plants, thus increasing soil acidity. Furthermore, because needles tend to decompose slowly, a thick layer of partly decayed organic material accumulates at the soil surface. This dramatic shift in conditions between broad-leaved and needle-leaved forests marks the edges of distributions of many shrub and herb species within each forest type. Similarly, at boundaries between grassland and shrubland or between grassland and forest, sharp changes in surface temperature, soil moisture, light intensity, and fire frequency result in many species replacements. Boundaries between grasslands and shrublands are often sharp because when one or the other vegetation type holds a slight competitive edge, it dominates the community. Grasses prevent the growth of shrub seedlings by reducing the moisture content of surface layers of soil; shrubs depress the growth of grass seedlings by shading them. Fire maintains a sharp edge between prairies and forests in the midwestern United States. Perennial grasses resist fire damage that kills tree seedlings outright, but fires cannot penetrate deeply into the moister forest habitats (see Chapter 22).

The continuum concept

Although distinct ecotones often form where physical conditions in the environment change abruptly, they are less

Figure 21.5 Ecotones are often associated with an abrupt change in the physical properties of adjacent habitats. In this section of the coast of the Bay of Fundy, New Brunswick, seaweed extends only to the high tide mark. Between the high tide mark and the spruce forest, waves wash soil from rocks and salt spray kills pioneering land plants, leaving the area devoid of vegetation. Photo by R. E. Ricklefs.

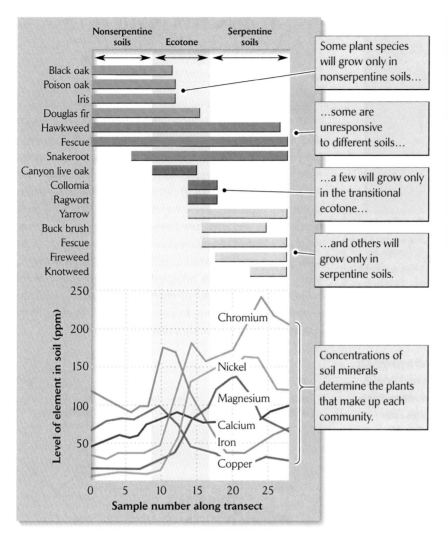

Some plant species will grow only in nonserpentine soils...

...some are unresponsive to different soils...

...a few will grow only in the transitional ecotone...

...and others will grow only in serpentine soils.

Concentrations of soil minerals determine the plants that make up each community.

▌Figure 21.6 Soil conditions reflecting underlying geology may cause prominent community boundaries. The diagram and graph show replacement of plant species (*above*) and changes in the concentration of elements in the soil (*below*) across a boundary between nonserpentine (samples 1–10) and serpentine soils (samples 18–28) in southwestern Oregon. This photograph of the edge of a serpentine barren in northern California shows the sharpness of the boundary. Data from C. D. White, *Vegetation-Soil Chemistry Correlations in Serpentine Ecosystems,* Ph.D. dissertation, University of Oregon (1971). Photo by R. E. Ricklefs.

likely to occur along gradients of gradual environmental change. The broad-leaved deciduous forests of eastern North America are bounded at conspicuous ecotones to the north, where they are replaced by cold-tolerant evergreen needle-leaved forests, to the west by drought- and fire-resistant grasslands, and to the southeast by fire-resistant pine forests. The broad-leaved forests themselves are not, however, homogeneous. Early botanical explorations clearly showed that different species of trees and other plants occurred in different areas within the deciduous forest biome. If Clements's closed community viewpoint were correct, the distinctive vegetation of each area would have represented a distinct community separated by sharp vegetational transitions from other communities. But as ecologists studied plant distributions in more detail, they found that plant associations fit the closed community concept less and less well: few species had closely overlapping geographic and ecological distributions, and sharp ecotones were not found. As Gleason had suggested, species tended to be dis-

tributed independently of one another over gradients of ecological conditions.

As the closed community concept lost support, ecologists became more interested in an open concept of community organization, referred to as the **continuum concept.** According to this concept, within broadly defined habitats, such as forest, grassland, or estuary, populations of plants and animals replace one another continuously along gradients of physical conditions. The environments of the eastern United States form a continuum, with a north–south temperature gradient and an east–west rainfall gradient. The species of trees found in any one region (for example, those native to eastern Kentucky) have different geographic ranges, suggesting independent evolutionary backgrounds and ecological relationships (**▌Figure 21.7**). Some species reach their northern limits in Kentucky, some their southern limits. Because few species have broadly overlapping geographic ranges, the assemblage of plant species that is found in any given spot does not represent a closed community.

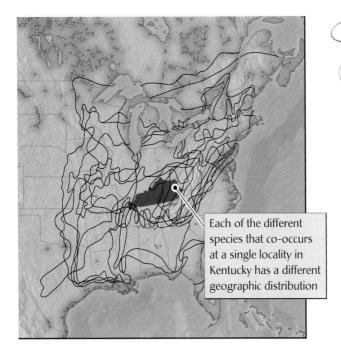

▌Figure 21.7 The species that occur together at a particular place may have different geographic distributions. None of the twelve tree species that occur together in plant associations in eastern Kentucky have the same geographic range. After H. A. Fowells, *Silvics of Forest Trees of the United States*, U.S. Department of Agriculture, Washington, D.C. (1965).

> Each of the different species that co-occurs at a single locality in Kentucky has a different geographic distribution

A more detailed view of the forests of Kentucky would reveal that many tree species segregate along local gradients of environmental conditions. Some grow along ridge tops, others along moist river bottoms; some on poorly developed, rocky soils, others on rich, organic soils. The species represented in each of these more narrowly defined associations might exhibit correspondingly closer distributions, but the open community concept would still better describe these associations of plants.

Gradient analysis

The validity of the continuum concept depends on the way in which species are distributed along ecological gradients. In a **gradient analysis,** closed community organization should reveal itself by the presence of sharp ecotones, as suggested in Figure 21.4. A gradient analysis is usually undertaken by measuring the abundances of species and the physical conditions at a number of locations and then plotting the abundances of each species as a function of the value of the physical condition. The range of conditions might embrace any number of physical variables, such as moisture, temperature, salinity, exposure, or light level.

ECOLOGISTS IN THE FIELD

How are species distributed along an ecological gradient?

Cornell University ecologist Robert Whittaker pioneered gradient analysis in North America, and his work was influential in putting to rest Clements's extreme view of closed communities. Whittaker conducted most of his work in mountainous areas, where moisture and temperature vary over short distances according to elevation, slope, and exposure. These variables in turn determine light, temperature, and moisture levels at a particular site. When Whittaker plotted abundances of species found at sites at the same elevation distributed over a range of soil moisture, he found that each species occupied a unique ecological distribution, with their peaks of abundance scattered along the environmental gradient (▌Figure 21.8). These findings were consistent with an open community structure.

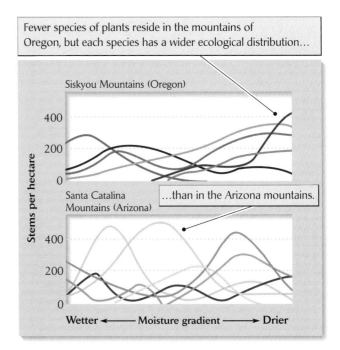

> Fewer species of plants reside in the mountains of Oregon, but each species has a wider ecological distribution…
>
> …than in the Arizona mountains.

▌Figure 21.8 Montane plant distributions indicate open community structure. Distributions of plant species along soil moisture gradients are shown for two locations: in the Siskyou Mountains of Oregon at 460–470 m elevation and in the Santa Catalina Mountains of southeastern Arizona at 1,830–2,140 m elevation. Each species has its greatest abundance at a different point on the moisture gradient. Fewer species are found in the mountains of Oregon, but each species has a wider ecological distribution, on average. After R. H. Whittaker, *Ecol. Monogr.* 30:279–338 (1960); R. H. Whittaker and W. A. Niering, *Ecology* 46:429–452 (1965).

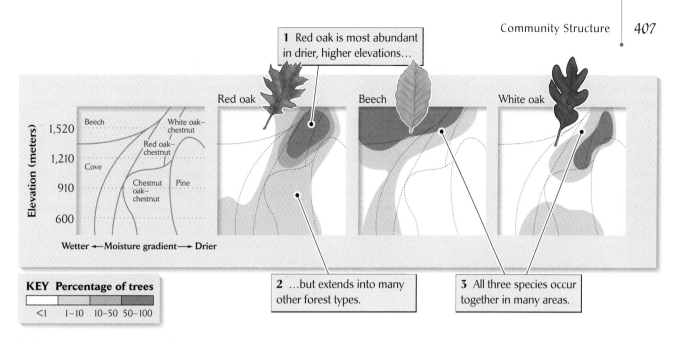

1 Red oak is most abundant in drier, higher elevations...

2 ...but extends into many other forest types.

3 All three species occur together in many areas.

▌Figure 21.9 Tree species of the Great Smoky Mountains show distinct but overlapping distributions. The approximate boundaries of major forest types with respect to altitude and soil moisture are shown at left for beech; red oak–chestnut; white oak–chestnut; cove; hemlock; chestnut oak–chestnut; pine. Distributions of red oak, beach, and white oak are not limited to the forest types bearing their names. Relative abundance is measured as the percentage of stems more than 1 cm in diameter of the focal species in samples of approximately 1,000 stems. After R. H. Whittaker, *Ecol. Monogr.* 26:1–80 (1956).

In the Great Smoky Mountains of Tennessee, Whittaker found that dominant species of trees occur widely outside the plant associations that bear their names (▌Figure 21.9). For example, red oak grows most abundantly in relatively dry sites at high elevations, but its distribution extends into forests dominated by beech, white oak, chestnut, and even hemlock (an evergreen conifer) and extends throughout the entire range of elevation in the Smoky Mountains. Beech prefers moister conditions than red oak, and white oak reaches its greatest abundance in drier conditions, but all three species occur together in many areas.

Whittaker's studies, and many others on plant and animal distributions since then, have found few cases of distinct ecotones between associations of species. Whittaker and others demonstrated that plant species are distributed more or less independently over ranges of ecological conditions. Few cases of consistent association between species were apparent, and these were overwhelmed by the predominately open structure of ecological communities.

Feeding relationships organize communities in food webs

When a community is viewed from an ecosystem perspective, with a focus on the flow of energy, one sees that species occur in functional groups whose members occupy similar trophic, or feeding, positions (see Chapter 6). Thus, plants can be lumped together as producers, all plant eaters (from ant to zebra) share the herbivore label, and so on. Yet this view is limited because it ignores the variation in the numbers and evolutionary histories of the species that make up the community.

When we apply a **food web** perspective to the community, we tend to emphasize diversity. Although food webs are also based on functional relationships, they stress connections between populations and recognize, for example, that not all herbivores consume all producers. Because food web analysis includes species-level information about a community, it has greater power than ecosystem analysis to differentiate community structure. Yet because community structure is difficult to define and measure, different food web analyses often produce different results. For instance, we can ask whether a more complex food web structure leads to increased dynamic stability. A reasonable answer might be: When predators have alternative prey, their population sizes depend less on fluctuations in numbers of a particular prey species; and where energy can take many routes through a system, disruption of one pathway merely shunts more energy through another. Both factors would contribute to increased community stability. Yet it is also reasonable to suggest that as communities become more diverse, species exert greater influence on one another through their various interactions; these biological links in turn may create pervasive time lags in population processes (see Chapters 15 and 18), which tend to destabilize diverse systems.

Feeding relationships may also affect species diversity within a community. For example, when a predator controls the population of a dominant competitor, it may allow inferior competitors to persist because they avoid predation. Thus, the diversity of a particular trophic level within a food web may depend on predation by populations at higher trophic levels.

ECOLOGISTS IN THE FIELD

Food web complexity in the rocky intertidal zone

The relationship between food web organization and community diversity was first demonstrated in an influential study by Robert T. Paine of the University of Washington. Paine compared food webs of rocky shore intertidal communities on the coast of Washington and in the Gulf of California, both of which are dominated by imposing predators, the sea stars *Pisaster* and *Heliaster* respectively (▌Figure 21.10). In the Gulf of California, however, there were abundant populations of other predators, such as snails, crabs, and fish. The herbivore and producer trophic levels were also more diverse in the Gulf of California, in spite of the lower primary productivity of the ecosystem.

This pattern suggested to Paine that predators might have a controlling influence on the diversity of lower trophic levels. To test this hypothesis, he conducted predator removal experiments in study areas on the coast of Washing-

ton. When Paine removed sea stars from these areas, their primary prey, the mussel (*Mytilus*), spread rapidly, crowding other organisms out of the experimental plots and reducing the diversity and complexity of local food webs, particularly the diversity of herbivores (see Chapter 19). Paine showed that the same principle applies to the diversity of primary producers. Removal of the urchin *Strongylocentrotus*, an herbivore, allowed a small number of competitively superior algae to dominate an area, crowding out many ephemeral or grazing-resistant species (see Chapter 18).

By manipulating competitive relationships among species at lower trophic levels in this way, Paine showed that consumers could promote diversity and thereby control the structure of a community. Such species are called keystone predators because when they are removed, the edifice of the community tumbles (▌Figure 21.11). Thus, maintenance of populations of **keystone predators** is an important component of the stability of a community.

Paine and others who followed his example stressed that the consumer–resource relationships represented in food webs hold a key to understanding community organization. Paine also distinguished different types of food webs, which describe different ways in which populations influence one another within communities. **Connectedness webs** emphasize feeding relationships among organisms, portrayed as links in a food web. **Energy flow webs** represent an ecosystem viewpoint, in which connections between species are quantified by flux of energy between a resource and its consumer. In **functional webs**, the importance of each population in maintaining the integrity of a community is reflected in its influence on the growth rates of other populations. This controlling role, which can

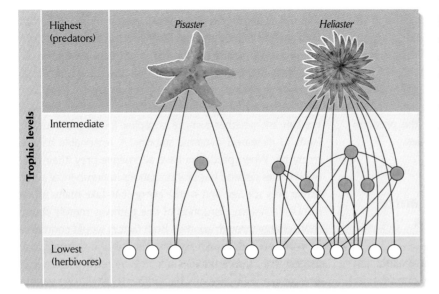

▌**Figure 21.10 Predators influence the diversity of lower trophic levels.** The sea stars *Pisaster*, on the coast of Washington, and *Heliaster*, in the northern Gulf of California, strongly influence the structure of the communities to which they belong. The *Heliaster*-dominated food web has numerous other predator species (shown as shaded symbols). The lowest trophic levels that are indicated on these food web diagrams include such herbivores as chitons, limpets, herbivorous gastropods, and barnacles. These are much more diverse in the *Heliaster*-dominated food web. After R. T. Paine, *Am. Nat.* 100:65–75 (1966).

▌Figure 21.11 Elimination of keystone predators shows their controlling influence on species diversity. The plot on the right side of the photograph was sprayed with insecticide for 8 years; the plot on the left is an unsprayed control plot. The insecticide kept populations of the chrysomelid beetle *Microrhopapla vittata* from reaching outbreak levels and defoliating the goldenrod *Solidago altissima*, its preferred food plant. Consequently, goldenrod came to dominate the sprayed plot and shaded out the many other species growing in the more diverse control plot. Courtesy of Walter Carson, from W. P. Carson and R. B. Root, *Ecol. Monogr.* 70:73–99 (2000).

be revealed only by experiments, need not correspond to the amount of energy flowing through a particular link in the food web, as shown dramatically for an intertidal zone food web in ▌Figure 21.12.

As Paine and others advocated the use of food web diagrams to depict the structure of biological communities, a few ecologists questioned whether differences in the structure of food webs could affect the dynamics, stability,

and persistence of communities. Is one particular arrangement of feeding relationships among species intrinsically more stable than a different arrangement among the same number of species? How important is food web stability to the structure of natural communities? Here, *stability* means the ability of the community to resist environmental perturbations or to return to equilibrium population sizes following perturbation. These qualities, which are

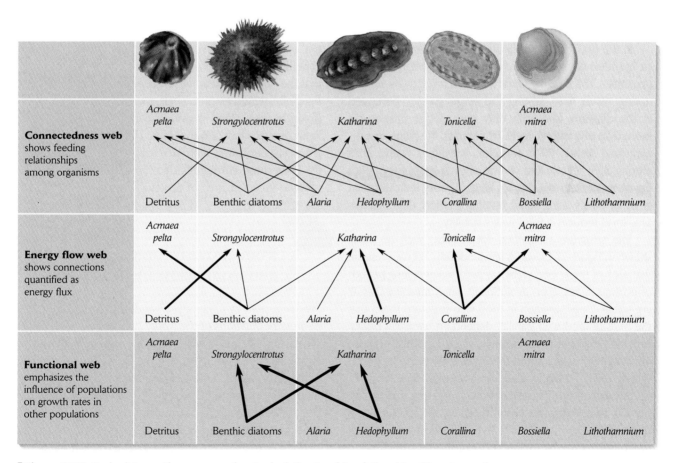

▌Figure 21.12 Ecologists use three approaches to depicting trophic relationships. Three types of food web diagrams, here applied to the species of a rocky intertidal habitat on the coast of Washington, depict different ways in which populations influence one another within communities. The thickness of an arrow reflects the strength of that relationship. From R. T. Paine, *J. Anim. Ecol.* 49:667–685 (1980).

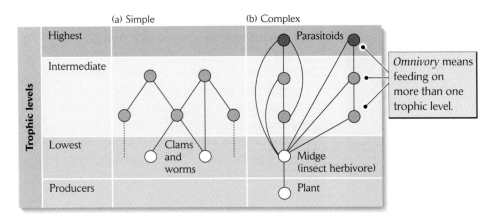

▌Figure 21.13 Different communities of similar diversity may have different food web structures. (a) A mudflat community containing intertidal gastropods, bivalves, and their prey has only one omnivorous species. (b) This food web, based on the plant *Baccharis,* its insect herbivores, and their parasitoids, is more complex, involving several omnivorous species. In food web diagrams such as these, lines connect the resources below to the consumer above. Not all prey species are depicted. After S. L. Pimm, *Food Webs,* Chapman & Hall, London and New York (1982).

best investigated by theoretical approaches, are also thought to influence the persistence of populations within communities, hence community diversity.

In the two food webs illustrated in ▌Figure 21.13, similar numbers of species are organized in strikingly different structures. The mudflat community (a) is relatively simple, having seven links among the seven species portrayed in the diagram, and with only one species preying on more than one trophic level. By contrast, the plant–insect–parasitoid system (b) is complex; it exhibits twelve links among eight species and several cases of **omnivory** (feeding on more than one trophic level). Theory indicates that, all other things being equal, feeding on more than one trophic level reduces a food web's stability.

The structure of natural food webs, such as those shown in Figure 21.13, varies tremendously. Yet we presume that each of these food webs has persisted over long periods of ecological and even evolutionary time, meaning that all are stable. Does variation in food web structure mean that the rules of food web stability depend on particular organisms and ecological circumstances? Or is stability not an important consideration in food web structure, which then merely reflects feeding relationships of individual species?

An important first step toward answering these questions is to characterize community organization in ways that capture its complexity. Two important attributes of communities are the total number of species and the average number of feeding links per species. Comparisons of a large number of food webs have shown that the number of feeding links per species is independent of the species richness of the community, as shown in ▌Figure 21.14 for

assemblages of invertebrates that form in the water trapped by pitcher plants. Thus, the number of interactions that each species has with other species is independent of the overall diversity of the community. Another generalization that has emerged from these comparisons is that the number of trophic levels, and the number of feeding guilds within trophic levels, increases with community diversity. This trend is also apparent in the food webs from pitcher-plant communities. Thus, increasing diversity is usually associated with increased community complexity. How these attributes of communities are determined remains an active area of ecological research.

Trophic levels are influenced from above by predation and from below by production

We have seen in Chapters 17 and 18 that predators can depress populations of their prey dramatically. This principle can apply equally well to entire trophic levels. In a classic paper published in 1960, three University of Michigan ecologists, Nelson Hairston, Frederick Smith, and Larry Slobodkin, suggested that the earth is green because carnivores depress the populations of herbivores that would otherwise consume most of the vegetation. This phenomenon, which emphasizes the indirect effects of consumer–resource interactions extended through additional trophic levels of the community, is called a **trophic cascade** (▌Figure 21.15). When higher trophic levels determine the size of the trophic levels below them, the situation is referred to

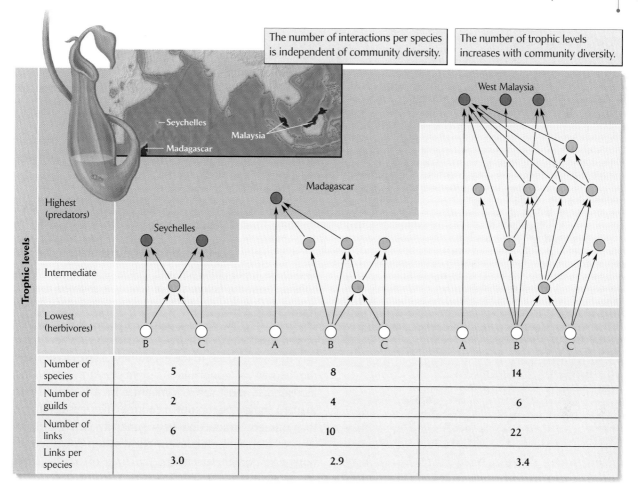

The number of interactions per species is independent of community diversity.

The number of trophic levels increases with community diversity.

Figure 21.14 Increasing species richness is associated with increased food web complexity.
Food webs are shown for invertebrates living in *Nepenthes* pitcher plants (*inset*) in different regions bordering on the Indian Ocean. The food web diagrams show increasing ecological diversity (more trophic levels and guilds within trophic levels) and longer food chains, but similar numbers of feeding links per species, with increasing species diversity. Sources of food are live insects (A), recently drowned insects (B), and older organic debris (C). From R. A. Beaver, *Ecol. Entomol.* 10:241–248 (1985).

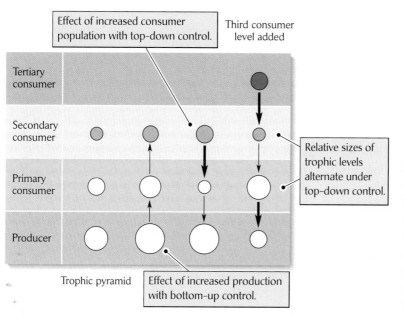

Effect of increased consumer population with top-down control.

Third consumer level added

Relative sizes of trophic levels alternate under top-down control.

Effect of increased production with bottom-up control.

Figure 21.15 The trophic structure of a community may be determined by bottom-up or top-down control. With bottom-up control, increased production results in greater productivity at all trophic levels. With top-down control, consumers depress the trophic level on which they feed, and this indirectly increases the next lower trophic level. This results in a trophic cascade linking all the trophic levels in a community.

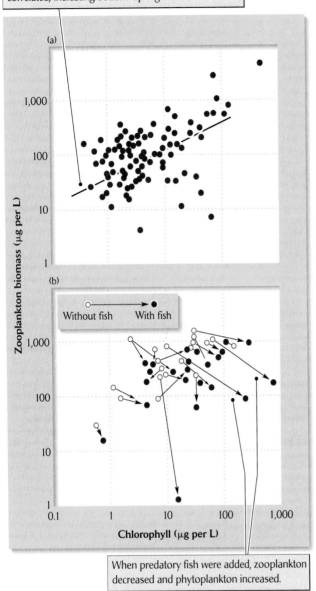

The relative sizes of the two trophic levels are positively correlated, indicating bottom-up regulation.

(a)

(b)

Without fish   With fish

Zooplankton biomass (μg per L)

Chlorophyll (μg per L)

When predatory fish were added, zooplankton decreased and phytoplankton increased.

Figure 21.16 Primary consumer biomass shows the effects of both bottom-up and top-down influences. (a) Relationship between zooplankton biomass and phytoplankton density in natural lakes sampled over a range of primary production. The relative sizes of the two trophic levels are positively correlated, indicating bottom-up regulation. (b) Introducing fish to lakes reduces zooplankton populations and results in an increase in phytoplankton density, indicating the effect of top-down regulation. After M. A. Leibold et al., *Annu. Rev. Ecol. Syst.* 28:467–494 (1997).

A survey of zooplankton and phytoplankton densities in natural lakes, assembled by Mathew Leibold and his colleagues at the University of Chicago, showed that the primary consumer trophic level varied in parallel with the producer trophic level, a pattern that is consistent with bottom-up control (**Figure 21.16**). However, when predatory fish were added to the experimental lakes to decrease the density of zooplankton, phytoplankton abundance increased in most cases, sometimes by a factor of more than 10, indicating top-down control. These results suggest that primary production very generally determines the sizes of higher trophic levels, but that top-down interactions can nonetheless adjust the sizes of trophic levels within a narrower range.

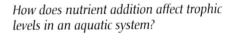

ECOLOGISTS IN THE FIELD

How does nutrient addition affect trophic levels in an aquatic system?

Lars–Anders Hansson and his colleagues at the University of Lund in Sweden investigated bottom-up control of trophic structure by adding inorganic nutrients (phosphorus and nitrogen) to experimental micro-cosm communities to boost their primary production. The experimental systems were established in hundreds of large cylindrical tanks in a greenhouse and stocked with either three trophic levels (heterotrophic bacteria, photosynthetic flagellates and algae, and zooplankton) or four trophic levels (adding fish as zooplankton predators) (**Figure 21.17a**). The results of the experiment revealed both bottom-up and top-down control of trophic level size. In both the three-level and four-level systems, adding inorganic nutrients increased the abundance of most of the trophic levels in the system. However, when fish were added (the fourth trophic level), zooplankton levels were reduced in both the low- and high-nutrient treatments, and numbers of producers were higher (**Figure 21.17b**).

as **top-down control.** When the size of a trophic level is determined by the rate of production of its food, the situation is referred to as **bottom-up control.**

Ecologists have debated the relative strengths of top-down and bottom-up control mechanisms for many years. For example, an alternative interpretation to top-down control of the abundance of vegetation in most terrestrial habitats is that plant parts resist consumption through various digestion inhibitors and toxic substances (see Chapter 17). Indeed, the best evidence for top-down control comes from aquatic ecosystems, in which plants and algae, especially phytoplankton, are highly digestible (see Chapter 6).

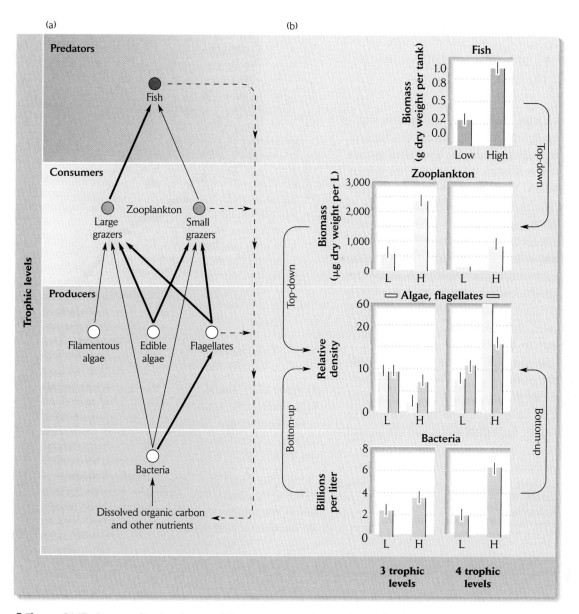

Figure 21.17 Community structure and its response to changes in productivity depends on the number of trophic levels. (a) Diagram of the interactions between the trophic levels. Thick arrows represent strong interactions; thin arrows, weak interactions. Dashed arrows show excretion. (b) Relative amounts of biomass in each trophic level in three-level and four-level experimental microcosms with low and high nutrient inputs. After L.-A. Hansson et al., *Proc. R. Soc. Lond.* B 265:901–906 (1998).

Thus, as in Leibold's comparative survey of natural and experimental lakes, Hansson et al.'s microcosm experiments provided evidence that increased primary production tends to augment all the overlying trophic levels. However, the experiments also showed that consumers could depress the size of the trophic level immediately below them and increase the abundance of organisms two levels below. Furthermore, zooplankton grazing shifted the dominance of organisms in the producer trophic level. With low nutrient addition, flagellates and algae were relatively more abundant than bacteria; with high nutrient addition, increasing zooplankton populations depressed flagellates and algae, and allowed bacterial densities to increase. When fish were added to the microcosm, they kept the zooplankton from increasing as much with nutrient addition, and the algae as well as the bacteria responded to the high nutrient levels.

Species in biological communities vary in relative abundance

Up to this point, we have talked about community structure without paying attention to differences among species. Yet even within a particular trophic level, each species occupies a distinctive ecological position and has unique ecological relationships. Differences in these ecological relationships often reveal themselves in the abundances of species. The Danish botanist Christen Raunkiaer noted early in the twentieth century that within a particular community, a few species attain high abundance—they are the **dominants** in the community—whereas most others are represented by relatively few individuals. When he plotted the number of individuals of several different species on the same trophic level, the result was a reverse-J-shaped curve, like that shown in ▌Figure 21.18.

Ecologists have devised many ways of portraying the relative abundances of species within communities. One common approach is to plot the abundances of species, usually on a logarithmic scale, ranked from the most common to the rarest (▌Figure 21.19). Such plots emphasize the general

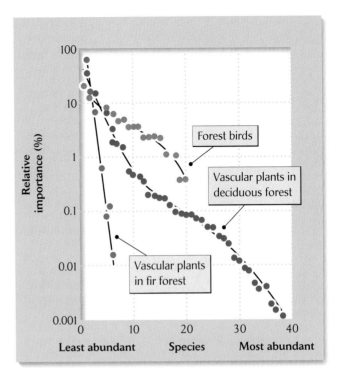

▌**Figure 21.19 Abundances of species can be plotted on a logarithmic scale.** Abundance curves are shown for three natural communities: birds in a deciduous forest in West Virginia; vascular plants in a subalpine fir forest in the Great Smoky Mountains, Tennessee; and vascular plants in a deciduous cove forest in the Great Smoky Mountains. Abundance is represented by number of individuals for birds and by net primary production for plants. From R. H. Whittaker, *Communities and Ecosystems* (2d ed.), Macmillan, New York (1975).

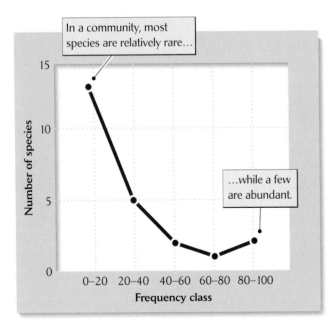

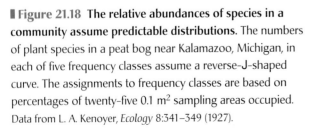

▌**Figure 21.18 The relative abundances of species in a community assume predictable distributions.** The numbers of plant species in a peat bog near Kalamazoo, Michigan, in each of five frequency classes assume a reverse-J-shaped curve. The assignments to frequency classes are based on percentages of twenty-five 0.1 m² sampling areas occupied. Data from L. A. Kenoyer, *Ecology* 8:341–349 (1927).

observation that few species are abundant in a community and many are rare. Additionally, they show that as the number of species on a trophic level increases, the dominance of the most abundant species—that is, the proportion it constitutes of all the individuals in the community—decreases. The abundance of each species appears to reflect the variety and abundance of resources available to it, as well as the influences of competitors, predators, and diseases.

As with many issues concerning the structure and dynamics of communities, ecologists have turned to quantitative models of patterns of community diversity with the hope of understanding the underlying processes that produce these patterns. Mathematics can serve two purposes here. On one hand, ecologists can use mathematics to describe data (species abundances, in this case) with simple equations and use values of variables in the equations to make comparisons among different communities. On the other hand, ecologists can use the logic of a mathematical model to investigate the processes that might produce the observed distributions. Without going into detail, let us simply say that models of relative abundance have served better as descriptive tools than they have as a way to elucidate the processes that determine relative abundances.

MORE ON THE WEB *The lognormal distribution.* The variation in relative abundance among species within a community can be described by a simple statistical relationship that shows how number of species increases with sample size.

Number of species increases with area sampled

As a rule, more species occur within large areas than within small areas. The botanist Olaf Arrhenius first formalized this **species–area relationship** in 1921. Since then, it has been common practice to characterize relationships between numbers of species (S) and area (A) with power functions of the form

$$S = cA^z,$$

where c and z are constants fitted to the data. Graphic portrayals of species–area relationships plot the logarithm of species number against the logarithm of area, as shown in ▌Figure 21.20. After log transformation, the species–area relationship becomes

$$\log S = \log c + z \log A,$$

which is the equation for a straight line.

Analyses of species–area relationships among many groups of organisms have revealed that most values of z fall within the range 0.20–0.35—that is, the number of species increases in proportion to the one-fifth to one-third power (fifth root to cube root) of area. This consistency of z suggests several possibilities. One is an artifact of sampling. As the area of a sample increases, the number of individuals included within it also increases; thus, as sample size increases, more and more species are discovered. Comparing samples of similar size can circumvent this problem. When this is done, species–area relationships persist, showing that these relationships are not simply artifacts. Furthermore, the slope of the species–area relationship, z, varies in predictable ways. For example, z values obtained for continental areas tend to be lower than those obtained for islands over a comparable size range. This difference occurs because rapid movement of individuals within continents, where barriers to dispersal are not strong, prevents local extinction of populations within small

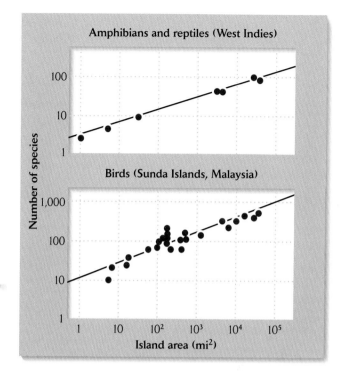

▌**Figure 21.20 The number of species increases with the area sampled.** Species–area curves for (*above*) amphibians and reptiles in the West Indies and (*below*) birds in the Sunda Islands, Malaysia, show this pattern. From R. H. MacArthur and E. O. Wilson, *Evolution* 17:373–387 (1963), and *The Theory of Island Biogeography*, Princeton University Press, Princeton, NJ (1967).

areas. Thus, small areas within continents have nearly the same species richness as large areas, and the species–area curve consequently is shallower.

Where a flora or fauna is perfectly known (that is, all species have been discovered), sampling by the investigator cannot be the reason that larger areas have more species. For example, we possess near-perfect knowledge of the land bird fauna of the West Indies, among which species increase with respect to area with a slope of about z = 0.24. Here, differences in diversity between large and small islands must signify differences in the intrinsic qualities of islands. Likely candidates include habitat heterogeneity, which undoubtedly increases with the size (and resulting topographic heterogeneity) of an island, and size per se, as larger islands make better targets for potential immigrants from the mainland. In addition, larger islands support larger populations, which may persist longer owing to their greater genetic diversity, broader distributions over habitats, and numbers large enough to prevent stochastic extinction.

Whether island size itself or habitat heterogeneity is more important to the species–area relationship can be

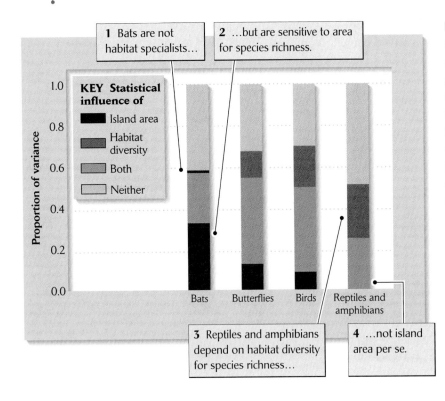

1 Bats are not habitat specialists...

2 ...but are sensitive to area for species richness.

KEY Statistical influence of
- Island area
- Habitat diversity
- Both
- Neither

Bats Butterflies Birds Reptiles and amphibians

Proportion of variance

3 Reptiles and amphibians depend on habitat diversity for species richness...

4 ...not island area per se.

▌Figure 21.21 **Area and habitat diversity both contribute to the species–area relationship.** The graph shows the proportion of variance in species richness of four types of animals among islands in the Lesser Antilles that is attributable to island area, habitat diversity, correlated variation between area and habitat diversity, and unexplained by either or both factors. After R. E. Ricklefs and I. J. Lovette, *J. Anim. Ecol.* 68:1142–1160 (1999).

determined by comparing island biotas in which these measures vary independently of one another, allowing their separate effects to be teased apart statistically. The Lesser Antilles exhibit a range of island size, but islands of any given size may be either high volcanic islands with varied habitats ranging from coastal mangroves to rain forest and cloud forest, or low areas of raised seabed dominated by dry forest and scrub. Analyses of species–area–habitat diversity relationships showed that the number of species of bats, which are not habitat specialists, was sensitive to island area and not to habitat diversity (▌Figure 21.21). At the other extreme, the diversity of reptiles and amphibians depended only on habitat diversity and was unrelated to island area per se. The reasons for these differences were not entirely clear, but they are probably related to differences among the taxa in habitat specialization and rates of movement between islands. This example emphasizes how important it is to study the underlying processes when one is interested in understanding the meaning of patterns.

Diversity indices weight species richness by relative abundance

How many species are found in a particular area? How does species richness vary from one place to another? These questions are important to conservationists and managers who need to know which areas support the greatest numbers of

species. Thus, ecologists are often faced with the problem of comparing the species diversity of different areas or habitats. However, differences in the abundances of species within communities present two practical problems. You have seen that the total number of species included in a sample varies with sample size because as more individuals are sampled, the probability of encountering rare species increases. Thus, ecologists cannot compare diversity among areas sampled at different intensities merely by comparing species counts. Also, not all species should contribute equally to our estimate of total diversity because their functional roles in a community vary in proportion to their overall abundance.

Ecologists have tackled the second problem by formulating **diversity indices** in which the contribution of each species is weighted by its **relative abundance,** meaning the proportion of the total number of individuals in a community that belong to that species. Two such indices are widely used in ecology: Simpson's index and the Shannon–Wiener index. Both indices are calculated from the proportions of each species (p_i) in the total sample of individuals. **Simpson's index,** D, is

$$D = \frac{1}{\sum p_i^2}$$

For any particular number of species in a sample (S), the value of D can vary from 1 to S, depending on the variation in species abundance. For example, when five species have equal abundances, each p_i is 0.20. Therefore, the value of

Table 21.2 Comparison of diversity indices D, H, and e^H for hypothetical communities of five species having different relative abundances

Proportion of sample represented by species					Diversity index		
A	B	C	D	E	D	H	e^H
0.20	0.20	0.20	0.20	0.20	5.00	1.609	5.00
0.25	0.25	0.25	0.25	0.00	4.00	1.386	4.00
0.24	0.24	0.24	0.24	0.04	4.30	1.499	4.48
0.25	0.25	0.25	0.25	0.001	4.02	1.393	4.03
0.50	0.30	0.10	0.07	0.03	2.81	1.229	3.42

each p_i^2 is 0.04, and $D = 1/(0.04 + 0.04 + 0.04 + 0.04 + 0.04) = 1/0.20 = 5$. Thus, the diversity index of this sample is 5, which is the number of species in the sample. When five species have unequal abundances, the diversity index is less than the total number of species, as shown in Table 21.2. Rarer species contribute less to the value of the diversity index than do common species.

The **Shannon–Wiener index** is calculated by the equation

$$H = -\Sigma\, p_i \log_e p_i,$$

where H is a logarithmic measure of diversity. As in the case of Simpson's index, higher values of H represent greater diversity. Also like Simpson's index, the Shannon–Wiener index gives less weight to rare species than to common ones. Because H is roughly proportional to the logarithm of number of species, it is sometimes preferable to express the index as e^H, which is proportional to the actual number of species. Table 21.2 presents values of e^H, which we may compare directly to Simpson's index.

As mentioned above, another problem in estimating species diversity is that the number of species in a sample tends to increase with the number of individuals sampled. If we wish to standardize measurements of diversity for comparison, we must base them on comparable sample sizes. When samples include different numbers of individuals, comparability can be achieved by a statistical procedure known as **rarefaction**, in which equal-sized subsamples of individuals are drawn at random from the total sample.

Rarefaction can be thought of as a means of portraying the relationship between number of species and sample size. Howard Sanders of the Woods Hole Oceanographic Institution applied this technique to samples of benthic marine organisms dredged from soft sediments at various localities in several marine environments. The number of specimens varied between samples because of differences

in densities of organisms in the substrate and unavoidable variation in sampling procedures. Therefore, it was impossible to tell whether differences in diversity between samples appeared as an artifact of sampling or reflected consistent differences between habitats. So the samples were rarefied to make them comparable (Figure 21.22).

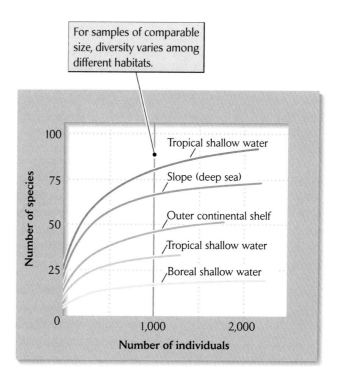

For samples of comparable size, diversity varies among different habitats.

Figure 21.22 Diversity in different-sized samples can be compared by rarefaction. The number of species of bivalves and polychaete worms found in different marine benthic habitats is shown as a function of sample size. Tropical and deep-sea faunas tend to be more diverse than faunas in less constant environments. After H. L. Sanders, *Brookhaven Symp. Biol.* 22:71–81 (1969).

The rarefaction curves allowed Sanders to compare diversity at each site. In this case, rarefaction curves clearly distinguished the sites, showing that for comparable samples, diversity varied among different habitats.

 ## Summary

1. A biological community is an association of interacting populations. Questions about communities address the evolutionary origins of community properties, relationships between community organization and stability, and the regulation of species diversity.

2. Ecologists characterize communities in terms of the number of species present, their relative abundances, their organization into guilds of species with similar feeding habitats, and food webs portraying feeding relationships among species.

3. Generally speaking, communities do not form discrete units separated by abrupt transitions in species composition. Species tend to distribute themselves over gradients of ecological conditions independently of the distributions of other species. Ecologists refer to this pattern as an open community structure.

4. Discontinuities between associations of plants and animals, called ecotones, sometimes occur at sharp physical boundaries or accompany changes in the growth forms that dominate a habitat. The aquatic–terrestrial transition provides an example of the first kind of ecotone, the prairie–forest transition an example of the second.

5. To analyze distributions of species with respect to environmental conditions and with respect to the distributions of other species, ecologists have devised various types of gradient analysis, in which they position sample localities on axes of physical conditions. The distributions of species along these environmental gradients emphasize the open structure of communities.

6. Community structure can be depicted by means of food webs showing the feeding relationships among species within a community. Food webs can be characterized by the number of feeding links per species and the average number of trophic levels on which a species feeds. The importance of a feeding link can be measured by the amount of energy flux through the link (energy flow web) or by the influence of variation in the prey population on variation in the predator population (functional web).

7. Experimental manipulations of trophic levels, particularly in aquatic systems, show that consumers can depress the sizes of the trophic level immediately below them, which indirectly increases populations two trophic levels below. This effect is called the trophic cascade and is also referred to as a top-down effect in biological communities. When the productivity of one trophic level affects the productivity of higher trophic levels, this is known as a bottom-up effect.

8. In any community, some species are common and others are rare. The most common species often are called dominants. The distribution of relative abundance within an ecological community can take on a characteristic pattern, but the meaning of such patterns is not well understood.

9. The number of species in a sample increases in direct proportion to the area sampled. This pattern results in part from larger areas giving rise to larger total samples. However, studies of well-known faunas and floras also indicate that larger areas are more heterogeneous ecologically, providing opportunities to sample more kinds of habitats, and that larger islands have more species because they are better targets for colonization and because larger populations better resist extinction.

10. Various indices of diversity, most notably Simpson's index and the Shannon–Wiener index, have been devised to account for variations in abundance when comparing diversity between samples. Because the number of species increases as sample size increases, ecologists have also devised rarefaction procedures and other statistical techniques to make samples of different sizes comparable.

 ## PRACTICING ECOLOGY

CHECK YOUR KNOWLEDGE

Plants upon Plants

Communities can sometimes occur nested within other communities. Epiphytes are a case in point. The word *epiphyte* literally means "upon plants," and thus epiphytes are plants that grow upon other plants, where they form a type of community. The term also applies to tiny marine algae that grow on large kelps, as well as to symbiotic algal-fungal lichens that grow on the bark of many tree species across North America. In the tropics, epiphytes are common on the trunks and branches of forest trees. They take root in the sparse lichens and organic debris that accumulate on the tops of branches, in the junctions between branches, and in other nooks and crannies of the forest canopy. Because of the thin soil layers and their position high in the forest canopy, epiphytes can be exposed to relatively high temperatures, temporary droughts, and bright light levels. Tropical epiphytes have evolved a variety of morphological and physiological adaptations to help them survive in this specialized microhabitat. They come in many shapes and sizes, and contain representatives of a wide array of plant taxonomic groups, including lichens, mosses, ferns, orchids, bromeliads, and even cacti!

The bromeliads (family Bromeliaceae, which includes the pineapple) exhibit a remarkable range of adaptations for survival in both terrestrial and epiphytic habitats. For example, members of the genus *Tillandsia* can absorb water from fog through their leaves via special valvelike structures. Tank bromeliads have their leaves arranged in a rosette, with an opening at the inner, central portion of the leaf bases. This configuration forms the "tank" for which they are named; the tank collects and stores water and organic debris. The tanks of these bromeliads can contain their own communities of frogs, other amphibians, aquatic insects, spiders, and a myriad of small invertebrates and microorganisms.

Nalini Nadkarni, of Evergreen State College, Olympia, Washington, has been studying the dynamics of tropical epiphyte communities. The habitats where epiphytes grow are subject to disturbance by wind, animals, waterlogging, branch breakage, and treefalls. How do epiphyte communities respond to such disturbances? Nadkarni used mountain-climbing gear to reach her research sites high above the ground, where she recorded the amount, composition, and location of epiphyte cover for a decade following experimental removal of epiphytes from mature trees in the cloud forest of Monteverde, Costa Rica. Colonization was so slow that no epiphytes returned for the first 5 years after removal. Also, the initial communities were quite different from the communities that were removed (Table 21.3). Re-colonization was expected to occur from the edge of plant material that remained on the branches. Instead, colonization began on the undersides of branches and moved upward over time. Nadkarni concluded that this pattern was a result of slower drying on the undersides of branches following precipitation.

CHECK YOUR KNOWLEDGE

1. What does the presence of communities within communities mean when one is attempting to calculate diversity indices?

2. Examine the data in Table 21.3, which is adapted from Nadkarni (2000). What is the likely sequence of species recovery based on these data?

3. Visit the Web site on epiphytes from the Royal Botanic Gardens at Kew, England, through *Practicing Ecology on the Web* at *http://www.*

MORE ON THE WEB

Table 21.3 Percent biomass or cover by epiphytes before and ten years after experimental removal

Component	Before removal	After removal
Bark (no epiphytes)	0	45
Crustose lichens	0	6
Foliose lichens	0	4
Bryophytes	20	35
Vascular plants	30	10
Dead organic matter	50	0

Data for components before removal are expressed as mean biomass, and for components after removal as mean cover. Data are adapted from Nadkarni (2000).

whfreeman.com/ricklefs. How do epiphytes interact with the bird and ant members of the forest canopy community?

4. Can you think of any other communities within communities?

 Suggested Readings

Brown, J. H. 1995. *Macroecology.* University of Chicago Press, Chicago.

Brown, J. H., and E. J. Heske. 1990. Control of a desert-grassland transition by a keystone rodent guild. *Science* 250:1705–1707.

Carson, W. P., and R. B. Root. 2000. Herbivory and plant species coexistence: Community regulation by an outbreaking phytophagous insect. *Ecological Monographs* 70:73–99.

Gee, J. H. R., and P. S. Giller. 1987. *Organization of Communities Past and Present.* Blackwell Scientific Publications, Oxford.

Hansson, L.-A., C. Brönmark, P. Nyström, L. Greenberg, P. Lundberg, P. A. Nilsson, A. Persson, L. B. Pettersson, P. Romare, and L. J. Tranvik. 1998. Consumption patterns, complexity and enrichment in aquatic food chains. *Proceedings of the Royal Society of London B* 265:901–906.

Jackson, J. B. C. 1994. Community unity? *Science* 264:1412–1413.

Leibold, M. A., J. M. Chase, J. B. Shurin, and A. L. Downing. 1997. Species turnover and the regulation of trophic structure. *Annual Review of Ecology and Systematics* 28:467–494.

MacArthur, R. H., and E. O. Wilson. 1967. *The Theory of Island Biogeography.* Princeton University Press, Princeton, NJ.

Magurran, A. E. 1988. *Ecological Diversity and Its Measurement.* Princeton University Press, Princeton, NJ.

Nadkarni, N. M. 2000. Colonization of stripped branch surfaces by epiphytes in a lower montane cloud forest, Monteverde, Costa Rica. *Biotropica* 32:358–363.

Paine, R. T. 1980. Food webs: Linkage, interaction strength and community infrastructure. *Journal of Animal Ecology* 49:667–685.

Palmer, M. W., and P. S. White. 1994. Scale dependence and the species–area relationship. *American Naturalist* 144:717–740.

Pimm, S. L. 1982. *Food Webs*. Chapman & Hall, London and New York.

Pimm, S. L. 1991. *The Balance of Nature?* University of Chicago Press, Chicago.

Power, M. E., D. Tilman, J. A. Estes, B. A. Menge, W. J. Bond, L. S. Mills, G. Daily, J. C. Castilla, J. Lubchenco, and R. T. Paine. 1996. Challenges in the quest for keystones. *BioScience* 46(8):609–620.

Reice, S. R. 1994. Nonequilibrium determinants of biological community structure. *American Scientist* 82:424–435.

Ricklefs, R. E., and I. J. Lovette. 1999. The roles of island area per se and habitat diversity in the species–area relationships of four Lesser Antillean faunal groups. *Journal of Animal Ecology* 68:1142–1160.

Ricklefs, R. E., and D. Schluter (eds.). 1993. *Species Diversity in Ecological Communities: Historical and Geographical Perspectives*. University of Chicago Press, Chicago.

Risser, P. G. 1995. The status of the science of examining ecotones. *BioScience* 45:318–325.

Terborgh, J. 1985. The role of ecotones in the distribution of Andean birds. *Ecology* 66:1237–1246.

Whittaker, R. H. 1953. A consideration of climax theory: The climax as a population and pattern. *Ecological Monographs* 23:41–78.

Whittaker, R. H. 1967. Gradient analysis of vegetation. *Biological Reviews* 42:207–264.

CHAPTER | 22

Community Development

The concept of the sere includes all the stages of successional change

Succession results in part from changes in the environment caused by colonists

Early and late successional species have different adaptations

Some climax communities are maintained by extreme environmental conditions

Transient and cyclic climaxes result from variable environments and unstable successional sequences

On August 27, 1883, the island of Krakatau in the Sunda Strait of present-day Indonesia exploded after months of volcanic activity. Most of the island was blown away, and all life was obliterated. Huge tidal waves swept the coasts of nearby Sumatra and Java, killing tens of thousands of people. Immense quantities of ash filled the atmosphere, dimming the sun and creating spectacular red sunsets all over the globe, and sending temperatures to the coldest levels in years.

Once the sobering effects of the enormous catastrophe had waned, scientists recognized the immense value of Krakatau as a laboratory to study the development of biological communities on a newly formed, raw terrain of volcanic ash (▌Figure 22.1). Expeditions were mounted and reports were made on the appearance and establishment of plants and animals over the ensuing century. The nearest sources of colonists for Krakatau were on Sumatra and Java, about 40 kilometers across the intervening ocean. As one might have expected, the first plants to show up on the island were sea-dispersed species that grow on tropical shores throughout the region (▌Figure 22.2). By 1886, 10 of the 24 species that had colonized Krakatau were sea-dispersed species. Most of the other pioneers were wind-dispersed plants such as grasses, whose seeds could be blown across the ocean. Thus, the first plant communities to develop away from the beach on Krakatau were dominated by grasses and ferns. Eventually, a few wind-dispersed species of trees began to arrive on the island. By the 1920s, closed forest had developed over most of Krakatau, and some of the pioneering species were pushed to marginal habitats or disappeared from the island. As forests developed on Krakatau, many birds and bats were attracted to the island. Some of these were fruit-eating species that brought the seeds of animal-dispersed trees and shrubs with them. Changes in the flora of Krakatau after the 1920s were

Figure 22.1 Volcanic terrains, such as this cinder slope in the Galápagos Islands, provide natural laboratories for the study of colonization and primary succession. Photo by R. E. Ricklefs.

dominated by animal-dispersed plants, which now outnumber sea- and wind-dispersed species.

The vegetation of Krakatau will continue to change for many years, as more plants invade the island and allow the development of distinct plant communities in different habitats. Moreover, the remaining island fragments that now make up Krakatau are constantly changing as a result of continuing volcanic eruptions,

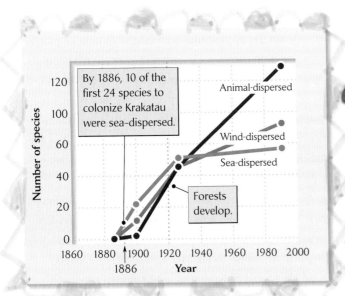

Figure 22.2 Plants dispersed by physical forces are the first to arrive in primary succession. This graph shows the number of plant species dispersed by sea, wind, and animals present on the island of Krakatau from 1883 through 1990. Data from R. J. Whittaker, *http://www.geog.ox.ac.uk/research/biogeography/krakatau-2.htm.*

erosion of soft ash deposits, and storms that pass through the region. Krakatau will continue to be an important laboratory for studying the dynamics of community change.

Communities exist in a state of continuous flux. Organisms die and others are born to take their places; energy and nutrients pass through the community. Yet the appearance and composition of most communities do not change appreciably over time. Oaks replace oaks, squirrels replace squirrels, and so on, in continual self-perpetuation. But when a habitat is **disturbed**—a forest cleared, a prairie burned, a coral reef obliterated by a hurricane, an island covered by volcanic ash—the community slowly rebuilds. Pioneering species adapted to disturbed habitats are successively replaced by other species as the community attains its former structure and composition (**Figure 22.3**).

The sequence of changes initiated by disturbance is called **succession,** and the ultimate association of species achieved is called a **climax community.** These terms describe natural processes that caught the attention of early ecologists, including Frederic Clements. By 1916, Clements had outlined the basic features of succession, supporting his conclusions with detailed studies of change in plant communities in a variety of environments. Since then, the study of community development has grown to include the processes that underlie successional change, the adaptations of organisms to the different conditions of early and late succession, and interactions between colonists and the species that replace them. Ecologists have come to realize that succession is a natural part of the dynamics of the community.

The concept of the sere includes all the stages of successional change

The creation of any new habitat—a plowed field, a sand dune at the edge of a lake, an elephant's dung, a temporary pond left by a heavy rain—attracts a host of species particularly adapted to be good pioneers. These colonizing species change the environment. Plants, for example, shade the earth's surface, contribute detritus to the soil, and alter soil moisture levels. These changes often inhibit the continued success of the pioneering species that caused them, but make the environment more suitable for the species that follow, which then exclude those responsible for the change. In this way, the character of the community changes with time.

The opportunity to observe succession presents itself conveniently in abandoned agricultural fields of various ages. On the Piedmont of North Carolina, bare fields are

(a) (b) (c)
(d) (e) (f)

▌**Figure 22.3 Species successively replace one another in the process of succession.**
Stages of succession in an oak–hornbeam forest in southern Poland are shown from
(a) immediately after clearing to (b) 7, (c) 15, (d) 30, (e) 95, and (f) 150 years thereafter.
Photographs by Z. Glowacinski, courtesy of O. Jarvinen. From Z. Glowacinski and O. Jarvinen,
Ornis Scand. 6:33–40 (1975).

quickly covered by a variety of annual plants (▌Figure 22.4).
Within a few years, herbaceous perennials and shrubs replace most of the annuals. Shrubs are followed by pines,
which eventually crowd out earlier successional species;
pine forests are in turn invaded and then replaced by a variety of hardwood species that constitute the last stage of the
successional sequence. Change comes rapidly at first. Crabgrass quickly enters an abandoned field, hardly allowing
time for the plow's furrows to smooth over. Horseweed and
ragweed dominate the field in the first summer after abandonment, aster in the second summer, and broomsedge in
the third. The pace of succession falls off as slower-growing
plants appear: the transition to pine forest requires 25 years,
and another century must pass before the developing hardwood forest begins to resemble the natural climax vegetation of the area.

The transition from abandoned field to mature forest is
only one of several successional sequences that may lead

▌**Figure 22.4 Abandoned fields undergo a series of
successional changes.** This old field on the Piedmont of North
Carolina is an example of the habitats that develop after
abandonment of agricultural land. Photo by R. E. Ricklefs.

to the same climax community within a given biome. In various regions of the eastern United States and Canada, a particular kind of forest is the end point of several different successional series. Each of these successional series, which are called **seres,** has a different beginning. For example, the sequence of species on newly formed sand dunes at the southern end of Lake Michigan in Indiana differs from the sere that develops on abandoned fields a few miles away. Sand dunes are first invaded by marram and bluestem grasses. Individuals of these species established in soils at the edge of a dune send out rhizomes (runners) under the surface of the sand, from which new shoots sprout (❙ Figure 22.5). These perennial grasses stabilize the dune's surface and add organic detritus to the sand. Numerous annuals follow these grasses onto the dunes, further enriching and stabilizing them and gradually creating conditions suitable for shrub species. Sand cherry,

dune willow, bearberry, and juniper form shrub layers before pines take hold. As in abandoned fields in North Carolina, pines on the Indiana dunes do not reseed themselves well after initial establishment and persist for only one or two generations. Eventually, the pines give way to the forests of beech, oak, maple, and hemlock that are characteristic of the region. In the same area, succession beginning in a marsh also ends in beech–maple forest as the wetland fills in with sediment and plant detritus and progressively dries out.

Primary succession begins in newly formed habitats

Beginning with Clements's classic work on succession, published in 1916, ecologists have classified seres into two types according to their origin. **Primary succession** is the

❙ **Figure 22.5 Primary succession on dunes begins with invasion of rhizomatous grasses from the edge.** Marram grass is the first invader of dunes, making use of underground rhizomes (a). Once the dunes are settled by marram grass (b) and organic nutrients begin to accumulate, shrubs can become established (c) and these are eventually replaced by trees (d). These scenes are from Indiana Dunes State Park, on the south shore of Lake Michigan. Photos by R. E. Ricklefs.

(a)

(b)

(c)

■ **Figure 22.6 Some ponds undergo bog succession.** In this bog forming behind a beaver dam (a) in Algonquin Provincial Park, Ontario, Canada, the open water in the center (b) is stagnant, poor in minerals, and low in oxygen. These conditions result in accumulation of detritus and lead to a gradual filling in of the bog, which passes through stages dominated by shrubs and, later, black spruce (c). Photos by R. E. Ricklefs.

establishment and development of plant communities in newly formed habitats previously lacking plants—sand dunes, lava flows, rock bared by erosion or landslides or exposed by receding glaciers. The regeneration of a climax community following a disturbance is called **secondary succession.** The distinction between the two blurs, however, because disturbances vary in the degree to which they destroy the fabric of a community and its physical support systems. A tornado that levels a large area of forest usually leaves intact the soil's bank of nutrients, seeds, and sproutable roots, and so succession follows quickly. In contrast, a severe fire may burn through organic layers of the soil, destroying hundreds or thousands of years of community development.

One of the most striking and best-studied examples of primary succession is the natural conversion of certain aquatic habitats in north temperate and boreal climates to dry land. Retreating glaciers left deep kettlehole ponds where large chunks of ice formed depressions and then melted; even today, new ponds are formed behind beaver dams. These ponds undergo a characteristic pattern of change known as bog succession. Bog succession begins when rooted aquatic plants become established at the

edge of a pond (■ Figure 22.6). Some species of sedges form mats on the water surface extending out from the shoreline. Occasionally these mats grow completely over a pond before it fills in with sediments, producing a more or less firm layer of vegetation over the water surface—a so-called "quaking bog." Detritus produced by the sedge mat accumulates as layers of organic sediments on the bottom of the pond, where the stagnant water contains little or no oxygen to sustain microbial decomposition. Eventually these sediments become peat, which is used by humans as a soil conditioner and sometimes as a fuel for heating (■ Figure 22.7). As a bog accumulates sediments and detritus, sphagnum moss and shrubs, such as Labrador tea and cranberry, become established along the edges, themselves adding to the development of a soil with progressively more terrestrial qualities. In northern peatlands, sphagnum is the major contributor to peat accumulation. At the edges of the bog, shrubs may be followed by black spruce and larch, which eventually give way to climax communities of forest trees, including birch, maple, and fir, depending on the locality. In this way, what started as an aquatic habitat is transformed over thousands of years through the accumulation of organic detritus until

Figure 22.7 Bog succession fills aquatic habitats with organic detritus. A 1-meter vertical section through a peat bed in a filled-in bog in Quebec, Canada, reveals accumulations of organic detritus from plants that successively colonized the bog as it was filled. The peat beds are several meters thick. Photo by R. E. Ricklefs.

the soil rises above the water table and a "terrestrial" habitat emerges.

Disturbance initiates secondary succession

Breaks in the canopy of a forest tend to close over as surrounding individuals take advantage of the new opportunities they provide. A small gap, such as that left by a falling limb, is quickly filled by growth of branches from surrounding trees. A big gap left by a fallen tree may provide saplings in the understory a chance to reach the canopy and claim a permanent place in the sun. A large area cleared by fire may have to be colonized anew by seeds blown or carried in from surrounding intact forest.

Even when reseeding initiates a secondary successional sequence, the size and type of disturbance influence which species become established first. Some plants require abundant sunlight for germination and establishment, and their seedlings are intolerant of competition from other species. These species usually have strong powers of dispersal; they often have small seeds that are easily blown about and can reach centers of large disturbances inaccessible to members of the climax community.

ECOLOGISTS IN THE FIELD

How does gap size influence succession on marine hard substrates?

The influence of gap size and isolation on succession has been investigated in marine habitats, where disturbance and recovery occur quickly. Off the coast of southern Australia, Michael Keough investigated the colonization of artificially created bare patches by various subtidal invertebrates that grow on hard surfaces. The major surface-growing taxa of the region vary considerably in their colonizing ability and competitive ability, which are generally inversely related (Table 22.1). Keough created bare patches ranging in size from 25 to 2,500 cm² (5–50 cm on a side). Some were cleared areas within larger areas of rock occupied by surface-growing invertebrates; others were hard substrates placed in sand, some distance from any source of colonists.

The gaps surrounded by intact communities were quickly filled in by such highly successful competitors as tunicates and sponges. In this case, gap size had little influence on community development, because the distances from the edges to the centers of the patches (less than 25 cm) were easily spanned by growth. The many bryozoan and polychaete larvae that attempted to colonize these patches were quickly overgrown. Among isolated patches, gap size had a much greater effect on the pattern of colonization. Tunicates and sponges, which do not disperse well, tended not to colonize small isolated patches, thereby giving bryozoans and polychaetes a chance to obtain a foothold. Because larger patches make bigger targets, many of these were settled by small numbers of tunicates and sponges, which then spread rapidly and eliminated other species that had colonized along with them. As a result, tunicates and sponges predominated on the larger isolated patches, but bryozoans and polychaetes—which, once established, can prevent the colonization of tunicate and sponge larvae—dominated many of the smaller patches.

In this system, bryozoans and polychaetes are disturbance-adapted species—what botanists call weeds.

Table 22.1   Life history attributes of the major surface-growing marine invertebrates at Edithburgh, South Australia

Taxon	Growth form	Colonizing ability	Competitive ability	Capacity for vegetative growth
Tunicates	Colonial	Poor	Very good	Very extensive, up to 1 m²
Sponges	Colonial	Very poor	Good	Very extensive, up to 1 m²
Bryozoans	Colonial	Good	Poor	Poor, up to 50 cm²
Serpulid polychaetes	Solitary	Very good	Very poor	Very poor, up to 0.1 cm²

Source: *M. J. Keough, Ecology 65:423–437 (1984).*

They get into open patches quickly, mature and produce off-spring at an early age, and then are often eliminated by more slowly colonizing but superior competitors. Such weedy species require frequent disturbances to stay in the system.

The size of a gap also influences whether predators and herbivores will be active there; these consumers can affect the course of succession as well as the trophic structure of the community. Some consumers feed in large gaps because large gaps are easy to find and require little travel time between them. Other consumers that are themselves vulnerable to predators may require the cover of intact habitat, from whose edges they venture to feed. These consumers are likely to graze small gaps more intensively than the centers of large gaps.

Wayne Sousa, of the University of California at Berkeley, showed this to be true of limpets (grazing mollusks), which live at the edges of mussel beds and feed on algae. In an intertidal rocky shore habitat in central California, he cleared patches of either 625 cm² (25 × 25 cm) or 2,500 cm² (50 × 50 cm) in mussel beds, then excluded limpets from half the patches of each size by applying a barrier of copper paint along their edges. Sousa monitored the colonization of the cleared patches by several species of algae during the following 3 years (Figure 22.8).

Limpets live in the crevices between mussels when they are not feeding; doing so enables them to avoid predation. Because this behavior limits their foraging range, densities of limpets in the small gaps (with a large edge-to-area ratio) exceeded those in the large gaps. Also, and not surprisingly, throughout the course of the experiment algae grew more densely in the large gaps. Where limpet grazing was prevented, total cover by all species of algae was high and did not differ between patches of different size.

In Sousa's experiment, limpet grazing depressed the establishment and growth of most species of algae, as we would have expected, but favored three rare, presumably competitively inferior, species: the brown alga *Analipus*, the green alga *Cladophora*, and the red alga *Endocladia*. These algae have low-lying, crustose growth forms that make them vulnerable to shading and overgrowth by other species, but help them to resist grazing. Grazing also affected the re-establishment of mussels. Where limpets occurred, colonizing mussels, which eventually crowd out all other space-occupying species, reached greater abundance in large gaps than in small ones. This difference resulted from the interaction of gap size and limpet grazing, rather than from gap size per se, because mussels colonized areas protected from grazing independently of patch size.

Figure 22.8 Consumers can affect the course of succession. A natural cleared patch in a bed of mussels (*Mytilus californianus*) on the central coast of California, about 1 meter across, has been colonized by a heavy growth of the green alga *Ulva*. Note the distinct grazed zone around the perimeter of the patch. It is created by limpets, which feed only short distances away from refuge in the mussel bed. Courtesy of W. P. Sousa, from W. P. Sousa, *Ecology* 65:1918–1935 (1984).

The climax community is the end point of succession

Ecologists have traditionally viewed succession as leading to the ultimate expression of community development, a climax community. Early studies of succession demonstrated that the many seres found within a region, each developing under a particular set of local environmental conditions, often progress toward the same climax. These observations led to a concept of mature communities as natural units—even as closed systems (see Chapter 21)—which was clearly stated by Frederic Clements in 1916:

> *The developmental study of vegetation necessarily rests upon the assumption that the unit or climax formation is an organic entity. As an organism the formation arises, grows, matures, and dies. Its response to the habitat is shown in processes or functions and in structures which are the record as well as the result of these functions. Furthermore, each climax formation is able to reproduce itself, repeating with essential fidelity the stages of its development. The life history of a formation is a complex but definite process, comparable in its chief features with the life history of an individual plant. (Carnegie Inst. Wash. Publ. 242:1–512.)*

Clements recognized fourteen climaxes in the terrestrial vegetation of North America, including two types of grassland (prairie and tundra), three types of scrub (sagebrush, desert scrub, and chaparral), and nine types of forest ranging from pine—juniper woodland to beech—oak forest. He believed that climate alone determined the nature of the local climax. Aberrations in community composition caused by soils, topography, fire, or animals (especially grazing) represented interrupted stages in the transition toward the local climax—immature communities.

In recent years, the concept of the climax as a closed system has been greatly modified—to the point of outright rejection by many ecologists—because it has become clear that communities are open systems whose composition varies continuously over environmental gradients. In addition, various factors, including the size of a disturbance and physical conditions during early succession, may result in alternative "climax" communities. Whereas in 1930 plant ecologists described the climax vegetation of much of Wisconsin as a sugar maple—basswood forest, by 1950 ecologists placed this forest type on an open continuum of climax communities. To the south, beech increased in prominence; to the north, birch, spruce, and hemlock were added to the climax community; in drier regions bordering prairies to the west, oaks became prominent. Locally, quaking

aspen, black oak, and shagbark hickory, long recognized as successional species on moist, well-drained soils, came to be accepted as climax species on drier upland sites.

Mature forest communities in southwestern Wisconsin, representing the end points of local seres, were ordered by J. T. Curtis and R. P. McIntosh along a **continuum index,** ranging from dry sites dominated by oak and aspen to moist sites dominated by sugar maple, ironwood, and basswood. The continuum index for Wisconsin forests was calculated from the species composition of each forest type, and its values varied between arbitrary extremes of 300 for a pure stand of bur oak to 3,000 for a pure stand of sugar maple. Although seral stages leading to the sugar maple climax have intermediate values, low and intermediate values may also represent local climax communities determined by topographic or soil conditions. Thus, the so-called climax vegetation of southwestern Wisconsin actually represents a continuum of forest (and, in some areas, prairie) types (▌Figure 22.9).

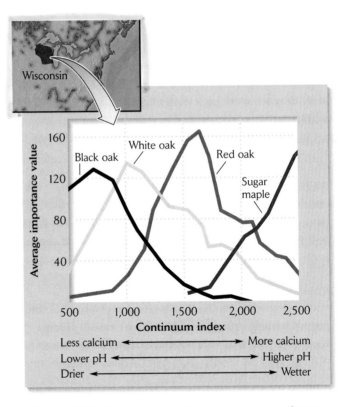

▌**Figure 22.9 Climax communities represent a continuum of vegetation types.** The relative importances (a measure of abundance) of several species of trees in forest communities of southwestern Wisconsin are arranged along a continuum index. Soil moisture, exchangeable calcium, and pH increase toward the right. After J. T. Curtis and R. P. McIntosh, *Ecology* 32:476–496 (1951).

Succession results in part from changes in the environment caused by colonists

Two factors determine when a species is established in a sere: how readily it invades a newly formed or disturbed habitat, and its response to changes that occur in the environment over the course of succession. Rapidly growing plants often produce many small seeds, which can be carried long distances by wind or by animals. Such plants have an initial advantage over species that disperse slowly, and they dominate the early stages of a sere. In a habitat that burns frequently, many species have fire-resistant seeds or root crowns that germinate or sprout soon after a fire and quickly reestablish their populations (see Chapter 9). Other species disperse slowly, or grow slowly once established, and therefore become established only late in the sere.

We have seen that early successional species sometimes modify environments in such a way as to allow later-stage species to become established. The growth of herbs on a cleared field shades the soil surface and helps the soil to retain moisture, providing conditions more congenial to the establishment of less drought-tolerant plants. Conversely, some colonizing species may inhibit the entrance of others into a sere, either by competing more effectively for limiting resources or by direct interference.

American ecologist Joseph Connell and Australian ecologist R. O. Slatyer classified this diverse array of processes governing the course of succession into three categories of mechanisms—facilitation, inhibition, and tolerance—that describe the effect of one species on the probability of establishment of a second, and whether that effect is positive, negative, or neutral.

Facilitation

Facilitation embodies Clements's view of succession as a developmental sequence in which each stage paves the way for the next, just as structure follows structure as an organism develops—or a house is built. Colonizing plants enable climax species to invade, just as wooden forms are essential to the pouring of a concrete wall, but have no place in the finished building. Alder trees (*Alnus*), which harbor nitrogen-fixing bacteria in their roots, provide an important source of nitrogen for soils developing on sandbars in rivers and in areas exposed by retreating glaciers (❚ Figure 22.10). Thus, alder facilitates the establishment and growth of nitrogen-limited plants, such as spruce, which eventually replace alder thickets on the developing soils.

❚ **Figure 22.10 Alder facilitates succession by adding nitrogen to soils.** An alder tree and (*inset*) alder cones. Main photo by K. Ward/Bruce Coleman; inset photo by Gilbert S. Grant/Photo Reseachers.

Soils do not develop in marine systems, but facilitation often occurs when one species enhances the quality of a site for the settling and establishment of another. Working with experimental panels placed below the tide level in Delaware Bay, T. A. Dean and L. E. Hurd found that for some species combinations, the presence of one species inhibited the establishment of a second, but that hydroids enhanced settlement of tunicates, and both hydroids and tunicates facilitated settlement of mussels. In southern California, early-arriving, fast-growing algae provide dense protective cover for reestablishment of kelp following disturbance by winter storms. In areas experimentally kept clear of early successional species of algae, grazing fish quickly removed settling kelp. In a parallel manner, establishment of the surfgrass *Phyllospadix scouleri* in rocky intertidal communities depends on the presence of certain early successional algae to which its seeds cling before

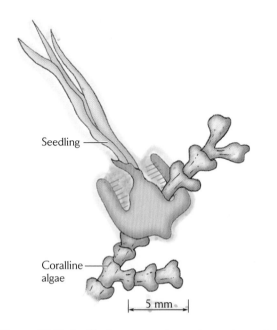

Seedling

Coralline algae

|← 5 mm →|

Figure 22.11 Facilitation occurs in marine systems. Seeds of the surfgrass *Phyllospadix* have barbs that enable them to become attached to certain types of early successional algae. After T. Turner, *Am. Nat.* 121:729–738 (1983).

germinating (**Figure 22.11**). In the absence of these algae, the seagrass cannot invade the community.

Inhibition

Inhibition of one species by the presence of another is a common phenomenon, which we have discussed in detail with respect to competition (Chapter 19) and predation (Chapter 18). One species may inhibit another by eating it, by reducing resources to a level below that at which it can subsist, or by confronting it with noxious chemicals or antagonistic behavior. In the context of succession, climax species, by definition, inhibit species characteristic of earlier stages: the latter cannot invade a climax community except following disturbance. Because inhibition is so intimately related to species replacement, it forms an integral part of the orderly succession from the early stages of a sere through the climax. When inhibition is acting, succession follows upon the establishment of one species or another only through the death and replacement of established individuals. Thus, just by chance, successional change moves toward predominance of longer-lived species.

Inhibition can give rise to an interesting situation when the outcome of an interaction between two species depends on which becomes established first. Colonists are often seeds or larvae, which are sensitive stages in the life history. Thus, it sometimes happens that neither of a pair of species can become established in the presence of competitively superior adults of the other. In this case, the course of succession depends on precedence. Precedence, in turn, may be strictly random, depending on which species reaches a disturbed site first, or it may depend on certain properties of a disturbed site—its size, location, the season, and so on. We have seen such a case in subtidal habitats of southern Australia, where bryozoans, when they become established first, can prevent the establishment of tunicates and sponges. Because of their stronger powers of dispersal, this is more likely to happen on small, isolated substrates than elsewhere.

Tolerance

Through the **tolerance** mechanism, a species can invade new habitat and become established independently of the presence or absence of other species, depending only upon its dispersal abilities and the physical conditions of the environment. Competitive exclusion then shapes the ensuing sere—that is, the life spans and competitive abilities of the colonists determine their position and dominance within the sere. Under the tolerance mechanism, early stages of succession are dominated by poor competitors that have short life cycles but become established quickly. Superior competitors constitute the climax species, but they may grow more slowly and may not express their dominance in the sere until the others have matured and reproduced.

ECOLOGISTS IN THE FIELD

Old-field succession on the Piedmont of North Carolina

Clearly, all three of Connell and Slatyer's mechanisms of succession—facilitation of establishment, inhibition of establishment, and tolerance of established populations—together with the life history characteristics of successional species, are important factors in every sere; none operates exclusively of the others. Duke University plant ecologist Henry J. Oosting observed during the 1940s and 1950s that these factors combine to influence the early stages of plant succession on old fields in the Piedmont region of North Carolina (see Figure 22.4). The first 3 to 4 years of old-field succession are dominated by a small number of species that replace one another in rapid sequence: crabgrass, horseweed, ragweed, aster, and broomsedge. The life cycle of each species partly determines its place in the succession (**Figure 22.12**).

Crabgrass, a rapidly growing annual, is usually the most conspicuous plant in a cleared field during the year in which

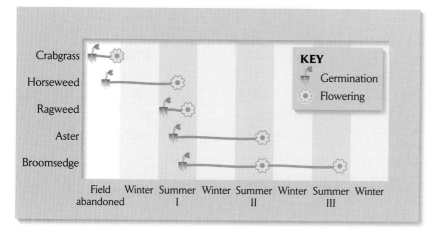

▌Figure 22.12 The life histories of plants influence their place in successional sequences. A schematic summary of the life histories of five early successional plants that colonize abandoned fields in North Carolina shows their place in the successional sequence. Photo by Stephen Collins/Photo Researchers.

the field is abandoned. Horseweed is a winter annual whose seeds germinate in autumn. Through winter, the plant exists as a small rosette of leaves; it blooms by the following midsummer. Because horseweed disperses well and develops rapidly, it usually dominates 1-year-old fields. But because its seedlings require full sunlight, horseweed is quickly replaced by shade-tolerant species. Thus, early succession is dominated by tolerance—colonizing species disperse readily and can cope with the harsh conditions of newly exposed ground—but rapidly shifts to inhibition.

Ragweed is a summer annual; its seeds germinate early in spring and the plants flower by late summer. In fields that are plowed under in late autumn, ragweed, not horseweed, dominates the first summer of succession. Aster and broomsedge are biennials that germinate in spring and early summer, exist through winter as small plants, and bloom for the first time in their second autumn. Broomsedge persists and flowers during the following autumn as well.

Horseweed and ragweed both disperse their seeds efficiently and, as young plants, tolerate desiccation. These abilities enable them to invade cleared fields rapidly and produce seed before competitors become established. Decaying horseweed roots stunt the growth of horseweed seedlings, so the species is self-limiting in the successional sequence. Such growth inhibitors presumably are by-products of other adaptations that increase the fitness of horseweed during the first year of succession. Regardless of how it arises, however, self-inhibition is common in early stages of succession.

Aster successfully colonizes recently cleared fields, but grows slowly and does not dominate old-field habitats until the second year. The first aster plants to colonize a field thrive in the full sunlight; their seedlings, however, are not shade-tolerant, and the adult plants shade their progeny out of existence. Furthermore, aster does not compete effectively with broomsedge for soil moisture. Catherine Keever observed this when she cleared a circular area, 1 m in radius, around several broomsedge plants and planted aster seedlings at various distances from them (▌Figure 22.13). Those closest to the broomsedge plants grew poorly because of reduced soil water availability.

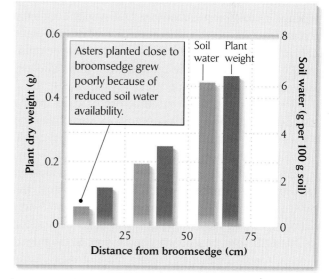

▌Figure 22.13 Some plants can inhibit others through competition for soil water and nutrients. The growth response of aster (dry weight) and soil water content is shown as a function of distance from broomsedge plants in an old field. From C. Keever, *Ecol. Monogr.* 20:230–250 (1950).

Approaching the climax

Succession continues until the addition of new species to the sere and the exclusion of established species no longer change the environment of the developing community. The progression from small to large growth forms modifies conditions of light, temperature, moisture, and soil nutrients. Conditions change more slowly, however, when the vegetation achieves the largest growth form that the environment can support. The final biomass dimensions of a climax community are limited by climate independently of events during succession.

Once forest vegetation establishes itself, patterns of light intensity and soil moisture do not change, except in the smallest details, with the introduction of new tree species. For example, beech and maple replace oak and hickory in northern hardwood forests because their seedlings are better competitors in the shade of the forest floor environment. But beech and maple seedlings develop as well under their parents as they do under the oak and hickory trees they replace. At this point, succession reaches a climax; the community growth form has come into equilibrium with its physical environment.

To be sure, the species composition of a community may change even after the climax growth form is reached. For example, a site near Washington, D.C., which was left undisturbed for nearly 70 years, developed a tall forest community dominated by oak and beech. Even though it had the appearance of a mature forest, the community had not reached equilibrium species composition at the time it was studied because the youngest individuals—the saplings in the forest understory that would eventually replace existing trees—included neither white nor black oak. In another century, this forest will probably be dominated by species with the most vigorous reproduction: red maple, sugar maple, and beech (❚ Figure 22.14).

The time required for succession to proceed from a disturbed habitat to a climax community varies with the nature of the climax and the initial quality of the soil. Obviously, succession is slower to gain momentum when it starts on bare rock than when it starts on recently exposed soil. A mature oak–hickory climax forest will develop within 150 years on an old field in North Carolina. Climax stages of western grasslands are reached in 20 to 40 years of secondary succession. In the humid Tropics, forest communities regain most of their climax elements within 100 years after clear-cutting, provided that the soil is not abused by farming or prolonged exposure to sun and rain. Primary succession usually proceeds much more slowly. Radiocarbon dating methods suggest that beech–maple climax forest requires up to 1,000 years to develop on Lake Michigan sand dunes.

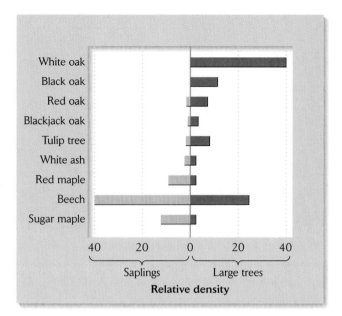

❚ **Figure 22.14 Species replacement can continue after the climax growth form is attained.** This graph shows the species composition of a forest undisturbed for 67 years near Washington, D.C. The relative predominance of beech and maple saplings in the understory foretells a gradual successional change in the community beyond the present oak–beech stage. After R. L. Dix, *Ecology* 30:663–665 (1957).

Ecologists generally agree that communities are most diverse and complex at intermediate stages of succession. However, we do not know whether this increase in the diversity of a community during its early stages of succession is related to increased production, greater constancy of physical characteristics of the environment, or greater structural heterogeneity of the habitat.

The biological properties of a developing community change as species enter and leave a sere. As a community matures, the ratio of biomass to productivity increases. The maintenance requirements of the community also increase until production can no longer meet demand, at which point net accumulation of biomass in the community stops. The end of biomass accumulation does not necessarily signal the attainment of a climax; species may continue to invade a community and replace others whether the biomass of the community increases or not. Attainment of steady-state biomass does mark the end of major structural change in the community, and further changes are usually limited to the adjustment of details.

As plant size increases with succession, a greater proportion of the nutrients available to a community come to reside in organic materials. Furthermore, because the plants of mature communities allocate much of their biomass to supportive tissue, which is less readily digestible than photosynthetic tissue, a larger proportion of the productivity enters detritus food chains rather than consumer food chains. Other aspects of the community change as well. The well-developed root systems of trees take up minerals more rapidly and store them to a greater degree than do the root systems of early successional plants. Forest soils hold nutrients more tightly because tree roots protect soils from erosion. The forest canopy protects the environment near the ground from extremes of heat and humidity, and conditions in the litter are more favorable to detritus-feeding organisms.

Early and late successional species have different adaptations

Succession in terrestrial habitats entails a regular progression of plant forms. Early inhabitants and late inhabitants use different strategies of growth and reproduction. Early-stage species capitalize on their dispersal ability to colonize newly created or disturbed habitats quickly. Climax species disperse and grow more slowly, but their shade tolerance as seedlings and their large size as mature plants give them a competitive edge over early successional species. Early successional species are adapted to colonize unexploited environments, whereas plants of climax communities are adapted to grow and prosper in the environments they create. The progression of successional species is therefore accompanied by a shift in the balance between adaptations promoting dispersal and adaptations enhancing competitive ability.

Some characteristics of early and late successional plants are compared in Table 22.2. To enhance their colonizing ability, early seral species produce many small seeds that are usually wind-dispersed (dandelion and milkweed, for example). Their seeds can remain dormant in

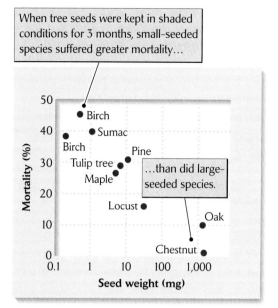

Figure 22.15 The survival of seedlings in shade is directly related to seed weight. The graph shows the relationship between seed weight and mortality of seedlings after 3 months under shaded conditions. After J. P. Grime and D. W. Jeffrey, *J. Ecol.* 53:621–642 (1965).

the soils of forest and shrub habitats for years, in what are called **seed banks,** until fires or treefalls create the bare-soil conditions required for their germination and growth. The seeds of most climax species are relatively large, providing their seedlings with ample nutrients to get started in the highly competitive environment of the forest floor. The survival of seedlings in shade is directly related to seed weight (Figure 22.15).

The ability of seedlings to survive the shady conditions of climax habitats is inversely related to their ability to grow rapidly in the direct sunlight of early successional habitats. When placed in full sunlight, early successional herbaceous

Table 22.2	General characteristics of early and late successional plants	
Characteristic	Early	Late
Number of seeds	Many	Few
Seed size	Small	Large
Dispersal	Wind, stuck to animals	Gravity, eaten by animals
Seed viability	Long, latent in soil	Short
Root : shoot ratio	Low	High
Growth rate	Rapid	Slow
Mature size	Small	Large
Shade tolerance	Low	High

species grew ten times more rapidly than shade-tolerant trees. Shade-intolerant trees, such as birch and red maple, had intermediate growth rates. Thus, plants must balance shade tolerance and growth rate against each other; each species must reach a compromise between those adaptations that best suits it for survival in a sere.

The rapid growth of early successional species results in part from their allocation of a relatively large proportion of seedling biomass to stems and leaves. A plant's allocation of tissue between roots and aboveground parts (shoots) influences its growth rate (see Chapter 3). Leaves sustain photosynthesis, and their productivity determines the net accumulation of plant tissue during growth. Thus the allocation of a large proportion of production to shoot biomass leads to rapid growth and production of large crops of seeds. Because annual plants must produce seeds quickly and abundantly, they mature early and never attain large size. Climax species allocate a larger proportion of their production to root and stem tissue to increase their competitive ability; thus they grow more slowly. In the seedlings of annual herbaceous species, the shoot typically accounts for 80–90% of the entire plant; in biennials, 70–80%; in herbaceous perennials, 60–70%; and in woody perennials, 20–60%.

Some climax communities are maintained by extreme environmental conditions

Many factors determine the composition of a climax community, among them soil nutrients, moisture, slope, and exposure. Fire is an important feature of many climax communities, favoring fire-resistant species and excluding species that otherwise would dominate. The vast southern pine forests along the Gulf coast and southern Atlantic coast of the United States are maintained by periodic fires. Pines have become adapted to withstand scorching that destroys oaks and other broad-leaved species (▌ Figure 22.16). Some species of pines do not even shed their seeds unless trig-

(a)

(b)   (c)

▌**Figure 22.16 Many species of plants are adapted to frequent fires.** (a) A stand of longleaf pine in North Carolina shortly after a fire. Although the seedlings are badly burned (b), the growing shoot is protected by the long, dense needles (c, shown on an unburned individual) and often survives. In addition, the slow-growing seedlings have extensive roots that store nutrients to support the plant following fire damage. Photos by R. E. Ricklefs.

gered by the heat of a fire passing through the understory below. After a fire, pine seedlings grow rapidly in the absence of competition from other understory species.

Any habitat that is occasionally dry enough to create a fire hazard but is normally wet enough to produce and accumulate a thick layer of plant detritus is likely to be influenced by fire. Chaparral vegetation in seasonally dry habitats in California is a fire-maintained climax that gives way to oak woodland in many areas when fire is prevented. The forest–prairie edge in the midwestern United States separates "climatic climax" and "fire climax" communities— terms that refer to the dominant physical factors that determine their species composition. Frequent burning kills seedlings of hardwood trees, but perennial prairie grasses sprout from their roots after a fire. The forest–prairie edge occasionally shifts back and forth across the countryside, depending on the intensity of recent drought and the extent of recent fires. After prolonged wet periods, the forest edge advances out onto the prairie as tree seedlings grow up and begin to shade out grasses. Prolonged drought followed by intense fire can destroy tall trees and permit rapidly spreading prairie grasses to gain a foothold. Once prairie vegetation establishes itself, fires become more frequent because of the rapid buildup of flammable litter. Reinvasion by forest species then becomes more difficult. By the same token, mature forests resist fires, and they are rarely damaged enough to allow encroachment of prairie grasses. Hence the forest–prairie ecotone remains fairly stable.

Grazing pressure also can modify a climax community. Grassland can be turned into shrubland by intense grazing. Herbivores may kill or severely damage perennial grasses and allow shrubs and cacti that are unsuitable for forage to invade. Most herbivores graze selectively, suppressing favored species of plants and bolstering competitors that are less desirable as food. On African plains, grazing ungulates make up a regular succession of species through an area, each using different types of forage. When wildebeests, the first of the succession, were experimentally excluded from some areas, the subsequent wave of Thomson's gazelles preferred to feed in areas previously used by wildebeests or other large herbivores (❚ Figure 22.17). Apparently, heavy grazing by wildebeests stimulates the growth of food plants that gazelles prefer and reduces cover within which predators of the smaller herbivores could conceal themselves. In western North America, grazing allows invasion of the alien cheatgrass (*Bromus tectorum*), which promotes fires and may lead succession to an alternative stable state—that is, a different climax community.

Transient and cyclic climaxes result from variable environments and unstable successional sequences

We usually view succession as a series of changes leading to a climax, whose character is determined by the local environment. Once established, a beech–maple forest perpetuates itself, and its general appearance does not change despite constant replacement of individuals within the community. Yet not all climaxes persist. Simple cases of **transient climaxes** include the development of animal and plant communities in seasonal ponds—small bodies of water that either dry up in summer or freeze solid in winter and thereby regularly destroy the communities that

(a)

(b)

❚ **Figure 22.17 Some grazers prefer to feed in areas previously used by others.**
Zebras (a) and Thomson's gazelles (b) both feed in the Serengeti ecosystem of East Africa, but eat different plants. The gazelles prefer to feed in areas previously grazed by wildebeests and other large herbivores. Photos by R. E. Ricklefs.

become established in them each year. Each spring the ponds are restocked from larger, permanent bodies of water or from resting stages left by plants, animals, and microorganisms before the habitat disappeared in the previous year, starting succession over again.

Succession recurs whenever a new environmental opportunity appears. Excreta and dead organisms, for example, provide resources for a variety of scavengers and detritus feeders. On African savannas, carcasses of large mammals are devoured by a succession of vultures (▮Figure 22.18). The first are large, aggressive species that gorge themselves on the largest masses of flesh. These are followed by smaller species that glean smaller bits of meat from the bones, and finally by a kind of vulture that cracks open bones to feed on marrow. Scavenging mammals, maggots, and microorganisms enter the sequence at different points and ensure that nothing edible remains. This succession has no climax because all the scavengers disperse when the feast concludes. We may, however, consider all the scavengers a part of a climax: the entire savanna community.

In simple communities, the particular life history characteristics of a few dominant species can create a **cyclic climax.** Suppose, for example, that species A can germinate only under species B, B only under C, and C only under A. This situation creates a regular cycle of species dominance in the order A, C, B, A, C, B, A, . . . , in which the length of each stage is determined by the life span of the dominant species. Cyclic climaxes usually follow such a scheme, often with one stage being bare substrate. Wind or frost heaving sometimes drives such a cycle (▮Figure 22.19).

When heaths and other types of vegetation in northern Scotland suffer extreme wind damage, shredded foliage and broken twigs create openings for further damage, and

▮**Figure 22.19 Cyclic succession is usually driven by stressful environmental conditions.** Waves of wind damage and regeneration in balsam fir forests move across the slopes of Mount Katahdin, Maine. Courtesy of D. G. Sprugel, from D. G. Sprugel and F. H. Bormann, *Science* 211:390–393 (1981).

the process becomes self-accelerating. Soon a wide swath is opened in the vegetation; regeneration occurs on the protected side of the damaged area while wind damage further encroaches upon exposed vegetation. Consequently, waves of damage and regeneration move through the community in the direction of the wind. If we watched the sequence of events at any one location, we would witness a healthy heath being reduced to bare earth by wind damage and then regenerating in repeated cycles (▮Figure 22.20).

Similar cycles occur in windy regions where hummocks, or small mounds of earth, form around the bases of clumps of grasses. As these hummocks grow, the soil becomes more exposed and better drained. As the soil dries out, shrubby lichens take over a hummock and exclude the grasses around which the hummock formed. However, the shrubby lichens are worn down by wind erosion, eventually giving way to prostrate lichens, which resist wind erosion but, lacking roots, cannot hold soil. Eventually the hummocks wear away completely, and grasses once more become established and renew the cycle.

Mosaic patterns of vegetation types typify any climax community where deaths of individuals alter the environment. Treefalls open a forest canopy and create patches of habitat that are dry, hot, and sunlit compared with the forest floor under the unbroken canopy. These openings are often invaded by early seral forms, which persist until the canopy closes again. Thus, treefalls create a shifting mosaic of successional stages within an otherwise uniform forest community. Indeed, adaptation by different species to growing in particular conditions created by different-sized openings in the canopy could enhance the overall diversity of a climax

▮**Figure 22.18 Scavengers may form a pattern of transient succession.** These vultures are feeding on a wildebeest carcass in Masai Mara Park, Kenya. Photo by R. E. Ricklefs.

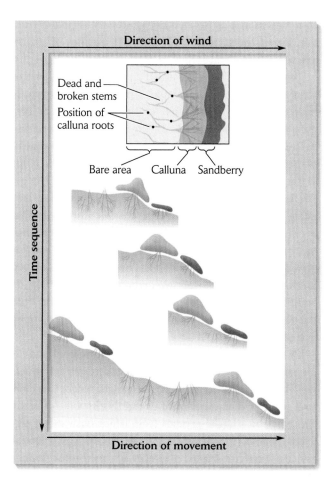

Figure 22.20 Cyclic succession may involve a sequence of damage and regeneration. This pattern is seen in dwarf heath communities of northern Scotland. At the top is a view of a heath from above, showing a band of sandberry growing in the protected lee of the calluna heath and wind damage to the heath on its upwind side (to the right). Below is a side view of the band of sandberry and heath as it appears to "migrate" downwind over time. After A. S. Watt, *J. Ecol.* 35:1–22 (1947).

community. Similar ideas have developed about intertidal regions of rocky coasts, where wave damage and intense predation continuously open new patches of habitat.

Our concept of the climax community must include cyclic patterns of changes, mosaic patterns of distribution, and alternative stable states. The climax is a dynamic state, self-perpetuating in composition, even if by regular cycles of change. Persistence is the key to the climax, and a persistent cycle defines a climax as well as an unchanging steady state does.

Succession emphasizes the dynamic nature of biological communities. By upsetting their natural balance, disturbance reveals to us the forces that determine the presence or absence of species within a community and the processes responsible for regulating community structure.

Succession also emphasizes the idea that communities often comprise patchwork mosaics of successional stages and that community studies must consider disturbance cycles on many scales of time and space.

Summary

1. Succession is community change following either habitat disturbance or the exposure of new substrate. The particular sequence of communities at a given location is referred to as a sere, and the ultimate stable association of plants and animals that is achieved is called a climax.

2. Succession on newly formed substrates—referred to as primary succession—involves substantial modification of the environment by early colonists. Moderate disturbances, which leave much of the physical structure of the ecosystem intact, are followed by secondary succession.

3. The initial stages of the sere depend on the intensity and extent of the disturbance, but its end point reflects climate and topography—that is, within a region, seres tend to converge on a single climax. However, variations in the area of a disturbance and in conditions during early stages of succession may lead to alternative stable states.

4. Especially in secondary succession, a species' entrance into and persistence in a sere depend on its colonizing and competitive abilities. Members of early stages tend to disperse well and grow rapidly; those of later stages tend to tolerate low resource levels or to dominate direct interactions with other species.

5. Joseph Connell and R. O. Slatyer categorized the mechanisms that govern succession as facilitation, inhibition, and tolerance. All three involve the effect of one established species on the probability of colonization by a second, potential invader.

6. Facilitation predominates in early stages of primary succession. Inhibition is a more common feature of secondary succession. It may be expressed in precedence effects—for example, resulting in competitive dominance of the first arrival.

7. The tolerance concept emphasizes the differing abilities of species to tolerate conditions of the environment as they change through succession, and downplays the effects of other species on their establishment. Tolerant species exclude others from the sere by competition.

8. Succession continues until a community is dominated by species capable of becoming established in their own and one another's presence. At this point the community becomes self-perpetuating.

9. In general, biomass increases during succession, whereas net production and diversity tend to be greatest in the intermediate stages.

10. Characteristics of species vary according to their place in a sere. Pioneering species tend to have many small seeds that are easily dispersed, have shade-intolerant seedlings that grow rapidly, and reach maturity quickly; late-stage species have the opposite features.

11. The character of the climax may be influenced profoundly by local conditions, such as fire and grazing, that alter interactions among seral species.

12. Transient climaxes develop on ephemeral resources and habitats, such as temporary pools and the carcasses of individual animals. In such cases, we may think of a regional climax as including transient successional sequences.

13. Cyclic local climaxes may develop where each species can become established only in association with some other species. Cyclic climaxes are often driven by harsh physical conditions, such as frost and strong winds. In this case, the regional climax includes any local cyclic seres.

PRACTICING ECOLOGY

CHECK YOUR KNOWLEDGE

Putting Succession to Work

Tropical forests are being cleared worldwide at a high rate; in the 1980s alone, an estimated 8% of tropical forests were cut. In Latin America, the main cause of forest clearing is the conversion to pasture for cattle grazing. Many factors must be addressed in efforts to conserve tropical forests, including slowing the rate of forest clearing, developing alternative agricultural practices, and finding sustainable uses for forest products. There is also a need to understand fundamental ecological processes within tropical forest systems, such as the ways forests undergo succession during recovery from disturbance. In particular, if the factors that limit forest recovery following disturbance are understood, then it should be possible to develop strategies to facilitate forest recovery in areas affected by human activities.

Karen Holl, of the University of California, Santa Cruz, has been studying the factors that limit the recovery of wet forests on abandoned cattle pastures in the mountains of southern Costa Rica. Her study site has been kept free of cattle since 1995; adjacent undisturbed forest shows what the pasture would have been like before clearing. At the time of abandonment, the pasture contained several species of non-native African grasses that spread clonally by underground rhizomes, as well as isolated trees that were remnants from the cleared forest. Since abandonment, patches of shrubs have established and spread to some degree.

Holl's research has identified a number of obstacles to natural recovery. For example, tropical seeds generally have a short period of viability during which they can germinate. For forest species to become established, seed dispersal into the pasture from the forest must be frequent. However, the rate of seed fall into experimental seed traps in the pasture was only 11% of that into traps in the forest. The seeds of forest trees that do reach the pasture then face being eaten by insects or other animals. For seeds that avoid the predation bottleneck and germinate, seedlings face a number of abiotic and biotic factors that may impede growth, including microclimatic conditions, lack of soil nutrients, absence of soil mycorrhizae, and herbivory. In addition, Holl's research has shown that competition from the non-native grasses strongly limits the growth of young tree seedlings in the pasture.

If movement of seeds from the forest into the pasture is the primary limiting factor in forest recovery, attracting animal seed dispersers into the pasture might reduce this limit. Holl conducted an experiment to test this hypothesis, using two types of artificial perches from which birds might distribute seeds to the pasture in their feces. One type (crossbar perches) were constructed of lumber, while the other (branch perches) were pieces of trees that were placed in the pasture to simulate young forest trees. The number of bird visitors suggested that branches were more effective than crossbar perches in attracting birds. The number of seeds caught below branch perches was higher than below crossbar perches and both were higher than the seed rain in open pasture. Holl concluded that bird perching structures did increase seed dispersal. Additional observation and experiments will be needed to determine whether or not other secondary factors, such as seed predation and low seed germination rates, would then begin to limit forest recovery.

CHECK YOUR KNOWLEDGE

1. Why is it important that tropical forest destruction be reduced, or that degraded forests be restored to a semblance of their former state?

2. Examine Table 22.3, adapted from Holl's article. What factors could explain the differences in number of seeds dispersed for each of the tree species? Why do the data show fractions of a seed in some cases?

 3. Visit the Web page "From Lava Flow to Forest: Primary Succession" through *Practicing Ecology on the Web* at *http://www.whfreeman.com/ricklefs*.

Table 22.3 Seed dispersal of several plant species caught in seed traps in open pasture, under crossbar perches, and under branch perches

Species	Open pasture	Crossbar perch	Branch perch
Cecropia polyphlebia	0.8	1.2	3.6
Dendropanax arboreus	0	0	25.6
Ficus spp.	0	0.4	20.8
Inga punctata	0	0	0.4
Ocotea whitei	0	0	0.4

Data are the average number of seeds per m² for 12 traps per crossbar or branch perch. Data are adapted from Holl (1998).

Although the page describes succession on lava in Hawaii, many tropical regions (including Costa Rica) have volcanoes. How does succession on lava differ from succession on cattle pastures? How can comparisons of succession in different ecosystems and under different disturbance types increase our understanding of how succession operates?

4. Can you think of any other ways that Holl might try to increase seed dispersal in disturbed forest systems?

Suggested Readings

Berkowitz, A. R., C. D. Canham, and V. R. Kelly. 1995. Competition vs. facilitation of tree seedling growth and survival in early successional communities. *Ecology* 76:1156–1168.

Callaway, R. M., and F. W. Davis. 1993. Vegetation dynamics, fire, and the physical environment in coastal central California. *Ecology* 74:1567–1578.

Christensen, N. L., and R. K. Peet. 1984. Convergence during secondary forest succession. *Journal of Ecology* 72:25–36.

Connell, J. H., and R. O. Slatyer. 1977. Mechanisms of succession in natural communities and their role in community stability and organization. *American Naturalist* 111:1119–1144.

Foster, B. L., and K. L. Gross. 1999. Temporal and spatial patterns of woody plant establishment in Michigan old fields. *American Midland Naturalist* 142:229–243.

Foster, B. L., and D. Tilman. 2000. Dynamic and static views of succession: Testing the descriptive power of the chronosequence approach. *Plant Ecology* 146:1–10.

Grubb, P. J. 1977. The maintenance of species diversity in plant communities: The importance of the regeneration niche. *Biological Reviews* 52:107–145.

Halpern, C. B., J. A. Antos, M. A. Geyer, and A. M. Olson. 1997. Species replacement during early secondary succession: The abrupt decline of a winter annual. *Ecology* 78:621–631.

Holl, K. D. 1998. Do bird perching structures elevate seed rain and seedling establishment in abandoned tropical pasture? *Restoration Ecology* 6:253–261.

Holl, K. D. 1999. Factors limiting tropical rain forest regeneration in abandoned pasture: Seed rain, seed germination, microclimate, and soil. *Biotropica* 31:229–242.

Keever, C. 1950. Causes of succession on old fields of the Piedmont, North Carolina. *Ecological Monographs* 20:230–250.

Keough, M. J. 1984. Effects of patch size on the abundance of sessile marine invertebrates. *Ecology* 65:423–437.

Knowlton, N. 1992. Thresholds and multiple stable states in coral reef community dynamics. *American Zoologist* 32:674–682.

Pickett, S. T. A., and P. S. White (eds.). 1985. *The Ecology of Natural Disturbance and Patch Dynamics.* Academic Press, Orlando, FL.

Prach, K., and P. Pysek. 1999. How do species dominating in succession differ from others? *Journal of Vegetation Science* 10:383–392.

Prach, K., P. Pysek, and P. Smilauer. 1997. Changes in species traits during succession: A search for pattern. *Oikos* 79:201–205.

Riggan, P. J., S. Goode, P. M. Jacks, and R. N. Lockwood. 1988. Interaction of fire and community development in chaparral of southern California. *Ecological Monographs* 58:155–176.

Sousa, W. P. 1984. Intertidal mosaics: Patch size, propagule availability, and spatially variable patterns of succession. *Ecology* 65:1918–1935.

Turner, M. G., V. H. Dale, and E. H. Everham. 1997. Fires, hurricanes, and volcanoes: Comparing large disturbances. *BioScience* 47:758–768.

Watt, A. S. 1947. Pattern and process in the plant community. *Journal of Ecology* 35:1–22.

Whittaker, R. J., M. B. Bush, and K. Richards. 1989. Plant recolonization and vegetation succession on the Krakatau Islands, Indonesia. *Ecological Monographs* 59:59–123.

Whittaker, R. J., S. H. Jones, and T. Partomihardjo. 1997. The rebuilding of an isolated rain forest assemblage: How disharmonic is the flora of Krakatau? *Biodiversity and Conservation* 6:1671–1696.

Zobel, D. B., and J. A. Antos. 1997. A decade of recovery of understory vegetation buried by volcanic tephra from Mount St. Helens. *Ecological Monographs* 67:317–344.

CHAPTER 23

Biodiversity

Large-scale patterns of diversity reflect latitude, habitat heterogeneity, and productivity

Diversity has both regional and local components

Local communities contain a subset of the regional species pool

Ecological release provides evidence for local interactions

Diversity can be understood in terms of niche relationships

Equilibrium theories of diversity balance factors that add and remove species

Explanations for high tree species diversity in the Tropics focus on forest dynamics

The great naturalist–explorers of the nineteenth century–Charles Darwin, Henry W. Bates, Alfred Russel Wallace, and others–discovered in the Tropics a great store of species unknown to European scientists. Many of these species had unusual forms and habits. This wealth of diversity still has not been fully described. So far, fewer than 2 million species have been catalogued worldwide. But by extrapolating rates of discovery of new insects and other life forms, some biologists have estimated that from 10 million to as many as 30 million species of animals and plants may inhabit the earth, most of them small insects in tropical forests.

Within most large groups of organisms–plant, animal, and perhaps microbial–numbers of species increase markedly toward the equator (▌Figure 23.1). For example, within a small region at 60° north latitude, we might find 10 species of ants; at 40°, between 50 and 100 species; and in a similar sampling area within 20° of the equator, between 100 and 200 species. By one count, Greenland is home to 56 species of breeding birds, New York to 105, Guatemala to 469, and Colombia to 1,395. Diversity in marine environments follows a similar trend: arctic waters harbor 100 species of tunicates, but over 400 species are known from temperate regions and more than 600 species from tropical seas. Latitudinal trends in diversity extend even to the greatest depths of the oceans, where conditions were once thought to be unvarying over the entire globe.

Why are there so many different kinds of organisms in the Tropics (and why are there so few toward the poles)? The factors that regulate the diversity of natural communities, and presumably provide answers to such questions, are the subject of this chapter.

Biologists hold two views on the issue of diversity. One maintains that diversity increases without limit over time; thus, tropical habitats, being much older than temperate and arctic habitats, have had time to accumulate more species. The second view holds that diversity reaches an equilibrium at which factors that remove species from a system

(a) Bivalve species diversity

(b) Ant species diversity

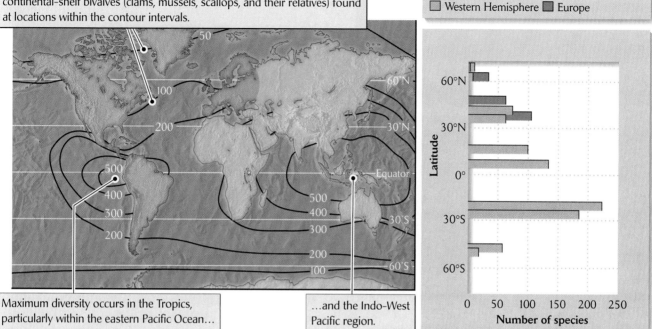

Contour lines on the map indicate the number of species of nearshore and continental-shelf bivalves (clams, mussels, scallops, and their relatives) found at locations within the contour intervals.

KEY
☐ Western Hemisphere ■ Europe

Maximum diversity occurs in the Tropics, particularly within the eastern Pacific Ocean…

…and the Indo-West Pacific region.

Figure 23.1 Global surveys reveal higher species diversity in the Tropics. (a) The contour lines on the map indicate the number of species of nearshore and continental-shelf bivalves (clams, mussels, scallops, and their relatives) found at locations within the contour intervals. Maximum diversity occurs in the Tropics, particularly within Australasia and the eastern Pacific Ocean. From F. G. Stehli, A. L. McAlester, and C. E. Helsley, *Geol. Soc. Am. Bull.* 78:455 (1967). (b) Number of ant species found within small sampling areas as a function of latitude. Peak diversity appears to be in subtropical South America, rather than at the equator. European localities support greater diversity than is found at similar latitudes in North America because of their generally warmer temperatures. After data in N. Kuznezov, *Evolution* 11:298 (1957).

balance those that add species. Accordingly, factors that add species would have to weigh more heavily in the balance, or factors that remove species would have to weigh less heavily, closer to the Tropics.

Throughout the first half of the twentieth century, the first viewpoint enjoyed the broader favor. Ecologists believed that tropical habitats had persisted since the beginning of life, whereas changes in climate (particularly during the last Ice Age) had occasionally destroyed most temperate and arctic habitats, resetting the diversity clock, so to speak. More recently, however, with the integration of population ecology into community theory, ecologists have come to consider diversity as an equilibrium state. Accordingly, at the equilibrium point, opposing diversity-dependent processes balance each other, just as equilibrium population size represents a balance between

opposing density-dependent birth and death processes. This viewpoint challenges ecologists to identify the processes responsible for adding species to, and removing species from, communities, and to discover why the balance between these processes differs systematically from place to place.

Large-scale patterns of diversity reflect latitude, habitat heterogeneity, and productivity

On a regional basis, numbers of species vary according to suitability of physical conditions, heterogeneity of habitats, isolation from centers of dispersal, and primary productivity. In North America, the number of species in most groups of animals and plants increases from north to south, but the influence of geographic heterogeneity and the isolation of peninsulas also is apparent. One way in

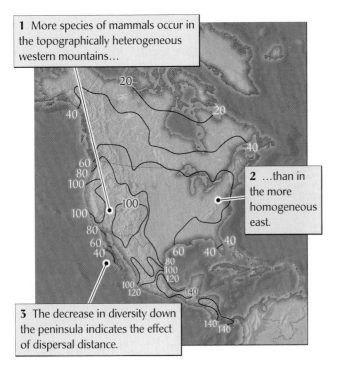

1 More species of mammals occur in the topographically heterogeneous western mountains...

2 ...than in the more homogeneous east.

3 The decrease in diversity down the peninsula indicates the effect of dispersal distance.

❙ Figure 23.2 Within North America, mammal species diversity increases toward the equator and in regions of high habitat heterogeneity. The contour lines on the map indicate the number of mammal species occurring in grid squares 150 miles on a side. From G. G. Simpson, *Syst. Zool.* 13:57–73 (1964).

which ecologists can describe such patterns is to tabulate the number of species occurring within large areas arbitrarily delimited by latitude and longitude. For example, within the area of North America extending to the Isthmus of Panama, the number of mammal species occurring in square blocks 150 miles on a side increases from 15 in northern Canada to more than 150 in Central America (❙ Figure 23.2). Within a narrow range of latitude across the middle of the United States, more species of mammals live in the topographically heterogeneous western mountains (90–120 species per block) than in the more uniform environments of the East (50–75 species per block). Presumably, the greater variety of environments in the West provides suitable conditions for a greater number of species. Notice also that diversity decreases toward the south along the peninsula of Baja California. Thus, fewer species occur at a progressively greater distance from the southwestern United States, indicating an effect of dispersal distance. The number of species of breeding land birds follows a similar pattern to that of mammals, but reptile and amphibian faunas do not. Reptiles are more diverse in the eastern half of the United States than in the mountainous western regions; amphibians are strikingly under-

represented in the deserts of the Southwest because most species require abundant water.

Vegetation structure overrides primary productivity in determining local diversity

Within a region, vegetation structure is an important determinant of diversity. Censuses of breeding birds in small areas (usually 5–20 ha) in the temperate zone reveal an average of about 6 species in grasslands, 14 in shrublands, and 24 in floodplain deciduous forests. There is some tendency for more productive habitats to harbor more species, but habitats with simple vegetation structure, such as grasslands and marshes, have fewer species than more complex habitats with similar productivity (❙ Figure 23.3).

The same principle applies to the primary producers themselves. Marshes, for example, are very productive but are structurally uniform, and have relatively few species of plants. Desert vegetation is less productive than marsh vegetation, but its greater variety of structure apparently makes room for more kinds of inhabitants (❙ Figure 23.4).

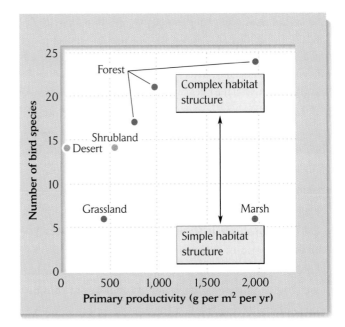

❙ Figure 23.3 Diversity is higher in complex habitats. Among seven habitats in temperate regions, the average number of bird species tends to increase with average net annual aboveground primary production. Diversity is lowest, however, in structurally simple marsh and grassland habitats, and highest in structurally complex forest habitats. From E. J. Tramer, *Ecology* 50:927–929 (1960); productivity data from R. H. Whittaker, *Communities and Ecosystems*, 2d ed., Macmillan, New York (1975).

(a)

(b)

Figure 23.4 Vegetation structure may be more important than primary productivity in determining diversity. (a) The Sonoran Desert of Baja California and (b) a salt marsh in Eastham, Massachusetts, illustrate extremes of productivity with an inverse relationship between productivity and species diversity. Photo (a) by R. E. Ricklefs; photo (b) by David Weintraub/Photo Researchers.

Structural complexity and diversity have always gone together in the minds of bird-watchers and other naturalists, but Robert and John MacArthur were the first to place this relationship in a quantitative framework that made it accessible to analysis. They did this simply by plotting the diversity of birds observed in different habitats according to diversity in foliage height, a measure of the structural complexity of the vegetation (**Figure 23.5**). Others were quick to demonstrate similar diversity relationships. Among web-building spiders, for example, species diversity varies in direct relation to heterogeneity in heights of the tips of vegetation to which spiders attach their webs. In desert habitats of the southwestern United States, lizard species diversity closely parallels total volume of vegetation per unit of area.

Diversity is correlated with overall energy input into the environment

The level of **potential evapotranspiration** (PET) has been found to be a good predictor of diversity over large regions (**Figure 23.6**). PET is the amount of water that could be evaporated from the soil and transpired by plants, given the average temperature and humidity. This measure integrates temperature and solar radiation and is thus an index to the overall energy input into the environment. The correlation between PET and diversity has become known as the energy–diversity hypothesis.

The mechanisms that relate diversity to PET are not understood. One idea is that a larger amount of energy input into an ecosystem can be shared by a larger number of species. Other mechanisms might also explain the species–energy relationship. For example, greater energy

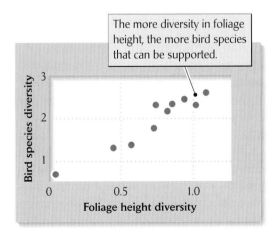

The more diversity in foliage height, the more bird species that can be supported.

Figure 23.5 Bird species diversity is correlated with foliage height diversity. This relationship is shown for areas of deciduous forest habitat in eastern North America. From R. H. MacArthur and J. MacArthur, *Ecology* 42:594–598 (1961).

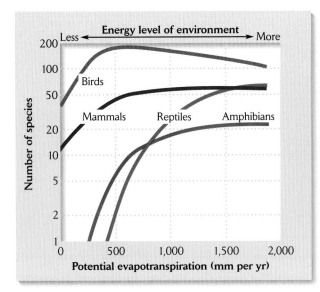

Figure 23.6 Species diversity is correlated with energy input into the environment. Relationship of potential evapotranspiration to species richness for birds, mammals, amphibians, and reptiles in North America. From D. J. Currie, *Am. Nat.* 137:27–49 (1991).

input might lead to larger population sizes and reduced rates of extinction, thereby allowing the persistence of species that could not maintain populations at a lower level of energy input. Although these ideas are attractive, none of these mechanisms has been verified experimentally, and there is some doubt that the energy–diversity relationship is a general pattern.

Diversity has both regional and local components

Diversity can be measured at a number of spatial levels. **Local diversity** (or alpha diversity) is the number of species in a small area of homogeneous habitat. Clearly, local diversity is sensitive to how one defines habitat and how intensively one samples the community.

Regional diversity (or gamma diversity) is the total number of species observed in all habitats within a geographic area that includes no significant barriers to dispersal of organisms. Thus, how we define a region depends on which organisms we are considering. The important point is that within a region, distributions of species should reflect selection of suitable habitats rather than inability to disperse to a particular locality.

If each species occurred in all habitats within a region, local and regional diversities would be the same. However, if each habitat had a unique flora and fauna, regional diversity would equal the sum of the local diversities of all the habitats in the region. Ecologists refer to the difference, or turnover, in species from one habitat to another as **beta diversity.** The greater the difference in species between habitats, the greater is beta diversity.

A useful measure of beta diversity is the number of habitats within a region divided by the average number of habitats occupied per species. Accordingly, if all species were habitat generalists, all species would occupy all habitats, and beta diversity would equal 1.0. As habitat specialization increased, the ratio of total habitats to average number of habitats per species—hence beta diversity—would increase.

According to this definition, regional diversity equals local diversity times beta diversity. For example, on the island of St. Lucia in the West Indies, each of 9 habitats (grassland, scrub, lowland forest, cloud forest, mangrove, and so on) had an average of 15.2 species of birds (local diversity). Each species occupied, on average, 4.15 of the 9 habitats, and thus beta diversity was

$$9 \text{ habitats}/4.15 \text{ habitats} = 2.17.$$

Regional diversity, which is the number of birds observed in all 9 habitats on the island, was

$$15.2 \text{ species} \times 2.17 = 33 \text{ species}.$$

As we saw in the last chapter, regional diversity on islands varies in relation to island area. The smaller island of St. Kitts has only 20 species, local diversity averages 11.9 species, and beta diversity is 1.68. The larger island of Jamaica has 56 species, and local diversity there averages 21.4 species, but beta diversity is 2.62, indicating greater differences in species between habitats, hence greater habitat specialization. We'll return to these components of diversity later in this chapter.

Local communities contain a subset of the regional species pool

The species that occur within a region are referred to as its **species pool.** All the members of the regional species pool are potential members of each local community. Yet not all species are found everywhere. A central concept of ecology is that membership in local communities is restricted to the species that can coexist together in the same habitat. Thus, each local community is a subset of the regional species pool.

The presence of a particular species within a local community signifies that the species can tolerate the conditions of the environment and find suitable resources for survival and reproduction. Ecologists say that these conditions occur within the **fundamental niche** of the species—the range of conditions and resources within which individuals of the species can persist. However, other species also compete for these resources, or are predators or pathogens. Competitors and predators may so reduce the population growth of a species within some parts of its fundamental niche that it cannot maintain its population in such places. Thus, other species may restrict the distribution of a particular species to those parts of its fundamental niche where it is most successful. This more restricted range of conditions and resources is referred to as the **realized niche** of the species. Species usually occur only in habitats that fall within their realized niche space.

The membership of a species in a local community is determined partly by its adaptations to conditions and resources and partly by competitive and other interactions with other species. The species present within the regional species pool are thus sorted into different communities based on their adaptations and interactions. This process is referred to as **species sorting**. The process of species sorting can be demonstrated experimentally by bringing together a large set of species from a regional species pool in a variety of habitats. Over time, competition and predation will cause the elimination of some species from these local communities, although which species disappear will vary from habitat to habitat depending on the particular adaptations of the species to local environmental conditions and resources.

ECOLOGISTS IN THE FIELD

Species sorting in wetland plant communities

Evan Weiher and Paul Keddy of the University of Ottawa established a large experiment to investigate the principles of species sorting. Working with wetland plants, they sowed seeds of 20 species into 120 wetland microcosms that differed with respect to soil fertility, water depth, fluctuations in water depth, soil texture, and organic leaf litter. These artificial communities were followed for 5 years. Over the duration of the experiment, the number of species in each of the communities decreased as poor competitors were excluded by dominants (▌Figure 23.7). After 5 years, several distinct communities had developed under different environmental conditions.

Weiher and Keddy thought of environmental variables as filters. If a species was unable to tolerate or compete effectively under a particular set of conditions, then it did not join communities having these conditions (▌Figure 23.8). One of the 20 species in the original species pool failed to germinate

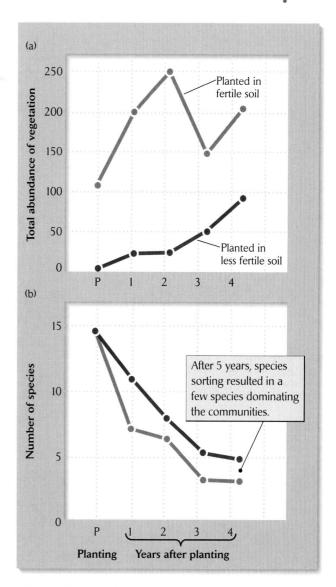

▌**Figure 23.7 Experimentally composed communities demonstrate species sorting.** (a) Total relative abundance of vegetation in fertile and infertile microcosms over the course of Weiher and Keddy's experiment. (b) Decrease in the average number of species in each type of microcosm with time. From E. Weiher and P. A. Keddy, *Oikos* 73:323–335 (1995).

under the conditions of the experiment, and 5 others were unable to persist under any of the combinations of conditions in the experimental microcosms. Important environmental filters sorting out the remaining 14 species were high versus low water levels and high versus low soil fertility. These environmental variables contributed the most to differentiating the particular assemblages of species within each of the microcosms. The experiment showed clearly the effect of species sorting from a regional species pool in relation to the particular habitat conditions of local communities.

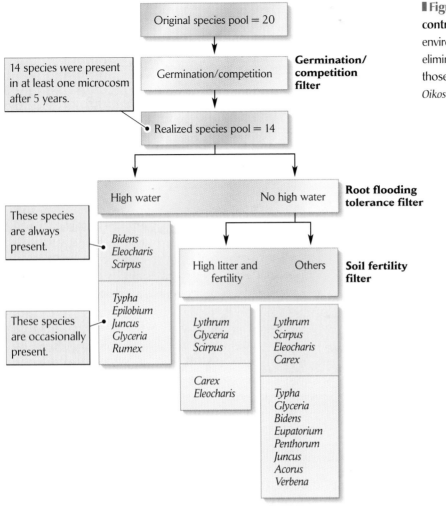

Figure 23.8 Environmental filters contribute to species sorting. Certain environmental conditions in a local community eliminate those species that cannot tolerate those conditions. From E. Weiher and P. A. Keddy, *Oikos* 73:323–335 (1995).

Ecological release provides evidence for local interactions

Competitive interactions between species play a major role in species sorting. For a given range of habitats, species sorting should be greatest where the regional pool contains the most species. In such a situation, each species should be able to maintain itself over only a narrow range of habitats—those to which it is best adapted—and beta diversity should be high.

This relationship has been most carefully noted in studies comparing islands and neighboring continental regions. In such studies, one can compare levels of diversity (resulting from different degrees of geographic isolation) among regions having a similar climate and range of habitats. Islands usually have fewer species than comparable mainland areas, but island species often attain greater densities than their mainland counterparts. Also, they ex-

pand into habitats that would normally be filled by other species on the mainland. Collectively, these phenomena are referred to as **ecological release.**

Ecological release can be seen in surveys of bird communities in seven continental areas and islands of various sizes within the Caribbean basin. These surveys show that where fewer species occur, each is likely to be more abundant and to live in more habitats (**Figure 23.9**). Thus, as the species pool decreases, the realized niche of each species becomes larger, as shown by greater habitat breadth and denser populations within each habitat. In this set of regions, local diversity and beta diversity contribute almost equally to variation in regional diversity. The numbers of individuals of all bird species added together were similar in each of the seven regions, although the numbers of species (regional diversity) differed by a factor of almost 7 between Panama and St. Kitts. In each habitat in Panama (mainland), about three times as many species (local diversity) were recorded, and populations of each species were about half as dense, as in corresponding habitats on St. Kitts (the smallest island). Beta diversity increased by a

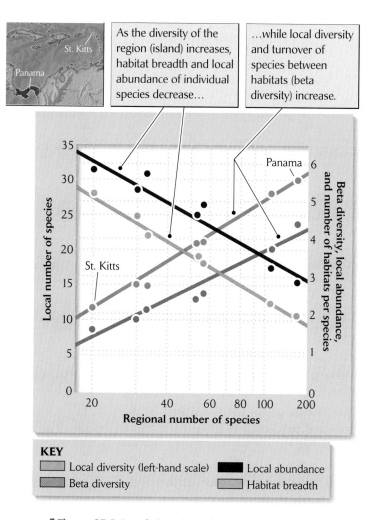

As the diversity of the region (island) increases, habitat breadth and local abundance of individual species decrease…

…while local diversity and turnover of species between habitats (beta diversity) increase.

KEY

Local diversity (left-hand scale)   Local abundance
Beta diversity   Habitat breadth

▌**Figure 23.9 Populations in regions with few species show ecological release.** From G. W. Cox and R. E. Ricklefs, *Oikos* 29:60–66 (1977); J. M. Wunderle, *Wilson Bull.* 97:356–365 (1985).

Diversity can be understood in terms of niche relationships

As we have seen, a niche represents the range of conditions and resource qualities within which an individual or species can survive and reproduce. Thus, for example, the boundaries of a particular species' niche might extend between temperatures of 10°C and 30°C, prey sizes of 4 and 12 mm, and daytime light levels between 10 and 50 W per m². Of course, the niche of any species would include many more variables than these three, and ecologists often cite the multidimensional nature of the niche to acknowledge the complexity of species–environment relationships.

The degree to which the niches of two species overlap determines how strongly the species might compete with each other. Thus, the niche relationships of species provide an informative measure of the structural organization of biological communities. Every ecological community can be thought of as having a total niche space within which all the niches of its member species must fit. Within this niche space, adding or removing species has certain consequences, particularly if the species' niches can expand or be compressed, as we saw in the case of ecological release.

Communities with different numbers of species may differ with respect to any one, or a combination, of three factors: total community niche space, niche overlap, and the niche breadth of individual species. In ▌Figure 23.10, species niches in their original condition, denoted as A, are portrayed as bell-shaped curves of resource use or activity along a single continuous axis that represents a resource quality or environmental condition. How might additional species be accommodated in such a community?

1. Without any change in niche relationships (breadth and overlap), the total niche space of a community would have to increase in direct proportion to the number of species (condition B). (Note that niche space refers to variety of resources, not their amounts.)

2. Without a change in niche breadth, increased diversity could be accommodated by increased niche overlap (condition C). In this case, the average productivity of each species would decline as a consequence of increased sharing of resources, all other things being equal.

3. Without an increase in niche overlap, increased specialization could accommodate additional species within a community's niche space (condition D). Here, too, average productivity would decline because each species would have access to a narrower range of resources.

Competition, diversity, and the niche

Both theory and experiment have shown that intense competition leads to exclusion of species from a community. Consequently, many ecologists have argued that in communities with high diversity, competition between species must be relatively weak to permit the coexistence of numerous species. This situation might be represented by narrower niches and reduced niche overlap (condition D in Figure 23.10). What mechanisms might lead to reduced interspecific competition? Greater ecological specialization,

factor of almost 3 between St. Kitts and Panama. Because the range of habitats was the same in each of the regions, the differences in abundance and habitat distribution of each species could be explained only by local interactions among species within the pool of each region.

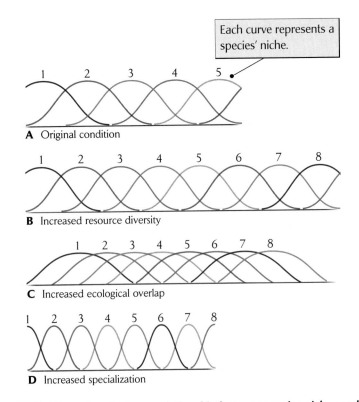

Each curve represents a species' niche.

A Original condition

B Increased resource diversity

C Increased ecological overlap

D Increased specialization

▌**Figure 23.10 Diversity reflects the relationship between species niches and total community niche space.** These schematic diagrams show how resource use along a continuum could be altered to accommodate more species. The horizontal axis represents some ecological resource or condition that defines the total niche space of the community, such as average size of prey items or height distribution within the intertidal zone. The height of each curve (vertical axis) represents the intensity of utilization of the resource or condition by each species.

greater resource availability, reduced resource demand, and intensified predation have all been cited as possibilities.

Most ecologists agree that the high diversity in the Tropics results at least in part from the presence of a greater variety of ecological roles there. That is, the total community niche space occupies greater volume near the equator, where there are many species, than it does toward the poles, where there are few. Condition B in Figure 23.10 illustrates this situation. However, greater niche space could also include an increase in the number of niche axes as well as the length of each. For example, part of the increase in the number of bird species toward the Tropics is related to an increase in fruit-feeding and nectar-feeding species and in insectivorous species that hunt by searching for their prey while quietly sitting on perches. These types of behavior are uncommon among birds in temperate regions and represent an increase in the number of axes of the community niche space. Among mammals, the Tropics are species-rich

primarily because of the many species of bats in tropical communities. Nonflying mammals are no more diverse at the equator than they are in temperate regions, although their variety does decrease as one goes farther north toward polar regions. Epiphytes and lianas (woody vines), which are generally absent from forests at higher latitudes, augment tropical plant diversity. Thus, variation in species diversity is generally paralleled by variation in the functional, or niche, diversity of species.

Species diversity parallels niche diversity

One way to assess niche diversity within a community is to use the morphology of a species as an indicator of its ecological role—that is, to assume that differences in morphology among related species reveal different ways of life. For example, size of prey captured varies in relation to the body size of the consumer, and different shapes of appendages can be related to different methods of loco-motion for hunting and escaping predators. Morphological analyses of communities have consistently revealed a rela-tively constant density of species packing in morphologi-cally defined niche space—in other words, the average size of species niches is independent of community diversity. Therefore, as species diversity increases, so does the total variety of morphology. This finding suggests that species added to a community increase the variety of ecological roles played by its members.

To illustrate this principle, we'll compare bat commu-nities in temperate and tropical localities. A morphological space defined by two axes captures many of the important ecological properties of bats. The first axis—the ratio of ear length to forearm length—is a measure of ear size relative to body size. This ratio is related to the bat's sonar system and thus to the type and location of its prey. The second axis—the ratio of the lengths of the third and fifth digits of the hand bones in the wing—describes whether the wing is long and thin or short and broad. Therefore, this axis determines a bat's flight characteristics, hence the types of prey it can pursue and the habitats within which it can capture prey efficiently.

When we plot each bat species in a community on a graph whose axes are these two morphological ratios, we can visualize niche relationships among species (**Figure 23.11**). In the less diverse community in Ontario, in southern Canada, all the bat species have similar morphology and all play similar ecological roles: all are small insectivores. The bats of the morphologically more diverse community in Cameroon, in tropical West Africa, play a greater variety of ecological roles. In addition to small insectivorous spe-cies, fruit eaters, nectar eaters, fish eaters, and large, preda-tory bat eaters make up this bat community.

Another example of the relationship between species diversity and ecological roles comes from freshwater fish communities. In streams and rivers, species diversity in most taxonomic groups increases from the headwaters to the mouths of rivers. One presumes that as a river increases in size, it offers a greater variety of ecological opportunities, more abundant resources, and more stable and therefore more reliable physical conditions. Local communities reflect these changes. For example, a headwater spring in the Rio Tamesi drainage of east central Mexico was found to sup-port only one species of fish, a detritus-feeding platyfish (*Xiphophorus*) (**Figure 23.12**). Farther downstream, three species occurred: the platyfish, a detritus-feeding molly (*Poe-cilia*) that prefers slightly deeper water, and a mosquito fish (*Gambusia*) that eats mostly insect larvae and small crus-taceans. Fish communities even farther downstream in-cluded additional carnivores—among them, fish eaters—and

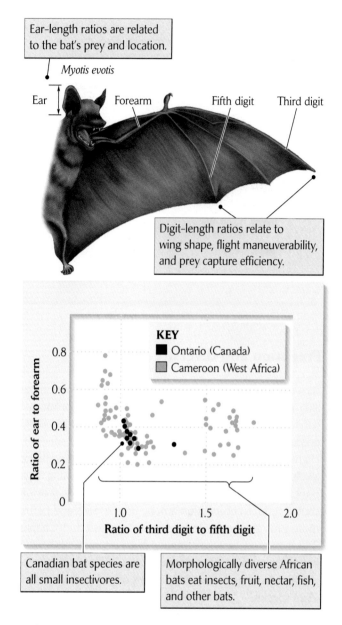

Figure 23.11 Niche relationships can be portrayed in morphological space. Distribution in morphological space differs between the aerial-feeding bat faunas of southeastern Ontario and Cameroon, West Africa. After M. B. Fenton, *Can. J. Zool.* 50:287–296 (1972).

other fish that feed primarily on filamentous algae and vas-cular plants. Downstream communities had all the species of upstream communities plus additional ones restricted to downstream localities. Thus species diversity increases as a stream becomes larger and presents more kinds of habitats and a greater variety and abundance of food items. A gen-eral rule evident in these examples is that species diversity is positively correlated with niche diversity.

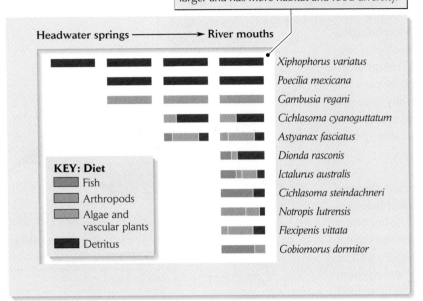

Species diversity increases as a stream becomes larger and has more habitat and food diversity.

Headwater springs ⟶ River mouths

Xiphophorus variatus
Poecilia mexicana
Gambusia regani
Cichlasoma cyanoguttatum
Astyanax fasciatus
Dionda rasconis
Ictalurus australis
Cichlasoma steindachneri
Notropis lutrensis
Flexipenis vittata
Gobiomorus dormitor

KEY: Diet
- Fish
- Arthropods
- Algae and vascular plants
- Detritus

Figure 23.12 Fish exhibit more ecological roles in more diverse communities. The bars show species of fish and their diets in four communities (vertical columns), from a headwater spring containing one species (leftmost column) to a downstream community with up to eleven species (rightmost column). The communities sampled were in the Rio Tamesi drainage of east central Mexico. From R. M. Darnell, *Am. Zool.* 10:9–15 (1970).

Predation and diversity

As we have seen, species diversity generally increases in parallel with habitat productivity. Although more productive habitats may have a greater variety of ecological resources, predation may also contribute to their higher diversity. Higher productivity results in more energy reaching higher levels in the trophic pyramid and therefore supporting larger populations of predators. Increased predation pressure should reduce competition among prey species and permit more prey to coexist. Indeed, as we discussed in Chapter 21, when predators are removed from a community experimentally, a common consequence is loss of prey species resulting from increased populations of competitively dominant species.

Ability to avoid predation is as important as competition to the persistence of populations, and avenues of escape from predators represent ways in which species may diversify. We would expect predators to be most efficient when they specialize on abundant prey species that share a common method of escape. Where many prey species use the same mechanisms to escape predation, predators with adaptations or learned behaviors that enable them to exploit such prey will prosper; thus, those prey populations will suffer increased mortality. Conversely, prey having unusual adaptations for escape should be strongly favored by natural selection. Consequently, predation pressure should lead to diversification of escape mechanisms among prey.

The variety of color patterns and resting positions exhibited by moths, which reflect the variety of backgrounds against which they hide during the day to avoid visually hunting predators, tends to be much higher in more diverse tropical communities than it is in moth communities in temperate latitudes (**Figure 23.13**). On one hand, tropical habitats may have a greater variety of potential resting backgrounds to which moth species can adapt. On the other hand, predation pressure to evolve unusual escape tactics may be stronger in tropical communities. In either case, it is evident that predators may play an important role in shaping niche relationships and regulating diversity within communities.

Figure 23.13 The diversity of adaptations for escaping predators is high among moths in the Tropics. These moths from the Amazon basin in Ecuador show some of the variety of concealing coloration evolved to match the resting background. Photos by R. E. Ricklefs.

Equilibrium theories of diversity balance factors that add and remove species

Our survey of diversity patterns suggests several general conclusions. At a global scale, there is a pronounced increase in diversity from high latitudes toward the equator. Within latitudinal belts, diversity appears to be correlated with topographic heterogeneity within a region and the complexity of local habitats. Islands exhibit species impoverishment. Everywhere, higher diversity is associated with greater niche variety.

How do we explain these patterns of diversity? Some biologists have claimed that diversity increases with time and depends on the age of the habitat, but the strong correlation between vegetation structure and diversity would seem to cast doubt on that hypothesis. Alternatively, most ecologists now believe that diversity achieves an equilibrium value at which processes that add species and those that subtract species balance each other. Production of new species within regions, combined with movement of individuals between habitats and regions, adds to the number of species in local communities. Species are removed by competitive exclusion, elimination by efficient predators, or the plain bad luck of succumbing to a regional disaster such as a major volcanic eruption. If competitive exclusion set limits on the ecological similarity of species (and thus on the intensity of competition between them), communities could become saturated with species. Accordingly, diversity should reach a steady state: new species added to the local community by regional diversification and migration should be balanced by local exclusion of close competitors. Physical conditions, variety of resources, predators, environmental variability, and, perhaps, other factors should affect the equilibrium point itself. Thus, conditions in the Tropics might allow greater numbers of species to coexist locally by reducing the intensity or consequences of competition.

Equilibrium theory of community diversity resembles the theory of density-dependent regulation of population size, discussed in Chapter 14. Births are analogous to the formation of new species or colonization by species from elsewhere; deaths are analogous to local extinctions of species. Each type of community has an equilibrium number of species, often referred to as the saturation number, just as a habitat has a carrying capacity for the population of a particular species. Ecologists were attracted to this view because it helped explain what was known about species diversity within local habitats. It also placed at least part of the problem of species diversity within the domain of ecology: present-day processes taking place within small areas.

Diversity on islands

One of the most influential developments of the equilibrium approach to diversity concerned the number of species on islands. During the 1960s, Robert MacArthur, then at the University of Pennsylvania, and E. O. Wilson, at Harvard University, developed their famous **equilibrium theory of island biogeography,** which states that the number of species on an island balances regional processes governing immigration against local processes governing extinction. This can be seen in the following simple model.

Consider a small offshore island. Its species diversity can increase only by immigration from other islands or from a continental landmass. The flora and fauna of the closest continental area make up the species pool of potential colonists. As the number of species on the island increases, the rate of immigration of new species to the island decreases; that is, as more potential mainland colonists establish themselves on the island, fewer immigrating individuals belong to new species. When all mainland species occur on the island, the immigration rate of new species must be zero. If species disappear from the island at random, the rate of extinction increases with the number of species present on the island. Where the immigration and extinction curves cross, the corresponding number of species on the island attains a steady state at the level $\hat{S}$ (**Figure 23.14**).

Immigration and extinction rates probably do not vary in strict proportion to the number of potential colonists and the number of species established on an island. Some species undoubtedly colonize more easily than others, and they reach the island first. Thus, the rate of immigration to an island initially decreases more rapidly with increasing island diversity than it would if all mainland species had equal

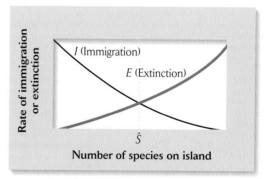

Figure 23.14 The equilibrium theory of island biogeography balances immigration against extinction. The steady-state number of species ($\hat{S}$) is determined by the intersection of the immigration (*I*) and extinction (*E*) curves. After R. H. MacArthur and E. O. Wilson, *Evolution* 17:373–387 (1963); R. H. MacArthur and E. O. Wilson, *The Theory of Island Biogeography,* Princeton University Press, Princeton, NJ (1967).

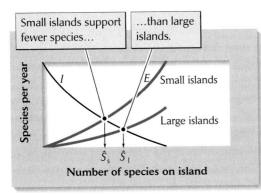

Figure 23.15 Smaller islands support fewer species because of higher extinction rates. $\hat{S}_s$ = equilibrium number of species for small islands; $\hat{S}_l$ = equilibrium number of species for large islands.

potential for dispersal. Consequently, immigration rate follows a curved line, as in Figure 23.14. As the number of species on the island increases, competition between species likely increases the probability of extinction by reducing population sizes of individual species (see Figure 23.9), so the extinction curve may rise progressively more rapidly as species diversity increases.

If probability of extinction increased as absolute population size decreased, extinction curves also would be higher for species on small islands than for those on large islands. Therefore, small islands would support fewer species than large islands (■ Figure 23.15). If the rate of immigration to islands decreased with increasing distance from mainland sources of colonists, the immigration curve would be lower for distant islands than for near islands. Therefore, the equilibrium number of species for distant islands would lie to the left of that for islands close to the mainland (■ Fig-

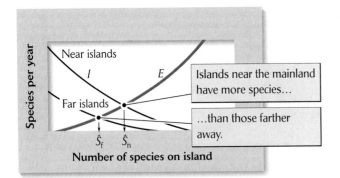

Figure 23.16 Islands close to the mainland support more species because of higher immigration rates. $\hat{S}_f$ = equilibrium number of species on far islands; $\hat{S}_n$ = equilibrium number of species on near islands.

ure 23.16). These predictions have been verified for islands throughout the world (■ Figure 23.17).

HELP ON THE WEB ▪ *Go to Living Graphs at http://www.whfreeman.com/ ricklefs and use the interactive tutorial to understand better the dynamics of species diversity on islands.*

ECOLOGISTS IN THE FIELD

Experimental manipulation of island faunas

As a corollary to its predictions concerning equilibrium diversity, the equilibrium theory of island biogeography also predicts that if some disaster reduced the diversity of a particular island, new colonists would, over time, restore diversity to its predisturbance equilibrium. This prediction was tested several years ago by Daniel Simberloff, at that time a graduate student at Harvard University and now at the University of Tennessee, and his advisor, E. O. Wilson. After first counting the number of species of arthropods present on each of four small mangrove islands in Florida Bay, Simberloff removed all the insects by fumigating the islands with methyl bromide. The islands were then censused at regular intervals for a year (■ Figure 23.18).

As predicted, the diversity of islands nearer to sources of colonists increased more rapidly than that of islands farther away. Numbers of species on both near and far islands began to level off before the end of the experiment, indicating that a steady state had been reached in each case. Also consistent with the predictions of the theory, the new equilibrium numbers of species were similar to the numbers of species present on the islands prior to defaunation. This confirmation of equilibrium theory suggested that one could understand patterns of variation in species diversity in terms of ecological processes.

Equilibrium theory in continental communities

We can also apply an equilibrium view of diversity to assemblages of species on continents. The major difference between islands and continental regions is that in continental regions, new species are more likely to form within the region, as well as arriving from outside the region by immigration. In a large region isolated from others by barriers to dispersal (an island continent such as Australia, for example), species are added to the regional pool primarily by formation of new species within the region. Curves relating rates of species production and extinction to regional

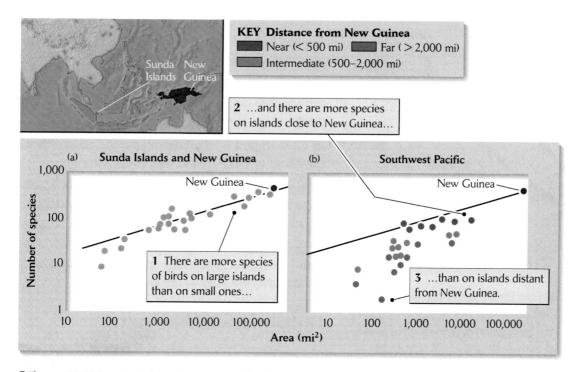

KEY **Distance from New Guinea**
Near (< 500 mi)   Far (> 2,000 mi)
Intermediate (500–2,000 mi)

2 ...and there are more species on islands close to New Guinea...

(a) **Sunda Islands and New Guinea**

New Guinea

1 There are more species of birds on large islands than on small ones...

(b) **Southwest Pacific**

New Guinea

3 ...than on islands distant from New Guinea.

Number of species

Area (mi²)

▌Figure 23.17 Species richness increases with island size and decreases with distance from the colonization source. Data for land and freshwater birds on (a) the Sunda Islands of Malaysia and Indonesia together with the Philippines and New Guinea, and (b) islands of the southwest Pacific demonstrate this pattern. The latter islands show the effect of distance from the major source of colonization (New Guinea) on the size of the avifauna. From R. H. MacArthur and E. O. Wilson, *Evolution* 17:373–387 (1963).

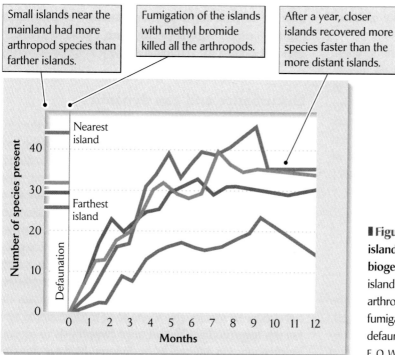

Small islands near the mainland had more arthropod species than farther islands.

Fumigation of the islands with methyl bromide killed all the arthropods.

After a year, closer islands recovered more species faster than the more distant islands.

Nearest island

Farthest island

Number of species present

Defaunation

Months

▌Figure 23.18 The re-colonization of four small islands supports the equilibrium theory of island biogeography. The entire faunas of four small mangrove islands in the Florida Keys, consisting almost solely of arthropods, were exterminated by methyl bromide fumigation. Estimated numbers of species present before defaunation are indicated at left. From D. S. Simberloff and E. O. Wilson, *Ecology* 50:278–296 (1970).

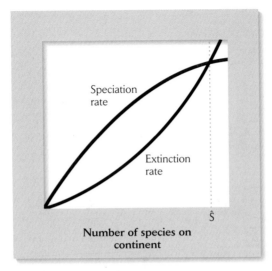

Figure 23.19 Equilibrium theory can be applied to continental regions. On a continent, new species are added to the regional pool by the evolutionary process of speciation as well as by immigration from elsewhere. After R. H. MacArthur, *Biol. J. Linn. Soc.* 1:19–30 (1969).

diversity might look like those in **Figure 23.19**. The probability of extinction per species would increase at higher diversity if competitive exclusion increased with diversity, whereas it would decrease if mutualistic relationships and alternative pathways of energy flow buffered processes that lead to extinction. The rate of speciation could level off if opportunities for further diversification were restricted by increasing diversity, whereas it could increase if diversity led to greater specialization and a higher probability of reproductive isolation of subpopulations.

Regardless of the particular shape of the immigration, speciation, and extinction curves, many biologically reasonable models can define an equilibrium level of diversity. In every case, increased rates of speciation, decreased rates of extinction, or both lead to greater equilibrium species diversity. Once this principle is understood, the challenge is to identify the factors responsible for variation in rates of species appearance and disappearance in different places.

Explanations for high tree species diversity in the Tropics focus on forest dynamics

Because the diversity of plant resources influences the potential diversity of animals in a straightforward manner, the most rigorous tests of general explanations for diversity lie in their application to plant communities. The question

"Why are there so many different kinds of trees in the Tropics?" has many plausible answers. These answers focus on a small number of mechanisms:

1. Environmental heterogeneity allows species to coexist because they can specialize on different parts of the niche space.

2. Gaps created by disturbances provide environmental conditions to which species may be specialized.

3. Herbivores and pathogens affect common species more than they do rare ones, and the resulting rare-species advantage allows many species to coexist.

4. Because species of trees are similar ecologically, competitive exclusion takes a long time, and species added to a community are likely to remain there.

Environmental heterogeneity

Many ecologists have argued that diversity of trees varies in proportion to the heterogeneity of the environment. Abundant evidence suggests that tropical forest trees may be specialized to certain soil and climatic conditions. Some species are found primarily on well-drained soils of slopes, and others are more abundant in low-lying wet soils. But could variation in the physical environment in the Tropics account for a tenfold (or more) greater diversity of plants in tropical than in temperate forests? It seems unlikely, unless plants recognize much finer habitat differences in the Tropics than they do in temperate regions, especially since temperate regions have greater heterogeneity in some climatic factors than do tropical regions. Thus, although many tropical plants do exhibit habitat specialization, this factor probably does not explain the latitudinal gradient in species diversity.

Disturbance and gap dynamics

Several ecologists, particularly Joseph Connell of the University of California at Santa Barbara, have related the high diversity of tropical rain forests to habitat heterogeneity created by disturbance. We have already touched upon the role of disturbance in Chapter 22: disturbances to communities caused by physical conditions, predators, or other factors open up space for colonization and initiate a cycle of succession by species adapted to colonizing disturbed sites. With a moderate level of disturbance, a community becomes a mosaic of habitat patches at different stages of succession; together, these patches contain the full variety of species characteristic of a successional sere. For this **intermediate disturbance hypothesis** to account satisfactorily for differences in diversity between regions, especially of the magnitude of latitudinal differences in tree species diversity, there would have to be comparable differences in the level of disturbance. Yet death rates of individual forest trees do not differ appreciably between temperate and tropical areas (▮ Figure 23.20). Nor is it likely that major disturbances such as storms and fires are more frequent in the Tropics.

Although disturbances may occur with similar frequency in tropical and temperate latitudes, their effect on the heterogeneity of environments, particularly for the germination and establishment of seedlings, may vary with latitude. In the Tropics, rains are heavier than in temperate regions, soils have less organic matter, and the sun beats down from directly overhead for much of the day. These factors are likely to create more heterogeneity between forest gaps and the rest of the environment, and thus provide more opportunity for habitat specialization, in the Tropics.

However, studies of the colonization of gaps in tropical forests suggest that gap specialists—species that can grow only in high light environments—are no more likely to establish themselves in gaps than are shade-tolerant

species that do not require gaps for germination and establishment. Steve Hubbell, of the University of Georgia, and his colleagues at the Smithsonian Tropical Research Institute have investigated recruitment of tree seedlings in gaps in a tropical rain forest on Barro Colorado Island, Panama, for several years. Within one 50-hectare plot, spatial and temporal variation in gap formation caused by disturbance did not explain variation in tree species richness. The number of species also was the same in gaps and in non-gap control sites. Furthermore, gaps were colonized primarily by shade-tolerant species.

These findings led Hubbell to suggest that even though species of trees may be specialized differently for germination sites, the species that actually invade a particular gap depend more on the vagaries of recruitment than on the particular ecological conditions in the gap. Accordingly, the level of competition for germination sites is reduced by dispersal limitation: not all species get to germination sites for which they can compete effectively. We should note, however, that recruitment limitation is partly a consequence of the high species diversity and low average density of species in the tropical forest. Thus, although this observation does not explain why tropical forests became so diverse, it does suggest that as they did so, competitive exclusion may have become a smaller factor in the maintenance of diversity. Thus, biological diversity may be self-accelerating in this regard.

Herbivore and pathogen pressure

When predators reduce prey populations below their carrying capacities, they may reduce competition among them and promote the coexistence of many prey species. Moreover, selective predation on superior competitors may allow competitively inferior species to persist in a system.

From Charles Darwin's time to the present, naturalists have believed that both selective and nonselective herbivory can influence the diversity of plant species. Daniel Janzen, of the University of Pennsylvania, suggested that herbivory could promote high diversity in tropical forests. He argued that herbivores feed on buds, seeds, and seedlings of abundant species so efficiently as to reduce their densities, allowing other, less common species to grow in their place. The key to this idea is that abundance per se, rather than any particular quality of individuals as resources, makes a species vulnerable to consumers. Consumers locate abundant resource species easily, and therefore their own populations grow to high levels.

Several lines of evidence support this "pest pressure" hypothesis. For example, attempts to establish plants in monoculture frequently fail because of infestations of

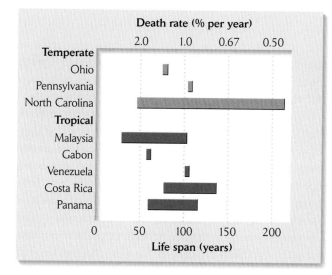

▮ Figure 23.20 The death rate of canopy trees does not differ between tropical and temperate forests. Data are presented for a number of plots in several tropical regions and in temperate North America. Average life span does not include the time required to grow into the canopy, which is estimated to be 54–185 years for various temperate zone species. Data from J. R. Runkle, in S. T. A. Pickett and P. S. White (eds.), *The Ecology of Natural Disturbance and Patch Dynamics*, Academic Press, Orlando, FL (1985), pp. 17–33; F. E. Putz and K. Milton, in E. G. Leigh, Jr., A. S. Rand, and D. M. Windsor (eds.), *The Ecology of a Tropical Forest*, Smithsonian Institution Press, Washington, D. C. (1982), pp. 95–100; N. V. L. Brokaw, in *The Ecology of Natural Disturbance and Patch Dynamics*, pp. 53–69.

herbivores. Dense plantations of rubber trees in their native habitats in the Amazon basin, where many species of herbivores have evolved to exploit them, have met with singular lack of success. But rubber tree plantations thrive in Malaysia, where specialist herbivores are not (yet) present. Attempts to grow many other commercially valuable crops in single-species stands in the Tropics have met the same disastrous end that befell the rubber plantations.

The pest pressure hypothesis predicts that seedlings should be less likely to establish themselves close to adults of the same species than at a distance from them. Adult individuals may harbor populations of specialized herbivores and pathogens that could readily infest nearby progeny. Furthermore, because most seeds fall close to their parent, herbivores may be attracted to the abundance of seedlings there while overlooking the few that disperse to a more distant location. The prediction that success in germination and establishment should increase with distance from the parent has been tested in a number of studies, which have yielded varied but generally supportive results (Figure 23.21). For example, detailed observations on a 50-hectare forest plot on Barro Colorado Island, Panama, have shown that seedling survival exhibits strong density dependence in most species, being much higher at greater distance from adult trees of the same species.

In the temperate zone, few seeds escape predation by squirrels and weevils, and herbivores and pathogens attack seedlings just as they do in the Tropics. If pest pressure does promote greater diversity in the Tropics, it must operate differently in different latitude belts. In particular, either tropical herbivores and plant pathogens must be more specialized with respect to species of host plant, or their populations must be more sensitive to the density and dispersion of host populations. Too few experimental studies have been conducted to determine whether such latitudinal differences in pest pressure exist and whether they are sufficient to explain differences in tree species diversity.

Reduced competitive exclusion

Pest pressure and recruitment limitation can reduce the consequences of interspecific competition for community membership. Steve Hubbell has argued that these factors make most species of tropical trees competitively equivalent. Accordingly, new species that invade a community are likely to remain there for long periods, if not indefinitely. Species disappear by extinction more or less at random, just as neutral alleles disappear from a population by random genetic drift. Large populations of trees in extensive areas of tropical forest would be relatively immune to such extinction, and so the numbers of species could build up to high levels through

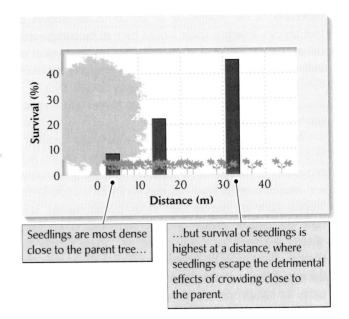

Figure 23.21 Seedling density and survival vary with distance from the parent tree. Seedlings of the Neotropical tree *Dipteryx panamensis* were followed to 18 months of age. From D. H. Janzen, *Am. Nat.* 104:501–528 (1970); D. A. Clark and D. B. Clark, *Am. Nat.* 124:769–788 (1984).

the production of new species within large geographic regions. If this were the case, the latitudinal gradient in species diversity would reflect regional species production more than the ability of species to coexist in local communities.

How well Hubbell's model explains patterns in diversity for trees, or for any other group, remains to be seen. However, his idea does emphasize the potential importance of large-scale regional processes for understanding ecological patterns. If rates of species production have been high throughout the region, and local processes have failed to remove new species from communities, then diversity has been accumulating within tropical rain forests for tens of millions of years. In the next chapter, we shall consider some of the basic issues that history and geography raise for the study of ecological systems.

Summary

1. A conspicuous pattern revealed by studies of biological communities is the tendency of species diversity in tropical regions to greatly exceed that at higher latitudes. Diversity is also high where habitats are heterogeneous, and where energy input into the environment is greatest.

2. The total diversity of species within a region containing many habitats is the regional diversity. In regions of high

species diversity, individual species are more likely to be habitat specialists than are their counterparts in regions of lower diversity. Hence, although more species occupy each habitat (local diversity), species also distinguish differences between habitats more finely (beta diversity).

3. Local communities contain a subset of the regional species pool. Membership in a local community is determined by the adaptations of species to environmental conditions and competitive relationships between them. The processes that determine local community composition are collectively referred to as species sorting.

4. Ecological release, which is an increase in population density and habitat distribution of species in less diverse communities, provides strong evidence that competition for resources structures ecological communities and limits diversity.

5. Diverse communities have either a larger total niche space, a smaller niche size per species, or greater niche overlap between species, than less diverse ones. Attempts to measure niche space consistently show that more diverse communities have larger total niche space and that the average size of the species niche is independent of community diversity.

6. Recently, thinking about diversity has been dominated by equilibrium theories, which state that diversity reflects a balance between processes that add species to a community and those that remove species. Thus, differences in diversity between communities reflect differences in the relative rates of these processes.

7. Differences in the numbers of species on islands emphasize the importance of regional processes—immigration from a continent or from other islands—to the maintenance of species diversity. On continents, the addition of species to local communities reflects, in part, the rate of production of new species, which is also a regional process.

8. Several explanations for high tree species diversity in tropical forests focus on the role of disturbance in creating mosaics of successional stages and in creating heterogeneous conditions for seedling establishment within gaps in the forest canopy. However, there is no evidence that these processes are more important in tropical than in temperate regions.

9. Predators may enhance diversity among their prey by reducing prey populations (and hence competition for resources), thereby making competitive exclusion less likely. Evidence that predators and pathogens may act in a density-dependent manner supports this pest pressure hypothesis. Density-dependent predation favors the persistence of rare species and enhances diversity.

PRACTICING ECOLOGY

CHECK YOUR KNOWLEDGE

Soil Pathogen Effects on Seedling Mortality

One of the most compelling explanations for the high diversity of trees in tropical forests is the pest pressure hypothesis. According to this idea, seedling survival near parent trees is low because specialist seed predators and herbivores focus on the high density of seeds and seedlings next to adults. Seeds dispersed away from the parent plant have a higher probability of survival because they are at a lower density and do not attract seed predators.

The pest pressure hypothesis has been experimentally tested, primarily in the Tropics. The results have been mixed, but generally supportive of the hypothesis. Occasionally, more seedlings become established directly under a parent tree because the large number of seeds overwhelms the ability of predators to eat them all. If indeed pest pressure accounts for the differences in tree diversity between the Tropics and temperate regions, then temperate species should generally not exhibit low seedling survival close to parent trees where seed density is greatest.

Alissa Packer and Keith Clay of Indiana University, Bloomington, recently put this hypothesis to the test in a population of black cherry trees (*Prunus serotina*) in Indiana, where they recorded seedling density and survival as a function of distance from the parent tree. Each parent tree was at least 50 meters from the next nearest black cherry tree. Surveys of more than 1,000 seedlings over a 2-year period showed that the density of germinated seeds was highest close to a parent tree, but that survival was highest at a distance.

Packer and Clay suspected that a soil pathogen was the primary cause of the seedling mortality. They tested this hypothesis by growing black cherry seedlings at a distance from parent trees but in soil that had been obtained from locations adjacent to parent trees. For half these plants, the soil was first sterilized by heating to 211°C for 4 hours. In a second set of experiments, the soil was inoculated with extracts containing the pathogenic species *Pythium* sp. taken from roots of dying seedlings. The results of this experiment are shown in Table 23.1.

CHECK YOUR KNOWLEDGE

1. In the experiment described above, seedlings were maintained in pots in greenhouses. Why might this be an important aspect of the method?

2. Examine the results in Table 23.1. How can you account for the fact that seedling survival was not 100% for control

Table 23.1	Percent seedling survival for *Prunus serotina* as a function of root inoculum treatment

Treatment	% survival
Control (potting mix)	73
Control + 5 ml sterile fungal growth medium	63
Pathogen isolate 1	22
Pathogen isolate 2	22
Pathogen isolate 3	24

Note: *The three different pathogen isolates were collected from dead seedlings. Survival was measured 19 days after treatment. Data are adapted from Packer and Clay (2000).*

plants? Why might the control + fungal growth medium produce a lower seedling survival?

MORE ON THE WEB

3. Visit the Web site highlighting the research of Charles L. Wilson through *Practicing Ecology on the Web* at *http://www.whfreeman.com/ricklefs*. Why is it important to study the effect of natural plant substances on soil pathogens? What implications does preservation of biological diversity have for this kind of research?

4. How do species-specific pests benefit overall biodiversity?

Suggested Readings

Brokaw, M., and R. T. Busing. 2000. Niche versus chance and tree diversity in forest gaps. *Trends in Ecology and Evolution* 15: 183–188.

Bush, M. B., R. J. Whittaker, and T. Partomihardjo. 1995. Colonization and succession on Krakatoa: An analysis of the guild of vining plants. *Biotropica* 27:355–372.

Case, T. J., and M. L. Cody. 1987. Testing island biogeographic theories. *American Scientist* 75:402–411.

Connell, J. H. 1978. Diversity in tropical rain forests and coral reefs. *Science* 199:1302–1310.

Cornell, H. V., and J. H. Lawton. 1992. Species interactions, local and regional processes, and limits to the richness of ecological communities: A theoretical perspective. *Journal of Animal Ecology* 61:1–12.

Currie, D. J. 1991. Energy and large-scale patterns of animal- and plant-species richness. *American Naturalist* 137:27–49.

Gaston, K. J. 2000. Global patterns in biodiversity. *Nature* 405 (11 May):220–227.

Givnish, T. J. 1999. On the causes of gradients in tropical tree diversity. *Journal of Ecology* 87:193–210.

Heywood, V. H. (ed.). 1996. *Global Biodiversity Assessment.* Cambridge University Press, Cambridge.

Hubbell, S. P., R. B. Foster, S. T. O'Brien, K. E. Harms, R. Condit, B. Wechsler, S. J. Wright, and S. L. de Lao. 1999. Light-gap disturbances, recruitment limitation, and tree diversity in a Neotropical forest. *Science* 283:554–557.

Janzen, D. H. 1970. Herbivores and the number of tree species in tropical forests. *American Naturalist* 104:501–528.

MacArthur, R. H. 1965. Patterns of species diversity. *Biological Reviews* 40:510–533.

MacArthur, R. H. 1972. *Geographical Ecology: Patterns in the Distribution of Species.* Harper & Row, New York.

Packer, A., and K. Clay. 2000. Soil pathogens and spatial patterns of seedling mortality in a temperate tree. *Nature* 404:278–281.

Pärtel, M., M. Zobel, K. Zobel, and E. Van der Maarel. 1996. The species pool and its relation to species richness: Evidence from Estonian plant communities. *Oikos* 75:111–117.

Purvis, A., and A. Hector. 2000. Getting the measure of biodiversity. *Nature* 405 (11 May):212–219.

Ricklefs, R. E. 1987. Community diversity: Relative roles of local and regional processes. *Science* 235:167–171.

Ricklefs, R. E., and D. Schluter (eds.). 1993. *Species Diversity in Ecological Communities: Historical and Geographical Perspectives.* University of Chicago Press, Chicago.

Rosenzweig, M. 1995. *Species Diversity in Space and Time.* Cambridge University Press, Cambridge.

Terborgh, J. 1992. *Diversity and the Tropical Rain Forest.* Scientific American Library, New York.

Terborgh, J., R. B. Foster, and P. Nuñez. 1996. Tropical tree communities: A test of the non-equilibrium hypothesis. *Ecology* 77: 561–567.

Weiher, E., and P. Keddy. 1995. The assembly of experimental wetland plant communities. *Oikos* 73:323–335.

Wills, C., R. Condit, R. B. Foster, and S. P. Hubbell. 1997. Strong density- and diversity-related effects help to maintain species diversity in a Neotropical forest. *Proceedings of the National Academy of Sciences USA* 94:1252–1257.

Zobel, M. 1997. The relative roles of species pools in determining plant species richness: An alternative explanation of species coexistence? *Trends in Ecology and Evolution* 12(7):266–269.

CHAPTER 24

History and Biogeography

The history of life can be gauged by the geologic time scale

Continental drift has changed the positions of landmasses

Biogeographic regions reflect the long-term evolutionary isolation of large areas

Changes in climate have shifted the distributions of plants and animals

Catastrophes have caused major changes in the direction of evolution

Organisms in similar environments converge in form and function

Communities in similar environments often include different numbers of species

Processes on many scales regulate biodiversity

In the last chapter, we saw that the number of species on a small island depends on the pool of potential colonists on distant continents as well as on processes occurring locally on the island. Thus, we cannot understand the structure and composition of ecological communities on islands without knowledge of their broader geographic context. We also accept that adaptation to environmental conditions is a historical process that occurs within a population over hundreds or thousands of generations. Thus, we cannot interpret the adaptations of organisms properly without understanding their past environments and ancestry.

The origin and maintenance of the earth's biodiversity is one of ecology's central issues. It is important to know whether the total diversity of the earth has been maintained in a steady state over a long period or whether it has varied, because this helps us to decide between taking an equilibrium or a nonequilibrium viewpoint. The data currently available from the fossil record suggest that during the past 600 million years, diversity has remained constant within some groups at some times, has increased in others—notably flowering plants, fishes, birds, and mammals—and has also decreased dramatically in many groups at times (▌Figure 24.1). The resolution of the fossil record is often poor, and many sources of bias can creep in. Nevertheless, it is hard to escape the conclusion that regional species pools have varied considerably in size over the earth's history, declining, sometimes precipitately, because of catastrophic events and growing because of biological diversification. To the degree that local communities reflect the regional species pool, we must question whether ecological systems ever truly achieve equilibrium.

One explanation for high species richness in the Tropics is that tropical conditions appeared on the earth's surface earlier than colder environments, allowing time for the evolution of a greater variety of

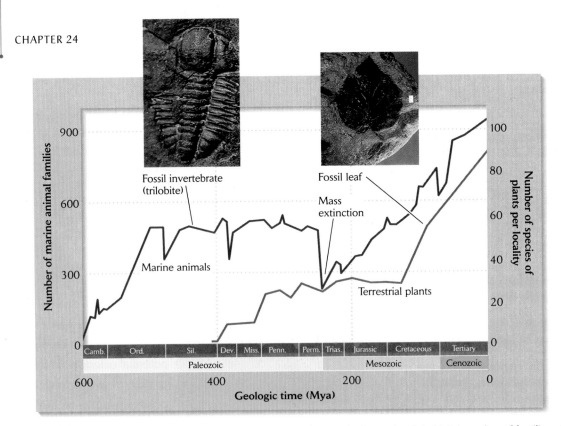

▌Figure 24.1 Diversity has generally increased over geologic time. This graph shows the global total number of families of marine animals since the beginning of the Paleozoic era and the average number of terrestrial plant species represented in local fossil floras since the mid-Paleozoic. Note the nearly constant diversity of marine animal families through the latter part of the Paleozoic era, the mass extinction at the end of the Permian period, and the continuous increase in diversity since then, coinciding with the proliferation of modern marine taxa (primarily fishes, mollusks, and crustaceans). Note also the nearly constant local diversity of plants throughout much of the Paleozoic and Mesozoic eras, the absence of a mass extinction of plants at the end of the Permian period, and the rapid rise in local diversity beginning with the appearance of flowering plants in the late Jurassic period. From J. J. Sepkoski, Jr., *Paleobiology* 10:246–267 (1984); A. J. Knoll, in J. Diamond and T. J. Case (eds.), *Community Ecology*, Harper & Row, New York (1986), pp. 126–141. Photo (*left*) by Newman & Flowers/Photo Researchers; photo (*right*) courtesy of Else Marie Friis.

plants and animals. This "time hypothesis" is not new. It was fully stated in 1878 by the English naturalist Alfred Russel Wallace, codiscoverer with Darwin of the theory of evolution by natural selection:

The equatorial zone, in short, exhibits to us the result of a comparatively continuous and unchecked development of organic forms; while in the temperate regions there have been a series of periodical checks and extinctions of a more or less disastrous nature, necessitating the commencement of the work of development in certain lines over and over again. In the one, evolution has had a fair chance; in the other, it has had countless difficulties thrown in its way. The equatorial regions are then, as regards their past and present life history, a more ancient world than that represented by the temperate zones, a world in which the laws which have governed the progressive development of life have operated with comparatively little check for

countless ages, and have resulted in those wonderful eccentricities of structure, of function, and of instinct— that rich variety of colour, and that nicely balanced harmony of relations which delight and astonish us in the animal productions of all tropical countries. (A. R. Wallace, Tropical Nature and Other Essays, Macmillan, New York and London.)

Because the tropical zone girdles the earth about its equator—the earth's widest point—tropical latitudes include more area, both land and sea, than temperate and arctic regions. For this reason alone, it is not surprising that the Tropics should harbor more species than temperate or arctic zones. During periods when the earth's climate was warmer, the area of the Tropics and Subtropics was even larger, and temperate and arctic zones were squeezed into smaller areas closer to the poles. During the last 25 million years, the climate of the earth has become cooler and drier, and the Tropics have contracted.

Both high and low latitudes have experienced drastic fluctuations in climate during the Ice Age of the last 2 million years. Temperate and arctic areas witnessed the expansion and retreat of glaciers, which caused major habitat zones to be displaced geographically and, possibly, to disappear. Periods of glacial expansion were coupled with low rainfall and reduced temperatures in the Tropics. The Amazonian rain forest, which today covers vast regions of the Amazon River's drainage basin, is thought by some biologists to have been repeatedly restricted to small, isolated refuges during dry periods correlated with glacial expansion in the north. Restriction and fragmentation of rain forest habitat could have driven many species to extinction; conversely, isolation of populations in patches of rain forest could have facilitated the formation of new species. This chapter shows how we can begin to evaluate such historical hypotheses.

The earth provides an ever-changing setting for the development of biological communities. Over millions of years of earth history, organisms have witnessed changes in climate and other physical conditions, rearrangements of continents and ocean basins, growth and wearing down of mountain ranges, evolution of new tactics by their predators and pathogens, and catastrophic impacts with extraterrestrial bodies. These changes have helped to direct the course of the evolution and diversification of organisms and have influenced the development of biological communities. The history of life reveals itself to us in the geochemical record of past environments, in fossil traces left by long-extinct taxa, and in the geographic distributions and evolutionary relationships of living species.

The most obvious consequence of this history is a nonuniform distribution of animal and plant forms over the surface of the earth. Australia, for example, has many unique forms—koalas, kangaroos, and eucalyptus trees—because of its long isolation as an island continent surrounded by ocean barriers to the dispersal of terrestrial organisms (▌Figure 24.2). Every part of the earth has its own distinctive fauna and flora. Even the major ocean basins, interconnected as they are by continuous corridors of water, have partly differentiated biotas, isolated by ecological barriers of temperature and salinity.

To study biological history, we must recognize that the structure and functioning of organisms are influenced as much by ancestry as by local environment. Thus, the morphology, physiology, and behavior of organisms reflect not

(a)

(b)

(c)

▌**Figure 24.2 Australia has many unusual terrestrial organisms.** Pictured here are (a) leaves and flowers of *Eucalyptus,* (b) the inflorescence of a species of *Banksia,* and (c) a red kangaroo. These life forms are highly distinctive and are found nowhere else on earth. Photos by R. E. Ricklefs.

only the conditions and resources of their environments, but also constraints imposed by the characteristics of their ancestors. For example, the marsupial mode of reproduction (involving, among other characteristics, early birth and subsequent development of young in a pouch) is uniquely a property of the marsupial line of mammalian evolution. It is not the result of unique ecological properties of the continent of Australia, where marsupials are now most diverse. Biologists refer to such characteristics shared by a lineage irrespective of environmental factors as **phylogenetic effects.** These effects reflect the inertia of evolution—the lack of change of some attributes in the face of change in the environment. Ecologists recognize that phylogenetic effects may influence structure and functioning in ecological systems, although this is difficult to demonstrate experimentally. Imagine that the plants and animals of Australia were replaced by a similar number of taxa from other regions with a similar climate. Would these new ecosystems function in the same manner, with similar levels of biological productivity and responses to environmental perturbation? Eucalyptus leaves have high concentrations of oils, which promote fire in the litter of Australian forests (see Figure 19.14). Would forests with different kinds of trees not be so susceptible to fire? What consequences would this have for ecosystem function?

As ecologists, we must also recognize that history and geography affect the diversification of species. To the extent that the histories of each region of the earth have differed, we might also expect biological diversity and the development of biological communities to differ. Because of this variation, it may be difficult to interpret patterns of diversity solely in terms of local environmental conditions.

In this chapter, we shall first briefly examine some historical processes that have shaped the distribution and development of ecological systems. Then we shall examine the principle of convergence, which states that inhabitants of similar environments with disparate historical origins often resemble one another because they adapt to similar ecological factors. This principle can also be applied to the diversity of biological communities. We shall see that history and biogeography have indeed influenced the character of local communities and have played an important role in the development of patterns of diversity.

The history of life can be gauged by the geologic time scale

The earth formed about 4.5 billion years ago, and life arose within its first billion years. For most of the history of the earth, life forms remained primitive. Physical conditions at the earth's surface, and the ecological systems that developed,

Figure 24.3 Many inhabitants of the Cambrian seas had hardened outer shells. This reconstruction shows a number of invertebrates representing sponges, segmented worms, and arthropods, as well as many forms that failed to leave descendants in later faunas. Painting by D. W. Miller; from D. Erwin, J. Valentine, and D. Jablonski, *American Scientist* 85(2):126–137 (1997).

were strikingly different from those of the present. The atmosphere had extremely little oxygen, and early microbes used strictly anaerobic metabolism. The oxygen in our present-day atmosphere was largely produced by photosynthetic microorganisms during this Archean eon of earth history.

At some point, the oxygen concentration of the atmosphere became high enough to sustain oxidative metabolism and made it possible for more complex life forms to evolve. The eukaryotic cell, which is the basic building block of all modern complex organisms, is a product of the last billion years of evolution. Little record of its development exists because most ancient life forms did not have hard skeletons or shells that form fossils. Much evidence of early complex life forms consists of tracks and burrows in the mud in which they lived.

All of this changed about 590 million years ago (Mya), when most of the modern phyla of invertebrate organisms suddenly appear in the fossil record. Echinoderms, arthropods, mollusks, and brachiopods rose to prominence in the oceans of that period, as did other life forms—evolutionary experiments, so to speak—that are no longer with us (█ Figure 24.3). No one knows for sure why animals began to protect themselves with hard shells or outer skeletons at that moment in history, but paleontologists regard the occasion as the beginning of life in its modern form. The interval between that critical moment and the

Table 24.1   The geologic time scale

Era	Period	Epoch	Distinctive features	Years before present
Cenozoic	Quaternary	Recent	Modern humans	11,000
		Pleistocene	Early humans	1,800,000
	Tertiary	Pliocene	Large carnivores	5,000,000
		Miocene	First abundant grazing animals	24,000,000
		Oligocene	Large running mammals	37,000,000
		Eocene	Many modern types of mammals	58,000,000
		Paleocene	First placental mammals	65,000,000
Mesozoic	Cretaceous		First flowering plants; extinction of dinosaurs and ammonites at end of period	144,000,000
	Jurassic		First birds and mammals; dinosaurs and ammonites abundant	213,000,000
	Triassic		First dinosaurs; abundant cycads and conifers	248,000,000
Paleozoic	Permian		Extinction of many kinds of marine animals, including trilobites	286,000,000
	Carboniferous	Pennsylvanian	Great coal-forming forests; conifers; first reptiles	320,000,000
		Mississippian	Sharks and amphibians abundant; large primitive trees and ferns	360,000,000
	Devonian		First amphibians and ammonites; fishes abundant	408,000,000
	Silurian		First terrestrial plants and animals	438,000,000
	Ordovician		First fishes; invertebrates dominant	505,000,000
	Cambrian		First abundant record of marine life; trilobites dominant, followed by massive extinction at end of period	590,000,000
	Precambrian		Fossils extremely rare, consisting of primitive aquatic plants	

Source: *J. H. Brown and M. V. Lomolino,* Biogeography, *2d ed., Sinauer Assoc., Sunderland, MA (1998).*

present, occupying about one-eighth of the total history of the earth, has been divided into a series of eras, periods, and epochs (Table 24.1). The first of these divisions is the Paleozoic ("old animals") era, and the Cambrian period is the first period within the Paleozoic era.

The divisions of geologic time coincide with changes in the fauna and flora of the earth that are easily perceived in the fossil record. Thus the end of the Cambrian period marks the disappearance of several prominent groups from the fossil record and their replacement in the subsequent Ordovician period by others not seen before. The major divisions at the end of the Paleozoic era and between the

Mesozoic ("middle animals," also known as the age of reptiles) and Cenozoic ("recent animals," also known as the age of mammals) eras also coincided with major extinctions of animal taxa: trilobites among others in the first case, and dinosaurs in the second. Thus, boundaries between the various periods of geologic time signal either major disruptions or less dramatic changes in the course of the development of life forms. In some cases, these boundaries reflect changes in the earth's crust. In at least one case, at the end of the Cretaceous period (Mesozoic era), disruption was caused by the explosive impact of an extraterrestrial object on the surface of the earth.

Continental drift has changed the positions of landmasses

The earth's surface has been restless over its history. The continents are islands of low-density rock floating on the denser material of the earth's interior. Giant convection currents in the semimolten material of the mantle carry continents along like gigantic logs on the surface of water. At some times in the past continents have coalesced, and at other times they have drifted apart. This movement of landmasses on the surface of the earth is called **continental drift.** The process has two important consequences for ecological systems: First, the positions of continents and major ocean basins profoundly influence climatic patterns. Second, continental drift creates and breaks down barriers to dispersal, alternately connecting and separating evolving biotas in different regions of the earth.

Because we are interested in the origin of modern ecological systems, we shall consider continental drift beginning with the early part of the Mesozoic era, about 200 Mya, when all the continents were joined in a giant landmass known as **Pangaea** (▌Figure 24.4). By 144 Mya, at the beginning of the Cretaceous period, the northern continents, which collectively made up **Laurasia,** had separated from the southern continents, which made up **Gondwana,** with the Tethys Sea forming in the space between them. In addition, Gondwana itself had begun to break up into three parts: West Gondwana, including Africa and South America; East Gondwana, including Antarctica and Australia; and India, which had separated from present-day Africa and was drifting toward the collision with Asia that occurred about 45 Mya (Table 24.2, ▌Figure 24.5).

By the end of the Mesozoic era, 65 Mya, South America and Africa were widely separated. The connection between Australia and South America through a temperate Antarctica finally dissolved about 50 Mya. At about the same time, in the Northern Hemisphere, a widening

Table 24.2   Estimated times of some of the major biogeographic events in earth history caused by continental drift

Period or epoch	Time (million years before present)	Event
Early Jurassic	200	The continental crust formed a single continent, Pangaea
Late Jurassic	180	West Laurasia (North America) ↔ Africa; West Gondwana (Africa + South America) ↔ India ↔ East Gondwana (Australia + Antarctica)
Early Cretaceous	135–125	South America ↔ Africa in far south because of rotational movement
Mid-Cretaceous	110–100	South America ↔ Africa at latitude of Brazil; Africa ↔ Madagascar ↔ India; Africa, India, and Australia all drifting northward
Late Cretaceous	80	North America ↔ (Europe + Greenland); (Antarctica + Australia) ↔ (New Zealand + New Caledonia)
Very late Cretaceous	70	Contact (Beringia) made between northwestern North America and northeastern Siberia
Very early Paleocene	63	Africa ↔ Europe (temporarily)
Eocene	49	Dispersal route between North America and Eurasia, from being predominantly via North Atlantic, switches to Beringia as North Atlantic becomes wider and Beringia becomes warmer
Eocene	~49	Australia ↔ Antarctica
Eocene	45	India drifts into contact with Asia
Oligocene	~30	Turgai Strait (east of Ural Mountains) finally dries up
Miocene	17	Europe and Africa rejoined
Miocene	15	The narrowing gap between Australia and Southeast Asia, and the appearance of stepping-stone islands, permits plant dispersal
Pliocene	3–6	North America and South America joined by land bridge

Source: E. C. Pielou, Biogeography, *Wiley, New York (1979).*
Note: *Double-headed arrows denote separations.*

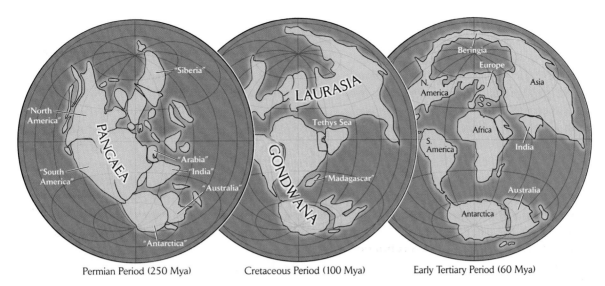

Figure 24.4 The positions of the continents have changed over geologic time. At the beginning of the Mesozoic era, the continents formed a single landmass, known as Pangaea. Subsequent drifting of continents to their present positions has isolated biotas in distinct biogeographic regions. After E. C. Pielou, *Biogeography*, Wiley, New York (1979).

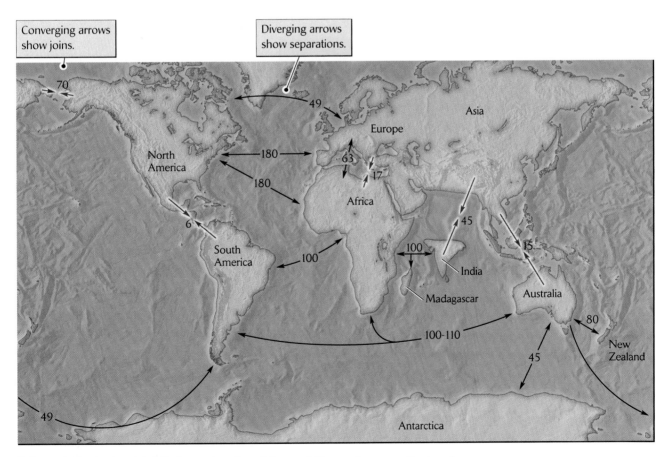

Figure 24.5 Continental drift changed routes of dispersal. The numbers are estimates of times (in millions of years before present) at which direct dispersal routes between landmasses were created or broken. The joining of Australia and Asia refers to the narrowing of a gap and the appearance of stepping-stone islands. After E. C. Pielou, *Biogeography*, Wiley, New York (1979).

Atlantic Ocean finally separated Europe and North America, but a land bridge had already formed by 70 Mya on the other side of the world between North America and Asia. More recent events of significance were the joining of Europe and Africa with the closure of the Tethys Sea about 17 Mya, and completion of a land bridge between North and South America 3–6 Mya.

Many details of continental drift have yet to be resolved, particularly in such complicated areas as the Caribbean Sea, Australasia, and the Mediterranean Sea–Persian Gulf region. Nevertheless, the history of connections between the continents endures in the distributions of animals and plants. We have only to look at the distribution of the flightless ratite birds to see the connection between the southern continents that made up Gondwana. Emus and cassowaries in Australia and New Guinea, rheas in South America, ostriches in Africa, and the extinct moas of New Zealand all descended from a common ancestor that inhabited Gondwana before its breakup (▌Figure 24.6). The splitting of a widely distributed ancestral population by continental drift is referred to as **vicariance.**

Biogeographic regions reflect the long-term evolutionary isolation of large areas

The distributions of animals he observed led Alfred Russel Wallace to recognize six major biogeographic regions (▌Figure 24.7). We now know that these regions correspond to landmasses isolated many millions of years ago by continental drift. Over the course of that isolation, animals and plants in each region developed distinctive characteristics independently of evolutionary changes in other regions.

The **Nearctic** and **Palearctic** regions, corresponding roughly to North America and Eurasia, maintained connections across either what is now Greenland or the Bering Strait between Alaska and Siberia throughout most of the past 100 million years. Consequently, these two areas share many groups of animals and plants. European forests seem familiar to travelers from North America, and

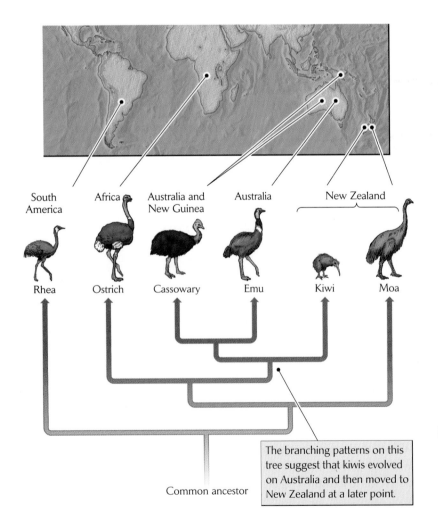

South America — Rhea
Africa — Ostrich
Australia and New Guinea — Cassowary
Australia — Emu
New Zealand — Kiwi
New Zealand — Moa

The branching patterns on this tree suggest that kiwis evolved on Australia and then moved to New Zealand at a later point.

Common ancestor

▌Figure 24.6 **Lineages of ratite birds were separated by the fragmentation of Gondwana.** The ancestors of these flightless birds once ranged over the continuous landmass of Gondwana.

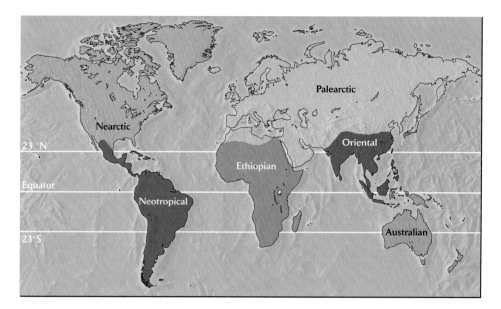

Figure 24.7 The major zoogeographic regions of the earth are based on the distributions of animals. This scheme, which is widely accepted today, originated with Alfred Russel Wallace in 1876. Biogeographic regions based on plant distributions are similar.
From J. H. Brown and M. V. Lomolino, *Biogeography* (2d ed.), Sinauer Associates, Sunderland, MA (1998).

vice versa; few species are the same, but both regions have representatives of many of the same genera and families.

The continents of the Southern Hemisphere, particularly Africa (**Ethiopian** region) and Australia (**Australian** region), experienced long histories of isolation from the rest of the terrestrial world, during which time many distinctive forms of life evolved in each. The **Oriental** region includes the biota of Southeast Asia, which was isolated from tropical areas of Africa and South America, in addition to contributions from the landmass of India. As one might expect, temperate (Palearctic) and tropical (Oriental) Asia have closer affinities than temperate North America (Nearctic) and tropical South America (**Neotropical** region) because of the continuous land connection between them. Indeed, the temperate forests of Asia contain a high percentage of tree species derived primarily from tropical forests, whereas those of temperate North America lack such species.

A land connection between North and South America through the Isthmus of Panama was finally re-formed during the Pliocene epoch, about 3 Mya. Although some taxa had island-hopped between the continents before this time, the land bridge allowed the exchange of many taxa, as shown for mammals in **Figure 24.8**. The exchange was uneven. More North American lineages entered South America than the reverse, and some of the North American groups diversified and possibly caused the extinction of many South American endemics, including a rich fauna of marsupial mammals.

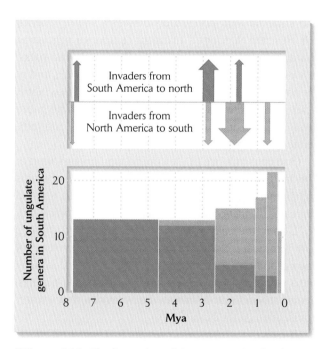

Figure 24.8 The formation of the Panamanian land bridge resulted in an interchange between North and South American biotas. The data shown here are for mammals, which have left a detailed fossil record. The initial migrations involved primarily South American mammals moving north, but later migration was primarily to the south. In South America, native ungulates were quickly replaced by lineages that colonized from North America. Data from D. S. Webb, *Paleobiology* 2:220–234 (1976).

Changes in climate have shifted the distributions of plants and animals

The climatic patterns of the earth ultimately depend on energy received from the sun, which warms the land and seas and evaporates water. Within this framework, the distribution of heat over the surface of the earth depends largely on the circulation of the oceans, which is driven by the rotation of the earth and constrained by the positions of the continents (see Chapter 4). When polar regions are occupied by landmasses or landlocked oceans, as they are at present, they can become very cold indeed. But this has not always been the case. In geologic periods when polar regions, which receive relatively little solar energy, were covered by oceans that extended to tropical areas, currents distributed heat rather evenly over the surface of the earth, and temperate climates extended quite close to the poles.

Fifty to 30 million years ago, large portions of North America and Europe were tropical. Tropical forests extended north to Washington State and into Canada, and warm temperate forests covered the Bering land bridge. The antarctic land connection between South America and Australia supported luxuriant temperate vegetation and animal life, as revealed by fossils in antarctic rocks. However, as Antarctica drifted over the South Pole, and as the northern polar ocean became trapped between North America and Eurasia, the earth's climate became more strongly differentiated into tropical (equatorial) and temperate (polar) zones.

One consequence of the cooling (and drying) trend at high latitudes was the retreat to lower latitudes of plants and animals that could not tolerate freezing; this resulted in greater distinction between temperate and tropical biotas. During the early part of the Tertiary period, what is now temperate North America supported a mixture of tropical and temperate forms growing side by side. Today these plants and animals occupy different climate zones, greater stratification of climate being matched by greater stratification of the biota.

These gradual changes in climate had profound effects on geographic distributions of plants and animals. During the past 2 million years, however, the gradual cooling of the earth gave way to a series of violent oscillations in climate that had immense effects on habitats and organisms in most parts of the world. This was the Ice Age, or Pleistocene epoch. Alternating periods of cooling and warming led to the advance and retreat of ice sheets at high latitudes over much of the Northern Hemisphere and caused cycles of cool, dry and warm, wet climates in the Tropics.

Ice came as far south as Ohio and Pennsylvania in North America and covered much of northern Europe, driving vegetation zones southward, possibly restricting tropical forests to isolated refuges with moist conditions, and generally disrupting biological communities all over the world.

One of the most striking and best-documented examples of this disruption concerns the migration of forest trees in response to the changing climate of eastern North America. Many tree species were restricted to southern refuges during the last glacial maximum, but after the glaciers began to retreat, about 18,000 years ago, a general pattern of reforestation ensued. Pollen grains deposited in the lakes and bogs left by retreating glaciers record the coming and going of plant species. These records show plainly that the composition of plant associations changed as species migrated over different routes across the landscape. Spruce forest, which dominated the area until about 10,000 years ago, was followed by associations of pine and birch, which were later replaced by more temperate species such as elm and oak.

The migrations of trees from their southern refuges since the height of the last glaciation are mapped for some representative species in ▌Figure 24.9. The distribution of spruce shifted northward just behind the retreating glaciers. Oaks expanded out of southern refuges to cover most of the eastern part of temperate North America, from southern Canada to the Gulf coast. Various species of pine had glacial refuges in the Carolinas, and their distributions shifted to the north and west, where they are currently centered near the Great Lakes. Hemlock had a more restricted refuge in the valleys of the Appalachian Mountains, and extended northward through the mountains into Pennsylvania, New York, and New England. Ironwood expanded out of small refuges in the Gulf states to cover most of eastern North America 12,000 to 10,000 years ago, and then contracted with further climatic warming to its present range in Michigan and southern Ontario. As a result of this movement in response to climatic change, the composition of forests during the past 18,000 years has included combinations of species that do not occur anywhere in eastern North America today, and has lacked combinations of species that do occur at present. For some species, the environment changed too rapidly during the Ice Age, and they disappeared altogether.

Catastrophes have caused major changes in the direction of evolution

Although the Ice Age brought dramatic changes in climate and spelled extinction for many forms of plants and animals, it pales beside the total disruption occasionally visited

Thousands of years before present

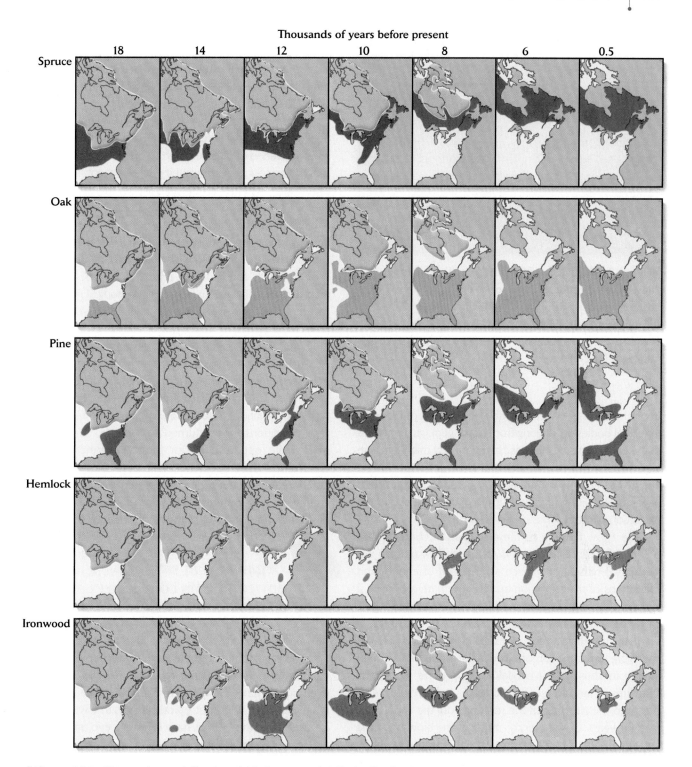

Figure 24.9 Climate change following glaciation caused shifts in distributions.
Migration of five types of trees in eastern North America from their Pleistocene refuges
18,000 years ago to their present distributions following the retreat of the glaciers. From G. L.
Jacobson, T. Webb, III, and E. C. Grimm, in W. F. Ruddiman and H. E. Wright, Jr. (eds.), *North America
during Deglaciation*, Geological Society of America, Boulder, CO (1987), pp. 277–288.

❚ Figure 24.10 An asteroid impact 65 million years ago caused the extinction of many forms of life. The impact left characteristic deposits of fine clay, like this one near Gubbio, Italy, in many parts of the world. This layer marks the boundary between Cretaceous and Tertiary deposits. Photo by Michael M. Follo.

upon the earth by collisions with asteroids and other extraterrestrial bodies or resulting from major geologic upheavals in the earth's crust. Collisions have occurred many times in earth history, with consequences in direct proportion to the energy liberated by the impact. On occasion, such impacts have caused widespread destruction of ecosystems and the extinction of many kinds of life on earth.

The most famous of these impacts occurred about 65 Mya. Evidence now points to the shallow seas off the Yucatán Peninsula of Mexico as the point of impact, and indications of the explosion and its aftermath appear as a layer of clay in geologic strata all over the world (❚ Figure 24.10). Scientists have estimated that an asteroid 10 km in diameter traveling 25 km per second may have been responsible. Such a collision would have released enough energy to cause massive tidal waves around the world, start fires on an unprecedented scale, and throw enough dust into the air to block the sun and cool the earth's surface for years. Consequently, much of the biomass of the earth would have been destroyed, either immediately by the direct effects of the impact or more slowly during the weeks and months afterward, and plant production in the oceans and on the land would have slowed to a standstill. The killing left a thin band of carbon preserved in sedimentary rocks of the time, along with thick deposits resulting from massive erosion in some areas.

One of the results of the impact was the extinction of a large proportion of the species on earth and of many higher taxa as well. Such episodes are referred to as **mass extinctions.** Not all plants and animals felt the impact

equally. All the dinosaurs disappeared, as did some large groups of marine organisms. Most of the larger evolutionary groups of plants survived—perhaps many of them as seeds in the soil—and mammals and birds survived to fill ecological vacancies left by the dinosaurs.

Catastrophes of such magnitude have happened infrequently—perhaps at intervals of tens or hundreds of millions of years—yet often enough to disrupt ecosystems and change the course of community development. Each major catastrophe, whether it originates within the earth through geologic upheavals or in outer space, brings about a period of extreme environmental stress. Geologists have found evidence in the geochemistry of sediments formed after such catastrophes that thousands of years may be required for environmental conditions to return to normal. The fossil record shows that some ecosystems—tropical reefs are a case in point—may disappear for millions of years, sometimes to be rebuilt by new kinds of organisms.

From the perspective of community development, catastrophes have several important consequences. They may eliminate species and thus greatly reduce diversity. They may foster rapid evolutionary responses to new types of conditions, and these changes many remain long after conditions have returned to "normal." Finally, they may create opportunities for the development of new types of biological associations. Although their effects cannot be easily identified or interpreted from present-day conditions and communities, such unique events in the past reach down through history to influence the present.

Organisms in similar environments converge in form and function

Just as long periods of isolation have led to unique life forms in many regions of the earth, similar environmental conditions in each of these regions have also led to the evolution of similar solutions to common problems. Plants inhabiting areas with subtropical climates in Central America and in eastern Africa have different evolutionary origins reflecting more than 100 million years of isolation, but they share similar growth forms and similar adaptations to arid conditions (see Figure 5.1). Thus, the different evolutionary histories and taxonomic affinities of plants and animals of the earth's regions are in part obliterated by convergence in form and function.

Convergence is the process whereby unrelated species living under similar ecological conditions come to resemble one another more than their ancestors did. There are many examples of convergent form and function. Where woodpeckers are absent from a fauna, as they are from many iso-

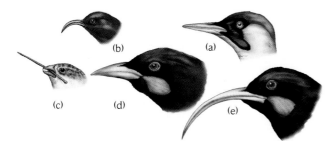

Figure 24.11 Several unrelated birds have become convergently adapted to extract insects from wood.
(a) European green woodpeckers excavate with their beaks and probe with their long tongues. (b) Hawaiian honeycreepers (*Heterorhynchus*) tap with their short lower mandibles and probe with their long upper mandibles. (c) Galápagos woodpecker-finches trench with their beaks and probe with cactus spines. New Zealand huias (now extinct) divided foraging between the sexes. Males (d) excavated with their short beaks, and females (e) probed with their long beaks.
After D. Lack, *Darwin's Finches*, Cambridge University Press, Cambridge (1947).

lated islands, other species have adapted to fill their role and have converged on the woodpecker lifestyle (**Figure 24.11**). Rain forests in Africa and South America are inhabited by plants and animals that have different evolutionary origins but are remarkably similar in appearance (**Figure 24.12**). Plants and animals of North and South American deserts resemble each other morphologically more than one would expect from their different phylogenetic origins. Similarities have also been noted in the behavior and ecology of Australian and North American lizards, despite the fact that they belong to different families and have evolved independently for perhaps 100 million years. Dolphins and penguins both evolved from terrestrial ancestors, but both have body shapes more closely resembling those of tuna, whose swimming lifestyle they share.

Convergence exists, and it reinforces our belief that adaptations of organisms to their environments obey certain general rules governing structure and function. However, detailed studies often turn up remarkable differences between the plants and animals in superficially similar environments. Despite striking convergence among desert-dwelling communities, for example, the ancient Monte Desert of South America lacks bipedal, seed-eating, water-independent rodents like the kangaroo rats of North America and the gerbils of Asia. Among frogs and toads, several South American forms have carried adaptation to desert environments a step further than their North American counterparts: they construct nests of foam to protect

their eggs from drying out. Differences between the Australian agamid lizard *Amphibolurus inermis* and its North American iguanid analog, *Dipsosaurus dorsalis,* include diet, optimal temperature for activity, burrowing behavior, and annual cycle, even though at first glance the species resemble each other closely.

Such differences are thought to reflect unique aspects of the evolutionary history of organisms in different regions, or perhaps subtle differences in the environments of the two regions. It has been suggested, for example, that

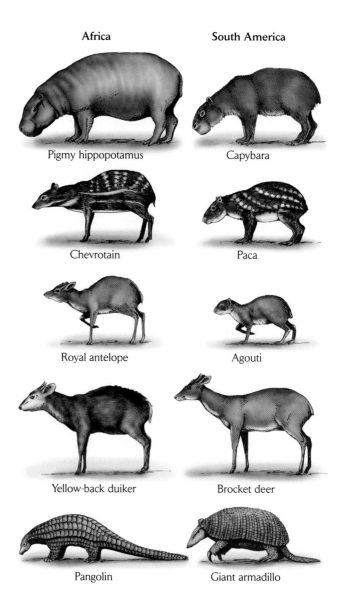

Figure 24.12 Unrelated African and South American rain forest mammals show striking convergence. Each pair is drawn to the same scale. After F. Bourliere, in B. J. Meggars, E. S. Ayensu, and W. D. Duckworth (eds.), *Tropical Forest Ecosystems in Africa and South America: A Comparative Review*, Smithsonian Institution Press, Washington, D.C. (1973). pp. 279–292.

many of the unique attributes of the reptile fauna of Australia, including the large number of species there, are related to poor soils. The logic of this idea is as follows. The Australian landmass is old and tectonically relatively quiet. Consequently, most of its soils are deeply weathered and have few nutrients. Plants, especially in the dry interior of Australia, have low nutrient content and high levels of toxic defensive compounds in their leaves, and consequently few insects can feed on them. Birds, which rely on arthropod food supplies, are thus not common in Australian deserts. Birds are major predators of lizards, and their absence has allowed a proliferation of lizard species and lifestyles not possible in other regions of the earth. Thus, some attributes of the unique reptile fauna of Australia may represent an accident of local geology rather than reflecting the unique ancestry of the fauna.

These issues have yet to be fully resolved. Certainly, ecologists must ensure that the comparisons they make involve habitats with closely matched physical characteristics. Otherwise, they cannot conclude that differences in structure or function have resulted from different histories rather than different contemporary environments. However, in general, convergence of form and function under similar environmental conditions is a broadly applicable principle of ecology and evolution. Difficulties with the concept in particular situations are more likely to reflect inadequate data about environment and evolutionary history than difficulties with the principle itself.

Communities in similar environments often include different numbers of species

The principle of convergence, when applied to communities, assumes that numbers of species and other aspects of community structure and function are determined primarily by local environmental conditions. Accordingly, we would expect independently derived communities that occupy similar habitats in different regions to have similar numbers of species, regardless of the number of species in the regional species pool. We can test this prediction by comparing diversity in similar habitats in different geographic regions having different regional diversities. If local diversity is the same, in spite of differing regional diversities, it is likely that local factors have predominated in determining the local coexistence of species. However, if local diversity varies, in spite of similarity among local environments, we must conclude that regional processes and the unique histories of different regions have left an imprint on local community diversity (❚ Figure 24.13).

We can also investigate whether local or regional processes determine community diversity by comparing the

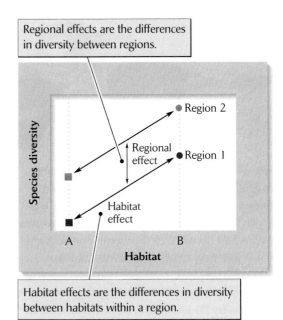

Regional effects are the differences in diversity between regions.

Habitat effects are the differences in diversity between habitats within a region.

❚ **Figure 24.13 Comparing the diversities of communities in similar habitats in different regions can reveal regional effects.** Local effects reflect the outcome of interactions of species with the environment and one another. Regional effects indicate that differences in the size of the regional species pool also influence the number of locally coexisting species.

relationship between local and regional species richness among areas with similar environments (❚ Figure 24.14). If local community structure is constrained by interactions among species, then community diversity should reach an upper limit, or saturation point, above which a further increase in the regional species pool should have no influence on local species diversity. If regional processes influence local communities, then local diversity should be higher where regional diversity is higher. That is, local communities should sample small regional pools and large regional pools in the same proportion. In the relationship between local (within-habitat) diversity and regional (island-wide) diversity in the West Indies, discussed in Chapter 23 (see Figure 23.9), local diversity increased continually with increasing regional diversity, although turnover of species between habitats (beta diversity) also increased. This pattern suggests that species can be added to local communities as the regional pool increases, but also that membership in a local community becomes more difficult as the number of species increases.

Studies of the relationship between local and regional diversity have generally favored the idea that communities are open to invasion at any level of diversity when additional species are produced within a region. For example, the number of fish species in short stretches of river (local

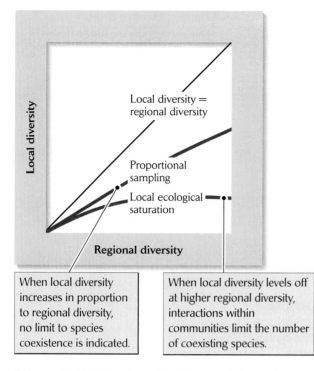

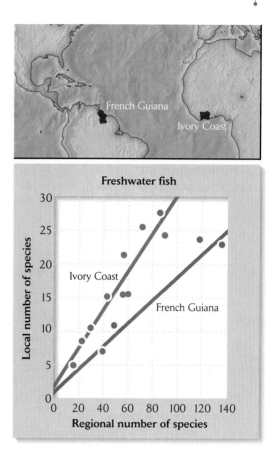

Figure 24.14 Saturation of local communities can be tested by relating local to regional diversity. If local interactions among species limit membership in the community, the number of species in the community should reach an upper limit, or saturation point, above which regional diversity has no effect. When species are added to a community independently of its existing diversity, the local community should sample the regional species pool proportionally. The line "Local diversity = regional diversity" would pertain if all species within a region were found in all habitats.

Figure 24.15 The diversity of tropical fish communities varies in proportion to regional species pools. Data from the Ivory Coast in West Africa and French Guiana in northern South America clearly indicate that these communities are not saturated. From B. Huegeny, L. T. Demorais, S. Merigoux, B. Demerona, and D. Ponton, *Oikos* 80:583–587 (1997).

communities) is directly proportional to the regional pool of species within entire river drainages (**Figure 24.15**) in northern South America and West Africa. That the regional species pool for various groups of organisms may differ among the major continental landmasses is not debated. However, the causes of these differences, and the degree to which they influence the structure of local communities, are not agreed upon.

ECOLOGISTS IN THE FIELD

Species diversity in temperate deciduous forests

The regional species pool for temperate deciduous forests of eastern North America includes 253 species of trees, more than twice the number (124) found in similar habitats in Europe. Temperate eastern Asia, whose climate is also similar to that of eastern North

America, has 729 tree species (**Figure 24.16**). Thus, although environment and forest structure—deciduous, broad-leaved trees—are uniform among these three regions, species diversity varies by a factor of nearly 6. These figures represent the total diversity of each region, but local diversity within small areas of uniform habitat exhibits parallel differences. Thus, regional diversity and local diversity appear to be closely related.

Although these diversity patterns are clearly related to the histories and unique biogeographic positions of the three regions, Roger Latham and I examined several kinds of data to interpret the differences in more detail. Analysis of the taxonomic relationships of trees shows that part of the greater diversity in Asia results from a greater proportion of species (32%) belonging to predominantly tropical genera. Over evolutionary time, the continuous corridor of forest habitat from the Tropics of Southeast Asia to the north has allowed tropical plants and animals to invade temperate ecosystems. In the Americas, the humid Tropics of Central

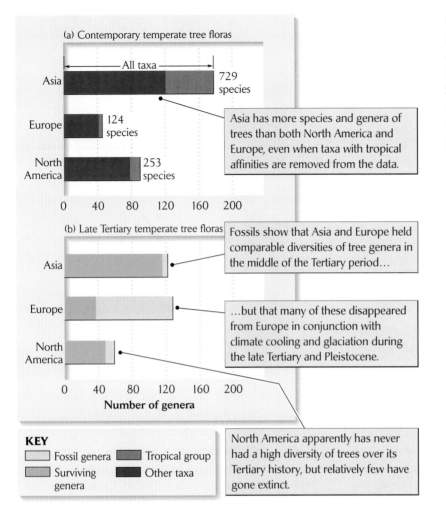

(a) Contemporary temperate tree floras

Asia — 729 species

Europe — 124 species

North America — 253 species

Asia has more species and genera of trees than both North America and Europe, even when taxa with tropical affinities are removed from the data.

(b) Late Tertiary temperate tree floras

Fossils show that Asia and Europe held comparable diversities of tree genera in the middle of the Tertiary period…

…but that many of these disappeared from Europe in conjunction with climate cooling and glaciation during the late Tertiary and Pleistocene.

Number of genera

North America apparently has never had a high diversity of trees over its Tertiary history, but relatively few have gone extinct.

KEY

Fossil genera
Surviving genera
Tropical group
Other taxa

Figure 24.16 Tree diversity varies among temperate forests on three continents. From R. E. Latham and R. E. Ricklefs, in R. E. Ricklefs and D. Schluter (eds.), *Species Diversity in Ecological Communities: Historical and Geographical Perspectives*, University of Chicago Press, Chicago (1993), pp. 294–314.

America are separated from the moist temperate areas of North America by a broad subtropical band of dry vegetation. In Europe, the Mediterranean Sea and arid North Africa effectively isolate temperate ecosystems from tropical Africa.

The fossil record suggests an ancient origin for the diversity difference among eastern North America, Europe, and eastern Asia. Almost twice as many genera of trees occur as fossils in eastern Asia as in North America (see Figure 24.16), paralleling the difference observed today. It is likely that the more complex geography of eastern Asia compared with eastern North America, as well as the consistent connection of temperate eastern Asia with tropical Southeast Asia, resulted in a higher rate of species production over the Tertiary period. But notice that the fossil record of Europe includes many more genera of trees than that of North America, in contrast with the present. A large proportion of the European genera became extinct in association with the climatic cooling leading to the Ice Age, while few genera of North American trees disappeared. As Europe cooled, the Alps and the Mediterranean Sea posed

effective barriers to southward movement (see Figure 25.11), and many cold-intolerant plant taxa died out. In North America, southward migration to areas bordering the Gulf of Mexico was always possible during cold periods.

Compared with temperate forests, the difference in species diversity between mangrove forests of the Caribbean region and of the Indo-West Pacific region is even more striking. Mangroves are tropical forests that occur within tidal zones along coastlines and river deltas (Figure 24.17). Mangrove trees tolerate high salt concentrations and anaerobic conditions in the water-saturated sediments in which they take root (see Chapter 3). Fifteen lineages of terrestrial trees have independently colonized the mangrove habitat, and several of these have subsequently diversified there.

At present, the mangrove flora of the Atlantic and Caribbean includes 7 species in 4 genera, 3 of which are cosmopolitan (occur worldwide). In contrast, the mangrove flora of the Indo-West Pacific includes at least 40 species in

(a)  (b)

Figure 24.17 Mangrove vegetation is less diverse in the Atlantic-Caribbean region than in the Indo-West Pacific region. (a) This mangrove forest in an estuary on the Pacific coast of Costa Rica contains few species. Note the prop roots of the *Rhizophora* trees at left and the buttressed trunks of *Pelliciera* at right. These trees are established on mud substrate within the tidal zone; hence the soil is flooded periodically with salt water. (b) The extensive prop roots of *Rhizophora* trap sediments and help to stabilize the coastline, as well as provide hard surfaces for growth of marine organisms. Photos by R. E. Ricklefs.

17 genera; 14 genera are endemic to the region (occur nowhere else). We cannot explain the much greater diversity of Indo-West Pacific mangroves on the basis of habitat availability, because both regions have roughly equal areas of a similar variety of mangrove habitats. Along with their greater species diversity, mangroves in the Indo-West Pacific occupy a greater proportion of the available habitat, having invaded farther down into the intertidal zone and farther upriver into estuaries than their Atlantic-Caribbean counterparts (**Figure 24.18**).

The large diversity anomaly in mangroves appears to have resulted from plant taxa invading mangrove habitats more frequently in the Indo-West Pacific than in the Caribbean region, although the reasons for this are not clear. In addition, fewer lineages may have suffered extinction in the Indo-West Pacific. Possibly, the terrestrial habitats fringing much of the Caribbean were arid during the latter part of the Tertiary. Consequently, wet terrestrial forest vegetation would have had little direct contact with mangrove habitat, and thus there would have been few opportunities for terrestrial taxa to adapt gradually to mangrove conditions. This has not been a limiting factor in Southeast Asia, where wet

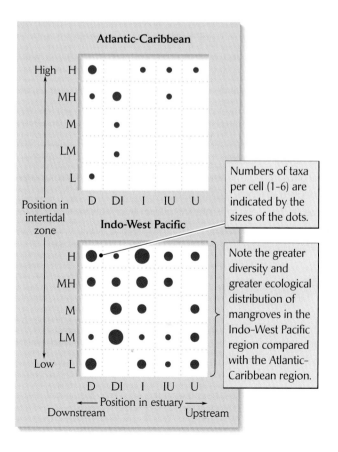

Figure 24.18 Mangrove vegetation occupies a greater ecological range where it is most diverse. The available mangrove habitat can be divided into cells representing categories of height in the intertidal zone (L, low; M, mid; H, high) and position within an estuary (D, downstream; I, intermediate; U, upstream). From R. E. Ricklefs and R. E. Latham, in R. E. Ricklefs and D. Schluter (eds.), *Species Diversity in Ecological Communities: Historical and Geographical Perspectives,* University of Chicago Press, Chicago (1993), pp. 215–229, after data in N. C. Duke, in A. I. Robertson and D. M. Alongi (eds.), *Tropical Mangrove Ecosystems,* American Geophysical Union, Washington, D.C. (1992), pp. 63–100.

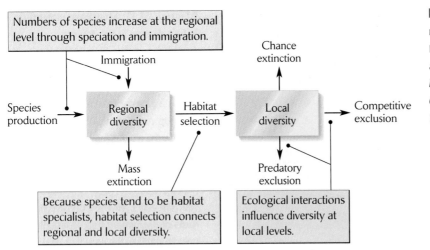

▌Figure 24.19 **Many factors influence regional and local species diversity.** From R. E. Ricklefs and D. Schluter, in R. E. Ricklefs and D. Schluter (eds.), *Species Diversity in Ecological Communities. Historical and Geographical Perspectives,* University of Chicago Press, Chicago (1993), pp. 350–363.

conditions have prevailed in tropical habitats throughout most of the Tertiary. In addition, much of Malaysia consists of islands of various size scattered on a shallow continental shelf, perhaps affording ideal conditions for isolation of populations in mangrove habitat and formation of new species of mangrove specialists.

Processes on many scales regulate biodiversity

Perhaps the example that we have just considered represents an extreme, although similar disparities between the more diverse Indo-West Pacific region and the less diverse Caribbean region are found in marine organisms (corals, reef fishes, mollusks, seagrasses), and numerous less striking examples have been reported. Nonetheless, these examples emphasize that the history and geographic position of a region may influence the diversity of both the entire region and its local habitats. Interactions of species within local habitats make up only half of the diversity equation (▌Figure 24.19).

Several processes are important to the regulation of biodiversity, each with a different characteristic scale of time and space. Scale in space varies from the activity ranges of individuals, through the geographic and ecological dispersal of individuals within populations, to the expansion and contraction of geographic ranges of populations. Scale in time varies from individual movements (behavior), through death and replacement of individuals in populations (demography and population regulation), interactions between populations (competitive exclusion), and selective replacement of genotypes within populations (evolution), to formation of new species after prolonged

isolation of subpopulations. Both local, contemporary processes and regional, historical processes shape community attributes. The fate of a local population depends partly on the tendencies of physical conditions, interspecific competition, and predation to influence local population size. Balancing these tendencies is the immigration of individuals from surrounding areas of population surplus. The persistence of a local population depends on the balance between these factors.

Local diversity of species depends on local rates of extinction—resulting from predators, disease, competitive exclusion, changes in the physical environment, and stochastic changes in small populations—and regional rates of species production and immigration. Every point on earth has limited accessibility, via dispersal, to sources of colonizing species. Local diversity depends not only on the capacities of environments to support a variety of species, but also on the accessibility of a region to colonists, the capacity of that region to generate new forms through speciation, and its ability to sustain taxonomic diversity in the face of environmental variation. Although ecology has traditionally focused on local, contemporary systems, it is now expanding its purview to embrace global and historical processes that have traditionally belonged to the disciplines of systematics, evolution, biogeography, and paleontology.

Summary

1. Life first appeared several billion years ago, but an abundant fossil record of modern life appeared about 590 Mya, a point that marks the beginning of the Paleozoic era of geologic history. The Mesozoic era, dominated on land by reptiles, began about 248 Mya; the age of mammals, the Cenozoic era, began 65 Mya.

2. The positions of the continents have changed continuously throughout the evolution of life, opening and closing different pathways of dispersal between continental landmasses and ocean basins and greatly altering climates on earth.

3. Animals and plants have evolved to some extent independently on different continents during prolonged periods of geographic isolation. Consequently, we can distinguish six major biogeographic regions: the Neotropical, Ethiopian, and Australian regions, derived from the former landmass of Gondwana, and the Oriental, Palearctic, and Nearctic regions, derived for the most part from the Northern Hemisphere landmass of Laurasia.

4. The climate of the earth cooled considerably during the Cenozoic era, causing tropical environments to contract to a narrower equatorial band and causing temperate and arctic environments to expand in extent.

5. The Cenozoic cooling trend culminated in the Ice Age (alternating periods of glacial advance and retreat in the Northern Hemisphere), which caused shifts in distributions and extinctions of many species of plants and animals.

6. Infrequent global catastrophes have punctuated the development of life when large extraterrestrial bodies have hit the earth. One of the best known of these events occurred 65 Mya, causing the extinction of the dinosaurs and other higher taxa of animals and bringing the Mesozoic era to a close. Such catastrophic changes in environments and their inhabitants have opened up new opportunities for evolution and have resulted in drastic reorganizations of biological communities.

7. The principle of convergence of form and function states that, despite their different histories of independent evolution, inhabitants of similar environments on different continents often resemble one another because they adapt to similar ecological factors.

8. If community diversity were regulated only by local interactions among species, whose outcome is determined primarily by environmental conditions, then biodiversity would also exhibit convergence between regions. Several examples of nonconvergence in the diversity of temperate forests and mangrove forests demonstrate that the unique histories and biogeographic settings of each continent also influence local species diversity.

9. Biodiversity reflects a broad array of local, regional, and historical processes and events operating on a hierarchy of temporal and spatial scales. Thus, understanding patterns of species diversity requires consideration of the history of a region and integration of ecological study with the

related disciplines of systematics, evolution, biogeography, and paleontology.

PRACTICING ECOLOGY

CHECK YOUR KNOWLEDGE

History of Diversity of North American Herbivorous Mammals

Assemblages of fossil mammals furnish an unusual opportunity to study the dynamics of biological communities over long periods. This work can provide information on communities over time scales encompassing climate change (i.e., glaciation), large-scale movements of species across continents, and evolutionary change among species. If one can sample enough fossils from many localities, then it should be possible to pose questions about changes in diversity over time in response to environmental change, as well as about the relationship between local and regional diversity patterns. Indeed, the mammalian fossil assemblage of North America is quite extensive and therefore ideal for such studies of historical changes in species diversity over space and time.

Blaire Van Valkenburgh, of the University of California, Los Angeles, and Christine Janis, of Brown University, studied 115 fossil mammal assemblages from different locations in North America, mostly in the western United States. They divided the fossil sites into eight geographical regions and tabulated the average number of species of herbivores and carnivores for each location, as well as the total number of species in each region. Their sampling covered the time period from the Middle Eocene (about 44 Mya) to the present. The Middle Eocene represents the height of warm, moist conditions in the Northern Hemisphere—a time when most of what is now the United States was covered by tropical forest. At the Oligocene–Miocene boundary (about 24 Mya) the climate of North America started to become colder and drier, a trend that has continued until recently.

The total numbers of herbivore and carnivore species in all the fossil assemblages show that herbivore diversity increased to a maximum during the mid-Miocene and then declined steadily through the Pliocene (■ Figure 24.20). A dramatic decrease in species diversity over the past 300,000 years then followed. How might one relate species diversity at the local and regional spatial scales? To do this, Van Valkenburgh and Janis plotted local species diversity as a function of the total continental diversity during the same period (■ Figure 24.21). From the pattern of this relationship, Van Valkenburgh and Janis concluded that increases in regional diversity reflecting the development of new

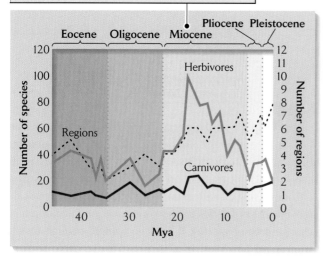

The middle of the Miocene was a period of warm, moist tropical conditions in the United States.

Figure 24.20 Herbivore and carnivore diversity increased to a maximum during the mid-Miocene epoch.

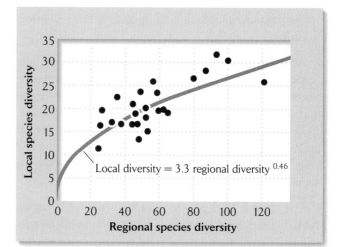

Local diversity = 3.3 regional diversity $^{0.46}$

Figure 24.21 Local species diversity is a function of the total continental diversity from Eocene to Pleistocene epochs.

communities, or exchange of species between communities, lead to an increase in the local species diversity of fossil assemblages.

CHECK YOUR KNOWLEDGE

1. What factors might have led to the decrease in large herbivorous mammal diversity in the second half of the Cenozoic era?

2. From ❙ Figure 24.21, what is the mathematical relationship between local and regional species diversity? What might account for the slope of the line for this equation?

MORE ON THE WEB
3. Visit the *Burgess Shale Fossils* Web site through *Practicing Ecology on the Web* at *http://www.whfreeman.com/ricklefs*. The Burgess Shale is located in Yoho National Park in the Rocky Mountains of British Columbia, Canada. What is it about this locality that accounts for its unusual assemblage of fossils? What can we learn about historical patterns of biodiversity from the Burgess Shale?

 Suggested Readings

Ben-Avraham, Z. 1981. The movement of continents. *American Scientist* 69:291–299.

Bennett, K. D. 1997. *Evolution and Ecology: The Pace of Life*. Cambridge University Press, Cambridge.

Brown, J. H. 1995. *Macroecology*. University of Chicago Press, Chicago.

Brown, J. H., and M. V. Lomolino. 1998. *Biogeography*. 2d ed. Sinauer Associates, Sunderland, MA.

Carlquist, S. 1981. Chance dispersal. *American Scientist* 69:509–516.

Farrell, B. D., C. Mitter, and D. J. Futuyma. 1992. Diversification at the insect–plant interface. *BioScience* 42:34–42.

Flessa, K. W. 1986. Causes and consequences of extinction. In D. M. Raup and D. Jablonski (eds.), *Patterns and Processes in the History of Life*, pp. 234–257. Springer-Verlag, Heidelberg and New York.

Marshall, L. G. 1988. Land mammals and the Great American Interchange. *American Scientist* 76:380–388.

Orians, G. H., and R. T. Paine. 1983. Convergent evolution at the community level. In D. J. Futuyma and M. Slatkin (eds.), *Coevolution*, pp. 431–458. Sinauer Associates, Sunderland, MA.

Paruelo, J. M., E. G. Jobbágy, O. E. Sala, W. K. Lauenroth, and I. C. Burke. 1998. Functional and structural convergence of temperate grassland and shrubland ecosystems. *Ecological Applications* 8(1):194–206.

Pielou, E. C. 1991. *After the Ice Age*. University of Chicago Press, Chicago.

Qian, H., and R. E. Ricklefs. 2000. Large-scale processes and the Asian bias in species diversity of temperate plants. *Nature* 407 (14 September):180–182.

Ricklefs, R. E., and D. Schluter (eds.). 1993. *Species Diversity in Ecological Communities: Historical and Geographical Perspectives*. University of Chicago Press, Chicago and London.

Stucky, R. K. 1990. Evolution of land mammal diversity in North America during the Cenozoic. *Current Mammalogy* 2:375–432.

Van Valkenburgh, B., and C. M. Janis. 1993. Historical diversity patterns in North American large herbivores and carnivores. In R. E. Ricklefs and D. Schluter (eds.), *Species Diversity in Ecological Communities: Historical and Geographical Perspectives*, pp. 330–340. University of Chicago Press, Chicago.

Vermeij, G. J. 1991. When biotas meet: Understanding biotic interchange. *Science* 253:1099–1104.

Extinction and Conservation

Biological diversity is incompletely described and catalogued

The value of biodiversity arises from social, economic, and ecological considerations

Extinction is natural but its present rate is not

Humans have caused extinctions by several mechanisms

Conservation plans for individual species must include adequate habitat for a self-sustaining population

Some critically endangered species have been rescued from the brink of extinction

Humans have an immense impact on the earth. There are so many of us (the year 2000 population of 6 billion is increasing at a rate of almost 2% per year), and each individual uses so much energy and so many resources, that our activities influence virtually everything in nature. Most of the land surface of the earth and, increasingly, the oceans have come under the direct control of humankind. Virtually all areas within temperate latitudes that are suitable for agriculture have been brought under the plow or fenced. Worldwide, fully 35% of the land area is used for crops or permanent pastures; countless additional hectares are grazed by livestock. Tropical forests are being felled at the alarming rate of 17 million hectares (almost 2% of the remaining primary stands) each year. Semiarid subtropical regions, particularly in sub-Saharan Africa, have been turned into deserts by overgrazing and collection of firewood. Rivers and lakes are badly polluted in many parts of the world. Our atmosphere reeks of gases produced by chemical industries and the burning of fossil fuels.

We are fouling our nest, and we are still rushing to exploit much of what remains to be taken. If unchecked, this deterioration of the environment will lead to a declining quality of life for all human inhabitants of the earth, as it already has for many. The animals and plants with which we share this planet, and on which we depend for all kinds of sustenance, are feeling the impact of the human population even more. They have been pushed aside as we have taken over land and water for our own living space and for the production of our food. We have poisoned their environments with our wastes. Entire species have succumbed to habitat destruction, hunting, and other forms of persecution.

This deterioration need not continue. Humans can live in a clean and sustaining world, but only by placing support for our own population into balance with preservation of other species and of the ecological processes that nurture us. Legislation in many countries has already led to cleaner air and water, more efficient use of energy and material resources, and the rescuing of endangered species from further decline.

The science of ecology has much to say about rational development and management of the natural world as a sustainable, self-replenishing system. What we have learned about the adaptations of organisms, the dynamics of populations, and the processes that occur in ecosystems suggests simple guidelines for living in reasonable harmony with the natural world.

First, environmental problems can never be brought under control as long as the human population continues to increase. Certainly the earth could support many more individuals than it does at present, but quality of life would be drastically reduced in the short term, and there would be little prospect for sustainability in the long term. Even the present-day human population cannot maintain itself on a sustainable basis. Reforestation cannot keep pace with growing demands for timber, paper, and fuelwood, and so vast amounts of previously uncut forest are being harvested each year. Most of the important fisheries of the Northern Hemisphere have collapsed and now yield only a fraction of their previous production. Large areas of deteriorated farmland are lost to agriculture every year. As the human population increases, such demands on the environment will only increase.

Studies of natural populations show that their control depends on factors that act in a density-dependent fashion; these factors (which include food shortage, disease, predation, and social strife) reduce fecundity, or increase mortality, or both, as populations grow. If the human population were to come under such external control, the toll in human suffering—disease, famine, warfare—would be enormous. Thus, maintaining individual quality of life at a high level will require, above all, that humans exhibit a reproductive restraint that defies the entire history of evolution, during which "fitness" has been measured in terms of reproductive success rather than quality of life. Only an appreciation of the negative economic and environmental consequences of overpopulation will cause humanity to value individual human experience over numbers of progeny as the two become increasingly incompatible.

Second, individual consumption of energy, resources, and food produced at higher trophic levels must be reduced. The earth could not sustain the resource and energy depletion that would result if everyone consumed at the level now exhibited by affluent citizens of developed countries. Efficiency can be increased and superfluous consumption reduced without impairing comfort or enjoyment of life. Each individual human can reduce her or his impact by eating lower on the food chain (reducing meat consumption, for example), investing in energy- and resource-efficient technologies, and living closer to equilibrium with the physical world (for example, lowering the thermostat setting in winter and raising it in summer).

Third, although it is inevitable that most of the world will come under human management, ecosystems should be maintained as close to their natural state as possible to keep natural ecosystem processes intact. As a general rule, the less we alter nature, the easier it will be to sustain the environment in a healthy condition. For example, many areas covered by tropical forests are unsuitable for grazing or agriculture because these activities upset natural nutrient regeneration and cause soils to deteriorate. Such areas should be left as forest preserves or recreation areas, or as sites for sustained exploitation of forest products. Deserts can be watered, and they often become tremendously productive for certain types of agriculture. But the costs of maintaining such managed systems can become extremely high as soils accumulate salts from irrigation water and aquifers become depleted. Living with nature is always preferable to, and less costly than, going against it.

In this chapter, we shall consider solutions to the problem of conserving species—that is, preventing their populations from dwindling to extinction. In Chapter 26, we shall discuss ways of maintaining natural populations and ecosystem processes so that our generation and future generations will benefit from them. The solutions to all of these problems can be understood by applying basic principles of ecology. We must remember, however, that although solutions can be proposed, implementing them will require concerted social, political, and economic action.

Biological diversity is incompletely described and catalogued

Nearly 1,500,000 species of plants and animals worldwide have been described and given Latin names (Figure 25.1). Insects account for about half of these. Many more species, particularly in poorly explored regions of the Tropics, await scientific discovery. Some experts have estimated that the final species count could reach between 10 and 30 million. Such estimates may be inflated, but it is incontestable that we share this planet with several million other kinds of plants, animals, and microbes.

Making lists of species names is one way of tabulating diversity, but such lists represent only part of the concept

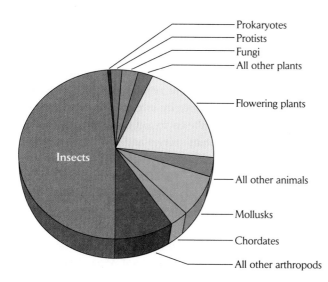
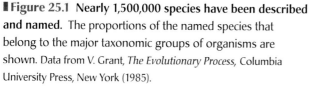

Prokaryotes
Protists
Fungi
All other plants

Flowering plants

Insects

All other animals

Mollusks

Chordates

All other arthropods

Figure 25.1 Nearly 1,500,000 species have been described and named. The proportions of the named species that belong to the major taxonomic groups of organisms are shown. Data from V. Grant, *The Evolutionary Process,* Columbia University Press, New York (1985).

of biological diversity, or **biodiversity,** which includes the many unique attributes of all living things. Although each species differs from every other species in the name that science has assigned to it, it also differs in the way its adapta-

tions define its place in the ecosystem. Different species of plants, for example, have dissimilar tolerances for soil conditions and water stress and disparate defenses against herbivores; they also differ in growth form and in strategies for pollination and seed dispersal. Animals, too, have adaptations that define their place in nature. These variations constitute **ecological diversity.**

Biodiversity results from genetic changes, or evolution. Because genetic variability is crucial to the evolutionary response of populations to changes in the environment, **genetic diversity,** both between and within species, is another important component of biodiversity.

Finally, biodiversity has a geographic component. Different regions have different numbers of species. If diversity were a contest, tropical rain forests and coral reefs would be the clear winners. Equally important, however, is the fact that some regions harbor unique species found nowhere else. Species whose distributions are limited to small areas are called **endemic species,** and regions with large numbers of endemic species are said to possess a high level of **endemism.** Clearly, conservation of global biodiversity is best served by directing efforts toward areas of high endemism as well as high diversity.

Oceanic islands are well known for harboring endemic forms; virtually all the birds, plants, and insects of such isolated islands as the Hawaiian and Galápagos archipelagoes occur nowhere else (**Figure 25.2**). As a result, when habitat

(a)

(b)

Figure 25.2 Many oceanic islands harbor endemic species. (a) The Hawaiian silversword is found only at high elevation on Haleakala Volcano on the island of Maui, Hawaii. (b) This tortoise is endemic to the Galápagos archipelago, where each island has a distinctive form. Photos by R. E. Ricklefs.

destruction, hunting, or the introduction of alien species results in a loss of local populations in such places, this loss is likely to signify worldwide extinction. Fossil-bearing deposits have shown that more than half of the avifauna of Hawaii has disappeared since human colonization of the islands. These birds occurred nowhere else; now they are gone forever. So are the giant moas (flightless birds) of New Zealand. Steller's sea cow (a giant relative of dugongs and manatees), which was endemic to the Bering Sea, became extinct by 1768, less than 30 years after it was discovered and first hunted by Europeans.

ECOLOGISTS IN THE FIELD

Identifying critical areas for biodiversity

Norman Myers, of Oxford University, and his colleagues have identified 25 biodiversity "hotspots" worldwide, which they have proposed for special conservation consideration (▮Figure 25.3). The boundaries of these hotspots correspond to the boundaries of major biogeographic regions. These boundaries are relatively easy to set for such island areas as the Caribbean, Madagascar, and New Caledonia. Within continents, the boundaries usually correspond to the edges of important biomes, such as the dry "cerrado" vegetation of Brazil and the Mediterranean climate region of southern Europe and northern Africa. To qualify as a hotspot, a region must have a high level of endemism. The remaining natural vegetation in all the hotspots identified by Myers occupies only 1.4% of the total land area of the earth, yet these hotspots hold as many as 44% of all plant species and 35% of all species of terrestrial vertebrates. They are also regions of rapid habitat destruction where a high proportion of species are threatened with declining populations or extinction. Within these hotspot areas, an average of 88% of the natural vegetation has already disappeared.

As Richard Cincotta and his colleagues at Population Action International have pointed out, Myers's biodiversity hotspots also tend to have above-average densities of human populations combined with high population growth rates. More than 1.1 billion people—nearly 20% of the human population—live in the 12% of the earth's land area included in the hotspots, and the average growth rate of these populations is 1.8% per year. Among the most densely populated are southern India and Sri Lanka, the Philippines, and the West Indies. Population growth rates are particularly high in Andean Colombia, Ecuador, and Peru, in Madagascar, and

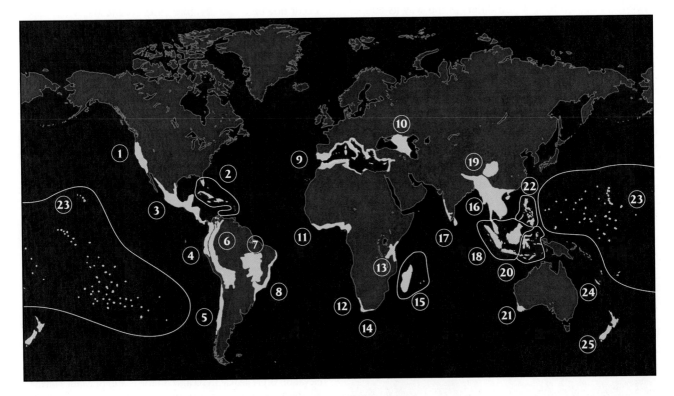

▮**Figure 25.3 Twenty-five biodiversity hotspots have been identified worldwide.**
These areas are receiving special conservation efforts. From N. Myers, R. A. Mittermeier,
C. G. Mittermeier, G. A. B. da Fonseca, and J. Kent, *Nature* 403:853–858 (2000).

in West Africa. Each of these areas presents special and urgent conservation challenges. The three major tropical wilderness areas—the Amazon basin of South America, the Congo basin of Africa, and New Guinea—have low population densities at present, but high population growth rates, due in large part to immigration. From the standpoint of preserving biodiversity, the hotspots identified by Myers and his colleagues are logical places to focus conservation and management efforts.

The value of biodiversity arises from social, economic, and ecological considerations

The rate of disappearance of certain kinds of species, particularly those most vulnerable to hunting, pollution, and destruction of habitat, is probably now at an all-time high in the history of the earth. Some estimates suggest the disappearance of more than one species each day, most of them tropical rain forest insects. This accelerated loss of species is directly linked to the growth and technological capacities of the human population.

Why do we care? What concern is it of ours if a species of beetle disappears from South America? Many species already are gone. Do we really miss them? In fact, extinction occurs normally in natural systems. Why should we try to stop it?

Moral responsibility

The rationale for conserving biodiversity depends on many values related to our own personal interest and involvement. For many people, extinction raises a moral issue. Some take the position that because humankind affects all of nature, it is our moral responsibility to protect nature. If morality derives from a natural law—that is, if morality is intrinsic to life itself—then we may presume that the rights of nonhuman individuals and species are as legitimate as the rights of individuals within human society. Of course, no species is guaranteed a right to perpetual existence, just as no human is guaranteed immortality. But extinction by unrestrained hunting, pollution, habitat destruction, and irresponsible spread of disease is considered by many to be like murder, manslaughter, genocide, and other infringements of individual human rights.

Economic benefits

In the absence of moral protection, the value of individual species can be argued from the standpoint of their economic and recreational benefits to humankind. Individual

Figure 25.4 Many species have economic value to humans. Public markets in tropical countries, such as this one in Nairobi, Kenya, offer hundreds of varieties of local plant products, such as fruits, fibers, and medicinals. Many of these products are harvested from natural ecosystems, others from species cultivated locally or throughout the world. Photo by R. E. Ricklefs.

species have obvious economic importance as food resources, game species, and sources of forest and other natural products, drugs, and organic chemicals (**Figure 25.4**). For example, more than a hundred important medicinal drugs (including codeine, colchicine, digitalin, L-dopa, morphine, quinine, strychnine, and vinblastine), which account for about one-fourth of all prescriptions filled in the United States, are extracted directly from flowering plants.

Some plant and animal species of economic importance have been cultivated or domesticated and then selectively bred to enhance their valuable qualities. These species are not in danger of extinction, but making room for their cultivation on a large scale has often endangered other species that are perceived as having lesser value. An example is the classic conflict between sheep ranchers and wolves, which occasionally kill sheep and other livestock. Because of the value of sheep farming, wolves were driven out of most of North America, often with bounties on their heads, and often with the result that herds of deer and other herbivores became so large as to damage the environment, including, ironically, its value for grazing sheep. The point is that assigning economic value to species favors some over others and often does not address the issue of conserving biodiversity in a general sense.

In many circumstances, the short-term gains of converting natural systems to human use (such as the conversion of forest to agriculture) or of overexploiting resources (such as

▌Figure 25.5 Ecotourism confers economic value on biodiversity. Ecotourists "on safari" in Kenya, East Africa, bring in millions of dollars in internationally traded hard currencies. Ecotourists also provide employment for many of the local people as guides and wardens, as well as in hotels and restaurants and the businesses that support them. Photos by R. E. Ricklefs.

the intensive fishing of Atlantic cod and swordfish populations) are assumed to outweigh any long-term value of conserving the natural system for sustained income. The value of conserved species and habitats usually becomes apparent only when the long-term costs of overexploitation or habitat conversion are properly accounted for.

High value may be placed on some individual species because they attract tourists to an area. The practice of visiting an area to see unspoiled habitats and the animals and plants that live in them is referred to as **ecotourism.** Many tropical countries have capitalized on this attraction by establishing parks and support services for tourists (▌Figure 25.5). In Latin America, quetzals, macaws, and monkeys draw tourists to many areas where these species are protected. Diversity itself is often the attraction in tropical rain forests and coral reefs, with their hundreds of different species of trees, birds, or fish.

In East Africa, lions, elephants, and rhinoceroses have great value because of the tourist dollars, pounds, francs, and yen they bring into countries that are badly in need of foreign currencies (▌Figure 25.6). Unfortunately, conflict exists between those few people who poach elephants for their ivory and rhinoceroses for their horn and the many who enjoy watching them. The intensity of poaching was revealed in a study in a national park in Zambia, where the fraction of tuskless female elephants increased from 10% in 1969 to 38% in 1989 as a direct result of selective illegal ivory hunting. Tusklessness in females is a genetic trait, and because poachers kill only individuals with tusks, poaching strongly favors tusklessness in a population. A change in the frequency of a trait from 10% to nearly 40% within one generation is strong selection indeed.

Ecotourism has been responsible for the development and maintenance of an increasing number of parks and preserves in many parts of the world. Its impact will expand as more people become aware of the gratification that comes from experiencing nature directly. The capacity of ecotourism to confer enough value on species to guarantee their protection is, however, finite. People have limited money to spend, and merely increasing preserve systems will not necessarily generate more tourism. Furthermore, some areas of immense biological importance, with high diversity and endemism, either are not attractive as destinations or are inaccessible to most tourists. Deserts, semiarid regions, many islands, and most marine ecosystems fall into these categories. Moreover, most species simply are not interesting or even perceptible to the general public. Their preservation will depend on their living in association with more highly valued species or habitats.

Indication of environmental quality

Individual species may have considerable value as indicators of broad and far-reaching environmental change. During the 1950s and 1960s, populations of many predatory and fish-eating birds in the United States (particularly the peregrine falcon, bald eagle, osprey, and brown pelican) declined drastically to the point that several of these species had disappeared from large areas, the peregrine from the entire eastern United States. The causes of these population declines were traced to pollution of aquatic habitats by breakdown products (residues) of DDT, a pesticide that was widely used to great immediate benefit after World War II. Unfortunately, the pesticide's residues resisted degradation

Figure 25.6 Biodiversity may attract tourists to an area. The inspiring diversity of wildlife in Africa is as much a part of its attraction as any particular species. Nowhere else can one see elephants, cheetahs, zebras, and dozens of other large mammals in their natural environments. Photos by R. E. Ricklefs.

and entered aquatic food chains, where they accumulated in the fatty tissues of animals and were concentrated with each step in the food chain. The high doses consumed by predatory birds interfered with their physiology and reproduction, causing overly thin eggshells and deaths of embryos (**Figure 25.7**). Breeding success plummeted, and populations followed.

The viability of the peregrine population was a sensitive indicator of the general health of the environment. Its demise sounded the alarm to environmentalists; Rachel Carson warned of a "silent spring" when no birds would be left to sing. The United States government responded by banning DDT and related pesticides, and chemical companies have since devised alternatives that have less drastic environmental effects. Bald eagles and ospreys are becoming familiar sights once again, and, thanks to the helping hands of dedicated biologists who reared birds obtained from other parts of the geographic range and released them

Figure 25.7 Predatory birds were important bioindicators of the effects of the pesticide DDT. Broken eggshells in the nest of a brown pelican; the shells were thinned by DDT. Photo by Betty Anne Schreiber/Animals Animals.

in the eastern United States, peregrine falcons have staged a comeback. This was a major victory, not only for the peregrine and the cause of species conservation, but also for the general quality of our own environment, because the peregrine population could not be saved until DDT, which affects many species, was banned from use. Sadly, this silver lining draws attention to a dark cloud: manufacture and export of DDT to foreign countries is still legal in the United States and remains a source of great profit to some chemical companies. DDT will bring a bleaker future to many people living in countries that have yet to ban the toxin in favor of safer alternatives.

Maintenance of ecosystem function

Species diversity in ecological systems may have intrinsic value for stabilizing ecosystem function. An increasing number of studies are showing that diverse systems are better able to maintain high productivity in the face of environmental variation. For example, David Tilman and J. A. Downing of the University of Minnesota, using experimental plots of Minnesota prairie containing differing numbers of species, demonstrated that biomass production was less affected by severe drought on high-diversity plots than on low-diversity plots. Such results can be explained by positing that higher-diversity systems are more likely to include some species that can withstand particular stresses. As the environment changes, different species can take over the roles of predominant producers in a system. Such switching among species is less likely to occur in less diverse systems. Controlled experiments with artificial communities established in small plots or in environmental chambers similarly show that increasing the number of guilds among primary producers or increasing the number of links in the food web increases ecosystem productivity and stability in the face of environmental variation (❚ Figure 25.8).

This general background illustrates why we should value biodiversity and conserve it. But what happens if our conflicting values are not resolved in favor of preserving biodiversity?

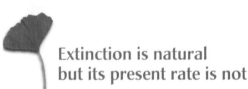

Extinction is natural but its present rate is not

Extinction is a major concern of conservationists because it represents the disappearance of evolutionary lineages that never can be recovered. We discussed the extinction of small populations in Chapter 15, and learned about major extinctions over the history of life in Chapter 24. Humans have already caused the extinction of many species, and the

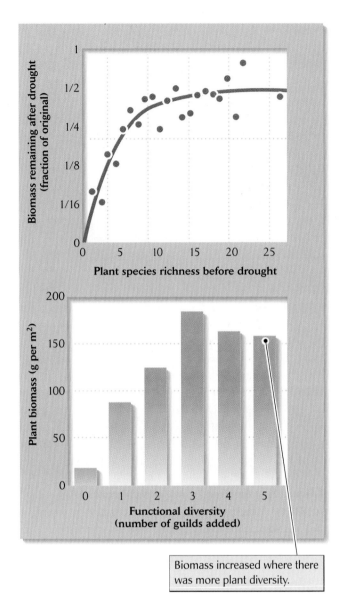

Biomass increased where there was more plant diversity.

❚ **Figure 25.8 Diverse ecosystems are more productive and better resist perturbation.** (*Above*) Decrease in plant biomass in prairie plots as a result of drought is related to plant species richness before the drought. (*Below*) Plant biomass in artificially seeded experimental plots increases in relation to the number of plant guilds (early-season annuals, late-season annuals, perennial bunchgrasses, perennial forbs, and nitrogen fixers). From D. Tilman and J. A. Downing, *Nature* 367:363–365 (1994) and D. Tilman, J. Knops, D. Wedin, P. Reich, M. Ritchie, and E. Siemann, *Science* 277:1300–1305 (1997).

rate of species loss is likely to accelerate. To design solutions to this problem, we need to understand the basic causes of extinction.

It will be useful for us to distinguish three types of extinction. As ecosystems change, some species disappear and others take their places. This turnover of species, at a relatively low rate, is known as **background extinction.** It appears to be a normal characteristic of natural systems.

Mass extinction refers to the dying off of large numbers of species because of natural catastrophes. Volcanoes, hurricanes, and meteor impacts happen occasionally. Some occur locally, others affect the entire globe, and species that happen to be in their way disappear.

Anthropogenic extinction—extinction caused by humans—is similar to mass extinction in the number of taxa affected and in its global dimensions and catastrophic nature. Anthropogenic extinction differs from mass extinction, however, in that its causes theoretically are under our control.

Most information on background extinction comes from the fossil record, which reveals appearances and disappearances of species through geologic time. The life spans of species in the fossil record vary according to taxon, but they generally fall within the range of 1 to 10 million years. Thus, on average, the probability that a particular species will go extinct in a single year is in the range of one in a million to one in 10 million. If, as conservative estimates have it, on the order of 1 to 10 million species inhabit the earth, this would amount to a background extinction rate of about one species extinction per year.

Mass extinctions occupy the other end of the spectrum. Natural catastrophes may cause the disappearance of a substantial proportion of species locally or globally, depending on the severity and geographic extent of the catastrophe. Local catastrophes may include prolonged drought, hurricanes of great force, and volcanic eruptions. When Krakatau, a volcanic island in the East Indies, exploded on August 26, 1883, not an organism was left alive; any that survived the initial explosion were buried under a thick layer of volcanic debris and ash (see Chapter 22). Whether any species disappeared in this catastrophe cannot be known because the biotic diversity of the island had not been well surveyed prior to the explosion, but any species endemic to the island certainly went extinct.

Some mass extinctions detectable in the fossil record are thought to have been caused by the impacts of large comets or asteroids (collectively referred to as bolides) Major bolide impacts with global effects have occurred at intervals of 10 to 100 million years over the history of life. As we have seen in the previous chapter, spectacular examples of mass extinctions occurred at the end of the Paleozoic era (Permian period) and the Mesozoic era (Cretaceous period). The first involved the disappearance of perhaps 95% of species and numerous higher taxa. The second is most famous for the extinction of the dinosaurs, but some other major groups, notably ammonites (predatory, nautilus-like mollusks), disappeared as well (▌Figure 25.9). Whatever

▌**Figure 25.9 The earth's second major mass extinction occurred at the end of the Cretaceous period.** Dinosaurs and other groups, including ammonites, of which a fossil is pictured above, were extinguished by the global effects of a bolide impact. Photo by Kerry Givens/Bruck Coleman.

their exact cause, these extinctions were associated with discrete, calamitous events.

And anthropogenic extinction? Are we to be regarded as a "human bolide" in terms of our impact on biodiversity? Well, not yet. Many extinctions have undoubtedly gone unrecorded, and rates of extinction in many groups (particularly among large animals hunted for food and among island forms) are far above background levels. Nevertheless, if humankind turns out to be a disaster for global biodiversity, the full force of its impact will come in the future. Most important, such a disaster is preventable. Examining the causes of extinction will enable us to see why this is so.

Humans have caused extinctions by several mechanisms

Populations disappear when deaths exceed births over a prolonged period. This much is obvious, but the statement also emphasizes that extinction may result from a variety of mechanisms that influence birth and death processes within a population. It has also been stated that extinction represents failure to adapt to changing conditions, whether because the changes occur too fast or because a population is unable to respond to them.

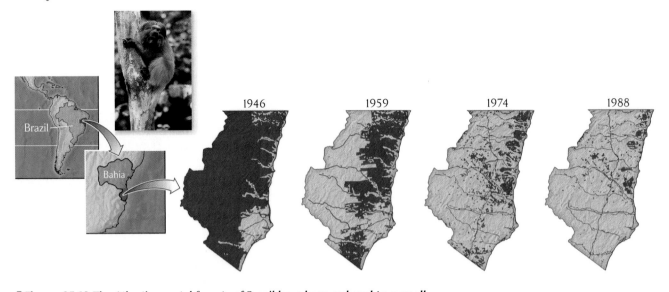

■ **Figure 25.10 The Atlantic coastal forests of Brazil have been reduced to a small fraction of their former extent.** These maps document the decimation of Atlantic coastal forests in Bahia, southern Brazil, during the past 50 years. Several unique, endemic species have disappeared from this area, and others, such as the golden lion tamarin, are gravely threatened. Maps by J. R. Mendonça, Projeto Mata Atlântica Nordeste, Convênio CEPLAC/New York Botanical Garden. Photo by Tom McHugh/Photo Researchers.

T. C. Foin and his colleagues recently surveyed the causes of population declines for endangered species in the United States. The primary causes they found were (1) habitat reduction and modification (67% of cases), (2) small population size, (3) introductions of exotic species, and (4) overexploitation.

Habitat reduction and fragmentation

Reduction of habitat and, especially, fragmentation of habitat into small remnants poses a tremendous threat to some kinds of wildlife. For example, the Atlantic coastal forests of Brazil have been reduced to only a small percentage of their former extent (■ Figure 25.10), critically endangering many endemic birds and mammals, such as the golden lion tamarin. The remaining fauna is now coming under intense conservation efforts, but possibly too late for many of the region's inhabitants. Even in North America, habitat fragmentation is causing population declines. Tall-grass prairie now exists only in a few isolated refuges, and many populations of prairie plants and animals are locally extinct. Our largest national parks have lost many of their mammal species in the past 50 years, suggesting that these preserves are insufficient in size to maintain viable populations.

The disappearance of many native songbird species from small fragments of temperate forest in North America is partly a consequence of small population size and stochastic local extinction. But fragmentation has also increased access to forest habitat by some predators and nest parasites that are more typical of fields and agricultural lands, with drastic consequences for songbird survival and reproductive success in some areas (see Figure 13.2). Thus, habitat fragmentation also causes a deterioration of habitat quality.

Over time, changes in climate cause the positions of the major biomes, and of habitats within each biome, to shift. Habitat fragmentation poses a serious threat to the ability of organisms to move with the changing climate. Over the long history of the earth, changes in global climate have been brought about by the drifting of continents and associated changes in oceanic circulation. Where physical barriers to dispersal prevented distributions of species from following shifts in climate belts, local populations have become extinct. For example, drastic changes in climate during the Ice Age, combined with barriers to dispersal in southern Europe, were responsible for today's impoverished European flora and fauna (■ Figure 25.11).

The effects of habitat fragmentation may be compounded by the increasing rate of climate change being caused by global warming. This anthropogenic change in temperature, which may amount to a rise of 2°–6°C over the next 50 years, could equal the warming of the earth's climate since the last glaciation, only 50 times faster (see Chapter 26). It is likely to cause the extinction of many species, particularly plants with narrow temperature tolerances, that are prevented from shifting their range distributions by habitat fragmentation.

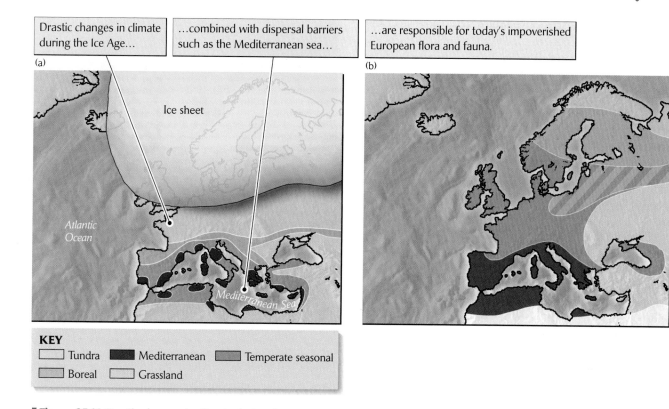

Drastic changes in climate during the Ice Age...

...combined with dispersal barriers such as the Mediterranean sea...

...are responsible for today's impoverished European flora and fauna.

(a)

(b)

Ice sheet

Atlantic Ocean

Mediterranean Sea

KEY
- Tundra
- Mediterranean
- Temperate seasonal
- Boreal
- Grassland

❚ **Figure 25.11 Drastic changes in climate during the Ice Age contributed to Europe's low species richness.** Maps of vegetation types in Europe (a) at the height of the last glaciation and (b) today show how the location of the Mediterranean Sea could have prevented some species from moving southward with the climate. After J. Blondel and J.-D. Vigne, in R. E. Ricklefs and D. Schluter (eds.), *Species Diversity in Ecological Communities,* University of Chicago Press, Chicago (1993), pp. 135–146.

Small population size

Just by chance, every population experiences variations in births and deaths during any particular period. These cause what is known as stochastic, or random, variation in population size (see Chapter 15). The magnitude of this variation varies inversely with the number of individuals in a population. Very small populations, such as those isolated in restricted fragments of habitat, may become extinct just by chance if they suffer a series of very unlucky years. This phenomenon is referred to as stochastic extinction, and although it is relatively unlikely except in the smallest populations, its probability increases with fragmentation of suitable habitat. It is a particular threat to species, such as large predators, that normally have low population densities.

Small population size may further increase the probability of extinction by reducing genetic variation in a population (see Chapter 16). A small population contains a smaller proportion of the species' gene pool than a larger population. Furthermore, inbreeding, or mating among close relatives, tends to reduce genetic variation. If a population goes through a population bottleneck and loses

genetic variation, it may not have the capacity to respond to rapid change in the environment. The collared lizard, for example, lives in small populations—generally 20–50 individuals—in isolated glades of xeric habitat on rocky outcrops in the Ozark Mountains of Missouri (see Figure 13.3). This lizard is a resident of Southwestern deserts that colonized the Ozarks during a period of hot, dry climate 4,000 to 8,000 years ago. Genetic surveys have shown that the Ozark lizards are genetically uniform within populations, but differ between populations. This is exactly the pattern expected to result from random loss of genetic diversity within small populations.

It is difficult to generalize about problems resulting from population bottlenecks because there are several cases of species that have been reduced to near extinction and have lost much of their genetic variability, but have recovered with spectacular growth when protected. The northern elephant seal is a case in point (see Figure 14.12). By 1890, hunting had reduced its once numerous population to about 20 individuals. Since then the population has increased explosively, passing 30,000 in 1970 and extending throughout much of the species' former range in California and

Mexico. Several years ago, investigators could not detect any genetic differences between individuals within the species, though they used tests that reveal ample genetic variation in other species of mammals. Similarly, one of Africa's large cats, the cheetah, has no detectable genetic variation within its population (see Figure 16.6). Nonetheless, although its reproductive success appears to be somewhat impaired in captivity, the cheetah population appears to be healthy and self-sustaining where it is not persecuted by humans.

Introductions of exotic species

Ultimately, of course, habitat reduction may cause extinction by wiping out suitable places to live. Animals of the forest will disappear when all the forest has been cut. Even where suitable habitat remains, however, conditions within the habitat may change, causing a population to begin a decline toward extinction. Frequently, a decrease in habitat quality can be traced to the introduction of predators, competitors, or disease organisms—that is to say, biological agents of change. Over the last 200 years, North America has been invaded by more than 70 species of fish, 80 species of mollusks, 2,000 species of plants, and 2,000 species of insects. These have arrived either accidentally—for example, in ship ballast—or as a result of deliberate introductions of crops, ornamentals, game species, or biological control agents.

Many of the areas most severely affected by exotics are isolated islands to which organisms have been introduced from more diverse continental biotas. The island organisms, having evolved in the absence of the new predators, competitors, or pathogens, are often poorly adapted to cope with them. The brown tree snake (*Boiga irregularis*), introduced from Asia, has literally eaten most of the endemic land birds of Guam to extinction. The Hawaiian Islands have also suffered greatly from introductions of exotics. The native Hawaiian tree snails have fallen prey to introduced predatory snails (❚ Figure 25.12). Major causes of mortality in Hawaiian birds have been malaria and pox virus—which would not have been a problem if the mosquito that transmits these diseases had not also been introduced to the islands. Native forests of Hawaii have also suffered from invasion by aggressive, weedy species, which crowd out native species.

Continental areas have also been vulnerable to exotics, which may escape the natural controls of predators, parasites, and herbivores that keep their populations in check within their native ranges. Purple loosestrife (*Lythrum salicaria*) and Japanese honeysuckle (*Lonicera japonica*), both introduced as ornamental plants, now dominate wetlands and forest understory vegetation in much of eastern North America. The subtropical climate of southern Florida is ideal for the introduced *Melaleuca* and Brazilian pepper

❚ **Figure 25.12 The introduction of exotic species often has negative effects on native biota.** Hawaii's native tree snails are being eliminated by introduced predatory snails like this one. Photo by Bob Gossington/Bruce Coleman.

(*Shinus terebinthifolius*), which cover immense areas and have crowded out native vegetation wherever they have spread. Exotic insects have also naturalized successfully. European honeybees were introduced to North America and elsewhere in the world to pollinate crops. They have been extremely beneficial to agriculture, but have also displaced native bees in many areas. Ironically, naturalized honeybee populations have declined dramatically in recent years owing to an introduced parasitic mite. Other insects, such as fire ants and gypsy moths, have caused drastic changes in local ecosystems, some of which were already altered early in the twentieth century by Dutch elm disease and chestnut blight.

Aquatic ecosystems seem to be particularly vulnerable to exotic species, perhaps because of the dominance of top-down control by predators and the high rate at which energy and nutrients are processed in aquatic food webs. In the first chapter of this book we saw the dramatic effect of the introduced Nile perch on Lake Victoria, in the Rift Valley of Africa. Countless numbers of endemic cichlid fish were driven to extinction. As the abundance of cichlids, many of which were zooplankton grazers, decreased, zooplankton abundance increased, and primary production by phytoplankton dropped to a low level, dramatically altering the entire ecosystem of the lake. A large share of the lake's nutrients are now taken up by introduced water hyacinth plants, which form continuous mats over thousands of hectares of the lake surface.

Opossum shrimp (*Mysis relicta*) were introduced to Flathead Lake in Montana to increase food supplies for salmon and other game fish. Unfortunately, the exotic shrimp ate so much zooplankton that the productivity of the lake actually decreased, and populations of salmon and the bald eagles that fed on them declined.

Zebra mussels (*Dreissena polymorpha*) arrived in the North American Great Lakes in 1988 from the Caspian Sea, apparently in ship ballast water. They have now become so abundant in lakes and rivers that they are crowding out native species, severely depleting phytoplankton food sources, and causing billions of dollars in damage to boats, dams, power plants, and water treatment facilities. North America has also supplied its share of exotics to the rest of the world. The comb jelly (*Mnemiopsis leidyi*) arrived in the Black Sea from North American coastal waters in 1982. By 1989, this exotic made up over 95% of the biomass of aquatic organisms in the Black Sea. There is no sign that introductions of exotic species are slowing down. Sadly, this homogenization of much of the world's biota is occurring at the expense of endemic species.

Overexploitation

Weapons and other harvesting tools, such as kilometer-long drift nets, have made humans such efficient hunters that many species have literally been hunted to extinction. Within recent history, North American losses have included the Steller's sea cow, great auk, passenger pigeon, and Labrador duck—all formerly abundant species, all prized for food, all vulnerable, and all slaughtered mercilessly until the last were gone. Long-range fleets and improved fishing technology have reduced catches of some fish and changed the composition of the catch on Georges Bank off New England and maritime Canada, once one of the richest fishing regions in the world (❚ Figure 25.13).

Extinction by overhunting and overfishing is not, however, a recent phenomenon. Wherever humans have colonized new regions, some elements of the fauna have

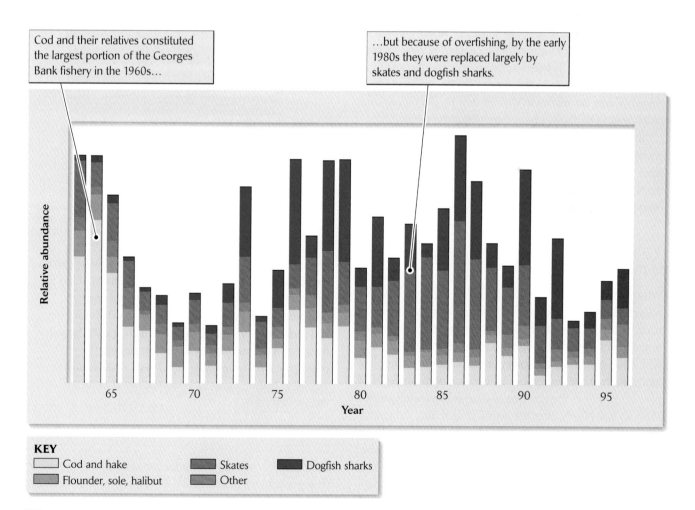

KEY

- Cod and hake
- Flounder, sole, halibut
- Skates
- Other
- Dogfish sharks

❚ **Figure 25.13 Overexploitation often changes the species composition of a community.** Whereas cod were abundant in catches on Georges Bank until the early 1960s, they have since been largely replaced by skates and dogfish sharks. From M. J. Fogarty, and S. A. Murawski, *Ecological Applications* 8(1) Supplement: S6–S22 (1998).

suffered. For example, early human populations in the Mediterranean region ate large quantities of tortoises and shellfish, which were easy to catch. As supplies of these foods were depleted, about 30,000 years ago in what is now Italy and about 15,000 years ago in what is now Israel, those populations were forced to switch to hunting hares, partridges, and other small mammals and birds (▌Figure 25.14).

Shortly after aboriginal people colonized Australia some 50,000 years ago, several large marsupial mammals, flightless birds, and a tortoise disappeared from the island continent. The advent of humans in the Americas about 12,000 years ago was accompanied by the rapid extinction of 56 species in 27 genera of large mammals, including horses, a giant ground sloth, camels, elephants, the saber-toothed tiger, a lion, and others. Madagascar, a large island off the southeastern coast of Africa, received its first human inhabitants only 1,500 years ago, yet this event brought the demise of 14 of 24 species of lemurs (mostly large species suitable for food) and between 6 and 12 species of elephant birds, flightless giants found only on Madagascar. A recent analysis suggested that fewer than 1,000 Polynesian colonists of New Zealand hunted 11 species of moas (large flightless birds) to extinction in less than a century at about the same time. Similar extinctions occurred widely on islands in the western Pacific Ocean and on Hawaii as humans spread throughout the region.

In each of these cases, a technologically superior species encountered populations unaccustomed to hunting pressures. Their lack of defenses and failure even to recognize the danger spelled disaster for these species; lack of restraint on the part of their hunters turned disaster into extinction.

Vulnerability to extinction is poorly understood

Why do some species seem more vulnerable to anthropogenic extinction than others? This question has been difficult to answer. Clearly, species that attract the attention of human exploiters are brought under great pressure. In addition, species that have evolved in the absence of hunting (particularly those on remote islands lacking most types of predators) and in the absence of diverse disease organisms seem to fare poorly after the arrival of humans. Vulnerability is also associated with limited geographic range, restricted habitat distribution, and small local population size.

But what makes one species rare and locally distributed when a close relative that exhibits superficially similar adaptations is abundant and widely distributed? The difference between success and failure in natural systems may hinge on very small differences in breeding success or longevity—perhaps too small for us to detect in studies of natural pop-

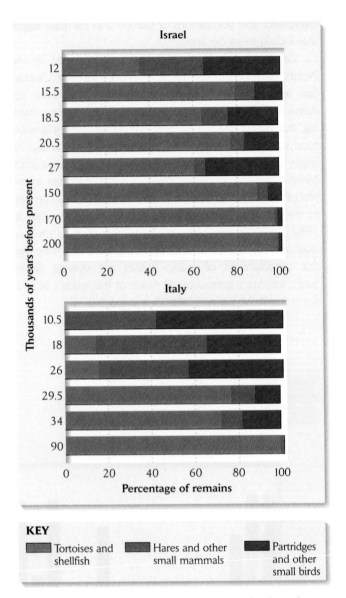

▌**Figure 25.14 The overexploitation of some food species has forced people to switch to others.** Remains of food items in Paleolithic archeological sites in Israel and Italy document a shift from easily caught prey, such as tortoises and shellfish, to prey requiring more hunting skill. The shift occurred as the former prey types were depleted by overexploitation. From data in M. C. Steiner, N. D. Munro, T. A. Surovell, E. Tchernov, and O. Bar-Yosef, *Science* (8 January) 283:190–194 (1999).

ulations. Most species persist for a million or more years, so their populations must be fully self-sustaining and capable of recovering from setbacks inflicted by a variable world. The forces that cause a population to embark on a decline to extinction may be very subtle. So far, ecologists have been able to say little on this point. If rarity is related to specialization, we must still ask what forces promote specialization, especially considering that close relatives of rare species may be abundant and widespread. What ecologists can address, however, is the problem of reversing the declining population trends of species that, in the absence of anthropogenic pressures, would be self-sustaining.

Conservation plans for individual species must include adequate habitat for a self-sustaining population

The straightforward way to maintain a population of a particular species is to guarantee the existence of a sufficient area of suitable habitat that can be kept free of exotic competitors, predators, and diseases. In practice, the design of nature preserves must take into account the ecological requirements of the species and the amount of space needed to support a **minimum viable population** (MVP)—the smallest population that can sustain itself in the face of environmental variation. The minimum viable population must be large enough to remain out of danger of stochastic extinction brought about by chance events. The population also must be distributed widely enough that local calamitous events, such as hurricanes and fires, cannot affect the entire species. At the same time, some degree of population subdivision may prevent the spread of disease from one part of a population to another.

Guaranteeing suitable habitat becomes more complex when a population has different habitat requirements during different seasons, or when it undertakes large-scale seasonal migrations. In the Serengeti ecosystem of East Africa, patterns of rainfall distribution and plant growth vary seasonally within the region. Huge populations of grazers, such as wildebeests, zebras, and gazelles, undertake long-distance seasonal movements in search of suitable grazing (see Figure 9.13). It would not be possible to isolate a part of this area as a preserve, because these populations need the entire area of the Serengeti ecosystem at different times of the year. Likewise, thundering herds of buffalo can never be restored to American prairies, because their migration routes are now blocked by miles of fencing and habitat converted to agriculture. Buffalo survive in a few small preserves in the American West—most notably the Greater Yellowstone ecosystem—but the natural environment of the buffalo has been irrecoverably lost (■ Figure 25.15).

Long-distance migration poses similar problems for the conservation of many types of birds. Wading birds, such as sandpipers, breed on arctic tundra, but maintenance of their populations also depends on conservation of the beaches and estuaries that they use during spring and fall migrations and as wintering grounds (■ Figure 25.16). Many American songbirds, whose populations have been declining during recent decades, spend their winters—wisely, it would seem—in forests of Central and South America. Their populations have been placed in double jeopardy by forest fragmentation throughout much of their breeding range in North America and by extensive clearing of forests and spraying of pesticides such as DDT in Latin America. Migration systems between Europe and Africa and between Siberia and Southeast Asia face the same problems, exacerbated by the toxic pesticides used throughout these regions.

When threats of extinction come from the dwindling of suitable habitat for a particular species, conservation strategy is relatively straightforward: the habitat should be preserved.

■ **Figure 25.15 Habitat changes caused by human land use limit conservation strategies for many species.** Prairies throughout the American West are now fenced and used for crops and cannot be set aside for buffalo and elk. The Greater Yellowstone ecosystem is one of the few places where these animals have access to large areas of grazing land. Photo by Fred Bruemmer/DRK Photo.

Figure 25.16 Populations that undergo long-distance and seasonal migrations present conservation challenges. Sandpipers and turnstones feed on eggs laid by horseshoe crabs during May along the shores of Delaware Bay. These eggs are a major food source for migrating shorebirds. This habitat is as necessary for the birds' survival as their breeding and wintering grounds. Photo by John Bova/Photo Researchers.

But this may be expensive and politically difficult to achieve. It is also impractical to develop a conservation strategy for every species, and the well-being of the majority will necessarily depend on conservation efforts directed toward a few of the most critically endangered or conspicuous species. As habitat preservation becomes more and more the focus of conservation efforts, however, it becomes especially important to identify habitats that are most critical to maintaining species diversity as a whole and to determine the area of habitat required to maintain minimum viable populations of most species. Each decision about a species or a habitat will depend on value judgments. What determines which species should be saved? How is their "value" measured?

Critical areas for conservation

What makes an area critical for conservation? The most valuable areas are those that provide havens for the largest number of species not represented elsewhere—such as the biodiversity hotspots described earlier in this chapter. Thus conservation value reflects a combination of local diversity and endemism. As a rule, endemism is highest on oceanic islands, in the Tropics, and in mountainous regions. Thus, such localities as Madagascar and the Hawaiian, Galápagos, and Canary islands are extremely critical ones for conservation. Extensive surveys of biodiversity resources in continental regions are beginning to identify critical areas there, but these efforts are continually hampered by lack of detailed information and by conflicting values attached to different components of biodiversity.

On a continent, preserves, whose number and area are necessarily limited by economic considerations, must target habitats and areas of special biological interest. From the standpoint of biodiversity, more is to be gained by setting aside several small preserves spread out over a variety of habitats and areas of high endemism than by preserving an equal area within a single habitat type. Values other than biodiversity may dictate the setting aside of large preserves, however, and we shall discuss some of them in the next chapter. One thing is certain: the cost of setting aside larger and larger amounts of habitat increases out of proportion to the area itself. This is so simply because land that is least expensive in terms of economic, social, and political values is set aside first. As more land is added to a preserve system, the cost of acquiring it, in terms of purchase price and potential resources forfeited, invariably increases. It is no accident that most parks and preserves are located in remote, underpopulated areas and that establishing a preserve becomes more difficult when it conflicts with economic interests.

One example of this conflict involved the setting aside in 1968 of a large tract of old-growth redwood forest in northern California as Redwood National Park, which was opposed by the timber industry. In this case, the uniqueness of the redwood habitat and its rapid conversion into managed tree farms greatly increased the value to society of setting aside a large area of this habitat for posterity. A similar controversy surrounds the old-growth Douglas fir forests of Washington and Oregon, where the fates of such unique inhabitants as spotted owls and marbled murrelets come into conflict with the local timber economy (**Figure 25.17**).

▌Figure 25.17 Conflicting interests and values present obstacles to conservation. In Washington and Oregon, the fate of the spotted owl has come into conflict with the timber industry. Main photo by Tom & Pat Leeson/Photo Researchers; inset photo by Janis Burger/Bruce Coleman.

Many tropical countries, particularly in Central and South America, are in the enviable position of having large tracts of uncut forest and relatively undisturbed tropical habitats of other kinds. These habitats have been protected in the past by their geographic remoteness and by the small size of local human populations. It is still possible to set aside large parks and preserves in such countries as Brazil, Ecuador, Peru, and Bolivia, and several governments have moved rapidly during the past decade to preserve tracts of what remains. The problem is complicated, however, by the rapid growth of the human population, by increased exploitation of forest products, and by the conversion of forests to agriculture. Such exploitation is justified by a legitimate need to feed people and generate export income for economic development. Thus, the price of conservation is rising rapidly in much of the world, and many developing

countries are unable to foot the bill. Even when lands are set aside "on paper," many countries cannot afford to protect them from squatters, poachers, and politicians who grant mining and logging concessions within protected lands in order to extract short-term profits. For this reason, conservation must be an international effort, and the wealth of the developed countries must be shared globally to protect the earth's biodiversity. One of the most successful recipes for conservation areas is to involve local people in their design and management so that the benefits of conservation become tangible and economically compelling.

Design of nature preserves

In some situations, those who design preserves have some latitude in deciding just how to draw the boundaries of a park or conservation area. Here, ecological principles derived from the theory of island biogeography (see Chapter 23) can help planners to arrive at the best design. There are three guiding principles: the species–area relationship, the avoidance of edge effects, and freedom to migrate. As we saw in Chapter 21, large areas support more species than small areas because large population sizes of individual species reduce the chances of stochastic extinction, promote genetic diversity within populations, and buffer populations against disturbances. In preserving a single type of habitat, edges should be minimized because the effects of habitat alteration extend for some distance beyond the areas directly altered (see Chapter 13).

According to these considerations, when a preserve is to be carved from an area of uniform habitat, such as a broad expanse of tropical rain forest, (1) larger is better than smaller, (2) one large area is better than several smaller areas that add up to the same total size, (3) corridors connecting isolated areas are desirable, and (4) circular areas are better than elongate ones with much edge. However, faced with choosing between a single large area of uniform habitat and several smaller areas, each in a different habitat, planners should remember that the smaller areas will often contain a greater total number of species among them because endemic species may be found in one habitat, but not the others.

As always, nature preserves must be designed in accordance with the habits of their inhabitants, and requirements for special features (such as nesting sites, water holes, and salt licks) must be taken into account. In mountainous areas, many species undertake altitudinal migrations over the seasonal cycle, and so preserves set aside at different elevations must be connected by suitable corridors for travel. Roads and pipelines set in the way of migratory movements or dispersal must be bridged in some manner to allow passage.

Some critically endangered species have been rescued from the brink of extinction

Many species have come so close to extinction that their preservation has required exceptional human intervention. Such efforts, which may cost millions of dollars, usually are directed toward species that appeal to the public. Some may question the wisdom of spending several million dollars (as happened recently) to free three gray whales trapped in arctic ice. But the incident dramatized the empathy that many humans feel for the plight of some other creatures. And many people, though perhaps less enthusiastic about spending so much to rescue individuals, are willing to devote considerable resources to the preservation of species.

In recent decades, zoological parks have become increasingly involved in maintaining viable, genetically diverse populations of species that are endangered or even extinct in the wild. Eventually, with the development of suitable preserves, many of these populations could be reintroduced into natural settings.

As the population of California condors in southern California dwindled below thirty and then below twenty individuals in the wild during the 1970s and early 1980s, management personnel made the difficult decision to bring the entire population into captivity (∎ Figure 25.18). In specially constructed breeding facilities located at the Los Angeles and San Diego zoos, the birds are protected from several mortality factors that were destroying the wild population: indiscriminate shooting, lead poisoning from slugs left in deer carcasses upon which condors fed, and poisons and traps set out for coyote and rodent control, to which condors were attracted by baits or poisoned carcasses. In addition, condors in captivity can be induced to lay up to three eggs a year, instead of the usual one, and most of the chicks are reared successfully. The objective of such a captive rearing program is to produce young that can be reintroduced into their native habitat. Such a program is costly, and its success ultimately depends on controlling those mortality factors that threatened the population in the first place, which often requires legislation, land purchases, and public education. In the particular case of the California condor, the program's success can be properly judged only after 30 or 40 years and the expenditure of tens of millions of dollars. At this time, however, the number of California condors has increased to more than 150 individuals, many of which have been released into the wild in California and Arizona.

∎ **Figure 25.18 Some endangered species have been brought back from the brink of extinction.** An intensive captive rearing program is intended to re-establish wild populations of the California condor in the western United States. Photo by Tom McHugh/Photo Researchers.

The California condor can be saved from extinction, as can the Hawaiian crow, the black-footed ferret, and many other species that are now the subject of captive breeding and release programs. The experience gained through these programs will be useful to similar efforts in the future. The California condor program, like other such programs, has heightened local residents' awareness of conservation issues and has resulted in the preservation of large tracts of habitat in mountainous regions of southern California and elsewhere. People have also come to understand that as long as care is taken, viable condor populations are compatible with other land uses, such as recreation (as long as human access to nesting sites is restricted), hunting (as long as steel rather than lead bullets are used), and ranching (as long as coyote and rodent control programs, if they are to persist at all, are condor-safe). Concessions to condors are neither difficult nor expensive. Making them simply depends on instilling values that acknowledge natural systems as an integral part of the environment of humankind.

Summary

1. Humankind has an immense impact on the earth, managing or otherwise affecting most of its land surface and waters. Human activities have caused deterioration in ecological systems and the extinctions of many species. The repercussions are accelerating as the human population grows beyond 6 billion individuals and the per capita consumption of energy and resources increases apace.

2. The environmental crisis cannot be fully resolved until human population growth is stopped, consumption of energy and resources declines, and economic development takes ecological values into consideration.

3. Of immediate concern is the preservation of biodiversity, which encompasses the variety of living beings—plants, animals, and microbes—on earth. The concept of biodiversity recognizes genetic diversity within and between populations and acknowledges the special value of areas that are inhabited by large numbers of endemic species.

4. The value of individual species is rooted in generalized moral considerations, in aesthetics, in the economic and recreational benefits we derive from them, and in their role as indicators of environmental deterioration. Diversity itself may also help to stabilize ecosystem function in the face of environmental variation.

5. Background extinction consists of natural extinctions resulting from environmental change and from the evolutionary turnover of species within communities. Mass extinctions, which appear episodically in the fossil record, reflect calamitous events in earth history, including impacts of extraterrestrial bodies. Anthropogenic extinction is the disappearance of species as a result of human activities: habitat destruction, overexploitation, introduction of exotic species, and pollution of various kinds.

6. Reduction of habitat area may hasten a population's decline toward extinction by making it more vulnerable to stochastic, or random, changes in population size or by causing reduced genetic variability and thereby impairing the capacity of the population to survive environmental change.

7. Introduction of exotic species has been a major factor in the decline and extinction of some native populations and in dramatic changes in some ecosystems.

8. Optimally designed nature preserves should include a high proportion of endemic species. For a given area of uniform habitat, preserves should consist of a single large area (rather than be dispersed in several small areas) to reduce the chances of stochastic extinction due to small population size, and they should be close to circular in shape to reduce edge effects.

9. In extreme cases, individual species can be rescued from the brink of extinction by massive recovery efforts that may include habitat restoration and captive breeding. Such costly programs, although they are focused on individual species, often highlight more general conservation problems and result in the conservation of habitat whose ecological value greatly exceeds that of the individual species it was preserved to save.

PRACTICING ECOLOGY

CHECK YOUR KNOWLEDGE

A Road Runs Through It

As we have read throughout this and previous chapters, factors that increase death rates over birth rates will lead to a decrease in population size, and can lead to localized extinction of populations. Human activities are increasing the death rates and thus the threat of extinction for many species. The impact of roads on plant and animal life—a major cause of this increase—has recently received increased attention from researchers. Roads, and road building, can affect plants and animals in many ways: by increasing mortality during construction; by occasions of collisions with vehicles; by changing the behavior of animals (including pollinators and seed dispersers); by enhancing the invasion of non-native species; and by increasing encroachment and disturbance by humans. Findings concerning the impacts of roads on ecosystems convinced the U.S. Forest Service in May 2000 to propose a limit on new construction of logging roads in roadless areas.

What is the actual area affected by roads in the United States? Richard Forman of Harvard University has investigated this question, and the short answer is that it is considerable. A recent study compared several ecological impacts of roads in The Netherlands and Massachusetts and identified a "road-effect zone." Forman estimates that approximately 20% of the land area of the United States is directly affected by public roads. One option for reducing the detrimental effects of roads on wildlife is to provide highway underpasses or corridors to avoid road crossings and to increase dispersal among fragmented populations. Anthony Clevenger, of the University of Alberta, and Nigel Waltho, of York University, Toronto, studied the effectiveness of such underpasses in Banff National Park in Alberta, Canada. They compared 14 different factors associated with the underpass structure, neighboring landscape features, and human activities in the area of 11 underpasses. The number of crossings by black bears, grizzly bears, cougars, wolves, deer, elk, and

Table 25.1 Factors affecting the use of underpasses by different species in Banff National Park, Alberta

Attribute	Carnivores	Ungulates
Noise level	+7	+2
Forest cover	−5	−1
Distance to town	+1	+13
Combined human uses	−3	−9
Bicycle traffic	−8	−11
Horse traffic	−4	−4
Foot traffic	−2	−6

The sign indicates a positive or negative influence, and the number indicates the relative strength of the influence. Data are adapted from Clevenger and Waltho (2000).

moose was determined by setting up "tracking sections," belts of sand, clay, and silt placed at the entry and exit of each underpass. After a few days, the tracking sections were checked and paw and hoof prints were identified to species and counted. These numbers were compared with the expected frequencies of such crossings estimated from radio telemetry, pellet counts, and habitat–suitability indices. Not surprisingly, species responded differentially to disturbance factors in the vicinity of these underpasses, as you can see in Table 25.1.

CHECK YOUR KNOWLEDGE

1. In addition to estimation and monitoring of species population dynamics, how can knowledge of the ecology of a species help in its conservation?

2. From the data shown in Table 25.1, what factors lead to the greatest use of underpasses by carnivores? By ungulates? How do you explain the strong negative effects of bicycle traffic?

MORE ON THE WEB | 3. Visit the U.S. Forest Service Web page on the "Roadless Area Initiative" through *Practicing Ecology on the Web* at *http://www.whfreeman.com/ricklefs*. What are the three aspects of the initiative? How will this proposal affect plant and animal conservation?

 Suggested Readings

Angermeier, P. L., and J. R. Karr. 1994. Biological integrity versus biological diversity as policy directives. *BioScience* 44:690–697.

Burney, D. A. 1993. Recent animal extinctions: Recipes for disaster. *American Scientist* 81:240–251.

Ceballos, G., and J. H. Brown. 1995. Global patterns of mammalian diversity, endemism, and endangerment. *Conservation Biology* 9:559–568.

Cincotta, R. P., J. Wisnewski, and R. Engelman. 2000. Human population in the biodiversity hotspots. *Nature* 404 (27 April): 990–992.

Clevenger, A. P., and N. Waltho. 2000. Factors influencing the effectiveness of wildlife underpasses in Banff National Park, Alberta, Canada. *Conservation Biology* 14:47–56.

Diamond, J., and T. J. Case. 1986. Overview: Introductions, extinctions, exterminations, and invasions. In J. Diamond and T. J. Case (eds.), *Community Ecology*, pp. 65–79. Harper & Row, New York.

Fogarty, M. J., and S. A. Murawski. 1998. Large-scale disturbance and the structure of marine systems: Fishery impacts on Georges Bank. *Ecological Applications* 8(1) Supplement: S6–S22.

Foin, T. C., S. P. D. Riley, A. L. Pawley, D. R. Ayres, T. M. Carlsen, P. J. Hodum, and P. V. Switzer. 1998. Improving recovery planning for threatened and endangered species. *BioScience* 48(3): 177–184.

Forman, R. T. T. 2000. Estimate of the area affected ecologically by the road system in the United States. *Conservation Biology* 14: 31–35.

Hadfield, M. G., S. E. Miller, and A. H. Carwile. 1993. The decimation of endemic Hawaiian tree snails by alien predators. *American Zoologist* 33:610–622.

Haig, S. M., J. R. Belthoff, and D. H. Allen. 1993. Population viability analysis for a small population of red-cockaded woodpeckers and an evaluation of enhancement strategies. *Conservation Biology* 7: 289–301.

Holdaway, R. N., and C. Jacomb. 2000. Rapid extinction of the moas (Aves: Dinornithiformes): Model, test, and implications. *Science* 287 (24 March): 2250–2254.

Houlahan, J. E., C. S. Findlay, B. R. Schmidt, A. H. Meyer, and S. L. Kuzmin. 2000. Quantitative evidence for global amphibian population declines. *Nature* 404 (13 April): 752–755.

Lawton, J. H., D. E. Bignell, B. Bolton, et al. 1998. Biodiversity inventories, indicator taxa and effects of habitat modification in tropical forest. *Nature* 391:72–76.

Leach, M. K., and T. J. Givnish. 1996. Ecological determinants of species loss in remnant prairies. *Science* 273:1555–1558.

Marmontel, M., S. R. Humphrey, and T. J. O'Shea. 1997. Population viability of the Florida manatee (*Trichechus manatus latiriostris*), 1976–1991. *Conservation Biology* 11:467–481.

Mills, E. L., H. H. Leach, J. T. Carlton, and C. L. Secor. 1994. Exotic species and the integrity of the Great Lakes. *BioScience* 44: 666–676.

Mills, L. S., M. E. Soulé, and D. F. Doak. 1993. The keystone-species concept in ecology and conservation. *BioScience* 43:219–224.

Myers, N., R. A. Mittermeier, C. G. Mittermeier, G. A. B. da Fonseca, and J. Kent. 2000. Biodiversity hotspots for conservation priorities. *Nature* 403:853–858.

Naeem, S. 1998. Species redundancy and ecosystem reliability. *Conservation Biology* 12:39–45.

Naeem, S., D. R. Hahn, and G. Schuurman. 2000. Producer–decomposer co-dependency influences biodiversity effects. *Nature* 403:762–764.

Noon, B. R., and K. S. McKelvey. 1996. Management of the spotted owl: A case history in conservation biology. *Annual Review of Ecology and Systematics* 27:135–162.

Pimentel, D., L. Lach, R. Zuniga, and D. Morrison. 2000. Environmental and economic costs of nonindigenous species in the United States. *BioScience* 50(1):53–65.

Pimm, S. L. 1991. *The Balance of Nature? Ecological Issues in the Conservation of Species and Communities*. University of Chicago Press, Chicago.

Primack, R. L. 1998. *Essentials of Conservation Biology*. 2d ed. Sinauer Associates, Sunderland, MA.

Redford, K. H. 1992. The empty forest. *BioScience* 42:412–422.

Robinson, S. K., et al. 1995. Regional forest fragmentation and the nesting success of migratory birds. *Science* 267:1987–1990.

Rolston, H. III. 1985. Duties to endangered species. *BioScience* 35:718–726.

Simons, T., S. K. Sherrod, M. W. Collopy, and M. A. Jenkins. 1988. Restoring the bald eagle. *American Scientist* 76:252–260.

Soulé, M. E. 1985. What is conservation biology? *BioScience* 35:727–734.

Soulé, M. E. (ed.). 1986. *Conservation Biology: The Science of Scarcity and Diversity.* Sinauer Associates, Sunderland, MA.

Spencer, C. N., B. R. McClelland, and J. A. Stanford. 1991. Shrimp stocking, salmon collapse and eagle displacement. *BioScience* 41:14–21.

Terborgh, J. 1989. *Where Have All the Birds Gone?* Princeton University Press, Princeton, NJ.

Terborgh, J. 1992. *Diversity and the Tropical Rain Forest.* Scientific American Library, W. H. Freeman, New York.

Tilman, D., and J. A. Downing. 1994. Biodiversity and stability in grasslands. *Nature* 367:363–365.

Tilman, D., J. Knops, D. Wedin, P. Reich, M. Ritchie, and E. Siemann. 1997. The influence of functional diversity and composition on ecosystem processes. *Science* 277:1300–1302.

Vitousek, P. M., C. M. D'Antonio, L. L. Loope, and R. Westbrooks. 1996. Biological invasions as global environmental change. *American Scientist* 84:468–478.

Wilson, E. O. (ed.). 1988. *Biodiversity.* National Academy Press, Washington, D.C.

CHAPTER 26

Economic Development and Global Ecology

Ecological processes hold the key to environmental policy

Human activities threaten local ecological processes

Toxins have accumulated in the environment

Atmospheric pollution threatens the environment on a global scale

Human ecology is the ultimate challenge

We have set aside, or have plans to set aside, large areas of the earth's natural environments as nature preserves and maintain their capacity for supporting species. But what of the 90% or so of the rest of the earth that has been, or soon will be, converted to supporting the human population—devoted to living space, food production, forestry, mineral production, hunting, and so on? What will the future of our own species be? Can the earth sustain an expanding human population indefinitely at a high quality of life? To what degree are human values compatible with natural values? Can managed ecosystems serve some of the same functions as natural ecosystems, or will nature preserves stand in stark contrast to completely altered environments dominated by humans and their domesticated species?

A sustainable biosphere is unlikely as long as the human population continues to grow. The earth offers no new regions to colonize. Except for portions of the humid Tropics, much of which cannot support dense human populations, most of the habitable areas of the earth have been filled. Further population increase will lead to further crowding, tearing not only the fabric of human society but also of the life-supporting systems of the environment.

Pessimism comes easily in the present environmental climate, but there is also plenty of room for optimism. Many programs for cleaning up the environment and protecting endangered species have been undeniable successes. In the United States, the Clean Air Act (1970), the Clean Water Act (1972), and the Endangered Species Act (1973) have resulted in considerable protection of the environment. Other countries have adopted similar legislation, and international agreements deal with such diverse issues as air pollution, trade in endangered species, and property rights for natural products. These successes have not been limited to the developed countries. Relatively straightforward ecological and engineering solutions exist for most environmental problems. Most important, people all over the globe share a deep concern for their environments. The challenge to ecologists is to provide the scientific information needed to develop social consensus, build political commitment, and inform decision making on issues concerning the environment.

Ecological processes hold the key to environmental policy

Throughout this book, we have discussed processes involved in biological production and in the regulation of communities and ecosystems. In the previous chapter, we saw how an understanding of ecological processes can contribute to the conservation of individual populations and species and of biodiversity more generally. Here we will consider the roles of ecological processes in sustaining the function of ecosystems. These processes occur in managed as well as natural ecosystems.

Two key aspects of ecosystem function are the harnessing of energy and the continual recycling of materials. In natural systems, the primary source of energy is sunlight; recycling is accomplished by a variety of regenerative processes, some of them physical or chemical, some of them biological. Any imbalance that leads to the accumulation or depletion of some component of an ecosystem is usually corrected by the ecosystem's self-maintaining dynamic processes. For example, when dead organic matter accumulates within a system, detritivores increase in numbers and consume the excess detritus. When herbivores increase to high numbers and begin to deplete their food resources, declining birth rates and increasing mortality check population growth and restore a sustainable relationship between consumer and resource.

These restorative processes may be physical, but more often they involve biological transformations. From the composition of the atmosphere to the most basic character of many environments, plants, animals, and microbes have greatly modified the condition of the earth's ecosystems and are responsible for maintaining their qualities. When natural processes are disrupted, ecosystems may not be able to maintain themselves. Nowhere is this more dramatically shown than in the changes that occur in the soil and streams following clearing of tropical rain forest (see Chapter 8). Thus, maintaining a sustainable biosphere re-

quires that we conserve the ecological processes responsible for its productivity.

Human activities threaten local ecological processes

All human activities have consequences for the environment. Fishing provides a good example. The goal is to harvest a food resource for human consumption. However, when we simply maximize short-term returns from a fishery, preferred fish stocks are reduced or even collapse, and our attention turns to other less desirable but nonetheless exploitable populations.

We can see this pattern in the commercial whale fishery, in which one species after another was hunted to near extinction. Humpback, right, bowhead, and gray whale populations were decimated during the nineteenth century. During the twentieth century, the whaling industry turned its attention to the now more profitable blue whales, the commercial catch of which declined to uneconomical levels during the middle of the century. Whalers then moved on to fin whales, which declined precipitously between 1965 and 1975, and progressed to less and less profitable species, such as sperm and, finally, sei whales (Figure 26.1). Under current laws, including a total moratorium on commercial whaling since 1985, some populations of whales, such as the gray, bowhead, and humpback, are increasing at rates of about 2% per year.

Other fisheries, such as the once immensely profitable sardine (*Sardinops*) fishery of western North America, have not recovered from overexploitation. It is possible that these fish stocks were pushed below the levels from which they could recover, and that the entire ecosystem has shifted so that it no longer includes the sardine as a prominent link in the food chain. Interestingly, following the collapse of the sardine fishery in California in the early 1950s, anchovies (*Engraulis*) have increased and are now the basis of an active fishery.

Often the consequences of human activities are less direct. To provide a relatively straightforward example, the clearing of watershed land for agriculture or timber frequently leads to erosion and deposition of silt downstream from the watershed over long periods. Thus, riverine habitats may be altered and reservoirs behind dams filled in. Erosion of logged land in Queensland, Australia, is causing damage to streams and to the Great Barrier Reef off the coast. We'll look at some more examples of both direct and indirect effects of human activities as we consider different kinds of threats to ecological processes.

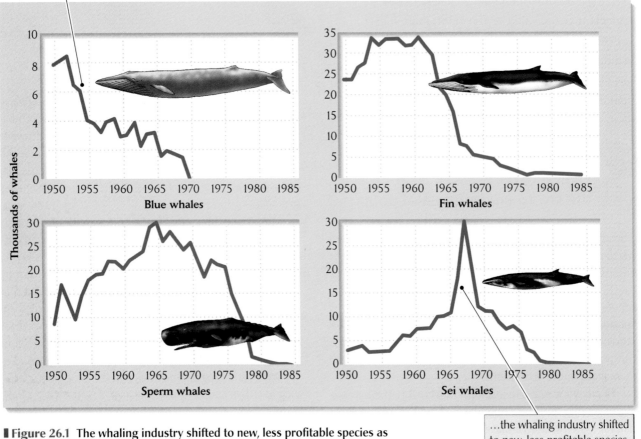

As numbers of large whales declined from overhunting...

...the whaling industry shifted to new, less profitable species.

▌Figure 26.1 The whaling industry shifted to new, less profitable species as populations of overhunted species dwindled. Commercial catches of four types of whales during the middle of the twentieth century illustrate this shift. After R. Payne, in W. Jackson (ed.), *Man and the Environment*, 2d ed., W. C. Brown, Dubuque, IA (1973), p. 143.

Overexploitation

Fishing, hunting, grazing, fuelwood gathering, logging, and the like are classic consumer-resource interactions. In most natural systems, such interactions achieve steady states because as a resource becomes scarce, efficiency of exploitation plummets. Consumer populations then begin to decline or seek alternative resources until consumers and their first resource are brought back into balance. Efficiency of exploitation and ability of resources to resist exploitation are characteristics of consumers and resources that have evolved over long periods of interaction (see Chapter 20).

In economic systems, consumer–resource interactions may also come into balance because as a resource becomes scarce and its price increases, demand for that resource drops; people either do without or find cheaper alternatives. However, because the human population's abil-

ity to exploit natural systems has been escalated out of all proportion by its ability to use tools, renewable resources may not become scarce until they are very nearly depleted and are unable to sustain even reduced exploitation. Technological skills have advanced too rapidly for nature to keep pace. Consequently, many ecosystems that historically supported the growth of the human population, such as the vast forests and prairies of North America, have been converted to other uses.

Where the fertility of the land has been exhausted (as in drier parts of Africa and increasingly elsewhere), the human population may lose the resource base it needs to support itself. Where the population can no longer move to other areas or shift to new food sources, the prospect of population control by starvation and associated diseases and social strife may become reality. Food shortages are already occurring in some parts of the world. In sub-Saharan Africa, over-

Figure 26.2 In many parts of the world, local land use has left the land unproductive. Grazing by goats has removed most of the groundlevel vegetation in this woodland in Madagascar. Photo by Walt Anderson/Visuals Unlimited.

grazing and fuelwood gathering have left little vegetation to support human or any other life. In other parts of the world, vast areas of formerly productive land have been laid waste by local land use practices (**Figure 26.2**).

Although much of the Tropics can sustain intensive agriculture, particularly in mountainous regions with volcanic soils, the high natural productivity of large portions of the Tropics depends on the presence of native vegetation (see Chapter 8). Because old lowland tropical soils contain few nutrients, the natural fertility of many tropical ecosystems is maintained by constant recycling of nutrients between detritus and living plants. Break the cycle by clear-cutting the forest and the nutrients are lost. Over much of the Amazon basin, forested land cleared for cattle grazing becomes so infertile that it must be abandoned after 3 years of ranching. After vegetation has been cleared and burned, harvesting the cattle takes away the last remaining nutrients from the system. To be sure, the forests will regenerate, but many decades or even centuries must pass before the natural fertility of the ecosystem is restored.

Many human populations in the Tropics maintain themselves by the practice of "shifting agriculture," in which small patches of forest are cut and burned to release nutrients into the soil, planted for 2 or 3 years, and then abandoned in favor of a new patch (**Figure 26.3**). A particular patch typically recovers sufficiently to repeat this process in 50–100 years. Accordingly, as long as only 1–2% of the forest is cut each year, and thus perhaps 2–6% is under cultivation at any one time, the land can sustain this

practice. This type of agriculture requires little input of labor, materials, or energy and takes advantage of the natural successional processes of tropical forests, but it supports only sparse human populations. When the land is cultivated more intensively by tilling, fertilization, watering, and weeding, productivity and long-term sustainability may increase greatly, but so do the inputs of labor and materials, and so does the cost of the agricultural products.

Many tropical regions lack abundant mineral resources, and their people must rely on agricultural exports to pay for more expensive staple foods and to import the manufactured goods that are considered essential to a high standard of living. Therefore, land has been cleared to grow cash crops for export to other parts of the world: coffee, sugar, bananas, and beef, to name a few. While this kind of agriculture is sustainable on some soils, for much of the Tropics it means a decline in the natural productivity of the environment and precludes alternative sustainable land uses.

There are many ways to reduce these problems. Among the most effective are limiting exploitation of resource populations to maximum sustainable yields, considering alternative sustainable uses of land, increasing agricultural intensity on land that will bear it, and improving distribution of food between areas of production and areas of need. Most of these solutions carry a price tag. Planning for sustained use cuts short-term returns, and these losses must be made

Figure 26.3 Shifting agriculture can be a sustainable practice. Small subsistence farms, such as this one in Dominica, can provide much of a family's food requirements without overstressing the environment, as long as population density is low. Photo by Tom Bean/Corbis.

up elsewhere to maintain the wealth of a country. Increasing agricultural intensity requires disproportionately greater inputs of energy, labor, and chemical fertilizers, each contributing its own problems. Relying on crops selected or genetically engineered for high food production often makes agriculture more vulnerable to outbreaks of pests and disease.

Providing for larger populations also has human costs. Segments of the human population that are forced by impoverished land to import food also must earn money to buy it; otherwise, they will be reduced to welfare status and impose a burden on other segments of the population. As long as local human population growth is not tied to the ability of local resources to support it, imbalances between human consumers and their resources will continue to worsen.

Introductions of exotic species

Both intentionally and unintentionally, humans have taken other species everywhere they have traveled (see Chapter 25). Aborigines brought dingoes (semidomesticated dogs) to Australia; Polynesians brought rats to Hawaii. Of course, the global movements of species by human agency have increased immensely since Europeans began colonizing most of the world some 500 years ago. It has been estimated that 50,000 nonindigenous species have been introduced to the United States, some examples of which are shown in ▌Figure 26.4. These have included edible and horticultural varieties of plants, and their pests; commercially valuable trees; domesticated animals for work or meat; familiar backyard animals, especially birds; mammals for sport hunting; disease organisms; and such frequenters of human habitation and transportation as the ubiquitous cockroach and dandelion. The result of this movement of species between continents is a globally distributed flora and fauna of alien species that have in some cases displaced or otherwise wreaked havoc with local biotas. Fully 40% of the species listed in the United States under the Endangered Species Act are threatened because of competition, predation, parasitism, and herbivory by non-native species.

An extreme example of a country with a predominantly alien flora and fauna is New Zealand, most of the area of which is occupied by introduced plants and animals (▌Figure 26.5). The native forest was cut long ago and replaced by pines from North America and eucalyptus from Australia. Moas (large, flightless birds that grazed vegetation and ate fruit) were killed off by Maori natives before Europeans arrived, and sheep now take their place. Most birds of the countryside are those that were transplanted from England to stave off the homesickness of early colonists.

(a)

(b)

▌**Figure 26.4 Approximately 50,000 species have been introduced into the United States from Europe in the last 500 years.** (a) Loosestrife (*Lythrum salicaria*), introduced from Europe, now dominates vast areas of wetland habitat, such as this meadow in Great Meadows National Wildlife Refuge, Massachusetts. (b) Introduced zebra mussels cover the shell of a native freshwater mussel. Photo (a) by Jeff Lepore/Photo Researchers; photo (b) by Grant Heilman/Runk/Schoenberger.

Only at the southern tip of New Zealand do native forests of southern beech persist in wet and remote fiordlands. Of the total New Zealand flora of 2,500 species, fully 500 are naturalized introductions, and they account for most of the present vegetation. These aliens prospered for a variety of reasons. Most of the natural habitat in New Zealand had been greatly disturbed by logging, farming, and ranching,

(a)

(b)

▌Figure 26.5 Very few native landscapes remain in New Zealand. Most native forests (a) have been replaced by agricultural landscapes featuring plants and animals introduced from Europe (b). Photos by R. E. Ricklefs.

perch introduced to Lake Victoria in East Africa and the peacock bass introduced to Gatun Lake in Panama, can virtually eliminate entire trophic levels of smaller planktivorous fish. Consequently, densities of zooplankton increase dramatically and algae are cleared from the water, reducing the overall productivity of the environment. In this way, a single predator—referred to as a keystone predator in this case because of its critical place in ecosystem function—can shift the character of a habitat from one state to a qualitatively different state.

Habitat conversion

Altering the basic nature of a habitat often upsets natural processes of regeneration and control. As we have already seen, cutting a tropical forest on impoverished soil breaks the tight cycling of nutrients that maintains forest productivity. It also greatly alters the physical structure of the soil by exposing it to increased leaching and sunlight. The productivity of the land decreases precipitately, and soil erosion may increase tenfold or more. In the Amazon basin, for example, erosion rates increased from 6–10 metric tons (T) per hectare per year in 1960 to 18–190 T per hectare per year by 1985, largely as a result of deforestation and overgrazing.

Problems associated with habitat conversion are not restricted to tropical forests. On the American prairies, plowing destroyed the dense root mats of perennial herbs that formerly held the soil together. A prolonged drought in the

which made invasion by weedy European species, accustomed to disturbance and intensely cultivated landscapes, relatively easy. Also, because of their comparatively low diversity and simple community structure, island ecosystems tend to be easier to invade than continental ecosystems—there are simply fewer native species to provide effective competition.

Although aliens may displace native plants and animals, they do not necessarily disrupt ecosystem function because they can assume the ecosystem roles of native species. However, the effects of introduced species are often difficult to predict. In aquatic systems in particular, introduced consumers at high levels in the food chain have seriously altered ecosystem function and have caused basic changes in community structure. Efficient predators, such as the Nile

Figure 26.6 Plowing of prairie land contributed to the creation of the Midwest's "dust bowl." This farmstead in the Midwest was abandoned during the height of the Dust Bowl period in 1937. Courtesy of the U.S. Department of Agriculture, Soil Conservation Service.

central United States during the 1920s and 1930s turned former prairies converted to agriculture into a devastated "dust bowl" of blowing soil (**Figure 26.6**). The global inventory of topsoil—the nutrient-rich upper layers of the soil profile—on croplands is about 6,500 gigatons (GT). Erosion from croplands is currently about 100 GT per year, or about five times faster than new topsoil is being formed. Thus, at the present rate, the earth's cropland topsoil reserves will be

depleted in an average of 65 years, forcing abandonment of now fertile agricultural regions and increasing dependence on chemical fertilizers. By some estimates, up to 1% of the earth's topsoil is lost to erosion every year (**Figure 26.7**).

Other kinds of environments are also being converted at a high rate. Mangrove forests provide natural protection for coastlines in many parts of the Tropics. Where they have been cleared for fuelwood, shrimp farming, and land reclamation, coasts have been laid bare to rampaging hurricane-driven floodwaters. Damming rivers brings the benefits of flood control, irrigation water, and power generation, but also increases silt transport, blocks fish migrations, alters downstream water conditions, and may even change the local weather.

Irrigation

Water makes the desert bloom. Humankind has employed various irrigation schemes to increase the productivity of land since the beginning of agriculture (**Figure 26.8**). Only recently, however, has irrigation been applied on immense scales to land that would otherwise be totally unsuitable for agriculture. The benefits are tremendous, but so are the costs, many of which surface only after years of profitable irrigation. The primary costs are the environmental effects of developing the dams, wells, canals, and dike work required to support irrigation; lowered water tables where wells are the source of irrigation water; reduction of groundwater quality through the introduction of pesticides and fertilizers or the concentration of naturally occurring toxic elements; the accumulation of salt in irrigated soils in arid zones; and transmission of diseases by aquatic organ-

(a)   (b)

Figure 26.7 Almost 1% of the earth's topsoil is eroded each year. Soil erosion and gully formation is likely to occur on plowed farmland (a, Whitman County, Washington) and heavily grazed pasture (b, Shelby County, Tennessee). Photos by Tim McCabe, courtesy of the U.S. Department of Agriculture, Soil Conservation Service.

(a)

(b)

▌**Figure 26.8** **Irrigation has both benefits and costs.** (a) Irrigation can turn desert into productive farmland, as furrow irrigation of cotton has done in the Imperial Valley of southern California. (b) However, the accumulation of salt in the soil that accompanies irrigation can damage crops, as it has in this irrigated alfalfa field in Colorado. Photos by Tim McCabe, courtesy of the U.S. Department of Agriculture, Soil Conservation Service.

isms. In most cases, the costs of delivering water to crops—including the burden of future environmental problems—are underwritten by the population at large through taxes and other subsidies.

Fertilization and eutrophication

Any substance that enhances the productivity of a habitat may be considered a fertilizer. We apply fertilizers to agricultural lands to increase crop production, but some portion of these chemicals makes its way into groundwater and from there into rivers, lakes, and eventually the ocean. Nitrates, phosphates, and other inorganic fertilizers have the same effect on rivers and lakes as they do on agricultural lands: they increase biological production (see Chapter 8). Overproduction as a consequence of this artificial fertilization, often called eutrophication, may cause a change in the species composition of a river or lake. Input of inorganic fertilizers may also turn clear, oligotrophic waters into turbid environments that are less attractive for recreation. Often, nutrient inputs upset seasonal cycles of nutrient use and regeneration in natural bodies of water, leading to accumulation of organic material, high rates of bacterial decomposition, and deoxygenation of the water. Under such conditions, fish may suffocate and contribute further to the load of organic material in the water (▌Figure 26.9).

Direct input of organic wastes, such as sewage and runoff from feedlots, poses a greater problem for water quality. Suspended or dissolved organic materials in water create what is known as biological oxygen demand, meaning that the decomposition of these materials by bacteria uses up

the oxygen present in the water. The organic inputs are unrelated to the natural productivity of the system and create conditions such that a stream or lake may become anoxic for long periods and unsuitable for many forms of life.

Before water pollution came under strict controls in North America and Europe, large sections of major rivers became completely anoxic, killing off local fish populations

▌**Figure 26.9** **Input of organic wastes can create anoxic conditions in aquatic environments.** This fish die-off in a stream near Colorado Springs, Colorado, was caused by oxygen deficiency related to organic pollution. Photo by Shane Anderson/Saturdaze.

and preventing the migration of other species, such as shad and salmon, between the ocean and their headwater spawning grounds. In general, cutting the off sources of organic nutrients, either by diverting the inputs to larger bodies of water that can absorb them or by improving treatment of sewage, can restore natural conditions. The costs of these solutions have been more than repaid in the long run by the benefits of enhanced water quality for fisheries, public health, and recreation.

Toxins have accumulated in the environment

Toxins are poisons that kill animals and plants by interfering with their normal physiological functions. Many toxins occur naturally, but human activities have increased their accumulation in the environment. Toxic substances can be divided into several classes: acids, heavy metals, organic compounds, and radiation are the most notable.

Acids

Anthropogenic sources of acids are primarily of two kinds. The first occurs in coal mining areas where reduced sulfur compounds associated with coal are exposed to atmospheric oxygen. Sulfur bacteria oxidize pure sulfur and thiol (reduced) forms of sulfur to sulfates, which may then be converted to sulfuric acid in streams that drain mining areas—hence the term **acid mine drainage.** In some places, the water becomes so acid that it sterilizes the aquatic environment (see Figure 7.15).

The second, and more widespread, problem is acid rain. Coal and oil are not pure hydrocarbons; they contain sulfur and nitrogen compounds as well. After all, these fossil fuels are the remains of plants and animals that, when living, contained nitrogen and sulfur in their proteins and other organic molecules. Burning coal and oil, in addition to producing carbon dioxide and water vapor, spews nitrous oxides and sulfur dioxide into the atmosphere. When these gases dissolve in raindrops, they are converted to acids and cause acid precipitation. In highly industrialized areas, the pH of rain may drop to between 3 and 4, which is 100 to 1,000 times the acidity of natural rain.

The consequences of acid rain have been severe in some regions, such as the northeastern United States, Canada, and Scandinavia. Rivers and lakes in these regions tend to be oligotrophic and thus do not contain dissolved bases to buffer acid inputs. Consequently, their pH may drop to as low as 4.0, acidic enough to stunt the growth of, or even kill, fish and other organisms. Acid rain may also lower the pH of soil, which increases the rate of leaching of soil nutrients and precipitates phosphorus compounds, making them unavailable for uptake by plant roots. The solutions to acid rain are primarily technological and economic: scrubbing offending gases from the effluents of power plants and automobiles, finding alternatives to burning fossil fuels for energy, and reducing total demand for energy. Even when sulfur and nitrogen emissions are reduced, however, ecosystems may not return to normal for decades or even centuries (see Figure 8.1).

Heavy metals

Even in low concentrations, mercury, arsenic, lead, copper, nickel, zinc, and other heavy metals are toxic to most forms of life. They are introduced into the environment in a variety of ways, principally as refuse from mining and mineral smelting (■ Figure 26.10), as waste products of manufacturing processes, as fungicides (such as lead arsenate), and through the burning of leaded fuel (although no longer sold in the United States, leaded gasoline is still used widely throughout the world). The effects of heavy metals are varied, but include interference with neurological function in vertebrates.

Many toxic heavy metals, including copper and nickel particulates released into the atmosphere by smelters, eventually accumulate in soils. The concentration of copper averages about 30 parts per million (ppm) in unpolluted temperate zone soils. Concentrations in excess of 100 ppm adversely affect mosses, lichens, and large fungi. Earthworm abundance drops off dramatically above 1,000 ppm, and most species of vascular plants cannot tolerate concentrations above 5,000 ppm (0.5%). As fungi die out, decomposition of organic matter and nitrification of organic nitrogen in the soil decrease. In one study in Sweden, in soils with copper levels of 2,000 ppm fungal populations had fallen to only 20–30% of their natural levels.

Concentrations of heavy metals above 1,000 ppm may extend 10–20 km from sites of metal smelting. Taller smokestacks, which distribute wastes over larger areas at lower concentrations, can mitigate these effects, but solving the problem will ultimately require a change in the technology of metal production to reduce toxic by-products.

Organic compounds

Some natural organic compounds, such as nicotine and pyrethrins, are used as agricultural pesticides, but most organic pesticides are far more deadly concoctions produced in the laboratory, to which pests have had no previous exposure and no opportunity to evolve resistance. The latter include organomercurials (such as methylmercury),

❚ **Figure 26.10 Toxic heavy metals are released into the environment by mining and mineral smelting.** High concentrations of metals in the waste material from this open-pit copper mine in Bingham, Utah, will prevent re-colonization by plants and the redevelopment of natural vegetation. Photo by Gene Ahrens/Bruce Coleman.

chlorinated hydrocarbons (DDT, lindane, chlordane, dieldrin), organophosphorus compounds (parathion, malathion), carbamate insecticides, and triazine herbicides. These compounds do their job in agriculture and pest management, but many accumulate in other parts of the ecosystem, where they adversely affect plant production and wildlife populations.

Modern pesticides and delivery systems are being designed for efficient use with minimal effects on the environment. Unfortunately, this progress is offset by increasingly widespread and more intense use of chemicals of all kinds in agriculture. Because insects and other pests may evolve resistance to pesticides, the benefits of pesticides are often short-lived, and their amounts must be increased to achieve continued results (see Chapter 9). Through ecological research, we can assess the vulnerability of natural systems to these pollutants, prescribe safe applications, and—perhaps most important—find suitable alternatives to humankind's chemical warfare with agricultural pests. Some microorganisms—some of them genetically engineered for special biochemical properties—can be used to metabolize pesticides and other toxic compounds to innocuous byproducts. This type of approach, using biological agents to clean up the environment and help restore habitats, is referred to as **bioremediation.**

Another type of toxic environmental pollution caused by organic compounds is that resulting from oil spills. Crude oil is a complex mixture of hydrocarbons, containing up to 1% nitrogen and 5% sulfur. Oil pollution occurs at the source in areas of oil production, rarely as a result of breaks in the nearly 100,000 km of oil pipeline in the world, and most frequently in the ocean due to offshore drilling and the wreckage of oil tankers (❚ Figure 26.11). These losses

(a)

(b)

❚ **Figure 26.11 Oil spills occur most frequently in the ocean.** (a) Oil slick from a damaged tanker approaching the Caribbean coast of Panama. (b) Fringe of dead mangroves killed by the oil spill. Photos by Carl C. Hansen, courtesy of the Smithsonian Tropical Research Institute.

amount to 3 to 6 million tons of oil annually, or about 0.1–0.2% of global oil production. Petroleum kills by coating the surfaces of organisms and, because hydrocarbons are organic solvents, by disrupting biological membranes. Over time, oil slicks disperse by evaporation of the lighter fractions, emulsion of other fractions in water, and weathering and microbial breakdown of the rest. But certain types of sensitive ecosystems, such as coral reefs, may take decades to recover fully.

Radiation

Radiation comes in a broad spectrum of energy intensities, ranging from generally harmless long-wavelength radio and infrared radiation, through damaging ultraviolet radiation of shorter wavelengths, to the extremely energetic cosmic rays and subatomic particles released by the disintegration of atomic nuclei (radioactive decay). Natural sources create an unavoidable background level of radiation. In some circumstances, natural radioactive substances, such as the radon gas present in soils in regions having granitic bedrock, can become concentrated and pose public health hazards. Such dangers are minor, however, compared with the extreme radiation hazards that could result from accidents at nuclear power plants, such as those that occurred at Three Mile Island, Pennsylvania, in 1979 and at Chernobyl, Ukraine, in 1986, from the waste products of nuclear power generation, and from nuclear war. The effects of intense radiation on life forms can be seen clearly in the results of experiments in which habitats were exposed to radiation sources.

The possibility of nuclear war is diminishing as the superpowers dismantle their nuclear arsenals and nations turn their attention to economic, social, and environmental problems. Still, radioactive wastes produced by peaceful uses of the atom pose daunting disposal problems. Depending on the waste product, radiation does not decline to harmless levels for thousands or even millions of years—far beyond the life span of waste containers, not to mention that of the institutions to which their care is entrusted. The waste disposal problem may ultimately limit the use of nuclear power.

Atmospheric pollution threatens the environment on a global scale

Because of the circulation of the atmosphere and oceans, certain types of pollution have global consequences: their effects extend far beyond the sources of the pollution itself. By far the most worrisome of these changes in the environment are the destruction of the ozone layer in the upper atmosphere and the increase in carbon dioxide and other "greenhouse" gases.

The ozone layer and ultraviolet radiation

Ozone (O_3) is a molecular form of oxygen that is highly reactive as an oxidant, capable of chemically oxidizing organic molecules and destroying their proper functioning. Consequently, ozone is toxic to animal and plant life even in small concentrations. Near the earth's surface, ozone is produced by the oxidation of molecular oxygen (O_2) in the presence of nitrous oxide (NO_2) and sunlight. Because NO_2 is a product of gasoline combustion, ozone can reach high levels in the exhaust fumes that pollute cities, particularly where there is strong sunlight. In Los Angeles, for example, which was well known for its smog before pollution control measures, ozone concentrations in the atmosphere at ground level sometimes reached 0.5 ppm, which is perhaps 20–50 times the normal level and is damaging to human health, crops, and natural vegetation.

Ozone is also produced in the upper atmosphere, but there it has the beneficial effect of shielding the surface of the earth from ultraviolet radiation by absorbing solar radiation of short wavelengths (especially in the range of 200–300 nm) (see Chapter 2). Unfortunately, certain substances, among them chlorine atoms, cause the breakdown of ozone. Levels of chlorine in the upper atmosphere have been increasing because of the release into the atmosphere of chlorofluorocarbons (CFCs), which are used as propellants in spray cans and as coolants in air conditioning and refrigeration systems. Decreases in stratospheric ozone of 50% or more—so-called **ozone holes**—have been observed at high latitudes in both hemispheres (❙ Figure 26.12).

The resulting elevation of ultraviolet radiation at the earth's surface will increase the incidence of skin cancer, because DNA also absorbs ultraviolet radiation between 280 and 320 nm in wavelength, and this absorbed energy causes damage to DNA molecules. Of greater concern is the fact that ultraviolet radiation also damages the photosynthetic apparatus of plants and could cause reductions in primary production—the base of the food chain for the entire ecosystem. Such reductions have already been observed in the oceans surrounding Antarctica. The threat is so great that the international community, through the Vienna Convention for the Protection of the Ozone Layer (1985) and the Montreal Protocol (1987), agreed to phase out the use of all CFCs by the end of the twentieth century, and by 1997 in the European Community. Even though it is too early to see the result, we can anticipate that this action will reverse the damage that has already been done and allow atmospheric ozone to return to its natural equilibrium level, perhaps within a century.

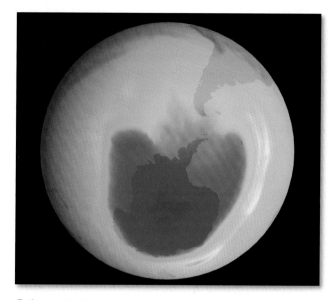

■ **Figure 26.12 Ozone holes are areas where the concentration of ozone has decreased by 50% or more.** This false-color satellite image, taken on October 3, 1999, shows the development of a large ozone hole over Antarctica. Courtesy of NASA.

Carbon dioxide and the greenhouse effect

Carbon dioxide (CO_2) occurs naturally in the atmosphere. Without it, the earth would be a very cold place, because most of the sunlight absorbed by the earth's surface would be reradiated into the cold depths of space. As it is now,

CO_2 forms an insulative blanket over the earth's surface that lets short-wavelength ultraviolet and visible light from the sun pass through, but retards loss of heat as longer-wavelength infrared radiation (see Chapter 2). Glass in greenhouses works on the same principle, and so the function of CO_2 in the atmosphere is known as the greenhouse effect.

At times in the distant past, the concentration of CO_2 in the atmosphere was far greater than during recent human experience, and the average temperature of the earth was correspondingly much warmer (see Chapter 7). As atmospheric CO_2 declined during the last 50–100 million years, the earth experienced a gradual cooling, resulting in an expansion of temperate and boreal climate zones and culminating in the Ice Ages of the past million years. The problem we face in global warming is not that the earth has never been so warm, but that the climate will change so quickly that ecological systems—not to mention the human population—will not be able to keep up.

Before 1850, the concentration of CO_2 in the atmosphere was on the order of 280 ppm (0.028%). During the last 150 years, which have witnessed tremendous increases in the burning of wood, coal, oil, and gas for energy production, it has increased to over 350 ppm (■ Figure 26.13). Half of this increase has occurred during the last 30 years, and the rate of increase appears to be rising. This change in the chemistry of the atmosphere has created fears of a major warming of the earth's climate. Current estimates of global warming in the next century suggest a rise in average global temperature of anywhere between 2°C and 6°C.

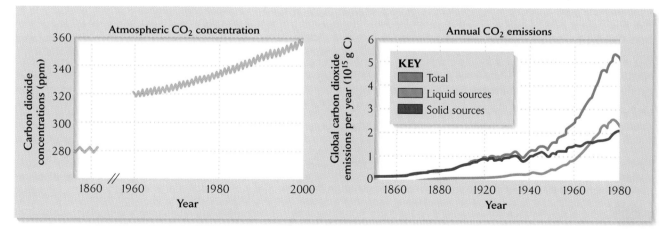

■ **Figure 26.13 The atmospheric concentration of carbon dioxide has increased about 25% in the last 150 years.** The concentration of carbon dioxide in the atmosphere, and global emissions of carbon dioxide from the combustion of solid (coal) and liquid (oil) fossil fuels, have increased since 1860. Accurate measurements of carbon dioxide concentration have been taken at Mauna Loa Observatory, Hawaii, which is far from regions of concentrated fossil fuel consumption, only since the late 1950s. Data from A. M. Solomon, J. R. Trabolka, D. E. Reichle, and C. M. Masters, in *Atmospheric Carbon Dioxide and the Global Carbon Dioxide Cycle*, U.S. Department of Energy, Washington, D.C. (1985), pp. 1–13 and 63–80.

The level of carbon dioxide in the atmosphere represents a balance between processes that add CO_2 and those that remove it. Before the Industrial Revolution, addition of CO_2 to the atmosphere by the respiration of terrestrial organisms (approximately 120 billion tons of carbon per year) was balanced by the gross primary production of terrestrial vegetation, and the total amount in the atmosphere was maintained in equilibrium. Humans add CO_2 to the atmosphere by burning fossil fuels and clearing and burning forests. At present, forest cutting accounts for the addition of about 2 billion tons of carbon to the atmosphere annually; burning fossil fuels adds about 5 billion tons. Thus, human activities have increased carbon flux into the atmosphere by about 6%—perhaps more, perhaps less, depending on whose figures we use.

The atmosphere exchanges carbon dioxide with the oceans, where excess carbon is precipitated as calcium carbonate sediments (see Figure 7.7). It has been estimated that at present, the oceans absorb more carbon than they release to the atmosphere by about 2.4 billion tons annually. In spite of the fact that the oceans are a net carbon dioxide sink, their capacity to absorb carbon is less than half of the anthropogenic input to the atmosphere, and part of that capacity may already be used to offset excess production of carbon dioxide by terrestrial systems. No matter how the arithmetic is done, it shows carbon dioxide concentrations in the atmosphere to be rising very rapidly.

Warmer temperatures caused by the greenhouse effect will have mixed effects on productivity. On the positive side, warmer temperatures lengthen the growing season and speed metabolism, and thereby tend to enhance production in moist environments. Because plants require carbon dioxide for photosynthesis, this effect is likely to be enhanced by the higher CO_2 concentration of the atmosphere. Balancing this benefit is the likelihood of increasing drought stress in arid environments, which may reduce agricultural production and accelerate the conversion of overused grazing lands and croplands to wasteland. Other potential problems include the inundation of coastal settlements by rising sea levels fed by melting polar ice caps and expansion of warmed ocean water.

Human ecology is the ultimate challenge

The human population continues to increase worldwide at a rate of almost 2% per year. The highest rates of population increase are in some of the poorest countries. Even if population growth were to stop today, staggering problems would remain. The present human population is consuming resources faster than new resources are being regenerated by the biosphere, all the while pouring forth so much waste that the quality of the environment in most regions of the earth is deteriorating at an alarming rate. If we are to leave a habitable world for future generations, our top priority must be to achieve a sustainable relationship with the rest of the biosphere. This will require putting an end to population growth, developing sustainable energy sources, providing for the regeneration of nutrients and other materials, and restoring deteriorated habitats.

ECOLOGISTS IN THE FIELD

Assessing the earth's carrying capacity for humankind

People have tried for many years to understand how many people the earth can support. Estimates of the sustainable human population vary between about 0.6 billion—roughly one-tenth the current population—to several tens of billions. These estimates depend on certain critical assumptions about the limits to human population size and the level of affluence for each individual. What is meant by level of affluence includes not only material goods, such as housing, televisions, recreational travel, and other measures of quality of life, but also the cost of producing and delivering that most basic human necessity—food. In the United States, for every kilojoule of food energy we consume, 10 kilojoules of fossil fuel energy have gone into the fertilizers, farm machinery, packaging, and transportation required to put that food on our tables. Most estimates of the earth's carrying capacity for humanity exclude the contributions of fossil fuels, which are nonrenewable resources with high environmental costs. Thus, to be sustainable, human energy needs must be met by renewable sources such as wind, solar, and hydroelectric power generation and ethanol production from crops.

Mathis Wackernagel, a member of the Task Force on Planning Healthy and Sustainable Communities at the University of British Columbia, developed the concept of the **ecological footprint** based on ethanol as a sustainable energy source. Making a generous assumption about how much ethanol can be produced from a hectare of fertile cropland, Wackernagel estimated that the citizens of The Netherlands, for example, would each require 5.3 hectares of fertile land to provide their energy requirements. The area of The Netherlands contains only 1.7 hectares of land for each of its 16,000,000 citizens. Clearly, to sustain their current standard of living, which is one of the highest in the world, the Dutch would have to import immense amounts of energy, as they currently do in the form of fossil fuels. The United States has the highest ecological footprint at 10.3 hectares per individual. Thus, even though there are 6.7 hectares of productive agricultural land available per person, the population at its present level is not sustainable by domestic renewable resources.

Globally, there are only about 1.7 hectares of land of average agricultural productivity available for each of the earth's 6 billion current inhabitants. This is clearly not enough to sustain the present population at a high standard of living according to Wackernagel's assumptions. What will happen when the human population increases to 10 billion, as it is projected to do by the middle of this century? Will the average standard of living decrease as sources of fossil fuel are used up, or will new technologies emerge to keep pace with the growing population? Cornell University ecologist David Pimentel has estimated that the earth can sustain 2 billion people at the standard of living currently enjoyed in the United States. Stanford University ecologist Paul Ehrlich puts the level much lower. According to many ecologists, larger populations, including the present 6 billion, can be sustained only at a lower average standard of living.

In many respects, humankind has stepped beyond the bounds of the usual ecological mechanisms of restraint and regeneration. Our ability to tap nonrenewable sources of energy in the form of coal, oil, and gas deposits has temporarily removed conventional food–energy limitations on our population growth. No longer is most of the human population supported by the land it occupies. Our technological and economic abilities to reach out for new land and resources have pushed density-dependent population feedbacks well into the future.

Our present course, however, leads in a predictable direction. It is not an inviting one: increasing energy, material, and food shortages; many people living in poverty and disease; a badly polluted environment; escalating social and political strife. These mechanisms of population control will inevitably come into play, as they do for every species.

The future need not be like this (▊ Figure 26.14). Where we have escaped natural restraint, we must substitute our own restraint. Where we produce waste products that cannot be regenerated by ecological systems, we must find ways to recycle them ourselves. Energy consumption must be scaled back, and production must be based increasingly on renewable energy sources, such as the sun, wind, and conversion of biomass to liquid fuels. Achieving these goals will require a consensus among social, economic, and political institutions. The hope for such a consensus lies in making people aware of the global deterioration of the quality of human life and educating them in the basic

(a)

(b)

(c)

▊ **Figure 26.14 Conservation efforts can help us take our proper place in the economy of nature.** (a) Well-maintained and productive farmland; (b) a well-planned industrial park; (c) an attractive urban setting with plenty of nature worked in. Photo (a) by Ron Nichols, courtesy of the U.S. Department of Agriculture, Soil Conservation Service; photo (b) by Macduff Everton/Corbis; photo (c) by Sotographs/The Stock Market.

ecological principles that must form the foundation of a self-sustaining system.

Above all, humankind has the choice of adopting a new attitude toward its relationship with nature. We are a part of nature, not apart from nature. To the extent that our intelligence, culture, and technology have given us the power to dominate nature, we must also use these abilities to impose self-regulation and self-restraint. This is our greatest challenge. We have succeeded famously in becoming the technological species. Our survival now depends on our becoming the ecological species and taking our proper place in the economy of nature.

Summary

1. The key to survival of the human population is development of sustainable interactions with the biosphere. This will require control of human population growth, an increasing reliance on renewable energy sources, and total recycling of material wastes.

2. Maintaining a sustainable biosphere means that we must conserve the ecological processes responsible for its productivity.

3. The principal local threats to natural processes are over-exploitation of resources, introductions of exotic species, habitat conversion, irrigation, enrichment with organic wastes (eutrophication), and production of toxic materials.

4. On a global scale, various airborne pollutants, especially chlorofluorocarbons, have reduced the concentrations of ozone (O_3) in the upper atmosphere, allowing more damaging ultraviolet radiation to reach the surface of the earth.

5. Increasing levels of carbon dioxide in the atmosphere, produced by the burning of fossil fuels, threaten to increase the average temperature of the earth by 2°C to 6°C, with potentially adverse consequences for natural ecosystems and agriculture. In addition, sea levels will rise owing to the melting of polar ice caps and the expansion of warmed ocean water.

6. Solutions to the environmental crisis will require new attitudes promoting sustainability and self-restraint.

PRACTICING ECOLOGY

CHECK YOUR KNOWLEDGE

Impacts of Decreasing Ozone

Atmospheric ozone occurs in two altitudinal zones. "Bad ozone" is a component of urban smog in the troposphere, close to the surface of the earth, where its high reactivity and oxidizing power can damage plant and animal tissues. "Good ozone" occurs in the stratosphere, at an altitude of about 25 km, where its capacity to absorb ultraviolet (UV) radiation protects organisms at the earth's surface from radiation damage to DNA and other biomolecules. On a global scale, the amount of good ozone is decreasing, while "bad" ozone concentrations in urban areas are increasing. Ozone concentration is measured in Dobson units (DU); average concentrations in the stratosphere are about 300 DU, but can vary from 100 to 500 DU globally. The lowest concentrations occur over the poles, particularly over the continent of Antarctica. These low concentrations are referred to as ozone holes. The size of the ozone hole over Antarctica, which varies seasonally, is greatest in September and October. Many scientists are seriously concerned because the size of the hole has been growing over the last several decades.

In September 2000 a NASA instrument determined that the Antarctic ozone hole was three times larger than the area of the United States—the largest area on record for this hole. The hole, about 28.3 million square kilometers, was measured on September 3, 2000, and was then about 1.1 million square kilometers bigger than the previous record, measured in September 1998. Because the concentration of stratospheric "good" ozone levels are presently near their lowest since measurements were first taken, it is thought that UV radiation levels at the surface of the earth (especially levels of the potent form, UV-B, which occurs at wavelengths of 280–315 nm) are close to their all-time maximum. Indeed, for every 1% decrease in ozone concentration, there is a 2% increase in UV radiation at the surface of the earth.

The impacts of ozone depletion have been studied mostly for aquatic organisms in Antarctica. Rousseaux et al. (1999), however, studied the effects of UV-B on the plant *Gunnera magellanica* as the ozone hole expanded during 1997 to expose Tierra del Fuego at the southern tip of South America (55°S). They showed that even though there was frequent cloud cover, the depleted stratospheric ozone hole caused greater intensity of UV radiation measured at ground level and large increases in DNA damage.

CHECK YOUR KNOWLEDGE

1. What measures can be taken to reduce tropospheric ozone levels? What can be done to affect stratospheric ozone levels?

2. Given the data from Rousseaux et al. (1999) shown in Table 26.1, what is the projected number of days with ozone levels less than 250 DU for the year 2050? From the data in Table 26.2, what is the amount of DNA damage expected for a UV dose of 1000 J per m²?

Table 26.1 Ozone depletion over time measured during spring (September–October), Tierra del Fuego National Park, Argentina

Year	Number of days with ozone < 250 DU
1983	2
1984	8
1985	3
1986	2
1987	2
1988	5
1989	4
1990	6
1991	15
1992	5
1994	16
1996	7
1997	19
1998	7

Ozone depletion was measured by a NASA satellite and is expressed as the number of days in a year when ozone concentration was less than 250 DU. Data for 1993 and 1995 were not available. Data are modified from Rousseaux et al. (1999).

Table 26.2 DNA damage units for leaves of *Gunnera magellanica* as a function of UV dose

UV (J per m^2)	DNA damage units
300	0.018
380	0.025
400	0.028
500	0.030
550	0.025
625	0.031
700	0.035
750	0.043

DNA damage was measured as cyclobutane pyrimidine dimer formation; methods are described in, and data are modified from, Rousseaux et al. (1999).

MORE ON THE WEB

3. Visit the Environmental Protection Agency's Web site for the Ultraviolet Monitoring Program through *Practicing Ecology on the Web* at *http://www.whfreeman.com/ricklefs*. Follow the links titled "Access Data" and "Graphs and Reports." Request a graph of daily UV for Riverside, California, and Denali National Park, Alaska, for the year 1998. By how much do the maximal levels of UV vary in mid-August for the two sites? If we assume that the data in Table 26.2 apply to plants in general (what's wrong with this assumption?), what would be the expected difference in DNA damage for the two sites?

 Suggested Readings

Arrow, K., et al. 1995. Economic growth, carrying capacity, and the environment. *Science* 268:520–521.

Ausubel, J. H. 1991. A second look at the impacts of climate change. *American Scientist* 79:210–221.

Bongaarts, J. 1994. Can the growing human population feed itself? *Scientific American* 270:36–42.

Bongaarts, J. 1994. Population policy options in the developing world. *Science* 263:771–776.

Browder, J. O. 1992. The limits of extractivism. *BioScience* 42:174–182.

Clark, W. C. 1989. Managing planet earth. *Scientific American* 261:46–64. (See also other articles in this issue.)

Cohen, A. N., and J. T. Carlton. 1998. Accelerating invasion rate in a highly invaded estuary. *Science* 279 (23 January): 555–557.

Daily, G. C., and P. R. Ehrlich. 1992. Population, sustainability, and Earth's carrying capacity. *BioScience* 42:761–771.

Davis, G. R. 1990. Energy for planet earth. *Scientific American* 263:54–62.

Goldemberg, J. 1995. Energy needs in developing countries and sustainability. *Science* 269:1058–1059.

Graedel, T. E., and P. J. Crutzen. 1995. *Atmosphere, Climate, and Change.* Scientific American Library, W. H. Freeman, New York.

Graham, R. L., M. G. Turner, and V. H. Dale. 1990. How increasing CO_2 and climate change affect forests. *BioScience* 40:575–587.

Hecht, S. B. 1993. The logic of livestock and deforestation in Amazonia. *BioScience* 43:687–695.

Houghton, R. A., D. L. Skole, C. A. Nobre, J. L. Hackler, K. T. Lawrence, and W. H. Chomentowski. 2000. Annual fluxes of carbon from deforestation and regrowth in the Brazilian Amazon. *Nature* 403 (20 January): 301–304.

Hughes, T. P. 1994. Catastrophes, phase shifts, and large-scale degradation of a Caribbean coral reef. *Science* 265:1547–1551.

Katzman, M. T., and W. G. Cale, Jr. 1990. Tropical forest preservation using economic incentives. *BioScience* 40:827–832.

Pimentel, D., et al. 1987. World agriculture and soil erosion. *BioScience* 37:277–283.

Pimentel, D., et al. 1991. Environmental and economic effects of reducing pesticide use. *BioScience* 41:402–409.

Pimentel, D., L. Lach, R. Zuniga, and D. Morrison. 2000. Environmental and economic costs of nonindigenous species in the United States. *BioScience* 50(1):53–65.

Rasmussen, P. E., K. W. T. Goulding, J. R. Brown, P. R. Grace, H. H. Janzen, and M. Körschens. 1998. Long-term agroecosystem experiments: Assessing agricultural sustainability and global change. *Science* 282 (30 October): 893–896.

Repetto, R. 1990. Deforestation in the Tropics. *Scientific American* 262:36–42.

Ricklefs, R. E., Z. Naveh, and R. E. Turner. 1984. Conservation of ecological processes. *The Environmentalist* 4 (Suppl. 8):1–16.

Rousseaux, M. C., et al. 1999. Ozone depletion and UVB radiation: Impact on plant DNA damage in southern South America. *Proceedings of the National Academy of Sciences USA* 96:15310–15315.

Rowland, F. S. 1989. Chlorofluorocarbons and the depletion of stratospheric ozone. *American Scientist* 77:36–45.

Schneider, S. H. 1989. The greenhouse effect: Science and policy. *Science* 243:771–782.

Stoddard, J. L., D. S. Jeffries, A. Lükewille, et al. 1999. Regional trends in aquatic recovery from acidification in North America and Europe. *Nature* 401 (7 October): 575–578.

Tibbetts, J. 1996. Farming and fishing in the wake of El Niño. *BioScience* 46:566–569.

Vitousek, P. M., P. R. Ehrlich, A. H. Ehrlich, and P. A. Matson. 1986. Human appropriation of the products of photosynthesis. *BioScience* 36:368–373.

Appendix A

International System of Units

The International System of Units (abbreviated SI after the French *Système international*) is described in detail in such publications as the National Bureau of Standards Special Publications 330, 1977 edition (U.S. Dept. of Commerce, Washington, D.C.), and M. H. Green, *Metric Conversion Handbook* (Chemical Publ. Co., New York, 1978).

The seven basic SI units are as follows:

Category of measurement	Name of SI unit	Abbreviation
Length	Meter	m
Mass	Kilogram	kg
Time	Second	s
Electric current	Ampere	A
Temperature	Kelvin	K
Amount of substance	Mole	mol
Luminous intensity	Candela	cd

The ampere, mole, and candela are not often used in ecology and will not be discussed here. One familiar unit of time is the second. The SI units of length and mass are metric, with simple conversions to and from the English system; for example, 1 meter equals 3.28 feet, and 1 kilogram equals 2.205 pounds. The kelvin is identical to the degree Celsius (°C) except that zero on the kelvin scale is absolute zero, the coldest temperature possible (−273°C). Most ecologists use °C instead of K, because of the convenient designations for the freezing and boiling points of water (0°C and 100°C).

In the SI system, all other measurements are derived from the basic seven. For example, the SI unit of flow is the cubic meter per second, which is derived from the base units meter and second. Some of these derived units frequently employed in ecology are shown in the table below.

There are also a number of other units that are not part of SI but nonetheless are widely used: the liter (L), a unit of volume equal to $1/1,000$ cubic meter; the are (a), a unit equal to 100 square meters, and the more often used hectare (ha), equal to 100 a and 10,000 m^2; and the minute, hour, and day, familiar units of time. The calorie (cal), a familiar unit of

Category of measurement	Name of SI unit	Abbreviation	Expressions in terms of	
			other units	base units
Area	Square meter	m^2		m^2
Volume	Cubic meter	m^3		m^3
Velocity	Meter per second	m per s		m per s
Flow	Cubic meter per second	m^3 per s		m^3 per s
Density	Kilogram per cubic meter	kg per m^3		kg per m^3
Force	Newton	N		kg m per s^2
Pressure	Pascal	Pa	N per m^2	kg per m per s^2
Energy, heat	Joule	J	N m	kg m^2 per s^2
Power	Watt	W	J per s	kg m^2 per s

heat energy, is not derived from SI base units and should be abandoned in favor of the joule (= 0.239 cal), even though ecologists continue to use the calorie and kilocalorie.

Multiples of 10 of the SI units are given special names by adding the prefixes below to the unit name.

Hence 1,000 meters becomes a kilometer (km) and $^1/_{1,000}$ meter is a millimeter (mm); 1,000 joules is a kilojoule (kJ) and 1 million watts is a megawatt (MW).

Factor	Prefix	Abbreviation	Number	Name
10^{12}	Tera	T	1,000,000,000,000	Trillion
10^9	Giga	G	1,000,000,000	Billion*
10^6	Mega	M	1,000,000	Million
10^3	Kilo	k	1,000	Thousand
10^2	Hecto	h	100	Hundred
10^1	Deka	da	10	Ten
10^{-1}	Deci	d	0.1	Tenth
10^{-2}	Centi	c	0.01	Hundreth
10^{-3}	Milli	m	0.001	Thousandth
10^{-6}	Micro	μ	0.000 001	Millionth
10^{-9}	Nano	n	0.000 000 0001	Billionth

*Milliard in the United Kingdom; the British billion is a million million (10^{12}), equivalent to the U.S. trillion, which is commonly used only in astronomy and the federal budget.

Appendix B

Conversion factors

Length
1 meter (m) = 39.4 inches (in)
1 meter = 3.28 feet (ft)
1 kilometer (km) = 3,281 feet
1 kilometer = 0.621 mile (mi)
1 micron (μ) = 10^{-6} meter
1 inch = 2.54 centimeters (cm)
1 foot = 30.5 centimeters
1 mile = 1,609 meters
1 angstrom (Å) = 10^{-10} meter
1 millimicron (mμ) = 10^{-9} meter

Area
1 square centimeter (cm^2) = 0.155 square inches (in^2)
1 square meter (m^2) = 10.76 square feet (ft^2)
1 hectare (ha) = 2.47 acres (a)
1 hectare = 10,000 square meters
1 hectare = 0.01 square kilometer (km^2)
1 square kilometer = 0.386 square mile
1 square mile = 2.59 square kilometers
1 square inch = 6.45 square centimeters
1 square foot = 929 square centimeters
1 square yard (yd^2) = 0.836 square meter
1 acre = 0.407 hectare

Mass
1 gram (g) = 15.43 grains (gr)
1 kilogram (kg) = 35.3 ounces (oz)
1 kilogram = 2.205 pounds (lb)
1 metric ton (T) = 2204.6 pounds
1 ounce = 28.35 grams
1 pound = 453.6 grams
1 short ton = 907 kilograms

Time
1 year (yr) = 8,760 hours (h)
1 day (d) = 86,400 seconds (s)

Volume
1 cubic centimeter (cc or cm^3) = 0.061 cubic inch (in^3)
1 cubic inch = 16.4 cubic centimeters
1 liter (L) = 1,000 cubic centimeters
1 liter = 33.8 U.S. fluid ounces (oz)
1 liter = 1.057 U.S. quarts (qt)
1 liter = 0.264 U.S. gallon (gal)
1 U.S. gallon = 3.79 liters
1 British imperial gallon = 4.55 liters
1 cubic foot (ft^3) = 28.3 liters
1 milliliter (mL) = 1 cubic centimeter
1 U.S. fluid ounce = 29.57 milliliters
1 British imperial fluid ounce = 28.4 milliliters
1 U.S. quart = 0.946 liter

Velocity
1 meter per second (m per s) = 2.24 miles per hour (mph)
1 foot per second (ft per s) = 1.097 kilometers per hour
1 kilometer per hour = 0.278 meter per second
1 mile per hour = 0.447 meter per second
1 mile per hour = 1.467 feet per second

Energy
1 joule (J) = 0.239 calorie (cal)
1 calorie = 4.184 joules
1 kilowatt-hour (kWh) = 860 kilocalories (kcal)
1 kilowatt-hour = 3,600 kilojoules
1 British thermal unit (Btu) = 252.0 calories
1 British thermal unit = 1,054 joules
1 kilocalorie = 1,000 calories

Power
1 kilowatt (kW) = 0.239 kilocalorie per second
1 kilowatt = 860 kilocalories per hour
1 horsepower (hp) = 746 watts
1 horsepower = 15,397 kilocalories per day
1 horsepower = 641.5 kilocalories per hour

Energy per unit area
1 calorie per square centimeter = 3.69 British thermal units per square foot
1 British thermal unit per square foot = 0.271 calorie per square centimeter
1 calorie per square centimeter = 10 kilocalories per square meter

Power per unit area
1 kilocalorie per square meter per minute = 52.56 kilocalories per hectare per year
1 footcandle (fc) = 1.30 calories per square foot per hour at 555 nm wavelength
1 footcandle = 10.76 lux (lx)
1 lux = 1.30 calories per square meter per hour at 55 nm wavelength

Metabolic energy equivalents
1 gram of carbohydrate = 4.2 kilocalories
1 gram of protein = 4.2 kilocalories
1 gram of fat = 9.5 kilocalories

Miscellaneous
1 gram per square meter = 0.1 kilogram per hectare
1 gram per square meter = 8.97 pounds per acre
1 kilogram per square meter = 4.485 short tons per acre
1 metric ton per hectare = 0.446 short ton per acre

Glossary

Acclimation. A reversible change in the morphology or physiology of an organism in response to environmental change; also called acclimatization.

Acid mine drainage. Water runoff from surface mining, usually coal strip mining, containing sulfuric acid, which forms when organic sulfur is oxidized on contact with the atmosphere.

Acid rain. Precipitation with high acidity (pH < 4) caused by the solution of certain atmospheric gases (sulfur dioxide and nitrous oxide) produced by combustion of fossil fuels.

Acidity. The concentration of hydrogen ions in a solution, often written on the logarithmic scale of pH.

Active transport. Movement of molecules or ions through a membrane against a diffusion gradient.

Activity space. The range of environmental conditions suitable for the activity of an organism.

Adaptation. A genetically determined characteristic that enhances the ability of an individual to cope with its environment; the evolutionary process by which organisms become better suited to their environments.

Adaptive radiation. The evolution of a variety of forms from a single ancestral stock; often occurs after organisms colonize an island group or enter a new adaptive zone.

Adiabatic cooling. The decrease in temperature with increasing elevation caused by the expansion of air in the lower atmospheric pressure.

Age class. The individuals in a population of a particular age.

Age structure. The distribution of individuals among age classes within a population.

Aging. Decrease in physiological functioning with age, usually associated with decreasing rates of survival and fecundity.

Alkaloids. Nitrogen-containing compounds, such as morphine and nicotine, that are produced by plants and are toxic to many herbivores.

Allele. One of several alternative forms of a gene.

Allelopathy. Direct inhibition of one species by another using noxious or toxic chemicals.

Allocation. The division of limited time, energy, or materials among competing functions or requirements.

Allochthonous. Referring to materials transported into a system, particularly minerals and organic matter transported into streams and lakes. *Compare with* Autochthonous.

Allometry. A relative increase in a part of an organism or a measure of its physiology or behavior in relation to some other measure, usually its overall size.

Allopatric. Occurring in different places; usually referring to geographic separation of populations.

Altruism. In an evolutionary sense, enhancing the fitness of another individual by acts that reduce the evolutionary fitness of the altruistic individual.

Ammonification. Metabolic breakdown of proteins and amino acids with ammonia as an excreted by-product.

Anaerobic. Without oxygen.

Anion. A part of a dissociated molecule carrying a negative electric charge.

Anoxic. Lacking oxygen; anaerobic.

Anthropogenic extinction. Extinction caused by the activities of humans, through either direct exploitation of a population or destruction of its habitat.

Aphotic zone. In lakes and oceans, the water layer below the depth to which light penetrates.

Aposematism. *See* Warning coloration.

Asexual reproduction. Reproduction without the benefit of the sexual union of gametes (fertilization).

Assimilation. Incorporation of any material into the tissues, cells, and fluids of an organism.

Assimilation efficiency. A percentage expressing the proportion of ingested energy that is absorbed into the bloodstream.

Assimilatory. Referring to a biochemical transformation that results in the reduction of an element to an organic form and hence its gain by the biological compartment of the ecosystem.

Association. A group of species living in the same place.

Assortative mating. Preferential mating between individuals having either similar appearance or genotypes (positive assortative mating) or dissimilar appearance or genotypes (negative assortative mating).

Asymmetric competition. An interaction between two species in which one exploits a particular resource more efficiently than the other; the second may persist by better avoiding predation or by subsisting on a different resource.

Australian region. The biogeographic region corresponding roughly to Australia, New Guinea, and nearby islands.

Autecology. The study of organisms in relation to their physical environment. *Compare with* Synecology.

Autochthonous. Referring to materials produced within a system, particularly organic matter produced and minerals cycled within streams and lakes. *Compare with* Allochthonous.

Autotroph. An organism that assimilates energy from either sunlight (green plants) or inorganic compounds (sulfur bacteria). *Compare with* Heterotroph.

Background extinction. Extinction of species or higher taxa during periods without rapid environmental change.

Barren. An area with sparse vegetation owing to some physical or chemical property of the soil.

Batesian mimicry. Resemblance of an unpalatable species (model) by an edible species (mimic) to deceive predators.

Benthic. On or within the bottom of a river, lake, or ocean.

Benthos. The environment on or within the bottom sediments of rivers, lakes, and oceans; also, the organisms that live there.

Beta diversity. The variety of organisms within a region arising from turnover of species between habitats.

Bet hedging. In a life history, reducing the risk of mortality or reproductive failure in a variable environment by adopting an intermediate strategy or several alternative strategies simultaneously, or by spreading one's risk over time and space (e.g., by perennial rather than annual reproduction).

Biodiversity. A measure of the variety of organisms within a local area or region, often including genetic variation, taxonomic uniqueness, and endemism. *See* Diversity.

Biological community. *See* Community.

Biological control. Use of natural enemies, particularly parasitoid insects, bacteria, and viruses, to control pest organisms.

Biological oxygen demand (BOD). The amount of oxygen required to oxidize the organic material in a water sample; high values in aquatic habitats often indicate pollution by sewage and other sources of organic wastes, or the overproduction of plant material resulting from overenrichment by mineral nutrients.

Biomass accumulation ratio. The ratio of weight to annual production.

Biome. A major type of ecological community (e.g., the grassland biome).

Bioremediation. Restoration of natural habitats or ecological conditions by use of biological agents (e.g., bacterial degradation of spilled oil or other pollutants).

Biosphere. All the environments and organisms of the earth.

Biota. Fauna and flora together.

Birth rate (b). The average number of offspring produced per individual per unit of time, often expressed as a function of age (x). *See also* Fecundity.

Boreal. Northern; often refers to the coniferous forest regions that stretch across Canada, northern Europe, and Asia.

Bottleneck. *See* Population bottleneck.

Bottom-up control. Influence of producers on the sizes of the trophic levels above them in the food web.

Boundary layer. A layer of still or slow-moving water or air close to the surface of an object.

Browsing. Consumption of a portion of a plant's tissues, generally applied to woody vegetation.

C_3 photosynthesis. Photosynthetic pathway in which carbon dioxide is initially assimilated into a three-carbon compound, phosphoglyceraldehyde (PGA), in the Calvin cycle.

C_4 photosynthesis. Photosynthetic pathway in which carbon dioxide is initially assimilated into a four-carbon compound, such as oxaloacetic acid (OAA) or malate.

Calcification. Deposition of calcium and other soluble salts in soils where evaporation greatly exceeds precipitation.

Calvin cycle. The basic assimilatory sequence of photosynthesis during which an atom of carbon is added to the five-carbon ribulose bisphosphate (RuBP) molecule to produce phosphoglyceraldehyde (PGA) and then glucose.

CAM photosynthesis. Photosynthetic pathway in which the initial assimilation of carbon dioxide into a four-carbon compound occurs at night; found in some succulent plants in arid habitats.

Canopy. The uppermost layer of vegetation in a forest.

Carbonate ion. An anion (CO_3^{2-}) formed by the dissociation of carbonic acid or one of its salts.

Carbonic acid. A weak acid (H_2CO_3) formed when carbon dioxide dissolves in water.

Carnivore. An organism that consumes mostly flesh.

Carrying capacity (K). The number of individuals in a population that the resources of a habitat can support; the asymptote, or plateau, of the logistic and other sigmoid equations for population growth.

Caste. Individuals within a social group sharing a specialized form or behavior.

Catastrophe. An unpredictable event that has a strong effect on a population.

Cation. A part of a dissociated molecule carrying a positive electrical charge.

Central place foraging. Foraging behavior in which acquired food is brought to a central place, such as a nest with young.

Chaos. Erratic change in the size of populations governed by difference equations and having high intrinsic rates of growth.

Character displacement. Divergence in the characteristics of two otherwise similar species where their ranges overlap, caused by the selective effects of competition between the species in the area of overlap.

Chemoautotroph. An organism that oxidizes inorganic compounds (often hydrogen sulfide) to obtain energy for synthesis of organic compounds (e.g., sulfur bacteria).

Clay. A fine-grained component of soil, formed by the weathering of granitic rock and composed primarily of hydrous aluminum silicates.

Climate zone. A region in Heinrich Walter's classification of the climates of the earth defined by temperature and precipitation.

Climax community. The end point of a successional sequence, or sere; a community that has reached a steady state under a particular set of environmental conditions.

Cline. Gradual change in population characteristics or adaptations over a geographic area.

Clone. A group of asexual individuals, all descended from the same parent and bearing the same genotype.

Closed community concept. The idea popularized by F. C. Clements that communities are distinctive associations of highly interdependent species.

Clumped distribution. A distribution of individuals in space indicating a tendency for association.

Coadaptation. Evolution of characteristics of each of two or more species in response to changes in the other(s), often to mutual advantage. *See* Coevolution.

Codominant. Referring to alleles that, in heterozygous form, produce a phenotype intermediate between the homozygous phenotypes.

Codon. A sequence of three nucleotides in DNA or RNA that specifies which amino acid will be placed at a particular position in a protein.

Coefficient of relationship (r). The probability that one individual shares with another a genetic factor inherited from a common ancestor.

Coevolution. The occurrence of genetically determined traits (adaptations) in two or more species selected by the mutual interactions controlled by these traits.

Coexistence. Occurrence of two or more species in the same habitat; usually applied to potentially competing species.

Cohort. A group of individuals of the same age recruited into a population at the same time.

Cohort life table. The age-specific survival and fecundity of a cohort of individuals in a population followed from birth to the death of the last individual. Also known as a dynamic life table.

Common garden experiment. Growing organisms from different habitats together in the same place to reveal genetic differences in the absence of different environmental influences.

Community. An association of interacting populations, usually defined by the nature of their interaction or the place in which they live.

Compartment model. A representation of a system in which the various parts are portrayed as units (compartments) that receive inputs from and provide outputs to other such units.

Compensation point. The depth of water or level of light at which respiration and photosynthesis balance each other; the lower limit of the euphotic zone.

Competition. Use or defense of a resource by one individual that reduces the availability of that resource to other individuals, whether of the same species (intraspecific competition) or other species (interspecific competition).

Competition coefficient (*a*). A measure of the degree to which one consumer uses the resources of another, expressed in terms of the population consequences of the interaction.

Competitive exclusion principle. The hypothesis that two or more species cannot coexist on a single resource that is scarce relative to demand for it.

Complementarity. The situation in which different food types or other entities each contain nutrients or other resources missing in the others.

Condition. A physical or chemical attribute of the environment that, while not being consumed, influences biological processes and population growth (e.g., temperature, salinity, acidity). *Compare with* Resource.

Conductance. The capacity of heat, electricity, or a substance to pass through a particular material.

Conduction. The ability of heat to pass through a substance.

Connectedness food web. A depiction of the feeding relationships among species within a community.

Constitutive defense. A defense that is always present. *Compare with* Induced defense.

Constraint. In a life history, a functional or genetic relationship between two or more characters that have opposing effects on evolutionary fitness and that therefore restrict evolutionary response.

Consumer. An individual or population that uses a particular resource.

Consumer chain. *See* Food chain.

Consumer–resource interaction. Any ecological or evolutionary interaction between species in which one preys upon or otherwise consumes the other.

Continental drift. Movement of continents on the surface of the earth over geologic time; rates of drift are on the order of centimeters per year.

Continuum. A gradient of environmental characteristics or of change in the composition of communities.

Continuum index. A scale of an environmental gradient based on changes in physical characteristics or community composition along that gradient.

Control. A treatment that reproduces all aspects of an experiment except the variable of interest.

Convection. Transfer of heat by the movement of a fluid (for example, air or water).

Convergence. Resemblance among organisms belonging to different taxonomic groups resulting from adaptation to similar environments.

Cooperation. Association or social interaction among individuals of the same or different species for mutual benefit.

Coral reef. In tropical oceans, a structure built of living corals in shallow water, often surrounding an island or ringing a submerged island, in which case the reef is an atoll.

Cost of meiosis. The fact that gamete production and fertilization results in a female parent's contributing only one-half of her genotype to each of her offspring.

Countercurrent circulation. Movement of fluids in opposite directions on either side of a separating barrier through which heat or dissolved substances can pass.

Crassulacean acid metabolism. *See* CAM photosynthesis.

Cross-resistance. Resistance or immunity to one disease organism resulting from infection by another, usually closely related, organism.

Crypsis. An aspect of the appearance of an organism whereby it avoids detection by others.

Cycle. Recurrent variation in a system periodically returning to its starting point.

Cyclic climax. A steady-state, cyclic sequence of communities, none of which by itself is stable.

Cyclic succession. Continual community change through a repeated sequence of stages.

Damped oscillation. Cycling with progressively smaller amplitude, as in some populations approaching their equilibria.

Death rate (*d_x*). The percentage of newborn individuals dying during a specified interval. *Compare with* Mortality.

Defensive mutualism. A relationship between two species in which one defends the other against some enemy and usually receives some type of nourishment or living space in return.

Demographic. Pertaining to populations, particularly their growth rate and age structure.

Demography. Study of the age structure and growth rate of populations.

Denitrification. Biochemical reduction, primarily by microorganisms, of nitrogen from nitrate (NO_3^-) eventually to molecular nitrogen (N_2).

Density. Referring to a population, the number of individuals per unit of area or volume; referring to a substance, the weight per unit of volume.

Density-dependent. Having an influence on the individuals in a population that varies with the degree of crowding within the population.

Density-independent. Having an influence on the individuals in a population that does not vary with the degree of crowding within the population.

Deoxyribonucleic acid (DNA). A long macromolecule whose sequence of subunits (nucleotides) encodes genetic information.

Deterministic. Referring to the outcome of a process that is not subject to stochastic (random) variation.

Detritivore. An organism that feeds on freshly dead or partially decomposed organic matter.

Detritus. Freshly dead or partially decomposed organic matter.

Developmental response. Acquisition of one of several alternative forms by an organism depending on the environmental conditions under which it grows.

Diapause. Temporary interruption of the development of insect eggs or larvae, usually associated with a dormant period.

Diffusion. Movement of particles of gas or liquid from regions of high concentration to regions of low concentration by means of their own spontaneous motion.

Dioecy. In plants, the occurrence of reproductive organs of the male and female sex on different individuals. *Compare with* Monoecy.

Diploid. Having two sets of chromosomes. *See* Haploid; Meiosis.

Directional selection. Differential survival or reproduction within a population favoring an extreme phenotype, resulting in an evolutionary shift in the population mean toward that phenotype.

Dispersal. Movement of organisms away from their place of birth or away from centers of population density.

Dispersion. The spatial pattern of distribution of individuals within populations.

Disruptive selection. Differential survival or reproduction within a population favoring two or more extreme phenotypes, tending to promote genetic polymorphism.

Dissimilatory. Referring to a biochemical transformation that results in the oxidation of the organic form of an element and hence its loss from the biological compartment of the ecosystem.

Distribution. The geographic extent of a population or other ecological unit.

Disturbance. Any marked change in a population or community caused by an outside (environmental) influence, usually thought of as displacing an ecological system from its equilibrium.

Diversity. The number of taxa in a local area or region. Also, a measure of the variety of taxa in a community that takes into account the relative abundance of each one.

Diversity index. A measure of the variety of taxa in a community that takes into account the relative abundance of each one.

DNA. *See* Deoxyribonucleic acid.

Dominance (species). The numerical superiority of a species over others within a community or association.

Dominance hierarchy. The orderly ranking of individuals in a group based on the outcome of aggressive encounters.

Dominant. Pertaining to an allele that masks the expression of another (recessive) allele of the same gene.

Dominant (species). Pertaining to species that are abundant or exert great ecological influence within an ecological system.

Donor. The individual that is the active party in a particular behavioral interaction.

Dormancy. An inactive state, such as hibernation, diapause, or seed dormancy, usually assumed during an inhospitable period.

Doubling time. The time required for a population to grow to twice its size.

Dynamic behavior. The change in ecological systems, especially applied to populations, over time.

Dynamic steady state. Condition in which fluxes of energy or materials into and out of a system are balanced.

Ecological diversity. A measure of diversity taking into account the varied ecological roles of different species.

Ecological efficiency. The percentage of energy in the biomass produced by one trophic level that is incorporated into the biomass produced by the next higher trophic level.

Ecological footprint. The amount of land required to sustain the energy needs of an individual human based on the conversion of crop plants to ethanol fuel.

Ecological release. Expansion of habitat and resource use by populations in regions of low species diversity resulting from low interspecific competition.

Ecological system. A regularly interacting or interdependent group of biological items forming a unified whole that functions in an ecological context.

Ecological tolerance. The range of conditions within which a species can survive.

Ecology. The study of the natural environment and of the relations of organisms to each other and to their surroundings.

Ecosystem. All the interacting parts of the physical and biological worlds.

Ecosystem ecology. The study of natural systems from the standpoint of the flow of energy and cycling of matter.

Ecotone. A habitat created by the juxtaposition of distinctly different habitats; an edge habitat; a zone of transition between habitat types.

Ecotourism. Travel for the recreational purpose of observing unusual species or ecological habitats and landscapes.

Ecotype. A genetically differentiated subpopulation that is restricted to a specific habitat

Ectomycorrhizae. Mutualistic associations of fungi with the roots of plants in which the fungus forms a sheath around the outside of the root.

Ectoparasite. A parasite that lives on or attached to the host's surface (e.g., a tick).

Ectothermy. The capacity to maintain body temperature by gaining heat from the environment, either by conduction or by absorbing radiation.

Edaphic. Pertaining to or influenced by the soil.

El Niño. A warm current from the Tropics that intrudes each winter along the west coast of northern South America. *Compare with* La Niña.

Emigration. Movement of individuals out of a population. *Compare with* Immigration.

Endemic. Confined to a certain region; with respect to disease, present at a low level within a local population.

Endemism. The quality of belonging to a particular region.

Endomycorrhizae. Mutualistic associations of fungi with the roots of plants in which part of the fungus resides within the root tissues.

Endoparasite. A parasite that lives within the tissues or bloodstream of its host.

Endothermy. The capacity to maintain body temperature by the metabolic generation of heat.

Energy. Capacity for doing work.

Energy flow food web. A depiction of the feeding relationships among species in a community based on the quantity of energy transferred as food.

ENSO. El Niño–Southern Oscillation: an occasional shift in winds and ocean currents, centered in the South Pacific region, that has worldwide consequences for climate and biological systems.

Environment. The surroundings of an organism, including the plants, animals, and microbes with which it interacts.

Epidemiology. The study of factors influencing the spread of disease through a population.

Epifaunal. Referring to animals living on the surface of a substrate.

Epilimnion. The warm, oxygen-rich surface layers of a lake or other body of water. *Compare with* Hypolimnion.

Epiphyte. A plant that grows on another plant and derives its moisture and nutrients from the air and rain.

Equilibrium. A state of balance between opposing forces.

Equilibrium isocline. A line on a population graph designating combinations of competing populations, or predator and prey populations for which the growth rate of one of the populations is zero.

Equilibrium theory of island biogeography. The idea that the number of species on an island exists as a balance between colonization by new immigrant species and extinction of resident species.

Estuary. A semi-enclosed coastal water body, often at the mouth of a river, having a high input of fresh water and great fluctuation in salinity.

Ethiopian region. The biogeographic region corresponding roughly to the continent of Africa.

Euphotic zone. The surface layer of water to the depth of light penetration at which photosynthesis balances respiration. *See* Compensation point.

Eusociality. The complex social organization of termites, ants, and many wasps and bees, dominated by an egg-laying queen that is tended by nonreproductive offspring.

Eutrophic. Rich in the mineral nutrients required by green plants; pertaining to an aquatic habitat or soil with high productivity.

Eutrophication. Enrichment of water by nutrients required for plant growth; often refers to overenrichment caused by sewage and runoff from fertilized agricultural lands and resulting in excessive bacterial growth and oxygen depletion.

Evaporation. The transformation of water from the liquid to the gaseous phase with the input of heat energy.

Evapotranspiration. The sum of transpiration by plants and evaporation from the soil. *See also* Potential evapotranspiration.

Evolution. Change in the heritable traits of organisms through the replacement of genotypes within a population.

Evolutionarily stable strategy (ESS). A strategy such that, if all members of a population adopt it, no alternative strategy can invade.

Excretion. Elimination from the body, by way of the kidneys, gills, and dermal glands, of excess salts, nitrogenous waste products, and other substances.

Experiment. A controlled manipulation of a system to determine the effect of a change in one or more factors.

Exploitation competition. Competition between individuals by way of their reduction of shared resources.

Exploitation efficiency. The proportion of production on one trophic level that is consumed by organisms on the next higher level.

Exponential growth. Continuous increase or decrease in a population in which the rate of change is proportional to the number of individuals at any given time. *See* Geometric growth.

Exponential rate of increase (r). The rate at which a population is growing at a particular time, expressed as a proportional increase per unit of time. *See* Geometric rate of increase.

External loading. Input of nutrients to a lake or stream from outside the system, especially sewage and runoff from agricultural lands. *Compare with* Internal loading.

Extinction. Disappearance of a species or other taxon from a region or biota.

Extra-pair copulation (EPC). Mating received by a female from a male that is not her mate.

Facilitation. Enhancement of a population of one species by the activities of another, particularly during early succession.

Facultative. Referring to the ability to adjust to a variety of conditions or circumstances; optional for the organism. *Compare with* Obligate.

Fall bloom. The rapid growth of algae in temperate lakes following the autumnal breakdown of thermal stratification and mixing of water layers.

Fall overturn. The vertical mixing of water layers in temperate lakes in autumn following breakdown of thermal stratification.

Fecundity. The rate at which an individual produces offspring. *See also* Birth rate.

Fertilization. In reproduction, the union of male and female gametes to form a zygote.

Field capacity. The amount of water that soil can hold against the pull of gravity.

Fitness. The genetic contribution by an individual's descendants to future generations of a population.

Fixation. An increase in the proportion of an allele to unity, resulting in the elimination of all alternative alleles.

Food chain. A representation of the passage of energy from a primary producer through a series of consumers at progressively higher trophic (feeding) levels.

Food chain efficiency. *See* Ecological efficiency.

Food web. A representation of the various paths of energy flow through populations in the community, taking into account the fact that each population shares resources and consumers with other populations.

Forbs. Herbaceous, broad-leaved vegetation (i.e., other than grasses) consumed by grazers.

Founder event. Colonization of an island or patch by a small number of individuals that possess less genetic variation than the parent population. *Compare with* Population bottleneck.

Frequency dependence. The condition in which the expression of a process varies with the relative proportions of phenotypes in a population.

Frequency-dependent selection. The condition in which the fitness of a genetic trait or phenotype depends on its frequency in the population.

Functional food web. A depiction of the relationships among species within a community based on the influence of consumers on the dynamics of resource populations.

Functional response. A change in the rate of exploitation of prey by an individual predator as a result of a change in prey density. Type I: Exploitation is directly proportional to prey density. Type II: Exploitation levels off at high prey density (predator satiation). Type III: As Type II, but exploitation is additionally low at low prey density owing to lack of a search image or to efficient prey escape mechanisms. *See also* Numerical response.

Fundamental niche. The range of conditions and resources within which individuals of a species can persist.

Game theory. Analysis of the outcomes of behavioral decisions where the outcomes depend on the behavior of other interacting individuals.

Gamete. A haploid cell that fuses with another haploid cell of the opposite sex during fertilization to form a zygote. In animals, the male gamete is called the sperm and the female gamete is called the egg or ovum.

Gene flow. The exchange of genetic traits between populations by movement of individuals, gametes, or spores.

Gene pool. The whole body of genes in an interbreeding population.

Generalist. A species with broad food or habitat preferences.

Generation time. The average age at which a female gives birth to her offspring, or the average time for a population to increase by a factor equal to the net reproductive rate.

Genetic code. The correspondence between the three-nucleotide combinations in DNA sequences (codons) and the amino acids they encode in the protein sequence.

Genetic diversity. A measure of the variety of genetic factors in the gene pool of a population.

Genetic drift. Change in allele frequencies due to random variations in fecundity and mortality in a population.

Genotype. All the genetic characteristics that determine the structure and functioning of an organism; often applied to a single gene locus to distinguish one allele or combination of alleles from another.

Genotype–environment interaction. Variation in the relative expression of alternative genetic factors (alleles) depending on the environment.

Geographic range. The distribution of a population in space.

Geometric growth. Periodic increase or decrease in a population in which the increment is proportional to the number of individuals at the beginning of the period, often the breeding season. *See* Exponential growth.

Geometric rate of increase (λ). The factor by which the size of a population changes over a specified period. *See* Exponential rate of increase.

Gonad. An organ that produces the male or female gametes.

Gondwana. A giant landmass in the Southern Hemisphere during the early Mesozoic era, made up of present-day South America, Africa, India, Australia, and Antarctica.

Gradient analysis. The portrayal and interpretation of the abundances of species along gradients of physical conditions.

Grazing. Consumption of a portion of a plant's tissues, generally applied to grasses and other herbaceous vegetation; removing plant or algal cover from a substrate, generally applied to aquatic systems.

Greenhouse effect. The warming of the earth's climate because of the increased concentration of carbon dioxide and certain other pollutants in the atmosphere.

Gross primary production. The total energy assimilated by plants through photosynthesis.

Gross production. The total energy or nutrients assimilated by an organism, a population, or an entire community. *See also* Net production.

Gross production efficiency. The percentage of ingested food used for growth and reproduction by an organism.

Growing season. The period of the year during which conditions are suitable for plant growth; in temperate regions, generally between the first and last frosts.

Growth form. One of several categories of the physical structure of plants, such as tree, herbaceous perennial, or liana.

Guild. A group of species that occupy similar ecological positions within the same habitat.

Habitat. The place where an animal or plant normally lives, often characterized by a dominant plant form or physical characteristic (that is, a stream habitat, a forest habitat).

Habitat matrix. In landscape models, the types of habitat surrounding suitable habitat patches for a particular species.

Habitat patch. An area of habitat with the necessary resources and conditions for a population to persist.

Habitat selection. Choice of or preference for certain habitats.

Hadley cell. Vertical and latitudinal circulation of air within the atmosphere driven by the warming effect of the sun.

Handicap principle. The idea that elaborate, sexually selected displays and adornments act as handicaps that demonstrate the generally high fitness of the bearer.

Haplodiploidy. A sex-determining mechanism by which females develop from fertilized eggs and males from unfertilized eggs.

Haploid. Having one set of chromosomes.

Hardy–Weinberg equilibrium. The proportions of genotypes in populations that obey the Hardy–Weinberg law.

Hardy–Weinberg law. The mathematical proposition that the frequencies of genes and genotypes within a population remain unchanged in the absence of selection, mutation, genetic drift, and assortative mating.

Hawk–dove game. A game-theory analysis of social behavior between sharing and nonsharing dominant individuals.

Heat. A measure of the kinetic energy of the atoms or molecules in a substance.

Heat budget. All the gains and losses of heat by an organism, including metabolism, evaporation, radiation, conduction, and convection.

Herbivore. An organism that consumes living plants or their parts.

Hermaphrodite. An organism that has the reproductive organs of both sexes.

Heterosis. The situation in which a heterozygote has higher fitness than either homozygote genotype.

Heterotroph. An organism that uses organic materials as a source of energy and nutrients. *Compare with* Autotroph.

Heterozygous. Containing two forms (alleles) of a gene, one derived from each parent.

Hibernation. A state of winter dormancy involving lowered body temperature and metabolism.

Holistic concept. The idea championed by F. E. Clements that the organisms in a community form a discrete, complex unit, analogous to a superorganism.

Homeostasis. The maintenance of constant internal conditions in the face of a varying external environment.

Homeothermy. The ability to maintain a constant body temperature in the face of a fluctuating environmental temperature; warm-bloodedness. *Compare with* Poikilothermy.

Homozygous. Containing two identical alleles at a gene locus.

Horizon. A layer of soil distinguished by its physical and chemical properties.

Host. The living organism on or within which a parasite resides.

Humus. Fine particles of organic detritus in soil.

Hydrological cycle. The movement of water throughout the ecosystem.

Hyperdispersion. A pattern of distribution in which distances between individuals are more even than expected from random placement; overdispersion.

Hyperosmotic. Having an osmotic potential (generally, salt concentration) greater than that of the surrounding medium.

Hyphae. The threadlike filaments that make up the mycelium, or major part of the body, of a fungus.

Hypolimnion. The cold, oxygen-depleted part of a lake or other body of water that lies below the zone of rapid change in water temperature (thermocline). *Compare with* Epilimnion.

Hypo-osmotic. Having an osmotic potential (generally, salt concentration) less than that of the surrounding medium.

Hypothesis. A conjecture about or explanation for a pattern or relationship embracing a mechanism for its occurrence.

Ideal free distribution. The distribution of individuals across resource patches of different intrinsic quality that equalizes the net rate of gain of each individual when competition is taken into account.

Identity by descent. The probability that homologous genes in different individuals are direct copies of the same gene in a common ancestor.

Immigration. Movement of individuals into a population. *Compare with* Emigration.

Inbreeding. Mating between closely related individuals.

Inbreeding depression. The decrease in fitness caused by homozygous deleterious genes exposed in the offspring of matings between close relatives.

Inclusive fitness. The fitness of an individual plus the fitnesses of its relatives, the latter weighted according to degree of relationship; usually applied to the consequences of social interaction between relatives.

Indeterminate growth. Continuing growth, usually at a decreasing rate, after maturity.

Individualistic concept. The idea espoused by H. A. Gleason that the distributions of species reflect their tolerances of physical factors, not interactions between species.

Induced defense. A defensive structure or compound that is produced in response to herbivory or predation.

Inducible response. Any change in the state of an organism caused by an external factor; usually reserved for the response of organisms to parasitism and herbivory.

Industrial melanism. The evolution of dark coloration by cryptic organisms in response to industrial pollution, especially by soot, in their environments.

Inflection point. The point at which a logistic or other sigmoid-shaped growth curve changes from its accelerating to its decelerating phase.

Infrared (IR) radiation. Electromagnetic radiation having a wavelength longer than about 700 nm.

Inhibition. The suppression of a colonizing population by another that is already established, especially during successional sequences.

Innate capacity for increase (r_0). The intrinsic growth rate of a population under ideal conditions without the restraining effects of competition.

Interference competition. Direct antagonistic interaction between competing individuals, usually by behavioral or chemical means.

Intermediate disturbance hypothesis. The idea that species diversity is greatest in habitats with moderate amounts of physical disturbance, owing to the coexistence of early and late successional species.

Internal loading. Regeneration of nutrients within a system, usually referring to the sediments of a lake or river. *Compare with* External loading.

Interspecific competition. Competition between individuals of different species.

Intertropical convergence. The region within which surface currents of air meet near the equator and begin to rise under the warming influence of the sun.

Intraspecific competition. Competition between individuals of the same species.

Intrinsic rate of increase (r_m). Exponential growth rate of a population with a stable age distribution, that is, under constant conditions.

Ion. One of the dissociated parts of a molecule, each of which carries an electric charge, either positive (cation) or negative (anion).

Iteroparity. The condition of reproducing repeatedly during the lifetime. *Compare with* Semelparity.

Joint equilibrium. The combination of population sizes at which two or more populations are in equilibrium with respect to each other.

Key factor. An environmental factor that is particularly responsible for change in the size of a population.

Key factor analysis. A statistical treatment of population data designed to identify factors most responsible for change in population size.

Keystone species. A species, often a predator, having a dominating influence on the composition of a community, which may be revealed when the keystone species is removed.

Kin selection. Differential reproduction among lineages of closely related individuals based on genetic variation in social behavior.

Kranz anatomy. An arrangement of tissues in the leaves of C_4 plants in which photosynthetic cells containing chloroplasts are grouped in sheaths around vascular bundles.

Lake. A body of fresh water in any kind of depression.

Landscape ecology. A branch of ecology that considers the mosaic structure of habitat patches and their influence on population and ecosystem processes.

Landscape model. A depiction of subpopulations in which movement between patches is influenced by the types of intervening habitats.

La Niña. A global climate condition characterized by strong winds and cool ocean currents flowing westward from the coast of South America into the tropical Pacific Ocean.

Laterite. A hard substance rich in oxides of iron and aluminum, frequently formed when tropical soils weather under alkaline conditions.

Laterization. Leaching of silica from soil, usually in warm, moist regions with an alkaline soil reaction.

Laurasia. A large landmass in the Northern Hemisphere during the Mesozoic era, consisting of what is presently North America, Europe, and most of Asia.

Leaching. Removal by water of soluble compounds from leaf litter or soil.

Liana. A climbing plant of tropical rain forests, usually woody, that roots in the ground.

Liebig's law of the minimum. The idea that the growth of an individual or population is limited by the essential nutrient present in the lowest amount relative to requirement.

Life history. The adaptations of an organism that more or less directly influence life-table values of age-specific survival and fecundity (e.g., reproductive rate, age at maturity, reproductive risk).

Life table. A summary by age of the survivorship and fecundity of individuals in a population.

Life zone. A more or less distinct belt of vegetation occurring within and characteristic of a particular latitude or range of elevation.

Limestone. A rock formed chiefly by the sedimentation of shells and the precipitation and sedimentation of calcium carbonate ($CaCO_3$) in marine systems.

Limit cycle. An oscillation of predator and prey populations that occurs when stabilizing and destabilizing tendencies of their interaction balance.

Limiting resource. A resource that is scarce relative to demand for it.

Limnetic. Of or inhabiting the open water of a lake.

Limnology. The study of freshwater habitats and communities, particularly lakes, ponds, and other standing waters.

Littoral. Pertaining to the shore of the sea, especially the intertidal zone, and often including waters to the depth limit of emergent vegetation.

Local diversity. The number of species in a small area of homogeneous habitat.

Local mate competition. Direct interactions between males that compete to mate with females, particularly at or near their place of birth, hence potentially involving mating with close relatives.

Logistic equation. The mathematical expression for a particular sigmoid growth curve in which the percentage rate of increase decreases in linear fashion as population size increases.

Mark–recapture method. A method of estimating the size of a population by the recapture of marked individuals.

Mass extinction. Abrupt disappearance of a large fraction of a biota, thought to be caused by such environmental catastrophes as meteor impacts; significant mass extinctions occurred at the end of the Permian and Cretaceous periods.

Mate choice. Selection of a mate based on the characteristics of its phenotype or territory.

Mate guarding. Close association of males with their female mates to prevent mating by other males.

Mating system. The pattern of matings between individuals in a population, including number of simultaneous mates, permanence of pair bond, and degree of inbreeding.

Matric potential. Referring to the water potential generated by soil.

Maturity. Acquisition of sexual function; often the age at which this occurs.

Maximum sustainable yield (MSY). The highest rate at which individuals may be harvested from a population without reducing the size of the population—that is, at which recruitment equals or exceeds harvesting.

Mediterranean climate. A pattern of climate found in middle latitudes, characterized by cool, wet winters and warm, dry summers.

Meiosis. A series of two divisions by cells destined to produce gametes, involving pairing and segregation of homologous chromosomes and a reduction of chromosome number from diploid to haploid.

Mesic. Referring to habitats with plentiful rainfall and well-drained soils.

Metamorphosis. An abrupt change in form during development that fundamentally alters the function of the organism.

Metapopulation. A population that is divided into subpopulations between which individuals migrate from time to time. Habitat fragmentation is causing many species to assume a metapopulation structure.

Microcosm. A small, simplified system, often maintained in a laboratory, that contains the essential features of a larger natural system.

Microenvironment. The conditions within a microhabitat, that is, those experienced by an individual at a particular time.

Microhabitat. The particular part of the habitat that an individual encounters at any particular time in the course of its activities.

Migration. The movement of individuals between one place and another or between subpopulations in a metapopulation.

Mimic. An organism adapted to resemble another organism or an object.

Mimicry. Resemblance of an organism to some other organism or an object in the environment, evolved to deceive predators or prey into confusing the organism with that which it mimics.

Mineralization. Transformation of elements from organic to inorganic forms, often by dissimilatory oxidations.

Minimum viable population (MVP). The minimum number of individuals necessary to prevent a population from losing genetic variation or suffering stochastic extinction over an acceptably long period.

Mitosis. The division of a cell into two identical daughter cells, involving replication and equal partitioning of the cell's DNA.

Mixed evolutionarily stable strategy. An evolutionarily stable strategy involving more than one phenotype within a population; generally the outcome of frequency-dependent fitnesses of the phenotypes.

Model. An organism, usually unpalatable or otherwise noxious to predators, upon which a mimic is patterned.

Model (mathematical). A quantitative representation of the relationships among the entities in a system (e.g., among the species in a community).

Monoecy. In plants, the occurrence of male and female reproductive organs in different flowers on the same individual. *Compare with* Dioecy.

Monogamy. A mating system in which each individual mates with only one individual of the opposite sex, generally involving a strong and lasting pair bond. *Compare with* Polygamy.

Mortality (m_x). Ratio of the number of deaths to the number of individuals at risk, often described as a function of age (x). *Compare with* Death rate.

Müllerian mimicry. Mutual resemblance of two or more conspicuously marked, unpalatable species to enhance predator avoidance.

Mutation. Any change in the genotype of an organism occurring at the gene, chromosome, or genome level; usually applied to changes in genes to new allelic forms.

Mutualism. A relationship between two species that benefits both.

Mycelium. The rootlike network of filaments (hyphae) making up the nonreproductive part of the body of a fungus.

Mycorrhizae. Close associations of fungi and tree roots in the soil that facilitate the uptake of minerals by trees.

Natural selection. Change in the frequency of genetic traits in a population through differential survival and reproduction of individuals bearing those traits.

Nearctic. The biogeographic region corresponding to temperate North America.

Negative assortative mating. Preferential mating between individuals having differing appearances or genotypes.

Negative feedback. The tendency of a system to counteract externally imposed change and return to a stable state.

Neighborhood size. The number of individuals in a population included within the dispersal distance of a single individual.

Neotropical region. The biogeographic region corresponding to South America and tropical Central America and the West Indies.

Neritic zone. The region of shallow water adjoining a seacoast.

Net aboveground productivity (NAP). Accumulation of biomass in aboveground parts of plants (trunks, branches, leaves, flowers, and fruits) over a specified period; usually expressed on an annual basis (NAAP).

Net primary production. Energy accumulated in the tissues of plants.

Net production. The total energy or nutrients accumulated as biomass by an organism, a population, or an entire community by growth and reproduction; gross production minus respiration.

Net production efficiency. The percentage of assimilated food used for growth and reproduction by an organism.

Net reproductive rate (R_0). The expected number of offspring of a female during her lifetime.

Niche. The ecological role of a species in the community; the ranges of many conditions and resource qualities within which the organism or species persists, often conceived as a multidimensional space.

Niche breadth. The variety of resources used and range of conditions tolerated by an individual, population, or species.

Niche overlap. The sharing of niche space by two or more species; similarity of resource requirements and tolerance of ecological conditions.

Nitrification. Breakdown of nitrogen-containing organic compounds by microorganisms, yielding nitrates and nitrites.

Nitrogen fixation. Biological assimilation of atmospheric nitrogen to form organic nitrogen-containing compounds.

Nonrenewable resource. A resource present in fixed quantity that can be used completely by consumers (e.g., space).

Nucleotide. Any of several chemical compounds forming the structural units of RNA and DNA and consisting of a purine or pyrimidine base, ribose (RNA) or deoxyribose (DNA) sugar, and phosphoric acid.

Null model. A set of rules for generating community patterns, presupposing no interaction between species, against which observed community patterns can be compared statistically.

Numerical response. A change in the population size of a predatory species as a result of a change in the density of its prey. *See also* Functional response.

Nutrient. Any substance required by organisms for normal growth and maintenance.

Nutrient cycle. The path of an element as it moves through the ecosystem, including its assimilation by organisms and its regeneration in a reusable inorganic form.

Nutrient use efficiency (NUE). Ratio of production to uptake of a required nutrient.

Obligate. Referring to a way of life or response to particular conditions without alternatives. *Compare with* Facultative.

Oceanic zone. Region of the ocean beyond the continental shelves.

Oligotrophic. Poor in the mineral nutrients required by green plants; pertaining to an aquatic habitat with low productivity.

Omnivore. An organism whose diet is broad, including both plant and animal foods; specifically, an organism that feeds on more than one trophic level.

Omnivory. In the sense of food-web analysis, feeding on more than one trophic level.

Open community. A local association of species having independent and only partially overlapping ecological distributions.

Open community concept. The idea advocated by H. A. Gleason and R. H. Whittaker that communities are the local expression of the independent geographic distributions of species.

Optimal foraging. A set of rules, including breadth of diet, by which organisms maximize food intake per unit of time or minimize the time needed to meet their food requirements; risk of predation may also enter the equation for optimal foraging.

Optimal outcrossing distance. The distance from which pollen is received by a plant that maximizes seed set.

Optimum. The narrow range of environmental conditions to which an organism is best suited.

Ordination. A set of mathematical methods by which communities are ordered along physical gradients or along derived axes over which distance is related to dissimilarity in species composition.

Organism. A living being; the most fundamental unit of ecology.

Organismal viewpoint. The idea that a community is a discrete, highly integrated association of species within which the function of each species is subservient to the whole.

Oriental region. The biogeographic region corresponding roughly to tropical southeast Asia and India.

Oscillation. Regular fluctuation through a fixed cycle above and below some mean value.

Osmosis. Diffusion of substances in aqueous solution across the membrane of a cell.

Osmotic potential. The attraction of water to an aqueous solution owing to its concentration of ions and other small molecules; usually expressed as a pressure.

Outcrossing. Mating with unrelated individuals within a population.

Overdispersion. *See* Hyperdispersion.

Overlapping generations. The co-occurrence of parents and offspring in the same population as reproducing adults.

Oxidation. Removal of one or more electrons from an atom, ion, or molecule. *Compare with* Reduction.

Ozone. A molecule consisting of three atoms of oxygen (O_3), which, in the upper atmosphere, blocks the penetration of ultraviolet light to the earth's surface.

Ozone hole. A region of severe ozone depletion in the upper atmosphere, usually at high latitude.

Palearctic. The biogeographic region corresponding roughly to temperate Asia and Europe.

Pangaea. A supercontinent existing at the end of the Paleozoic era that included practically all the earth's landmasses, including the future Laurasia and Gondwana.

Parasite. An organism that consumes part of the blood or tissues of its host, usually without killing the host.

Parasite-mediated sexual selection. The hypothesis that females choose mates on the basis of their ability to resist parasite infection.

Parasitoid. Any of a number of insects whose larvae live within and consume their host, usually another insect.

Parent–offspring conflict. The situation arising when the optimum level of parental investment in a particular offspring differs from the viewpoints of the parent and that offspring. This conflict derives from the fact that offspring are genetically equivalent from the point of view of their parents, but siblings carry identical copies of only half of one another's genes.

Parity. The age-specific pattern of giving birth.

Parthenogenesis. Reproduction without fertilization by male gametes, usually involving the formation of diploid eggs whose development is initiated spontaneously.

Partial pressure. The proportional contribution of a particular gas to the total pressure of a mixture.

Pathogen. A parasitic organism that causes a disease in its host.

Patch. An area of habitat with the necessary resources and conditions for a population to persist.

Pelagic. Pertaining to the open sea.

Per capita. Expressed on a per individual basis.

Perennial. Referring to an organism that lives for more than one year; lasting throughout the year.

Perfect flower. A flower having both male and female sexual organs (anthers and pistils).

Periodic cycle. Fluctuation in population size with regular intervals between high and low numbers.

Permafrost. A layer of permanently frozen ground in very cold regions, especially the Arctic and Antarctic.

Pest pressure hypothesis. The idea that individuals are vulnerable to pests and pathogens when crowded in the vicinity of their parents, permitting the coexistence of many different types of species that are attacked by different pests.

pH. A scale of acidity or alkalinity; the logarithm of the concentration of hydrogen ions.

Phenolics. Aromatic hydrocarbons produced by plants, many of which exhibit antimicrobial properties.

Phenology. Study of the timing of biological activity over the course of a year, particularly in relation to climate.

Phenotype. The physical expression in an organism of the interaction between its genotype and its environment; the outward appearance and behavior of the organism.

Phenotypic plasticity. Variation among individuals produced by the influence of the environment on form and function.

Photic. Pertaining to surface waters to the depth of light penetration.

Photoautotroph. An organism that uses sunlight as its primary energy source for the synthesis of organic compounds.

Photoperiod. The length of the daylight period each day.

Photorespiration. Oxidation of carbohydrates to carbon dioxide and water by the enzyme responsible for CO_2 assimilation, in the presence of bright light.

Photosynthesis. Use of the energy of light to combine carbon dioxide and water into simple sugars.

Photosynthetic efficiency. Percentage of light energy assimilated by plants, based either on net production (net photosynthetic efficiency) or on gross production (gross photosynthetic efficiency).

Phylogenetic effect. Resemblance of the morphology or ecology of species resulting from their common ancestry.

Phylogeny. The evolutionary relationships among species or other taxa.

Phytoplankton. Microscopic floating aquatic plants.

Plankton. Microscopic floating aquatic plants (phytoplankton) and animals (zooplankton).

Podsolization. Breakdown and removal of clay particles from the acidic soils of cold, moist regions.

Poikilothermy. Inability to regulate body temperature; cold-bloodedness. *Compare with* Homeothermy.

Poisson distribution. A statistical description of the random distribution of items among categories, often applied to the distribution of individuals among sampling plots.

Polyandry. A mating pattern in which a female mates with more than one male at the same time or in quick succession.

Polygamy. A mating system in which a male pairs with more than one female (polygyny) or a female pairs with more than one male (polyandry) at the same time. *Compare with* Monogamy.

Polygyny. A mating pattern in which a male mates with more than one female at the same time or in quick succession.

Polygyny threshold. The difference between the intrinsic values of territories or males such that the realized values of an unmated male on a poorer territory and a mated male on a better territory are equal in the eyes of an unmated female.

Polymorphism. The occurrence of more than one distinct form of individual or genotype in a population.

Pool. A stretch of slow-flowing, often deep water in a stream between riffles.

Population. The group of organisms of a particular species that inhabit a particular area.

Population bottleneck. A period of extremely small population size during which a population is vulnerable to the loss of genetic diversity through genetic drift. *Compare with* Founder event.

Population genetics. The study of changes in the frequencies of genes and genotypes within a population.

Population structure. Referring to the density and spacing of individuals within a suitable habitat and the proportions of individuals in each age and sex class.

Positive assortative mating. Preferential mating between individuals having similar appearance or genotypes.

Potential evapotranspiration (PET). The amount of transpiration by plants and evaporation from the soil that would occur, given the local temperature and humidity, if water were not limited.

Prairie. An extensive area of level or rolling, almost treeless grassland in central North America.

Preadaptation. A trait evolved for one purpose that becomes useful for another purpose in a changed environment.

Precipitation. Rainfall or snowfall. Also, the change of a compound from a dissolved form to a solid form.

Predator. An animal (rarely, a plant) that kills and eats animals.

Prediction. A logical consequence of a hypothesis or outcome of a model describing some aspect of a system.

Primary consumer. An herbivore, the lowermost consumer on the food chain.

Primary producer. A green plant or other autotroph that assimilates the energy of light to synthesize organic compounds.

Primary production. Assimilation (gross primary production) or accumulation (net primary production) of energy and nutrients by green plants and other autotrophs.

Primary productivity. The rate at which primary production occurs within an ecosystem.

Primary succession. The sequence of communities developing in a newly exposed habitat devoid of life.

Production. Accumulation of energy or biomass.

Programmed death. Death as part of an adaptation to maximize a single, terminal, reproductive episode.

Promiscuity. Mating with many individuals within a population, generally without the formation of strong or lasting pair bonds.

Proximate factor. An aspect of the environment that the organism uses as a cue for behavior (e.g., day length). Proximate factors often are not directly important to the organism's well-being. *Compare with* Ultimate factor.

Pyramid of energy. The concept that the energy flux through any given link in the food chain decreases with progressively higher trophic levels.

Pyramid of numbers. Charles Elton's concept that the sizes of populations decrease with progressively higher trophic levels, that is, as one progresses along the food chain.

Queen. In a bee, wasp, ant, or termite colony, a fertile, fully developed female whose function is to lay eggs and who is the mother of all the members of the colony.

Radiant flux. Light intensity, expressed in energy per unit of area per unit of time.

Radiation. Energy emitted in the form of electromagnetic waves.

Rain shadow. A dry area on the leeward side of a mountain range.

Random distribution. The condition in which each individual occurs without regard to the position of other individuals

Rarefaction. A method of determining the relationship between species diversity and sample size by randomly deleting individuals from a sample.

Reaction norm. The relationship between the appearance of a phenotype and variations in the environment produced by a particular genotype.

Realized niche. The range of conditions and resources within which a population can persist in the presence of competitors and predators.

Recessive. Pertaining to an allele whose expression is masked by an alternative (dominant) allele in a diploid, heterozygous state.

Recipient. The individual that is the passive party in a particular behavioral interaction.

Reciprocal altruism. The exchange of altruistic acts between individuals.

Reciprocal transplant experiment. An exchange of individuals of the same species between two different habitats or regions to determine the relative contributions of genotype and environment to the phenotype.

Recombination. The formation of offspring with combinations of genes that did not occur in either parent by crossing over between chromosomes and independent assortment and mixing of maternal and paternal chromosomes.

Recruitment. Addition of new individuals to a population by reproduction; often restricted to the addition of breeding individuals.

Red Queen hypothesis. The idea that biological change, especially in predators and pathogens, applies continual selective pressure on populations.

Reduction. Addition of one or more electrons to an atom, ion, or molecule. *Compare with* Oxidation.

Refugium. A place where a species or community can persist in the face of environmental change over the remainder of its distribution.

Region. A geographic area without significant internal barriers to dispersal but generally much larger than the dispersal distance of an individual.

Regional diversity. The number of species or other taxa occurring within a large geographic area encompassing many biomes or habitats.

Relative abundance. Proportional representation of a species in a sample or a community.

Renewable resource. A resource that is continually supplied to the system such that it cannot be fully depleted by consumers.

Reproductive effort. The allocation of time or resources or the assumption of risk in order to increase fecundity.

Rescue effect. Prevention of the extinction of a local population by immigration of individuals from elsewhere, often from a more productive habitat.

Residence time. The ratio of the size of a compartment to the flux through it, expressed in units of time; thus, the average time spent by energy or a substance in the compartment.

Resource. A substance or object required by an organism for normal maintenance, growth, and reproduction. If the resource is scarce relative to demand, it is referred to as a limiting resource. Nonrenewable resources (such as space) occur in fixed amounts and can be fully used; renewable resources (such as food) are produced at a rate that may be partially determined by their use.

Respiration. Use of oxygen to break down organic compounds metabolically to release chemical energy.

Rhizome. An underground, usually horizontal stem of a plant that produces both roots and aboveground shoots and that may be modified to store carbohydrate nutrient reserves.

Riffle. A shallow stretch of fast-moving, often rough water between quieter pools in a stream.

Riparian. Along the bank of a river or lake.

Risk-sensitive foraging. The situation in which the rate or place of foraging is influenced by the presence of predators or risk of predation.

River continuum concept. The idea that a river system encompasses a continuum of conditions from the headwaters to the mouth, characterized by increasing streambed size and water flow and interconnected by the movement of nutrients and organisms with downstream currents.

RuBP. Ribulose bisphosphate, a five-carbon carbohydrate to which a carbon atom is attached during the assimilatory step of the Calvin cycle in photosynthesis.

RuBP carboxylase. An enzyme in the Calvin cycle of photosynthesis responsible for the reaction of ribulose bisphosphate and carbon dioxide to form two molecules of phosphoglyceraldehyde.

Ruderal. Pertaining to or inhabiting highly disturbed sites. *See* Weed.

Runaway sexual selection. The situation in which females persistently choose the most extreme male phenotypes in a population, leading to continuous elaboration of secondary sexual characteristics.

Saturation point. With respect to primary production, the amount of light that causes photosynthesis to attain its maximum rate.

Scale. The dimension in time or space over which variation is perceived.

Sclerophyllous. Referring to the tough, hard, often small leaves of drought-adapted vegetation.

Secondary plant compounds. Chemical products of plant metabolism that are produced specifically for the purpose of defense against herbivores and disease organisms.

Secondary sexual characteristics. Traits, other than the sexual organs themselves, that distinguish the sexes, including, for example, the beard of human males.

Secondary succession. Progression of communities in habitats where the climax community has been disturbed or removed.

Sedimentation. The settling of particulate material to the bottom of an ocean, lake, or other body of water.

Seed bank. Seeds remaining viable in the soil that can germinate when conditions are favorable.

Selection. Differential survival or reproduction of individuals in a population owing to phenotypic differences among them.

Self-incompatible. Unable to mate with oneself because of structural, biochemical or timing factors.

Selfing. Mating with oneself; applicable, of course, only to individuals (usually plants) having both male and female reproductive organs (hermaphrodites).

Selfishness. Behavior that conveys an advantage to the individual without regard for others.

Self-thinning curve. In populations of plants limited by space or other resources, the characteristic relationship between average plant weight and density.

Semelparity. The condition of having only one reproductive episode during the lifetime. *Compare with* Iteroparity.

Semipermeable. Referring to a membrane that blocks the passage of some molecules, usually large ones, but not other, smaller molecules.

Senescence. Gradual deterioration of function in an organism with age, leading to increased probability of death; aging.

Sequential hermaphroditism. The condition in which an individual is first one sex and then another.

Sere. A series of stages of community change in a particular area leading toward a stable state.

Serpentine. An igneous rock rich in magnesium that forms soils toxic to many plants.

Sex ratio. The ratio of the number of individuals of one sex to that of the other sex in a population.

Sexual dimorphism. The condition in which the males and females of a species differ in appearance.

Sexual reproduction. Reproduction by means of the union of two gametes (fertilization) to form a zygote.

Sexual selection. Selection by one sex for specific characteristics in individuals of the opposite sex, usually exercised through courtship behavior.

Shannon–Wiener index (*H*). A logarithmic measure of the diversity of species weighted by the relative abundance of each.

Simpson's index (*D*). A measure of the diversity of species weighted by the relative abundance of each.

Simultaneous hermaphroditism. The condition in which an individual has both male and female sexual organs at the same time.

Sink population. A population that would continually decrease in size owing to high mortality, low reproduction, or both if it were not maintained by immigration from other populations.

Social behavior. Any direct interaction among distantly related individuals of the same species; usually does not include courtship, mating, parent–offspring, and sibling interactions.

Soil. The solid substrate of terrestrial communities, resulting from the interaction of weather and biological activities with the underlying geologic formation.

Soil horizon. A layer of soil formed at a characteristic depth and distinguished by its physical and chemical properties.

Soil profile. A characterization of the structure of soil vertically through its various horizons, or layers.

Soil skeleton. The physical structure of mineral soil, referring principally to sand grains and silt particles

Solar constant. The intensity of solar radiation reaching the outer limit of the earth's atmosphere, approximately 1,400 W per m^2.

Solar equator. The parallel of latitude that lies directly under the sun at any given season.

Solute. Any substance dissolved in a solvent.

Source population. A population that produces an excess of individuals over the number needed for self-maintenance and from which there is net emigration.

Source–sink model. A depiction of population structure in which some subpopulations produce excess offspring (sources) and others would decline in size without immigration (sinks).

Spaced distribution. A distribution in which each individual maintains a minimum distance between itself and its neighbors.

Specialist. A species that uses a restricted range of habitats or resources.

Specialization. An adaptation of form or function that suits an individual particularly well to a restricted range of habitats, resources, or environmental conditions; the evolutionary process of such restriction.

Speciation. The division of a single species into two or more reproductively incompatible daughter species.

Species. A group of actually or potentially interbreeding populations that are reproductively isolated from all other kinds of organisms.

Species–area relationship. The pattern of increase in the number of species with respect to the area of the habitat, island, or region being considered.

Species diversity. *See* Diversity.

Species pool. The entire group of species within a source area from which colonists of an island or habitat are drawn.

Species richness. A simple count of the number of species.

Species sorting. Restriction of species from a regional pool in local communities owing to their tolerance of conditions, requirements for resources, or interactions with competitors and predators.

Spitefulness. A social interaction in which the donor of a behavior incurs a cost in order to reduce the fitness of a recipient.

Spring overturn. The vertical mixing of water layers in temperate lakes in spring as surface ice disappears.

Stabilizing selection. Differential survival or reproduction of phenotypes closest to the mean value of a population; selection against extreme phenotypes tending to maintain the population mean.

Stable age distribution. The proportion of individuals in various age classes in a population that has been growing at a constant rate.

Stable equilibrium. The particular state to which a system returns if displaced by an outside force.

Standard deviation (*s* or σ). A measure of the variability among items in a sample, such as individuals in a population; the square root of the variance and hence the square root of the average squared deviation from the mean.

Static life table. The age-specific survival and fecundity of individuals of different age classes within a population at a given time; a time-specific life table.

Steady state. The condition of a system in which opposing forces or fluxes are balanced.

Steppe. Usually treeless plains, especially in southeastern Europe and Asia in regions of extreme temperature range and sandy soil.

Stochastic. Referring to patterns resulting from random effects.

Stochastic extinction. Decrease of a population's size to zero resulting from random fluctuations in births and deaths.

Stomate. The opening in a leaf surface through which gas exchange with the atmosphere takes place.

Storage. Accumulation of resources during favorable conditions for use during less favorable conditions.

Stratification. The establishment of distinct layers of temperature or salinity in bodies of water based on the different densities of warm and cold water or saline and fresh water.

Stream. Water running over the surface of the land.

Subpopulation. A subdivision within a population having restricted exchange of individuals with the remainder of the population.

Subtropical high-pressure belts. Regions of high atmospheric pressure and dry air centered approximately 30° north and south of the equator.

Succession. Replacement of populations in a habitat through a regular progression to a stable state.

Survival (*s$_x$*). The probability of living from one age or time period (*x*) to the next.

Survivorship (*l$_x$*). Proportion of newborn individuals alive at age *x*.

Switching. A change in diet to favor items of increasing suitability or abundance.

Symbiosis. An intimate and often obligatory association of two species, usually involving coevolution. Symbiotic relationships can be parasitic or mutualistic.

Sympatric. Occurring in the same place; usually referring to areas of overlap in species distributions.

Taiga. A moist coniferous forest bordering the arctic zone, dominated by spruce and fir trees.

Tannins. Polyphenolic compounds produced by most plants that bind proteins, thereby impairing digestion by herbivores and inhibiting microbes.

Temperature profile. The relationship of temperature to depth below the surface of the water or soil, or height above the ground.

Tension–cohesion theory. The idea that the force required to draw water from the soil and roots is generated by the evaporation of water from the leaves of plants.

Terpenoids. Plant secondary compounds, including essential oils, latex, and plant resins, used to resist herbivory.

Territory. Any area defended by one or more individuals against intrusion by others of the same or different species.

Theory of island biogeography. *See* Equilibrium theory of island biogeography.

Thermal conductance. The rate at which heat passes through a substance.

Thermal stratification. Sharp delineation of layers of water by temperature, the warmer layer generally lying over the top of the colder layer.

Thermocline. The zone of water depth within which temperature changes rapidly between the upper warm water layer (epilimnion) and lower cold water layer (hypolimnion).

Thermodynamic. Relating to heat and motion.

$-3/2$ (three-halves) power law. A generalization proposing that the relationship between the logarithms of the biomass and density of a population of plants has a slope of $-3/2$.

Time lag. A delay in the response of a population or other system to change in the environment; also called time delay.

Time-specific life table. *See* Static life table.

Tolerance. In reference to succession, the indifference of establishment of one species to the presence of others.

Top-down control. Influence of predators on the sizes of the trophic levels below them in the food web.

Torpor. Loss of the power of motion and feeling, usually accompanied by a greatly reduced rate of respiration.

Trade-off. A balancing of factors, all of which are not attainable at the same time, especially traits that tend to increase the evolutionary fitness or performance of the individual.

Transient climax. Climax community that develops in a temporary habitat, such as a vernal pool or carcass of a dead animal.

Transit time. The average time that a substance or energy remains in the biological realm or any compartment of a system; ratio of biomass to productivity.

Transpiration. Evaporation of water from leaves and other parts of plants.

Transpiration efficiency. The ratio of net primary production to transpiration of water by a plant, usually expressed as grams per kilogram of water; water use efficiency.

Trophic. Pertaining to food or nutrition.

Trophic cascade. Influence of consumers or producers on populations that are two or more trophic levels removed.

Trophic level. Position in the food chain, determined by the number of energy-transfer steps to that level.

Trophic mutualism. A symbiotic relationship between two species based on complementary methods of obtaining energy and nutrients.

Trophic structure. The organization of a community based on the feeding relationships of populations.

Types I, II, and III functional responses. *See* Functional response.

Understory. A layer of vegetation under the canopy of a forest.

Ultimate factor. An aspect of the environment that is directly important to the well-being of an organism (e.g., food). *Compare with* Proximate factor.

Ultraviolet (UV) radiation. Electromagnetic radiation having a wavelength shorter than about 400 nm.

Upwelling. Vertical movement of water, usually near coasts and driven by offshore winds, that brings nutrients from the depths of the ocean to surface layers.

Vertical mixing. Exchange of water between deep and surface layers.

Vicariance. The breaking up of a widely distributed ancestral population by continental drift or some other barrier to dispersal.

Wahlund effect. An excess of homozygotes compared with heterozygotes in a population, relative to Hardy–Weinberg frequencies, caused by the mixing of two populations with different equilibrium allele frequencies.

Warning coloration. Conspicuous patterns or colors adopted by noxious organisms to advertise their noxiousness or dangerousness to potential predators; aposematism.

Water potential. The force by which water is held in the soil by capillary and hygroscopic attraction.

Watershed. The drainage area of a stream or river.

Water use efficiency. *See* Transpiration efficiency.

Weathering. Physical and chemical breakdown of rock and its component minerals at the base of the soil.

Weed. A plant or animal, generally having high powers of dispersal, capable of living in highly disturbed habitats.

Wilting coefficient. The minimum water content of the soil at which plants can obtain water. Also called the wilting point.

Xeric. Referring to habitats in which plant production is limited by the availability of water.

Xerophyte. A plant that tolerates dry (xeric) conditions.

Zonation. The distribution of organisms in bands or regions corresponding to changes in ecological conditions along a continuum, for example, intertidal zonation and elevational zonation.

Zooplankton. Tiny floating aquatic animals.

Zygote. A diploid cell formed by the union of male and female gametes during fertilization.

Illustration Acknowledgments

The staff at J. B. Woolsey is to be thanked for their artistic talents and beautiful work on the line illustrations.

I am grateful to the following individuals and institutions for providing many of the fine photgraphs used in this book: The American Museum of Natural History, Scott Bauer, F. Bennet, Rudy Boonstra, Otis Brown, Alfred Brousseau, J. Burgett, Ron Burton, Mark Carle, Walter Carson, Jack K. Clark, Dale H. Clayton, The Cornell Laboratory of Ornithology, The Cornell University Integrated Pest Management Program, Paul Dayton, Natalia Demong, Jim Ehleringer, Thomas Eisner, Robert Evans, Ola Fincke, Michael M. Follo, Larry Jon Friesen, Elsa Marie Friis, Z. Glowacinski, Marcos Guerra, Antonio Guillen, D. Habek, Eric Hanauer, Carl C. Hansen, W. H. Haseler, L. Higley, Norman Hodgkin, Hudson's Bay Company Archives, C. B. Huffaker, Terry P. Hughes, Rick Karban, Uwe Kils, J. Kimmel, Jonathan B. Losos, Bob Marquis, Michele McCauley, Andy McGregor, D. W. Miller, C. H. Muller, Gary Munkvold, National Aeronautics and Space Administration, National Oceanic and Atmospheric Administration (NOAA), National Undersea Research Program, David Pimentel, David W. Schindler, Anthony R. E. Sinclair, Thomas R. Sinclair, Barry Sinervo, Thomas B. Smith, Smithsonian Tropical Research Institute, Soil Conservation Service, Wayne P. Sousa, Douglas G. Sprugel, R. B. Suter, David Tilman, The U.S. Department of Agriculture, The U. S. Forest Service, The University of California Digital Library Project, The University of California Statewide Integrated Pest Management Program, C. Van Dover, Eduardo Venticinque, Chris Whelan, Lon Wilkens, Gerald S. Wilkinson. and Truman P. Young.

The chapter-opening images were provided by the following photographers: Chapter 1, Luiz C. Marigo/Peter Arnold; Chapter 2, Bruce Sidell, University of Maine; Chapter 3, François Gohier/Photo Researchers; Chapter 4, John Barr/Gamma Liaison; Chapter 5, Alan & Linda Detrick/Photo Researchers; Chapter 6, R. E. Ricklefs; Chapter 7, Tom & Pat Leeson/Photo Researchers; Chapter 8, Will McIntyre/Photo Researchers; Chapter 9, Tom Bean/DRK Photo; Chapter 10, Michel Roggo-Bios/Peter Arnold; Chapter 11, Simon D. Pollard/Photo Researchers; Chapter 12, John R. MacGregor/Peter Arnold; Chapter 13, E. R. Degginger/Photo Researchers; Chapter 14, Rainer Grosskopf/Tony Stone; Chapter 15, Ernst Haas/Tony Stone; Chapter 16, Tui de Ray/Bruce Coleman; Chapter 17, D. Habeck and F. Bennet, University of Florida; Chapter 18, Tom Brakefield/DRK Photo; Chapter 19, Stuart Westmorland/Photo Researchers; Chapter 20, Andy Rouse/DRK Photo; Chapter 21, Tom Bean/DRK Photo; Chapter 22, R. E. Ricklefs; Chapter 23, Joshua Singer; Chapter 24, John Cancalosi/DRK Photo; Chapter 25, Holt Confer/DRK Photo; and Chapter 26, William Campbell/Peter Arnold. The photograph of the horse-chestnut leaf used as an icon at the beginning of each chapter was contributed by Grant Heilman/Runk/Schoenberger.

Index

Please note that **boldfaced** page numbers indicate references to boldfaced words or phrases in the text. *Italicized* page numbers indicate pages with illustrations.

Conversion Factors

Length

1 meter (m) = 39.4 inches (in)
1 meter = 3.28 feet (ft)
1 kilometer (km) = 3,281 feet
1 kilometer = 0.621 mile (mi)
1 micron (μ) = 10^{-6} meter
1 inch = 2.54 centimeters (cm)
1 foot = 30.5 centimeters
1 mile = 1,609 meters
1 angstrom (Å) = 10^{-10} meter
1 millimicron (mμ) = 10^{-10} meter

Area

1 square centimeter (cm^2) = 0.155 square inch (in^2)
1 square meter (m^2) = 10.76 square feet (ft^2)
1 hectare (ha) = 2.47 acres (a)
1 hectare = 10,000 square meters
1 hectare = 0.01 square kilometer (km^2)
1 square kilometer = 0.386 square mile
1 square mile = 2.95 square kilometers
1 square inch = 6.45 square centimeters
1 square foot = 929 square centimeters
1 square yard (yd^2) = 0.836 square meter
1 acre = 0.407 hectare

Mass

1 gram (g) = 15.43 grains (gr)
1 kilogram (kg) = 35.3 ounces (oz)
1 kilogram = 2.205 pounds (lb)
1 metric ton (T) = 2,204.6 pounds
1 ounce = 28.35 grams
1 pound = 453.6 grams
1 short ton = 907 kilograms

Time

1 year (yr) = 8,760 hours (h)
1 day = 86,400 seconds (s)

Volume

1 cubic centimeter (cc or cm^3) = 0.061 cubic inch (in^3)
1 cubic inch = 16.4 cubic centimeters
1 liter (L) = 1,000 cubic centimeters
1 liter = 33.8 U. S. fluid ounces (oz)
1 liter = 1.057 U. S. quarts (qt)
1 liter = 0.264 U. S. gallon (gal)
1 U.S. gallon = 3.79 liters
1 British imperial gallon = 4.55 liters
1 cubic foot (ft^3) = 28.3 liters
1 milliliter (mL) = 1 cubic centimeter
1 U. S. fluid ounce = 29.57 milliliters
1 British imperial fluid ounce = 28.4 milliliters
1 quart = 0.946 liter

Velocity

1 meter per second (m per s) = 2.24 miles per hour (mph)
1 foot per second (ft per s) = 1.097 kilometers per hour
1 kilometer per hour = 0.278 meter per second
1 mile per hour = 0.447 meter per second
1 mile per hour = 1.467 feet per second

Energy

1 joule (J) = 0.239 calorie (cal)
1 calorie = 4.184 joules
1 kilowatt-hour (kWh) = 860 kilocalories (kcal)
1 kilowatt-hour = 3,600 kilojoules
1 British thermal unit (Btu) = 252.0 calories
1 British thermal unit = 1,054 joules
1 kilocalorie (kcal) = 1,000 calories

Power

1 kilowatt (kW) = 0.239 kilocalorie per second
1 kilowatt = 860 kilocalories per hour
1 horsepower (hp) = 746 watts
1 horsepower = 15,397 kilocalories per day
1 horsepower = 641.5 kilocalories per hour

Energy per unit area

1 calorie per square centimeter = 3.69 British thermal units per square foot
1 British thermal unit per square foot = 0.271 calorie per square centimeter
1 calorie per square centimeter = 10 kilocalories per square meter

Power per unit area

1 kilocalorie per square meter per minute = 52.56 kilocalories per hectare per year
1 footcandle (fc) = 1.30 calories per square foot per hour at 555 nm wavelength
1 footcandle = 10.76 lux (lx)
1 lux = 1.30 calories per square meter per hour at 555 nm wavelength

Metabolic energy equivalents

1 gram carbohydrate = 4.2 kilocalories
1 gram of protein = 4.2 kilocalories
1 gram of fat = 9.5 kilocalories

Miscellaneous

1 gram per square meter = 0.1 kilogram per hectare
1 gram per square meter = 8.97 pounds per acre
1 kilogram per square meter = 4.485 short tons per acre
1 metric ton per hectare = 0.446 short tons per acre